Mössbauer Spectroscopy and Transition Metal Chemistry

Philipp Gütlich • Eckhard Bill •
Alfred X. Trautwein

Mössbauer Spectroscopy and Transition Metal Chemistry

Fundamentals and Applications

with electronic supplementary material at
extras.springer.com

Springer

Prof. Dr. Philipp Gütlich
Universität Mainz
Inst. Anorganische und
Analytische Chemie
Staudingerweg 9
55099 Mainz
Germany
guetlich@uni-mainz.de

Dr. Eckhard Bill
MPI für Bioanorganische
Chemie
Stiftstr. 34-36
45470 Mülheim
Germany
bill@mpi-muelheim.mpg.de

Prof. Dr. Alfred X. Trautwein
Universität zu Lübeck
Inst. Physik
Ratzeburger Allee 160
23538 Lübeck
Germany
trautwein@physik.uni-luebeck.de

Additional material to this book is available on CD-ROM; in subsequent printings this material can be downloaded from http://extras.springer.com using the following password: 978-3-540-88427-9

ISBN 978-3-540-88427-9 e-ISBN 978-3-540-88428-6
DOI 10.1007/978-3-540-88428-6
Springer Heidelberg Dordrecht London New York

Library of Congress Control Number: 2010927411

Cover image: The picture shows a roman mask after restoration at the Roman-German Museum of Mainz. The mask originates from the roman aera and had fallen into many fragments which were found mixed together with fragments from other iron-containing items such as weapons. With the help of Mössbauer spectroscopy (^{57}Fe) the fragments belonging to the mask could be identified for successful restoration (P. Gütlich, G. Klingelhöfer, P. de Souza, unpublished).

Cover design: Kuenkel Lopka GmbH, Heidelberg, Germany

Printed on acid-free paper

Springer is part of Springer Science+Business Media (www.springer.com)

Preface

More than five decades have passed since the young German physicist Rudolf L. Mössbauer discovered the *recoilless nuclear resonance (absorption) fluorescence of γ-radiation*. The spectroscopic method based on this resonance effect – referred to as Mössbauer spectroscopy – has subsequently developed into a powerful tool in solid-state research. The users are chemists, physicists, biologists, geologists, and scientists from other disciplines, and the spectrum of problems amenable to this method has become extraordinarily broad. Up to now, more than 60,000 reports have appeared in the literature dealing with applications of the Mössbauer effect in the characterization of a vast variety of materials. Besides many workshops, seminars, and symposia, a biannual conference series called The International Conference on the Applications of the Mössbauer Effect (ICAME) started in 1960 (Urbana, USA) and regularly brings together scientists who are actively working on fundamental – as well as industrial – applications of the Mössbauer effect. Undoubtedly, Mössbauer spectroscopy has taken its place as an important analytical tool among other physical methods of solid-state research. By the same token, high-level education in solid-state physics, chemistry, and materials science in the broadest sense is strongly encouraged to dedicate sufficient space in the curriculum to this versatile method. The main objective of this book is to assist the fulfillment of this purpose.

Many monographs and review articles on the principles and applications of Mössbauer spectroscopy have appeared in the literature in the past. However, significant developments regarding instrumentation, methodology, and theory related to Mössbauer spectroscopy, have been communicated recently, which have widened the applicability and thus, merit in our opinion, the necessity of updating the introductory literature. We have tried to present a state-of-the-art book which concentrates on teaching the fundamentals, using theory as much as needed and as little as possible, and on practical applications. Some parts of the book are based on the first edition published in 1978 in the Springer series "Inorganic Chemistry Concepts" by P. Gütlich, R. Link, and A.X. Trautwein. Major updates have been included on practical aspects of measurements, on the computation of Mössbauer parameters with modern quantum chemical techniques (special chapter by guest-authors F. Neese and T. Petrenko), on treating magnetic relaxation phenomena (special chapter by guest-author S. Morup), selected applications in coordination chemistry, the use of synchrotron radiation to observe nuclear forward scattering (NFS) and inelastic scattering (NIS), and on the miniaturization of a Mössbauer spectrometer for mobile spectroscopy in space and on earth (by guest-authors G. Klingelhöfer and Iris Fleischer).

The first five chapters are directed to the reader who is not familiar with the technique and deal with the basic principles of the recoilless nuclear resonance and essential aspects concerning measurements and the hyperfine interactions between nuclear moments and electric and magnetic fields. Chapter 5 by guest-authors F. Neese and T. Petrenko focuses on the computation and interpretation of Mössbauer parameters such as isomer shift, electric quadrupole splitting, and magnetic dipole splitting using modern DFT methods. Chapter 6, written by guest-author S. Morup, describes how magnetic relaxation phenomena can influence the shape of (mainly ^{57}Fe) Mössbauer spectra. Chapter 7 presents an up-to-date summary of the work on all Mössbauer-active transition metal elements in accordance with the title of this book. This chapter will be particularly useful for those who are actively concerned with Mössbauer work on noniron transition elements. We are certainly aware of the large amount of excellent Mössbauer spectroscopy involving other Mössbauer isotopes, for example, ^{119}Sn, ^{121}Sb, and many of the rare earth elements, but the scope of this volume precludes such extensive coverage. We have, however, decided to describe and discuss some special applications of ^{57}Fe Mössbauer spectroscopy in Chap. 8. This is mainly based on work from our own laboratories and we include these to give the reader an impression of the kind of problems that can be examined by Mössbauer spectroscopy. In Chap. 8, we give examples from studies of spin crossover compounds, systems with biological relevance, and the application of a miniaturized Mössbauer spectrometer in NASA missions to the planet Mars as well as mobile Mössbauer spectroscopy on earth. Finally, Chap. 9 is devoted to the most recent developments in the use of synchrotron radiation for nuclear resonance scattering (NRS), both in forward scattering (NFS) for measuring hyperfine interactions and inelastic scattering (NIS) for recording the density of local vibrational states.

A CD-ROM is attached containing a teaching course of Mössbauer spectroscopy (ca. 300 ppt frames), a selection of examples of applications of Mössbauer spectroscopy in various fields (ca. 500 ppt frames), review articles on computation and interpretation of Mössbauer parameters using modern quantum-mechanical methods, list of properties of isotopes relevant to Mössbauer spectroscopy, appendices refering to book chapters, and the first edition of this book which appeared in 1978. In subsequent printruns files are available via springer.extra.com (see imprint page).

The authors wish to express their thanks to the *Deutsche Forschungsgemeinschaft*, the *Bundesministerium für Forschung und Technologie, the Max Planck-Gesellschaft* and the *Fonds der Chemischen Industrie* for continued financial support of their research work in the field of Mössbauer spectroscopy. We are very much indebted to Dr. M. Seredyuk, Dr. H. Paulsen, and Mrs. P. Lipp, for technical assistance, and to Professor Frank Berry, for critical reading of the manuscript, and to Dr. B.W. Fitzsimmons for assistance in the proof-reading.

Mainz, Mülheim, Lübeck, November 2010

Philipp Gütlich, Eckhard Bill,
Alfred X. Trautwein

Contents

Chapter 1
Introduction

Some 50 years ago, Rudolf L. Mössbauer, while working on his doctoral thesis under Professor Maier-Leibnitz at Heidelberg/Munich, discovered *the recoilless nuclear resonance absorption (fluorescence) of γ rays*, which subsequently became known as the *Mössbauer effect* [1–3]. Some three decades before this successful discovery, Kuhn had speculated on the possible observation of the nuclear resonance absorption of γ-rays [4] similar to the analogous optical resonance absorption which had been known since the middle of the nineteenth century. The reason that nuclear resonance absorption (or fluorescence) was so difficult to observe was clear. The relatively high nuclear transition energies on the order of 100 keV impart an enormous recoil effect on the emitting and absorbing nuclei, such recoil energies being up to five orders of magnitude larger than the γ-ray line width. As a consequence, the emission and absorption lines are shifted away from each other by a very large distance, that is, some 10^5 times the line width, and the resonance overlap between the emission and absorption line is therefore no longer possible. In optical resonance absorption, the electronic transition energies are much smaller, typically of only a few electron volts, and hence the resultant recoil energy loss is negligibly small such that the emission and absorption lines are hardly shifted and can readily overlap. Several research groups had tried to compensate for the nuclear recoil energy loss by making use of the Doppler effect. Moon mounted the radioactive source on a centrifuge and moved it with suitably high velocity towards the absorber [5]. Malmfors heated both the source and the absorber in order to broaden the line widths, leading to a higher degree of overlap of emission and absorption lines [6]. In both cases, a very small but measurable resonance effect was observed. Mössbauer's procedure, in order "to attack the problem of recoil-energy loss at its root in a manner which, in general, ensures the complete elimination of this energy loss," as he said [7], was fundamentally different from the methods described by Moon and Malmfors. The basic feature of his method was that the resonating nuclei in the source and absorber were bound in crystals. He employed radioactive sources which emitted 129 keV γ-quanta leading to the ground state of ^{191}Ir. His first experiment aimed at measuring the lifetime of the 129 keV state of ^{191}Ir using an experimental arrangement similar to that of Malmfors. However, instead of

P. Gütlich et al., *Mössbauer Spectroscopy and Transition Metal Chemistry*,
DOI 10.1007/978-3-540-88428-6_1, © Springer-Verlag Berlin Heidelberg 2011

increasing the temperature as in the Malmfors experiment, Mössbauer decided to decrease the temperature because he believed that chemical binding in the crystal could play a decisive role in absorbing the recoil effect, particularly at lower temperatures. The results of his experiments were most spectacular: the nuclear resonance effect increased tremendously on lowering the temperature [1–3]. Not only was the *recoilless nuclear resonance absorption (fluorescence)* experimentally established but also theoretically rationalized on quantum mechanical grounds. The key features in his interpretation are twofold: (1) part of the nuclear recoil energy is imparted onto the whole crystal instead of a freely emitting and absorbing atom; this part becomes negligibly small because of the huge mass of a crystal in comparison to a single atom; (2) the other part of the recoil energy is converted into vibrational energy. Due to the quantization of the lattice phonon system, there is a certain probability, which is high for hard materials such as metals and lower for soft materials such as chemical compounds, that lattice oscillators do not change in vibrational energy (*zero-phonon-processes*) on the emission and absorption of γ-rays. For this probability, known as the Lamb–Mössbauer factor, the emission and absorption of γ-rays takes place entirely radiationless. Rudolf Mössbauer received the Nobel Prize in Physics at the age of 32 for this brilliant achievement.

The nuclear resonance phenomenon rapidly developed into a new spectroscopic technique, called Mössbauer spectroscopy, of high sensitivity to energy changes on the order of 10^{-8} eV (ca. 10^{-4} cm^{-1}) and extreme sharpness of tuning (ca. 10^{-13}). In the early stages, Mössbauer spectroscopy was restricted to low-energy nuclear physics (e.g., determination of excited state lifetimes and nuclear magnetic moments). After Kistner and Sunyar's report on the observation of a "chemical shift" in the quadrupolar perturbed magnetic Mössbauer spectrum of α-Fe_2O_3 [8], it was immediately realized that the new spectroscopic method could be particularly useful in solid state research, solving problems in physics, chemistry, metallurgy, material- and geo-sciences, biology, archeology, to name a few disciplines. It now transpires that the largest portion of the more than 60,000 papers on Mössbauer spectroscopic studies published to date deal with various kinds of problems arising from, or directly related to, the electronic shell of Mössbauer active atoms in metals and nonconducting materials, for example, magnetism, electronic fluctuations, relaxation processes, electronic and molecular structure, and bond properties. Such properties are characteristic of different materials (compounds, metals, alloys, etc.) and are the basis for nondestructive chemical analysis. The method, therefore, serves as a kind of "fingerprint" technique.

Up to the present time, the Mössbauer effect has been observed with nearly 100 nuclear transitions in about 80 nuclides distributed over 43 elements (cf. Fig. 1.1). Of course, as with many other spectroscopic methods, not all of these transitions are suitable for actual studies, for reasons which we shall discuss later. Nearly 20 elements have proved to be suitable for practical applications. It is the purpose of the present book to deal only with Mössbauer active transition elements (Fe, Ni, Zn, Tc, Ru, Hf, Ta, W, (Re), Os, Ir, Pt, Au, Hg). A great deal of space will be devoted to the spectroscopy of ^{57}Fe, which is by far the most extensively used Mössbauer nuclide of all. We will not discuss the many thousands of reports on ^{57}Fe

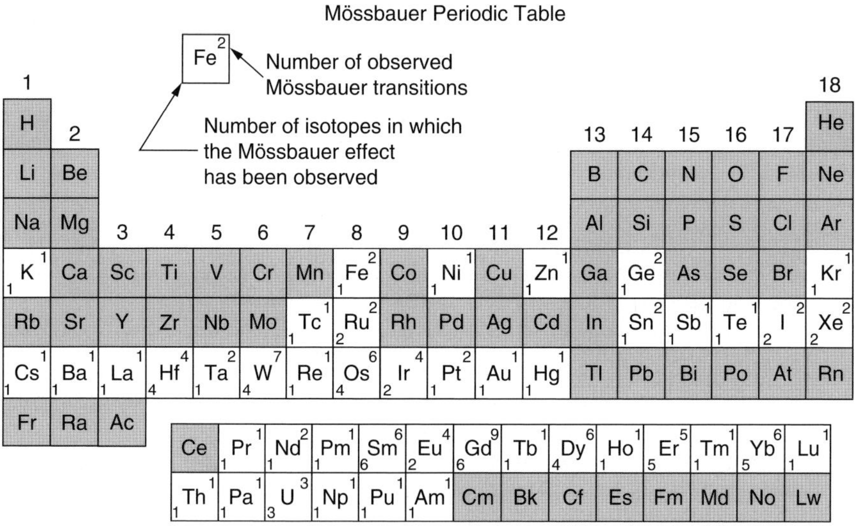

Fig. 1.1 Periodic table of the elements; those in which the Mössbauer effect has been observed are marked appropriately. (Taken from the 1974 issue of [10])

spectroscopy that have been published so far. Instead, we endeavor to introduce the reader to the various kinds of chemical information one can extract from the electric and magnetic hyperfine interactions reflected in the Mössbauer spectra. Particular emphasis will be put on the interpretation of bonding and structural properties in connection with electronic structure theories.

A CD-ROM (arranged in power-point format) is attached to the book. The first part of it contains lecture notes by one of the authors (P.G.) covering the fundamentals of Mössbauer spectroscopy, the hyperfine interactions and selected applications in various fields. This part of the CD (ca. 300 frames) is primarily arranged for teaching purposes. The second part of the CD (nearly 500 frames) contains examples of the applications of Mössbauer spectroscopy in physics, chemistry, biology, geoscience, archeology, and industrial applications. These examples are contributions from different laboratories and describe Mössbauer effect studies which, from our point of view, demonstrate the usefulness of the relatively new method. Those who are further interested in using Mössbauer spectroscopy in their research work may consult the many original reports as compiled in the Mössbauer Effect Data Index [9, 10], the Mössbauer Effect Reference and Data Journal [11], and relevant books [12–44].

References

1. Mössbauer, R.L.: Z. Physik. **151**, 124 (1958)
2. Mössbauer, R.L.: Naturwissenschaften **45**, 538 (1958)
3. Mössbauer, R.L.: Z. Naturforsch. **A 14**, 211 (1959)

4. Kuhn, W.: Philos. Mag. **8**, 625 (1929)
5. Moon, P.B.: Proc. Phys. Soc. (London) **64**, 76 (1951)
6. Malmfors, K.G.: Ark. Fys. **6**, 49 (1953)
7. Mössbauer, R.L.: Nobel Lecture, December 11, (1961)
8. Kistner, O.C., Sunyar, A.W.: Phys. Rev. Lett. **4**, 229 (1960)
9. Muir Jr., A.H., Ando, K.J., Coogan, H.M.: Mössbauer Effect Data Index (1958–1965). Interscience, New York (1966)
10. Stevens, J.G., Stevens, V.E.: Mössbauer Effect Data Index 1965–1975. Adam Hilger, London
11. Stevens, J.G., Khasanov, A.M., Hall, N.F., Khasanova, I.: Mössbauer Effect Reference and Data Journal. Mössbauer Effect Data Center, The University of North Carolina at Asheville, Asheville, NC (2009) (up to 2009)
12. Frauenfelder, H.: The Mössbauer Effect. Benjamin, New York (1962)
13. Wertheim, G.K.: Mössbauer Effect: Principles and Applications. Academic, New York (1964)
14. Wegener, H.: Der Mössbauer Effekt und seine Anwendung in Physik und Chemie. Bibliographisches Institut, Mannheim (1965)
15. Goldanskii, V.I., Herber, R. (eds.): Chemical Applications of Mössbauer Spectroscopy. Academic, New York (1968)
16. May, L. (ed.): An Introduction to Mössbauer Spectroscopy. Plenum, New York (1971)
17. Greenwood, N.N., Gibb, T.C.: Mössbauer Spectroscopy. Chapman and Hall, London (1971)
18. Bancroft, G.M.: Mössbauer Spectroscopy: An Introduction for Inorganic Chemists and Geochemists. McGraw-Hill, London, New York (1973)
19. Gonser, U. (ed.): Mössbauer Spectroscopy, in Topics in Applied Physics, vol. 5. Springer, Berlin (1975)
20. Gruverman, I.J. (ed.): Mössbauer Effect Methodology, vol. 1. Plenum, New York (1965). 1965 and annually afterwards
21. Gibb, T.C.: Principles of Mössbauer Spectroscopy. Wiley, New York (1976)
22. Cohen, R.L. (ed.): Applications of Mössbauer Spectroscopy, vol. 1. Academic, London (1976)
23. Shenoy, G.K., Wagner, F.E.: Mössbauer Isomer Shifts. North Holland, Amsterdam (1978)
24. Gütlich, P., Link, R., Trautwein, A.X.: Mössbauer Spectroscopy and Transition Metal Chemistry. Inorganic Chemistry Concepts Series, vol. 3, 1st edn. Springer, Berlin (1978)
25. Vertes, A., Korecz, L., Burger, K.: Mössbauer Spectroscopy. Elsevier, Amsterdam (1979)
26. Cohen, R.L.: Applications of Mössbauer Spectroscopy, vol. 2. Academic, New York (1980)
27. Barb, D.: Grundlagen und Anwendungen der Mössbauerspektroskopie. Akademie Verlag, Berlin (1980)
28. Gonser, U.: Mössbauer Spectroscopy II: The Exotic Side of the Effect. Springer, Berlin (1981)
29. Thosar, V.B., Iyengar, P.K., Srivastava, J.K., Bhargava, S.C.: Advances in Mössbauer Spectroscopy: Applications to Physics, Chemistry and Biology. Elsevier, Amsterdam (1983)
30. Long, G.J.: Mössbauer Spectroscopy Applied to Inorganic Chemistry, vol. 1. Plenum, New York (1984)
31. Herber, R.H.: Chemical Mössbauer Spectroscopy. Plenum, New York (1984)
32. Cranshaw, T.E., Dale, B.W., Longworth, G.O., Johnson, C.E. (eds.): Mössbauer Spectroscopy and its Applications. Cambridge University Press, Cambridge (1985)
33. Dickson, D.P.E., Berry, F.J. (eds.): Mössbauer Spectroscopy. Cambridge University Press, Cambridge (1986)
34. Long, G.J.: Mössbauer Spectroscopy Applied to Inorganic Chemistry. Modern Inorganic Chemistry Series, vol. 2. Plenum, New York (1989)
35. Long, G.J., Grandjean, F.: Mössbauer Spectroscopy Applied to Inorganic Chemistry. Modern Inorganic Chemistry Series, vol. 3. Plenum, New York (1989)
36. Vertes, A., Nagy, D.L.: Mössbauer Spectroscopy of Frozen Solutions. Akademiai Kiado, Budapest (1990)
37. Mitra, S.: Applied Mössbauer Spectroscopy: Theory and Practice for Geochemists and Archeologists. Elsevier, Amsterdam (1993)

38. Long, G.J., Grandjean, F. (eds.): Mössbauer Spectroscopy Applied to Magnetism and Materials Science, vol. 1. Plenum, New York (1993)
39. Belozerskii, G.N.: Mössbauer Studies of Surface Layers. Elsevier, Amsterdam (1993)
40. Long, G.J., Grandjean, F.: Mössbauer Spectroscopy Applied to Magnetism and Materials Science, vol. 2. Plenum, New York (1996)
41. Vertes, A., Hommonay, Z.: Mössbauer Spectroscopy of Sophisticated Oxides. Akademiai Kiado, Budapest (1997)
42. Stevens, J.G., Khasanov, A.M., Miller, J.W., Pollak, H., Li, Z.: Mössbauer Mineral Handbook. Mössbauer Effect Data Center, The University of North Carolina at Asheville, Asheville, NC (1998)
43. Ovchinnikov, V.V.: Mössbauer Analysis of the Atomic and Magnetic Structure of Alloys. Cambridge International Science, Cambridge (2004)
44. Murad, E., Cashion, J.: Mössbauer Spectroscopy of Environmental Materials and Their Industrial Utilization. Kluwer, Dordrecht (2004)

Chapter 2
Basic Physical Concepts

Mössbauer spectroscopy is based on recoilless emission and resonant absorption of γ-radiation by atomic nuclei. The aim of this chapter is to familiarize the reader with the concepts of nuclear γ-resonance and the Mössbauer effect, before we describe the experiments and relevant electric and magnetic hyperfine interactions in Chaps. 3 and 4. We prefer doing this by collecting formulae without deriving them; comprehensive and instructive descriptions have already been given at length in a number of introductory books ([7–39] in Chap. 1). Readers who are primarily interested in understanding their Mössbauer spectra without too much physical ballast may skip this chapter at first reading and proceed directly to Chap. 4. However, for the understanding of some aspects of line broadening and the preparation of optimized samples discussed in Chap. 3, the principles described here might be necessary.

2.1 Nuclear γ-Resonance

Most readers are familiar with the phenomenon of resonant absorption of electromagnetic radiation from the observation of light-induced electronic transitions. Visible light from a white incident beam is absorbed at exactly the energies of the splitting of *d*-electrons in transition metal ions or at the energies corresponding to metal-to-ligand charge transfer transitions in coordination compounds. These are the most common causes of color in inorganic complexes. Only when the quantum energy of the light matches the energy gap between the electronic states involved does such resonant absorption occur.

An analogous process is possible for γ-radiation, for which nuclear states are involved as emitters and absorbers. In such experiments, the emission of the γ-rays is mostly triggered by a preceding decay of a radioactive precursor of the resonance nuclei with Z protons and N neutrons (Fig. 2.1). The nuclear reaction (α-, or β-decay, or K-capture) yields the isotope (Z, N) in the excited state (e) with energy E_e. The excited nucleus has a limited mean lifetime τ and will undergo a transition to its ground state (g) of energy E_g, according to the exponential law of decay. This leads,

P. Gütlich et al., *Mössbauer Spectroscopy and Transition Metal Chemistry*,
DOI 10.1007/978-3-540-88428-6_2,

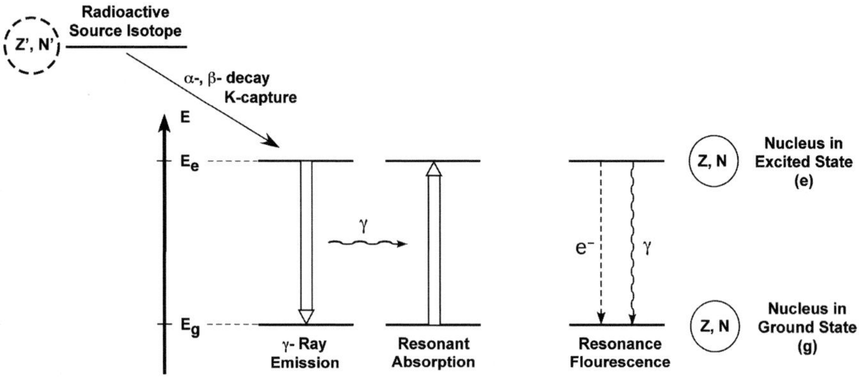

Fig. 2.1 Nuclear resonance absorption of γ-rays (Mössbauer effect) for nuclei with Z protons and N neutrons. The top left part shows the population of the excited state of the emitter by the radioactive decay of a mother isotope (Z', N') via α- or β-emission, or K-capture (depending on the isotope). The right part shows the de-excitation of the absorber by re-emission of a γ-photon or by radiationless emission of a conversion electron (thin arrows labeled "γ" and "e^-", respectively)

with a certain probability, to the emission of a γ-photon,[1] which has the quantum energy $E_0 = E_e – E_g$ if the process occurs without recoil. Under this, and certain other conditions which we shall discuss below, the γ-photon may be reabsorbed by a nucleus of the same kind in its ground state, whereby a transition to the excited state of energy E_e takes place. The phenomenon, called *nuclear resonance absorption of γ-rays*, or *Mössbauer effect*, is described schematically in Fig. 2.1.

Resonant γ-ray absorption is directly connected with *nuclear resonance fluorescence*. This is the re-emission of a (second) γ-ray from the excited state of the absorber nucleus after resonance absorption. The transition back to the ground state occurs with the same mean lifetime τ by the emission of a γ-ray in an arbitrary direction, or by energy transfer from the nucleus to the K-shell via internal conversion and the ejection of conversion electrons (see footnote 1). *Nuclear resonance fluorescence* was the basis for the experiments that finally led to R. L. Mössbauer's discovery of nuclear γ-resonance in ^{191}Ir ([1–3] in Chap. 1) and is the basis of Mössbauer experiments with synchrotron radiation which can be used instead of γ-radiation from classical sources (see Chap. 9).

In order to understand the Mössbauer effect and the importance of recoilless emission and absorption, one has to consider a few factors that are mainly related to the fact that the quantum energy of the γ-radiation used for Mössbauer spectroscopy ($E_0 \approx$ 10–100 keV) is much higher than the typical energies encountered, for instance, in optical spectroscopy (1–10 eV). Although the absolute widths of the

[1]Not all nuclear transitions of this kind produce a detectable γ-ray; for a certain portion, the energy is dissipated by internal conversion to an electron of the K-shell which is ejected as a so-called conversion electron. For some Mössbauer isotopes, the total internal conversion coefficient α_T is rather high, as for the 14.4 keV transition of ^{57}Fe ($\alpha_T = 8.17$). α_T is defined as the ratio of the number of conversion electrons to the number of γ-photons.

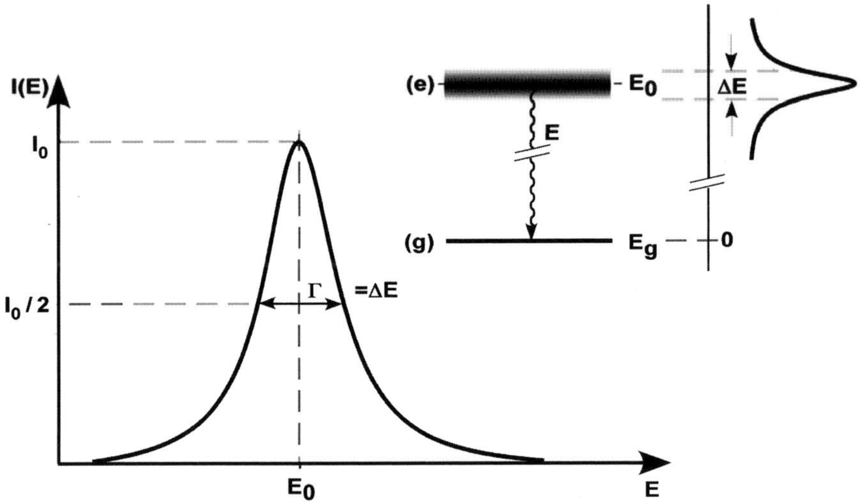

Fig. 2.2 Intensity distribution $I(E)$ for the emission of γ-rays with mean transition energy E_0. The Heisenberg natural line width of the distribution, $\Gamma = \hbar/\tau$, is determined by the mean lifetime τ of the excited state (e)

energy levels involved in both spectroscopies are rather similar ([15] in Chap. 1), the *relative* widths of the nuclear levels are very small because of the high mean energies ($\Delta E/E_0 \approx 10^{-13}$ or less, see Fig. 2.2). Consequently, the *recoil* connected with any emission or absorption of a photon is a particular problem for nuclear transitions in gases and liquids, because the energy loss for the γ-quanta is so large that emission and absorption lines do not overlap and nuclear γ-resonance is virtually impossible. Thermal motion and the resulting *Doppler* broadening of the γ-lines are another important aspect. R. L. Mössbauer showed that, for nuclei fixed in solid material, a substantial fraction *f* of photons, termed the Lamb–Mössbauer factor, are emitted and absorbed without measurable recoil. The corresponding γ-lines show natural line widths without thermal broadening.

2.2 Natural Line Width and Spectral Line Shape

The energy E_0 of a nuclear or electronic excited state of mean lifetime τ cannot be determined exactly because of the limited time interval Δt available for the measurement. Instead, E_0 can only be established with an inherent uncertainty, ΔE, which is given by the Heisenberg uncertainty relation in the form of the conjugate variables energy and time,

$$\Delta E\ \Delta t \geqslant \hbar, \tag{2.1}$$

where $h = 2\pi\hbar$ = Planck's constant.

The relevant time interval is on the order of the mean lifetime, $\Delta t \approx \tau$. Consequently, ground states of infinite lifetime have zero uncertainty in energy.

As a result, the energy E of photons emitted by an ensemble of identical nuclei, rigidly fixed in space, upon transition from their excited states (e) to their ground states (g), scatters around the mean energy $E_0 = E_e - E_g$. The intensity distribution of the radiation as a function of the energy E, the *emission line*, is a Lorentzian curve as given by the Breit–Wigner equation [1]:

$$I(E) = \frac{\Gamma/(2\pi)}{(E - E_0)^2 + (\Gamma/2)^2}. \tag{2.2}$$

The emission line is centered at the mean energy E_0 of the transition (Fig. 2.2). One can immediately see that $I(E) = 1/2\, I(E_0)$ for $E = E_0 \pm \Gamma/2$, which renders Γ the full width of the spectral line at half maximum. Γ is called the *natural width* of the nuclear excited state. The emission line is normalized so that the integral is one: $\int I(E)\mathrm{d}E = 1$. The probability distribution for the corresponding absorption process, the *absorption line,* has the same shape as the emission line for reasons of time-reversal invariance.

Weisskopf and Wigner [2] have shown that the natural width of the emission and the absorption line is readily determined by the mean lifetime τ of the excited state because of the relation (note the equal sign):

$$\Gamma\tau = \hbar. \tag{2.3}$$

The ratio Γ/E_0 of width Γ and the mean energy of the transition E_0 defines the precision necessary in nuclear γ-absorption for "tuning" emission and absorption into resonance. Lifetimes of excited nuclear states suitable for Mössbauer spectroscopy range from $\sim 10^{-6}$ s to $\sim 10^{-11}$ s. Lifetimes longer than 10^{-6} s produce too narrow emission and absorption lines such that in a Mössbauer experiment they cannot overlap sufficiently because of experimental difficulties (extremely small Doppler velocities of $<\mu\mathrm{m\ s}^{-1}$ are required). Lifetimes shorter than 10^{-11}s are connected with transition lines which are too broad such that the resonance overlap between them becomes smeared and no longer distinguishable from the base line of a spectrum. The first excited state of ^{57}Fe has a mean lifetime of $\tau = t_{1/2}/\ln 2 = 1.43 \cdot 10^{-7}$ s; and by substituting $\hbar = 6.5826 \cdot 10^{-16}$ eV s, the line width Γ evaluates to $4.55 \cdot 10^{-9}$ eV.

2.3 Recoil Energy Loss in Free Atoms and Thermal Broadening of Transition Lines

In the description of nuclear γ-resonance, we assume that the photon emitted by a nucleus of mean energy $E_0 = E_e - E_g$ carries the entire energy, $E_\gamma = E_0$. This is not true for nuclei located in free atoms or molecules, because the photon has

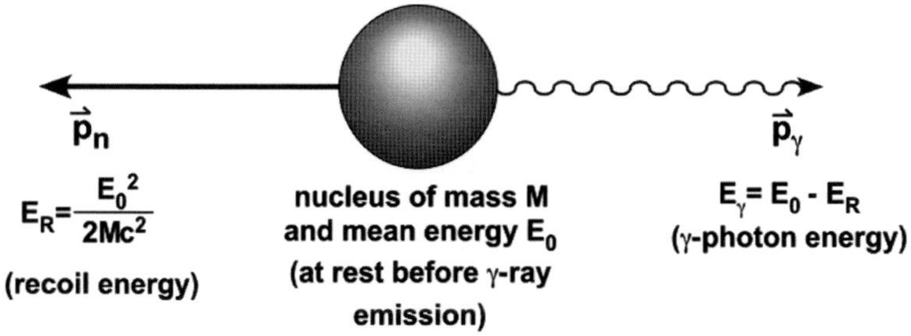

Fig. 2.3 Recoil momentum $\vec{p}_n$ and energy E_R imparted to a free nucleus upon γ-ray emission

momentum. When a photon is emitted from a nucleus of mass M, recoil is imparted to the nucleus and consequently the nucleus moves with velocity υ in a direction opposite to that of the γ-ray propagation vector (see Fig. 2.3).

Suppose the nucleus was at rest before the decay, it takes up the recoil energy

$$E_R = \frac{1}{2}M\upsilon^2. \tag{2.4}$$

Momentum conservation requires that

$$p_n = -p_\gamma, \tag{2.5}$$

where p_n ($= M\upsilon$) and p_γ are the linear momenta of the nucleus and the γ-photon, respectively, and c is the velocity of light. The momentum p_γ of the (mass-less) photon is given by its quantum energy:

$$p_\gamma = -E_\gamma/c, \quad \text{with} \tag{2.6}$$

$$E_\gamma = E_0 - E_R. \tag{2.7}$$

Because of the large mass of the nucleus and the low recoil velocity involved, we may use the nonrelativistic approximation

$$E_R = \frac{1}{2}M\upsilon^2 = \frac{(M\upsilon)^2}{2M} = \frac{p_n^2}{2M} = \frac{E_\gamma^2}{2Mc^2}. \tag{2.8}$$

Since E_R is very small compared to E_0, it is reasonable to assume that $E_\gamma \approx E_0$, so that we may use the following elementary formula for the recoil energy of a nucleus in an isolated atom or molecule:

$$E_R = \frac{E_0^2}{2Mc^2}. \tag{2.9}$$

By substituting numerical values for c and $M = m_{n/p} \cdot A$ one obtains

$$E_R = 5.37 \cdot 10^{-4} \frac{E_0^2}{A} \text{ eV}, \tag{2.10}$$

where $m_{n/p}$ is the mass of a nucleon (proton or neutron), A is the mass number of the Mössbauer isotope, and E_0 is the transition energy in keV. For example, for the Mössbauer transition between the first excited state and the ground state of ^{57}Fe ($E_0 = E_e – E_g = 14.4$ keV), E_R is found to be $1.95 \cdot 10^{-3}$ eV. This value is about six orders of magnitude larger than the natural width of the transition under consideration ($\Gamma = 4.55 \cdot 10^{-9}$ eV).

The recoil effect causes an energy shift of the emission line from E_0 to smaller energies by an amount E_R, whereby the γ-photon carries an energy of only $E_\gamma = E_0 - E_R$. However, a recoil effect also occurs in the absorption process so that the photon, in order to be absorbed by a nucleus, requires the total energy $E_\gamma = E_0 + E_R$ to make up for the transition from the ground to the excited state and the recoil effect (for which $\vec{p}_n$ and $\vec{p}_\gamma$ will have the same direction).

Hence, nuclear resonance absorption of γ-photons (the Mössbauer effect) is not possible between free atoms (at rest) because of the energy loss by recoil. The deficiency in γ-energy is two times the recoil energy, $2E_R$, which in the case of ^{57}Fe is about 10^6 times larger than the natural line width Γ of the nuclear levels involved (Fig. 2.4).

In real gases and liquids, however, atoms are never at rest. If γ-emission takes place while the nucleus (or atom) is moving at velocity υ_n in the direction of the γ-ray propagation, the γ-photon of energy E_γ is modulated by the Doppler energy E_D [3]:

$$E_D = \frac{\upsilon_n}{c} E_\gamma, \tag{2.11}$$

which adds to E_γ:

$$E_\gamma = E_0 - E_R + E_D. \tag{2.12}$$

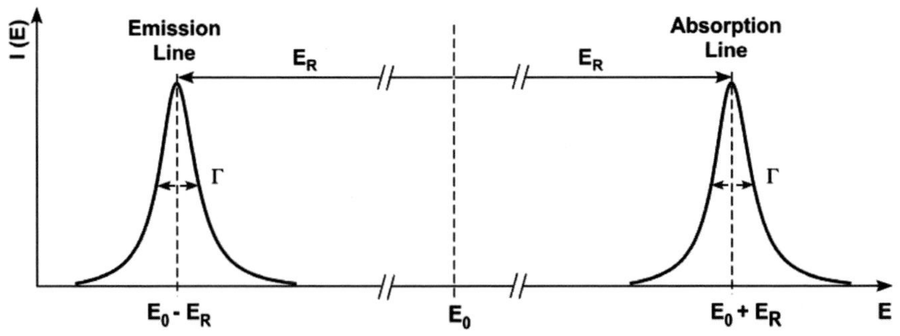

Fig. 2.4 Energy separation of γ-emission and absorption lines caused by recoil of resting free nuclei ($2E_R \approx 10^6\Gamma$, note the three separate sections of the energy scale). Since there is virtually no overlap between emission and absorption line, resonant absorption is not possible

Kinetic gas theory predicts a broad variation of velocities in gases, which for an ideal gas obeys the classical Maxwell distribution [3]. At normal temperature and pressure, the average time between collisions of the gas particles is so long (10^{-9}–10^{-10} s) that a typical Mössbauer isotope during its mean lifetime of about 10^{-9} s hardly experiences changes in motion. Relevant to the Doppler shift imparted to a certain γ-emission therefore is the component of the nuclear motion parallel or antiparallel to $\vec{p}_\gamma$. The large variety of possible results, with E_D possibly being positive or negative, leads to a wide statistical scattering of γ-energies, the so-called *Doppler broadening* of the transition line. The distribution has its maximum at $E_D = 0$, which is plausible since all motions with a net velocity vector close to $\vec{v}_n \perp \vec{p}_\gamma$ contribute virtually nothing to the energy of the photon, $E_D \approx 0$. For sufficiently high γ-energies (large recoil energies) and fast kinetic motion, the shape of the recoil-shifted and Doppler-broadened emission and absorption lines for free atoms and molecules can be approximated by a Gaussian distribution. A derivation of the subject is given in ([15] in Chap. 1). There it is also shown that the width of the Doppler broadening can be given in terms of the recoil shift E_R of the γ-energy and the average kinetic energy ½kT of the gas particles at temperature T:

$$\Gamma_D = 2\sqrt{E_R \cdot kT}. \tag{2.13}$$

The Doppler broadening Γ_D is of the order of E_R or larger. For ^{57}Fe with $E_0 = 14.4$ keV and $E_R = 1.95 \cdot 10^{-3}$ eV, for instance, it exceeds E_R by a factor of 5: $\Gamma_D \approx 10^{-2}$ eV at 300 K. Thus, there must be a finite probability for nuclei in gases and liquids to compensate for the recoil loss E_R of the photon by the Doppler shift E_D. The strength of the absorption and the shape of the absorption line are mathematically obtained from folding the overlapping emission and absorption lines. Since the line amplitudes are very small (the area of the broadened lines is the same as that of the sharp natural lines), the absorption probability is very small. Experimentally, it is difficult to detect nuclear γ-resonance in gases and liquids at all, except for very viscous fluids. For practical applications, it is more important that the usual Doppler modulation of the γ-radiation, as it is used as the "drive" system in classical Mössbauer spectroscopy (see Fig. 2.6), does not affect γ-absorption in gases and liquids. Motions with velocities of a few millimeters per second are negligible because of the extreme line broadening in nonsolid samples, which is more than 10^6 times the natural width Γ.

2.4 Recoil-Free Emission and Absorption

The arguments seen in section 2.3 suggest that resonant γ-absorption should decrease at very low temperatures because the Doppler broadening of the γ-lines decreases and may even drop below the value of the recoil energy. In his experiments with solid sources and absorbers, however, R.L. Mössbauer ([1] in Chap. 1) observed on the contrary a dramatic increase in resonant absorption when the temperature approached that of liquid nitrogen. The correct explanation of this effect is found in the quantized nature of vibrations in solids ([1–3] in Chap. 1) [4]. In the following, we shall briefly illustrate the corresponding principles by means of a simple model. More information on this topic is found in Chap. 9 on Mössbauer spectroscopy with synchrotron radiation and nuclear inelastic scattering.

In the solid state, the Mössbauer active nucleus is more or less rigidly bound to its environment and not able to recoil freely, but it can vibrate within the framework of the chemical bonds of the Mössbauer atom. The effective vibration frequencies are of the order of $1/\tau_{\text{vib}} \approx 10^{13}\text{s}^{-1}$ ([15] in Chap. 1). Since, under this condition, the mean displacement of the nucleus essentially averages to zero during the time of the nuclear transitions, $\tau \approx 10^{-7}$ s, there is, firstly, no Doppler broadening of the γ-energy and, secondly, the recoil momentum can only be taken up by the "crystallite" as a whole: $p = M_{\text{crystal}}\upsilon$. The induced velocity υ of the emitter in this case is vanishing because of the large mass of the system (even the finest "nano"-particles may contain 10^{14} atoms or molecules) and, hence, the corresponding recoil energy $E_R = 1/2M_{\text{crystal}}\upsilon^2$ of translational motion is negligible.

Instead, part of the energy E_0 of the nuclear transition can be transferred to the lattice vibrational system if the recoil excites a lattice vibration, a so-called phonon. Alternatively, a phonon can also be annihilated by the nuclear event. In either case, the corresponding energy deficit or excess of the emitted γ-quantum, E_{vib}, is again orders of magnitude larger than the natural line width Γ of the nuclear levels. Nuclear γ-resonance absorption is therefore not possible if phonon excitation or annihilation is involved. However, a quantum mechanical description of the nucleus and its vibrational environment includes a certain finite probability f for so-called zero-phonon processes. The factor f, also known as the Lamb–Mössbauer factor, denotes the fraction of γ-emissions or absorptions occurring without recoil, the *recoil-free fraction*. It is in fact equivalent to the Debye–Waller factor for Bragg X-ray scattering by periodic lattices.[2] Characteristic f-values are, e.g., 0.91 for the 14.4 keV transition of ^{57}Fe in metallic iron at room temperature, and 0.06 for the 129 keV transition of ^{191}Ir in metallic iridium.

Independent of specific theoretical models for the phonon spectrum of a solid matrix, the recoil-free fraction can be given in terms of the γ-energy E_γ and the mean *local* displacement of the nucleus from its equilibrium position ([2] in Chap. 1) [5]:

$$f = \exp\left[-\langle x^2\rangle E_\gamma^2/(\hbar c)^2\right], \tag{2.14}$$

[2]The Bragg scattering of X-rays by a periodic lattice in contrast to a Mössbauer transition is a collective event which is short in time as compared to the typical lattice vibration frequencies. Therefore, the mean-square displacement $\langle x^2\rangle$ in the Debye–Waller factor is obtained from the average over the ensemble, whereas $\langle x^2\rangle$ in the Lamb–Mössbauer factor describes a time average. The results are equivalent.

where $\langle x^2 \rangle$ is the expectation value of the squared vibrational amplitude in the direction of γ-propagation, known as the mean-square displacement. The f factor depends on the square of the γ-energy, similar to recoil in free atoms. This in fact limits the choice of isotopes for Mössbauer spectroscopy; nuclei with excited state energies beyond 0.2 MeV are found to be impracticable because of prohibitively small f factors for the transitions.

The recoil-free fraction is temperature-dependent, as one would expect from the introductory remarks. Higher temperatures yield larger mean-square displacements $\langle x^2 \rangle$, and according to (2.14), lower values for the f factor. A thorough description of the temperature dependence would require a detailed and comprehensive description of the phonon spectrum of the solid matrix, which is virtually unavailable for most Mössbauer samples. Since the nucleus is a local probe of the lattice vibrations, sophisticated approaches are often not necessary. For most practical cases, the simple *Debye model* for the phonon spectrum of solids yields reasonable results, although in general it is not adequate for chemical compounds and other complex solids. This model is based on the assumption of a continuous distribution of phonon frequencies ω ranging from zero to an upper limit ω_D, with the density of states being proportional to ω^2 [3]. The highest phonon energy $\hbar\omega_D$ at the Debye frequency limit ω_D depends on the elastic properties of the particular material under study, and often it is given in terms of the corresponding *Debye temperature* $\Theta_D = \hbar\omega_D/k$ representing a measure of the strength of the bonds between the Mössbauer atom and the lattice. The following expression is obtained for the temperature dependence of $f(T)$:

$$f(T) = \exp\left[\frac{-3E_\lambda^2}{k_B\Theta_D Mc^2}\left\{\frac{1}{4} + \left(\frac{T}{\Theta_D}\right)^2 \int_0^{\Theta/T} \frac{x}{e^x - 1}dx\right\}\right], \tag{2.15}$$

where $E_\gamma^2/2Mc^2$ is the free-atom recoil energy E_R, and k is the Boltzmann factor. Appropriate approximations of the integral yield a T^2 dependence for $f(T)$ in the low temperature limit ($T \ll \Theta_D$):

$$f(T) = \exp\left[-\frac{E_\gamma^2}{k_B\Theta_D 2Mc^2}\left(\frac{3}{2} + \frac{\pi^2 T^2}{\Theta_D^2}\right)\right], \tag{2.16}$$

whereas at high temperatures ($T > \Theta_D$), it approaches a linear regime

$$f(T) = \exp\left[-\frac{3E_\gamma^2 T}{k_B\Theta_D^2 Mc^2}\right]. \tag{2.17}$$

The Debye temperature Θ_D is usually high for metallic systems and low for metal–organic complexes. For metals with simple cubic lattices, for which the model was developed, Θ_D is found in the range from 300 K to well above 10^3 K. The other extreme may be found for iron in proteins, which may yield Θ_D as low as 100–200 K. Figure 2.5a demonstrates how sharply $f(T)$ drops with temperature for such systems. Since the intensity of a Mössbauer spectrum is proportional to the

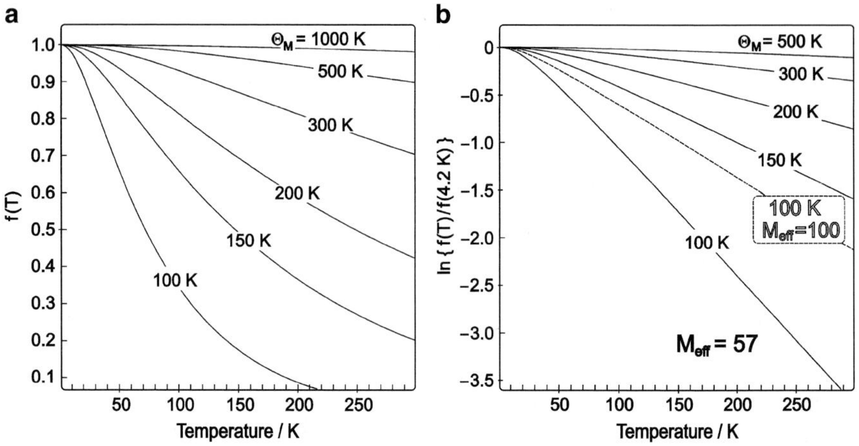

Fig. 2.5 (**a**) Temperature dependence of the recoil-free fraction $f(T)$ calculated on the grounds of the Debye model as a function of the Mössbauer temperature Θ_M, using $M_{eff} = 57$ Da. (**b**) Plot of $\ln(f(T)/f(4.2\text{ K}))$ for the same parameters, except for the dotted line ($M_{eff} = 100$ Da, $\Theta_M = 100$ K)

recoil-free fraction and since the measuring time necessary to obtain a good signal-to-noise ratio depends on the square of the intensity (see Sect. 3.2), the performance of a corresponding experiment can be significantly improved by cooling the sample to liquid nitrogen or liquid helium temperature.

The recoil-free fraction f is an important factor for determining the intensity of a Mössbauer spectrum. In summary, we notice from inspecting (2.14)–(2.17) and Fig. 2.5a that

1. f increases with decreasing transition energy E_γ;
2. f increases with decreasing temperature;
3. f increases with increasing Debye temperature Θ_D.

For a more detailed account of the recoil-free fraction and lattice dynamics, the reader is referred to relevant textbooks ([12–15] in Chap. 1).

The Lamb–Mössbauer factor as spectroscopic parameter

For practical applications of (2.15), i.e., the exploration of the elastic properties of materials, it is convenient to eliminate instrumental parameters by referring the measured intensities to a base intensity, usually measured at 4.2 K. The corresponding expression is then [6]:

$$\ln\left(\frac{f(T)}{f(4.2\,\mathrm{K})}\right) = \frac{-3E_\gamma^2}{k_B \Theta_M^3 M_{eff} c^2}\left(T^2\int_0^{\Theta_M/T}\frac{x}{e^x-1}\mathrm{d}x - 29K^2\right). \tag{2.18}$$

The logarithm expression can be obtained from measured Mössbauer intensities, or areas $A(T)$ of the spectra, $\ln(f(T)/f(4.2\,\mathrm{K})) = \ln(A(T)/A(4.2\text{ K}))$. The mass M of the nucleus is replaced in (2.18) by a free adjustable *effective mass*, M_{eff}, to take into account also collective motions of the Mössbauer atom together with its ligands. These may vibrate together as large entities, particularly in "soft" matter like proteins or inorganic compounds with large organic ligands. Similarly, the Debye temperature is replaced by the *Mössbauer temperature* Θ_M, which is also specific to the local environment sensed by the Mössbauer nucleus. These modifications are consequences of the limitations of the Debye model. The parameters Θ_M and M_{eff} are not to be considered as universal quantities; they rather represent effective local variables that are specific to the detection method. Nevertheless, their values can be very helpful for the characterization of Mössbauer samples, particularly when series of related compounds are considered. A series of simulated traces for $\ln(f(T)/f(4.2\,K))$, calculated within the Debye model as a function of different Mössbauer temperatures Θ_M and for two effective masses M_{eff}, is plotted in Fig. 2.5b.

The recoil-free fraction depends on the oxidation state, the spin state, and the elastic bonds of the Mössbauer atom. Therefore, a temperature-dependent transition of the valence state, a spin transition, or a phase change of a particular compound or material may be easily detected as a change in the slope, a kink, or a step in the temperature dependence of $\ln f(T)$. However, in fits of experimental Mössbauer intensities, the values of Θ_M and M_{eff} are often strongly covariant, as one may expect from a comparison of the traces shown in Fig. 2.5b. In this situation, valuable constraints can be obtained from corresponding fits of the temperature dependence of the *second-order-Doppler shift* of the Mössbauer spectra, which can be described by using a similar approach. The formalism is given in Sect. 4.2.3 on the temperature dependence of the isomer shift.

2.5 The Mössbauer Experiment

From Sect. 2.3 we learnt that the recoil effect in free or loosely bound atoms shifts the γ-transition line by E_R, and thermal motion broadens the transition line by Γ_D, the Doppler broadening (2.13). For nuclear resonance absorption of γ-rays to be successful, emission and absorption lines must be shifted toward each other to achieve at least partial overlap. Initially, compensation for the recoil-energy loss was attempted by using the Doppler effect. Moon, in 1950, succeeded by mounting the source to an ultracentrifuge and moving it with high velocities toward the absorber [7]. Subsequent experiments were also successful showing that the Doppler effect compensated for the energy loss of γ-quanta emitted with recoil.

The real breakthrough in nuclear resonance absorption of γ-rays, however, came with Mössbauer's discovery of recoilless emission and absorption. By means of an experimental arrangement similar to the one described by Malmfors [8], he intended to measure the lifetime of the 129 keV state in ^{191}Ir. Nuclear resonance absorption was planned to be achieved by increasing overlap of emission and absorption lines via heating and increased thermal broadening. By lowering the temperature, it was expected that the transition lines would sharpen because of less effective Doppler broadening and consequently show a smaller degree of overlap.

However, in contrast, the resonance effect increased by cooling both the source and the absorber. Mössbauer not only observed this striking experimental effect that was not consistent with the prediction, but also presented an explanation that is based on zero-phonon processes associated with emission and absorption of γ-rays in solids. Such events occur with a certain probability f, the recoil-free fraction of the nuclear transition (Sect. 2.4). Thus, the factor f is a measure of the recoilless nuclear absorption of γ-radiation – the Mössbauer effect.

In an actual Mössbauer transmission experiment, the radioactive source is periodically moved with controlled velocities, $+\upsilon$ toward and $-\upsilon$ away from the absorber (cf. Fig. 2.6). The motion modulates the energy of the γ-photons arriving at the absorber because of the Doppler effect: $E_\gamma = E_0(1 + \upsilon/c)$. Alternatively, the sample may be moved with the source remaining fixed. The transmitted γ-rays are detected with a γ-counter and recorded as a function of the Doppler velocity, which yields the Mössbauer spectrum, $T(\upsilon)$. The amount of resonant nuclear γ-absorption is determined by the overlap of the shifted emission line and the absorption line, such that greater overlap yields less transmission; maximum resonance occurs at complete overlap of emission and absorption lines.

2.6 The Mössbauer Transmission Spectrum

In the following, we consider the shape and the width of the Mössbauer velocity spectrum in more detail. We assume that the source is moving with velocity υ, and the *emission* line is an unsplit Lorentzian according to (2.2) with natural width Γ. If we denote the total number of γ-quanta emitted by the source per time unit toward the detector by N_0, the number $N(E)\mathrm{d}E$ of recoil-free emitted γ-rays with energy E_γ in the range E to $E + \mathrm{d}E$ is given by ([1] in Chap. 1)

$$N(E,\upsilon) = f_s N_0 \frac{\Gamma/(2\pi)}{[E - E_0(1 + \upsilon/c)]^2 + (\Gamma/2)^2}, \tag{2.19}$$

where f_S is the recoil-free fraction of the source, E_0 is the mean energy of the nuclear transition, and $E_0(1 + \upsilon/c)$ is the Doppler-shifted center of the emission line.

In a Mössbauer transmission experiment, the absorber containing the stable Mössbauer isotope is placed between the source and the detector (cf. Fig. 2.6). For the absorber, we assume the same mean energy E_0 between nuclear excited and ground states as for the source, but with an additional intrinsic shift ΔE due to chemical influence. The *absorption* line, or resonant absorption cross-section $\sigma(E)$, has the same Lorentzian shape as the emission line; and if we assume also the same half width Γ, $\sigma(E)$ can be expressed as ([1] in Chap. 1)

$$\sigma(E) = \sigma_0 \frac{(\Gamma/2)^2}{(E - E_0 - \Delta E)^2 + (\Gamma/2)^2}. \tag{2.20}$$

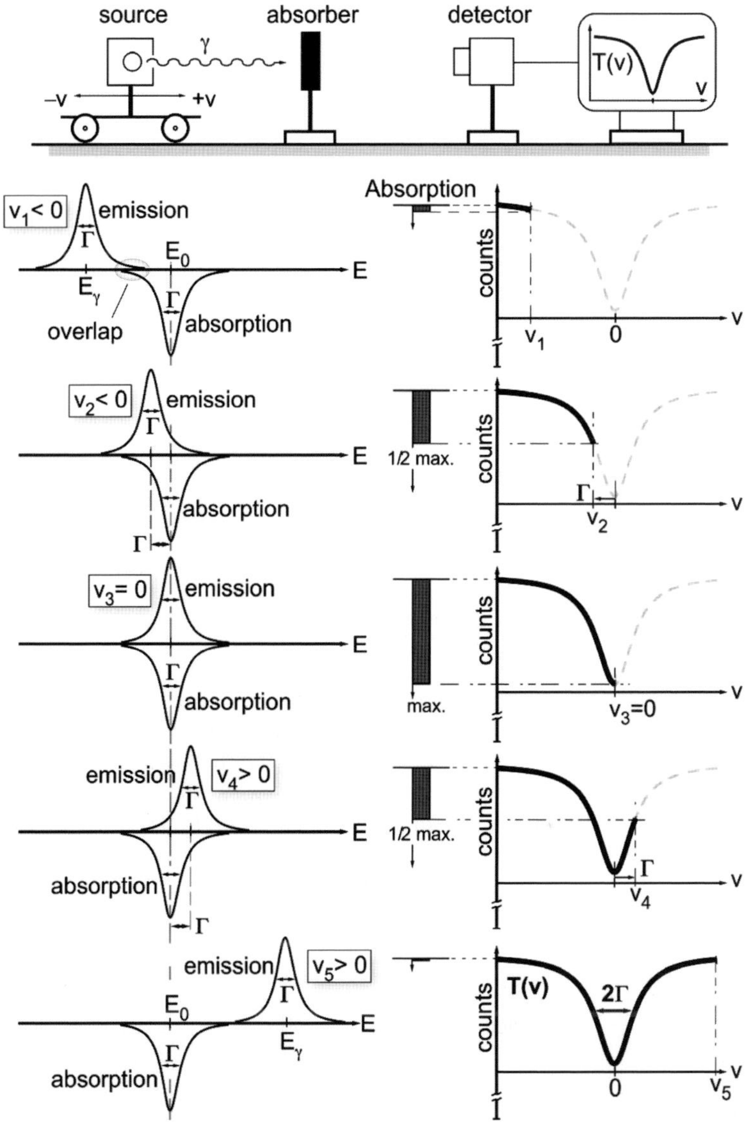

Fig. 2.6 Schematic illustration of a Mössbauer transmission experiment in five steps. The "Absorption" bars indicate the strength of recoilless nuclear resonant absorption as determined by the "overlap" of emission and absorption lines when the emission line is shifted by Doppler modulation (velocities $\upsilon_1, \ldots, \upsilon_5$). The transmission spectrum $T(\upsilon)$ is usually normalized to the transmission $T(\infty)$ observed for $\upsilon \rightarrow \infty$ by dividing $T(\upsilon)/T(\infty)$. Experimental details are found in Chap. 3

The absorption line is normalized to the maximum cross section[3] σ_0 at resonance, $E = E_0 + \Delta E$, which depends on the γ-energy E_γ, the spins I_e and I_g of the

[3]The maximum resonant cross section in cm^2 is: $\sigma_0(\mathrm{cm}^2) = \frac{2.446 \times 10^{-15}}{(E_\gamma\,\mathrm{keV})^2} \frac{2I_e+1}{2I_g+1} \frac{1}{1+\alpha}$.

nuclear excited and ground states, and the internal conversion coefficient α for the nuclear transition (footnote 1). In the case of ^{57}Fe, for example, the maximum cross section is $\sigma_0 = 2.56 \cdot 10^{-18}\ \mathrm{cm}^2$.

We are interested in the transmission of γ-quanta through the absorber as a function of the Doppler velocity. The radiation is attenuated by resonant absorption, in as much as emission and absorption lines are overlapping, but also by mass absorption due to photo effect and Compton scattering. Therefore, the number $T_\mathrm{M}(E)\mathrm{d}E$ of recoilless γ-quanta with energies E to $E + \mathrm{d}E$ traversing the absorber is given by

$$T_\mathrm{M}(E,\upsilon) = N(E,\upsilon)\exp\{-[\sigma(E)f_\mathrm{abs}n_\mathrm{M} + \mu_\mathrm{e}]t'\}, \tag{2.21}$$

where t' is the absorber thickness (area density) in gcm^{-2}, f_abs is the probability of recoilless absorption, n_M is the number of Mössbauer nuclei per gram of the absorber, and μ_e is the mass absorption coefficient[4] in $\mathrm{cm}^2\mathrm{g}^{-1}$.

The total number of recoil-free photons arriving at the detector per time unit is then obtained by integration over energy

$$T_\mathrm{M}(\upsilon) = \int_{-\infty}^{+\infty} N(E,\upsilon)\exp\{-[\sigma(E)f_\mathrm{abs}n_\mathrm{M} + \mu_\mathrm{e}]t'\}\mathrm{d}E. \tag{2.22}$$

So far we have considered only the recoil-free fraction of photons emitted by the source. The other fraction $(1 - f_\mathrm{S})$, emitted with energy loss due to recoil, cannot be resonantly absorbed and contributes only as a nonresonant background to the transmitted radiation, which is attenuated by mass absorption in the absorber

$$T_\mathrm{NR} = (1 - f_\mathrm{S})N_0\ \exp\{-\mu_\mathrm{e}t'\}, \tag{2.23}$$

so that the total count rate arriving at the detector is[5]

$$C(\upsilon) = T_\mathrm{NR} + T_\mathrm{M}(\upsilon) \quad \text{or} \tag{2.24}$$

$$C(\upsilon) = N_0\mathrm{e}^{-\mu_\mathrm{e}t'}\left[(1 - f_\mathrm{S}) + \int_{-\infty}^{+\infty} N(E,\upsilon)\cdot\exp\{-f_\mathrm{abs}n_\mathrm{M}\sigma(E)t'\}\mathrm{d}E\right]. \tag{2.25}$$

[4]Mass absorption is taken as independent of the velocity υ, because the Doppler shift is only about 10^{-11} times the γ-energy, or less.

[5]This expression holds only for an ideal detection system, which records only Mössbauer radiation. Practical problems with additional nonresonant background contributions from γ-ray scattering and X-ray fluorescence are treated in detail in Sects. 3.1 and 3.2.

The expression is known as the *transmission integral* in the actual formulation, which is valid for ideal thin sources without self-absorption and homogeneous absorbers assuming equal widths Γ for source and absorber [9]. The transmission integral describes the experimental Mössbauer spectrum as a convolution of the source emission line $N(E,\upsilon)$ and the absorber response $\exp\{-\sigma(E) f_{\mathrm{abs}} n_{\mathrm{M}} t'\}$. The substitution of $N(E,\upsilon)$ and $\sigma(E)$ from (2.19) and (2.20) yields in detail:

$$C(\upsilon) = N_0 \mathrm{e}^{-\mu_e t'} \Bigg[(1 - f_{\mathrm{S}}) + \int_{-\infty}^{+\infty} \frac{f_{\mathrm{S}} \Gamma/(2\pi)}{[E - E_0(1+\upsilon/c)]^2 + (\Gamma/2)^2} \cdot \exp\left\{ \frac{-t(\Gamma/2)^2}{(E - E_0 - \Delta E)^2 + (\Gamma/2)^2} \right\} \mathrm{d}E \Bigg], \tag{2.26}$$

by which we have introduced the *effective absorber thickness t*. The dimensionless variable summarizes the absorber properties relevant to resonance absorption as

$$t = f_{\mathrm{abs}} n_{\mathrm{M}} t' \sigma_0 \quad \text{or} \quad t = f_{\mathrm{abs}} N_{\mathrm{M}} \sigma_0, \tag{2.27}$$

where $N_{\mathrm{M}} = n_{\mathrm{M}} t'$ is the number of Mössbauer nuclei per area unit of the absorber (in cm^{-2}). Note that the expressions hold only for *single lines*, which are actually rare in practical Mössbauer spectroscopy. For spectra with split lines see footnote.[6]

2.6.1 *The Line Shape for Thin Absorbers*

For thin absorbers with $t \ll 1$, the exponential function in the transmission integral can be developed in a series, the first two terms of which can be solved yielding the following expression for the count rate in the detector:

$$C(\upsilon) = N_0 \mathrm{e}^{-\mu_e t'} \left(1 - f_{\mathrm{S}} \frac{t}{2} \frac{\Gamma^2}{[E_0(\upsilon/c) - \Delta E]^2 + \Gamma^2} \right). \tag{2.28}$$

[6] Most Mössbauer spectra are split because of the hyperfine interaction of the absorber (or source) nuclei with their electron shell and chemical environment which lifts the degeneracy of the nuclear states. If the hyperfine interaction is static with respect to the nuclear lifetime, the Mössbauer spectrum is a superposition of separate lines (i), according to the number of possible transitions. Each line has its own effective thickness $t(i)$, which is a fraction of the total thickness, determined by the relative intensity w_i of the lines, such that $t(i) = w_i t$.

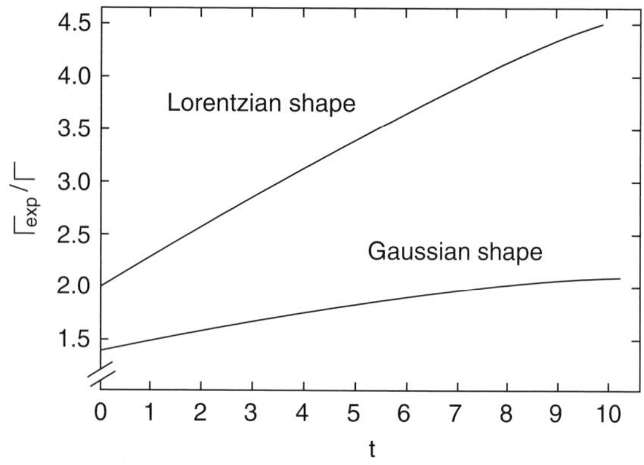

Fig. 2.7 Dependence of the experimental line width Γ_{exp} on the effective absorber thickness t for Lorentzian lines and inhomogenously broadened lines with quasi-Gaussian shape (from [9])

Since the resonance absorption vanishes for $\upsilon \rightarrow \infty$, the count rate off resonance is given entirely by mass absorption

$$C(\infty) = N_0 \mathrm{e}^{-\mu_{\mathrm{e}} t'}, \tag{2.29}$$

so that the absorption spectrum for a thin absorber is

$$\frac{C(\infty) - C(\upsilon)}{C(\infty)} = f_{\mathrm{s}} \frac{t}{2} \frac{\Gamma^2}{\left[E_0(\upsilon/c) - \Delta E\right]^2 + \Gamma^2}. \tag{2.30}$$

Thus, the experimental Mössbauer spectrum of a thin single-line absorber is a Lorentzian line, with full-width at half maximum twice the natural line width of the separate emission and absorption lines:[7] $\Gamma_{\mathrm{exp}} = 2\Gamma$.

Broadening of Mössbauer lines due to saturation.

Although Lorentzian line shapes should be strictly expected only for Mössbauer spectra of thin absorbers with effective thickness t small compared to unity, Margulies and Ehrman have shown [9] that the approximation holds reasonably well for moderately thick absorbers also, albeit the line widths are increased, depending on the value of t (Fig. 2.7). The line broadening is approximately

$$\begin{aligned} &\Gamma_{\mathrm{exp}}/2 = (1 + 0.135t)\Gamma_{\mathrm{nat}} && \text{for } t \le 4 \quad \text{and} \\ &\Gamma_{\mathrm{exp}}/2 = (1.01 + 0.145t - 0.0025t^2)\Gamma_{\mathrm{nat}} && \text{for } t > 4. \end{aligned} \tag{2.31}$$

Exact analyses of experimental spectra from thick absorbers, however, have to be based on the transmission integral (2.26). This can be numerically evaluated following the procedures described by Cranshaw [10], Shenoy et al. [11] and others.

In many cases, the actual width of a Mössbauer line has strong contributions from *inhomogeneous* broadening due to the distribution of unresolved hyperfine splitting in the source or absorber. Often a Gaussian distribution of Lorentzians,

[7] The experimental line width is 2Γ because an emission line of the same width scans the absorption line; see Fig. 2.6.

which approaches a true Gaussian envelope when the width of the distribution substantially exceeds the natural line width and thickness broadening, can give the shape of such Mössbauer lines. The problem has been discussed in detail by Rancourt and Ping [12]. They suggest a fit procedure, which is based on *Voigt* profiles. The depth of such inhomogeneously broadened lines with quasi-Gaussian shapes depends to some extent also on the effective thickness of the absorber; an example is shown in Fig. 2.7. Additional remarks on line broadening due to diffusion processes and line narrowing in coincidence measurements are found in the first edition of this book, Sect. 3.5, p. 38 (cf. CD-ROM).

2.6.2 Saturation for Thick Absorbers

The broadening of the experimental Mössbauer line for thick, saturating absorbers can be regarded as a consequence of line shape distortions: saturation is stronger for the center of the experimental line, for which the emission and absorption lines are completely overlapping, than for the wings, where the absorption probability is less. Thus, the detected spectral line appears to be gradually more compressed from base to tip. Consequently, there must also be a nonlinear dependence of the maximum absorption on the effective thickness t. An inspection of the transmission integral, (2.26), yields for the maximum *fractional absorption* of the resonant radiation the expression [13]

$$\varepsilon(t) = [1 - e^{-t/2} J_0(i\tfrac{t}{2})], \tag{2.32}$$

such that the depth of the Mössbauer line is $f_s\varepsilon(t)$ (cf. Fig. 2.8b). Here $J_0(it/2)$ is the zeroth-order Bessel function and f_s is the recoil-free fraction of the source. From the

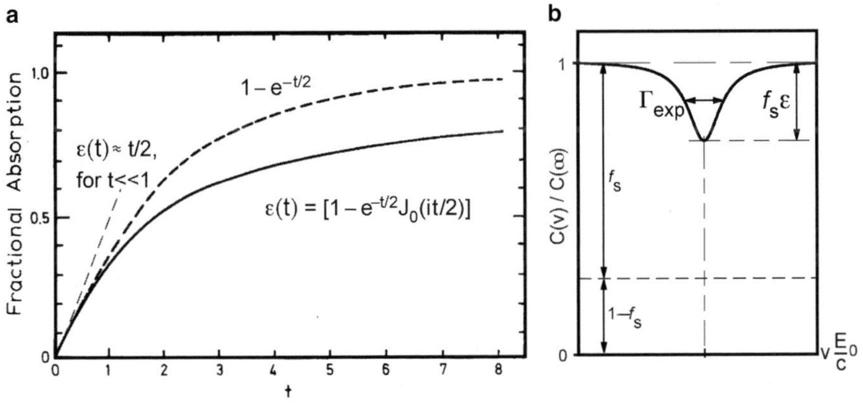

Fig. 2.8 (**a**) Fractional absorption of a Mössbauer absorption line as function of the effective absorber thickness t. (**b**) The depth of the spectrum is determined by $f_s\varepsilon$. The width Γ_{exp} for thin absorbers, $t \ll 1$, is twice the natural line width Γ of the separate emission and absorption lines (see (2.30)). ΔE is the shift of the absorption line relative to the emission line due to chemical influence

plot of $\varepsilon(t)$ shown in Fig. 2.8a, it is clear that the amplitude of the Mössbauer line is proportional to the effective absorber thickness only when $t \ll 1$, where an expansion of the fractional absorption yields $\varepsilon(t) \approx t/2$.

References

1. Breit, G., Wigner, E.: Phys. Rev. **49**, 519 (1936)
2. Weisskopf, V., Wigner, E.: Z. Physik **63**, 54 (1930); **65**, 18 (1930)
3. Atkins, P., De Paula, J.: Physical Chemistry. Oxford University Press, Oxford (2006)
4. Visscher, W.M.: Ann. Phys. **9**, 194 (1960)
5. Lamb Jr., W.E.: Phys. Rev. **55**, 190 (1939)
6. Shenoy, G.K., Wagner, F.E., Kalvius, G.M.: Mössbauer Isomer Shifts. North Holland, Amsterdam (1978)
7. Moon, P.B.: Proc. Phys. Soc. **63**, 1189 (1950)
8. Malmfors, K.G.: Arkiv Fysik **6**, 49 (1953)
9. Margulies, E., Ehrmann, J.R.: Nucl. Instr. Meth. **12**, 131 (1961)
10. Cranshaw, T.E.: J. Phys. **E 7**, 122; **7**, 497 (1974)
11. Shenoy, G. K., Friedt, J. M., Maletta, H., Ruby, S. L.: Mössbauer Effect Methodology, vol. 9 (1974)
12. Rancourt, D.G., Ping, J.Y.: Nucl. Instr. Meth. **B 58**, 85 (1991)
13. Mössbauer, R.L., Wiedemann, E.: Z. Physik **159**, 33 (1960)

Chapter 3
Experimental

In this chapter, we present the principles of conventional Mössbauer spectrometers with radioactive isotopes as the light source; Mössbauer experiments with synchrotron radiation are discussed in Chap. 9 including technical principles. Since complete spectrometers, suitable for virtually all the common isotopes, have been commercially available for many years, we refrain from presenting technical details like electronic circuits. We are concerned here with the functional components of a spectrometer, their interaction and synchronization, the different operation modes and proper tuning of the instrument. We discuss the properties of radioactive γ-sources to understand the requirements of an efficient γ-counting system, and finally we deal with sample preparation and the optimization of Mössbauer absorbers. For further reading on spectrometers and their technical details, we refer to the review articles [1–3].

3.1 The Mössbauer Spectrometer

Mössbauer spectra are usually recorded in transmission geometry, whereby the sample, representing the absorber, contains the stable Mössbauer isotope, i.e., it is not radioactive. A scheme of a typical spectrometer setup is depicted in Fig. 3.1. The radioactive Mössbauer *source* is attached to the electro-mechanical velocity transducer, or *Mössbauer drive*, which is moved in a controlled manner for the modulation of the emitted γ-radiation by the Doppler effect. The Mössbauer drive is powered by the electronic *drive control unit* according to a reference voltage (V_R), provided by the digital *function generator*. Most Mössbauer spectrometers are operated in constant-acceleration mode, in which the drive velocity is linearly swept up and down, either in a saw-tooth or in a triangular mode.[1] In either case,

[1]Most Mössbauer spectrometers use triangular velocity profiles. *Saw-tooth* motion induces excessive ringing of the drive, caused by extreme acceleration during fast fly-back of the drive rod. *Sinusoidal* operation at the *eigen* frequency of the vibrating system is also found occasionally and

DOI 10.1007/978-3-540-88428-6_3,

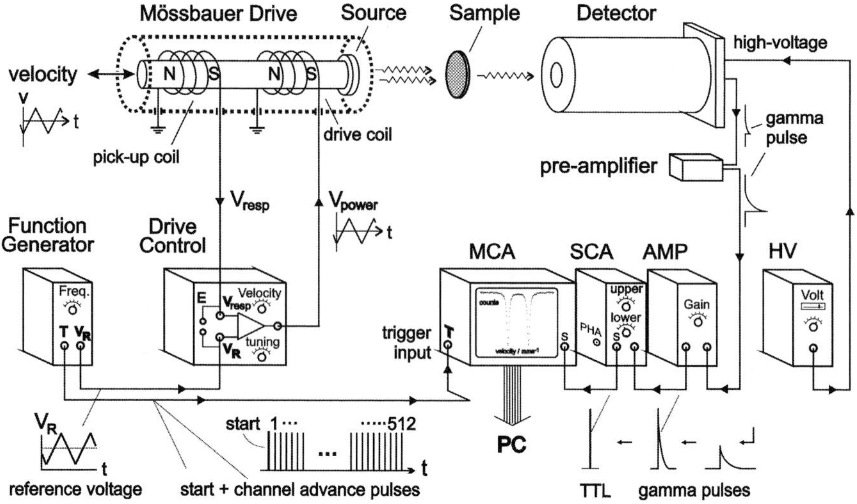

Fig. 3.1 Diagrammatic representation of a Mössbauer spectrometer for transmission measurements. MCA, multi-channel analyzer; SCA, single-channel analyzer; AMP, pulse amplifier; HV, high-voltage supply; VR, reference voltage provided by the function generator for the drive control unit; V_{resp}, output response voltage from the drive (proportional to the actual velocity); E, error signal (output on the drive control unit to monitor the difference of the reference voltage VR and the actual voltage V_{resp}); "start" and "channel advance" pulses, trigger pulses provided by the function generator to the MCA to initiate and control data acquisition (trigger signal "T"); TTL, shaped needle pulses suitable for digital networks (transistor–transistor logic), provided by the SCA upon receiving "gamma pulses" from the detector/amplifier chain; PC indicates the connection to a personal computer for data visualization and final data transfer. (The unit and signal labels are generic and may be different in any spectrometer)

the source moves periodically back and forth. The emitted γ-photons arriving at the *γ-detector* are converted into electric pulses which are then amplified, shaped and selected by a chain of electronic components, comprising a *preamplifier*, a main *amplifier* and a *single-channel analyzer* (SCA). The SCA is a pulse discrimination device which can be tuned to the Mössbauer radiation such that pulses of resonance energy are passed on to the *multi-channel analyzer* (MCA) for acquisition, and pulses from unwanted nonresonant background radiation are rejected. The MCA is the central device of the Mössbauer spectrometer for data acquisition and storage. It essentially consists of an array of digital counters with input logics for discrete electric pulses. In the following text, we shall describe the functional principles of the spectrometer components in detail.

---

requires the least effort for accurate control. However, it has the disadvantage that the source lingers a relatively longer time around zero velocity and passes fast through the high-velocity regimes yielding highly nonlinear baselines. For special applications, it may be very useful to choose the constant velocity mode (switching only between $+\upsilon$ and $-\upsilon$), or to scan a limited *region-of-interest* (focused sweeps through short linear regimes, $+\upsilon_{\max}$ to $+\upsilon_{\min}$, and $-v_{\max}$ to $-\upsilon_{\min}$).

3.1.1 The Mössbauer Drive System

The motion of a Mössbauer source must be controlled with an accuracy of 10^{-2} mm s^{-1} or better to resolve sharp Mössbauer spectral lines, such as those of ^{57}Fe which have full width $2\Gamma_{nat} \approx 0.2$ mm s^{-1}. Other isotopes are less demanding, e.g., ^{197}Au, for which the lines are ten times wider. Most spectrometers are equipped with electromechanical Mössbauer velocity transducers of the "loudspeaker" type. This technique is suitable for velocity variations ranging from less than 1 mm s^{-1} full scale up to several cm s^{-1} and covers the whole reach of hyperfine splitting for most of the common isotopes. Kalvius, Kankeleit, Cranshaw, and others [1–5] have been pioneers in the field, who laid foundations for the development of high-precision drives with feedback amplifiers for proper linear velocity scales with high stability and low hum. Other techniques for Doppler modulation have been developed for isotopes with extremely narrow hyperfine lines, e.g., ^{67}Zn. For such isotopes, piezo-electric transducers are mostly used [6, 7], more details of which are found in Sect. 7.2.1.

3.1.1.1 Setup and Function

All reference and trigger signals necessary for the basic function of a Mössbauer spectrometer are provided by the digital *function generator*. The unit provides (a) an oscillating reference voltage to the *drive control unit* (output V_R), and (b) a series of trigger pulses to the MCA for the timing and synchronization of data recording (output T). The reference voltage V_R is applied to one of the two inputs of a difference amplifier in the drive control unit (input V_R). The amplified output, in turn, energizes the drive coil of the electro-mechanical *Mössbauer drive* to afford the source motion. The Mössbauer drive is a linear motor particularly designed for high-definition oscillating motion.

The moving part of the Mössbauer drive is a rod (or a tube), which carries the radioactive Mössbauer source at one end. The parts are elastically suspended by two disc springs (not shown) that allow only one-dimensional movements in axial direction. The driving elements for the motion are two firmly attached magnets (or two magnet sections, as shown here), which are positioned within the *drive coil* and the so-called *pick-up* coil.[2] The magnetic field of the energized drive coil leads to an acceleration of the appending magnet, whereas the pick-up magnet *senses* the actual motion via induction of an output voltage (V_{resp}) in the pick-up coil.

---

[2]This is a highly schematic description. Commercial Mössbauer spectrometers have fixed magnets and moving coils in the style of acoustic loudspeaker systems. Drive and pick-up coils are attached to the moving rod or tube. Both are mounted inside the narrow gap of strong yoke magnets made of a neodymium–iron–boron alloy or another strong magnetic material.

The pick-up coil is connected to the second input (V_{resp}) of the differential amplifier in the drive control unit. If at any time the actual pick-up voltage, according to the source motion, does not match the reference voltage (V_R) because the source motion deviates from the nominal values, the difference between (V_{resp}) and (V_R) affords a correction signal at the amplifier output which pushes the drive coil to oppose the deviation. The amplifier and the motor form a tightly coupled electro-mechanical negative *feedback loop* that forces the source velocity to follow the objective defined by the oscillating reference voltage signal (V_R) with high accuracy. For a properly set system, the remaining error and deviation from linearity can be as small as 1‰ of the nominal amplitude (Fig. 3.2).

3.1.1.2 Tuning the Drive Performance

The fundamental frequency of the (mostly triangular) drive motion should be set slightly above the first resonance of the mechanical unit, which may be in the range of 10–30 Hz. The frequency determines the repetition rate of the Mössbauer measurement by which the source sweeps up and down the whole velocity scale for spectra accumulation until the end of the measurement. The *amplitude* of the velocity sweep is selected by scaling the reference voltage (V_R) at the drive control unit. The *quality* of the velocity control can be tuned by adjusting the characteristics of the feedback loop, i.e., the *gain* of the differential amplifier and eventually some *frequency–response* parameters. The actual performance of the drive can be judged from the size of the so-called *error* signal, which is the voltage difference at the inputs (V_R) and (V_{resp}) of the difference amplifier of the drive control unit. The error signal can be monitored with an oscilloscope connected to output (E). To optimize the performance, the corresponding dials on the front panel of the control unit are adjusted to minimize the error signal while observing the oscilloscope (Fig. 3.2).

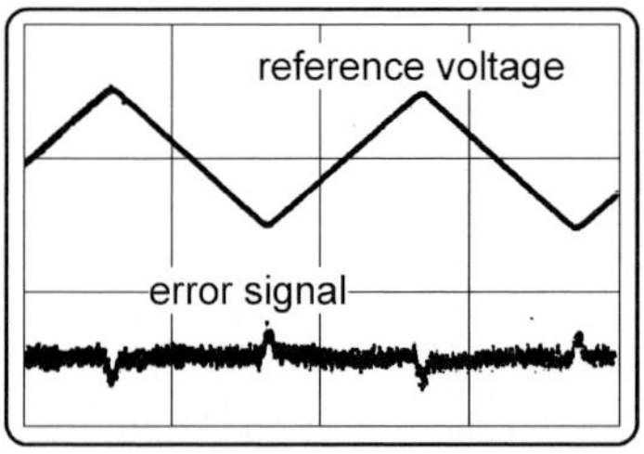

Fig. 3.2 Triangular velocity reference signal (*top*) and drive error signal (*bottom*) of a Mössbauer drive operating in constant acceleration mode. The error signal is taken from the monitor output "*E*" of the drive control unit (see Fig. 3.1). Usually it is internally amplified by a factor of 100. Here, the deviations, including hum, are at the ≈2‰ level of the reference. The peaks at the turning points of the triangle are due to ringing of the mechanical component, induced by the sudden change in acceleration (there should be no resonance line at the extremes of the velocity range)

3.1.2 Recording the Mössbauer Spectrum

To accumulate a Mössbauer spectrum, the electric pulses delivered by the γ-detection system must be recorded synchronously to the sweep of the source velocity. This is achieved by operating the MCA in the so-called multichannel scaling (MCS) mode, in which the function generator triggers the digital counters (channels) of the MCA sequentially, one by one. The channels accumulate the incoming γ-pulses by incrementing their digital storage value.

A memory sweep is initialized when the MCA receives a "start" pulse from the function generator at the trigger input (*T*). The "start" pulse is synchronized with the sweep of the reference voltage (V_R) for the Mössbauer drive. It opens the first MCA channel when the source velocity passes through the minimum (cf. Fig. 3.3). After this "start" trigger, a train of 512 "channel advance" pulses follows with exact delay times of about 100 μs each.[3] On receiving such a "channel advance" pulse, the

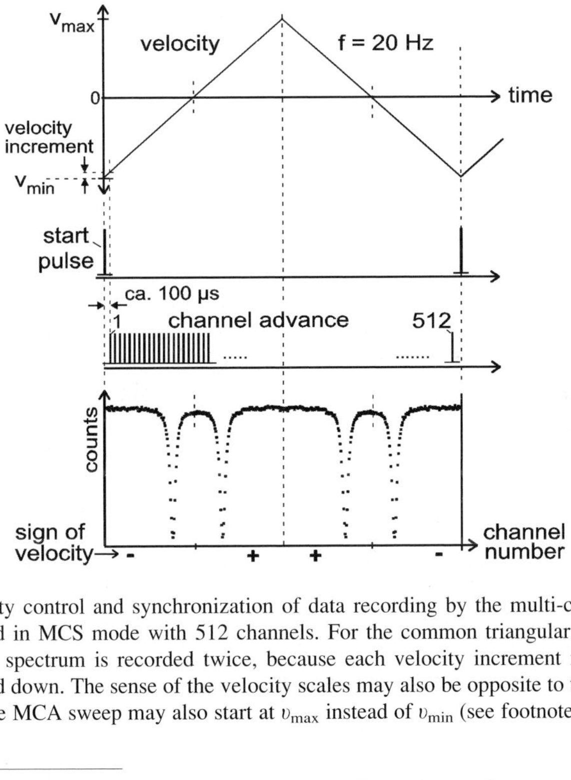

Fig. 3.3 Velocity control and synchronization of data recording by the multi-channel analyzer (MCA) operated in MCS mode with 512 channels. For the common triangular velocity profile shown here the spectrum is recorded twice, because each velocity increment is reached upon sweeping up and down. The sense of the velocity scales may also be opposite to that shown here, which means the MCA sweep may also start at υ_{max} instead of υ_{min} (see footnote 4)

[3]The numbers are generic for our example and may be different for a real system, depending on the number of channels used in the MCA and the frequency chosen for the drive system. The channel numbers are usually a certain power of two; like $512 = 2^9$. The value of ≈100 μs applies to a spectrometer with 512 channels operated at 20 Hz, which means a period time of 50 ms. The exact dwell time is obtained from the ratio of period time to the number of channels, which is 50 ms/512 = 97.7 μs per channel.

MCA first closes the open channel, then increments the channel address by one, advances to the next channel, and opens that for γ-pulse recording. When the last channel is closed (address $=$ 512 in our example), the MCA is reset and is ready to start the next cycle. The process is repeated with the spectrometer frequency ($\approx$20 Hz) for every velocity cycle until the measurement is stopped. By this procedure the channels are sequentially opened and closed for identical dwell times, which in total match the period of the velocity sweep. Hence, γ-recording is tightly synchronized with the source motion such that each counting channel corresponds to exactly one distinct increment of the velocity range. At the end, the counts stored in each channel of the MCA array represent the Mössbauer spectrum given in counts as a function of channel number or more commonly in counts as a function of Doppler velocity. Typical acquisition times for such a Mössbauer spectrum, for instance for a nonenriched iron compound, may range from several hours to several days, depending mostly on the iron concentration and absorber thickness. When the sample is prepared with enriched Mössbauer isotope, the time for running a Mössbauer spectrum with sufficient signal-to-noise ratio (SNR) may be reduced considerably, even down to seconds.

3.1.2.1 "Folding" of Raw Spectra

After measurement, the content of the MCA memory is transferred to a computer and stored as an array of N numbers for further processing and analysis, where N is the number of channels used to record the spectrum (usually 512). If the spectrometer is operated in the constant acceleration mode with triangular velocity profile, the raw numbers code for two data sets, one recorded in the "up" branch and one in the "down" branch of the velocity sweep (Fig. 3.3). The two parts are antisymmetric mirror images of the same spectrum showing opposite velocity scales. Therefore, "folding" of the raw spectra is usually the first step of data reduction to obtain the average spectrum of the "up" and "down" branches. This also makes it possible to detect perturbing deviations from linearity in the motion of the transducer. There are different algorithms for folding, but essentially it is the pair-wise addition of channels with corresponding velocity in the left and right halves of the full data set. Folding reduces the length of the data set to $N/2$ channels and increases the signal-to-noise ratio (SNR) by a factor of $\sqrt{2}$.

Since the actual motion of the Mössbauer drive, as for any frequency transmission system, can show phase shifts relative to the reference signal, the ideal folding point (FP) of the raw data in terms of channel numbers may be displaced from the center at channel number $(N - 1)/2$ ($=$ 255.5 in the example seen earlier). The folding routine must take this into account. Phase shift and FP depend on the settings of the feedback loop in the drive control unit. Therefore, any change of the spectrometer velocity tuning requires the recording of a new calibration spectrum.

One should note that the "polarity" of a Mössbauer drive may be unknown in advance. Depending on the internal polarity of the feedback loop and on which side of the drive the source is mounted, the actual source motion may be toward the absorber or away from it at the time of the "START" signal for the MCA sweep. If the MCA sweep starts at υ_{max} instead of υ_{min}, the sense of the velocity scale of the raw data is opposite to that shown in Fig. 3.3. Some commercial Mössbauer motors can even carry two sources, one on either side, to record two spectra at the same time and having opposite velocity scales.[4] The folding routine must also take this into account and fold either "right to left" or "left to right." The correct sense of folding can be recognized from the features of a calibration spectrum as is shown in the following section.

3.1.3 Velocity Calibration

The calibration of the energy scale of a Mössbauer spectrometer is an essential step of data analysis as in any other spectroscopic method. The simplest and cheapest way is to use the spectrum of a standard absorber with known hyperfine splitting. In the past, a variety of substances were used, e.g., sodium nitroprusside $Na_2[Fe(CN)_5NO]\cdot 2H_2O$, and potassium ferrocyanide $K_4[Fe(CN)_6]\cdot 3H_2O$. For ^{57}Fe spectroscopy, it has now become common practice to calibrate with foils of pure metallic iron (<0.02% carbon) at ambient temperature. The pure metal has a very stable crystalline phase with body-centered cubic lattice (α-iron) and shows at room temperature an accurately known splitting of 10.657 mm s^{-1} between the two outermost lines of the magnetic six-line-pattern (cf. Fig. 3.4). Calibration foils should be no more than 5–25 μm thick; otherwise substantial line broadening occurs because of thickness effects (see Sect. 3.2.1). Line widths of about 0.20–0.23 mm s^{-1} can be achieved with good spectrometers, which is close to the physical limit of the natural line width of $2\Gamma_{nat}$ (= 0.19 mm s^{-1}).

3.1.3.1 Velocity Range and Calibration Factor

Calibration spectra must be measured at defined temperatures (ambient temperature for α-iron) because of the influence of second-order Doppler shift (see Sect. 4.2.1) for the standard absorber. After folding, the experimental spectrum should be simulated with Lorentzian lines to obtain the exact line positions in units of channel numbers which for calibration can be related to the literature values of the hyperfine splitting. As shown in Fig. 3.4, the velocity increment per channel, υ_{step}, is then obtained from the equation: $\upsilon_{step} = D_i(\text{mm s}^{-1})/D_i(\text{channel numbers})$. Different

[4]The polarity of both the drive coil and pick-up coil of the Mössbauer motor can be changed together without changing the performance.

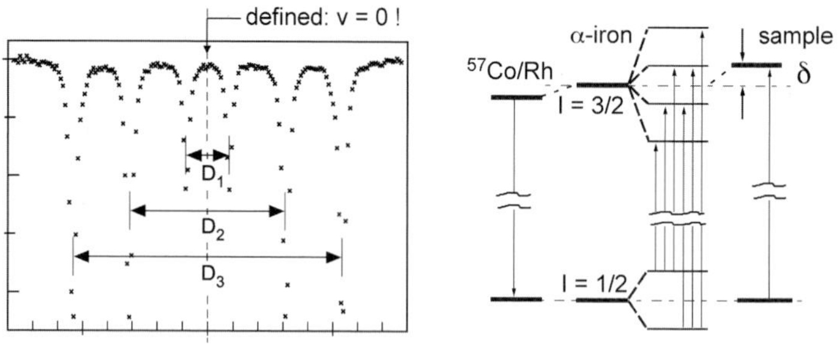

Fig. 3.4 Calibration spectrum of metallic iron and magnetic hyperfine splitting of the nuclear levels. The values of the hyperfine splitting in α-iron are $D_1 = 1.677$ mm s^{-1}, $D_2 = 6.167$ mm s^{-1}, $D_3 = 10.657$ mm s^{-1}. The center of the calibration spectrum is defined as velocity zero (*left*). The isomer shift of a specific sample with respect to metallic iron is indicated as δ (*right*)

results, eventually found for D_1, D_2, and D_3 due to statistical noise, can be averaged; significant deviations ($>0.1\Gamma$), however, would stem from nonlinearity in the source motion and require a refurbishment of the system. Instead of using υ_{step} as the calibration factor for the subsequent Mössbauer measurements, it is also common to use the velocity span in terms of the maximum velocity $\upsilon_{\max}$, which is given by the equation $\upsilon_{\max} = \upsilon_{\text{step}} \cdot (N - 1)/2$.

The calibration procedure is sufficiently accurate for Doppler velocities in the regime 0 to ± 10 mm s^{-1}; beyond this range, laser calibration is more suitable. Calibration with α-iron, as described, can also be used for Mössbauer measurements with other isotopes, e.g., ^{61}Ni, ^{197}Au, and ^{193}Ir, for which suitable standard absorbers are not available (provided that the Doppler velocity range of interest is not significantly greater than ± 10 mm s^{-1}). This, of course, requires that the spectrometer is temporarily equipped with a ^{57}Co source and an α-iron absorber.

3.1.3.2 Velocity Zero and Isomer Shift References

The point of true zero velocity in a folded Mössbauer spectrum recorded in the constant acceleration mode with triangular wave form is theoretically found at the channel corresponding to half the FP (FP/2) (cf. Fig. 3.3). Isomer shifts given with respect to this point refer to the nuclei in the employed Mössbauer source material. Such reference scales are commonly found in older literature, when isomer shifts are quoted relative to Pt, Pd, Cu, or other metals that are used as source matrix. For ^{57}Fe, however, it is now common practice to re-adjust the velocity scale according to the calibration absorber and *define* velocity zero as the center of the α-iron calibration spectrum. The corresponding channel number is again taken from the average of line positions found by simulation. The advantage of this procedure is that isomer shifts are referenced to α-iron, irrespective of the specific source used for the Mössbauer measurement (Fig. 3.4). Table 3.1 provides the isomer shifts of

Table 3.1 Isomer shifts[a] of reference materials for ^{57}Fe (14.4 keV)

Material	δ (mm s^{-1})	Material	δ (mm s^{-1})
$Na_2[(CN)_5NO]\cdot 2H_2O$	−0.2576(14)	Pd	+0.1798(12)
Cr	−0.146(3)	Cu	+0.2242(10)
$K_4[Fe(CN)_6]\cdot 3H_2O$	−0.042(3)	Pt	+0.3484(24)
Rh	+0.1209(22)	α-Fe_2O_3	+0.365(3)

[a]Given relative to α-Fe at 300 K; taken from Mössbauer Effect Data Journal [8]

some alternative reference materials and can be used to convert previously referenced data into values relative to that of α-iron (by addition of the tabulated values).

Interestingly, the correct *polarity of the Mössbauer drive* can be checked by using the isomer shift of α-iron with respect to the materials in Table 3.1. After folding of the raw data, the center of the calibration spectrum without further correction must be at −0.12 mm s^{-1}, relative to the ^{57}Co/Rh source material.

3.1.3.3 Laser Calibration

Mössbauer spectra of ^{161}Dy, ^{129}I, ^{121}Sb, and other isotopes may show resonances at Doppler velocities of up to a few centimeters per second. Extrapolation of the calibration factors derived from standard spectra to such high velocities, assuming linearity of the transducer, may not be accurate and is often questionable. In this case, the source velocity can be accurately measured by using a Michelson interferometer with a He/Ne-laser as reported by Fritz and Schulze [9], Cranshaw [10], and Viegers [11]. The principle of the interferometer is shown in Fig. 3.5. A laser beam is split into two light paths, one of which passes straight through the splitter and is reflected at a fixed mirror, whereas the other one is directed toward the Mössbauer drive and is reflected at the moving velocity transducer, usually on the back site of the drive. Both reflected beams superimpose in the direction of the laser

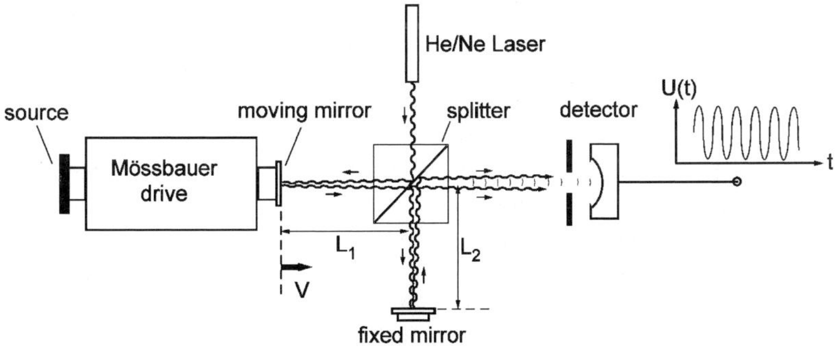

Fig. 3.5 Principle of a laser interferometer for absolute calibration of the transducer velocity. L_1 and L_2 denote the lengths of the two light paths of the split laser beam, giving a path difference $\Delta S = 2(L_1 - L_2)$

detector and interfere according to their phase difference, as determined by the difference in light paths $\Delta S = 2(L_1 - L_2)$. When the *position* of the moving mirror changes, a fringing pattern passes the entrance window of the detector, so that the output signal of the detector photodiode is modulated by a cosine function according to $U(\Delta S) \propto \cos(4\pi\Delta S/\lambda)$, where λ is the wave length of the laser light (here 632.8 nm). Moreover, moving the Mössbauer source and the drive mirror with a *velocity* υ generates a periodic output signal $U(t) \propto \cos(2\pi(2\upsilon/\lambda)t)$, the frequency $f = 2\upsilon/\lambda$ of which is proportional to the source velocity [9]. Substituting λ yields the output frequency $f = 3{,}160$ Hz for a source velocity of $\upsilon = 1$ mm s^{-1}.

The number of beatings in the output signal $U(t)$ can be recorded as a function of the channel number simultaneously to the Mössbauer spectrum (usually in a second part of the MCA memory). Accordingly, the number of beats stored in each channel per velocity sweep divided by the dwell time of the MCA channels yields the specific average velocity for each detection channel.

3.1.4 The Mössbauer Light Source

For nuclear γ-resonance absorption to occur, the γ-radiation must be emitted by source nuclei of the same isotope as those to be explored in the absorber. This is usually a stable isotope. To obtain such nuclei in the desired excited meta-stable state for γ-emission in the source, a long-living radioactive parent isotope is used, the decay of which passes through the Mössbauer level. Figure 3.6a shows such a transition cascade for ^{57}Co, the γ-source for ^{57}Fe spectroscopy. The isotope has a half-life time $t_{1/2}$ of 270 days and decays by K-capture, yielding ^{57}Fe in the 136 keV excited state (^{57}Co nuclei capture an electron from the K-shell which reduces the

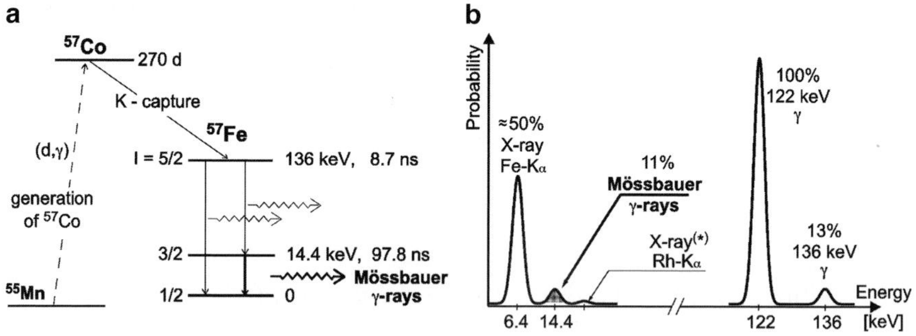

Fig. 3.6 (**a**) Decay scheme of ^{57}Co and (**b**) ideal emission spectrum of ^{57}Co diffused into rhodium metal. The nuclear levels in (**a**) are labeled with spin quantum numbers and lifetime. The dashed arrow up indicates the generation of ^{57}Co by the reaction of ^{55}Mn with accelerated deuterons (d in; γ out). Line widths in (**b**) are arbitrarily set to be equal. The relative line intensities in (%) are given with respect to the 122-keV γ-line. The weak line at 22 keV, marked with (*), is an X-ray fluorescence line from rhodium and is specific for the actual source matrix

nuclear charge and converts cobalt into iron). Within about 10 ns, the 136 keV state of ^{57}Fe undergoes a transition to the ground state either directly (15%) by the emission of a 136 keV γ-photon or (85%) via the 14.4 keV Mössbauer excited state by the emission of a 122 keV photon first. However, of all the ^{57}Fe nuclei produced in the source, only about 11% will emit the desired 14.4 keV Mössbauer radiation because of high probability of the Mössbauer level to decay by radiationless internal conversion ($\alpha \approx 0.9$, see Sect. 3.2.1).

The emission spectrum of ^{57}Co, as recorded with an ideal detector with energy-independent efficiency and constant resolution (line width), is shown in Fig. 3.6b. In addition to the expected three γ-lines of ^{57}Fe at 14.4, 122, and 136 keV, there is also a strong X-ray line at 6.4 keV. This is due to an after-effect of K-capture, arising from electron–hole recombination in the K-shell of the atom. The spontaneous transition of an L-electron filling up the hole in the K-shell yields Fe–K_α X-radiation. However, in a practical Mössbauer experiment, this and other soft X-rays rarely reach the γ-detector because of the strong mass absorption in the Mössbauer sample. On the other hand, the sample itself may also emit substantial X-ray fluorescence (XRF) radiation, resulting from photo absorption of γ-rays (not shown here). Another X-ray line is expected to appear in the γ-spectrum due to XRF of the carrier material of the source. For rhodium metal, which is commonly used as the source matrix for ^{57}Co, the corresponding K_α line is found at 22 keV.

The complexity of the ^{57}Co emission spectrum and the low fraction of the desired 14.4 keV radiation require an efficient Mössbauer counting system that is able to discriminate photons of different energies and reject the unwanted events. Otherwise a huge nonresonant background would add to the counting statistics of the spectra and fatally increase the noise of the spectrometer.

Mössbauer sources are technically prepared by diffusing the radioactive long-lived precursor of the Mössbauer isotope into a matrix of a nonmagnetic metal with cubic crystal lattice. The acquisition of narrow Mössbauer emission lines without hyperfine splitting requires that the matrix impose neither a magnetic moment nor an electric field gradient to the source nuclei. For experiments with ^{57}Fe, the source is kept at room temperature and usually consists of approximately 0.5–1.85 GBq (1.85 GBq $\triangleq$ 50 mCi) of ^{57}Co diffused into 6 μm thick rhodium foil which is glued to a disc for mounting on the velocity transducer. The recoil-free fraction of a Mössbauer source of ^{57}Co in rhodium or palladium at room temperature is about $f_s \approx 0.75$ [12].

3.1.5 Pulse Height Analysis: Discrimination of Photons

The γ-detector of a Mössbauer spectrometer converts the incident γ-photons into electric output pulses of defined charge (see Sect. 3.1.6). The detector signals are electronically amplified and shaped by an amplifier network to obtain strong needle pulses with well-defined rise time, so that the pulse *height* is proportional to the energy of the incident photon. The amplifiers are usually adjusted to obtain

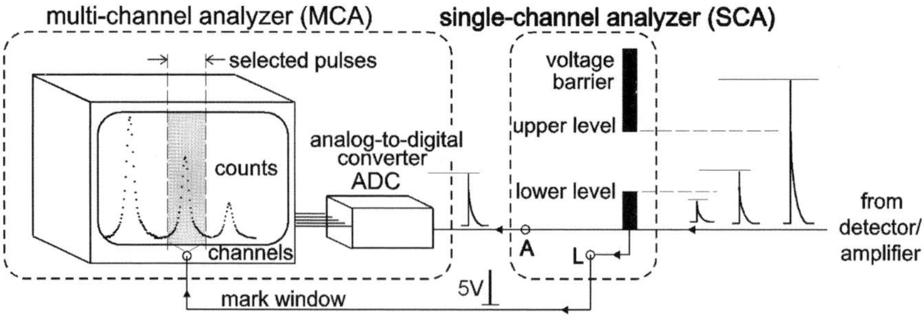

Fig. 3.7 Pulse-height analysis (PHA): Function of the single-channel analyzer (SCA) and data recording by the *multi-channel analyzer* (MCA). The output (L) of the SCA yields a 5 V square-shaped pulse, a so-called TTL pulse for each γ-pulse matching the voltage selection window. The SCA is set to select the Mössbauer pulses for the subsequent measurement

pulse amplitudes of 1–10 V for the most relevant part of the respective γ-spectrum, which is around 14 keV for iron.

The SCA discriminates the incoming pulses by comparing their height with the tunable lower and upper levels of a voltage selection window (cf. Fig. 3.7). Pulses matching the discrimination window trigger a so-called "logical" TTL pulse at output (L) of the SCA (a short square-shaped 5 V needle pulse). The TTL pulses are used downstream by the MCA for the registration of the Mössbauer spectrum in the MCS experiment, as described in Sect. 3.1.2. This effectively suppresses the unwanted pulses from nonresonant γ- and X-radiation if the discriminator window of the SCA has been properly tuned to the Mössbauer line.

3.1.5.1 Tuning the SCA

The setting of the SCA discrimination window can be monitored by using the MCA in PHA mode. In this mode, an *analog-to-digital converter* (ADC) in the input circuits of the MCA converts the height of the incoming pulses into a channel address in the range of the MCA memory size (mostly 1–1,024 channels) and triggers the incrementing of the channel content. Thus, γ-pulses are recorded at high or low channels according to their height and the energy of the respective photons. Data accumulation yields the *pulse height spectrum*, which represents the γ-spectrum or the selected region of interest as sketched on the screen of the MCA[5] in Fig. 3.7.

For the visualization of the SCA discrimination window on the MCA screen, the TTL pulses, which map the coincidently incoming γ-pulses, can be used to mark the range of channel numbers corresponding to the selected voltage range. The marks allow instantaneous control while setting the dials for lower and upper limits to the

[5]This is meant schematically; modern MCA modules usually do not have a screen but are controlled by a front-end PC for parameter setting and data visualization.

Mössbauer line.[6] The appearance of the pulse height spectrum including the overall pulse height and the resolution of lines and noise level also aids the setup of the Mössbauer detection system, via the adjustment of the high voltage for the detector, the amplifier gains, and the shaping parameters of the main amplifier. These need to give a narrow Mössbauer line in the PHA-spectrum with minimum nonresonant background contribution.

In summary, pulse-height analysis (PHA) prior to a Mössbauer measurement is an essential step in tuning a Mössbauer spectrometer. PHA allows the adjustment of the γ-detection system to the Mössbauer photons and the reduction of noise by rejecting nonresonant background radiation.

3.1.6 Mössbauer Detectors

Detectors for Mössbauer γ-radiation convert the photons transmitted through the sample into electric pulses. According to the properties of the Mössbauer source, the process must be energy-sensitive to enable subsequent discrimination of the specific Mössbauer γ-line from the other nonresonant radiation. Preferably a detector should have high sensitivity for the Mössbauer radiation, but mostly ignore other photons, to keep the counter dead time short and to protect the downstream electronic components from unnecessary load. Such features help to achieve fast counting events with strong pulses and high count rates.

3.1.6.1 Proportional Counters

Mössbauer spectrometers for transmission measurements are normally equipped with gas-filled *proportional counters*. Their energy resolution is sufficiently high, they are suitable for high count rates, and their sensitivity can be adapted to the specific Mössbauer radiation by selecting the counting gasses and the dimensions of the counting chamber defining the absorber path length. In addition they are cheap and robust. Proportional counters are filled either with argon, krypton, or xenon as the counting gas, with a small amount of a so-called quench gas such as 10% methane. Argon has little sensitivity to the unwanted 122 and 136 keV radiation of ^{57}Fe, but a reasonably good sensitivity is achieved for the 14 keV Mössbauer line with larger path lengths for the photon absorption. Heavier atoms, like krypton, are more efficient and thus allow smaller counter dimensions, but they also respond to high-energy quanta. The typical overall efficiency of proportional counters is 60–80% for the 14.4 keV of ^{57}Co.

[6]The true energy scale of the γ-spectrum in units of keV usually cannot be derived directly from the pulse height spectrum because the overall amplification of the detection system is not known. Therefore, the γ-lines eventually have to be identified by trial and error when a new system is set up by checking for the occurrence of the Mössbauer effect.

The dominating process for the interaction of Mössbauer γ-quanta with the gas of a proportional counter is absorption by the photo effect and the emission of a photoelectron. Since the ionization energy of the gas atoms is much lower than the γ-energy, the released photoelectron has sufficient kinetic energy to ionize a certain number of other counting gas atoms by collision. High voltage is applied to separate the primary charges by directing the positive gas ions to the counter wall and the electrons to the central counting wire. The voltage is adjusted such that the accelerated electrons ionize more atoms by collision to cause a charge avalanche of electrons rushing to the counting wire without achieving a full discharge breakdown as in a Geiger–Müller counter. The effective "gas gain" is kept in the so-called proportional range for which the total charge of the finally collected electric pulse is proportional to the energy of the initially absorbed γ-quantum. The function of the quench gas in this process is to suppress random motions of the electrons which yield shorter signal rise times at the counter output and higher signal amplitude.

3.1.6.2 Other γ-Detectors

Some properties of the detectors most commonly used for transmission experiments are summarized in Table 3.2. Alternative counters are scintillation detectors based on NaI or plastic material that is attached to a photomultiplier, and solid-state detectors using silicon- or germanium-diodes.

Scintillation counters based on NaI show better sensitivity than gas counters particularly for higher γ-energies, but their energy resolution is low such that they can be hardly used for 14.4 keV. Moreover, Compton scattering of high-energy photons may contribute an unfavorable broad nonresonant background to the counting statistics of the Mössbauer radiation. However, the thickness of the scintillation crystal can be adapted to higher photon energy. For some critical Mössbauer experiments with isotopes with γ-energies >50 keV and with low count rates and strong scattering from the absorber material, scintillation counters are often the best choice.

Table 3.2 Properties of detectors commonly used in Mössbauer spectroscopy

Detector	Resolution	Efficiency	Maximum count rate in single-pulse technique (kHz)
Proportional counter	2–4 keV at 14 keV	80% at 14 keV	50
NaI	8–10 keV at 60 keV	100% at 60 keV	100–200
Si (Li)	~400 eV at 10 keV	100% at 10 keV	10–40
Ge (Li) Ge (pure)	~600 eV at 100 keV	80–100%	10–40

The count rates are integral values for the incident total radiation

Solid-state detectors based on silicon- or germanium-diodes possess better resolution than gas counters, particularly when cooled with liquid nitrogen, but they allow only very low count rates. PIN diodes have also recently become available and have been developed for the instruments used in the examination of Martian soils (Sects. 3.3 and 8.3). A very recent development is the so-called silicon-drift detector (SDD), which has very high energy resolution (up to ca. 130 eV) and large sensitive detection area (up to ca. 1 cm^2). The SNR is improved by an order of magnitude compared to Si-PIN detectors. Silicon drift detectors may also be used in X-ray florescence spectroscopy, even in direct combination with Mössbauer spectroscopy (see Sects. 3.3 and 8.3).

Current integration. In special Mössbauer experiments with very strong sources, for which single-pulse detection is not possible because of exceedingly high radiation intensity, a detection scheme may be used which is called *current integration* technique. For ^{61}Ni, ^{197}Au, ^{199}Hg Mössbauer spectroscopy, for instance, it is desirable and possible to produce sources with activities up 18–800 GBq to compensate for the small recoilless fraction resulting from high energetic nuclear transitions, for the low Debye temperatures, or for low densities of Mössbauer active nuclei in the absorber. Since single pulse counting fails in this case because of "overload," the counter current is monitored.

Details of the method including a wiring diagram are provided in the first edition of this book, Sect. 4.1.3, p. 45 (cf. CD-ROM).

3.1.6.3 Detectors for Conversion Electrons and Scattered Radiation

So far, we have discussed only the detection of γ-rays transmitted through the Mössbauer absorber. However, the Mössbauer effect can also be established by recording scattered radiation that is emitted by the absorber nuclei upon de-excitation after resonant γ-absorption. The decay of the excited nuclear state proceeds for ^{57}Fe predominantly by internal conversion and emission of a conversion electron from the K-shell ($\approx$90%). This event is followed by the emission of an additional (mostly K_α) X-ray or an Auger[7] electron when the vacancy in the K shell is filled again. Alternatively, the direct transition of the resonantly excited nucleus causes re-emission of a γ-photon (14.4 keV).

Since conversion electrons, Auger electrons, X- and γ-rays are emitted by the absorber nuclei in full 4π solid angle, the detection of these events is predominantly used for recording Mössbauer spectra from opaque or excessively thick absorbers for which resonantly back-scattered radiation can be detected on the side of the sample exposed to the moving Mössbauer source. Such spectra appear as emission spectra with the Mössbauer lines pointing "upward" (Fig. 3.8, right). The background can be very low, depending on the quality of the detector and nonresonant

[7]The Auger effect is the phenomenon involving electron–hole recombination in an inner-shell vacancy causing the emission of another electron.

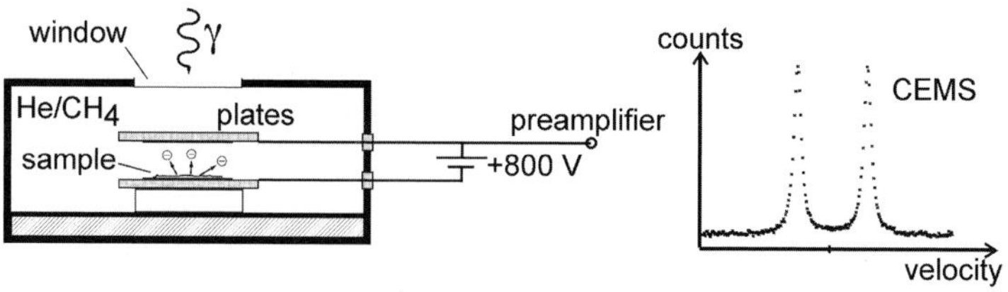

Fig. 3.8 Scheme of a PPAC counter for CEMS and a typical CEMS spectrum. The electrodes of the PPAC are parallel plates made of Perspex coated with carbon (adapted from [19])

scattering in the sample. Conversion electron Mössbauer spectroscopy (CEMS), using conversion and Auger electrons, has gained increasing importance in the study of surface layers, surface reactions, or implantation reactions because of the small escape depth of low energy electrons in solids (about 1,000 Å in iron metal for 7.2 keV Auger electrons [13]).

The simplest detector for conversion and Auger electrons is a flat proportional counter filled with a mixture of He/methane. Since these light gasses are virtually insensitive to γ-radiation, the sample can be mounted inside the detector on its back wall. The Mössbauer radiation enters through a front window and passes through a grid of thin counting wires spanned as cathode above the sample surface (Fig. 3.8, left). Resonantly emitted electrons are efficiently detected in 2π solid angle with a very low background of the 6.4 keV X-rays (<1%). Swanson and Spijkerman published the first description of such an electron detector [14]; it is currently widely used in the study of corrosion and surface reactions. An improved version of such a CEMS detector [15] is described in Sect. 4.1.3 of the first edition of this book (cf. CD-ROM). More recent developments for low-temperature applications, for instance, are found in [16, 17].

An interesting variant of a CEMS counter is the parallel-plate avalanche counter (PPAC) [18, 19], which carries the sample between parallel electrodes made of Perspex coated with graphite (Fig. 3.8, left). A counter gas is used to amplify the low conversion-electron current emitted by the sample, with an avalanche effect taking place between the plates. Compared with the CEMS proportional counters, PPAC gives a larger signal-to-background ratio, faster time response, simpler construction, and better performance at low temperatures.

Due to electron–electron collisions in the sample material, the energy of conversion- and Auger-electrons leaving the sample surface depends on the depth of their origin [20]. This establishes the interesting possibility to record *depth-selective* conversion electron Mössbauer spectra (DCEMS) for the study of layers of thin films or other surfaces. To this end, the electron energy has to be discriminated for two-dimensional data recording. However, the energy resolution of conventional CEMS and PPAC detectors is not sufficient for this purpose. Some improvement has been achieved with micro-channel plates [21] but well-resolved DCEMS require genuine electric or magnetic electron spectrometers for the electron detection [22–24]. The energy discrimination of these complicated and expensive devices can be better than ~1% with an efficiency of about 15% [25]. A high-resolution electron spectrometer with orange-shape geometry is described in [26]. Very thin layers of films and other surface samples can be studied by ILEEMS, integral low-energy electron Mössbauer spectroscopy, for which electrons of $E < 20$ eV are detected. The methodology and a detector based on a channeltron have been described by de Grave [27] (cf. CD-ROM, Part II).

3.1.6.4 Limits of Counter Resolution

The resolution of a detector for ionizing radiation is principally limited by the mean number N_{m} of ion–electron pairs released per absorbed photon. Since energy dissipation after photo absorption occurs via stochastic collision processes, the result obeys a Poisson or Gaussian distribution of width $\Delta N = \sqrt{N_{\mathrm{m}}}$. Accordingly, the *relative* resolution of the photon-to-charge conversion is $\Delta N/N = 1/\sqrt{N_{\mathrm{m}}}$, if noise from other sources, such as the signal amplifiers, is neglected. Creation of an ion–electron pair in argon for instance requires about 30 eV, so that a 14.4 keV photon cannot yield about 480 primary electron–ion pairs. Thus, the theoretical statistical limit of the counter resolution $(1/\sqrt{480})$ is about 5%. Even with >10% resolution, as is often encountered in practice, it is readily possible to discriminate the Mössbauer line from other radiation coming from the source or the absorber. Better values are obtained with solid-state detectors (2–3% resolution) because semiconductors have energy gaps of only 1–2 eV for charge excitation, which causes the release of many more electrons than in gas-filled counters.

It is notable that the energy resolution of a γ-detection system does not reach the natural width of the Mössbauer lines, such as the value $\Delta E/E = 10^{-13}$ for ^{57}Fe. Some 10^{26} primary electrons would be necessary to overcome the statistical uncertainty of the detection process that leads to the broadening of the lines observed in the γ-spectrum (i.e., in the pulse-height spectrum). Fortunately, this is not applicable in Mössbauer spectroscopy because the spectrometer resolution does not depend at all on the energy resolution of the detector. Instead, the shape of the Mössbauer absorption line is probed by convenient *intensity* measurements of the radiation transmitted through the absorber, while the energy modulation is tuned through the Mössbauer resonance. The resolution of this process is given by the accuracy of the Doppler modulation; the properties of the counting system decide the statistical quality of the data and the SNR.

3.1.7 Accessory Cryostats and Magnets

An important accessory in many applications of Mössbauer spectroscopy is a cryostat for low temperature and temperature-dependent measurements. This may be necessary to keep samples frozen or to overcome small Debye–Waller factors of the absorbers at room temperature in the case of an isotope with high γ-energy. Paramagnetic samples are measured at liquid-helium temperatures to slow down spin relaxation and to determine electronic ground states. Phase transitions and spin crossover phenomena have to be explored over wide temperature ranges.

In a simple helium bath or flow cryostat, the temperature can be controlled in the range 1.2–500 K. For source-cooled experiments (e.g., ^{61}Ni, ^{197}Au), it is desirable to independently vary the temperature of the source and the absorber. Very low temperatures (down to ~0.03 K) can be obtained with ^{3}He/^{4}He refrigerators which, however, are complicated to handle and expensive. Interesting recent developments are closed-cycle or even cryogen-free cryo-cooler systems for Mössbauer spectroscopy [28] with or without cryo-magnets included. There has been significant progress in vibration decoupling of the cooler unit and the sample insert which made this relatively inexpensive cryogenic system interesting and competitive to conventional helium-cooled cryostats. For basic information on Mössbauer cryostats we refer to references ([10] in Chap. 2, [1, 29]).

Figure 3.9 shows a schematic sectional drawing of a liquid-helium Mössbauer cryostat with a superconducting magnet optionally included. The layout is kept generic to highlight a few issues that are essential for applications in transition metal chemistry:

(a) The system is top-loading, which allows fast sample change and, even more importantly, mounting of frozen samples without warming.

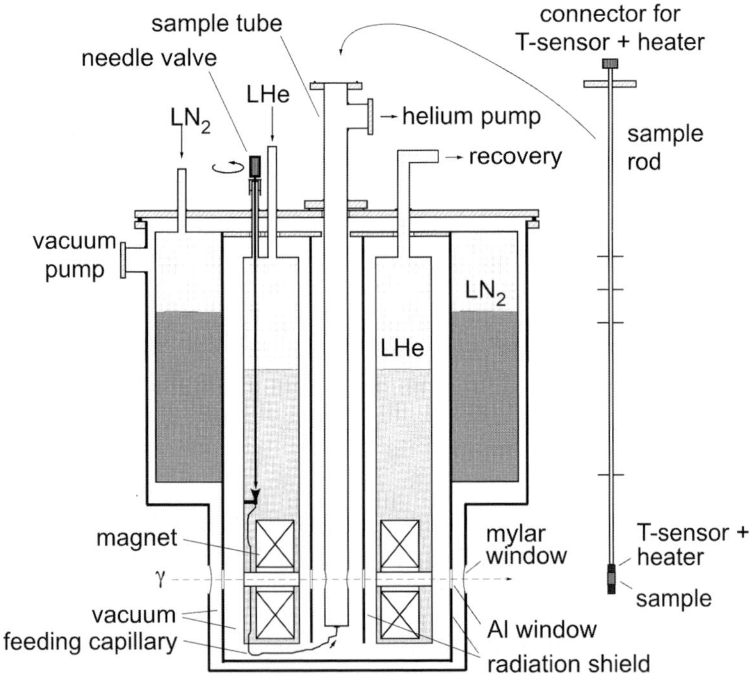

Fig. 3.9 Schematic view of a Mössbauer cryostat with a superconducting cryo-magnet for horizontal transmission experiments with the field parallel to the γ-rays. Without a magnet the helium reservoir may end above the optical axis, such that the cryostat can have a narrow tail with much shorter window-to-window distance

(b) The sample tube is a variable temperature (VT) insert, the temperature of which can easily be changed from about 2 to 300 K independent of the temperature of the main bath. Sample cooling is achieved by a helium flow through the feeding capillary into the sample tube. The speed of the helium pump and the setting of the needle valve control the cooling power. An advantage of such a system is that the coolant consumption can be reduced considerably at elevated temperatures. Absorber temperatures below 4.2 K can be reached by pumping on the coolant liquid helium.

(c) Cryostats for *liquid nitrogen* should not have VT inserts of this type because liquid nitrogen may accumulate in the sample cavity and then strongly absorbs the γ-radiation. This is a severe problem for ^{57}Fe spectroscopy. In this case, the VT insert should have a heat exchanger which prevents liquid nitrogen from getting into the Mössbauer light path. Alternatively, a simple bath system may be employed in which the sample tube is cooled by thermal contact with the outer main reservoir. The latter is a very economical procedure if measurements at base temperature (temperature of the liquid coolant) are predominantly performed.

(d) Cryostat windows for γ-radiation must be free of iron. This is not so much a problem for aluminized mylar used for vacuum-tight outer or inner windows, but more for solid beryllium metal and for plain aluminum (Al) foils, which are often applied as pressure-free inner thermal shields. Products from the supermarket or from the ordinary lab store may contain several percent of iron.

(e) Magnet cryostats should not impose stray fields stronger than 20–50 mT at the source to prevent substantial line broadening. Similar limits hold for the Mössbauer drive, which can eventually be put into a remote position using an extended drive rod for mounting the source. Magnets for longitudinal fields ($B \parallel \gamma$) may need an integrated compensation coil to shield the source. In split-pair magnets for transversal fields (for instance: B vertical, γ-rays horizontal through the gap of the magnet, sample insert from top along the magnet axis), the source may be positioned inside the gap of the magnet in the zone of zero field between field "up" inside the coil and field "down" outside.

3.1.8 Geometry Effects and Source–Absorber Distance

In a Mössbauer transmission experiment, one may be tempted to keep the source–detector distance as short as possible to obtain a high count rate and consequently good statistical accuracy of the spectrum in a short measuring time. The corresponding large solid angle of γ-acceptance in such an arrangement may however lead to considerable distortion of the baseline and even of the shape of the spectrum. In the following discussion of these geometry effects, we assume that the bore of the absorber holder determines the *aperture* α of the spectrometer, which is the ratio of the collimator diameter to the source–collimator distance (cf. Fig. 3.10). The following arguments also hold for any other layout of the γ-ray path.

Baseline distortions: The relative motion of source and absorber causes periodic variations of the solid angle θ and consequently of the incident count rate as a

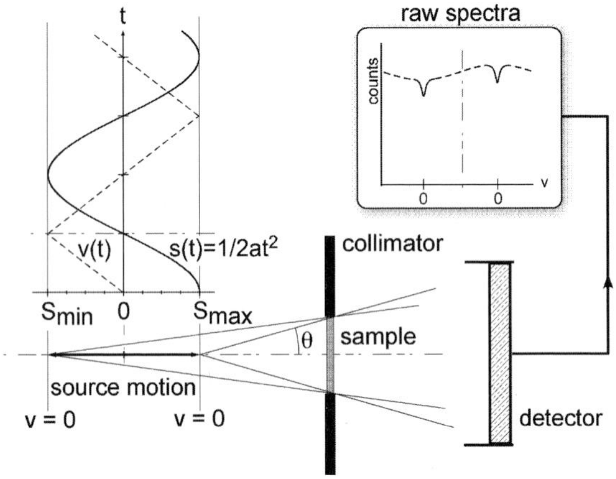

Fig. 3.10 Variation of the spectrometer aperture as a function of the source motion for Mössbauer spectrometers operated in constant acceleration mode with triangular velocity profile, and the resulting nonlinear baseline distortion of the unfolded raw spectra. For simplicity a point-source is adopted, in contrast to most real cases (≈6 mm active spot for ^{57}Co in Rh)

function of the velocity, as indicated in Fig. 3.10. The result is a *nonlinear baseline distortion* of the unfolded Mössbauer spectrum. In the constant acceleration mode, the two Mössbauer spectra taken in the "up" and "down" branch of the velocity sweep are antisymmetric in the sense of their velocity scales and the baseline distortions are virtually opposite, as long as θ remains sufficiently small. Thus, folding of the two "mirror images" will basically eliminate this geometric effect (cf. Sect. 3.1.2).

Line broadening. A serious consequence of a large aperture is the so-called *cosine smearing* of the velocity. Photons, arriving at the absorber at an angle θ with respect to the direction of the actual velocity υ between the source and the absorber, have a reduced Doppler energy shift because the effective velocity υ' along their propagation vector is only $\upsilon' = \upsilon \cos\theta$ (see inset in Fig. 3.11). Thus, the distribution of photon directions within the solid angle of acceptance given by the aperture leads to a spread of effective velocities υ' for any source velocity υ. The intensity distribution of this smearing effect is shown in Fig. 3.11 for different apertures α. The lines of a Mössbauer spectrum measured at large aperture will therefore be asymmetrically broadened, depending on υ, and even the center of the lines will be displaced. Thus, curve fitting with symmetric Lorentzian lines will yield incorrect results. In principle, the cosine effect can be included in the fitting procedure but, since the experimental resolution is reduced for large apertures in general, it should be kept small by refraining from apertures beyond $\alpha \approx 0.2$.

General remarks. The γ-beam must be properly collimated around the sample holder to prevent radiation from reaching the detector without passing through the sample. It is also important that the effective aperture does not change during the source–absorber motion at some critical distance. Particularly, the detector should not be positioned such that the effective collimation can "jump" at some source position from the bore in the sample holder to the entrance window of the γ-counter. Such variations would induce rather unpredictable intensity changes and baseline distortions that cannot be eliminated by folding.

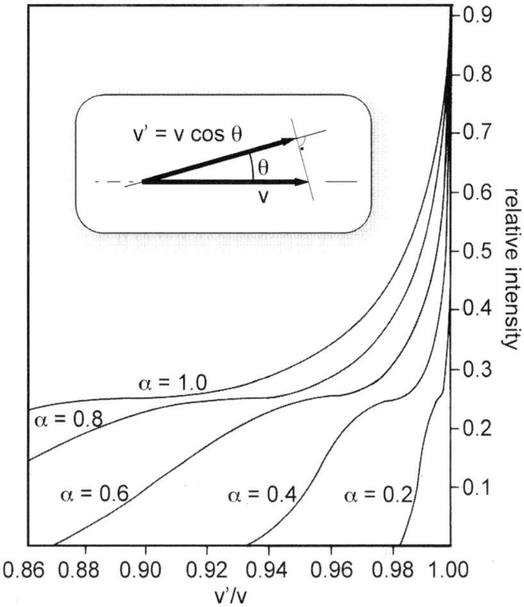

Fig. 3.11 Velocity distribution (cosine smearing effect) in the case of identical source and collimator radius, for different aperture α, the ratio of collimator radius to source–collimator separation (adapted from [30])

Deviating from the setup discussed earlier, the γ-ray beam can also be consistently collimated by structures other than the absorber holder. If this is the entrance window of the detector, the counter should have a lead shield, and the absorber must be sufficiently large to prevent radiation from passing by. For Mössbauer scattering experiments, the same arguments have to be considered.

Scattered radiation. In a transmission experiment, the Mössbauer sample emits a substantial amount of scattered radiation, originating from XRF and Compton scattering, but also γ-radiation emitted by the Mössbauer nuclei upon de-excitation of the excited state after resonant absorption. Since scattering occurs in 4π solid angle, the γ-detector should not be positioned too close to the absorber so as not to collect too much of this unwanted scattered radiation. The corresponding pulses may not only unnecessarily overload the detector and increase the counting dead time, but they may also affect the γ-discrimination in the SCA and increase the nonresonant background noise.

3.2 Preparation of Mössbauer Sources and Absorbers

Most Mössbauer experiments are currently performed with commercially available radioactive sources. For some applications, however, a so-called *source experiment* may be useful, in which the sample is labeled with the radioactive parent-isotope of the Mössbauer nucleus such as ^{57}Co. The γ-radiation of the radioactive sample is then analyzed by moving a single-line absorber for Doppler modulation in front of the detector.

If radioactive sources are to be prepared for specific Mössbauer isotopes, which are commercially not available, there are a number of criteria to be considered. First of all, if there are several different nuclear transitions leading to the excited nuclear level of interest, one should preferentially choose the one that leads to the highest intensity of Mössbauer quanta and has the longest half-life of the precursor nucleus. The chemical composition of the source material should be such as to obtain narrow and intense emission lines with low background from Compton scattering and XRF. Any electric quadrupole or magnetic hyperfine perturbation would split or at least broaden the emission lines, which in turn reduces the spectral resolution and renders the evaluation cumbersome. The Debye–Waller factor should be as large as possible (corresponding to a transition energy as small as possible) to obtain the highest possible resonance absorption. The source material should be chemically inert during the lifetime of the source and resistant against autoradiolysis. Various methods of source preparation are outlined in Chap. 7.

3.2.1 Sample Preparation

The sample for a Mössbauer absorption measurement can be a plate of solid material, compacted powder or a frozen solution, if it contains the Mössbauer isotope in sufficient concentrations and if the γ-radiation can penetrate the material. The first condition is almost trivial, low concentrations of the resonance nuclei cause low signals and exceedingly long acquisition times, but the second condition may be more severe than suspected, because the Mössbauer γ-rays are soft and get strongly absorbed in many materials by nonresonant mass absorption, particularly when heavy atoms (like chlorine or higher) are present in appreciable amounts. The electronic absorption can be neglected only for samples with a high content of the Mössbauer isotope and sufficiently light other atoms such as C, N, O, and H. (Then the optimum absorber thickness is limited only by the onset of spectral distortions when samples are too thick). High mass absorption of the 14.4 keV radiation for ^{57}Fe spectroscopy is already an issue when samples are made of glass or enclosed in glass because of the absorption coefficient of silicon ($N = 14$, $\mu_e = 14\ \mathrm{cm}^2\ \mathrm{g}^{-1}$). The choice of a good absorber thickness in such cases is a compromise between affording a strong signal and avoiding low count rate due to nonresonant γ-attenuation. Optimization of this ratio is the topic of this section.

3.2.1.1 Basic Considerations

The strength of a Mössbauer signal is determined by the *effective thickness* of the absorber, $t = f_A N_M \sigma_0$ (2.27). This dimensionless factor includes the number N_M of Mössbauer nuclei per square centimetre, the Debye–Waller factor f_A of the absorber material, and the resonance cross-section σ_0 of the Mössbauer isotope. For a multiline spectrum, the result must be split into separate values for each line, which are obtained by weighting t with the relative transition probability of each line.

The *relative absorption depth* of the Mössbauer line is determined by the product of the recoil-free fraction f_S of the Mössbauer source and the *fractional absorption* $\varepsilon(t)$ of the sample, $abs_r = f_S \cdot \varepsilon(t)$, where $\varepsilon(t)$ is a zeroth-order Bessel function ((2.32) and Fig. 2.8). Since $\varepsilon(t)$ increases linearly for small values of t, the "thin absorber approximation," $\varepsilon(t) \approx t/2$, holds up to $t \approx 1$. On the other hand, values as small as $t = 0.2$ may cause already appreciable *thickness broadening* of the Mössbauer lines, according to (2.31), $\Gamma_{exp} \approx 2\Gamma_{nat}(l + 0.135t)$. In practice, therefore the sample thickness may be limited to values of $t \approx 0.2$–0.5 to keep the resulting distortions of line areas, heights, widths, and shape at a tolerable level.[8]

Using the value $t = 0.2$ for the effective thickness, the amount of resonance nuclei (^{57}Fe) for a good "thin" absorber can be easily estimated according to the relation $N_M = t/(f_A \cdot \sigma_0)$. For a quadrupole doublet with two equal absorption peaks of natural width and a recoil-free fraction of the sample $f_A = 0.7$ one obtains for the concentration of a good thin absorber: $N_M{}^{thin} = 2.23 \cdot 10^{17}$ atoms per cm^2 or 21.1 mg cm^{-2} (the resonance cross-section for ^{57}Fe is $\sigma_0 = 2.56 \cdot 10^{-18}$ cm^2). One milligram of natural iron, having 2.18% of ^{57}Fe, corresponds to $2.35 \cdot 10^{17}$ resonance nuclei, or 0.386 μmol ^{57}Fe.

Thus, one may state as a rule of thumb that an absorber should have about 21 μg cm^{-2} of ^{57}Fe (0.37 μmol cm^{-2}) or ca. 1 mg cm^{-2} of natural iron (17.9 μmol cm^{-2}) for a symmetric quadrupole doublet with natural line width.

The corresponding absorption depth for $t = 0.2$ is 0.074, or 7.4% of the baseline counts ($abs_r = 0.7 \cdot t/2$). In practice, this value can be hardly accomplished because of inhomogeneous line broadening arising from various sources and because of increased background due to the contributions from nonresonant radiation reaching the detector. The experimental absorption depth for a quadrupole doublet recorded with an absorber having 1 mg Fe per cm^2 will not be more than ≈ 5% if the intrinsic line width is 0.26 – 0.28 mm s^{-1} as is often found for inorganic compounds, whereas $2\Gamma_{nat} = 0.194$ mm s^{-1}.

One can also infer in turn from these arguments that the relative absorption depth of a Mössbauer line should not exceed 10–15%, because of the increasing thickness broadening and the related line distortions.

3.2.1.2 Counting Statistics and Acquisition Time

The Mössbauer spectrometer will typically divide the velocity scale into 256 channels. For a 0.93 GBq source (25 mCi), the total count rate of photons arriving at the detector and having the proper pulse-height is usually about $C = 20{,}000$ counts s^{-1}. Only about 85% of these will be 14.4 keV radiation; the others are

[8] It is difficult to give an exact limit because the impact of thickness broadening depends on the intrinsic width of experimental lines [31], which often exceeds the natural width $2\Gamma_{nat}$ by 0.05–0.1 mm s^{-1} for ^{57}Fe as studied in inorganic chemistry. This inhomogeneous broadening, which is due to heterogeneity and strain in the sample, causes a reduction of the effective thickness. Rancourt et al. have treated this feature in detail for iron minerals [32].

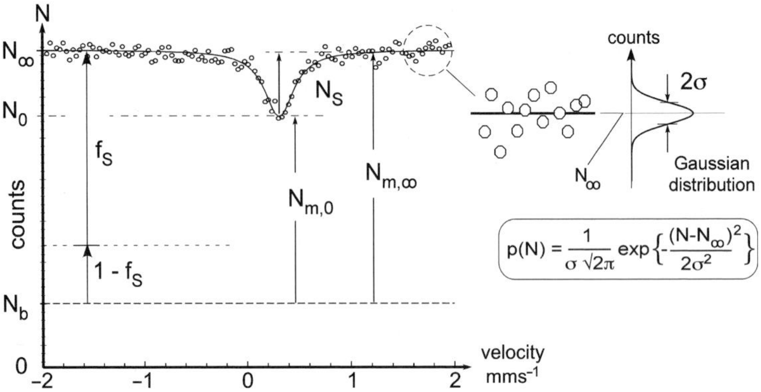

Fig. 3.12 Contributions to a Mössbauer spectrum given in counts per channel. N_b is the nonresonant background from scattered γ-radiation and X-ray fluorescence in source and absorber

nonresonant background from Compton scattering, etc. If the spectrometer efficiency is 66%, the counting rate of 14.4-keV Mössbauer pulses per channel will be typically $C' \approx 40$ counts per second per channel ($= C \cdot 0.85 \cdot 0.66/256$). Thus, when the mean numbers of counts collected in and off resonance after a certain acquisition time Δt are denoted N_0 and N_∞, respectively (see Fig. 3.12), the absorption depth of the spectrum will be $N_s = (N_\infty - N_0)$, or $N_s \approx \text{abs}_r \cdot C' \cdot \Delta t$.

Since γ-emission is a stochastic process, the collected data will scatter around their mean values. The statistics of this result are described by the Poisson distribution which for large numbers N approaches a Gaussian of half-width $\sigma = \sqrt{N} \equiv \Delta N$. Whenever the baseline at N_∞ is exactly known (from a line fit, etc.), the statistical error or "noise" of the Mössbauer signal is essentially given by the distribution of counts at resonance. Since for sufficiently small values of abs_r the distribution width $\sqrt{N_0}$ is nearly the same as $\sqrt{N_\infty}$, we can adopt the approximation $\Delta N_S \approx \sqrt{N_\infty} = \sqrt{C' \cdot \Delta t}$. A common and convenient expression for the quality of the data is the SNR, which we can define here by the simple approximation $\text{SNR} = N_S/\Delta N_S$. Substitution of the relations for N_S and ΔN_S yields $\text{SNR} = \text{abs}_r \cdot \sqrt{C' \cdot \Delta t}$, and resolving for the Δt finally gives us the acquisition time.

$$\Delta t = \text{SNR}^2 / \left(\text{abs}_r^2 \cdot C'\right). \tag{3.1}$$

Hence, if we require a SNR of 40 for the spectrum, the acquisition time Δt will be 16,000 s or about $4 \cdot 3/4$ h for the aforementioned conditions ($\text{abs}_r = 0.05$, $C' = 40$ counts s^{-1} per channel).

3.2.1.3 Minimal Thickness of a Mössbauer Sample

The fact that according to (3.1) the run time for a spectrum is inversely proportional to the square of the peak absorption abs_r and that the "noise" improves only with the square root of time implies also that there is a *lower limit* for the effective thickness of a Mössbauer absorber. If, for instance, the iron content of a sample is ten times less than that of a "good" sample as discussed earlier ($t = 0.2$), the acquisition time will be 100 times longer, which means $\Delta t \approx 445$ h. Since this may be at the limit of acceptable acquisition time, a sample for ^{57}Fe spectroscopy should have an effective thickness of at least $t_{min} \approx 0.02$, corresponding to 0.1 mg cm^{-2} of natural iron or 0.039 μmol of ^{57}Fe cm^{-2}.

3.2.2 Absorber Optimization: Mass Absorption and Thickness

The thickness of a Mössbauer sample affects not only the strength of the Mössbauer signal but also the intensity of the radiation arriving at the detector because the γ-rays are inherently attenuated by the sample because of nonresonant mass absorption caused by the photo effect and Compton scattering as mentioned earlier. The counting rate C in the detector decreases exponentially with the density of the absorber,

$$C = C_0 \exp(-t' \cdot \mu_e), \tag{3.2}$$

where μ_e is the total *mass absorption coefficient* of the sample given in cm^2 g^{-1}, and t' is the *thickness* of the sample, or the *area density* given in g^{-1} cm^2. The attenuation limits the statistical uncertainty of the number of counts N collected in each channel after a certain time Δt because of the relation $\Delta N = \sqrt{N}$ (seen earlier). Since the Mössbauer signal increases almost linearly with the thickness of the sample, whereas the count rate decays exponentially, there must exist an ideal absorber thickness for which the measuring time is minimal and the SNR for a certain acquisition time is maximal.

Particularly for "thin" Mössbauer absorbers with a low concentration of the resonance nuclide and high mass absorption, it may be problematic to apply the recommendation for sample preparation ($t \approx 0.2$), because the resulting electronic absorption may be prohibitively high. In such a case, it may pay well to optimize the absorber thickness, i.e., the area density t'. To this end, following the approach of Long et al. [33], we adopt the general expression:

$$\mathrm{SNR} = N_S \Big/ \sqrt{\Delta N_\infty{}^2 + \Delta N_0{}^2}, \tag{3.3}$$

where $N_S = N_\infty - N_0$ represents the signal amplitude in counts and N_0 and N_∞ are the counts per channel in and off resonance (Fig. 3.12). A relative simple solution to the problem can be obtained for absorbers with low signal ($t \leq 1$) when the nonresonant background N_b can be neglected. Long et al. have shown [33] that the variations of SNR then obey the plain exponential.

$$\mathrm{SNR}(t') \propto t'\, \mathrm{e}^{-t'\mu_e/2}. \tag{3.4}$$

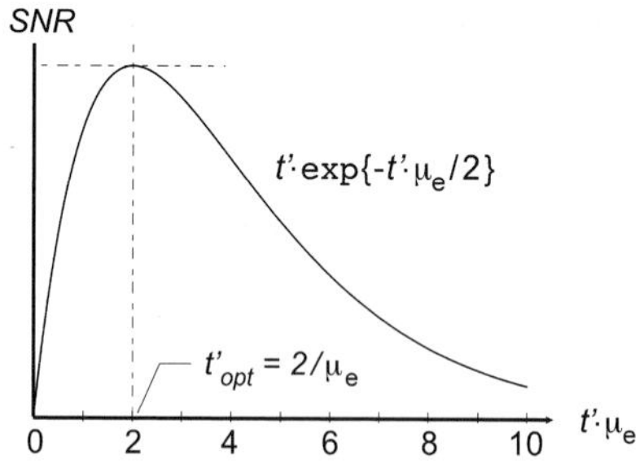

Fig. 3.13 Signal-to-noise ratio of Mössbauer spectra as a function of the area density t' of the sample for thin absorbers ($t \leq 1$) and negligible nonresonant background, $N_b \ll N_\infty$

Interestingly, the curve shape of SNR(t') does not depend explicitly on the concentration of Mössbauer nuclei in the sample. The function has a maximum at $t'\mu_e = 2$ (see Fig. 3.13), as one can easily verify from the root of the first derivative. The ideal thickness of a Mössbauer absorber is therefore given by [33]:

$$t'_{opt} = 2/\mu_e \text{ (for low nonresonant background, } N_b \ll N_\infty) \tag{3.5}$$

When N_b cannot be neglected, the analysis yields [33]

$$t'_{opt} = 1/\mu_e \text{ (for high nonresonant background } (N_b > N_\infty/2). \tag{3.6}$$

The derivation of the expressions (3.3)–(3.6) is found in Appendix A (cf. CD-ROM). Since for most ^{57}Fe-spectra the level of nonresonant background counts, N_b, may be in the range of 10–30% of the total counts, the absorber thickness is usually best adjusted to a value between the limits given above. The maximum of SNR(t') is naturally rather broad, such that deviations from t'_{opt} of even ±50% are fairly immaterial.

Conclusions. A Mössbauer sample with a low content of the resonance nuclide has ideal thickness when it attenuates the incident radiation by ca. 63–85% ($\mu_e \cdot t' = 1$–2, $C'/C_0 \approx e^{-1} - e^{-2}$). However, the optimization should be subordinated to the requirement of a "thin absorber" having an effective thickness $t < 1$ to avoid excessive line broadening.

3.2.2.1 Mass Absorption Coefficients

The mass absorption coefficient μ_e used in (3.5)–(3.6) is obtained from tabulated mass absorption coefficients (given in cm^2 g^{-1}) and the *mass fractions* c_i for the elements in the sample:

$$\mu_e = \sum c_i \mu_i. \tag{3.7}$$

Mass fractions for molecules are easily obtained from the relation $c_i = M_i/M$, where M_i is the product of the atomic mass of the ith element and its abundance in

Table 3.3 Mass fractions and mass absorption coefficients for $C_{1610}H_{3624}N_{110}FeP_{101}F_{606}$

	M_i [a]	c_i	$\mu_{e,i}$ (cm^2 g^{-1})	$c_i\mu_{e,i}$ (cm^2 g^{-1})
C	12 · 1,610	0.493	0.87	0.429
H	1 · 3,624	0.093	0.387	0.036
N	14 · 110	0.039	1.4	0.055
Fe	55.85 · 1	0.001	64	0.064
P	31 · 101	0.079	14.2	1.122
F	19 · 606	0.294	2.7	0.794
				$\Sigma = 2.499$ cm^2 g^{-1}

[a] Atomic mass times abundance per formula unit; the total molar mass is $M = 39{,}228$ g mol^{-1}

the formula unit, and M is the total molar mass of the compound. Atomic mass absorption coefficients for the 14.4 keV radiation of ^{57}Fe are summarized in Appendix B (cf. CD-ROM) and can also be found in [33, 34].

An Example

For illustration, we shall demonstrate the thickness optimization of the iron(III) bis-azide complex, [(cyclam)$Fe^{III}(N_3)_2$]PF_6, mixed with a 100-fold excess of (TBA)PF_6 (tetrabutylammonium–hexafluorophosphate), which is often used as a conducting electrolyte in electrochemistry. The iron compound has the chemical composition $C_{10}H_{24}N_{10}FePF_6$ and a molar mass of $M = 485.2$ g mol^{-1}. (TBA)PF_6 has the composition $C_{16}H_{36}NPF_6$ and a mass of $M = 387.428$ g mol^{-1}. We define an effective composition of the mixture by summing up the composition of the compound and one hundred times that of (TBA)PF_6, yielding $C_{1610}H_{3624}N_{110}FeP_{101}F_{606}$ with an effective mass of $M = 39{,}228$ g mol^{-1}. The data for the participating atoms are compiled in Table 3.3. The total mass absorption coefficient is found to be $\mu_e = 2.499$ g cm^{-2}, which according to (3.5) yields the ideal absorber thickness $t'_{opt} = 2/2.499$ cm^2 g^{-1} $= 0.89$ g cm^{-2}, corresponding to about 1.3 mg of natural iron per cm^2!

3.2.2.2 Solvents, Solutions, and Powders

Mass absorption increases strongly with the atomic number Z. For the 14.4 keV radiation of ^{57}Fe, the coefficient follows approximately the relation $\mu_e \approx 0.003 \cdot Z^3$ from oxygen to krypton. Therefore, organic solvents containing sulfur or chlorine are virtually opaque to the Mössbauer radiation. The sulfur component of a 2 mm layer of dimethylsulfoxide (DMSO) absorbs ≈70% of the Mössbauer radiation ($t' = 1.1$ g cm^{-2}) [35]. Even worse is dichloromethane (CH_2Cl_2), having an absorption coefficient of 16.83 cm^2 g^{-1}. A layer of 0.1 g cm^{-2}, which is only 0.75 mm thick ($\rho = 1.33$ g cm^{-3}), absorbs about 82% of 14.4 keV radiation.[9] For the same reason, chlorinated polymers (PVC) or glass should not be used for

[9] The mass absorption of a solution of an iron complex in CH_2Cl_2 is usually entirely dominated by the solvent. With $\mu_e = 16.83$ cm^2 g^{-1}, the optimized sample thickness t'_{opt} is between 0.059 and 0.119 g cm^{-2}, or 45–89 μl cm^{-2}, which corresponds to a layer thickness of 0.45–0.89 mm.

making sample holders. Water has an absorption coefficient of $\mu_e = 1.7\ cm^2\ g^{-1}$, whereas a typical protein has $\mu_e = 1.3\ cm^2\ g^{-1}$ [35]. The ideal thickness of protein samples with water as the solvent is about 0.8 g cm^{-2}, which corresponds to a layer thickness of about 8 mm.

Very thick absorbers may be required for *applied-field measurements* to achieve reasonable absorption depths and measuring times because the Mössbauer spectra are usually split into several hyperfine components. Here the iron content may be as large as ≈100 μg ^{57}Fe per cm^2 (1.75 μmol ^{57}Fe per cm^2), which would correspond to $t \approx 1$ for a two-line spectrum. For studies of frozen solutions, ^{57}Fe concentrations of 1 mM are desirable for each nonequivalent iron site [35].

Mössbauer absorbers should be reasonably homogeneous. Holes must be avoided, particularly when particles or frozen liquids are mounted. A mixture of grains having large thickness with spots of vanishing iron content may yield thickness broadening and extremely small absorption. Powder material must be fixed to avoid sliding. Moreover, minor "roughness" of the absorber layer hardly affects the performance of the spectrometer because the γ-beam does not pass through any optical component.

3.2.2.3 Isotope Enrichment

Compounds with natural iron can be measured in nonchlorinated organic solvents or water when the concentration is at least 35–50 mM (sample thickness 6–8 mm). Since the natural abundance of ^{57}Fe is only 2.18%, the Mössbauer isotope can be enriched up to 40 times to lower the sample concentration if necessary. The stable isotope is available as metal sheets, metal powder, or oxide with about 95% of the isotope ^{57}Fe. Proteins are usually prepared by in vivo enrichment up to 90% of ^{57}Fe; in vitro enrichment is also possible if the iron site of the natural protein can be metal-depleted and reconstituted with the Mössbauer isotope ions in the original oxidation state. An interesting aspect of both methods is the possibility of selective enrichment of certain iron sites if multicomponent enzymes are to be studied. For the enrichment of synthetic compounds, it might be favorable not to use the highest isotope concentration, but rather to synthesize a larger amount of the substance with only 30–50% enrichment. Handling is facilitated, and good absorbers can be easily prepared, if solubility of the compound is sufficiently high.

3.2.3 *Absorber Temperature*

The recoil-free fraction f_A of transition metal complexes or proteins in frozen solution can be as small as 0.1–0.3, when measured just below the melting point, but the *f*-factor increases strongly when the temperature is lowered to liquid nitrogen temperatures (77 K), and at liquid helium temperatures (4.2 K) it may reach values of 0.7–0.9 [35]. This makes a substantial difference to the acquisition time of the spectra because of the square dependency on the signal (3.1).

Paramagnetic compounds are measured at liquid helium temperatures to slow-down spin relaxation when magnetic hyperfine parameters are to be evaluated, and to increase the difference in Boltzmann population of the magnetic sublevels. For half-integer spin systems, the application of fields of only a few milli-Tesla may induce sizable magnetic hyperfine splitting in the static limit of slow relaxation without line broadening (see Sect. 4.6 and Chap. 6).

Mössbauer spectra can yield valuable information about the abundance of different Mössbauer sites in a sample from the relative intensities of the corresponding subspectra if the *f*-factors are known. These, however, depend critically on temperature. Differences for the individual sites must be expected at ambient temperatures; however, they all vanish at 4.2 K because $f(T)$ approaches one for $T \rightarrow 0$ (see (2.15)). Often measurements at 80 K are sufficient for reliable estimates of the true intensity ratio of Mössbauer subspectra.

3.3 The Miniaturized Spectrometer MIMOS II

Göstar Klingelhöfer[10]

3.3.1 Introduction

Iron-57 Mössbauer spectroscopy has been used in earth-based laboratories to study the mineralogy of iron-bearing phases in a variety of planetary samples, including lunar samples returned to earth by American Apollo astronauts and Soviet (Russian) robotic missions, and meteorites that have asteroidal and Martian origin (see [36, 40] and references therein). In the early 1990s, the development of a miniaturized Mössbauer spectrometer was initiated by E. Kankeleit at the Technical University of Darmstadt, originally for the Russian Mars mission Mars-92/94. This development was led by Klingelhöfer, who brought the project to completion for the NASA Mars Exploration Rover 2003 mission and the ESA Mars-Express Beagle 2 mission at the Institute of Inorganic Chemistry, University of Mainz. The Beagle 2 mission failed, but the Mars Exploration Rovers (MER) (Fig. 3.14; see also Sect. 8.3) Spirit and Opportunity landed successfully on the Red Planet in January 2004. The goal of the mission was to search for signatures of water possibly from the past. Both Rovers, Spirit and Opportunity [37–39, 53–55, 60, 62], carry the miniaturized Mössbauer spectrometer MIMOS II [36] as part of their scientific payload. For more details, see Sect. 8.3. In the following section we describe the technical details and specialities of the miniaturized Mössbauer spectrometer MIMOS II.

[10]Institut für Anorganische Chemie und Analytische Chemie, Johannes Gutenberg-Universität Mainz, Staudingerweg 9, 55099 Mainz, Germany; e-mail: klingel@uni-mainz.de

Fig. 3.14 *Left*: NASA Mars-Exploration-Rover (artist view; courtesy NASA, JPL, Cornell). On the front side of the Rover the robotic arm carrying the Mössbauer spectrometer and other instruments can be seen in stowed position. *Right*: robotic arm before placement on soil target at Victoria crater rim, Meridiani Planum, Mars. The Mössbauer instrument MIMOS II with its circular contact plate can be seen, pointing towards the rover camera. See also Sect. 8.3

3.3.2 Design Overview

The MIMOS instrument II is extremely miniaturized compared to standard laboratory Mössbauer spectrometers and is optimized for low power consumption and high detection efficiency [36, 40, 41]. All components were selected to withstand high acceleration forces and shocks, temperature variations over the Martian diurnal cycle, and cosmic ray irradiation. Because of restrictions in data transfer rates, most instrument functions and data processing capabilities, including acquisition and separate storage of spectra as a function of temperature, are performed by an internal dedicated microprocessor (CPU) and memory. The dedicated CPU is also required because many Mössbauer measurements are done at night when the rover CPU is turned off to conserve power. High detection efficiency is extremely important to minimize experiment time. Experiment time can also be minimized by using a strong ^{57}Co/Rh source. Instrument internal calibration is accomplished by a second, less intense radioactive source mounted on the end of the velocity transducer opposite to the main source and in transmission measurement geometry with a reference sample (see also Sects. 3.1.3 and 3.3.6). The spectrometer can also be calibrated using an external target (e.g., Fe-metal or magnetite) on the rover spacecraft.

Physically, the MIMOS II Mössbauer spectrometer has two components that are joined by an interconnect cable: the sensor head (SH) and electronics printed-circuit board (PCB). On MER, the SH is located at the end of the Instrument Deployment Device (IDD) and the electronics board is located in an electronics box inside the rover body. On Mars-Express Beagle-2, a European Space Agency (ESA) mission in 2003, the SH was mounted also on a robotic arm integrated to the Position

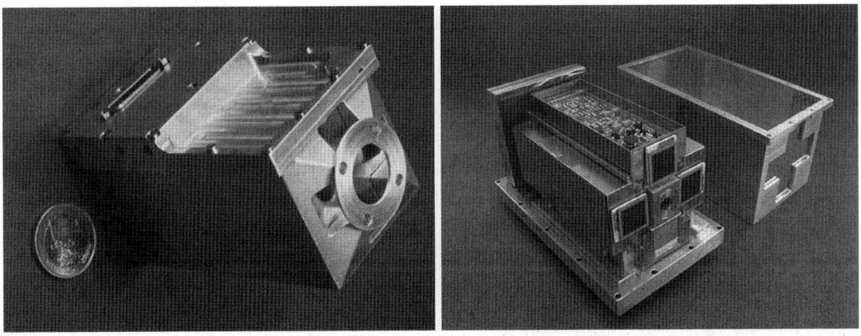

Fig. 3.15 *Left*: External view of the MIMOS II sensor head (SH) with pyramid structure and contact ring assembly in front of the instrument detector system. The diameter of the one Euro coin is 23 mm; the outer diameter of the contact-ring is 30 mm, the inner diameter is 16 mm defining the field of view of the instrument. *Right*: Mimos II SH (without contact plate assembly) with dust cover taken off to show the SH interior. At the front, the end of the cylindrical collimator (with 4.5 mm diameter bore hole) is surrounded by the four Si-PIN detectors that detect the radiation re-emitted by the sample. The metal case of the upper detector is opened to show its associated electronics. The electronics for all four detectors is the same. The Mössbauer drive is inside (in the center) of this arrangement (see also Fig. 3.16), and the reference channel is located on the back side in the metal box shown in the photograph

Adjustable Workbench (PAW) instrument assembly. The SH shown in Figs. 3.15 and 3.16 contains the electromechanical transducer (mounted in the center), the main and reference ^{57}Co/Rh sources, multilayered radiation shields, detectors and their preamplifiers and main (linear) amplifiers, and a contact plate and sensor. The contact plate and contact sensor are used in conjunction with the IDD to apply a small preload when it places the SH holding it firmly against the target. The electronics board contains power supplies/conditioners, the dedicated CPU, different kinds of memory, firmware, and associated circuitry for instrument control and data processing. The SH of the miniaturized Mössbauer spectrometer MIMOS II has the dimensions $\sim (5 \times 5.5 \times 9.5)$ cm^3 and weighs only ca. 400 g. Both 14.4 keV γ-rays and 6.4 keV Fe X-rays are detected simultaneously by four Si-PIN diodes. The mass of the electronics board is about 90 g [36].

3.3.2.1 Mössbauer Sources, Shielding, and Collimator

To minimize experiment time, the highest possible source activity is desirable, with the constraint that the source line width should not increase significantly (maximum by a factor of 2–3) over the ~9–12 months' duration of the mission. Calculations and tests indicate an optimum specific activity for ^{57}Co at 1 Ci per cm^2 [42, 43]. Sources of ~350 mCi ^{57}Co/Rh with a specific activity close to this value and extremely narrow source line width (<0.13 mm s^{-1} at room temperature), given

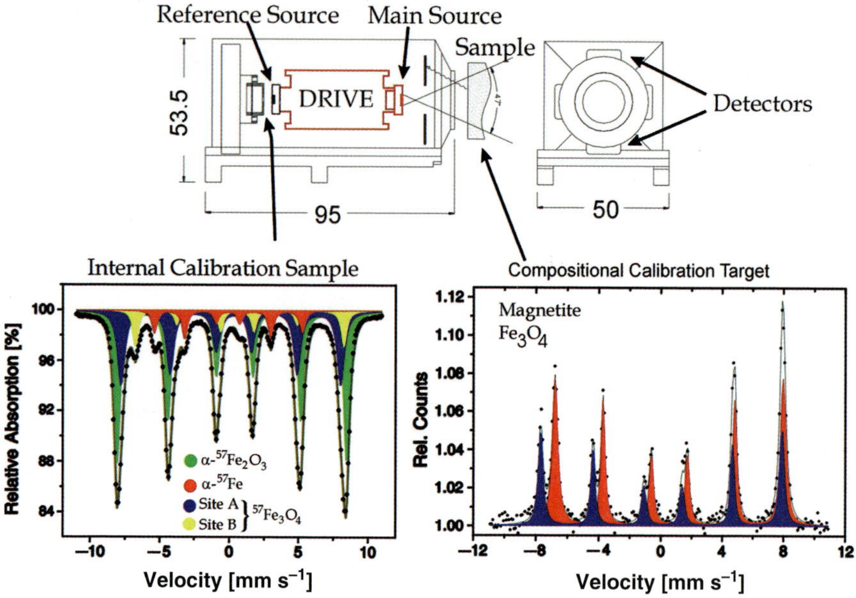

Fig. 3.16 Schematic drawing of the MIMOS II Mössbauer spectrometer. The position of the loudspeaker type velocity transducer to which both the reference and main $^{57}Co/Rh$ sources are attached is shown. The room temperature transmission spectrum for a prototype internal reference standard shows the peaks corresponding to hematite (α-Fe_2O_3), α-Fe, and magnetite (Fe_3O_4). The internal reference standards for MIMOS II flight units are hematite, magnetite, and metallic iron. The backscatter spectrum for magnetite (from the external CCT (Compositional Calibration Target) on the rover) is also shown

the high activity, were produced by Cyclotron Co. Ltd. (Russia) in custom-made space-qualified Ti-holders, tested successfully, and mounted in flight instruments approximately 90 days prior to launch. The rhodium matrix precludes additional line broadening at lower temperatures on Mars.

The effective shielding of the detector system from direct and cascade radiation from the $^{57}Co/Rh$ source is also very important. A graded shield consisting of concentric tubes of brass, tantalum, and lead was selected. The thickness and the shape of different parts of the shielding were optimized so that nearly zero direct 122 and 136 keV radiation (emitted by the ^{57}Co source) was in a direct line with the detectors (see Fig. 3.16).

The shielding also acts as the collimator which fixes the diameter of the target that is illuminated by γ-rays. As discussed previously, this diameter is as large as possible to minimize experiment time within the constraint of acceptable cosine smearing [44, 50, 51]. The measure used for acceptable cosine smearing was the ability to reliably resolve the strongly overlapping spectra of hematite and maghemite in a 1:1 mixture of those oxides. A series of experiments with this mixture and the pure oxides were conducted at constant source intensity and variable collimator diameters between 4.5 and 7.1 mm, which correspond to illumination diameters of

∼12 and ∼17 mm, respectively. The analysis of the mixture of spectra reproduced the Mössbauer parameters for and relative proportions of the individual oxide components for collimator diameters of 4.5 and 5.6 mm. For collimator diameters of 6.2 and 7.1 mm, the calculated and actual relative proportions of the oxides differed significantly. Therefore, a collimator diameter of 5.6 mm was selected for MIMOS II flight instruments. This diameter limits the maximum γ-ray emission angle to ∼25° and gives a "beam diameter" of ∼14 mm on the sample (see also Sect. 3.3.3).

3.3.2.2 Drive System

The simplest way to meet volume and weight constraints was to scale down drive systems we have built for laboratory instruments for many years [4, 5]. We constructed a drive system that had about one fifth the size of a standard laboratory system. It had a diameter of 22 mm, a length of 40 mm, and ∼50 g mass [45] (see Fig. 3.17; left). The MIMOS II design is based on a rigid Al tube connection between drive and the velocity pick-up coils in the double-loudspeaker arrangement with good electrical and magnetic shielding between the two coils to avoid cross-talk. The intense main ^{57}Co/Rh source is mounted at one end of the Al tube, and the weaker source for the reference absorber is mounted at the other end. The short tube guarantees a fast transfer of information with the velocity of sound in the Al and thus a minimum phase lag and a high feedback gain margin. Fortunately, and despite the increase of unwanted crosstalk resulting from the smaller distance between the coils, the crosstalk relative contribution is still less than 0.01% in the frequency domain of the triangular waveform and is therefore negligible. The system is equipped with SmCo permanent magnets and was optimized to give a

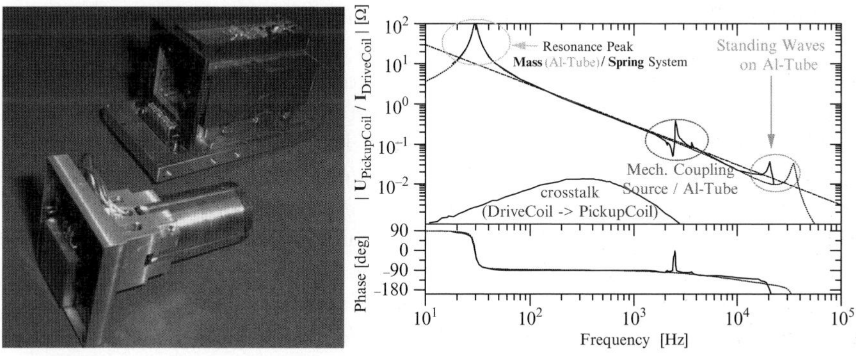

Fig. 3.17 *Left/bottom*: Picture of the drive (cylindrical piece; length 40 mm; diameter 22 mm) and its control unit (quadratic box holding the cylindrical drive unit); this whole unit slides into the detector system unit (top part in the left picture); *Right*: Transfer function of MIMOS II velocity transducer (see also Sect. 3.1.1). For details of the drive unit and the model describing its behavior see [36, 45]

homogeneous and high magnetic field in the coil gaps. The drive operates at a frequency of ~25 Hz, which is also its main resonance (Fig. 3.17; right). This low frequency allows a broad bandwidth for the closed loop system, and good performance with a triangular reference signal, but requires rather soft springs. As a consequence, rotation of the drive from horizontal to vertical position in Earth gravity leads to a shift of about 0.4 mm from the equilibrium position of the tube. However, the resulting nonlinearity between velocity and pickup voltage remains <0.1% at room temperature. In order for the drive to operate at expected Martian surface temperatures (~170–280 K), it was necessary to degrade the velocity linearity slightly by modification of the feedback circuit, which after calibration is correctable by software. The design provides mechanical limiters to avoid destruction of the springs during the large accelerations associated with launch and landing. Vibration and shock tests of the drive system with levels up to several 100 g, slightly exceeding the specifications for both MER and Beagle-2 missions, were successfully performed.

3.3.2.3 Detector System and Electronics

The main disadvantage of the backscatter measurement geometry employed by MIMOS II is the secondary radiation caused by primary 122 and 136 keV radiation from the decay of the ^{57}Co source. To reduce the background at the energies of the 14.4 keV γ-ray and the 6.4 keV X-ray lines, a detector with good energy resolution is required. In addition, a detector system covering a large solid angle is needed to minimize data acquisition time. Good resolution is even more important should it prove possible to use these detectors for elemental analysis with the X-ray fluorescence (XRF) technique (i.e., using the PHA spectra that are also acquired as a part of our measurement procedure). For this reason, four Si-PIN-diodes with a 10×10 mm^2 active area were selected as detectors, instead of gas-counters as considered by others. A detector thickness of about 400–500 μm was a good choice according to our calculations and experience. The energy resolution is ~1.0–1.5 keV at room temperature and improves at lower temperatures. The efficiencies at 6.4 and 14.4 keV are nearly 100% and about 70%, respectively [46]. The 100V DC bias voltage for the detector diodes is generated by high-frequency cascade circuitry with a power consumption of less than 5 mW. Noise contributions are minimized by incorporating a preamplifier–amplifier–SCA (single channel analyzer) system for each individual detector. Spectra of ^{57}Co/Rh radiation backscattered from an Al and a stainless steel (SS) plate (same recording times) are shown in Fig. 3.18. A continuum is seen above 122 keV resulting from the few 692 keV γ-rays which are not completely absorbed in the shielding. Although no photo peak appears at 122 keV, this radiation shows up as a broad Compton distribution, being more intense for the lower Z Al. A second Compton distribution originates in the detector itself as seen in the rising slope starting below 40 keV. The peak at 22.1 keV results from the silver (Ag) backing of the detector. Below this energy, the 14.4 keV Mössbauer resonance line and

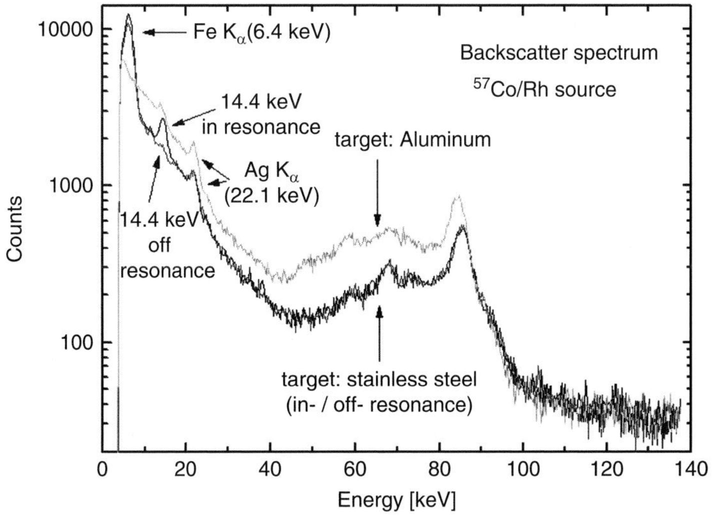

Fig. 3.18 Pulse-height analysis (PHA) spectrum (or energy spectrum) for ^{57}Co/Rh Mössbauer source radiation backscattered nonresonantly and/or resonantly from aluminum and stainless steel plates. Data were obtained with Si-PIN diodes with sensitive area of 1 cm^2 per diode and a thickness of $\sim$400 μm (from [36, 46])

also the 6.4 keV X-ray line dominate in SS at zero velocity, in contrast to Al, where only the $((14.4 - 0.4) = 14.0)$ keV Compton scattered line appears.

In addition to the four detectors used to detect backscattered radiation from the sample, there is a fifth detector to measure the transmission spectrum of the reference absorber (α-^{57}Fe, α-$^{57}Fe_2O_3$, Fe_3O_4; see Fig. 3.16). Sample and reference spectra are recorded simultaneously, and the known temperature dependence of the Mössbauer parameters of the reference absorber can be used to give a measurement of the average temperature inside the SH, providing a redundancy to measurements made with the internal temperature sensor (see Sect. 3.3.4).

3.3.3 Backscatter Measurement Geometry

Because of the complexity of sample preparation, backscatter measurement geometry (see Fig. 3.19) is the choice for an in situ planetary Mössbauer instrument [36, 47–49]. No sample preparation is required, because the instrument is simply presented to the sample for analysis. On MER, the MIMOS II SH is mounted on a robotic arm that places it in physical contact with the analysis target (e.g., rock or soil) [36, 37].

As discussed in Sects. 3.1.1–3.1.3, successful acquisition of Mössbauer spectra depends on accurate knowledge of the relative velocity of the source and sample. External vibrations that impart differential velocity components to the source and sample would degrade the quality of the Mössbauer spectrum. This degradation

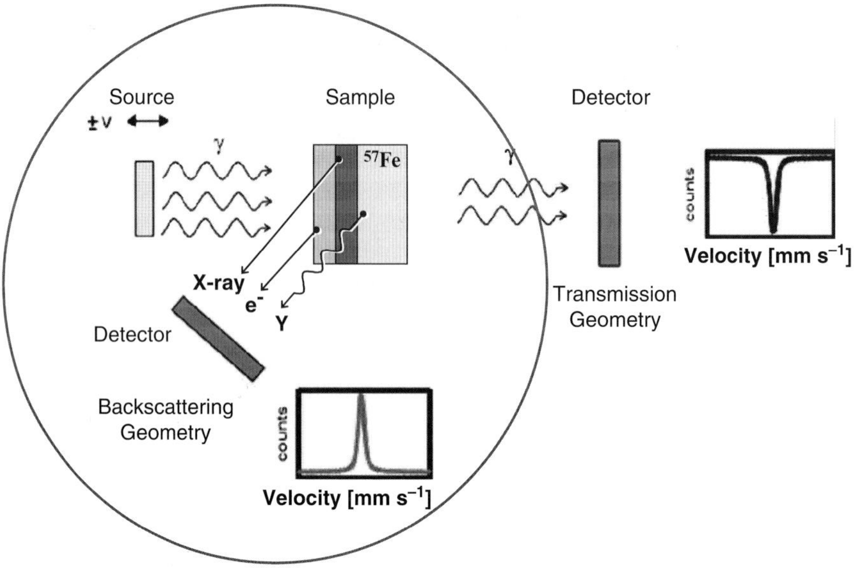

Fig. 3.19 Schematic illustration of the measurement geometry for Mössbauer spectrometers. In transmission geometry, the absorber (sample) is between the nuclear source of 14.4 keV γ-rays (normally ^{57}Co/Rh) and the detector. The peaks are negative features and the absorber should be "thin" with respect to absorption of the γ-rays to minimize nonlinear effects. In emission (backscatter) Mössbauer spectroscopy, the radiation source and detector are on the same side of the sample. The peaks are positive features, corresponding to recoilless emission of 14.4 keV γ-rays and conversion X-rays and electrons. For both measurement geometries Mössbauer spectra are counts per channel as a function of the Doppler velocity (normally in units of mm s^{-1} relative to the mid-point of the spectrum of α-Fe^0 in the case of Fe Mössbauer spectroscopy). MIMOS II operates in backscattering geometry (*circle*), but the internal reference channel works in transmission mode

ranges from slight line broadening in mild cases to complete obliteration of the Mössbauer spectrum in severe cases. External vibrations are not generally a problem in laboratory settings because the sample and velocity transducer are rigidly held. On Mars, wind-induced vibrations are an obvious environmental factor that might degrade the quality of the Mössbauer spectra. However, backscatter spectra obtained for hematite with a prototype MIMOS II instrument during field tests in the Mojave desert, California, did not show detectable line broadening. The MER IDD has been designed to assure that velocity noise at the MIMOS II SH will not exceed 0.1 mm s^{-1} [36, 37].

3.3.3.1 Cosine Smearing

Because instrument volume and experiment time must both be minimized for a planetary Mössbauer spectrometer, it is desirable in backscatter geometry to illuminate as much of the sample as possible with source radiation. However, this requirement at some point compromises the quality of the Mössbauer spectrum because of an effect known as "cosine smearing" [36, 44, 50] (see also Sects. 3.1.8 and 8.3). Examples for scattering geometry are shown in Fig. 3.20. The energy of the 14.4 keV γ-rays incident on the sample is a function of both the velocity of the source and the angle (θ) between the axis of source motion and the emission angle of source γ-rays. The energy is changed by a factor of cos θ, which gives the effect its name. The effect of cosine smearing on the Mössbauer spectrum is to increase

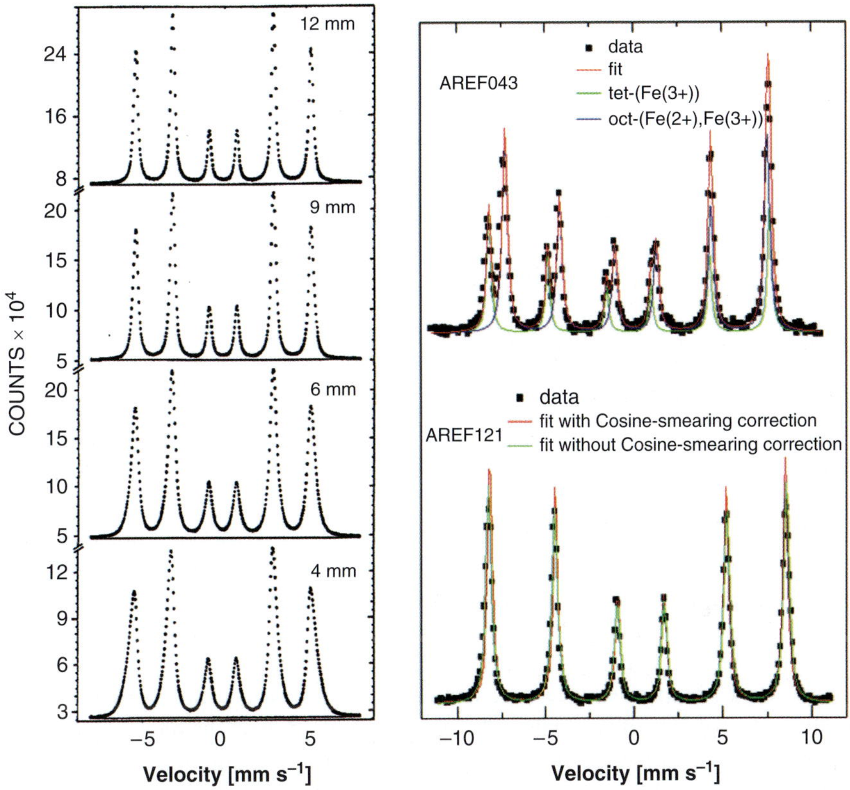

Fig. 3.20 *Left*: Backscattering Mössbauer spectra measured with conversion electrons as a function of the distance (4–12 mm) between sample and source; sample and source diameters are 4 and 3 mm, respectively; the cosine smearing effect is seen (from [44]). *Right*: Mössbauer spectra (512 data channels folded to 256 channels) obtained at 293 K using an engineering-model of MIMOS II spectrometer for two reference rock slabs named AREF043 (*top*) and AREF121 (*bottom*). The experiment time and source intensity for AREF043 were 49 h and 25 mCi, which corresponds to a 12 h integration at the Martian source intensity (~100 mCi). The Mössbauer parameters derived from the fits indicate that the iron-bearing phases in AREF043 and AREF121 are magnetite and hematite respectively. The spectrum AREF043 of magnetite was fitted with correction for cosine smearing. The spectrum AREF121 of hematite shows the difference between a fit with and without correction for cosine smearing (*red and green lines,* respectively). (From [36])

the linewidth of Mössbauer peaks (lower resolution) and shift their centers outward (affects the values of Mössbauer parameters). Therefore, the diameter of the source "γ-ray beam" incident on the sample, which is determined by a collimator, is a compromise between the acceptable experiment time and the acceptable velocity resolution. For coaxial symmetry, cosine smearing can be neglected when the ratio of the collimator radius to the distance between the source and the collimator is ≤ 0.1 [36]. Note that the distortion in peak shape resulting from cosine smearing can be accounted for mathematically in spectral fitting routines, but the reduction in resolution can lead to irretrievable loss of information. Riesenmann et al. [50] derived a formula for the peak shape when cosine smearing is present for coaxial symmetry in transmission measurement geometry. Geometrical considerations are more complicated for MIMOS II, where backscatter measurement geometry is employed and the symmetry, although coaxial for the excitation process, is more complex for detection. Held [51] solved this problem mathematically and implemented the result in the fitting routines we use.

3.3.4 Temperature Dependence and Sampling Depth

3.3.4.1 Temperature Dependence

For measurements on planetary surfaces, the instrument has to meet special requirements. In the case of Mars, MIMOS II is able to work at temperatures as low as -140°C and an average atmospheric pressure of 7 mbar of CO_2 [36, 40, 47, 52–55].

MIMOS II has three temperature sensors: one on the electronics board and two on the SH. One temperature sensor in the SH is mounted near the internal reference absorber, and the measured temperature is associated with the reference absorber and the internal volume of the SH. The other sensor is mounted outside the SH at the contact ring assembly. It gives the approximate analysis temperature for the sample on the Martian surface. This temperature is used to route the Mössbauer data to the different temperature intervals (maximum of 13, with the temperature width software selectable) assigned in memory areas. Shown in Fig. 3.21 are the data of the three temperature sensors taken on Mars (rover Opportunity at Meridiani Planum) in January 2004 between 12:10 PM on Sol 10 (10 Martian days after landing) and 11:30 AM on Sol 11. The temperature of the electronics board inside the rover is much higher than the temperatures inside the SH and the contact plate sensor, which are nearly identical and at ambient Martian temperature.

An example of a simulated overnight experiment on Mars is shown in Fig. 3.22 for eight temperature intervals using the Compositional Calibration Target (CCT; made out of magnetite rock slab) on the rover as the target. In the case of contact-ring temperature sensor failure, the internal temperature sensor would be used (software selectable).

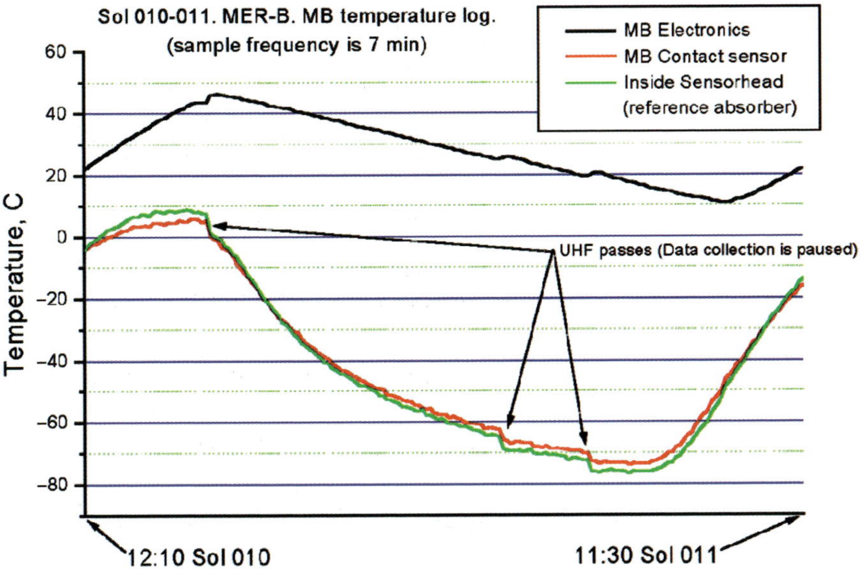

Fig. 3.21 Example of temperature variation as measured by MIMOS II temperature sensors on MER: (i) inside the rover body at MIMOS electronics board (*black curve*); (ii) outside the rover, at the MIMOS II SH (*green and red curves*), which is at ambient Martian temperature: (a) inside the sensor-head, at the reference absorber position (*green*); (b) outside the SH at the sample's contact plate (*red*). Temperatures at the two SH positions are nearly identical (difference less than ~2 K). During data transmission between the rover and the Earth (or the relay satellite in Mars orbit) the instrument is switched off resulting in immediate small but noticeable temperature changes (see figure above)

3.3.4.2 Sampling Depth

In addition to there being no requirement for sample preparation, backscatter measurement geometry has another important advantage (see also Sect. 8.3). Emission of internal conversion electrons, Auger electrons, and X-rays, which occur along with the recoilless emission and absorption of the 14.4 keV γ-ray of ^{57}Fe, can also be used for Mössbauer measurements. For ^{57}Fe, internal conversion X-rays have an energy of 6.4 keV. Because the penetration depth of radiation is inversely proportional to energy, the average depth from which 14.4 keV γ-rays emerge in emission measurements is greater than that for 6.4 keV X-rays. The importance of this difference in emission depths for an in situ Mössbauer spectrometer is that mineralogical variations that occur over the scale depths of the 14.4 and 6.4 keV radiations can be detected and characterized. Such a situation on Mars has been found for thin alteration rinds and dust coatings on the surfaces of otherwise unaltered rocks [56, 62] (see also Fig. 3.23; right).

Mössbauer spectra of layered samples are influenced by various parameters, such as the thickness of the layers, the density of the sample, and its elemental

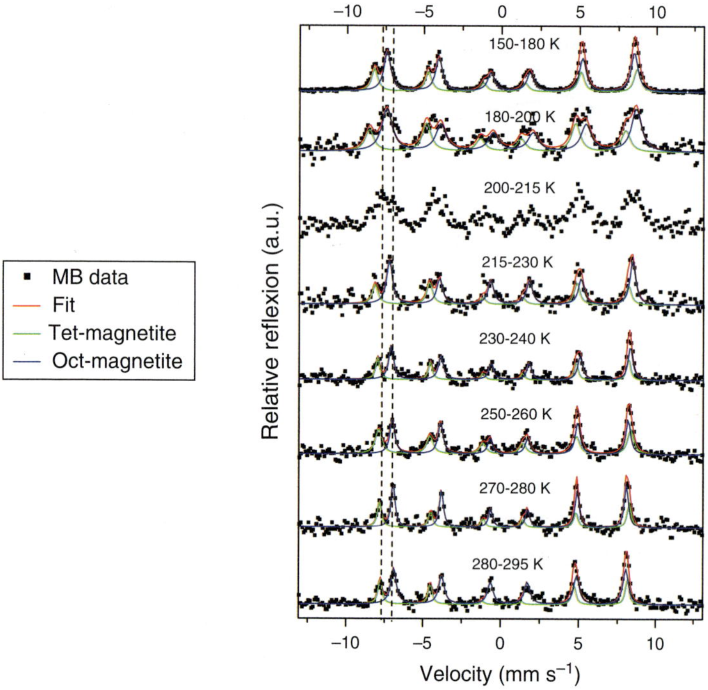

Fig. 3.22 Backscatter MIMOS II spectra collected in eight temperature intervals on the CCT target (magnetite rock) during a simulated overnight Mössbauer experiment on Mars

composition, especially the iron content. A Monte Carlo simulation allows for the independent variation of these parameters to study their influence on 6.4 and 14.4 keV spectra. The comparison of simulated and measured spectra can then be used to estimate the thickness of a surface layer.

The simulation described in [56] models the geometry of the MIMOS II instrument [36]. A sample composed of two distinct homogeneous layers, each containing up to ten different Mössbauer subspectra (singlets, doublets, and sextets), can be modeled. For every photon its emission from the Mössbauer source, interaction processes in the sample, and detection are simulated, updating its energy and direction of propagation after each interaction according to the source velocity. The depth selectivity in Mössbauer spectra has been demonstrated experimentally by analyzing samples composed of two distinct homogeneous layers of a well-known composition, using a laboratory version of the MIMOS II instrument [56]. An olivine thin section with a thickness of 60 μm or iron foils with thicknesses of 10 and 50 μm, respectively, were combined with substrates of pyrite (FeS_2) and hematite (α-Fe_2O_3). These samples were chosen because they differ significantly in their iron content, and their different hyperfine parameters render them easy to distinguish in spectra. Figure 3.23 compares the measured and the corresponding

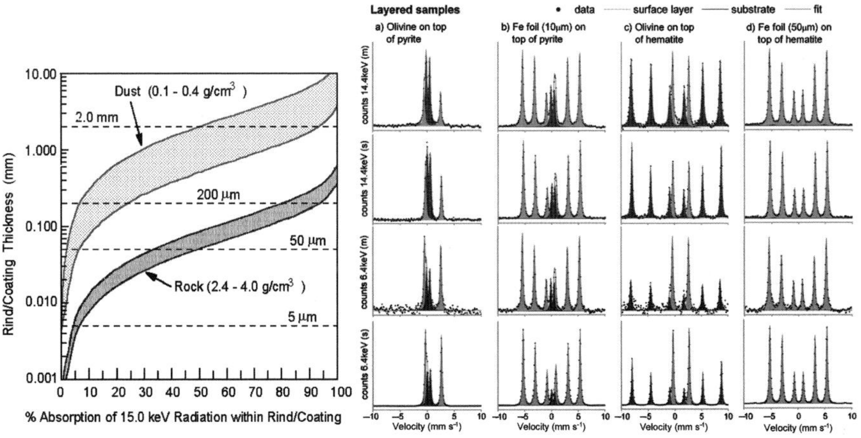

Fig. 3.23 *Left*: Calculated relationship between the thickness of an alteration rind and/or dust coating on a rock and the amount of 15.0-keV radiation absorbed in the rind/coating for densities of 0.4, 2.4, and 4.0 g cm^{-3} [57]. The bulk chemical composition of basaltic rock was used in the calculations, and the 15.0 keV energy is approximately the energy of the 14.4 keV γ-ray used in the Mössbauer experiment. The stippled area between densities of 2.4 and 4.0 g cm^{-3} is the region for dry bulk densities of terrestrial andesitic and basaltic rocks [58]. The stippled area between densities of 0.1 and 0.4 g cm^{-3} approximates the range of densities possible for Martian dust. The density of 0.1 g cm^{-3} is the density of basaltic dust deposited by air fall in laboratory experiments [59]. *Right*: Measured spectra obtained on layered laboratory samples and the corresponding simulated spectra, from top to bottom: 14.4 keV measured (m); 14.4 keV simulated (s); 6.4 keV measured (m); and 6.4 keV simulated (s). All measurements were performed at room temperature. Zero velocity is referenced with respect to metallic iron foil. Simulation was performed using a Monte Carlo-based program (see [56])

simulated spectra obtained from these layered samples at 6.4 and 14.4 keV, demonstrating the depth sensitivity of MIMOS II in backscattering geometry. The results show that the Fe-mineralogical composition of a surface layer and its substrate can be determined by comparing 6.4- and 14.4-keV Mössbauer spectra if the surface layer is not too thick. With the help of a Monte Carlo simulation, it is possible to estimate the thickness of a surface layer such as a weathering rind (see also Sect. 8.3).

3.3.5 Data Structure, Temperature Log, and Backup Strategy

One Mössbauer spectrum consists of 512 velocity channels (3 bytes per channel). One temperature interval consists of five Mössbauer spectra (one for each detector). There are 13 temperature intervals with selectable width. Thus, MIMOS II can accumulate simultaneously up to 65 Mössbauer spectra during one experiment session on Mars. All Mössbauer, energy, engineering, and temperature data taken during this session are stored in a volatile SRAM (Static Random Access

Memory) (128 KB) on the MIMOS II electronics board. Firmware parameters and the instrument logbook are stored in the nonvolatile memory ferroelectric RAM (FRAM) on the electronics board. There are three individual FRAMs on the MIMOS II electronics board with three identical copies of these parameters to ensure parameter integrity. The copies are compared with each other from time to time to verify that they are identical. If one copy deviates from the other two, it is replaced by a copy of the other two identical parameter sets. All parameters can be adjusted during mission operations.

To minimize the risk of data loss because of power failure or other reasons, the Mössbauer data are copied to a nonvolatile EEPROM (Electrically Erasable Programmable Read-Only Memory) every 9 minutes (software selectable). As the size of the EEPROM is smaller than the SRAM, the EEPROM can accumulate only up to ten Mössbauer spectra as a subset of the data from the SRAM. These spectra are obtained from the SRAM according to a pre-defined summation strategy.

3.3.6 Velocity and Energy Calibration

3.3.6.1 Velocity Calibration

The interpretation of acquired Mössbauer spectra is impossible without knowing the drive velocity and in particular the maximum velocity value precisely at any given time. Mössbauer drive velocity calibration for MIMOS II is rather straightforward and can be done in three different ways, thus ensuring redundancy. Prior to flight, each individual drive system was calibrated as a function of the software velocity settings by measuring in the backscattering mode an α-iron foil standard at different software velocity range settings. For each setting, the maximum drive velocity was preset by firmware and the corresponding Mössbauer spectrum acquired. Fitting the acquired Mössbauer spectrum using the well-known parameters of the α-iron foil then yielded the real velocity for that particular setting. In that way, a calibration curve for the real "maximum velocity" value as a function of the internal software velocity setting was obtained for each Mössbauer unit. This procedure was repeated at different temperatures within the Martian ambient range to account for the temperature dependence of the SmCo magnets in the drive and the drive control electronics.

During the mission, the magnetite CCT was measured in several runs to verify the functionality of MIMOS II. The well-known Mössbauer parameters of magnetite were used for velocity calibration, as shown in Fig. 3.22 for different temperatures. This kind of measurement was done in the laboratory with the flight units as a function of temperature to be used as a reference for the measurements on Mars. Figure 3.22 shows the Mössbauer spectra of the CCT at different mean temperatures.

The primary method for velocity calibration is the internal reference channel with reference target and detector configured in transmission geometry (Fig. 3.16).

The reference target is a mixture of α-Fe^0 (metallic iron, 30% enriched ^{57}Fe), α-Fe_2O_3 (hematite), and Fe_3O_4 (magnetite), both Fe-oxides 95% enriched in ^{57}Fe, and its Mössbauer spectrum is measured automatically during each backscattering experiment. Each component of the reference target has well-known Mössbauer parameters so that fitting of reference spectra enables velocity calibration for each individual measurement done in backscattering geometry ensuring that the actual drive velocity is always well-defined and known regardless of the prevailing environmental conditions.

3.3.6.2 Detector Calibration

Careful energy calibration on each detector was done to achieve optimal detection rate. Each SH was temperature cycled (153–293 K). During cycling energy spectra were measured. As a result of the analysis of these spectra, optimal firmware parameters were calculated for each detector and each temperature window. During operation instrument firmware automatically adjusts those parameters depending on temperature and ensures best detector performance.

3.3.7 The Advanced Instrument MIMOS IIa

Mössbauer spectroscopy has been proved by the MER project to be a valuable tool for the in situ exploration of extraterrestrial bodies and for the study of Fe-bearing samples. The MIMOS instrument operates in backscattering geometry with a ^{57}Co source irradiating a sample area at a distance of ~10 mm from the detector surface. MIMOS II has four square-shaped PIN diodes with a sensitive area of 1 cm^2 each. The advanced version MIMOS IIa will be equipped with a ring of Silicon Drift Detectors (SDD) (see Fig. 3.24) developed by PN Sensor GmbH together with the semiconductor laboratory of the Max–Planck-Institute in Munich, Germany. The SDD ring is divided into four individual chips each with two SDD cells of 45 mm^2 sensitive area and an active silicon thickness of 450 μm.

The improved version of the MIMOS II instrument is lighter with a total mass of ~320 g. The new ring detector system of SDDs [61, 63] with its integrated FET does significantly improve the energy resolution of the instrument and the area fill factor around the collimator. The actual SDDs with a sensitive area of 2×45 mm^2 (= 1 quarter of the ring) show at 5.9 keV (^{55}Fe source) an energy resolution <280 eV at room temperature (~+20°C) approaching the theoretical limit of ~130 eV at moderate cooling (~−30°C) (Fig. 3.24). This results in an increase of the SNR (Signal to Noise Ratio) by a factor of more than 7 (Fig. 3.25).

In addition to the Mössbauer data, SDDs allow the simultaneous acquisition of the XRF spectrum, thus providing data on the sample's elemental composition. A new control- and readout-electronics for MIMOS IIA allows spectra acquisition at highest possible count rates available at a total detector area of 360 mm^2.

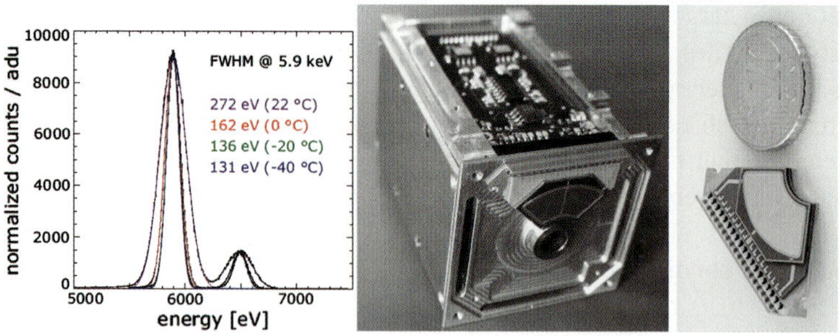

Fig. 3.24 *Left*: Energy resolution of the silicon-drift-detectors (SDD) at 5.9 keV (^{55}Fe radioactive source) as a function of temperature. *Middle*: One segment (out of four) of SDDs composed of two individual chips with front-end electronics. The dimensions of the Al housing are ~4 cm × 4 cm × 7 cm. In the center the opening of the radiation collimator can be seen. *Right*: MIMOS IIA SDD segment (1/4 of the complete detector ring; a 10 €-cent coin is shown for comparison)

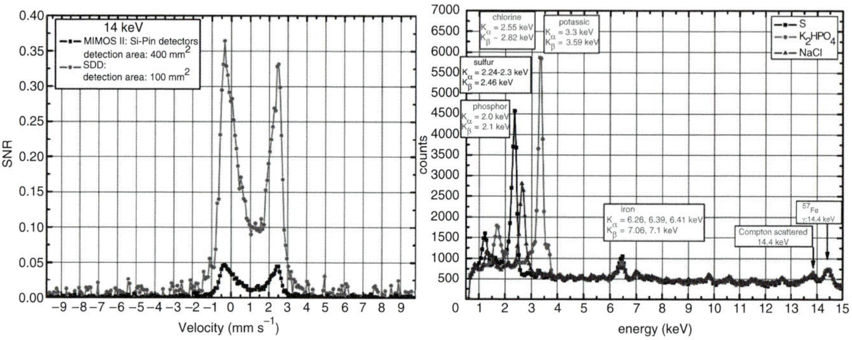

Fig. 3.25 *Left*: signal-to-noise ratio (SNR) of the Mössbauer spectra of a basalt taken with MIMOS II (full Si-PIN detector system; *black data-points*) and MIMOS IIA (1/4 of full SDD system; *red data-points*) respectively. *Right*: XRF spectra of low *Z* elements measured with MIMOS IIA (SDDs) at −20°C. The Compton scattered 14.4 keV "line" (at ~13.8 keV) and the resonant 14.4 keV Mössbauer line are well separated

This is possible because of digital filtering and pulsed reset technique. Readout of the SDD ring will be done using eight separate charge-sensitive preamplifiers. In the final configuration, four double ADCs sample the preamplifier signals. Parallel signal processing chains perform AC deconvolution and signal shaping in a single space qualified FPGA. All XRF and MB histograms are built inside the FPGA and transferred to an onboard memory. A microcontroller IP in the FPGA eliminates the need for an additional space qualified controller part. The FPGA also operates the Mössbauer drive that generates the Doppler velocity of the ^{57}Co source. The whole instrument runs autonomously at a power consumption of 3 W, which is the same consumption as the precursor instrument but at a significantly increased performance.

In Fig. 3.25 (left), the SNR for the 6.4 keV K Fe-X-rays and the 14.4 keV Mössbauer radiation is shown for simultaneous measurement of the Mössbauer spectra of a basalt sample in backscattering geometry. The use of the SDD gives a better SNR by a factor of 4 for 6.4 keV and by a factor of more than 7 for 14.4 keV radiation in comparison with MIMOS II. The use of the SDD ring with a total sensitive area of 360 mm^2 in MIMOS IIA therefore reduces the acquisition time by a factor of at least 7. Furthermore, measurements with the SDD show a separation of the resonant 14.4 keV peak (from the ^{57}Co source) and its Compton scattered peak (below about 13.8 keV). For the 14.4 keV Mössbauer measurement with the SDD an optimal choice of the lower and upper threshold is possible removing Compton scattered events. The result is again a significant increase of the SNR.

The X-ray-spectra of low *Z* elements clearly show (Fig. 3.25) that detection of XRF-lines down to 1 keV is possible with MIMOS IIA.

References

1. Kalvius, G. M., Kankeleit, E. In: Mössbauer Spectroscopy and its Applications Proceedings, p. 9. IAEA, Vienna (1972)
2. Cohen, R.L., Wertheim, G.K.: Methods Exp. Phys. **11**, 307 (1974)
3. Cranshaw, T. E. In: Mössbauer Spectroscopy Applied to Inorganic Chemistry, p. 27. Plenum, New York (1984)
4. Kankeleit, E.: Rev. Sci. Instrum. **35**, 194 (1964)
5. Kankeleit, E.: In: International Conference on Mössbauer Spectroscopy Proceedings, vol. 2, 43. Cracow (Poland) (1975)
6. Potzel, W.: In: Mössbauer Spectroscopy Applied to Magnetism and Material Science, p. 305. Plenum, New York, (1993)
7. Helisto, P., Katila, T., Potzel, W., Riski, K.: Phys. Rev. Lett. **A 85**, 177 (1981)
8. Stevens, J.G.: In: Stevens, J.G., Stevens, V.E., White, R.M., Gibson, J.L. (eds.): Mössbauer Effect Reference and Data Journal, p. 99. Mössbauer Effect Data Center, NC (1980)
9. Fritz, R., Schulze, D.: Nucl. Instrum. Methods **62**, 317 (1968)
10. Cranshaw, T.E.: J. Phys. **E 6**, 9 (1976)
11. Viegers, M.P.A.: Doctoral Dissertation p. 13, University of Nijmegen (1976)
12. Amersham International Limited, Technical Catalog RS16-8, Mössbauer Sources (1981)
13. Shigematsu, T., Pfannes, H.-D., Keune, W.: Phys. Rev. Lett. **45**, 1206 (1980)
14. Swanson, K.R., Spijkerman, J.J.: J. Appl. Phys. **41**, 3155 (1970)
15. Hanzel, D., Griesbach, P., Meisel, W., Gütlich, P.: Hyperfine Interact. **71**, 1441 (1992)
16. Ruskov, T., Passage, G., Rastanawi, A., Radev, R.: Nucl. Instrum. Methods **B 94**, 565 (1994)
17. Sawicki, J.A., Sawicka, B.D., Stanek, J.: Nucl. Instrum. Methods **138**, 565 (1976)
18. Weyer, G.: In: Mössbauer Effect Methodology, p. 301. Plenum, New York (1976)
19. Mantovan, R., Fanciulli, M.: Rev. Sci. Instrum. **78**, 063902 (2007)
20. Liljequist, D., Ismail, M.: Phys. Rev. **B 31**, 4131 (1985)
21. Sato, H., Mitsuhashi, M.: Hyperfine Interact. **58**, 2535 (1990)
22. Liljequist, D., Ismail, M., Saneyoshi, K., Debusmann, K., Keune, W., Brand, R.A., Kiauka, W.: Phys. Rev. **B 31**, 4137 (1985)
23. Macedo, W.A.A., Pelloth, D., Toriyama, T., Nikolov, O., Kruijer, S., Scholz, B., Brand, R.A., Keune, W.: Hyperfine Interact. **92**, 1221 (1994)
24. Shigematsu, T., Pfannes, H.-D., Keune, W.: In: Mössbauer Spectroscopy and its Chemical Applications. American Chemical Society, Washington, DC (1981)

25. Bokemeyer, H., Wohlfahrt, K., Kankeleit, E., Eckhardt, D.: Z. Phys. **A 274**, 305 (1975)
26. Stahl, B., Kankeleit, E.: Nucl. Instr. Sec. **B 122**, 149 (1997)
27. De Grave, E., Vandenberghe, R.E., Dauwe, C.: Hyperfine Interact. **161**, 147 (2005)
28. Alonzo, G., Consiglio, M., Gionfriddo, N., Bardi, G.: Hyperfine Interact. **46**, 703 (1989)
29. Kalvius, G.M., Katila, T.E., Lounasmaa, O.V.: In: Mössbauer Effect Methodology 5, p. 231. Plenum, New York (1970)
30. Spijkerman, J.J., Ruegg, F.C., de Voe, J.R.: In: Mössbauer Effect Methodology 1, p. 119. Plenum, New York (1965)
31. Ping, J., Rancourt, D.: Hyperfine Interact. **71**, 1433 (1992)
32. Rancourt, D.G., McDonald, A.M., Lalonde, A.E., Ping, J.Y.: Am. Mineral. **78**, 1 (1993)
33. Long, G.J., Cranshaw, T.E., Longworth, G.: In: Mössbauer Effect Reference and Data Journal, p.42. Mössbauer Effect Data Center, NC (1983)
34. Storm, E., Gilbert, E., Israel, H.: In: Gamma-Ray Absorption Coefficients for Elements 1 Through 100, Los Alamos Scientific Laboratory LA-2237. University of California, New Mexico (1958)
35. Münck, E.: In: Methods in Enzymology, vol. LIV, p. 346. Academic, New York (1978)
36. Klingelhöfer, G., Morris, R.V., Bernhardt, B., Rodionov, D., de Souza Jr., P.A., Squyres, S.W., Foh, J., Kankeleit, E., Bonnes, U., Gellert, R., Schröder, C., Linkin, S., Evlanov, E., Zubkov, B., Prilutski, O.: J. Geophys. Res. **108**, 8067 (2003)
37. Squyres, S.W., Arvidson, R.E., Baumgartner, E.T., Bell III, J.F., Christensen, P.R., Gorevan, S., Herkenhoff, K.E., Klingelhöfer, G., Madsen, M.B., Morris, R.V., Rieder, R., Romero, R.A.: J. Geophys. Res. **108**, 8062 (2003)
38. Knudsen, J.M., Madsen, M.B., Olsen, M., Vistisen, L., Koch, C.B., Moerup, S., Kankeleit, E., Klingelhöfer, G., Evlanov, E.N., Khromov, V.N., Mukhin, L.M., Prilutskii, O.F., Zubkov, B., Smirnov, G.V., Juchniewicz, J.: Hyperfine Interact. **68**, 83 (1992)
39. Knudsen, J.M.: Hyperfine Interact. **47**, 3 (1989)
40. Klingelhöfer, G.: Hyperfine Interact. **113**, 369 (1998)
41. Klingelhöfer, G.: In: Garcia, M., Marco, J.F., Plazaola, F. (eds.): Industrial Applications of the Mössbauer Effect, p. 369. American Institute of Physics, MD (2005)
42. Gummer, A.W.: Nucl. Instrum. Methods **B 34**, 224 (1988)
43. Evlanov, E.N., Frolov, V.A., Prilutski, O.F., Rodin, A.M., Veselova, G.V., Klingelhöfer, G.: Lunar Planet Sci. **24**, 459 (1993)
44. Klingelhöfer, G., Imkeller, U., Kankeleit, E., Stahl, B.: Hyperfine Interact. **71**, 1445 (1992)
45. Teucher, R.: Miniaturisierter Mössbauerantrieb, Diploma Dissertation, University Darmstadt, Inst. f. Nuclear Physics, (1994)
46. Held, P., Teucher, R., Klingelhöfer, G., Foh, J., Jäger, H., Kankeleit, E.: Lunar Planet Sci. **24**, 633 (1993)
47. Klingelhöfer, G., Fegley Jr., B., Morris, R.V., Kankeleit, E., Held, P., Evlanov, E., Priloutskii, O.: Planet Space Sci. **44**, 1277 (1996)
48. Klingelhöfer, G., Held, P., Teucher, R., Schlichting, F., Foh, J., Kankeleit, E.: Hyperfine Interact. **95**, 305 (1995)
49. Klingelhöfer, G.: In: Miglierini, M., Petridis, D. (eds.): Mössbauer Spectroscopy in Materials Science, p. 413. Kluwer, Dordrecht (1999)
50. Riesenmann, R., Steger, J., Kostiner, E.: Nucl. Instrum. Methods **72**, 109 (1969)
51. Held, P.: MIMOS II – Ein miniaturisiertes Mößbauerspektrometer in Rückstreugeometrie zur mineralogischen Analyse der Marsoberfläche, PhD Dissertation, TU Darmstadt, Inst. Nuclear Physics (1997)
52. Schröder, C., Klingelhöfer, G., Morris, R.V., Rodionov, D.S., Fleischer, I., Blumers, M.: Hyperfine Interact. **182**, 149 (2008)
53. Klingelhöfer, G., Morris, R.V., Bernhardt, B., Schröder, C., Rodionov, D.S., de Souza Jr., P.A., Yen, A., Gellert, R., Evlanov, E.N., Zubkov, B., Foh, J., Bonnes, U., Kankeleit, E., Gütlich, P., Ming, D.W., Renz, F., Wdowiak, T., Squyres, S.W., Arvidson, R.E.: Science **306**, 1740 (2004)

54. Morris, R.V., Klingelhöfer, G., Schröder, C., Fleischer, I., Ming, D.W., Yen, A.S., Gellert, R., Arvidson, R.E., Rodionov, D.S., Crumpler, L.S., Clark, B.C., Cohen, B.A., McCoy, T.J., Mittlefehldt, D.W., Shmidt, M.E., de Souza, P.A., Squyres, S.W.: J. Geophys. Res. **113**, E12S42 (2008)
55. Morris, R.V., Klingelhöfer, G.: In: Bell, J. (ed.): The Martian Surface, p. 339. Cambridge University Press, Cambridge (2008)
56. Fleischer, I., Klingelhöfer, G., Schröder, C., Morris, R.V., Hahn, M., Rodionov, D., Gellert, R., de Souza, P.A.: J. Geophys. Res. **113**, E06S21 (2008)
57. Morris, R.V., Golden, D.C., Bell, J.F., Shelfer, T.D., Scheinost, A.C., Hinman, N.W., Furniss, G., Mertzman, S.A., Bishop, J.L., Ming, D.W., Allen, C.C., Britt, D.T.: J. Geophys. Res. **105**, 1757 (2000)
58. Johnson, G.R., Olhoeft, G.R.: In: Carmichael, R.S. (ed.): Handbook of Physical Properties of Rocks, vol. III, p. 1. CRC, Boca Raton (1984)
59. Morris, R.V., Graff, T.G., Shelfer, T.D., Bell III, J.F.: Lunar Planet Sci. **32**, 1912 (2001)
60. Morris, R.V., Klingelhöfer, G., Schröder, C., Rodionov, D.S., Yen, A., de Souza Jr., P.A., Ming, D.W., Wdowiak, T., Gellert, R., Bernhardt, B., Evlanov, E.N., Zubkov, B., Foh, J., Bonnes, U., Kankeleit, E., Gütlich, P., Renz, F., Squyres, S.W., Arvidson, R.E.: J. Geophys. Res. – Planets **111**, E02S13 (2006)
61. Strüder, L., Lechner, P., Leutenegger, P.: Naturwissenschaften **11**, 539 (1998)
62. Fleischer, I., Klingelhöfer, G., Schröder, S., Rodionov, D.: Hyperfine Interact. **186**, 193 (2008)
63. Blumers, M., Bernhardt, B., Lechner, P., Klingelhöfer, G., d'Uston, C., Soltau, H., Strüder, L., Eckhardt, R., Brückner, J., Henkel, H., Girones Lopez, J., Maul, J. Nucl. Instrum. Methods A, (2010), doi: 10.1016/j.nima.2010.04.07

Chapter 4
Hyperfine Interactions

In Chap. 2, for the sake of simplicity, we dealt with transitions between unperturbed energy levels of "bare" nuclei with the mean transition energy E_0. In reality, however, nuclei are exposed to electric and magnetic fields created by the electrons of the Mössbauer atom itself and by other atoms in the neighborhood. These fields generally interact with the electric charge distribution and the magnetic dipole moment of the Mössbauer nucleus and perturb its nuclear energy states. The perturbation, called *nuclear hyperfine interaction*, may be such that it shifts the nuclear energy levels, as is the case in the *electric monopole interaction* ($e0$), or such that it splits degenerate states, as afforded by the electric *quadrupole interaction* ($e2$) and the magnetic *dipole interaction* ($m1$). Only these three kinds of interaction must be considered in practical Mössbauer spectroscopy.

A Mössbauer spectrum, in general, reflects the nature and the strength of the hyperfine interactions. The $e0$ interaction alters the position of the resonance lines on the energy scale in units of the Doppler velocity and gives rise to the so-called *isomer shift* (chemical shift) δ, whereas the $e2$ and the $m1$ interactions induce splitting of the resonance lines according to the allowed transitions from the ground to the excited state. Most valuable information regarding the chemical and physical characteristics of the sample under study can be extracted from these hyperfine interactions. In the following paragraphs, multipole expansion of the electrostatic interaction is described and the various contributions to electric and magnetic hyperfine interactions are discussed in detail.

4.1 Introduction to Electric Hyperfine Interactions

The total energy of the electrostatic interaction between a nucleus with charge Ze and charges around this nucleus may be expressed in classical terms as [1], [9] in Chap. 1:

P. Gütlich et al., *Mössbauer Spectroscopy and Transition Metal Chemistry*,
DOI 10.1007/978-3-540-88428-6_4, © Springer-Verlag Berlin Heidelberg 2011

$$E_{\text{el}} = \int \rho_{\text{n}}(\vec{r})V(\vec{r})\text{d}\tau. \tag{4.1}$$

The variable $\rho_{\text{n}}(\vec{r})$ denotes the nuclear charge density at a point $\vec{r}$ with coordinates $\vec{r} = (x_1, x_2, x_3)$, and $V(\vec{r})$ is the Coulomb potential set up at that point by all other charges (the Coulomb constant $k = 1/(4\pi\varepsilon_0)$ is dropped in this description). The integration variable in (4.1) is the volume element $\text{d}\tau = \text{d}x_1\,\text{d}x_2\,\text{d}x_3$. The origin of the coordinate system is chosen to coincide with the center of the nuclear charge. A more convenient expression can be obtained by expanding $V(\vec{r})$ at $\vec{r} = (0,0,0)$ in a Taylor series, that is,

$$V(\vec{r}) = V_0 + \sum_{i=1}^{3}\left(\frac{\partial V}{\partial x_i}\right)_0 x_i + \frac{1}{2}\sum_{i,j=1}^{3}\left(\frac{\partial^2 V}{\partial x_i \partial x_j}\right)_0 x_i x_j + \cdots. \tag{4.2}$$

The first derivative of the potential V at $\vec{r} = (0,0,0)$, taken as negative value, represents the electric field $\vec{E}$, and the second derivative represents the *electric field gradient* tensor $\overline{\overline{V}}$ at the nucleus,

$$E_i \equiv -\left(\frac{\partial V}{\partial x_i}\right)_0 \quad \text{and} \quad V_{ij} \equiv \left(\frac{\partial^2 V}{\partial x_i \partial x_j}\right)_0. \tag{4.3}$$

With these notations, we obtain for the electrostatic energy

$$E_{\text{el}} = V_0 \cdot \int \rho_{\text{n}}\text{d}\tau - \sum_{i=1}^{3} E_i \cdot \int \rho_{\text{n}} x_i \text{d}\tau + \frac{1}{2}\sum_{i,j=1}^{3} V_{ij} \cdot \int \rho_{\text{n}} x_i x_j \text{d}\tau + \cdots. \tag{4.4}$$

The first term in (4.4) is of no further interest here, because it represents only the total electrostatic energy of the nucleus (considered as a point charge). Since the nucleus always keeps its stable position in the center of the atom, it cannot experience an electric field $\vec{E}$; hence, the second term in (4.4) can also be dropped (moreover, nuclei with defined parity do not have an electric dipole moment, i.e. $\int \rho_{\text{n}} x_i \text{d}\tau = 0$). Given that terms higher than second order are negligible because the corresponding nuclear moments and interaction energies are vanishingly small, the only relevant term in (4.4) is

$$E_{\text{el}}^{(2)} = \frac{1}{2}\sum_{i,j=1}^{3} V_{ij} \int \rho_{\text{n}}(\vec{r}) x_i x_j \text{d}\tau. \tag{4.5}$$

The superscript (2) marks the quadratic approximation of the electrostatic energy.

4.1.1 Nuclear Moments

The integral over the nuclear charge distribution in (4.5) can be split into an isotropic and an anisotropic part by adding and subtracting $r^2 = \sum x_i^2$ which yields

$$\int \rho_n(\vec{r})x_i x_j \mathrm{d}\tau = \frac{1}{3}\int \rho_n(\vec{r})r^2 \mathrm{d}\tau + \frac{1}{3}\int \rho_n(\vec{r})(3x_i x_j - \delta_{ij}r^2)\mathrm{d}\tau. \tag{4.6}$$

The first term in (4.6), $\int \rho_n(r)r^2\mathrm{d}\tau$, depends only on the radial distribution of the nuclear charge. This term represents the so-called *nuclear monopole moment*; note that it is related to the extended finite size of the nucleus.[1]

In contrast, the second term in (4.6) comprises the full orientation dependence of the nuclear charge distribution in 2nd power. Interestingly, the expression has the appearance of an irreducible (3 × 3) second-rank tensor. Such tensors are particularly convenient for rotational transformations (as will be used later when nuclear spin operators are considered). The term here is called the *nuclear quadrupole moment* $\overline{\overline{Q}}$. Because of its inherent symmetry and the specific "cylindrical" charge distribution of nuclei, the quadrupole moment can be represented by a single scalar, Q (*vide infra*).

4.1.2 Electric Monopole Interaction

The isotropic part of the nuclear charge distribution, as described by the monopole moment $\int \rho_n(r)r^2\mathrm{d}\tau$ in (4.6), cannot be evaluated exactly because the precise nuclear charge distribution is not known. Instead, for simplicity, the nucleus may be approximated by a uniformly charged sphere of total charge $+Ze$. The radius of the sphere, R_u, is selected such[2] that the electric potential at the surface ($r = R_u$) has the same value and gradient as the potential of an equivalent hypothetical point charge Ze at $r = 0$. Following the concept of "equivalent uniform distribution" [2], R_u can be obtained from the experimental value of the root mean square nuclear radius as $R_u = \sqrt{5/3} \cdot \langle r^2 \rangle^{1/2}$ [3]. The charge density of the sphere can be given by using R_u as $\rho_n(r) = 3Ze/(4\pi {R_u}^3)$. Inserting into the integral yields for the nuclear monopole moment

$$\int \rho_n(\vec{r})r^2\mathrm{d}\tau = Ze \cdot \frac{3}{4\pi R_u^3}\int r^2\mathrm{d}\tau. \tag{4.7}$$

[1] The Coulomb interaction of the (point) nucleus with the potential V_0, which is also part of the monopole interaction, was neglected in (4.5) because it yields only an offset of the total energy.

[2] The subscript "u" in R_u is introduced to distinguish the radius of the uniformly charged sphere from the usual mean square radius R^2 which can be obtained from scattering experiments.

The integral can be solved by conversion from Cartesian to spherical coordinates. Then, the integration variable takes the convenient form $\mathrm{d}\tau = r^2 \mathrm{d}r \sin\theta \, \mathrm{d}\theta \, \mathrm{d}\phi$, which yields

$$\int r^2 \mathrm{d}\tau = \int_0^{R_\mathrm{u}} r^4 \mathrm{d}r \int_0^{\pi} \sin\theta \, \mathrm{d}\theta \int_0^{2\pi} \mathrm{d}\phi = \frac{4\pi}{5} R_\mathrm{u}^5. \tag{4.8}$$

Finally, a very simple expression is obtained when R_u is replaced by the mean square radius as given above, and the notation $\langle r^2 \rangle \equiv R^2$ is introduced, yielding

$$\int \rho_\mathrm{n}(\vec{r}) r^2 \mathrm{d}\tau = Ze \cdot \frac{3}{4\pi R_\mathrm{u}^3} \frac{4\pi}{5} R_\mathrm{u}^5 = \frac{3}{5} Ze \frac{5}{3} \langle r^2 \rangle = ZeR^2. \tag{4.9}$$

When inserting into (4.5), the term ZeR^2 will be multiplied with the elements of the electric field gradient tensor $\bar{\bar{V}}$. Fortunately, the procedure can be restricted to diagonal elements V_{ii}, because $\bar{\bar{V}}$ is symmetric and, consequently, a principal axes system exists in which the nondiagonal elements vanish, $V_{i \neq j} = 0$. The diagonal elements can be determined by using Poisson's differential equation for the electronic potential at point $r = 0$ with charge density $\rho_\mathrm{e}(0)$, $\Delta V = 4\pi\rho_\mathrm{e}$, which yields

$$\left(\sum_{i=1}^{3} V_{ii} \right)_0 = \left(\vec{\nabla}\vec{\nabla} V \right)_0 = (\Delta V)_0 = -4\pi e |\psi(0)|^2, \tag{4.10}$$

where $-e$ is the charge of the electron. The electronic charge density $\rho_\mathrm{e}(0) = -e|\psi(0)|^2$ essentially arises from the ability of s-electrons to penetrate the nucleus with a finite probability density $|\psi|^2$ at $r = 0$. With the above conversions, the energy of the electrostatic monopole interaction between a finite nucleus and the electrons is obtained by inserting the first term of (4.6), together with (4.9) and (4.10), into (4.5):

$$E_\mathrm{I} = -\frac{2\pi}{5} Ze^2 R^2 |\psi(0)|^2 \equiv \delta E. \tag{4.11}$$

The term causes a uniform shift of the nuclear energy states which, however, is different for the ground and excited state because the nuclear volume and, therefore, also the entity $\langle R^2 \rangle$ are different for ground and excited states. This gives rise to the *isomer shift* δ of the Mössbauer spectrum. The notation $\delta E \equiv E_\mathrm{I}$ is introduced to emphasize the very small change in energy ($\approx 10^{-8}$ eV), which is only a fraction (about 10^{-12}) of the transition energy. The isomer shift will be discussed in detail in Sect. 4.2.

4.1.3 Electric Quadrupole Interaction

The tensor of the nuclear quadrupole moment $\bar{\bar{Q}}$ has nine elements

$$Q_{ij} = \int \rho_\mathrm{n}(\vec{r})(3x_i x_j - \delta_{ij} r^2)\mathrm{d}\tau \quad \text{with } i, j = 1, 2, 3, \tag{4.12}$$

of which only five are independent, because $\bar{\bar{Q}}$ is symmetric and its trace vanishes,[3] that is, $Q_{ij} = Q_{ji}$ and $\Sigma Q_{ii} = 0$. The elements of the electric field gradient, $V_{ij} = (\partial^2 V/\partial x_i \partial x_i)_0$, as defined in (4.3), also form a symmetric (3 × 3) second-rank tensor (the second derivatives can be reversed, $V_{ij} = V_{ji}$). Inserting (4.12) into (4.6) and (4.5) yields the so-called quadrupole interaction energy

$$E_\mathrm{Q} = \frac{1}{6} \sum_{i,j=1}^{3} V_{ij} \cdot Q_{ij}. \tag{4.13}$$

Quadrupole interaction lifts the degeneracy of nuclear states with spin quantum numbers $I > 1/2$, and is manifested in the Mössbauer spectrum as quadrupole splitting ΔE_Q (as will be further discussed in Sect. 4.3). According to (4.5), the classical electric monopole and quadrupole interaction energies E_I and E_Q are additive, that is, $E_\mathrm{el}^{(2)} = E_\mathrm{I} + E_\mathrm{Q}$.

4.1.4 Quantum Mechanical Formalism for the Quadrupole Interaction

In quantum mechanics, the classical charge density ρ_n of the nucleus is replaced by the operator [4].

$$\hat{\rho}_\mathrm{n}(\vec{r}) = \sum_k e(\vec{r} - \vec{r}_k) = e \sum_k \delta(\vec{r} - \vec{r}_k), \tag{4.14}$$

for which the summation[4] runs over all protons (charge +e). The corresponding operator of the quadrupole moment is obtained by inserting (4.14) into (4.12), which yields

$$\begin{aligned} \hat{Q}_{ij} &= \int \hat{\rho}_\mathrm{n}(\vec{r})(3x_i x_j - \delta_{ij} r^2)\mathrm{d}\tau = e \cdot \sum_{\text{protons}} \int (3x_i x_j - \delta_{ij} r^2)\, \delta(\vec{r} - \vec{r}_k)\mathrm{d}\tau \\ &= e \cdot \sum_{\text{protons}} (3x_{ik} x_{jk} - \delta_{ij} r_k^2), \end{aligned} \tag{4.15}$$

so that the Hamiltonian operator for nuclear quadrupole interaction becomes

$$\hat{H}_\mathrm{Q} = \frac{1}{6} \sum_{i,j=1}^{3} V_{ij} \cdot \hat{Q}_{ij}. \tag{4.16}$$

[3] $\int \rho_\mathrm{n}(\vec{r})(3x_1^2 - (x_1^2 + x_2^2 + x_3^2) + 3x_2^2 - (x_1^2 + x_2^2 + x_3^2) + 3x_3^2 - (x_1^2 + x_2^2 + x_3^2))\mathrm{d}\tau = 0$.

[4] The operator $\hat{\vec{r}}$ and its components $\hat{y}$, $\hat{y}$, $\hat{z}$ have the same effect as in classical mechanics (just multiplication), so that the operator symbol ^ can be omitted.

Equation (4.15) would be extremely onerous to evaluate by explicit treatment of the nucleons as a many-particle system. However, in Mössbauer spectroscopy, we are dealing with eigenstates of the nucleus that are characterized by the total angular momentum with quantum number I. Fortunately, the electric quadrupole interaction can be readily expressed in terms of this momentum I, which is called the nuclear spin; other properties of the nucleus need not to be considered. This is possible because the transformational properties of the quadrupole moment, which is an irreducible 2nd rank tensor, make it possible to use Clebsch–Gordon coefficients and the Wigner–Eckart theorem to replace the awkward operators $3x_ix_j-\delta_{ij}r^2$ (in spatial coordinates) by angular momentum operators $\hat{I}^2, \hat{I}_z, \hat{I}_x, \hat{I}_y$ of the total spin.[5] With this elegant transformation, a convenient spin Hamiltonian operator is obtained for the electric quadrupole interaction [6]:

$$\hat{H}_Q = \frac{eQ}{6I(2I+1)} \sum_{i,j=1}^{3} V_{ij} \cdot \left[\frac{3}{2}(\hat{I}_i\hat{I}_j + \hat{I}_j\hat{I}_i) + \delta_{ij}\hat{I}^2\right]. \tag{4.17}$$

Remarkably, only one nuclear constant, Q, is needed in (4.17) to describe the *quadrupole moment of the nucleus*, whereas the full quadrupole tensor $\overline{\overline{Q}}$ has five independent invariants. The simplification is possible because the nucleus has a definite angular momentum (I) which, in classical terms, imposes cylindrical symmetry of the charge distribution. Choosing $x_i = z$ as symmetry axis, the off-diagonal elements Q_{ij} are zero ($i \neq j$), and the energy change caused by nuclear reorientation will depend only on the difference of the charge distribution along the z- and the x-axis (or y-axis). Since these are given by $\int \rho_n(r) z^2 d\tau$ and $\int \rho_n(r) x^2 d\tau$, respectively, the classical expression[6] arises

$$\begin{aligned} Q &\equiv \frac{1}{e}\int \rho_n(r)(z^2 - x^2)d\tau = \frac{1}{e}\int \rho_n(r)(3z^2 - r^2)d\tau \\ &= \frac{1}{e}\int \rho_n(r) r^2(3\cos^2\theta - 1)d\tau, \end{aligned} \tag{4.18}$$

where θ is the polar angle between the symmetry axis (z) and the vector $\vec{r}$, and $z = r\cos\theta$. According to this equation, the nuclear quadrupole moment Q is *positive* for elongated nuclei and *negative* for oblate nuclei (and zero for spherical charge distribution). Nuclear states with spin quantum number $I = 0, 1/2$ do not possess an observable quadrupole moment.

---

[5] C. P. Slichter's textbook on magnetic resonance [4] may be recommended for further reading. It presents a very educational introduction into this issue of operator equivalence. A comprehensive, elaborate article on quadrupole interaction in Mössbauer spectroscopy is provided by H. Spiering in [5].

[6] Recall: $z^2 - x^2 = (2z^2 - x^2 - y^2)/2$, $x^2 = y^2$, and $r^2 = x^2 + y^2 + z^2$.

4.2 Mössbauer Isomer Shift

The electric monopole interaction between a nucleus (with *mean square* radius R^2) and its environment is a product of the nuclear charge distribution ZeR^2 and the electronic charge density $e|\psi(0)|^2$ at the nucleus, $\delta E = \text{const} \cdot R^2|\psi(0)|^2$ (4.11). However, nuclei of the same mass and charge but different nuclear states (*isomers*) have different charge distributions ($ZeR_g^2 \neq ZeR_e^2$), because the nuclear volume and the mean square radius depend on the state of nuclear excitation ($R_g^2 \neq R_e^2$). Therefore, the energies of a Mössbauer nucleus in the ground state (g) and in the excited state (e) are shifted by different amounts $(\delta E)_e$ and $(\delta E)_g$ relative to those of a bare nucleus. It was recognized very early that this effect, which is schematically shown in Fig. 4.1, is responsible for the occurrence of the Mössbauer isomer shift [7].

The energy of a γ-photon emitted by an excited Mössbauer nucleus in the source is given by the transition energy

$$E_S = E_0 - \left[(\delta E)_e - (\delta E)_g\right]_{\text{source}} = E_0 + \frac{2\pi}{5} Ze^2 |\psi(0)|_S^2 (R_e^2 - R_g^2), \tag{4.19}$$

where E_0 is the transition energy of a bare nucleus, and $|\psi(0)|_s^2$ is the electronic charge density for the source material. The corresponding expression for the absorber nucleus, $E_A = E_0 - \left[(\delta E)_e - (\delta E)_e\right]_{\text{abs}}$, includes the electron density $|\psi(0)|_A^2$ for the absorber material, that is,

$$E_A = E_0 + \frac{2\pi}{5} Ze^2 |\psi(0)|_A^2 (R_e^2 - R_g^2). \tag{4.20}$$

Provided the electron densities $|\psi(0)|_S^2$ and $|\psi(0)|_A^2$ at the Mössbauer nuclei in the source and the absorber material are different because of the different chemical

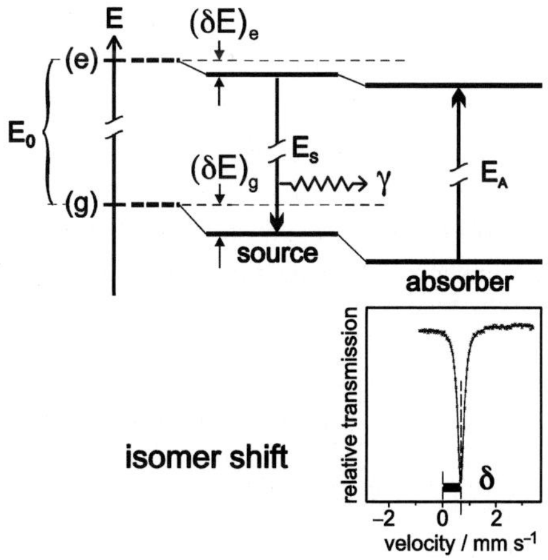

Fig. 4.1 The electric monopole interaction between the nuclear charge and the electron density at the nucleus shifts the energy of the nuclear states and gives the Mössbauer isomer shift

compositions, the transition energies E_S and E_A are different. Hence, the γ-resonance in the Mössbauer experiment does not appear at Doppler velocity zero, but is shifted by an amount δ, which is called the *isomer shift*:

$$\delta = E_A - E_S = \frac{2\pi}{5} Ze^2 \left\{ |\psi(0)|_A^2 - |\psi(0)|_S^2 \right\} \left(R_e^2 - R_g^2 \right). \tag{4.21}$$

With the convention $R_e - R_g = \Delta R$ and $R_e + R_g \approx 2R$ (R_e and R_g differ only slightly from each other) (4.21) takes the form

$$\delta = \frac{4\pi}{5} Ze^2 \left\{ |\psi(0)|_A^2 - |\psi(0)|_S^2 \right\} \cdot R^2 \left(\frac{\Delta R}{R} \right), \tag{4.22}$$

or, if the source properties may be taken as constant $\left(|\psi(0)|_S^2 = C \right)$,

$$\delta = \text{const} \; (\Delta R/R) \left\{ |\psi(0)|_A^2 - C \right\}. \tag{4.23}$$

The isomer shift of a resonance line (or the centroid of a line multiplet) in an experimental Mössbauer spectrum in terms of the Doppler velocity (mm s^{-1}) necessary to achieve resonance absorption is given by

$$\delta = \alpha \left\{ |\psi(0)|_A^2 - C \right\}, \quad \text{with } \alpha = \left(\frac{3Ze^2 cR^2}{5\varepsilon_0 E_0} \right) \frac{\Delta R}{R}. \tag{4.24}$$

The entity α is the so-called isomer shift calibration constant, c is the speed of light, ε_0 is the electric constant, and E_0 is the nuclear transition energy. (The Coulomb constant $k = 1/(4\pi\varepsilon_0)$, which was dropped in (4.1), is re-inserted here.) A comprehensive derivation of this expression is found in [8, 9].

The factor $\Delta R/R$, which describes the relative change of the nuclear radius during nuclear transition, has been determined for many Mössbauer nuclei either experimentally from electron capture lifetime variations [10, 11] or by internal-conversion-electron spectroscopy [12], or theoretically – from isomer shifts. The latter method is usually based on the determination of the calibration constant α from correlating experimental isomer shifts and the theoretically calculated values of $|\psi(0)|^2$ (using DFT or other methods [9, 13–18]; see also Chap. 5 and Appendix C). Apparently, the result depends on the quality (and type) of the electron density calculation in the vicinity of the nucleus. Although systematic errors basically cancel, because only differences of energy shifts are evaluated, α is usually regarded as an adjustable parameter which serves to calibrate the quantum chemical methods for the calculation of isomer shifts. Recently, a new physically interesting approach has been suggested in which the isomer shift is treated as a derivative of the total electronic energy with respect to the radius of a finite nucleus [19].

A calibration of the popular B3LYP and BP86 density functionals for the prediction of ^{57}Fe isomer shifts from DFT calculations [16], using a large number of complexes with a wide range of iron oxidation states and a span of about 2 mm s^{-1} for the isomer shifts, yielded a value for the calibration constant $\alpha = -0.3666$ mm s^{-1} a.u.^{3} (see Chap. 5). Note the negative sign, which indicates that a positive isomer shift of a certain compound relative to a reference material reveals a lower electron density at the nuclei in that compound as compared to nuclei in the reference material.

4.2.1 Relativistic Effects

Relativistic quantum mechanics yields the same type of expressions for the isomer shift as the classical approach described earlier. Relativistic effects have to be considered for the calculation of the electron density. The corresponding contributions to $|\psi(0)|^2$ may amount to about 30% for iron, but much more for heavier atoms. In Appendix D, a few examples of correction factors for nonrelativistically calculated charge densities are collected. Even the nonrelativistically calculated $\rho(0)$ values accurately follow the chemical variations and provide a reliable tool for the prediction of Mössbauer properties [16].

4.2.2 Isomer Shift Reference Scale

For a comparison of experimental Mössbauer isomer shifts, the values have to be referenced to a common standard. According to (4.23), the results of a measurement depend on the type of source material, for example, ^{57}Co diffused into rhodium, palladium, platinum, or other metals. For ^{57}Fe Mössbauer spectroscopy, the spectrometer is usually calibrated by using the known absorption spectrum of metallic iron (α-phase). Therefore, ^{57}Fe isomer shifts are commonly reported relative to the centroid of the magnetically split spectrum of α-iron (Sect. 3.1.3). Conversion factors for sodium nitroprusside dihydrate, $Na_2[Fe(CN)_5NO]\cdot 2H_2O$, or sodium ferrocyanide, $Na_4[Fe(CN)]_6$, which have also been used as reference materials, are found in Table 3.1. Reference materials for other isotopes are given in Table 1.3 of [18] in Chap. 1.

4.2.3 Second-Order Doppler Shift

The experimentally observed isomer shift, δ_{exp}, includes a relativistic contribution, which is called second-order Doppler shift, δ_{SOD}, and which adds to the genuine isomer shift δ,

$$\delta_{\mathrm{exp}} = \delta + \delta_{\mathrm{SOD}}. \tag{4.25}$$

The effect results from a relativistic shift of the energy of the γ-photon due to the thermal motion of the emitting and absorbing nuclei, which is proportional to the mean square *velocity* of the Mössbauer nuclei. Assuming that the emitting or absorbing atom is vibrating with mean squared velocity $\langle v^2 \rangle$, the second-order Doppler shift is given by (see also Appendix E):

$$\delta_{\mathrm{SOD}} = -E_\gamma \frac{\langle v^2 \rangle}{2c^2}. \tag{4.26}$$

Experimental isomer shifts, δ_{exp}, should be corrected for the contribution of δ_{SOD}, in order to avoid misinterpretations. The value of δ_{SOD} drops with temperature and becomes vanishingly small at liquid helium temperature, because $\langle v^2 \rangle$ is proportional to the mean kinetic energy of the Mössbauer atom. In practice, δ_{SOD} may already be negligible at liquid nitrogen temperature; it rarely exceeds -0.02 mm s^{-1} at 77 K. At room temperature, δ_{SOD} may be as large as -0.1 mm s^{-1} or more (Fig. 4.2).

The temperature dependence of δ_{SOD} is related to that of the recoil-free fraction $f(T) = \exp[-\langle x^2 \rangle {E_\gamma}^2/(\hbar c)^2]$, where $\langle x^2 \rangle$ is the mean square displacement (2.14). Both quantities, $\langle x^2 \rangle$ and $\langle v^2 \rangle$, can be derived from the Debye model for the energy distribution of phonons in a solid (see Sect. 2.4). The second-order Doppler shift is thereby given as [20]

$$\delta_{\mathrm{SOD}} = -\frac{9k_{\mathrm{B}}E_\gamma}{16M_{\mathrm{eff}}c^2}\left(\Theta_{\mathrm{M}} + 8T\left(\frac{T}{\Theta_{\mathrm{M}}}\right)^3 \int_0^{\Theta_{\mathrm{M}}/T} \frac{x^3}{\mathrm{e}^x - 1}\mathrm{d}x\right), \tag{4.27}$$

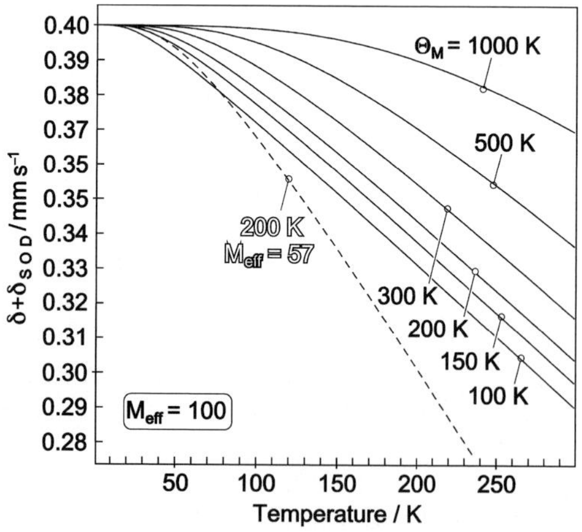

Fig. 4.2 Temperature dependence of the isomer shift due to the second-order Doppler shift, δ_{SOD}. The curves are calculated for different Mössbauer temperatures Θ_{M} by using the Debye model whereby the isomer shift was set to $\delta = 0.4$ mm s^{-1} and the effective mass to $M_{\mathrm{eff}} = 100$ Da, except for the dashed curve with $M_{\mathrm{eff}} = 57$ Da

where M_{eff} is an effective parameter for the vibrating mass which accounts for the Mössbauer atoms taking part in collective motions together with the surrounding ligands. Correspondingly, the *Debye temperature* Θ_D is replaced by the *Mössbauer temperature* Θ_M, which is specific for the environment as "sensed" by the Mössbauer nucleus [21]. (The corresponding expression for the Lamb–Mössbauer factor $f(T)$ is given in (2.15)). Figure 4.2 shows the temperature dependence of $\delta_{exp}(T) = \delta + \delta_{SOD}(T)$ simulated using the Debye model for ^{57}Fe ($E_0 = 14.41$ keV) and assuming different Mössbauer temperatures Θ_M and effective masses M_{eff}; the isomer shift δ was arbitrarily set to 0.4 mm s^{-1}.

In proteins, the value of Θ_M is usually in the low temperature range 100–300 K; inorganic compounds may show values of 150–500 K, whereas metals have Θ_M as high as 1,000 K or more.

4.2.4 Chemical Information from Isomer Shifts

The electron density $|\psi(0)|^2$ at the nucleus primarily originates from the ability of s-electrons to penetrate the nucleus. The core-shell 1s and 2s electrons make by far the major contributions. Valence orbitals of *p*-, *d*-, or *f*-character, in contrast, have nodes at $r = 0$ and cannot contribute to $|\psi(0)|^2$ except for minor relativistic contributions of *p*-electrons. Nevertheless, the isomer shift is found to depend on various chemical parameters, of which the oxidation state as given by the number of valence electrons in *p*-, or *d*-, or *f*-orbitals of the Mössbauer atom is most important. In general, the effect is explained by the contraction of inner *s*-orbitals due to shielding of the nuclear potential by the electron charge in the valence shell. In addition to this *indirect* effect, a *direct* contribution to the isomer shift arises from valence *s*-orbitals due to their participation in the formation of molecular orbitals (MOs). It will be shown in Chap. 5 that the latter issue plays a decisive role. In the following section, an overview of experimental observations will be presented.

4.2.4.1 Isomer Shift Correlations

The isomer shift is considered the key parameter for the assignment of oxidation states from Mössbauer data. The early studies, following the first observation of an isomer shift for Fe_2O_3 [7], revealed a general correlation with the (formal) oxidation state[7] of iron. However, isomer shifts have also been found to depend on the spin state of the Mössbauer atom, the number of ligands, the σ-donor and the

[7]The concepts of *formal oxidation state, spectroscopic oxidation state, oxidation number* are often subject to controversial discussion. Regarding the definitions, the reader is referred to Appendix F and [22, 23].

π-acceptor strengths of the ligands, and other parameters. Therefore, the study of a single compound may not be very informative unless data from similar compounds are available for comparison. However, systematic series of related compounds may yield close and revealing correlations of the isomer shift with one or more features of the electronic structure.

4.2.4.2 Oxidation State and Spin

A typical example of a correlation diagram for ^{57}Fe is given in Fig. 4.3. It summarizes the isomer shifts for a great variety of iron complexes with oxidation states (I) to (VI) in the order of the respective high-spin, intermediate-spin, and low-spin configurations. The plot of the corresponding values marked by grey, hatched and open bars demonstrates three major trends:

- *High-spin* iron compounds: the lower the oxidation state the more positive is the isomer shift. Note that the allocation of high-spin iron(II) is unique for δ-values >1 mm s^{-1}.
- *Low-spin* iron compounds: isomer shifts are rather similar. It is, for example, not possible to distinguish between low-spin iron(II) and low-spin iron(III) configurations from δ-values alone.
- *Low-spin compounds exhibit lower isomer shifts than high-spin compounds*, whereas the isomer shifts of intermediate-spin compounds often resemble those of the corresponding low-spin compounds.
- In addition to the variation in electronic configuration, the geometric details of the coordination sphere and the properties of iron–ligand bonds (different σ- or π-donor strength) also influence the isomer shift as observed for a series of compounds:
- *Covalent (soft) ligands induce lower isomer shifts than ionic (hard) ligands.* Compounds with sulfur ligands for example, yield lower δ values than those with oxygen or nitrogen ligands, mainly because of stronger participation of valence-s electrons in the bonds due to shorter bond lengths. In contrast to chemical intuition, the effect of "electron-accepting" and "electron-donating" ligands on δ is similar, because σ-donation of electrons from ligands to the metal and d_π–p_π backdonation of electrons from metal t_{2g} orbitals to empty p-orbitals of ligands both lead to an increase in s-electron density at the nucleus and give rise to more negative isomer shifts.
- *Four-coordinate complexes exhibit lower isomer shifts than six-coordinate compounds.* Metal–ligand bonds are shorter and more covalent if the coordination number is smaller because of less steric hindrance and less overlap with antibonding t_{2g} orbitals in the case of four as compared to six bonds.

Similar dependencies and trends are observed for other Mössbauer isotopes, for which more information is found in Chap. 7. It should be pointed out again that the nuclear parameter $\Delta R/R$ is negative for ^{57}Fe in contrast to many other nuclei. The sign of the isomer shift correlations is inverted for nuclei with $\Delta R/R > 0$.

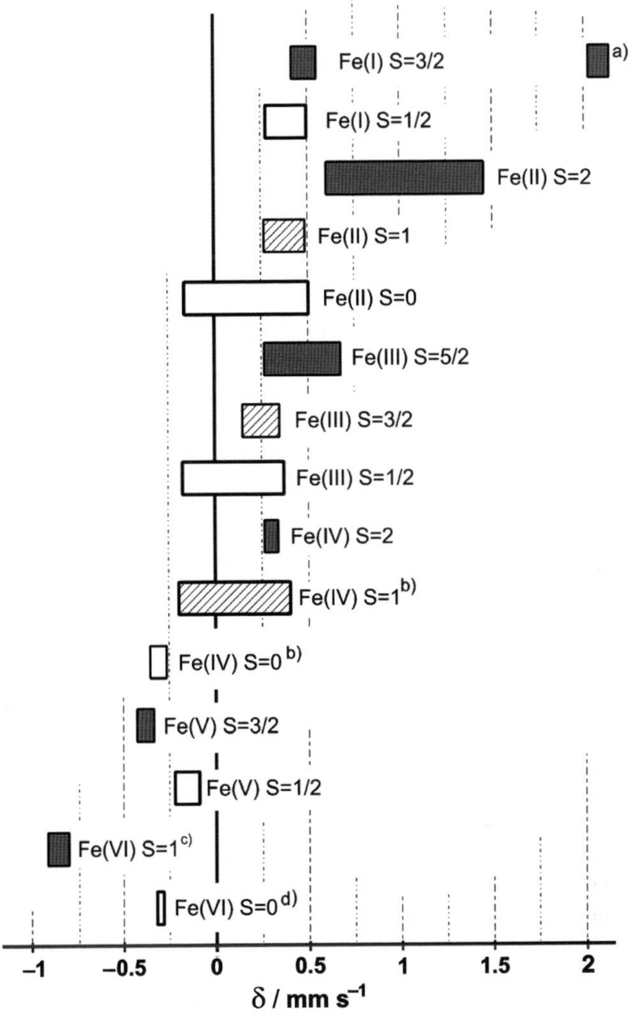

Fig. 4.3 Ranges of isomer shifts observed for ^{57}Fe compounds relative to metallic iron at room temperature (adapted from [24] and complemented with recent data). [a)]The high values above ~1.4–2 mm s^{-1} were obtained from ^{57}Co emission experiments with insulators like NaCl, MgO or TiO_2 [25–28], which yielded complex multi-component spectra. However, the assignment of subspectra for Fe(I) to Fe(III) in different spin states has never been confirmed by applied-field measurements, or other means. More recent examples of structurally characterized molecular Fe (I)-diketiminate and tris(phosphino)borate complexes with three-coordinate iron show values around ~0.45–0.57 mm s^{-1} [29–31]. [b)]The usual low-spin state for Fe(IV) with 3d^4 configuration is $S = 1$ for quasi-octahedral or tetrahedral coordination. The "low–low-spin" state with $S = 0$ is found for distorted trigonal-prismatic sites with three strong ligands [30, 32]. [c)]Occurs only in ferrates. [d)]There is only one example of a molecular iron(VI) complex; it is six-coordinate and has spin $S = 0$ [33]

4.2.4.3 Applications of Isomer Shift Correlations

The unusual oxidation state (+VI) of iron in a molecular iron(VI)–nitrido complex, generated in frozen solution at 77 K by the photolytic cleavage of an iron(IV)–azide precursor, was to be demonstrated on the basis of the Mössbauer isomer shift [33]. A series of $[(\text{cyclam-ac})\text{FeX}]^{+/2+}$ compounds were synthesized, where cyclam-ac is a macrocyclic planar N_4-donor ligand with a tethered acetate as a fifth ligand and X is a halide or azide (N_3^-) or a nitrido group (≡N). The presumed electron configurations of iron range within the series from d^2 to d^6. This assignment of the unprecedented iron(VI) state in a molecule was readily supported by the (linear) correlation with the measured isomer shifts within the series of compounds (see Sect. 8.2.3, Fig. 8.26 and Chap. 5).

Helpful isomer shift correlations have also been established for *iron–sulfur clusters*. This class of inorganic iron centers is found in proteins and synthetic analogs. Most FeS clusters are oligo-nuclear and include one to four tetrahedral high-spin $[FeS_4]$-sites, which are redox-active and show a variety of charge-delocalized mixed-valence states, as has been demonstrated by Mössbauer spectroscopy [34]. Isomer shifts have been empirically related to the average oxidation number x of the iron centers according to

$$\delta(x) = (1.43 - 0.40 \cdot x)\ \text{mm s}^{-1}, \tag{4.28}$$

with x ranging from 2.0 to 3.0 depending on how many iron sites and valence electrons are available [35].

4.2.4.4 Covalent Bonding Properties

Properties of the ligands directly bound to the Mössbauer atom can have a similarly strong influence on the isomer shifts as the oxidation state of the metal. One such property is the *electronegativity* (EN) of the ligands. Indeed, a positive, almost linear correlation has been established by using Pauling's EN scale for the closely related group of high-spin ferrous halides FeX_2 and $FeX_2(H_2O)_n$ (X = F, Cl, Br, I and $n = 1, 2, 4$), the bonds of which are governed by σ-interaction [24]. The experimental isomer shifts reach from $\delta = 1.03$ mm s^{-1} for FeI_2 to $\delta = 1.4$ mm s^{-1} for FeF_2 (relative to α-iron). The higher δ-value for the fluoride is due to the higher EN of fluorine as compared to iodine. This causes a lower 4*s* population in the case of FeF_2 and thus, a lower electron density at the nucleus and, in turn, a more positive isomer shift than for the iodide. The extension of this approach to other more complex systems is not feasible because of the ambiguities of EN.

Another example of the correlation between the isomer shift and covalent bonding properties is *π-backbonding*. The observed isomer shift of ferrous cyanides $[\text{Fe(II)(CN)}_5\text{X}^n]^{(3+n)-}$ [24] becomes more *negative* with increasing

d_π–p_π-backdonation in the order of X being $H_2O < NH_3 < NO_2^- < SO_3^{2-} < CO \ll NO^+$. Water is a rather weak σ-donor with negligible π-acceptor capabilities, whereas CO and NO^+ are known to be strong π-acceptor ligands. The actual values of the isomer shift found for the ferrous low-spin cyanides range from $\delta = 0.31$ mm s^{-1} for X = H_2O to $\delta = 0.15$ mm s^{-1} for X = CO and $\delta = 0.0$ mm s^{-1} for X = NO^+.

Backdonation also explains why the isomer shifts of $K_4[Fe(CN)_6]\cdot 3H_2O$ and $K_3[Fe(CN)_6]$ are nearly equal [36]; the degree of π-backbonding changes upon oxidation or reduction of the metal site and compensates for the change in the number of valence 3*d* electrons.

Remarkable variations of the isomer shift are also found for low-spin iron(II)-nitrosyl complexes, which are usually best classified by using the Enemark–Feltham notion [37], $\{FeNO\}^n$, where *n* is the total number of valence electrons on iron and NO. Almost irrespective of the other ligands, the isomer shift for six-coordinate nitrosyl complexes changes by about −0.2 mm s^{-1} on going from $n = 8$ to 7 and again by above −0.2 mm from 7 to 6. MO calculations revealed that all the compounds are best described as Fe(II) low-spin complexes [38]. This means, vice versa, that the nitrosyl ligand changes from NO to NO^- and from NO to NO^+ upon reduction and oxidation of the stable $\{FeNO\}^7$ species. Apparently, the significant changes in the isomer shift essentially reflect the actual π-acceptor strength of NO^-, NO,˙ and NO^+, in the respective cases.

4.2.4.5 Basic Interpretation

Mössbauer isomer shifts of iron-containing compounds are traditionally explained

1. By the shielding of the nuclear potential for core-*s* and valence-*s* electrons by the charge of the 3*d* electrons and
2. By the population of 4*s* orbitals due to participation in bond formation.

Shielding causes relaxation of *s*-orbitals and leads to a decrease of $|\psi(0)|^2$. The strength of shielding apparently depends on the 3*d* population. This effect and its inherent importance for the isomer shift were first theoretically verified by Watson using Hartree–Fock atomic structure calculations for *free atoms* and ions [39, 40] (see also Table 5.2). However, quantum chemical calculations have revealed that the total charge on the metal atom in complexes varies very little throughout a redox series; for the examples shown in Fig. 4.3, the change is effectively less than 1 electron on going from Fe(VI) to Fe(II). Accordingly, the classical shielding effect is found to be less important for the variation of the isomer shift of iron in coordination compounds, as will be shown in Chap. 5. Other effects (such as variations of 4*s* population) are equally or even more important.

An explanation of the direct 4*s* contribution to the charge density at the nucleus requires MO calculations. A simple MO diagram for octahedral complexes is shown in Fig. 4.4. The σ-interaction of metal *d*-orbitals and symmetry-adapted ligand orbitals usually yields the major part to the stability of the bonds. According

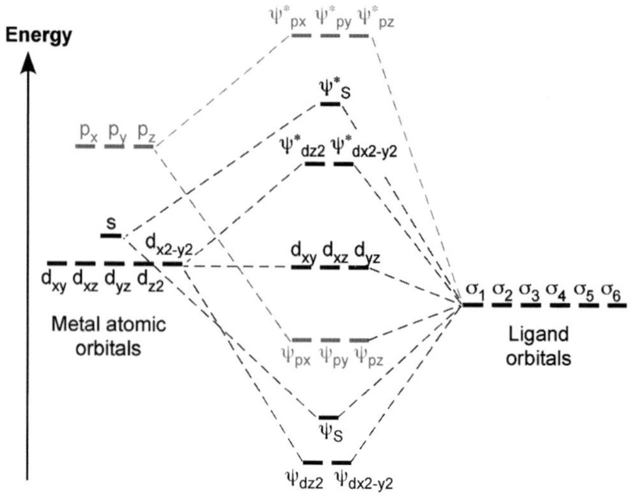

Fig. 4.4 Molecular orbital diagram for octahedral complexes (σ-interaction only)

to their symmetries, *p*- and *d*-orbitals form three sets of MOs which are bonding, nonbonding and antibonding in nature. The 4*s* atomic orbital of the metal is empty, but it is energetically rather close to the 3*d* shell and has the appropriate symmetry to interact with the ligand σ-orbitals. Therefore, the resultant bonding MO will be a linear combination of the metal orbital and ligand orbitals, $|\psi_s\rangle = a|4s\rangle + b\sum|\sigma_i\rangle$. The amount of mixing of the atomic orbitals in the MO is directly proportional to the magnitude of the atomic orbital overlap divided by their energy difference. The bonding MO, ψ_S, will have a predominant ligand character and a smaller amount of 4*s* character (and vice-versa for the antibonding combination $\psi_S{}^*$). Thus, the occupation of ψ_S by electrons will cause a small but not negligible contribution of iron 4s electron density to $|\psi(0)|^2$.

Apart from the direct 4*s* contribution to $|\psi(0)|^2$ caused by participation of the orbital in the bonding MO ψ_S, indirect contributions which are related to changes in iron–ligand bond lengths and coordination numbers also play a significant role in the variation of the isomer shift. Ligand coordination causes a compression of the radial distribution of the 3*s*- and 4*s*-iron orbitals and thus alters the electron density at the nucleus. This influence determines about 2/3 of the variations of $|\psi(0)|^2$ found for different iron complexes (Fig. 5.3 and Table 5.3). Based on theoretical perceptions presented in Chap. 5, the iron–ligand *bond lengths* play a decisive role for the isomer shift of a compound. This is also in accord with the experimental data shown in Fig. 4.3. Compounds with short and highly covalent bonds exhibit low (more negative) isomer shift, whereas long ionic bonds correlate with large positive isomer shifts. Therefore, for example, low-spin compounds have lower isomer shifts than high-spin compounds. Similarly, oxidation (e.g. from the ferrous high-spin to the ferric high-spin state) causes shorter iron–ligand bond lengths and lower isomer shifts. Also, strong σ-acceptor or π-donor interaction causes short iron–ligand bonds and low isomer shifts.

4.3 Electric Quadrupole Interaction

Quadrupole splitting is a salient feature of Mössbauer spectra. However, other techniques, such as perturbed-angular correlation spectroscopy, nuclear quadrupole resonance (NQR), solid-state nuclear magnetic resonance (NMR), and electron-paramagnetic or electron-nuclear double resonance (EPR/ENDOR) spectroscopy, can also detect electric quadrupole interaction. The basic theory is similar for all these techniques [4, 5, 41]; it deals with the rotational conformations that a non-spherical nucleus can take in the in-homogeneous electric field generated by a noncubic charge distribution of the surrounding electrons. In Sect. 4.1, the *electric field gradient* (EFG), (4.3), was used to describe this property of the electronic environment of the nucleus. An example for the rotational conformations of an elongate nucleus in the EFG generated by four model charges is sketched in Fig. 4.5.

The shape of the nucleus is best described by a power series, the relevant term of which yields the nuclear *quadrupole moment*. In Cartesian coordinates, this is represented by a set of intricate integrals of the type $\int \rho_{\mathrm{n}}(r)(3x_i x_j - \delta_{ij} r^2)\mathrm{d}\tau$, where $x_i = x, y, z$, and $\rho_{\mathrm{n}}(r)$ is the nuclear charge distribution (4.12). The evaluation of $\rho_{\mathrm{n}}(r)$ for any real nucleus would be very challenging.

Fortunately, in quantum mechanics, the corresponding spatial operations for the individual nucleons (4.15) can be replaced by convenient angular momentum operators that act on the *total spin I* of the nucleus [4]. The corresponding

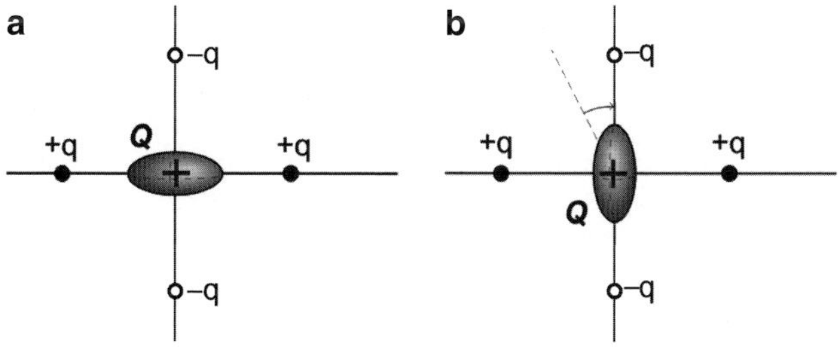

Fig. 4.5 Rotational conformations of a cigar-shaped nucleus with quadrupole moment Q in the center of two positive charges ($+q$) and two negative charges ($-q$). The picture visualizes quadrupole coupling, or quadrupole interaction, whereby the choice of outer charges is generic to model a simple EFG at the center of the nucleus. The configuration (**b**) is energetically more favorable than (**a**) because the positive tips of the elongate nucleus are closer to the negative charges. The rotational energy of the system depends on the strength of Q, the EFG and the rotation angle. The rotation angle and the energy of the system are quantized; they can take only discrete values

spin-Hamiltonian operator (4.17) commonly used for the computation of quadrupole interactions has the form [6]

$$\hat{H}_{\mathrm{Q}} = \frac{eQV_{zz}}{4I(2I-1)}\left[3\hat{I}_z^2 - I(I+1) + \frac{\eta}{2}(\hat{I}_+^2 + \hat{I}_-^2)\right], \tag{4.29}$$

where I is the nuclear spin quantum number, $\hat{I}_\pm = \hat{I}_x \pm \mathrm{i}\hat{I}_y$ are shift operators, and $\hat{I}_x$, $\hat{I}_y$, $\hat{I}_z$ are the operators of the nuclear spin projections onto the principal axes. Q denotes the quadrupole moment of the nucleus, and V_{zz} and η represent the main component and the asymmetry parameter of the EFG at the nucleus (*vide infra*). We note in passing that the product eQV_{zz} is also called the *nuclear quadrupole coupling constant* (NQCC); other notations of which are e^2qQ, eqQ, or even B, b, or υ^X, all in energy units (and $\hbar$) [42]. Note that NQCCs, like all hyperfine coupling constants, are products of a nuclear and an electronic property.

4.3.1 Nuclear Quadrupole Moment

Only one nuclear constant, Q, is needed in (4.29) to parameterize the properties of the nucleus, although the quadrupole moment, in general, is an irreducible 2nd rank tensor with five independent invariants (4.12) [5]. The simplification is mainly due to the fact that the nuclear charge distribution has cylindrical symmetry, because nuclei exhibit well-defined angular momenta [4]. Adopting z as the corresponding symmetry axis, Q corresponds in classical terms to the difference between the charge distribution parallel and transverse to z: $eQ = \int \rho_{\mathrm{n}} z^2 \mathrm{d}\tau - \int \rho_{\mathrm{n}} x^2 \mathrm{d}\tau$. The type of the nuclear deformation determines the sign of Q; a *positive* quadrupole moment indicates that the nucleus is elongate, like a cigar, whereas for a *negative* sign, the nucleus is oblate, or flattened like a pan-cake. Only states with $I > 1/2$ exhibit an electric quadrupole moment different from zero. Values of Q for the ground and excited states of some typical Mössbauer isotopes are given in Table 4.1, and a more complete summary for all Mössbauer isotopes is presented in Part IV of the attached CD-ROM. A comprehensive review of nuclear quadrupole moments is found in [43] and the IUPAC compilations [44, 45].

4.3.2 Electric Field Gradient

The electrons at the Mössbauer atom and the surrounding charges on the ligands cause an electric potential $V(\vec{r})$ at the nucleus (located at $\vec{r} = (0, 0, 0)$. The negative value of the first derivative of the potential represents the electric field, $\vec{E} = -\vec{\nabla}V$, which has three components in Cartesian coordinates, $\vec{E}$ $-(\partial V/\partial x,\ \partial V/\partial y,\ \partial V/\partial z)$.

Table 4.1 Quadrupole moments for the ground state (g) and the excited state (e) of some Mössbauer nuclei quoted in millibarn (1 mb = 10^{-31} m^2)

Isotope	I_g	Q_g (mb)	I_e	Q_e (mb)	Ref.
^{57}Fe	1/2	–	3/2	154.9(2)	[43]
^{61}Ni	3/2	162(15)	5/2	−200(30)[a]	[42]
^{99}Ru	5/2	79(4)	3/2	231(12)[a]	[43]
^{119}Sn	1/2	–	3/2	94(4)	[43]
^{127}I	5/2	−696(12)[b]	7/2	−710[a]	[42]

I is the quantum number of the nuclear spin
[a]Taken from Mössbauer Effect Data Center (MEDC), Prof. John Stevens, University of North Carolina, Asheville, NC, USA, September 2009; for a full list of the nuclear properties for all known Mössbauer isotopes see the MEDC web address: http://orgs.unca.edu/medc/Resources.html, or the corresponding pdf file in the CD-ROM of this book
[b]An older report [46] states −720 mb; the value reported by MEDC is −789 mb

The EFG at the nucleus corresponds to the second derivative of the electric potential $V(\vec{r})$ at $r = 0$, that is,

$$\text{EFG} = \left[-\vec{\nabla}\vec{E}\right] = \left[\vec{\nabla}\vec{\nabla}V\right] = \begin{bmatrix} V_{xx} & V_{xy} & V_{xz} \\ V_{yx} & V_{yy} & V_{yz} \\ V_{zx} & V_{zy} & V_{zz} \end{bmatrix}, \tag{4.30}$$

where $V_{ij} = (\partial^2 V/\partial x_i \partial x_j)_0$ are the nine components of the 3 × 3 second-rank EFG tensor.[8] Only five of these components are independent because of the symmetric form of the tensor, that is, $V_{ij} = V_{ji}$, and because of Laplace's equation which requires that the EFG be a traceless tensor,

$$(\Delta V)_0 = (\vec{\nabla}\vec{\nabla}V)_0 = V_{xx} + V_{yy} + V_{zz} = 0. \tag{4.31}$$

The trace vanishes because only *p*- and *d*-electrons contribute to the EFG, which have zero probability of presence at $r = 0$ (i.e. Laplace's equation applies as opposed to Poisson's equation, because the nucleus is external to the EFG-generating part of the electronic charge distribution). As the EFG tensor is symmetric, it can be diagonalized by rotation to a *principal axes system* (*PAS*) for which the off-diagonal elements vanish, $V_{i \neq j} = 0$. By convention, the principal axes are chosen such that the principal tensor components are denoted V_{zz}, V_{yy}, and V_{xx}, in order of decreasing modulus,

$$|V_{zz}| \geqslant |V_{yy}| \geqslant |V_{xx}|, \tag{4.32}$$

(also denoted as *eigenvalues* or *main components* of the EFG: $|V_{\max}| \geqslant |V_{\text{mid}}| \geqslant |V_{\min}|$). Given the traceless character, only two of the principal components are

[8]Each component of the electric field $\vec{E}$ vector has three derivatives, i.e. in *x*, *y* and *z* direction.

independent.[9] Typically, these are chosen to be the *largest eigenvalue* V_{zz} and the *asymmetry parameter*, η. The latter is defined by the ratio

$$\eta = \frac{V_{xx} - V_{yy}}{V_{zz}}, \quad \text{for which we find } 0 \leqslant \eta \leqslant 1. \tag{4.33}$$

Any EFG can be specified by these two invariants of the tensor, that is,

- V_{zz} also denoted as eq, (e = proton charge), and
- the asymmetry parameter η.

For a three- or fourfold axis of symmetry passing through the Mössbauer nucleus as the center of symmetry, one can show that the EFG is symmetric, that is, $V_{xx} = V_{yy}$ and therefore, $\eta = 0$. In a system with two mutually perpendicular axes of threefold or higher-symmetry, the EFG must be zero.

4.3.3 Quadrupole Splitting

The Hamiltonian operator for the electric quadrupole interaction, $\hat{H}_{\mathrm{Q}}$, given in (4.29), connects the spin of the nucleus with quantum number I with the EFG. In the simplest case, when the EFG is axial ($V_{xx} = V_{yy}$, i.e. $\eta = 0$), the Schrödinger equation can be solved on the basis of the spin functions $|I, m_I\rangle$, with magnetic quantum numbers $m_I = I, I-1, \ldots, -I$. The Hamilton matrix is diagonal, because only matrix elements of the type $\langle m_I, I|\hat{I}_z^2|m_I, I\rangle = m_I^2$ and $\langle m_I, I|\text{const}|\text{m}_I, I\rangle =$ const occur. The resulting energies for $\eta = 0$ are

$$E_{\mathrm{Q}}(m_I) = \frac{eQV_{zz}}{4I(2I-1)}\left[3m_I^2 - I(I+1)\right]. \tag{4.34}$$

The electric quadrupole interaction causes a splitting of the $(2I + 1)$ magnetic substates without shifting the mean energy of the nuclear spin manifold; substates with the same absolute value of $|m_I|$ remain degenerate for $\eta = 0$.

The effect of electric quadrupole interaction for ^{57}Fe is exemplified in Fig. 4.6. The ground state remains unsplit because of the lack of quadrupole moment for $I = 1/2$. The excited state with $I = 3/2$ splits into two doubly degenerate substates $|3/2, \pm 3/2\rangle$ and $|3/2, \pm 1/2\rangle$ due to the m_I^2 dependence of the quadrupole energies:

$$\begin{aligned} E_{\mathrm{Q}}(\pm 3/2) &= 3eQV_{zz}/12 \\ E_{\mathrm{Q}}(\pm 1/2) &= -3eQV_{zz}/12 \end{aligned} \quad (\text{for } I = 3/2). \tag{4.35}$$

[9] Three more parameters are implicitly included which are the Euler angles that describe the orientation of the principal axes system.

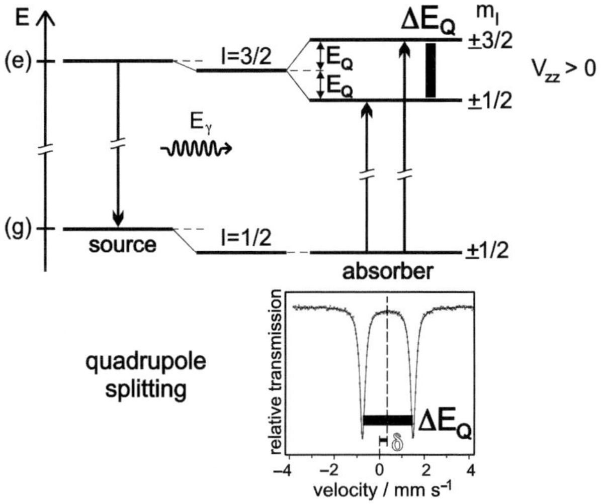

Fig. 4.6 Quadrupole splitting of the excited state of ^{57}Fe with $I = 3/2$ and the resulting Mössbauer spectrum. Quadrupole interaction splits the spin quartet into two degenerate sublevels $|I, \pm m_I\rangle$ with energy separation $\Delta E_\mathrm{Q} = 2E_\mathrm{Q}$. The ground state with $I = 1/2$ remains unsplit. The nuclear states are additionally shifted by electric monopole interaction giving rise to the isomer shift δ

The energy difference ΔE_Q between the two substates is

$$\Delta E_\mathrm{Q} = E_\mathrm{Q}(\pm 3/2) - E_\mathrm{Q}(\pm 1/2) = \mathrm{e}QV_{zz}/2. \tag{4.36}$$

In a conventional ^{57}Fe Mössbauer experiment with a powder sample, one would observe a so-called *quadrupole doublet* with two resonance lines of equal intensities.[10] The separation of the lines, as given by (4.36), represents the *quadrupole splitting* ΔE_Q. The parameter ΔE_Q is of immense importance for chemical applications of the Mössbauer effect. It provides information about bond properties and local symmetry of the iron site. Since the quadrupole interaction does not alter the mean energy of the nuclear ground and excited states, the isomer shift δ can also be derived from the spectrum; it is given by the shift of the center of the quadrupole spectrum from zero velocity.

The quadrupole interaction becomes more sophisticated when the EFG lacks axial symmetry, $\eta \neq 0$, because the shift operators $\widehat{I}_\pm^2$ connected to η introduce

[10]In practice, quadrupole doublets may be asymmetric, when the sample is a single crystal or an imperfect powder with partial ordering (texture). The related issue of line intensities will be treated in Sect. 4.6. Another, though more exotic reason for asymmetric intensities might be the presence of an anisotropic recoilless fraction *f*, the so-called Goldanskii–Karyagin effect [73]. Inhomogeneous *line broadening* also causes asymmetric doublets with different widths for the low- and the high-energy lines but with equal intensities (areas).

off-diagonal elements into the Hamilton matrix of the type $\langle m_I|\hat{H}_Q|m_I \pm 2\rangle$ The exact solution of the secular equation for the energies is available only for $I = 3/2$:

$$E_Q(I = 3/2, m_I) = \frac{eQV_{zz}}{4I(2I - 1)}\left[3m_I^2 - I(I + 1)\right]\sqrt{1 + \frac{\eta^2}{3}}. \quad (4.37)$$

The comparison of (4.34) and (4.37) shows that the value of η can modify the quadrupole splitting at most by $\approx$15%, as η is limited to the range 0–1.

Quadrupole interaction lifts the degeneracy of *integer-spin states* completely, except when the EFG is axial. The corresponding Schrödinger equation may be solved for nuclear spins different from $I = 3/2$ by perturbation treatment of the η-dependent terms. Exact solutions are obtained by numerical matrix diagonalization. The excited state of ^{182}W with $I = 2$, for instance, will be split into five separate "m_I" – states[11] when it is exposed to a nonaxial EFG ($\eta \neq 0$). Since the ground state of ^{182}W is a singlet ($I = 0$), the corresponding Mössbauer quadrupole spectrum will show five resonance lines. Only in the axial limit $\eta = 0$ do the states with $m_I = \pm1$ and ±2 remain degenerate, in which case, the spectrum will exhibit a three-line pattern and the quadrupole energies can be taken from (4.34).

Nuclear states with *half-integer* spin, in contrast, are never completely split by pure quadrupole interaction. Pairs of $|\pm m_I\rangle$ states, so-called Kramers doublets, remain doubly degenerate for any value of the asymmetry parameter, according to Kramers' degeneracy theorem, which states that the energy levels of half-integer spin systems remain at least doubly degenerate in the presence of purely electric fields [51]. The twofold degeneracy can be removed only by magnetic perturbation, which will be discussed in Sects. 4.4 and 4.5.

Quadrupole interaction can be positive or negative, depending on the sign of Q and of the main component V_{zz} of the EFG. Negative quadrupole coupling for the $I = 3/2$ excited state of ^{57}Fe (Fig. 4.6) would invert the order of the $|\pm1/2\rangle$ and the $|\pm3/2\rangle$ states. However, this cannot be easily recognized in the experiment, because the lines of the symmetric quadrupole doublet obtained from a powder sample cannot be assigned to the corresponding $\pm1/2 \rightarrow \pm1/2$ and the $\pm1/2 \rightarrow \pm3/2$ transitions. Zimmermann [52] suggested a method to reveal the sign of V_{zz} and the asymmetry parameter η from the intensity variations of single-crystal measurements [5]. In Sect. 4.5, it will be demonstrated how perturbations of the nuclear states by externally applied magnetic fields can solve the problem. However, for nuclear spin systems with multiplicities higher than 4, the sign of V_{zz} and the value of η may be determined from a zero-field spectrum if the number of observed resonance lines is sufficient to assign the splitting pattern for the ground and for the excited state of the Mössbauer nucleus [53].

[11]The eigenfunctions for nonaxial nuclear quadrupole interaction are mixtures of the $|I, m_I\rangle$ basis functions and thus do not possess well-defined magnetic quantum numbers. Strictly speaking, the states should not be labeled with pure quantum numbers "m_I."

A traceless tensor has always positive and negative principal elements. The sign of the EFG tensor is defined by the sign of the main component V_{zz}, which is meaningful at least for $\eta = 0$ (positive V_{zz} indicates an oblate charge distribution, negative V_{zz} means elongation). However, when η increases from 0 to 1, the sign loses its significance progressively, because $|V_{zz}|$ approaches $|V_{yy}|$, while the two components have opposite signs (see (4.32–4.33)). Finally, at the limit of $\eta = 1$, the EFG tensor has the diagonal elements $(0, -V_{zz}, V_{zz})$, for which a sign of the EFG is without any physical meaning. This emphasizes the importance of η for the parameterization of the electric field gradient.

4.3.4 *Interpretation and Computation of Electric Field Gradients*

The experimentally observed quadrupole splitting ΔE_Q for ^{57}Fe in inorganic compounds, metals, and solids reaches from 0 to more than 6 mm s^{-1} [30, 32]. The range of ΔE_Q for other Mössbauer isotopes may be completely different because of the different nuclear quadrupole moment Q of the respective Mössbauer nucleus, and also because the EFG values may be intrinsically different due to markedly different radial distributions of the atomic orbitals (vide infra). As Q is constant for a given isotope, variations in the quadrupole coupling constants eQV_{zz} can only arise from V_{zz}.

Although the EFG of a given system can be easily determined from a Mössbauer spectrum, it may be rather difficult to relate it to the electronic structure of the Mössbauer atom. In order to visualize a few typical cases, the computation of the EFG is described in the following for some selected charge distributions. A comprehensive quantum chemical interpretation of the quadrupole splitting will be given in Chap. 5.

4.3.4.1 EFG from Point Charges

The influence of a noncubic electronic charge distribution interacting with a Mössbauer nucleus may be exemplified by using *point charges*, for which the EFG is easy to calculate. A point charge q at a distance $r = (x^2 + y^2 + z^2)^{1/2}$ from a nucleus located at the origin of the coordinate system causes a potential $V_0 = q/r$ at the nucleus (Fig. 4.7a). The electric field $\vec{E}$ is the negative gradient of the potential, $-\vec{\nabla} V$, and the gradient of the electric field (EFG) is the second derivative of the potential, $V_{ij} = (\partial^2 V/\partial x_i \partial x_j)_0$, where x_i and x_j denote the Cartesian coordinates (4.30). Taking derivatives of the potential at point (0, 0, 0) in z-direction yields the EFG component V_{zz},

$$V_{zz} = \frac{\partial^2 V}{\partial z \partial z} = q r^{-5}(3z^2 - r^2), \qquad (4.38)$$

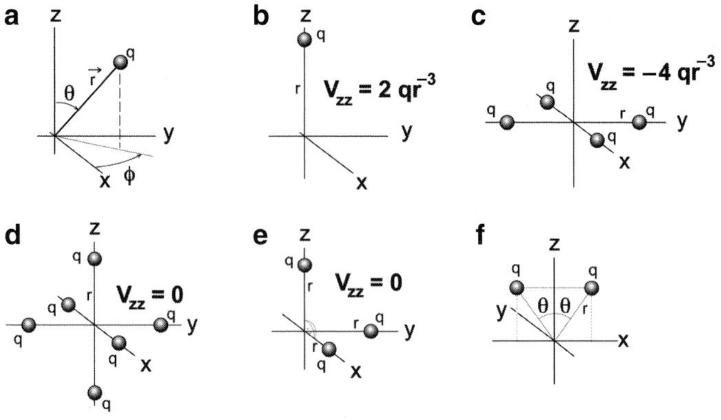

Fig. 4.7 Electric field gradients at the origin of the coordinate system for six different arrangements of identical point charges q. Panel (**a**): A charge positioned at an arbitrarily chosen point with polar coordinate (θ, ϕ, r) induces an EFG with components as given by (4.39). The tensor is not diagonal in the coordinate system (x, y, z). Diagonalization requires two Euler rotations of the coordinate system, one around z by an angle ϕ and one around the new axis y by an angle θ, to achieve the new z-axis passing through q. Panel (**b**) depicts the situation after rotation of the coordinate system; the EFG is diagonal and V_{zz} and η (=0) are easily obtained from (4.41)–(4.42). The arrangements (**a**) and (**b**) generate the same quadrupole splitting in the corresponding Mössbauer spectra. Panel (**c**): Four charges in the xy-plane generate an axial EFG with $V_{zz} = -4qr^{-3}$. Panel (**d**) and (**e**): Six charges located on a cube (**d**) or three charges in a "half-cube" arrangement (**e**) yield zero EFG. Panel (**f**): The EFG for two charges in the xz-plane at the points $(x, 0, z)$ and $(-x, 0, z)$ depend notably on the angle θ, according to $V_{zz} = 2qr^{-3}(3\cos^2\theta - 1)$ and $\eta = 3\sin^2\theta/(3\cos^2\theta - 1)$. The asymmetry parameter exceeds unity at $\sin\theta = 1/\sqrt{3}$ (i.e. $\theta \approx 35.2°$), whereas for $\theta = 90°$ the asymmetry parameter becomes negative, $\eta = -3$, and the zz-component is $V_{zz} = -qr^{-3}$. In both cases the (x, y, z) axes do not mark a "proper" PAS because (4.32), $|V_{zz}| \geqslant |V_{yy}| \geqslant |V_{xx}|$, is violated. The values of V_{zz} and η can be converted to those obtained in a "proper" reference frame $(\hat{x}, \hat{y}, \hat{z})$ by re-sorting the three diagonal elements of the EFG, such that $|V_{\hat{z}\hat{z}}|$ is the largest component and $|V_{\hat{x}\hat{x}}|$ is the smallest. Finally, we would point out that, when q is an electron with charge $-e$, the EFG main component V_{zz} is negative for cases (**a**) and (**b**) and positive for case (**c**). The asymmetry parameter is zero throughout, except for (**f**)

and the corresponding general expression for x_i and $x_j = x, y, z$,

$$V_{ij} = \frac{\partial^2 V}{\partial x_i \partial x_j} = q(3x_i x_j - r^2\delta_{ij})r^{-5}. \tag{4.39}$$

Conversion to spherical coordinates (see Fig. 4.7a) yields

$$V_{zz} = qr^{-3}(3\cos^2\theta - 1). \tag{4.40}$$

Thus, a charge q located on the z-*axis* at point (0, 0, r) produces an EFG at the nucleus (Fig. 4.7b) which has the from

$$V_{zz} = 2qr^{-3}. \tag{4.41}$$

The off-diagonal elements $V_{i \neq j}$ are zero for this arrangement and $V_{xx} = V_{yy} = -1/2V_{zz}$. As expected, the adopted coordinate system is a principal axes system (PAS) of the EFG, and the asymmetry parameter η is zero.

Since the EFG component V_{zz} and the asymmetry parameter $\eta = (V_{xx} - V_{yy})/V_{zz}$ in the PAS are invariants of the EFG, the two similar arrangements of the charge q shown in Fig. 4.7a, b must produce the same quadrupole splitting (because energies do not depend on the choice of the coordinate system).

Electric field gradients are additive; an EFG arising from a number of charges q_i located at different distances r_i from the nucleus is the sum of the contributions from the individual charges. The tensor components for each charge are computed by using (4.39), using a *common* coordinate system. The components are then added and the resulting tensor is diagonalized to obtain V_{zz} and η. However, if the PAS is known in advance (due to symmetry arguments for instance, as a three- or fourfold axis passing through the Mössbauer atom, marking the z-axis), V_{zz} and η can be obtained in the PAS from the formulae

$$V_{zz} = \sum_{i=1}^{n} q_i r_i^{-3}(3\cos^2\theta_i - 1), \tag{4.42a}$$

$$\eta = \frac{1}{V_{zz}} \sum_{i=1}^{n} q_i r^{-3} 3\sin^2\theta_i \cos 2\phi_i. \tag{4.42b}$$

Figure 4.7c–f summarizes a set of examples with different geometrical arrangements of point charges to elucidate some typical aspects of EFGs arising from symmetric or asymmetric charge distributions.

4.3.4.2 The "Lattice Contribution" to the EFG

The EFG parameters V_{zz} and η described by (4.42a) and (4.42b) do not represent the actual EFG felt by the Mössbauer nucleus. Instead, the electron shell of the Mössbauer atom will be distorted by electrostatic interaction with the noncubic distribution of the external charges, such that the EFG becomes *amplified*. This phenomenon has been treated by Sternheimer [54–58], who introduced an "anti-shielding" factor $(1-\gamma_\infty)$ for computation of the so-called *lattice contribution* to the EFG, which arises from (point) charges located on the atoms surrounding the Mössbauer atom in a crystal lattice (or a molecule). In this approach,[12] the actual lattice contribution is given by

$$(V_{zz})_{\text{lat}} = (1 - \gamma_\infty) \cdot V_{zz} \quad \text{and} \tag{4.43a}$$

$$(\eta)_{\text{lat}} = (1 - \gamma_\infty) \cdot \eta. \tag{4.43b}$$

[12] Sternheimer factors do not appear in quantum chemical calculations that include all electrons of a system This is in contrast to Sternheimer's early approach in which core electrons were taken as 'frozen' for the sake of saving computing time.

For iron compounds, $(V_{zz})_{\text{lat}}$ and η_{lat} are amplified by approximately $(1-\gamma_\infty) \approx 10$ as compared to the point charge contributions V_{zz} and η obtained from (4.42a) and (4.42b). Nevertheless, the lattice contribution is usually least significant for most iron compounds because it is superseded by a strong EFG from the valence electrons; details will be found in Chap. 5.

4.3.4.3 Local Contribution from Valence Electrons

The $1/r^3$ dependency of (4.38)–(4.43) indicates that charges close to the nucleus have the most important effect on the quadrupole interaction. Hence, the valence electrons belonging to the Mössbauer atom usually make the strongest contribution to the EFG if their charge distribution deviates from spherical symmetry. This *valence contribution*, $(V_{zz})_{\text{val}}$ and η_{val}, is determined by the MOs formed between the central Mössbauer atom and the ligands. A strong $(\text{EFG})_{\text{val}}$ arises when the oxidation state of the metal ion and the chemical bonds afford significant differences in the population of the local atomic valence orbitals of the metal. This is usually the case for open-shell complexes. (For example, the iron 3d population for a hypothetical compound might be $3d(xy)^{1.9}(xz,yz)^{2.1}(x^2-y^2)^{1.2}(z^2)^{0.9}$, which would be close to a $3d^6$ high-spin configuration for iron(II) in a crystal field description).

For ionic compounds, crystal field theory is generally regarded a sufficiently good model for qualitative estimates. Covalency is neglected in this approach, only metal *d*-orbitals are considered which can be populated with zero, one or two electrons. To evaluate $(V_{zz})_{\text{val}}$ at the Mössbauer nucleus, one may simply take the expectation value[13] of the expression $-e(3\cos^2\theta-1)r^{-3}$ for every electron in a valence orbital $|\psi_i|$ of the Mössbauer atom and sum up,

$$(V_{zz})_{\text{val}} = -e\sum\nolimits_i \langle\psi_i|(3\cos^2\theta-1)r^{-3}|\psi_i\rangle. \tag{4.44}$$

The expectation value of r^{-3} is given by the radial part $R(r)$ of the orbitals,

$$\langle r^{-3}\rangle = \int R(r)r^{-3}R(r)r^2\mathrm{d}r. \tag{4.45}$$

Numerical values may be obtained from experiments [59]; for *d*-orbitals of iron, a good estimate is $\langle r^{-3}\rangle \approx 5a_0{}^{-3}$. One can factorize out $\langle r^{-3}\rangle$ in (4.44) to have,

$$(V_{zz})_{\text{val}} = -e\sum_i \langle\psi_i|3\cos^2\theta-1|\psi_i\rangle\langle r_i^{-3}\rangle. \tag{4.46a}$$

[13] See (4.42a) for the spatial operator and footnote 4 for the notation.

The asymmetry parameter η_{val} can be obtained from the corresponding results for the tensor components in the x- and y-directions,

$$(V_{xx})_{val} = -e\sum_i \langle\psi_i|3\sin^2\theta\cos^2\phi - 1|\psi_i\rangle\langle r_i^{-3}\rangle \tag{4.46b}$$

$$(V_{yy})_{val} = -e\sum_i \langle\psi_i|3\sin^2\theta\sin^2\phi - 1|\psi_i\rangle\langle r_i^{-3}\rangle, \tag{4.46c}$$

which afford

$$\eta_{val} = -\frac{e}{(V_{zz})_{val}}\sum_i \langle\psi_i|3\sin^2\theta\cos 2\phi - 1|\psi_i\rangle\langle r_i^{-3}\rangle. \tag{4.46d}$$

The axes used for the EFG computation are usually those of the canonical coordinate system of the crystal field, for which a perturbed octahedral (or tetrahedral) symmetry may be adopted. Using (4.46a–d), one can calculate the tensor elements of $(\mathrm{EFG})_{val}$ for any number of electrons residing in the common (real) p- and d-orbitals. The individual tensors are diagonal in the canonical axes system (but a proper value of η is not obtained for every orbital), as summarized in Table 4.2. Note that s-electrons do not contribute to the EFG because of their spherical charge distribution, and f-electrons of higher transition elements are neglected here because they are well shielded from ligand influence by the outer s-, p- and d-electrons of the respective ions and do not participate in chemical bonds to a significant extent.

The expectation values given in Table 4.2 for p- and d-orbitals can be converted into quadrupole coupling energies (in mm s^{-1}) by inserting in (4.36) and (4.46) using actual values for the quadrupole moment and for $\langle r^{-3}\rangle$. A good estimate of $(\Delta E_Q)_{val}$ for iron is obtained with the conversion factor of 4.5 mm s^{-1}/(4/7) $e\langle r^{-3}\rangle$ (obtained for $\langle r^{-3}\rangle = 5a_0^{-3}$ and $Q = 0.16b$). Accordingly, a single electron in a hypothetically pure $d_{x^2-y^2}$-orbital is expected to yield a quadrupole splitting of +4.5 mm s^{-1} when covalency effects and lattice contributions are neglected.

Table 4.2 Expectation values of $(\mathrm{EFG})_{val}$ tensor elements for p- and d-electrons in units of $e\langle r^{-3}\rangle$, where e is the proton charge

Orbital	$\frac{(V_{xx})_{val}}{e\langle r^{-3}\rangle}$	$\frac{(V_{yy})_{val}}{e\langle r^{-3}\rangle}$	$\frac{(V_{zz})_{val}}{e\langle r^{-3}\rangle}$	η_{val} [a]
p_x	–4/5	+2/5	+2/5	–3
p_y	+2/5	–4/5	+2/5	+3
p_z	+2/5	+2/5	–4/5	0
d_{xy}	–2/7	–2/7	+4/7	0
d_{xz}	–2/7	+4/7	–2/7	+3
d_{yz}	+4/7	–2/7	–2/7	–3
$d_{x^2-y^2}$	–2/7	–2/7	+4/7	0
d_{z^2}	+2/7	+2/7	–4/7	0

[a] Values of $|\eta|$ larger than 1 indicate that the octahedral axes do no represent a "proper" axis system for the EFG in the sense of (4.32). To achieve $0 \le \eta \le 1$, the axes may be just interchanged

Figure 4.8 exemplifies the computation of EFG_{val} in the crystal field model for a few cases of different crystal-field splitting for iron(II) and iron(III). First, the corresponding tensor elements for all electrons of a given configuration are summed up, from which $(V_{zz})_{val}$ and η_{val} can then be obtained using (4.32) and (4.33). Part (b) presents the results in comparison with the reduced values of the quadrupole splitting, when spin–orbit coupling (SOC) causes mixing of close-lying orbital states. In practice, however, symmetry distortions and covalency effects may influence the actual value of ΔE_Q values to a similar extent. The deviation of the experimental values of a series of ferric- and ferrous-cyclam complexes given in part (c) from the crystal-field limits exemplifies these effects. If SOC states show appropriate splitting, then temperature-dependent quadrupole splitting can also occur, as mentioned in part (b).

In the case of covalent compounds, crystal-field theory is a poor model for estimating electric field gradients because of the extensive participation of ligand atomic orbitals in the chemical bonds. MO calculations are a much better choice, since the corresponding interactions are considered, and realistic (noninteger) population numbers are obtained for the central metal as well as the ligand atomic orbitals.

The general influence of covalency can be qualitatively explained in a very basic MO scheme. For example, we may consider the μ-oxo Fe(III) dimers that are encountered in inorganic complexes and nonheme iron proteins, such as ribonucleotide reductase. In spite of a half-filled *d*-shell (affording $(\Delta E_Q)_{val} = 0$ in the crystal-field model), the ferric high-spin ions show quadrupole splittings as large as 2.45 mm s^{-1} ($V_{zz} < 0$, $\delta = 0.53$ mm s^{-1}, 4.2–77 K) [61, 62]. This is explained by *anisotropy* of the covalent bonds induced by the noncubic surroundings. The iron centers are, in general, six-coordinate, including a particularly short Fe–O bond to the bridging oxygen atom ($d \approx 1.64-1.68$ Å [62]). The bonds to the other O- and N-ligand atoms are "normal" and less covalent. Therefore, the bonding and antibonding MOs between iron and the bridging oxygen will be particularly strong mixtures of those metal *d*- and oxygen *p*-orbitals, which are suitable for σ- and π-interaction along that particular direction. As a result, some iron *d*-electrons will be much more affected by the short Fe–O bond than others, and the population of the metal *d*-orbitals will significantly deviate from a half-filled shell situation in a purely ionic compound signifying distinct charge anisotropy. The (negative) sign of the resulting EFG, which indicates an *elongate* charge distribution around iron, may be rationalized by assuming strong σ-donation from the oxygen atom into the half-filled d_{z^2} orbital (if we assume the iron-oxo bond in z-direction). This is further enforced by π-donation into the empty d_{xz}, d_{yz} orbitals (see values in Table 4.2).

Another typical example for anisotropic covalency is found in five-coordinate ferric compounds with intermediate spin $S = 3/2$ (also discussed in Sect. 8.2). Crystal field theory predicts a vanishing valence contribution to the EFG, whereas large quadrupole splittings up to more than 4 mm s^{-1} are experimentally found.

The aforementioned examples demonstrate that qualitatively correct estimates of EFG values and signs may be obtained from crystal field arguments for compounds

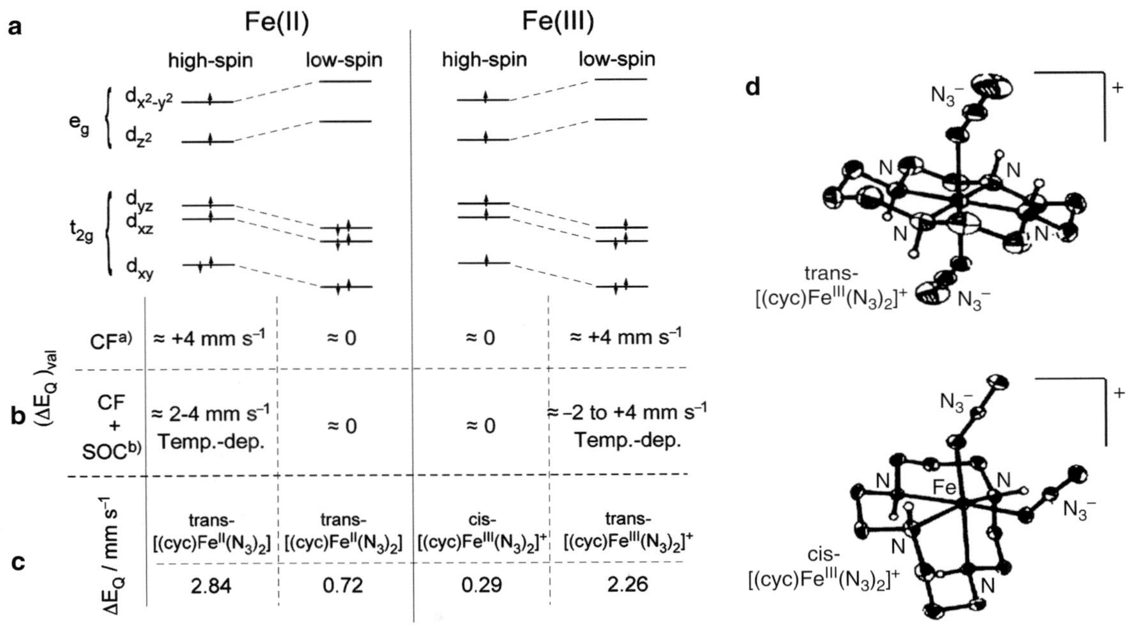

Fig. 4.8 (**a**) Electron configurations of ferrous and ferric compounds in perturbed octahedral symmetry and (**b**) valence contribution to the electric quadrupole splitting $(\Delta E_Q)_{val}$ derived from the expectation values of Table 4.2, and (**c**) experimental quadrupole splittings for a series of ferrous and ferric trans- and cis-[(cyclam)Fe(N$_3$)$_2$]$^{n+}$ complexes, where cyclam is a N$_4$-donating macrocycle [60], and (**d**) schematic view of the ferric complexes. [a]Calculated using crystal-field theory; [b]Calculated using crystal-field theory and spin–orbit coupling (SOC) which eventually mixes low-lying orbital states and causes temperature-dependent quadrupole splitting observed in the range 4.2–300 K if the splitting of the resulting SOC states is appropriate

with basic symmetry and "well-behaving" ligands. Such estimates of $(\mathrm{EFG})_{\mathrm{val}}$ can be helpful in many cases to achieve an elementary, qualitative understanding of experimentally obtained quadrupole splittings. However, reliable and predictive interpretations of the quadrupole splitting require thorough calculation of the electronic structure. A detailed introduction into the corresponding quantum chemical techniques will be given in Chap. 5.

4.4 Magnetic Dipole Interaction and Magnetic Splitting

A nucleus in a state with spin quantum number $I > 0$ will interact with a magnetic field by means of its magnetic dipole moment $\vec{\mu}$. This magnetic dipole interaction or nuclear Zeeman effect may be described by the Hamiltonian

$$\hat{H}_{\mathrm{m}} = -\hat{\vec{\mu}} \cdot \hat{\vec{B}} = -g_{\mathrm{N}}\mu_{\mathrm{N}}\hat{\vec{I}} \cdot \hat{\vec{B}}, \tag{4.47}$$

where the field $\vec{B}$ represents the magnetic induction, g_{N} is the nuclear Landé factor, and $\mu_{\mathrm{N}} = e\hbar/2M_{\mathrm{p}}c$ is the nuclear magneton (M_{p}: proton mass). Diagonalization of the Hamilton matrix yields the eigenvalues E_{M}:

$$E_{\mathrm{M}}(m_I) = -\mu B m_I/I = -g_{\mathrm{N}}\mu_{\mathrm{N}} B m_I. \tag{4.48}$$

The nuclear Zeeman effect splits the nuclear state with spin quantum number I into $2I + 1$ equally spaced and nondegenerate substates $|I, m_I\rangle$, which are characterized by the sign and the magnitude of the nuclear magnetic spin quantum number m_I. Figure 4.9 shows schematically the effect of magnetic dipole interaction for ^{57}Fe, where the excited state with $I = 3/2$ is split into its four magnetic substates and the ground state with $I = 1/2$ into the two substates. The allowed gamma transitions between the sublevels of the excited state and those of the ground state are found following the selection rules for magnetic dipole transitions: $\Delta I = 1$, $\Delta m = 0, \pm 1$. The six allowed transitions for ^{57}Fe are shown in Fig. 4.9. In a Mössbauer experiment with a single-line source, one would usually observe a resonance sextet, the centroid of which may be shifted from zero velocity by electric monopole interaction (isomer shift). The relative line intensities taken from left to right exhibit a 3:2:1:1:2:3 pattern for an isotropic distribution of the magnetic field with respect to the γ-ray propagation. The line intensities are determined by the squares of the relevant Clebsch–Gordan coefficients (Sect. 4.6, Table 4.3). When the hyperfine field $\vec{B}$ at the nucleus is perpendicular to the γ-direction, the intensity ratio is 3:4:1:1:4:3, whereas for $\vec{B}$ parallel to the γ-direction, the ratio is 3:0:1:1:0:3. Thus, the relative intensities of the $\Delta m_I = 0$ transitions (lines "2" and "4") are indicative of the field orientation.

The magnetic hyperfine splitting allows the determination of the effective magnetic field acting on the nucleus, which may be a superposition of an applied

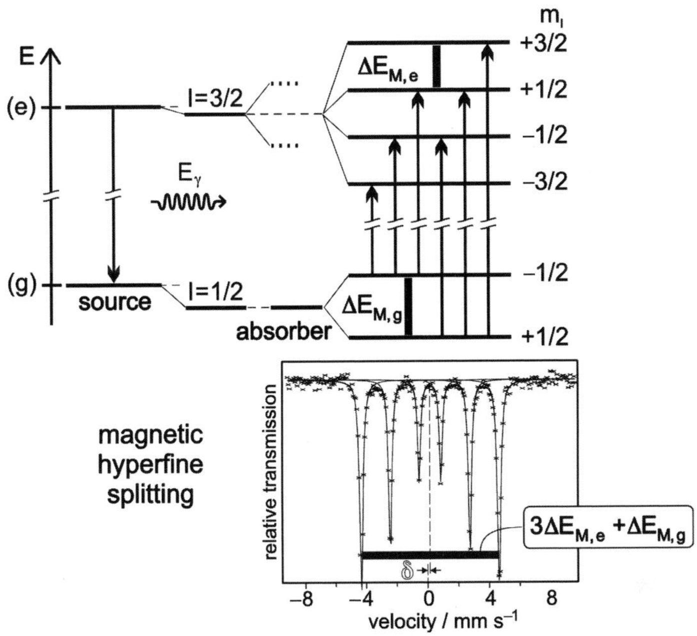

Fig. 4.9 Magnetic dipole splitting (nuclear Zeeman effect) in ^{57}Fe and resultant Mössbauer spectrum (schematic). The mean energy of the nuclear states is shifted by the electric monopole interaction which gives rise to the isomer shift δ. $\Delta E_{M,g} = g_g \mu_N B$ and $\Delta E_{M,e} = g_e \mu_N B$ refer to the Zeeman energies of the ground and the excited states, respectively. The splitting of the $I = 3/2$ state into two dotted lines in the middle panel indicates the effect of quadrupole splitting discussed earlier

field $\vec{B}_{ext}$ and an internal field $\vec{B}_{int}$ arising from a magnetic moment of the valence electrons. The various sources contributing to the internal field are

- The isotropic Fermi contact field B^C, which arises from a net spin-up or spin-down *s*-electron density at the nucleus as a consequence of spin-polarization of *s*-electrons by unpaired valence electrons [63];
- An anisotropic contribution B^L from the orbital motion of valence electrons with the quantum number L for the total orbital moment;
- An anisotropic spin-dipolar contribution B^d arising from nonspherical distribution of the electronic spin density.

The magnetic field contributions will be discussed in further detail in Chap. 5.

4.5 Combined Electric and Magnetic Hyperfine Interactions

Pure nuclear magnetic hyperfine interaction without electric quadrupole interaction is rarely encountered in chemical applications of the Mössbauer effect. Metallic iron is an exception. Quite frequently, a nuclear state is perturbed simultaneously by all three types of hyperfine interaction – electric monopole, magnetic dipole, and electric quadrupole interaction,

$$\hat{H} = \delta E + \hat{H}_{\mathrm{Q}} + \hat{H}_{\mathrm{M}}. \tag{4.49}$$

The monopole interaction δE, which yields the isomer shift, is easy to treat; it is just additive to all transition energies. Thus, the recorded spectrum has uniform shifts of all resonance lines with no change in their relative separations.

Magnetic dipole interaction $\hat{H}_{\mathrm{M}}$ (4.47) and electric quadrupole interaction $\hat{H}_{\mathrm{Q}}$ (4.29) both depend on the magnetic quantum numbers of the nuclear spin. Therefore, their combined Hamiltonian may be difficult to evaluate. There are closed-form solutions of the problem [64], but relatively simple expressions exist only for a few special cases [65]. In Sect. 4.5.1 it will be shown which kind of information can be obtained from a perturbation treatment if one interaction of the two is much weaker than the other and will be shown below. In general, however, if the interactions are of the same order of magnitude, $g_{\mathrm{N}}\mu_{\mathrm{N}}B \approx eQV_{zz}/2$, and particularly if the magnetic field takes arbitrary orientations with respect to the principal axes system of the EFG, the full nuclear Hamiltonian must be evaluated by numerical diagonalization. Appropriate computer routines are available for many nuclear systems and for many experimental conditions such as single-crystal and powder measurements.

4.5.1 *Perturbation Treatment*

Competing hyperfine interactions cannot occur for the ground state of ^{57}Fe because there is no quadrupole moment for $I = 1/2$. For the excited state ($I = 3/2$), the solution of the Schrödinger equation can be considerably simplified whenever one of the two contributing hyperfine interactions is weak and can be treated as a perturbation. Although the approach is scarcely used for quantitative analyses of actual Mössbauer spectra, it is very educational and may often help to rationalize the origin of a complex Mössbauer spectrum. For instance, with a first-order energy diagram of the nuclear states in mind, the sign of the EFG can usually be deduced from the spectra without numerical simulation. Several of the arguments outlined below will be used again in Sect. 4.7 in spin-Hamiltonian analyses of paramagnetic systems, for which strong or weak internal fields occur with pronounced anisotropy.

4.5.2 *High-Field Condition: $g_N\mu_N B \gg eQV_{zz}/2$*

The combined effect of strong nuclear magnetic (Zeeman) interaction and weak electric quadrupole interaction for the excited state of ^{57}Fe is demonstrated in

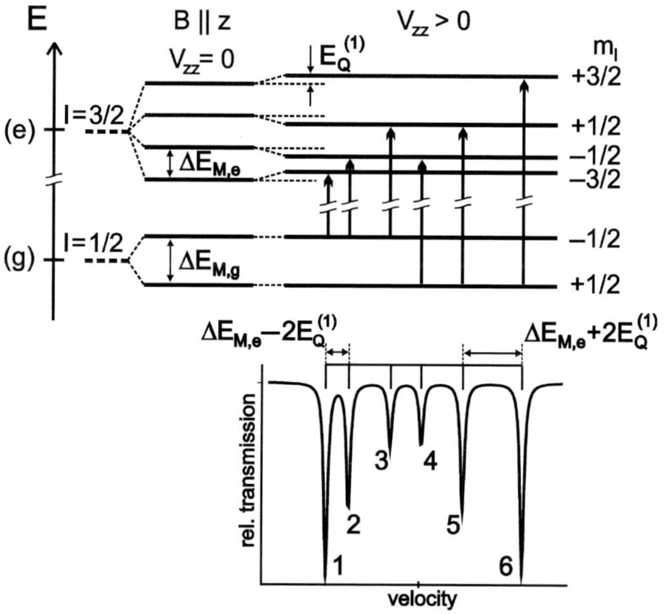

Fig. 4.10 Combined hyperfine interaction with strong magnetic field. *Top*: Level scheme for ^{57}Fe. Pure magnetic dipole splitting of the ground (g) and excited state (e) are shown on the left side. First-order quadrupole shifts, $E_Q^{(1)}$, are given on the right side (positive main component V_{zz} of EFG along the magnetic field, $\eta = 0$). *Bottom*: Magnetically split Mössbauer spectrum with line shifts due to first-order quadrupole interaction. The quadrupole splitting can be obtained from the energies L_i of the lines 1, 2, and 5, 6 as: $\Delta E_Q = 2E_Q^{(1)} = ½\,(L_6-L_5)-(L_2-L_1)$. The intensity distribution of the spectrum owes its origin to a powder distribution of the EFG with respect to the γ-beam, for which the magnetic field was kept parallel to the direction of V_{zz}. The quadrupole shifts indicated above are reversed if the main component of the EFG takes the opposite direction, $V_{zz} < 0$; in this case the measured spectrum is the mirror image of the one shown above

Fig. 4.10. The left side of the scheme represents the "starting situation" of pure Zeeman splitting, as described by (4.48) and shown before in Fig. 4.9. In this example, the field $\vec{B} = (0, 0, B)$, which defines the quantization axis, is chosen as the z-direction. The additional quadrupole interaction, as shown on the right side of Fig. 4.10, leads to a pair-wise shift of the Zeeman states with $m_I = \pm 3/2$ and $m_I = \pm 1/2$ up- and down-wards in opposite sense. In first order, all lines are shifted by the same energy $E_Q^{(1)}$ as expected from the m_I^2-dependence of the electric quadrupole interaction (see (4.34), whereby the superscript (1) used here denotes the order of the perturbation).

The value of $E_Q^{(1)}$ is given by the component of the EFG tensor *along* the main quantization axis. Therefore, in this example where the EFG is axial ($\eta = 0$) with the main component V_{zz}, the quadrupole shift $E_Q^{(1)}$ is $eQV_{zz}/4$. This is just half the quadrupole splitting that would be observed in an unperturbed quadrupole spectrum without a magnetic field at the nucleus.

4.5.2.1 Quadrupole Shifts in High-Field Magnetic Spectra

The perturbation of the four substates of the excited $I = 3/2$ manifold by $\pm E_Q^{(1)}$ induces a typical asymmetry of the resulting magnetically split Mössbauer spectrum as pictured at the bottom of Fig. 4.10 for positive V_{zz}; the inner four lines, 2–5, are shifted to lower velocities, whereas the outer two lines, 1 and 6, are shifted to higher velocities by equal amounts. In first order, the line intensities are not affected. For negative V_{zz}, the line asymmetry is just inverted, as the quadrupole shift of the nuclear $\pm 1/2$ and $\pm 3/2$ states is opposite. Thus, *the sign and the size of the* EFG *component along* the field can be easily derived from a magnetic Mössbauer spectrum with first-order quadrupole perturbation.

4.5.2.2 Angular Dependence of the Effect of Quadrupole Interaction in High-Field Spectra

Arbitrary orientations of the magnetic field that are not along the EFG main component yield basically the same splitting pattern for the nuclear excited state shown in Fig. 4.10 but the quadrupole shifts $E_Q^{(1)}$ for the $|3/2, \pm m_I\rangle$ states are determined by the *strength of the* EFG *along the field* (and not by the main component V_{zz}). The general first-order expression for the eigenvalues of the nuclear $I = 3/2$ manifold for an arbitrary field orientation (with respect to which the principal axes system of the EFG is defined by the polar and azimuthal angles θ and ϕ) is:

$$E_{\mathrm{M,Q}}(I = 3/2, m_I)^{(1)} = -g_{\mathrm{N}} B m_I + \mu_{\mathrm{N}} E_{\mathrm{Q}}(m_I, \theta, \phi)^{(1)}, \tag{4.50}$$

with the quadrupole shift $E_Q(m_I, \theta, \phi)^{(1)}$ represented by:

$$E_{\mathrm{Q}}(m_I, \theta, \phi)^{(1)} = -1^{|m_I|+1/2}(eQV_{zz}/8) \cdot (3\cos^2\theta - 1 + \eta \cdot \sin^2\theta \cos 2\phi). \tag{4.51}$$

This expression is the same as that for the first-order quadrupole effect[14] in high-field NMR spectra [4, 66]. The angular variation of (4.51) is visualized in Fig. 4.11. The graph on the left shows the contour of the function $(3\cos^2\theta - 1 + \eta\sin^2\theta \cos 2\phi)$ in the (x, z) plane for an axial EFG tensor ($\eta = 0$), whereas the other three graphs show the contour for a fully rhombic EFG tensor ($\eta = 1$) in three different planes. Figure 4.12 presents surface plots of the angular variation for three different values of the asymmetry parameter η to further visualize the tensorial character of the EFG.

The strong variation on θ and ϕ reflects the fact that the EFG is a traceless tensor with positive and negative components. For $V_{zz} > 0$ and $\eta = 0$, the shift $E_Q^{(1)}$ is

[14] A nice derivation is given by Dr. Pascal Man, Directeur de recherche, CNRS, Université Pierre et Marie Curie-Paris 6 at his web site: http://www.pascal-man.com/tensor-quadrupole-interaction/V20-static.shtml. The Mathematica-5 script is also given and can be used for solving the first order Hamiltonian to explain quadrupole effects in high-field NMR spectra.

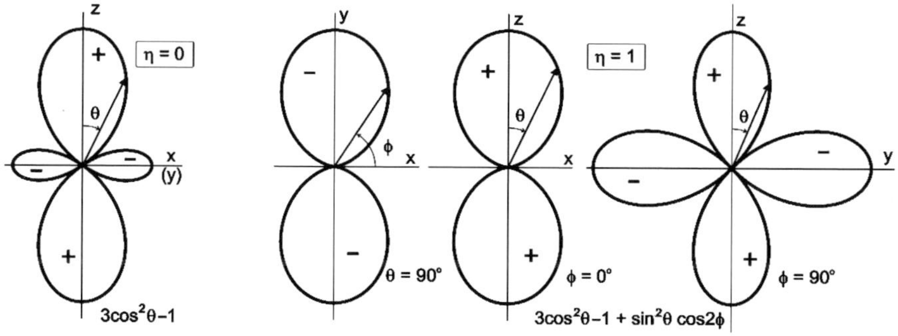

Fig. 4.11 Angular dependence ($3\cos^2\theta - 1 + \eta\sin^2\theta\cos2\phi$) of the first-order quadrupole shift, $E_Q^{(1)}$, in high-field Mössbauer spectra. *Left side*: $V_{zz} > 0$ and $\eta = 0$. *Right side*: $\eta = 1$ viewed in the (xy), (xz), and (yz) planes (*the sign* of the EFG is undefined in this case). The angles θ and ϕ describe the orientation of the PAS of the EFG with respect to the magnetic field at the nucleus. The length of the arrows correspond to the value of the apparent quadrupole shift $E_Q^{(1)}$ as shown in Fig. 4.10. The labels in the lopes denote the sign of the shifts. Note, the apparent similarity with real d-orbitals is not accidental but results from the underlying equivalence of the operators for quadrupole interaction and the angular momentum with $L = 2$

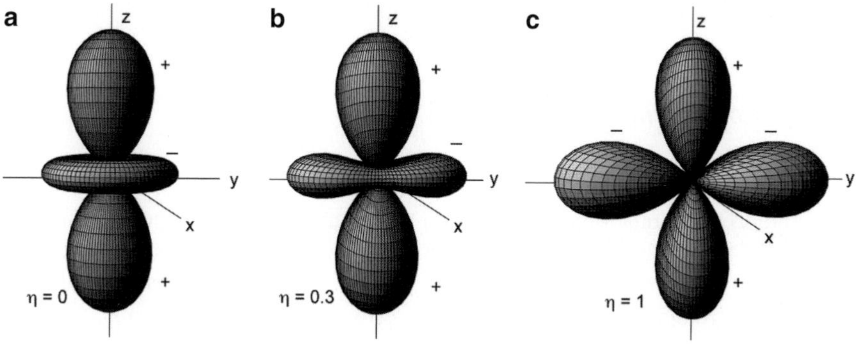

Fig. 4.12 Surface plots of the EFG tensor as determined from the angular dependence of the first-order quadrupole shift, $E_Q^{(1)}$, of high-field magnetic Mössbauer spectra. The plots visualize the value of the function ($3\cos^2\theta - 1 + \eta\sin^2\theta\ \cos2\phi$) for $V_{zz} > 0$ and $\eta = 0$ (**a**), $\eta = 0.3$ (**b**), and $\eta = 1$ (**c**)

positive and amounts to $\Delta E_Q/2$, if the field is at $\theta = 0$. It is negative and has half the size for $\theta = 90°$ because the field is along V_{xx}, which is $-0.5V_{zz}$. The quadrupole effect can also vanish in a magnetically split Mössbauer spectrum. This is most obvious for $\eta = 0$, if the field orientation is at the magic angle $\theta = 54.7°$, for which $3\cos^2\theta = 1$! Thus, quadrupole shifts in high-field magnetic Mössbauer spectra have to be interpreted with care, particularly when the relative orientation of the field and the EFG axes is not known.

Distinct quadrupole shifts do occur as well in magnetically split spectra of single-crystals, polycrystalline powder or frozen solution samples. In all three cases, the line shifts obey the simple first-order expression at high-field condition, if the EFG tensor has a fixed orientation with respect to the magnetic field. The relative line intensities of the magnetically split hyperfine lines of powder and frozen solution samples generally differ from those of single-crystals, but the relative line shifts are the same and depend solely on the EFG component along the field. Typical examples are oxides with strong magnetic anisotropy; the internal fields are fixed within magnetic domains, which strictly correlate with the EFG axes, even if weak external fields are applied to the powder samples. Other cases are paramagnetic systems with large zero-field splitting which may also cause pronounced magnetic anisotropy and strong internal fields pointing along an "easy-axis" of magnetization. In both cases, the spectra resemble that shown in Fig. 4.10.

In contrast, "soft" magnetic solids and paramagnetic systems with weak anisotropy may be completely polarized by an applied field, that is, the effective field at the Mössbauer nucleus is along the direction of the applied field, whereas the EFG is powder-distributed as in the case of crystallites or molecules. In this case, first-order quadrupole shifts cannot be observed in the magnetic Mössbauer spectra because they are symmetrically smeared out around the unperturbed positions of hyperfine lines, as given by the powder average of $E_Q(m_I, \theta, \phi)^{(1)}$ in (4.51). The result is a symmetric broadening of all hyperfine lines (however, distinct asymmetries arise if the first-order condition is violated).

4.5.3 Low-Field Condition: $eQV_{zz}/2 \gg g_N\mu_N B$

In the low-field condition, the energy-splitting of the $I = 3/2$ manifold of ^{57}Fe is dominated by the quadrupole interaction which affords a quadrupole splitting of $2E_Q$ (Fig. 4.13). The perturbing magnetic dipole interaction lifts the remaining degeneracy of the $|\pm m_I\rangle$ nuclear doublets of the $I = 3/2$ excited state as well as of the $I = 1/2$ ground state. Since the Zeeman effect depends linearly on m_I, the first-order magnetic splitting $\Delta E_{M,e}{}^{(1)}$ in the excited state is three times larger for $m_I = \pm 3/2$ than for $m_I = \pm 1/2$. Moreover, the ground state experiences pure Zeeman splitting $\Delta E_{M,g}$ as given by (4.48) (recall, the nuclear g factor of the $I_g = 1/2$ ground state is different from that of the $I_e = 3/2$ excited state).

The corresponding low-field Mössbauer spectrum shows a weak magnetic splitting superimposed on a dominating quadrupole splitting (Fig. 4.13, bottom). This is different for the low-energy and the high-energy parts of the spectrum, that is, the $|I = 1/2\rangle \rightarrow {}''|I = 3/2, \pm 1/2\rangle''$ and the $|I = 1/2\rangle \rightarrow {}''|I = 3/2, \pm 3/2\rangle''$ transitions.[15] The magnetic splittings are given by the sum of $\Delta E_{M,g}$ in the ground state and $\Delta E_{M,e}{}^{(1)}$ of the respective excited state. In our example, a typical four-line pattern is expected for the low-energy quadrupole line (left), and a two-line pattern for the high-energy line (right). The Mössbauer spectrum shown in Fig. 4.13 is

[15]The labels of the excited states are set in quotation marks because the true-wave functions are mixtures of the basis functions $|I = 3/2, \pm m_I\rangle$.

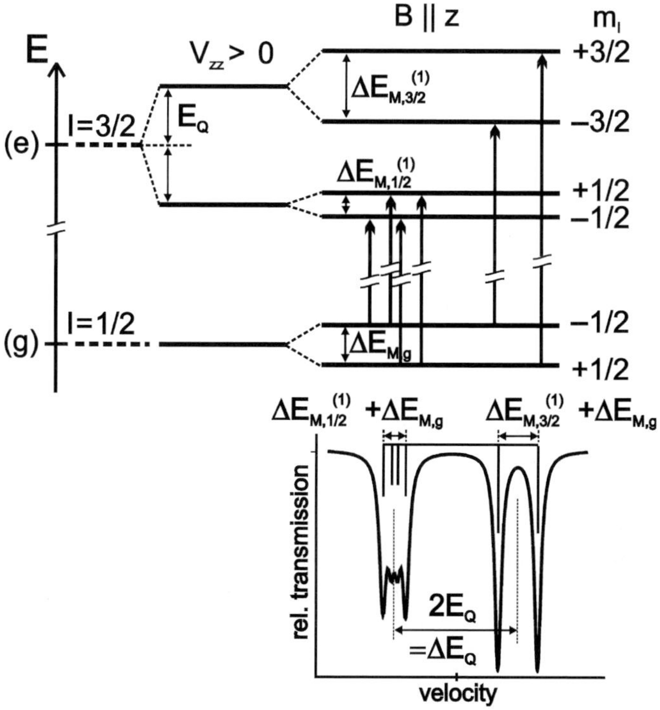

Fig. 4.13 Combined magnetic hyperfine interaction for ^{57}Fe with strong electric quadrupole interaction. *Top left*: electric quadrupole splitting of the ground (g) and excited state (e). *Top right*: first-order perturbation by magnetic dipole interaction arising from a "weak" field along the main component $V_{zz} > 0$ of the EFG ($\eta = 0$). *Bottom*: the resultant Mössbauer spectrum is shown for a single-crystal type measurement with $\vec{B}$ fixed perpendicular to the γ-rays and $\vec{B}$ oriented along V_{zz}. The magnetic splitting energies $\Delta E_{\mathrm{M,g}}$ and $\Delta E_{\mathrm{M,e,3/2}}{}^{(1)}$, $\Delta E_{\mathrm{M,e,1/2}}{}^{(1)}$ for ground and excited states are given by (4.48)

characteristic of $V_{zz} > 0$ and $\eta = 0$ (with $\vec{B}\|z$); for $V_{zz} < 0$, the spectrum would be inverted. This feature shows that *the sign of the EFG can also be derived from magnetically perturbed quadrupole spectra.*

In the low-field condition, the quantization axis is defined by the EFG main component V_{zz}. In this situation, V_{zz} and η can both be determined from *powder spectra* when recorded in an externally applied field. Figure 4.14 shows simulated spectra as is often encountered in practice such as in applied-field measurements of diamagnetic compounds or fast-relaxing paramagnetic compounds at high temperatures. The simulated traces differ in detail from a single-crystal spectrum as shown in Fig. 4.13, but their features still correlate in a unique manner with η and the sign of V_{zz}.

For *nonaxial* EFG tensors, mixing of the nuclear m_I basis functions occurs even for $\vec{B}$ being oriented along V_{zz}. This results from the contributions of the shift operators I_+^2 and I_-^2 in the Hamiltonian described by (4.29). These contributions change the eigenvalues such that the splittings $\Delta E_{\mathrm{M,e,1/2}}$ and $\Delta E_{\mathrm{M,e,3/2}}$

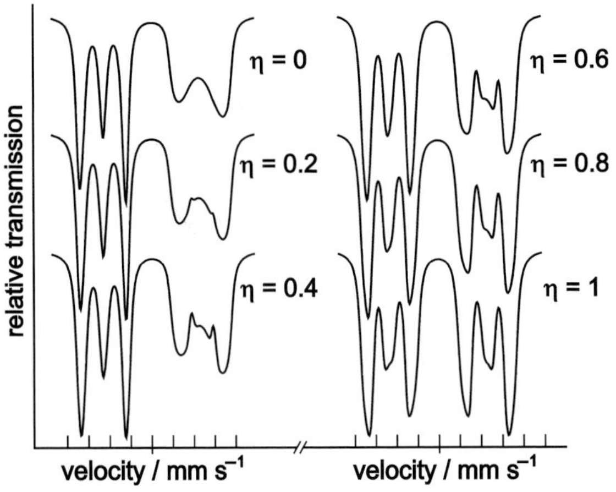

Fig. 4.14 Magnetically perturbed quadrupole spectra simulated for *powder distributions* of the EFG ($V_{zz} > 0$) with an applied field $B = 4$ T which is fixed in the laboratory system perpendicular to the γ-beam.[16] The value of the quadrupole splitting is kept constant at $\Delta E_Q = +4$ mm s^{-1}. For negative quadrupole splitting ($V_{zz} < 0$), the spectra would be inverted on the velocity scale. Note the difference in relative intensities for the spectrum for $\eta = 0$ and the "single-crystal" type spectrum given in Fig. 4.13. Similar patterns are obtained for $\vec{B} \parallel \gamma$

approach each other with increasing values of η. As a result, the asymmetry of the spectrum *decreases* with increasing asymmetry of the EFG so that the spectrum is fully symmetric for $\eta = 1$. This effect is demonstrated in Fig. 4.14 for powder spectra.

Powder simulations for the spectra of diamagnetic compounds recorded with applied field are rather straightforward because only the size and sign of V_{zz} and η (and the isomer shift δ) are unknown parameters. Often the absolute value of the quadrupole splitting, ΔE_Q, and δ can be taken as a constraint from a corresponding zero-field measurement. Therefore, the full EFG can be determined from a single magnetically split Mössbauer spectrum. Although the accuracy for η may not be as high as that obtained from perturbed angular correlation measurements, the information is usually sufficiently precise for electronic structure analyses.

Powder spectra of paramagnetic compounds measured with applied fields are generally more complicated than those shown in Fig. 4.14. Large internal fields at the Mössbauer nucleus that are temperature- and field-dependent give rise to this complication. If, however, the measurement is performed at sufficiently high temperature, which is above ca. 150 K, the internal magnetic fields usually collapse due to fast relaxation of the electronic spin system (*vide infra,* Chap. 6). Under

[16]The spectra are calculated by numerical diagonalization of the nuclear Hamiltonian (4.49) and powder summation over the distribution of the EFG for different values of the asymmetry parameter η. The line width was 0.25 mm s^{-1}.

these circumstances, the magnetically perturbed Mössbauer spectra show the same pattern as those obtained for diamagnetic compounds measured at the same field so that the EFG can be easily determined as described earlier.

4.5.4 Effective Nuclear g-Factors for $eQV_{zz}/2 \gg g_N\mu_N B$

If the electric quadrupole splitting of the $I = 3/2$ nuclear state of ^{57}Fe is larger than the magnetic perturbation, as shown in Fig. 4.13, the $m_I = |\pm 1/2\rangle$ and $|\pm 3/2\rangle$ states can be treated as independent doublets and their Zeeman splitting can be described independently by effective nuclear g factors and two effective spins $I^{\text{eff}} = 1/2$, one for each doublet [67]. The approach corresponds exactly to the spin-Hamiltonian concept for electronic spins (see Sect. 4.7.1). The nuclear spin Hamiltonian for each of the two Kramers doublets of the ^{57}Fe nucleus is:

$$\hat{H}_{\text{m}}^{\text{eff}} = -\mu_{\text{N}}\vec{I}^{\text{eff}} \cdot \bar{\bar{g}}_{\text{N}}^{\text{eff}}\vec{B}, \tag{4.52}$$

where $\bar{\bar{g}}_{\text{N}}^{\text{eff}}$ the respective effective nuclear g-matrix, which is diagonal in the PAS of the EFG. The principal values $g_{\text{N},i}^{\text{eff}}$ parameterize the magnetic Zeeman splitting according to the usual angular dependence:

$$\Delta E_{\text{M}} = -\mu_{\text{N}}\sqrt{(g_{\text{N},x}^{\text{eff}} \cdot \sin\theta\cos\phi)^2 + (g_{\text{N},y}^{\text{eff}} \cdot \sin\theta\sin\phi)^2 + (g_{\text{N},z}^{\text{eff}} \cdot \cos\theta)^2} \cdot B, \tag{4.53}$$

where B is the field strength, and θ and ϕ are the polar and azimuthal angles as used above to define the field orientation in the principal axes system of the EFG. The effective nuclear values $g_{\text{N},i}^{\text{eff}}$ can be obtained in second-order perturbation from a comparison of the matrix elements of (4.52) with those for the full $I = 3/2$ system described by $\hat{H} = \hat{H}_{\text{Q}} + \hat{H}_{\text{M}}$ ((4.29) and (4.47)). The concept is a useful simplification for low-field Mössbauer spectra with combined hyperfine interaction, because the effective values which are different for the $|3/2, m_I = \pm 1/2\rangle$ and the $|3/2, m_I = \pm 3/2\rangle$ doublet depend only on the asymmetry parameter η. The corresponding diagrams for the two sub-doublets of $I = 3/2$ are shown in Fig. 4.15.

Effective nuclear g values can be helpful in understanding (para) magnetic systems with moderately strong internal fields but very large magnetic anisotropy, because the orientation of the internal fields can be related to the EFG axes. Such conditions are frequently encountered for integer spin systems, for which strong zero-field interaction of the electronic spin system often lead to well isolated magnetic m_S substates yielding only weak internal fields with strong anisotropy. The corresponding Mössbauer spectra measured with weak or moderately strong applied fields are perturbed quadrupole spectra which show the typical features of a predominant “easy plane” or an “easy axis” of magnetization. This can be rationalized and assigned to molecular directions (i.e. the EFG principal axes system) by comparing the magnetic Mössbauer line splittings with the diagrams for effective

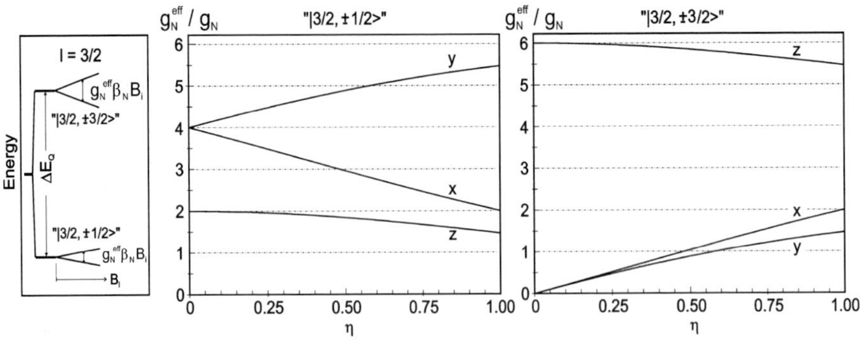

Fig. 4.15 Effective nuclear g values for the excited $I = 3/2$ state of ^{57}Fe in units of the corresponding nuclear g factor ($g_{N,e} = -0.10317$). The left panel shows the Zeeman splitting of the $I = 3/2$ manifold with large quadrupole slitting under the influence of a weak field, and the two panels on the right show the η-dependence of the corresponding effective nuclear g values for the $|m_I = \pm 1/2\rangle$ and $|m_I = \pm 3/2\rangle$ doublets with the field oriented along the x, y, and z principal axes of the EFG

nuclear g factors. This concept can be applied to paramagnetic systems (see Sect. 4.7).

4.5.5 *Remarks on Low-Field and High-Field Mössbauer Spectra*

The nuclear Zeeman effect is not a very strong interaction as compared to electric quadrupole splitting because of the relatively weak nuclear magneton. A field of $B = 1$ T at the nucleus of ^{57}Fe causes an energy splitting $\Delta E_M = g_{N,e}\mu_N B$ of the $|3/2, m_I = \pm 1/2\rangle$ nuclear doublet of just 0.78589 MHz, or 0.0676 mm s^{-1} in Mössbauer energy units.[17] The overall magnetic splitting of the corresponding six-line Mössbauer spectrum without quadrupole interaction is $4.75\Delta E_M$ or 0.321 mm s^{-1} (or 10.6 mm s^{-1} for $B = 33$ T as in metallic α-iron). Therefore, a quadrupole splitting $\Delta E_Q = eQV_{zz}/2$ of 1 mm s^{-1} is as strong a nuclear interaction as the Zeeman splitting obtained with a field of about 14.8 T! Since such strong fields are at the limit of what is technically available, Mössbauer measurements of *diamagnetic* compounds with externally applied fields usually fall in the low-field range of Mössbauer spectra: $g_N\mu_N B < eQV_{zz}/2$.

On the other hand, internal magnetic fields at the iron nucleus arising from the magnetic moments of unpaired valence electrons can be much stronger than any applied field and their effect can easily exceed the quadrupole interaction. For instance, Mössbauer nuclei in magnetic materials such as metals or oxides may experience fields of 30–50 T even without applied field. Similarly, the typical

[17]The number is derived from (4.48) by substituting the nuclear g factor of the excited state, $g_e = -0.103267$, and the nuclear Bohr magneton, $\mu_N = 7.622593$ MHz T^{-1} or 0.6557 mm s^{-1} T^{-1}.

(internal) magnetic hyperfine field of iron(III) in *paramagnetic* coordination compounds with "hard" ionic ligands can be as large as 52 T. To our knowledge, the "world-record" for ^{57}Fe is at 152 T, which was observed by W. Reiff et al. for the rigorously linear two-coordinate iron(II) complex bis(tris(trimethylsilyl)methyl)Fe (II) [68]. This high magnetic field arises from an extraordinarily large (positive) orbital contribution, which outweighs by far the negative Fermi contact contribution to the effective internal field. Later on, Kuzmann et al. confirmed this observation and measured a value of even 157.5 T [69]. Large orbital contributions to the magnetic moment have also been observed for square-planar coordinated Fe(II) in ferromagnetically ordered a-Fe(II) octa-ethyltetraazaporphyrin with a hyperfine field of $B_{\mathrm{hf}} = 62.4$ T [70]. Equally strong fields in the range $B_{\mathrm{hf}} = 62$–82 T are reported for a series of three-coordinated Fe(II) complexes [71]. The magnetic Mössbauer spectra of such systems are clearly in the high-field regime, $g_N \mu_N B \gg eQV_{zz}/2$.

4.6 Relative Intensities of Resonance Lines

4.6.1 Transition Probabilities

The transition probabilities or line intensities of the hyperfine components in a Mössbauer spectrum are determined by the properties of the nuclear transition. The spin and the parity of the excited and the ground states of the nucleus under consideration are important factors as are the transition- and geometric-configurations, that is, the direction of the wave vector $\vec{k}$ of the emitted γ-quanta with respect to the quantization axis, for example, the direction of the magnetic field or the principal axes system of the field gradient tensor. It will be shown that the geometric arrangement of the experimental set-up may result in additional information concerning the hyperfine interaction parameter and the anisotropy of the chemical bond (cf. Sect. 4.6.2).

For the sake of simplicity and a more instructive description, we shall restrict ourselves to the case of unpolarized single line sources of $I = 3/2 \leftrightarrow I = 1/2$ magnetic dipole transitions (M1) as for example in ^{57}Fe, which has only a negligible electric quadrupole (E2) admixture. It will be easy to extend the relations to arbitrary nuclear spins and multipole transitions. A more rigorous treatment has been given in [76, 78] and [14] in Chap. 1. The probability P for a nuclear transition of multipolarity M1 (L=1) from a state $|I_1, m_1\rangle$ to a state $|I_2, m_2\rangle$ is equal to

$$P\left(3/2\; m_{3/2},\; Lm \middle| 1/2\; m_{1/2},\; \theta,\; \phi\right) = \left| \left\langle 3/2\; m_{3/2},\; Lm \;\middle|\; 1/2\; m_{1/2} \right\rangle \right|^2 F_{Lm_{1/2}}^{Lm_{3/2}}(\theta, \phi) \langle I_1 \| L \| I_2 \rangle^2, \tag{4.54}$$

where θ, ϕ are the polar and azimuthal angles of the z-direction (defined by, e.g., the direction of the magnetic field) and the direction of the γ-ray emission (Fig. 4.16). $\langle |I_1 m_1, Lm | I_2 m_2\rangle$ are the Clebsch–Gordan coefficients [131] coupling together the

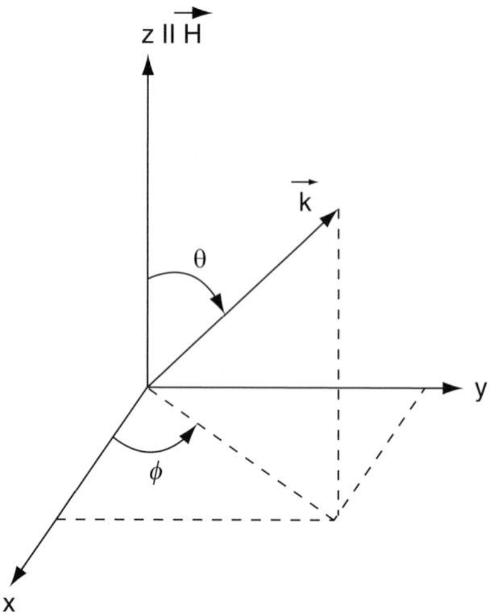

Fig. 4.16 Definition of the polar angles θ, ϕ. $\vec{k}$ is the wave vector of the emitted γ-ray. The z-axis may be defined by the direction of a magnetic field

Table 4.3 Values of the Clebsch–Gordan coefficients $<3/2\,m_{3/2}, Lm|1/2\,m_{1/2}>$ in the case of a $I_e = 3/2$, $I_g = 1/2$ magnetic dipole transition ($L = 1$) and electric quadrupole transition ($L = 2$) (see, e.g., [47, 77] and [14] in Chap. 1)

$m_{3/2}$	$m_{1/2}$	m	$<3/2\,m_{3/2},\,1m|1/2\,m_{1/2}>$	$<3/2\,m_{3/2},\,2m|1/2\,m_{1/2}>$
3/2	1/2	–1	$\sqrt{3/6}$	$\sqrt{1/10}$
1/2	1/2	0	$-\sqrt{2/6}$	$-\sqrt{2/10}$
–1/2	1/2	1	$+\sqrt{1/6}$	$\sqrt{1/10}$
–3/2	1/2	2	0	$-\sqrt{4/10}$
3/2	–1/2	–2	0	$+\sqrt{4/10}$
1/2	–1/2	–1	$+\sqrt{1/6}$	$-\sqrt{3/10}$
–1/2	–1/2	0	$-\sqrt{2/6}$	$+\sqrt{2/10}$
–3/2	–1/2	–1	$+\sqrt{3/6}$	$-\sqrt{1/10}$

three vectors $\vec{I}_1, \vec{L}, \vec{I}_2$, and $\langle I_1 \| L \| I_2 \rangle$ is the reduced matrix element, which does not depend on the magnetic quantum numbers. For M1 ($L = 1$) and E2 ($L = 2$) transitions and $I_1 = 3/2$, $I_2 = 1/2$, the Clebsch–Gordan coefficients are tabulated in Table 4.3. The angular-dependent terms $F_{Lm}^{L'm'}(\theta, \phi)$ do not depend on quantum numbers I, m_I, and are of more general validity.

The angular functions $F_{Lm}^{L'm'}$ are given in Table 4.4. Let us first consider the case of a magnetically split spectrum from a powder sample. Here, we have pure m_I-states $|Im_1\rangle$ with the z-axis which is parallel to the direction of the internal magnetic field being randomly distributed. Therefore, we have to integrate over the polar (θ) and azimuthal (ϕ) angles, yielding for L = 1

$$P(3/2\,m_{3/2}, 1m|1/2\,m_{1/2}) = \int_0^{\pi}\int_0^{2\pi} P(3/2\,m_{3/2}, 1m|1/2\,m_{1/2}, \theta, \phi)\sin\theta\,\mathrm{d}\theta\,\mathrm{d}\phi \tag{4.55}$$

Table 4.4 Direction-dependent terms $F_{Lm}^{L'm'}(\theta)$ for M1 and E2 nuclear γ-transitions (see, e.g., [47])

m \ m′	2	1	0	–1	–2
a)	$F_{1m}^{1m'} / 3e^{-i(m'-m)\phi}$				
1		$1/2\,(1+\cos^2\theta)$	$1/4\sqrt{2}\sin 2\theta$	$1/2\,\sin^2\theta$	
0		$1/4\sqrt{2}\sin 2\theta$	$\sin^2\theta$	$1/4\sqrt{2}\sin 2\theta$	
–1		$1/2\,\sin^2\theta$	$1/4\sqrt{2}\sin 2\theta$	$1/2(1+\cos^2\theta)$	
b)	$F_{2m}^{2m'} / 5e^{-i(m'-m)\phi}$				
2	$1/2\,(\sin^2\theta + 1/4\sin^2 2\theta)$	$-1/4\,(\sin 2\theta + 1/2\sin 4\theta)$	$-1/4\sqrt{3/2}\sin^2 2\theta$	$-1/4\,(\sin 2\theta - 1/2\sin 4\theta)$	$-1/2\,(\sin^2\theta - 1/4\sin^2 2\theta)$
1	$-1/4\,(\sin 2\theta + 1/2\sin 4\theta)$	$1/2\,(\cos^2\theta + \cos^2 2\theta)$	$1/4\sqrt{3/2}\sin 4\theta$	$1/2\,(\cos^2\theta - \cos^2 2\theta)$	$1/4\,(\sin 2\theta - 1/2\sin 4\theta)$
0	$-1/4\sqrt{3/2}\sin^2 2\theta$	$1/4\sqrt{3/2}\sin 4\theta$	$1/4\,\sin^2 2\theta$	$-1/4\sqrt{3/2}\sin 4\theta$	$-1/4\sqrt{3/2}\sin^2\theta$
–1	$-1/4\,(\sin 2\theta - 1/2\sin 4\theta)$	$1/2\,(\cos^2\theta - \cos^2 2\theta)$	$-1/4\sqrt{3/2}\sin^2 4\theta$	$1/2\,(\cos^2\theta + \cos^2 2\theta)$	$1/4\,(\sin 2\theta - 1/2\sin 4\theta)$
–2	$-1/2\,(\sin^2\theta - 1/4\sin^2 2\theta)$	$1/4\,(\sin 2\theta - 1/2\sin 4\theta)$	$-1/4\sqrt{3/2}\sin^2 2\theta$	$1/4\,(\sin 2\theta - 1/2\sin 4\theta)$	$1/2\,(\sin^2\theta + 1/2\sin^2 2\theta)$
c)	$F_{1m}^{2m'} / \sqrt{15}\cdot 2\cos(m'-m)\phi$				
1	$-1/2\sin 2\theta$	$1/2\,(\cos^2\theta + \cos 2\theta)$	$1/2\sqrt{3/2}\sin 2\theta$	$1/2\,(\cos^2\theta - \cos 2\theta)$	0
0	$-1/2\sqrt{2}\sin^2\theta$	$1/4\sqrt{2}\sin 2\theta$	0	$1/4\sqrt{2}\sin 2\theta$	$1/2\sqrt{2}\sin 2\theta$
–1	0	$-1/2\,(\cos^2\theta - \cos 2\theta)$	$1/2\sqrt{3/2}\sin 2\theta$	$-1/2\,(\cos^2\theta + \cos 2\theta)$	$-1/2\sin 2\theta$

or

$$P(3/2\ m_{3/2}, 1m|1/2\ m_{1/2}) \propto \left|\langle 3/2\ m_{3/2}, 1m \mid 1/2\ m_{1/2}\rangle\right|^2. \tag{4.56}$$

From (4.56) and Table 4.3, we derive the relative intensity ratios 3:2:1:1:2:3 for the hyperfine components of a Zeeman pattern of a powder sample. The transition probability for the case of the polar angle $\theta = \theta_0$ can readily be calculated by integrating (4.56) only over the azimuthal angle ϕ. One obtains a factor $(1 + \cos^2\theta_0)/2$ and $\sin^2\theta_0$ for $m = \pm 1$ and $m = 0$, respectively, which are multiplied by the square of the Clebsch–Gordan coefficients. As a consequence of the angular correlation of the transition probabilities, the second and fifth hyperfine components (Fig. 4.17) disappear if the direction $\vec{k}$ of the γ-rays and the magnetic field $\vec{H}$ are parallel ($\theta_0 = 0$).

Let us now consider a pure quadrupole spectrum which leads to the hyperfine splitting as shown schematically in Fig. 4.18. For the intensity ratio I_2/I_1, we obtain

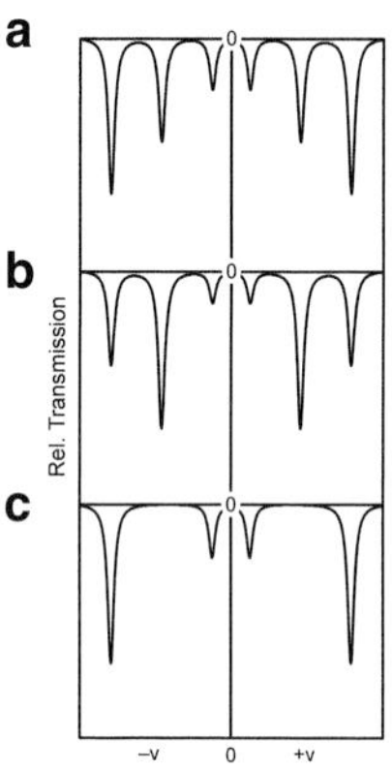

Fig. 4.17 Magnetic hyperfine pattern of a powder sample with randomly distributed internal magnetic field (**a**), and with (**b**) an applied magnetic field ($\theta_0 = 90°$), and (**c**) an applied magnetic field ($\theta_0 = 0°$)

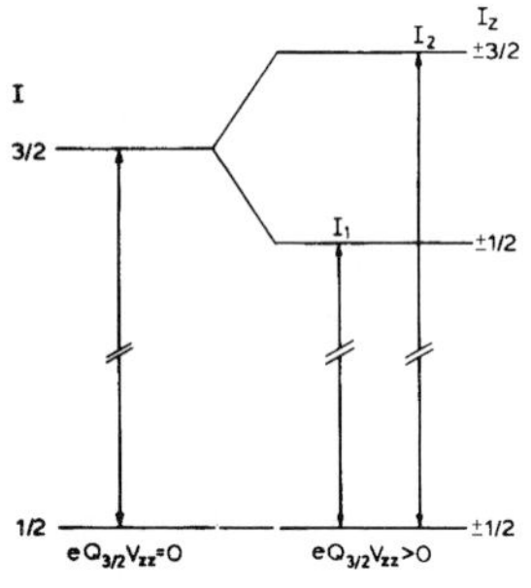

Fig. 4.18 Quadrupole hyperfine splitting for ^{57}Fe with $I_e = 3/2$ and $I_g = 1/2$; the quadrupole interaction parameter $eQ_{3/2}V_{zz}$ is assumed to be positive

from (4.54) and Table 4.4 for the case of an axially symmetric field gradient tensor (integration over ϕ).

$$I_2/I_1 = \frac{\int_0^{\pi} 3(1+\cos^2\theta)h(\theta)\sin\theta\,\mathrm{d}\theta}{\int_0^{\pi} (5-3\cos^2\theta)h(\theta)\sin\theta\,\mathrm{d}\theta}. \tag{4.57}$$

This relation is only valid for a crystal with isotropic *f*-factor. The effect of crystal anisotropy will be treated in Sect. 4.6.2. The function $h(\theta)$ describes the probability of finding an angle θ between the direction of the z-axis and the γ-ray propagation. In a powder sample, there is a random distribution of the principal axes system of the EFG, and with $h(\theta) = 1$, we expect the intensity ratio to be $I_2/I_1 = 1$, that is, an asymmetric Mössbauer spectrum. In this case, it is not possible to determine the sign of the quadrupole coupling constant eQV_{zz}. For a single crystal, where $h(\theta) = \frac{1}{\sin\theta}\delta(\theta-\theta_0)$($\delta$: delta-function), the intensity ratio takes the form

$$I_2/I_1 = \frac{3(1+\cos^2\theta_0)}{5-3\cos^2\theta_0}, \tag{4.58}$$

with values ranging from 3 for $\theta_0 = 0$ to 0.6 for $\theta_0 = 90°$. Therefore, in studies using single crystals, the sign of the quadrupole coupling constant eQV_{zz} can be determined. This in turn yields the sign of V_{zz} (EFG) if the sign of the electric quadrupole moment is known (which is the case for ^{57}Fe, viz., $Q(^{57}\text{Fe}) > 0$).

The situation becomes somewhat more complicated if both magnetic dipole and electric quadrupole interactions are present. Then, the states are no longer pure m_I states $|I, m_I\rangle$, but linear combinations of these, for example, for ^{57}Fe

$$\begin{aligned} \left|\psi_{3/2}\, i\right\rangle &= \sum_{m_{3/2}=-3/2}^{+3/2} C_{m_{3/2}}^{3/2\,i}\left|3/2\; m_{3/2}\right\rangle,\; i = 1,2,3,4, \\ \left|\psi_{1/2}\, j\right\rangle &= \sum_{m_{1/2}=-1/2}^{+1/2} C_{m_{1/2}}^{1/2\,j}\left|1/2\; m_{1/2}\right\rangle,\; j = 1,2. \end{aligned} \tag{4.59}$$

The coefficients $C_{m_{3/2}}^{3/2i}$, $C_{m_{1/2}}^{1/2j}$ depend on the strength of the magnetic dipole and electric quadrupole interactions and are calculated by diagonalizing the appropriate Hamiltonian $\hat{H} = \hat{H}_\mathrm{M} + \hat{H}_\mathrm{Q}$. The transition probability will then be given by [47, 76]

$$\begin{aligned} P(3/2i, 1/2j;\theta,\phi) \propto &\sum_{m_{3/2}}\sum_{m'_{3/2}}\sum_{m_{1/2}}\sum_{m'_{1/2}} C_{m_{1/2}}^{1/2j} C_{m_{3/2}}^{3/2i} C_{m'_{3/2}}^{3/2i} C_{m'_{1/2}}^{1/2j} \\ &\times \left\langle 3/2\; m_{3/2},\; 1m \middle| 1/2\; m_{1/2}\right\rangle \left\langle 3/2\, m'_{3/2},\; 1m' \middle| 1/2\; m'_{1/2}\right\rangle F_{1m}^{1m'}(\theta,\phi). \end{aligned} \tag{4.60}$$

In those cases where we are dealing with nuclear transitions which are a mixture of multipolarity M1 and E2 with a mixing parameter δ defined by $\delta = (\langle I_1||E2||I_2\rangle)/(\langle I_1||M1||I_2\rangle)$ (positive or negative), one obtains the extended relation

$$\begin{aligned} P(3/2\,i, 1/2\,j; \theta, \phi) &\propto (4.60) + \delta^2[(4.60) \quad \text{with } L = 2] \\ &-\delta \sum_{m_{3/2}} \sum_{m'_{3/2}} \sum_{m_{1/2}} \sum_{m'_{1/2}} C^{1/2j}_{m_{1/2}} C^{3/2i}_{m_{3/2}} C^{3/2i}_{m'_{3/2}} C^{1/2j}_{m'_{1/2}} \langle 3/2\; m_{3/2},\; 1m|1/2\; m_{1/2}\rangle \\ &\times \left\langle 3/2\; m'_{3/2},\; 2m'|1/2\; m'_{1/2} \right\rangle F^{2m'}_{1m}(\theta, \phi). \end{aligned} \tag{4.61}$$

For a poly crystalline absorber, where the z-axis is randomly distributed, (4.61) reduces to

$$\begin{aligned} P(3/2\,i, 1/2\,j) \propto &\left[\sum_{m_{3/2}} \sum_{m'_{3/2}} \sum_{m_{1/2}} \sum_{m'_{1/2}} C^{1/2j}_{m_{1/2}} C^{3/2i}_{m_{3/2}} C^{3/2i}_{m'_{3/2}} C_{m'_{1/2}}\, 1/2j \right. \\ &\left. \times \langle 3/2\; m_{3/2},\; 1m|1/2m_{1/2}\rangle \left\langle 3/2m'_{3/2},\; 1m'|1/2m_{1/2} \right\rangle \right] \\ &+ \delta^2 \cdot \; [\text{with } L = 2]. \end{aligned} \tag{4.62}$$

4.6.2 *Effect of Crystal Anisotropy on the Relative Intensities of Hyperfine Splitting Components*

In anisotropic crystals, the amplitudes of the atomic vibrations are essentially a function of the vibrational direction. As has been shown theoretically by Karyagin [72] and proved experimentally by Goldanskii et al. [48], this is accompanied by an anisotropic Lamb–Mössbauer factor f which in turn causes an asymmetry in quadrupole split Mössbauer spectra, for example, in the case of $I_e = 3/2$, $I_g = 1/2$ nuclear transitions in polycrystalline absorbers. A detailed description of this phenomenon, called the Goldanskii–Karyagin effect, is given in [73]. The Lamb–Mössbauer factor is given by

$$f = \exp\left(-\langle(\vec{k}\vec{r})^2\rangle\right), \tag{4.63}$$

$\vec{k}$: wave vector of the emitted γ-quantum,
$\vec{r}$: radius vector with the origin in the center of the vibrating atom.

According to Fig. 4.16, $\langle(\vec{k}\vec{r})^2\rangle$; can be calculated as a function of the angles θ and ϕ

$$\langle(kr)^2\rangle = k^2\left[\left(\langle r_x^2\rangle\cos^2\phi + \langle r_y^2\rangle\sin^2\phi\right)\sin^2\theta + \langle r_z^2\rangle\cos^2\theta\right]. \tag{4.64}$$

For an axially symmetric crystal, $\langle r_x^2 \rangle = \langle r_y^2 \rangle = \langle r_\perp^2 \rangle$ and $\langle r_z^2 \rangle = \langle r_\parallel^2 \rangle$. Using this notation, one obtains the following expression for f:

$$f(\theta) = \exp(-k^2 \langle r_\perp^2 \rangle) \exp(-k^2 \langle r_\parallel^2 - r_\perp^2 \rangle \cos^2\theta). \qquad (4.65)$$

Inserting (4.65) into (4.57), we find a modified relation for the relative intensity ratio I_2/I_1 for crystals with anisotropic (but axially symmetric) Lamb–Mössbauer factor f

$$I_2/I_1 = \frac{\int_0^\pi 3(1+\cos^2\theta)h(\theta)f(\theta)\sin\theta\,\mathrm{d}\theta}{\int_0^\pi (5-3\cos^2\theta)h(\theta)f(\theta)\sin\theta\,\mathrm{d}\theta} \quad \text{or}$$

$$I_2/I_1 = \frac{\int_0^\pi 3(1+\cos^2\theta)h(\theta)\exp(-N\cos^2\theta)\sin\theta\,\mathrm{d}\theta}{\int_0^\pi (5-3\cos^2\theta)h(\theta)\exp(-N\cos^2\theta)\sin\theta\,\mathrm{d}\theta}, \qquad (4.66)$$

$$N = k^2\left(\langle r_\parallel^2 \rangle - \langle r_\perp^2 \rangle\right).$$

With $h(\theta) = (1/\sin\theta)\delta(\theta - \theta_0)$, one obtains the same result as given by (4.58), which implies that the anisotropy of the f factor cannot be derived from the intensity ratio of the two hyperfine components in the case of a single crystal. It can, however, be evaluated from the absolute f value of each hyperfine component. However, for a poly-crystalline absorber ($h(\theta) = 1$), (4.66) leads to an asymmetry in the quadrupole split Mössbauer spectrum. The ratio of I_2/I_1, as a function of the difference of the mean square amplitudes of the atomic vibration parallel and perpendicular to the γ-ray propagation, is given in Fig. 4.19.

The anisotropic f factor may also manifest itself in the relative line intensities of Zeeman split hyperfine spectra in a poly crystalline absorber. Expanding $f(\theta)$ in a power series

$$f(\theta) = \sum_i a_i P_i(\theta), \quad (P_i(\theta):\ \text{Legendre polynomials}). \qquad (4.67)$$

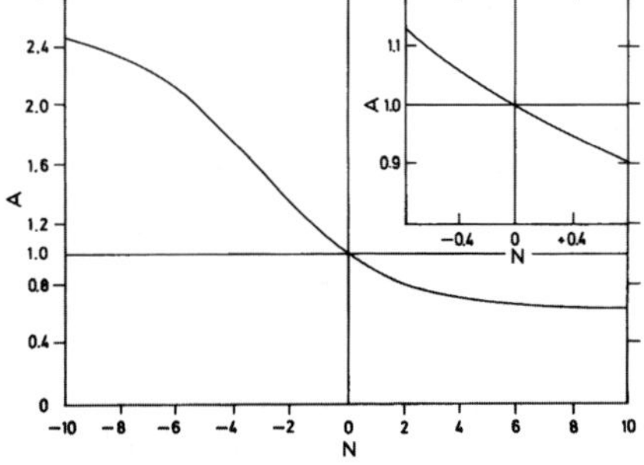

Fig. 4.19 The ratio of $A = I_2/I_1$ as function of the difference of the mean square of the vibrating amplitudes, $N = k^2\left(\langle r_\parallel^2 \rangle - \langle r_\perp^2 \rangle\right)$ (from [73])

Cohen et al. [49] obtained the following relations for the transition intensities

$$I_{m_{3/2}-m_{1/2}}:$$

$$\begin{aligned} I_{\pm 3/2-\pm 1/2} &= \frac{3(120-48N)}{120-40N} \\ I_{\pm 1/2-\pm 1/2} &= \frac{2(120-24N)}{120-40N} \\ I_{\pm 1/2-\pm 1/2} &= \frac{120-48N}{120-40N}\,. \end{aligned} \tag{4.68}$$

In [49, 76], the line intensities for electric quadrupole and Zeeman (magnetic dipole) splitting and including the anisotropy of the *f*-factor are also given for $I_e = 2 \leftrightarrow I_g = 0$ transitions (even–even isotopes, e.g., in the rare earth region or in W, Os).

We wish to conclude this section with a warning. Asymmetry in the intensities of observed quadrupole split spectra (3/2–1/2) is not always caused by an anisotropic Lamb–Mössbauer factor. There are other effects which may give rise to an asymmetric spectrum, for example, the line broadening due to the cosine smearing effect for spectra with nonzero isomer shift (cf. Sect. 3.1.8), and texture effects, that is, preferred orientation of crystallites in the absorber introduced during the preparation (packing) of the absorber, which cause $h(\theta)$ not to be isotropic. One has to measure the Mössbauer spectra as a function of the angle between the normal of the absorber plane and the direction of the γ-ray propagation as well as the temperature dependence of the asymmetry, which should increase with increasing temperature, in order to evaluate the effect of crystal anharmonicities on the Lamb–Mössbauer factor and find the correct source of line asymmetry.

4.7 ^{57}Fe-Mössbauer Spectroscopy of Paramagnetic Systems

The known spectroscopic oxidation states of iron in molecular systems range from (0) to (+VI). Most of them can occur with either low-spin, high-spin, or even intermediate-spin configuration. They all exhibit characteristic structural features and reactivity. Many iron compounds are paramagnetic because of the large number of open-shell configurations. The magnetic properties of paramagnetic compounds originating from the open-shell valence electrons allow the spectroscopist to probe the related electronic structure in detail. Among the techniques used are magnetic susceptibility measurements, MCD, EPR, ENDOR, NMR, and Mössbauer spectroscopy, with applied magnetic field. The latter has the virtue that the ^{57}Fe nucleus represents a *local* sensor, which connects microscopic *magnetic* information with the parameters for *electronic* charge density (isomer shift) and charge asymmetry (quadrupole splitting). Therefore, applied-field Mössbauer spectroscopy is an extremely useful technique in the (bio)inorganic spectroscopy of iron.

The spin-Hamiltonian formalism is a useful concept for the interpretation of paramagnetic data obtained from various types of measurements. The approach provides a common theoretical "clamp" for different techniques and assists their use as complementary tools. Its basic principle is the description of an energetically well-isolated electronic ground state of a paramagnetic molecule under the influence of a magnetic field by using an effective Hamiltonian that contains only variables of a "fictitious" electron spin S. The spin multiplicity is chosen to match the number of spin states under investigation. S must not necessarily be equal to the physical spin, but often it is. In the most general scheme, the spin Hamiltonian has a number of terms of the order $\vec{S}^n$ that correspond to the symmetry elements of the paramagnetic site under study. The expressions are obtained from operator equivalence for the symmetry-dependent terms of the electrostatic interaction [1]. Although this is "a priori" a rather formal approach, the method provides insight into the electronic structure of molecules because, in many cases, reasonable expressions can be derived for the leading terms in $\vec{S}$ up to $n = 2$ ($\vec{S}^2$) by perturbation treatment from ligand-field theory.

The spin-Hamiltonian concept, as proposed by Van Vleck [79], was introduced to EPR spectroscopy by Pryce [50, 74] and others [75, 80, 81]. H. H. Wickmann was the first to simulate paramagnetic Mössbauer spectra [82, 83], and E. Münck and P. Debrunner published the first computer routine for magnetically split Mössbauer spectra [84] which then became the basis of other simulation packages [85].[18] Concise introductions to the related modern EPR techniques can be found in the book by Schweiger and Jeschke [86]. Magnetic susceptibility is covered in textbooks on molecular magnetism [87–89]. An introduction to MCD spectroscopy is provided by [90–92]. Various aspects of the analysis of applied-field Mössbauer spectra of paramagnetic systems have been covered by a number of articles and reviews in the past [93–100].

4.7.1 The Spin-Hamiltonian Concept

In *ligand field theory* (LFT), the influence of the chemical environment on the electronic states of transition metal ions with d^n electron configuration is described by symmetry-dependant LF splittings of the free-ion states. The approach [101, 102] is an advancement of crystal field theory (CFT) in which all interactions between the metal ion and its environment are treated as pure electrostatic interactions between point charges. Instead, LFT adopts the idea of covalent π- and σ bonds between the metal valence and ligand electrons as the origin of term splitting. The orbitals used in LFT are still labeled as metal d-orbitals, although they are linear combinations of the metal valence and the ligand orbitals involved in that particular bonds. Other electrons are neglected. If covalency of the bonds plays only

[18] Another routine can be obtained from the authors. Write to: bill@mpi-muelheim.mpg.de.

a minor role, then the expressions found for the energies in LFT closely resemble those derived from CFT. In particular, the symmetry properties of the expressions remain the same, even for strongly covalent systems.

4.7.1.1 Ground State Properties and Zero-Field Splitting

Meaningful spin-Hamiltonian expressions can be derived only for orbitally *nondegenerate* ground states. This is the case for most of the coordination compounds found in inorganic chemistry and biochemistry because sterically demanding ligands impose strong symmetry distortions on the metal site and lift most of the degeneracies that are otherwise often encountered in small, highly symmetric molecules. In this case, the ground state with total spin S gives rise to $2S + 1$ substates with magnetic quantum numbers $M_s = -S, \ldots, S$, which (in the Born–Oppenheimer approximation)[19] are not split by electrostatic interactions. Direct spin–spin interaction and particularly *spin–orbit coupling* (SOC) can lift their degeneracy even in the absence of a magnetic field, due to mixing of ground and excited orbital states. The effect, referred to as *zero-field splitting* (ZFS), depends on the strength of the ligand field splitting between the orbital states and thus on the chemical bonds of the compound under study.

Figure 4.20 represents the energy diagram for the $3d^5$ electron configuration of high-spin iron(III) ($S = 5/2$) and the splitting of the ^6S ground state, as qualitatively obtained from LFT and SOC. The splitting pattern is schematically built up in five "tiers" with decreasing strength of the interactions from left to right.

In tier (1) of the diagram (for the electronic structure of iron(III)), only the total energy of the five metal valence electrons in the potential of the nucleus is considered. Electron–electron repulsion in tier (2) yields the free-ion terms (Russel–Saunders terms) that are usually labeled by term[20] symbols $^m\Gamma$. (The numbers given in brackets at the energy states indicate the spin- and orbital-multiplicities of these states.)

The ligand field of a quasi-octahedral coordination sphere acting on the *d*-orbitals lifts the degeneracy of the excited spin quartet (and others) as sketched in tier (3). The nature of the resulting states and their energy depends on the strength

[19]The Born-Oppenheimer approximation (BOA or adiabatic approximation) is a simplification of the Schrödinger equation for molecules with at least two nuclei and several electrons (or similar systems comprising a few heavy and many light particles); this problem cannot be solved exactly even for the lightest molecule, H_2^+. The simplification is based on the assumption that the motions of nuclei and electrons can be regarded as decoupled. Because of the huge difference of moments of inertia, the Coulomb forces between electrons and nuclei cause much stronger acceleration of the electrons than the nuclei. The motions therefore occur on different time scales. In the BOA, the nuclei are literally considered being at rest, such that a static potential exists for the electrons (BOA is named after M. Born and J. R. Oppenheimer).

[20]The term symbols $^m\Gamma$ used in panels (2) and (3) of the figure follow the usual convention, where $m = 2S + 1$ is the multiplicity of the spin, and Γ = S, P, D, ... denotes the orbital momentum; an S-term has $L = 0$; a P-term has $L = 1$; D has $L = 2, \ldots$

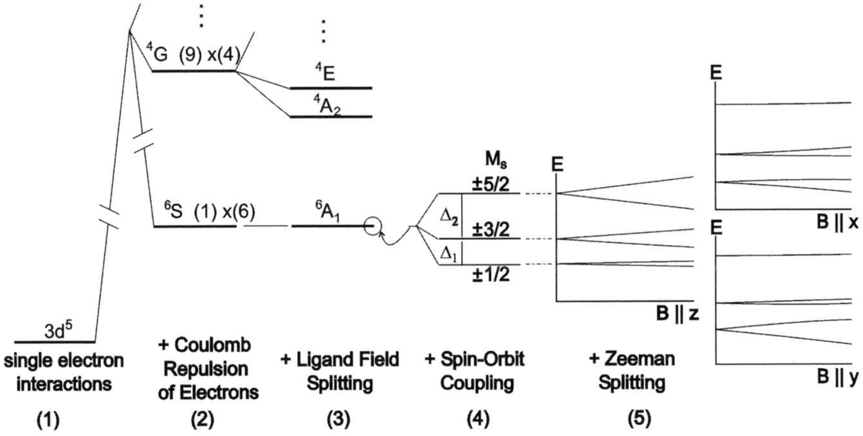

Fig. 4.20 A schematic view of the ground state terms of a transition metal ion with $3d^5$ electron configuration and spin $S = 5/2$, such as in high-spin iron(III), under various perturbations with decreasing interaction strength from left to right

and the symmetry of the LF. The LF splitting contains "chemical information," since it reflects the metal–ligand bond strength. The orbital momentum of the ground state at this stage is completely quenched. This is true not only for the S-type ground state with $L = 0$ in this particular example, but it holds also for any other nondegenerate orbital state, for example, resulting from the splitting of orbitally degenerate ground states by axial distortion.

Spin–orbit coupling (SOC) in tier (4) mixes the wave functions of ground and excited orbital states, which leads to partial restoration of the orbital momentum, and also to lifting of the orbital degeneracy of the ground state, which is called *zero-field splitting*. The spin sextet is split into three Kramers doublets,[21] which may be labeled by magnetic quantum numbers $M_S = \pm 5/2, \pm 3/2, \pm 1/2$. (The labels have only a symbolic meaning here; they would be strictly correct only if a weak, "polarizing" field would be applied along the z-direction of the LF system in order to lift the degeneracy of the $\pm M_S$ pairs. Otherwise, these pairs are strictly nonmagnetic, because the spin-wave functions have equal contributions from corresponding $+M_S$ and $-M_S$ components). The restored orbital contribution to the magnetic moment of the ground state of iron(III) can be measured by EPR or magnetic susceptibility, since it contributes to the (electronic) Zeeman interaction for the spin $S = 5/2$, which is seen as a *deviation of the g value* from the free electron value $g_e = 2.0023$.

[21]Kramers' degeneracy theorem states that the energy levels of systems with an odd number of electrons remain at least doubly degenerate in the presence of purely electric fields (i.e. no magnetic fields). This is a consequence of the time-reversal invariance of electric fields, and follows from an application of the antiunitary T-operator to the wavefunction of an odd number of electrons [51].

Finally in tier (5), the applied field lifts the remaining degeneracy of "M_S" states and yields a complex magnetic splitting pattern as shown on the right side of the diagram. The three different panels illustrate the effect of a field applied along the three principal directions x, y, and z of the LF system. The different splitting pattern demonstrates the magnetic anisotropy of the spin manifold which is related to zero-field splitting. These properties will be parameterized by a spin-Hamiltonian description in the following section.

4.7.2 The Formalism for Electronic Spins

The effective *spin Hamiltonian* for energetically well isolated and orbitally nongenerate ground states of transition metal ions with spin S is generally given by [1, 41, 103]:

$$\hat{H}_S = \hat{\vec{S}} \cdot \bar{\bar{D}} \cdot \hat{\vec{S}} + \mu_B \hat{\vec{S}} \cdot \bar{\bar{g}} \cdot \vec{B}, \tag{4.69}$$

where μ_B is the Bohr magneton and $\vec{B}$ the magnetic flux density, which is related to the magnetic field $\vec{H}$ by $\vec{B} = \mu_0 \vec{H}$. The symmetric 3×3 tensors $\bar{\bar{D}}$ and $\bar{\bar{g}}$ represent the zero-field splitting (ZFS) and the g-matrix of the anisotropic electronic Zeeman interaction. The Hamiltonian $\hat{H}_S$ operates on basis spin functions $|SM\rangle$ with $M = S$, $S-1, \ldots -S$. In a "proper" coordinate system, for which $\bar{\bar{D}}$ is diagonal, the ZFS can be written with the diagonal elements D_{ii} as:

$$\hat{\vec{S}} \cdot \bar{\bar{D}} \cdot \hat{\vec{S}} = D_{xx}\hat{S}_x^2 + D_{yy}\hat{S}_y^2 + D_{zz}\hat{S}_z^2. \tag{4.70}$$

If $\bar{\bar{D}}$ is taken as a traceless tensor, $\mathrm{Tr}(\bar{\bar{D}}) = \sum D_{ii} = 0$, there remain only two independent components for $\bar{\bar{D}}$ (neglecting the three Euler angles for orientation in a general coordinate system). Usually, these are the parameters D and E, for the *axial* and *rhombic* contribution to the ZFS:

$$D = D_{zz} - {}^1\!/_2(D_{xx} + D_{yy}), \quad E = {}^1\!/_2(D_{xx} - D_{yy}). \tag{4.71a, b}$$

The ratio E/D is the so-called *rhombicity parameter*. With some algebra, and recalling that $\hat{\vec{S}}^2 = \hat{S}_x^2 + \hat{S}_y^2 + \hat{S}_z^2$, one obtains

$$\hat{H}_S = D\left[\hat{S}_z^2 - {}^1\!/_3 S(S+1) + {}^E\!/_D\left(\hat{S}_x^2 - \hat{S}_y^2\right)\right] + \mu_B \hat{\vec{S}} \cdot \bar{\bar{g}} \cdot \vec{B}. \tag{4.72}$$

Although $\bar{\bar{D}}$ is in general not necessarily traceless,[22] (4.71) holds for the present purpose, because the corresponding term $\mathrm{Tr}(\bar{\bar{D}})\hat{\vec{S}}^2$, that is omitted here, would add only a uniform shift to all energy terms which would not be detectable. Thus, it can

[22]The trace of D vanishes when dipole coupling between paramagnetic centers determines the ZFS, since dipole interaction is traceless. A typical example is the ZFS of triplets arising from coupled radical pairs, for which SOC is negligible. For transition metal ions in contrast, SOC is the leading contribution to ZFS and the trace of $\bar{\bar{D}}$ in general has finite values.

be ignored by redefining energy zero. (However, one should be aware of exceptions: for exchange-coupled systems of oligo-nuclear compounds the trace of $\bar{D}$ principally does contribute to the total splitting of spin manifolds.)

The values of the rhombicity parameters are conventionally limited to the range $0 \leq E/D \leq 1/3$ without loss of generality. This corresponds to the choice of a proper coordinate system, for which $|D_{zz}|$ (in absolute values) is the largest component of the $\bar{\bar{D}}$ tensor, and $|D_{xx}|$ is smaller than $|D_{yy}|$. Any value of rhombicity outside the proper interval, obtained from a simulation for instance, can be "projected" back to $0 \leq E/D \leq 1/3$ by appropriate 90°-rotations of the reference frame, that is, by permutations of the diagonal elements of $\bar{D}$. To this end, the set of nonconventional parameters D' and E/D' has to be converted to the components of a traceless 3×3 tensor $\bar{\bar{D}}'$ using the relationships

$$D'_{xx} = E - 1/3D', \quad D'_{yy} = -E - 1/3D', \quad \text{and} \quad D'_{zz} = 2/3D', \qquad (4.73\ \text{a, b ,c})$$

(4.73a)–(4.73c) as can be obtained from (4.71a) and (4.71b) with the condition $\mathrm{Tr}(\bar{\bar{D}}') = 0$. The resulting tensor elements have to be rearranged in the conventional order $|D_{x'x'}| \leqslant |D_{y'y'}| \leqslant |D_{z'z'}|$ and, thus, the parameters D and E/D are obtained in the proper range $0 \leq E/D \leq 1/3$ by again using (4.71a) and (4.71b). With this transformation, the "nonconventional" parameter set $D' = -6\ \text{cm}^{-1}$, $E/D' = 0.66$, yields, for example, the values $D = +9\ \text{cm}^{-1}$, $E/D = 0.11$. The choice of a proper coordinate system is most important for any comparison of data sets to avoid the sign of D' being misleading.

The electronic spin Hamiltonian as given in (4.69) is widely used for the parameterization of magnetic susceptibility, EPR, MCD, and Mössbauer data. Representations of the Hamiltonian matrix in the basis of the usual M_S functions can be found in the literature for many quantum numbers S (see [1] and the remarks in Appendix G). Diagonalization of the matrix yields the eigenvalues and the eigenfunctions of the ground state manifold in terms of the ZFS parameters D, E/D and the $\bar{\bar{g}}$ matrix. From that, EPR transition energies as well as transition probabilities can be readily derived. Magnetic moments $\vec{m}_i$ for each of the magnetic sub-states $|i\rangle$ can be calculated from the energies E_i according to the relation $\vec{m}_i = -\partial E_i/\partial B$. For the simulation of Mössbauer spectra, however, the hyperfine interactions of the ^{57}Fe nucleus and the electronic charge and spin S also have to be considered as described in the following.

4.7.3 *Nuclear Hamiltonian and Hyperfine Coupling*

For the evaluation of magnetically split Mössbauer spectra within the spin-Hamiltonian formalism, the purely S-dependent Hamiltonian must be extended by an appropriate *nuclear Hamiltonian* for the nuclear spin I:

$$\hat{H}_{\text{nuc}} = \hat{\vec{I}} \cdot \bar{\bar{A}} \cdot \hat{\vec{S}} - g_N \mu_N \hat{\vec{I}} \cdot \vec{B} + \hat{H}_Q \,. \qquad (4.74)$$

The leading term in $\hat{H}_{\text{nuc}}$ is usually the magnetic hyperfine coupling $\vec{I} \cdot \overline{\overline{A}} \cdot \hat{\vec{S}}$ which connects the electron spin S and the nuclear spin I. It is parameterized by the hyperfine coupling tensor $\overline{\overline{A}}$. The I-dependent nuclear Zeeman interaction and the electric quadrupole interaction are included as 2nd and 3rd terms. Their detailed description for ^{57}Fe is provided in Sects. 4.3 and 4.4. The total spin Hamiltonian for electronic and nuclear spin variables is then:

$$\hat{H}_{\text{SH}} = \hat{H}_{\text{S}} + \hat{H}_{\text{nuc}}. \tag{4.75}$$

To calculate Mössbauer spectra, which consist of a finite number of discrete lines, the nuclear Hamiltonian, and thus also $\hat{H}_{\text{SH}}$, has to be set up and solved independently for the nuclear ground and excited states. The electric monopole interaction, that is, the isomer shift, can be omitted here since it is additive and independent of M_{I}. It can subsequently be added as an increment δ to the transition energies of each of the obtained Mössbauer lines.

4.7.3.1 Separation of *I*- and *S*-Dependent Contributions

The obvious basis for the diagonalization of $\hat{H}_{\text{SH}}$ are the product functions $|S, M_s\rangle \otimes |I, M_I\rangle$, which yield $(2S+1)(2I+1)$ eigenstates and eigenvalues. The interpretation of Mössbauer spectra is facilitated if the nuclear spin is decoupled from the electronic spin by the application of an external field. This may be achieved by the application of a weak field of 20 mT, which is sufficient for the electronic Zeeman term to exceed the magnetic hyperfine coupling, $g\mu_B B \gg A$ (recall that the nuclear magneton is about 2,000 times smaller than the Bohr magneton of electrons). Thus, the hyperfine coupling term, $\hat{\vec{I}} \cdot \overline{\overline{A}} \cdot \hat{\vec{S}}$, which is essential for the nuclear eigenvalues, needs not to be considered in the determination of the purely S-dependent electronic Hamiltonian $\hat{H}_{\text{S}}$. Instead, it can be treated as a part of the nuclear Hamiltonian for which the spin operator $\vec{S}$ is replaced by its expectation values $\langle\vec{S}\rangle_{\text{i}}$ for each of the $2S+1$ eigenstates $|\varphi_i\rangle$ of $\hat{H}_{\text{S}}$. With this approximation, the nuclear Hamiltonian can be re-written as:

$$\hat{H}_{\text{nuc},i} = \hat{\vec{I}} \cdot \overline{\overline{A}} \cdot \langle\vec{S}\rangle_i - g_{\text{N}}\mu_{\text{N}}\hat{\vec{I}} \cdot \vec{B} + \hat{H}_{\text{Q}}, \quad \text{for } i = 1, \ldots, 2S+1. \tag{4.76}$$

Quotations of the hyperfine coupling in terms of energy (or frequency) are not unique for Mössbauer spectra, unless the values are explicitly referenced to the nuclear ground or the excited state. Therefore, it is common to introduce the *internal field* $\vec{B}_i^{\text{int}}$ at the nucleus,

$$\vec{B}_i^{\text{int}} = -\overline{\overline{A}} \cdot \langle\vec{S}\rangle_i / g_{\text{N}}\mu_{\text{N}}, \tag{4.77}$$

which is independent of the nuclear state as can be seen by re-writing the nuclear Hamiltonian as

$$\begin{aligned}\hat{H}_{\text{nuc},i} &= -g_N\mu_N\hat{\vec{I}}\left(\frac{-\overline{\overline{A}}\langle\vec{S}\rangle_i}{g_N\mu_N}+\vec{B}\right)+\hat{H}_Q \\ &= -g_N\mu_N\hat{\vec{I}}\cdot(\vec{B}_i^{\text{int}}+\vec{B})+\hat{H}_Q.\end{aligned} \tag{4.78}$$

Because of the different properties of the nuclear ground and excited states, the hyperfine coupling constants (A-values) for Mössbauer nuclei are often quoted in units of the internal field, $A/g_N\mu_N$, where A represents an energy and μ_N is the energy per field unit.

4.7.4 Computation of Mössbauer Spectra in Slow and Fast Relaxation Limit

By decoupling nuclear and electronic spins, the evaluation of Mössbauer spectra is reduced to the independent diagonalization of the $2S + 1$ eigenvalues E_i of $\hat{H}_S$ and the $2I + 1$ eigenvalues for the nuclear ground state ($I = 1/2$) and excited state ($I_e = 3/2$). This procedure is numerically much faster than the diagonalization of the coupled system, which would have the matrix dimension $(2S+1)(2I+1) \times (2S+1)(2I+1)$. The eigenfunctions of $\hat{H}_S$ yield the electronic spin expectation values $\langle\vec{S}\rangle_i$ which represent the microscopic magnetic moments $\vec{m}_i$ of the spin states. If the electronic spin states are stationary due to slow spin relaxation, separate Mössbauer spectra have to be evaluated for each of the electronic states, and then the 2S + 1 spectra are superimposed using appropriate Boltzmann factors $\rho_i = \exp(-E_i/kT)/\sum_j \exp(-E_j/kT)$ for the actual temperature T. If, moreover, the Mössbauer sample is a powder or a frozen solution, which is usually the case, it is not sufficient to calculate the Mössbauer spectrum for a single orientation of the applied field. Instead, the procedure of evaluating subspectra has to be repeated in a loop for a large number of field orientations uniformly distributed on a unit sphere, and the results have to be summed to obtain the final powder spectrum.

Often the electronic spin states are not stationary with respect to the Mössbauer time scale but fluctuate and show transitions due to coupling to the vibrational states of the chemical environment (the "lattice" vibrations or phonons). The rate $1/T_1$ of this *spin–lattice relaxation* depends among other variables on temperature and energy splitting (see also Appendix H). Alternatively, spin transitions can be caused by spin–spin interactions with rates $1/T_2$ that depend on the distance between the paramagnetic centers. In densely packed solids of inorganic compounds or concentrated solutions, the *spin–spin relaxation* may dominate the total spin relaxation $1/T = 1/T_1 + 1/T_2$ [104]. Whenever the relaxation time is comparable to the nuclear Larmor frequency ($\langle S\rangle A/\hbar$) or the rate of the nuclear decay ($\approx 10^7 \text{s}^{-1}$), the stationary solutions above do not apply and a dynamic model has to be invoked to calculate the Mössbauer spectra from $\hat{H}_S$. An overview of the evaluation of such magnetic Mössbauer spectra with *intermediate spin relaxation* and their interpretation is presented in Chap. 6. However, for most iron centers, the experimental conditions can be chosen in such a way that either stationary conditions are achieved, called *slow spin relaxation*, or that the fluctuation rate exceeds the Mössbauer time scale, called *fast spin relaxation*. In the latter case, the thermal average $\langle\vec{S}\rangle_T$ of the $2S + 1$ spin expectation values

$$\langle\vec{S}\rangle_T = \frac{\sum_i \langle\vec{S}\rangle_i \exp(-E_i/kT)}{\sum_j \exp(-E_j/kT)} \quad (4.79)$$

can be substituted in (4.77), and a single Mössbauer spectrum is obtained from all electronic states together instead of $2S + 1$ separate subspectra. Averaging over an angular grid on the unit sphere again performs the powder summation.

4.7.5 *Spin Coupling*

The active site of many metalloproteins contains more than one paramagnetic transition metal ion. Moreover, during the reaction cycle of some metalloenzymes, intermediates are formed which comprise organic radicals in contact with the transition metal ion. Compounds with two or more paramagnetic centers are extensively investigated in inorganic chemistry because of the great variety of electronic structures and reactivity. The salient difference between the metal ions in such paramagnetic clusters and the corresponding mononuclear species is a coupling of the electronic spins, which affords a series of new energy states with splitting energies comparable to thermal energy kT up to ambient temperatures. Boltzmann population of these states causes interesting magnetic properties with non-Curie temperature dependence. Although the fundamental interaction of two paramagnetic metal ions *a* and *b*, or a metal ion and a radical connected by one or more diamagnetic bridging atoms, is of electrostatic origin and involves orbital states, the resulting spin coupling for systems without first-order orbital momentum can be adequately described by the effective spin operator (according to Heisenberg, Dirac, and Van Vleck, HDVV)

$$\hat{H}_{ex} = -2J\hat{\vec{S}}_a \cdot \hat{\vec{S}}_b, \quad (4.80)$$

The two spins $\hat{\vec{S}}_a$ and $\hat{\vec{S}}_b$ are assumed to be local, associated with the two sites a and b, respectively. The parameter J is the so-called exchange-coupling constant,[23] which expresses the strength of the (super)exchange interaction between the

[23]Note that also other forms of the HDVV operator are used in the literature, e.g. $J\hat{\vec{S}}_a \cdot \hat{\vec{S}}_b$ and $-J\hat{\vec{S}}_a \cdot \hat{\vec{S}}_b$; therefore, it is important to check the Hamiltonian when comparing J values.

paramagnetic sites (usually, the interaction between metal ions bridged by one or more diamagnetic atoms, is called superexchange). The values of J for most iron compounds encountered in (bio)inorganic chemistry are of the order +10 to $-300\ \text{cm}^{-1}$, depending on the nature of the bridging ligands and the cluster geometry. Interpretations of the exchange coupling constants in terms of the related chemical information are found in many textbooks and articles on magnetochemistry [87, 88, 105–107].

The dominating contribution to exchange interaction is isotropic as expressed by the scalar parameter used in (4.80); anisotropic and antisymmetric contributions [87] are mostly ignored. Equation (4.80) can be evaluated in the coupled presentation using the total spin S, that is, $\vec{S} = \vec{S}_a + \vec{S}_b$. The product $\hat{\vec{S}}_a \cdot \hat{\vec{S}}_b$ can be substituted according to the relation $\hat{\vec{S}}^2 = (\hat{\vec{S}}_a + \hat{\vec{S}}_b)^2 = \hat{\vec{S}}_a^2 + \hat{\vec{S}}_b^2 + 2\hat{\vec{S}}_a \cdot \hat{\vec{S}}_b$, achieving a simple operator of S^2,

$$\hat{H}_{ex} = -J\left[\hat{\vec{S}}^2 - \hat{\vec{S}}_a^2 - \hat{\vec{S}}_b^2\right], \tag{4.81}$$

which is diagonal in the usual $|S, M_s\rangle$ basis of the total spin. From this expression, the energies of the eigenstates are immediately obtained as

$$E(S) = -J[S(S+1) - S_a(S_a+1) - S_b(S_b+1)]. \tag{4.82}$$

The allowed values of S are restricted to $|S_a - S_b| \leqslant S \leqslant |S_a + S_b|$. The constant terms with $S_i(S_i + 1)$ can be eliminated by redefining zero energy. Thus, for any given pair of spins S_a and S_b, a certain sequence of energy states is obtained, which is also called a "spin ladder" (Fig. 4.21). The state with minimal spin is lowest in energy, if $J < 0$ (antiferromagnetic coupling), whereas ferromagnetic coupling ($J > 0$) inverts the scheme.

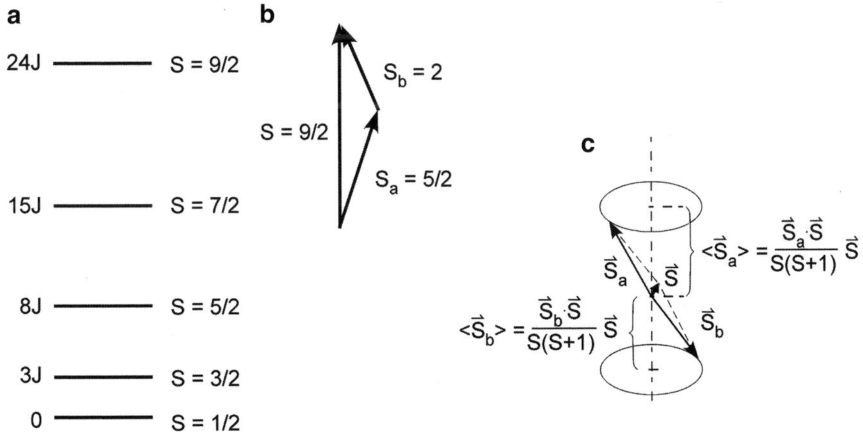

Fig. 4.21 The spin-coupling scheme for $S_a = 5/2$ and $S_b = 2$: (**a**) spin ladder according to (4.82), (**b**) vector coupling for $S = 9/2$, (**c**) spin projection for $S = 1/2$

A classical example of spin coupling in a metalloprotein is the $[2Fe–2S]^+$ cluster of reduced (plant-type) ferredoxins [108, 109]. Iron–sulfur proteins have attracted particular attention in the recent decade because of new functions discovered for these molecules [110, 111]. The ground state of the $[2Fe–2S]^+$ cluster is $S = 1/2$, whereby Mössbauer spectroscopy has shown the concurrent presence of an Fe(III) site with $S_a = 5/2$ and an Fe(II) site with $S_b = 2$. The ferric- and the ferrous-ions are tetrahedrally coordinated and they are bridged by two diamagnetic sulfur ligands which mediate strong antiferromagnetic coupling ($|J|>300\,\mathrm{cm}^{-1}$) [112].

If the coupling between the spins $S_a = 5/2$ and $S_b = 2$ is sufficiently strong , that is, if $H_{ex} \gg H_{zfs}, H_{Zeeman}$, both spins precess coherently together about an applied field yielding the resultant total spin S. (Fig. 4.21c). The magnetic moments or internal fields that can be sensed locally by a Mössbauer nucleus are proportional to the expectation values $\langle S_a \rangle$ and $\langle S_b \rangle$of the local spins, given by the projection of S_a and S_b onto the precession axis (i.e. the magnetic quantization axis)

$$\langle S_a \rangle = \frac{\vec{S}_a \cdot \vec{S}}{S} = \frac{\vec{S}_a \cdot \vec{S}}{S(S+1)} \langle S \rangle = 7/3 \langle S \rangle \tag{4.83a}$$

$$\langle S_b \rangle = \frac{\vec{S}_b \cdot \vec{S}}{S} = \frac{\vec{S}_b \cdot \vec{S}}{S(S+1)} \langle S \rangle = -4/3 \langle S \rangle. \tag{4.83b}$$

This means the internal fields, $\vec{B}_i^{\mathrm{int}} = -\bar{\bar{A}} \langle \vec{S}_i \rangle / g_N \mu_N$, at the nuclei $i = a, b$ are reduced with respect to the values expected for the uncoupled ions, and they will have opposite signs! The latter can be probed by the field dependence of spectra recorded at liquid helium temperatures, at which spin relaxation is in the slow limit. This means the magnetic Mössbauer spectrum of the $S = 1/2$ ground state of the Fe (III)–Fe(II) dimer is represented by the superposition of four subspectra, namely, the separate contributions from (1) the ferric- and (2) the ferrous-site. Each of both is additionally affected by the different effective fields $B^{\mathrm{eff}} = B + B^{\mathrm{int}}$ at the nuclei, arising (3) from spin-up ($\langle S \rangle = +1/2$) and (4) from spin down ($\langle S \rangle = -1/2$) expectation values. Corresponding simulations of Mössbauer spectra can be found, for instance, in [99, 113]. The effective field for the ground state level ($\langle S \rangle = -1/2$) decreases with increasing applied field at the ferric site (carrying the majority spin) and increases at the ferrous site (carrying the minority spin). These distinct properties can be used for assigning the subspectra of spin-coupled clusters.

The different magnetic hyperfine coupling of the spins a and b in this example as heuristically derived for the coupled presentation from above can also be expressed by using the hyperfine coupling constants

$$\bar{\bar{A}}_{a,t} = 7/3 \bar{\bar{A}}_a, \quad \text{and} \quad \bar{\bar{A}}_{b,t} = -4/3 \bar{\bar{A}}_b, \tag{4.84a, b}$$

where the subscript t indicates the total spin S, that is, $A_{a,t}$ is the A tensor for site (a) with respect to the total spin, and A_a is the A-tensor with respect to the local spin of that particular iron. The analogous assignment holds for site (b). Note, that only the local values should be compared with those of other systems or noncoupled iron compounds, even when the actual simulation of the spectra have been performed with the total spin in the coupled presentation.

Spin projection coefficients as given in (4.84) can be obtained in a general and mathematically proper manner by using the Wigner–Eckhart theorem. A detailed description of this topic is found in the textbook by Bencini and Gatteschi [106].

Finally, it is noteworthy that in addition to the isotropic part of spin coupling as treated above, there may also be an anisotropic contribution due to the presence of anisotropic exchange [114] or dipole interaction [106]. In this case, the exchange coupling constant is replaced by a tensor

$$\hat{H}_{ex} = -2\hat{\vec{S}}_a \cdot \bar{\bar{J}} \cdot \hat{\vec{S}}_b. \tag{4.85}$$

However, both anisotropic contributions are in generally negligibly small in molecular systems. In some cases, however, appreciably strong contributions of antisymmetric exchange have been observed, particularly for asymmetric homovalent ferric dimers. This contribution can mix total-spin manifolds more efficiently than zero-field splitting and induce a field-dependent magnetic moment for the singlet ground state of such dimers. Actual examples are found in [115–122].

4.7.6 Interpretation, Remarks and Relation with Other Techniques

The major contribution to the components of the D tensor as well as the deviations of the g values from 2.0023 arises from the mixing of ligand field states by SOC; other contributions to D result from direct spin–spin coupling, which mixes states of the same spin S. The $\bar{\bar{D}}$ tensor and the $\bar{\bar{g}}$ matrix both carry "chemical information" as they are related to the strength and symmetry of the LF, which is competing and counteracting to the effects of SOC. Details on the chemical interpretation of the parameters by quantum chemical means is found in Chap. 5.

The spin-Hamiltonian formalism is a "crutch" in the sense that it is a parameterized theory, but it provides a common theoretical frame for the various experimental techniques with a minimum number of adjustable parameters that describe the essential physics of the system under investigation. Even more important is the fact that the same parameters can be derived relatively easily from quantum chemical calculations. Therefore, theoreticians appreciate the concept as a "convenient place to rest in the analysis of experimental data by theoretical means" [123, 124].

Orbitally degenerate ground states, in general, cannot be treated in the spin-Hamiltonian approach. In this case, SOC has to be evaluated explicitly on an extended basis of spin–orbit functions. However, in coordination chemistry and bioinorganic chemistry, this is only of marginal importance, because the metal centers of relevant compounds are usually of sufficiently low symmetry in order to lift orbital degeneracy. Even highly symmetric geometries, as in the Fe(II)-hexaquo complex, may be lowered by vibronic coupling so that the spin-Hamiltonian approach is sufficient for the interpretation of magnetic Mössbauer, susceptibility, and inelastic neutron scattering data [125, 126].

Magnetic hyperfine coupling constants derived from Mössbauer spectra are often given in units of the internal field [in Tesla (T)] at the nucleus. To obtain such a value from the coupling energy, the energy value is divided by the nuclear g-factor and by the nuclear magneton [given in energy/Tesla]. For ^{57}Fe values up to the order of 10^2, Tesla are obtained for the corresponding entity $A/g_N\mu_N$. The convention of using the magnetic hyperfine field instead of hyperfine coupling energy is physically meaningful, because the Mössbauer transitions occur between nuclear ground and excited states, which have different g-factors. Quotations of the hyperfine coupling energy would remain ambiguous without specifying the corresponding nuclear state, whereas quotations of $A/g_N\mu_N$ in Tesla are unique. A comparison of field values with energy data from NMR and EPR spectroscopy for the same type of hyperfine interaction (given in energy units for the ground state splitting of e.g. the ^{57}Fe nucleus) requires that they have to be multiplied with the factor $g_N\mu_N$. The values of $A/g_N\mu_N$ should not be confused with A values obtained in EPR spectroscopy from the splitting of hyperfine lines in a field-swept EPR spectrum, which are given in Gauss or milli-Tesla. These numbers depend on the actual electronic g-value and have to be converted to energy units, for instance, by using the relation $A(\mathrm{cm}^{-1}) = A(\mathrm{mT}) \cdot g \cdot \mu_B(\mathrm{cm}^{-1}\mathrm{mT}^{-1})$. More information on conversion factors is given in Appendix I.

References

1. Abragam, A., Bleaney, B.: Electron Paramagnetic Resonance of Transition Ions. Dover, New York (1986)
2. Bodmer, A.R.: Proc. Phys. Soc. London **A 66**, 1041 (1953)
3. Stacey, D.N.: Rep. Prog. Phys. **29**, 171 (1966)
4. Slichter, C.P.: Principles of Magnetic Resonance, 3rd edn. Springer, Berlin (1990)
5. Spiering, H.: The Electric Field Gradient and Quadrupole Interaction. In: Long, G. (ed.) Mössbauer Spectroscopy Applied to Inorganic Chemistry, p. 79. Plenum, New York (1984)
6. Abragam, A.: The Prinicples of Nuclear Magnetism. Oxford University Press, London (1961)
7. Kistner, O.C., Sunyar, A.W.: Phys. Rev. Lett. **4**, 229 (1960)
8. Shirley, D.A.: Rev. Mod. Phys. **36**, 339 (1964)
9. Wdowik, U.D., Ruebenbauer, K.: J. Chem. Phys **129**, 104504 (2008)
10. Kündig, W., Müller, P.E.: Helv. Phys. Acta **52**, 555 (1979)
11. Ladriere, J., Cogneau, M., Meykens, A.: J. Physique **41**, C1–C132 (1980)
12. Spijkerveet, W.J.J., Pleiter, F.: Hyperfine Interact **7**, 285 (1979)
13. Nieuwpoort, W.C., Post, D., Duijnen, P.T.V.: Phys. Rev **B 17**, 91 (1978)
14. Marathe, V.R., Trautwein, A.: Calculation of charge density, electric field gradient and internal magnetic field at the nuclear site using molecular orbital cluster theory. In: Thosar,

B.V., Srivasta, J.K., Iyengar, P.K., Bhargava, S.C. (eds.) Advances in Mössbauer Spectroscopy, p. 398. Elsevier, Amsterdam (1985)

15. Lovell, T., Han, W.G., Liu, T., Noodelman, L.: J. Am. Chem. Soc. **124**, 5890 (2002)
16. Neese, F.: Inorg. Chim. Acta **337**, 181 (2002)
17. Nemykin, V.N., Hadt, R.C.: Inorg. Chem. **45**, 8297 (2006)
18. Wdowik, U.D., Ruebenbauer, K.: Phys. Rev. **B 76**, 155118 (2007)
19. Filatov, M.: J. Chem. Phys. **127**, 084101 (2007)
20. Herber, R.H.: Structure, bonding and the Mössbauer lattice temperature. In: Herber, R.H. (ed.) Chemical Mössbauer Spectroscopy. Plenum, New York (1984)
21. Yousif, A.A., Winkler, H., Toftlund, H., Trautwein, A.X., Herber, R.H.: J. Phys. Condens. Matter **1**, 7103 (1989)
22. IUPAC. Compendium of chemical terminology, 2nd ed. Compiled by McNaught, A.D. and Wilkinson A. Blackwell Scientific Publications, Oxford (1997). XML on-line corrected version: http://goldbook.iupac.org (2006) created by Nic, M., Jirat, J., Kosata, B.; updates compiled by Jenkins, A. ISBN 0-9678550-9-8 (2006)
23. Jörgensen, C.K.: Oxidation Numbers and Oxidation States. Springer, Berlin (1969)
24. Greenwood, N.N., Gibb, T.C.: Mössbauer Spectroscopy. Chapman and Hall, London (1971)
25. De Coster, M., Amelinckx, S.: Phys. Lett. **1**, 245 (1962)
26. Mullen, J.G.: Phys. Rev. **131**, 1410 (1963)
27. Chappert, J., Frankel, R.B., Misetich, A., Blum, N.A.: Phys. Rev. **179**, 578 (1969)
28. Wdowik, U.D., Ruebenbauer, K.: Phys. Rev. **B 63**, 125101 (2001)
29. Stoian, S.A., Yu, Y., Smith, J.M., Holland, P.L., Bominaar, E.L., Münck, E.: Inorg. Chem. **44**, 4915 (2005)
30. Hendrich, M.P., Gunderson, W., Behan, R.K., Green, M.T., Mehn, M.P., Betley, T.A., Lu, C.C., Peters, J.C.: Proc. Natl. Acad. Sci. USA **103**, 17107 (2006)
31. Stoian, S.A., Vela, J., Smith, J.M., Sadique, A.R., Holland, P.L., Münck, E., Bominaar, E.L.: J. Am. Chem. Soc. **128**, 10181 (2006)
32. Vogel, C., Heinemann, F.W., Sutter, J., Anthon, C., Meyer, K.: Angew. Chem. Int. Ed **47**, 2681 (2008)
33. Berry, J.F., Bill, E., Bothe, E., George, S.D., Mienert, B., Neese, F., Wieghardt, K.: Science **312**, 1937 (2006)
34. Beinert, H., Holm, R.H., Münck, E.: Science **277**, 653 (1997)
35. Hoggins, J.T., Steinfink, H.: Inorg. Chem. **15**, 1682 (1976)
36. Kerler, W., Neuwirth, W.: Z. Physik **167**, 176 (1962)
37. Enemark, J.H., Feltham, R.D.: Coord. Chem. Rev. **13**, 339 (1974)
38. Li, M., Bonnet, D., Bill, E., Neese, F., Weyhermüller, T., Blum, N., Sellman, D., Wieghardt, K.: Inorg. Chem. **41**, 3444 (2002)
39. Watson, R.E.: Phys. Rev. **118**, 1036 (1960)
40. Watson, R.E.: Phys. Rev. **119**, 1934 (1960)
41. Weil, J.A., Bolton, J.R., Wertz, J.E.: Electron Paramagnetic Resonance. Wiley, New York (1994)
42. Pyykkö, P.: Mol. Phys. **106**, 1965 (2008)
43. Raghavan, P.: Atomic Data Nucl. Data Tables **42**, 189 (1989)
44. Mills, I., Cvitas, T., Homann, K., Kallay, N.: Quantities, Units and Symbols in Physical Chemistry, 2nd edn. Blackwell, Oxford (1993)
45. Cohen, E.R., Cvitaš, T., Frey, J.G., Holmström, B., Kuchitsu, K., Marquardt, R., Mills, I., Pavese, F., Quack, M., Stohner, J., Strauss, H.L., Takami, M., Thor, A.J.: Quantities, Units and Symbols in Physical Chemistry, 3rd edn. RSC, Cambridge (2007)
46. Bieron, J.P., et al.: Phys. Rev. **A 64**, 052507 (2001)
47. Viegers, T.: Dissertation, University of Nijmegen, Netherlands (1976)
48. Goldanskii, V.I., Makarov, E.F., Khrapov, V.V.: Zh. Eksperim. Theor. Fiz. **44**, 752 (1963); Phys. Lett. **3**, 344 (1963)
49. Cohen, S.G., Gielen, P., Kaplow, R.: Phys. Rev. **141**, 423 (1966)

50. Abragam, A., Pryce, M.H.L.: Proc. R. Soc. **A 250**, 135 (1951)
51. Kramers, H.A.: Proceedings of the Koninklijke Akademie Van Wetenschappen, vol. 35, p. 1272. Amsterdam (1932)
52. Zimmermann, R.: Nucl. Instrum. Methods **128**, 537 (1975)
53. Shenoy, G.K., Dunlop, B.D.: Nucl. Instrum. Methods **71**, 285 (1969)
54. Sternheimer, R.M.: Phys. Rev. **80**, 102 (1950)
55. Sternheimer, R.M.: Phys. Rev. **84**, 244 (1951)
56. Foley, H.M., Sternheimer, R.M., Tycko, D.: Phys. Rev. **93**, 734 (1954)
57. Sternheimer, R.M., Foley, H.M.: Phys. Rev. **102**, 731 (1956)
58. Sternheimer, R.M.: Phys. Rev. **130**, 1423 (1963)
59. Barnes, R.G., Smith, W.V.: Phys. Rev. **93**, 95 (1954)
60. Meyer, K., Bill, E., Mienert, B., Weyhermüller, T., Wieghardt, K.: J. Am. Chem. Soc. **121**, 4859 (1999)
61. Atkin, C.I., Thelander, L., Reichhard, P., Lang, G.: J. Biol. Chem. **248**, 7464 (1973)
62. Vincent, J.B., Olivier-Lilley, G.L., Averill, B.A.: Chem. Rev. **90**, 1447 (1990)
63. Freeman, A.J., Watson, R.E.: Phys. Rev. **B 131**, 2566 (1963)
64. Blaes, N., Fischer, H., Gonser, U.: Nucl. Instrum. Methods Phys. Res **B 9**, 201 (1985)
65. Wertheim, G.K.: Mössbauer Effect: Principles and Applications. Academic, New York (1964)
66. Bonhomme, C., Livage, J.: J. Phys. Chem. **A 102**, 375 (1998)
67. Champion, P.M.: Disertation, University of Illinois, Urbana-Champaign (1975)
68. Reiff, W.M., LaPointe, A.M., Witten, E.H.: J. Am. Chem. Soc. **126**, 10206 (2004)
69. Kuzmann, E., Szalay, R., Vértes, A., Homonnay, Z., Pápai, I., de Châtel, P., Klencsár, Z., Szepes, L.: Struct. Chem. **20**, 453 (2009)
70. Reiff, W.M., Frommen, C.M., Yee, G.T., Sellers, S.P.: Inorg. Chem. **39**, 2076 (2000)
71. Andres, H., Bominaar, E.L., Smith, J.M., Eckert, N.A., Holland, P.L., Münck, E.: J. Am. Chem. Soc. **124**, 3012 (2002)
72. Karyagin, S.V.: Dokl. Akad. Nauk. SSSR **148**, 1102 (1963)
73. Goldanskii, V.I., Makarov, E.F.: Chemical Applications of Mössbauer Spectroscopy, p. 102. Academic, New York (1968)
74. Pryce, M.H.L.: Proc. Phys. Soc. **A 63**, 25 (1950)
75. Bleaney, B., Stevens, K.H.W.: Rep. Prog. Phys **16**, 108 (1953)
76. Gedikli, A., Winkler, H., Gerdau, E.: Z. Phys. **267**, 61 (1974)
77. Edmonds, A.R.: Drehimpulse in der Quantenmechanik. Bibliographisches Institut, Mannheim (1964)
78. Housley, R.M., Grant, R.W., Gonser, U.: Phys. Rev. **178**, 514 (1969)
79. Vleck, J.H.V.: Phys. Rev. **57**, 426 (1940)
80. Griffith, J.S.: Mol. Phys. **3**, 79 (1960)
81. Stevens, K.H.W.: Spin Hamiltonians. In: Rado, G.T., Suhl, H. (eds.) Magnetism, p. 1. Academic, New York (1963)
82. Wickman, H.H.: Mössbauer Effect Methodology, 2nd edn. Plenum, New York (1966)
83. Wickman, H.H., Klein, M.P., Shirley, D.A.: Phys. Rev. **152**, 345 (1966)
84. Münck, E., Grover, J.L., Tumolillo, T.A., Debrunner, P.G.: Comput. Phys. Commun. **5**, 225 (1973)
85. The simulation package WMOSS is available from http://www.wmoss.org/
86. Schweiger, A., Jeschke, G.: Principles of Pulse Electron Paramagnetic Resonance. Oxford University Press, Oxford (2001)
87. Kahn, O.: Molecular Magnetism. VCH, Weinheim (1993)
88. Girerd, J.J., Journaux, Y.: Molecular Magnetism in Bioinorganic Chemistry. In: Que, L. (ed.) Physical Methods in Bioinorganic Chemistry, p. 321. University Science Books, Sausalito (2000)
89. Blundell, S.: Magnetism in Condensed Matter. Oxford University Press, New York (2001)
90. Praneeth, V.K.K., Neese, F., Lehnert, N.: Inorg. Chem. **44**, 2570 (2005)
91. Stich, T.A., Buan, N.R., Brunold, T.C.: J. Am. Chem. Soc. **126**, 9735 (2004)

92. Quintanar, L., Gebhard, M., Wang, T.P., Kosman, D.J., Solomon, E.I.: J. Am. Chem. Soc. **126**, 6579 (2004)
93. Münck, E., Champion, P.M.: J. Physique Colloq. **C6**(Suppl. 12), C6–C33 (1974)
94. Oosterhuis, W.T.: In: Dunitz, J.D. et al. (eds.) Structure and Bonding, vol. 20, p. 59. Springer, Berlin (1974)
95. Johnson, C.E.: Mössbauer spectroscopy in biology. In: Gonser, U. (ed.) Mössbauer Spectroscopy, p. 139. Springer, Berlin (1975)
96. Debrunner, P.G.: Mössbauer spectroscopy of iron porphyrins. In: Lever, A.B.P., Gray, H.B. (eds.) Iron Porphyrins Part III, p. 137. VCH, Weinheim (1989)
97. Trautwein, A.X., Bill, E., Bominaar, E.L., Winkler, H.: Struct. Bonding **78**, 1 (1991)
98. Debrunner, P.G.: Mössbauer spectroscopy of iron proteins. In: Berliner, L.J., Reuben, J. (eds.) EMR of Paramagnetic Molecules, p. 59. Plenum, New York (1993)
99. Münck, E.: Aspekts of ^{57}Fe Mössbauer spectroscopy. In: Que, L. (ed.) Physical Methods in Bioinorganic Chemistry, p. 287. University Science Books, Sausalito (2000)
100. Schünemann, V., Winkler, H.: Rep. Prog. Phys. **63**, 263 (2000)
101. Schläfer, H.L., Gliemann, G.: Basic Principles of Ligand Field Theory. Wiley, New York (1969)
102. Cotton, F.A.: Chemical Applications of Group Theory. Wiley, New York (1990)
103. Atherton, N.M., Winscom, C.J.: Inorg. Chem. **12**, 383 (1973)
104. Pilbrow, J.R.: Transition Ion Electron Paramagnetic Resonance. Clarendon, Oxford (1990)
105. Anderson, P.W.: In: Rado, G.T., Suhl, H. (eds.) Magnetism, vol. 1. Academic, New York (1963)
106. Bencini, A., Gatteschi, D.: EPR of Exchange Coupled Systems. Springer, Berlin (1990)
107. Hotzelmann, R., Wieghardt, K., Floerke, U., Haupt, H.J., Weatherburn, D.C., Bonvoisin, J., Blondin, G., Girerd, J.J.: J. Am. Chem. Soc. **114**, 1681 (1992)
108. Gibson, J.F., Hall, D.O., Thornley, H.M., Whatley, F.R.: Proc. Natl. Acad. Sci. USA **56**, 987 (1966)
109. Rao, K.K., Cammack, R., Hall, D.O., Johnson, C.E.: Biochem. J. **122**, 257 (1971)
110. Johnson, M.K.: Curr. Opin. Chem. Biol. **2**, 173 (1998)
111. Rao, P.V., Holm, R.H.: Chem. Rev. **104**, 527 (2004)
112. Beardwood, P., Gibson, J.F., Bertrand, P., Gayda, J.-P.: Biochim. Biophys. Acta – Protein Struct. **742**, 426 (1983)
113. Johnson, C.E.: Mössbauer spectroscopy in biology. In: Gonser, U. (ed.) Mössbauer Spectroscopy, p. 139. Springer, Berlin (1975)
114. Meiklejohn, W.H., Bean, C.P.: Phys. Rev. **105**, 904 (1957)
115. Moriya, T.: Phys. Rev. **120**, 91 (1960)
116. Kent, T.A., Huynh, B.H., Münck, E.: Proc. Natl. Acad. Sci. USA **77**, 6574 (1980)
117. Kauffmann, K.E., Popescu, C.V., Dong, Y., Lipscomb, J.D., Que, L., Münck, E.: J. Am. Chem. Soc. **120**, 8739 (1998)
118. Sanakis, Y., Macedo, A.L., Moura, I., Moura, J.J.G., Papaefthymiou, V., Münck, E.: J. Am. Chem. Soc. **122**, 11855 (2000)
119. Yoo, S.J., Meyer, J., Achim, C., Peterson, J., Hendrich, M.P., Münck, E.: J. Biol. Inorg. Chem. **5**, 475 (2000)
120. Sanakis, Y., Power, P.P., Stubna, A., Münck, E.: Inorg. Chem. **41**, 2690 (2002)
121. Sanakis, Y., Yoo, S.J., Osterloh, F., Holm, R.H., Münck, E.: Inorg. Chem. **41**, 7081 (2002)
122. de Oliveira, F.T., Bominaar, E.L., Hirst, J., Fee, J.A., Münck, E.: J. Am. Chem. Soc. **126**, 5338 (2004)
123. Griffith, J.S.: The Theory of Transition-Metal Ions. Cambridge University Press, Cambridge (1961)
124. Sinnecker, S., Neese, F.: Top. Curr. Chem. **268**, 47 (2007)
125. Tregenna-Piggott, P.L.W., Andres, H.-P., McIntyre, G.J., Best, S.P., Wilson, C.C., Cowan, J. A.: Inorg. Chem. **42**, 1350 (2003)
126. Carver, G., Dobe, C., Jensen, T.B., Tregenna-Piggott, P.L.W., Janssen, S., Bill, E., McIntyre, G.J., Barra, A.L.: Inorg. Chem. **45**, 4695 (2006)

Chapter 5
Quantum Chemistry and Mössbauer Spectroscopy

Frank Neese* and Taras Petrenko**

5.1 Introduction

The underlying physics and analysis of Mössbauer spectra have been explained in detail in Chap. 4. In that chapter, the principles of how a spectrum is parameterized in terms of spin-Hamiltonian (SH) parameters and the physical origin of these SH parameters have been clarified. Many Mössbauer studies, mainly for ^{57}Fe, have been performed and there is a large body of experimental data concerning electric- and magnetic-hyperfine interactions that is accessible through the Mössbauer Effect Database.

Measured SH parameters are traditionally interpreted in terms of models that are derived from ligand field theory. Hence, attention is strongly focused on the absorber atom and its immediate environment is treated in a phenomenological way. Without any doubt, such ligand field models have been highly valuable. Many trends in experimental data can be understood on the basis of ligand field pictures. In particular, the changes in the Mössbauer parameters as a function of d^N electron configuration and distribution (e.g. as a function of oxidation- and spin-state) become intelligible.

However, useful as it is, ligand field theory is not a predictive first principles theory. Thus, it *cannot* be used to predict a priori the Mössbauer parameters of a given compound. Yet, the need to do so arises frequently in Mössbauer spectroscopy. For example, if a reaction intermediate or some other unstable chemical species has been characterized by freeze quench Mössbauer spectroscopy and its SH parameters become available, then the question arises as to the structure of the unstable species. Mössbauer spectroscopy in itself does not provide enough information to answer this question in a deductive way. However, the more modest question "which structures are compatible with the observed Mössbauer parameters" can be answered if one is able to reliably predict Mössbauer parameters

*Lehrstuhl für Theoretische Chemie, Universität Bonn, Germany: e-mail:neese@thch.uni-bonn.de
**Max-Planck Institut für Bioanorganische Chemie, Mülheim an der Ruhr, Germany

P. Gütlich et al., *Mössbauer Spectroscopy and Transition Metal Chemistry*,
DOI 10.1007/978-3-540-88428-6_5, © Springer-Verlag Berlin Heidelberg 2011

for arbitrary molecules. This is very valuable because under many, if not most, experimental circumstances there will be a finite number of "candidate structures" for the observed species.

There may, however, be a number of other reasons to pursue a predictive first principles theory of Mössbauer spectroscopy. For example, one may want to elucidate structure/spectroscopy correlations in the cleanest way. To this end one may construct in the computer a number of models with systematic variations in oxidation states, spin states, coordination numbers, and identity of ligands to name only a few "chemical degrees of freedom." In such studies it is immaterial whether these molecules have been made or could be made: what matters is that one can find out which structural details the Mössbauer parameters are most sensitive to. This can provide insight into the effects of geometry or covalency that are very difficult to obtain by any other means.

Fortunately, such first principles approaches to Mössbauer parameters became available in the late 1990s and during the 2000s. These approaches are deeply rooted in the methods of theoretical chemistry which in turn are based on the microscopic physics of the (relativistic) quantum mechanical N-electron problem. Clearly, the problem to solve the (relativistic) N-particle wave equation is a hopeless endeavor for any realistic system of practical relevance to Mössbauer spectroscopy. Hence, the present chapter is devoted to the description of approximate approaches. The underlying field of molecular quantum mechanics is a very large one and in a single chapter we can only provide a flavor of methods which are in large-scale use, their leading ideas, and how they may be applied to Mössbauer spectroscopy. Special emphasis will be given to density functional theory (DFT) since this is the methodology that has been most useful and is most widely used so far. Books [1, 2] and reviews [3–11] that describe the technical apparatus and an in-depth coverage of the underlying theory are available.

5.2 Electronic Structure Theory

5.2.1 The Molecular Schrödinger Equation

The problem is perhaps most transparently approached starting from the time-independent Born–Oppenheimer (BO) Schrödinger equation. This is the fundamental equation that describes the electronic structure of arbitrary molecules in the nonrelativistic limit. In this Hamiltonian, the nuclei are represented as positive point charges with charge Z. The only additional approximation is the "adiabatic" approximation which assumes the nuclei to be at rest for the sake of calculating the electronic wavefunction. Physically speaking, this means that the electrons are assumed to follow any nuclear motion instantaneously. If this is the case, the BO Schrödinger equation can be solved pointwise for each nuclear configuration. The nuclear degrees of freedom ($3M-6(5)$ for nonlinear (linear) molecules where M is the number of nuclei) then define a multidimensional "potential energy surface"

(PES), which is different for each electronic state of the system (i.e. each eigenfunction of the BO Schrödinger equation). Based on these PESs, the nuclear Schrödinger equation is solved to define, for example, the possible nuclear vibrational levels. This approach will be used below in the description of the nuclear inelastic scattering (NIS) method.

The BO Schrödinger equation reads:

$$H_{BO}\Psi(\mathbf{x}_1, ..., \mathbf{x}_N) = E\Psi(\mathbf{x}_1, ..., \mathbf{x}_N) \tag{5.1}$$

with

$$H_{BO} = V_{\mathrm{NN}} + T_{\mathrm{e}} + V_{\mathrm{eN}} + V_{\mathrm{ee}} \tag{5.2}$$

$$\begin{aligned} &= \sum_{A<B} \frac{Z_A Z_B}{|\mathbf{R}_A - \mathbf{R}_B|} - \frac{1}{2}\sum_i \nabla_i^2 - \sum_A \sum_i \frac{Z_A}{|\mathbf{R}_A - \mathbf{r}_i|} \\ &+ \frac{1}{2}\sum_i \sum_{j \neq i} \frac{1}{|\mathbf{r}_i - \mathbf{r}_j|}. \end{aligned} \tag{5.3}$$

In this equation, "atomic units" have been used in which $\hbar = 4\pi\varepsilon_0 = 1$ and consequently energy is measured in "Hartree" (1 Eh $=$ 27.2107 eV) and length in "Bohrs" (1 Bohr = 0.529. . .10^{-10} m). The solutions to the BO Schrödinger equation are the many-electron wavefunctions $\Psi_I(\mathbf{x}_1, ..., \mathbf{x}_N)$ (I = 0, 1, . . ., ∞) and their associated energies E_I. The square of the many-electron wavefunction provides the probability density for finding the electrons at positions $\mathbf{r}_1, ..., \mathbf{r}_N$ with spins $\sigma_1, ..., \sigma_N(\mathbf{x}_i = (\mathbf{r}_i, \sigma_i))$.

The individual terms in (5.2) and (5.3) represent the nuclear–nuclear repulsion, the electronic kinetic energy, the electron–nuclear attraction, and the electron–electron repulsion, respectively. Thus, the BO Hamiltonian is of treacherous simplicity: it merely contains the pairwise electrostatic interactions between the charged particles together with the kinetic energy of the electrons. Yet, the BO Hamiltonian provides a highly accurate description of molecules. Unless very heavy elements are involved, the exact solutions of the BO Hamiltonian allows for the prediction of molecular phenomena with "spectroscopic accuracy" that is defined by $\sim$1 cm^{-1} ($\sim$$10^{-6}$ Eh or $\sim$$10^{-4}$ eV). Unfortunately, exact solutions to the BO Schrödinger equations cannot be obtained, even for the simplest two-electron systems. Nevertheless, very accurate approximate solutions can be generated by the powerful apparatus of ab initio quantum chemistry for small molecules and these methods are computationally too demanding to become useful for Mössbauer (MB) spectroscopy in the foreseeable future.

5.2.2 *Hartree–Fock Theory*

The term that prevents an exact solution of the BO eigenvalue problem is the electron–electron repulsion that couples the motion of pairs of electrons. Without this term, the electronic motions would be independent and the Schrödinger equation would be solved by a simple product Ansatz:

$$\Psi_{product}(\mathbf{x}_1, ..., \mathbf{x}_N) = \psi_1(\mathbf{x}_1)....\psi_N(\mathbf{x}_N), \tag{5.4}$$

where the single-electron wavefunctions are called "orbitals" and would satisfy the one-electron Schrödinger equation:

$$h\psi_i(x) = \varepsilon_i\psi_i(x), \tag{5.5}$$

where the term $h = T_{\mathrm{e}} + V_{\mathrm{eN}}$ is the one-electron Hamiltonian. Obviously, this is a much more manageable problem than the N-particle problem itself and could be numerically solved with high-precision with modern computers. Unfortunately, the simple product wavefunction has the major flaw of not satisfying the Pauli principle, which states that the N-particle wavefunction must be antisymmetric with respect to particle interchanges, e.g. $\Psi(\mathbf{x}_1, ..., \mathbf{x}_i, ..., \mathbf{x}_j, ...\mathbf{x}_N) = -\Psi(\mathbf{x}_1, ..., \mathbf{x}_j, ..., \mathbf{x}_i, ...\mathbf{x}_N)$. However, this is easy to rectify by passing on to an antisymmetrized product wavefunction that is most elegantly written in terms of a "Slater determinant":

$$\Psi_{SD}(\mathbf{x}_1, ..., \mathbf{x}_N) = \frac{1}{\sqrt{N!}} \begin{vmatrix} \psi_1(\mathbf{x}_1) & \psi_2(\mathbf{x}_1) & \cdots & \psi_N(\mathbf{x}_1) \\ \psi_1(\mathbf{x}_2) & \psi_2(\mathbf{x}_2) & \cdots & \psi_N(\mathbf{x}_2) \\ \vdots & \vdots & \ddots & \vdots \\ \psi_1(\mathbf{x}_N) & \psi_2(\mathbf{x}_N) & \cdots & \psi_N(\mathbf{x}_N) \end{vmatrix}. \tag{5.6}$$

HF (HF) theory is based on the idea that one takes an antisymmetrized product wavefunction and uses the variational principle to obtain the best possible approximation to the N-particle wavefunction that *cannot* be represented by such a single determinant. Thus, one inserts the single determinant into the Rayleigh–Ritz functional and performs a constraint variation of the orbitals. The results of the variational process are the famous HF equations that are satisfied by each of the orbitals:

$$\hat{F}\psi_i(\mathbf{x}) = \varepsilon_i\psi_i(\mathbf{x}), \tag{5.7}$$

where the matrix elements of the Fock-operator between two arbitrary orbitals are:

$$\langle\psi_p|F|\psi_q\rangle = \langle\psi_p|h|\psi_q\rangle + \sum_i \langle\psi_i\psi_p||\psi_i\psi_q\rangle, \tag{5.8}$$

with the one- and two-electron integrals:

$$\langle\psi_i|h|\psi_i\rangle = \int \psi_i^*(x)\, h(x)\psi_i(x)\, \mathrm{d}x, \tag{5.9}$$

$$\left\langle \psi_i \psi_j || \psi_i \psi_j \right\rangle = \left\langle \psi_i \psi_j | \psi_i \psi_j \right\rangle - \left\langle \psi_i \psi_j | \psi_j \psi_i \right\rangle$$
$$= \int \int \psi_i^*(\mathbf{x}_1) \psi_j^*(\mathbf{x}_2) r_{12}^{-1} \left[\psi_i(\mathbf{x}_1) \psi_j(\mathbf{x}_2) - \psi_i(\mathbf{x}_2) \psi_j(\mathbf{x}_1) \right] dx_1 \, dx_2 \tag{5.10}$$

(we will also simply write: $\langle pq||rs \rangle = \langle \psi_p \psi_q || \psi_r \psi_s \rangle$). The total energy is then given by:

$$\langle \Psi_{SD} | H_{BO} | \Psi_{SD} \rangle = V_{NN} + \sum_i \langle \psi_i | h | \psi_i \rangle + \tfrac{1}{2} \sum_{i,j} \left\langle \psi_i \psi_j || \psi_i \psi_j \right\rangle. \tag{5.11}$$

Only at first glance do the HF equations represent single-particle equations – the Fock operator F depends, via the sum over occupied orbitals (second term), on its own eigenfunctions. Hence, unlike the original BO Schrödinger equation, the HF equations are highly nonlinear and need to be solved by iteration. Due to the nonlinearity, the total energy is *not* the sum of the orbital energies. Physically, HF theory is a mean-field theory that describes the motion of a single electron in the electrostatic field of the nuclei and the average field created by the remaining $N-1$ electrons. Note also that the HF orbitals are not unique: a unitary transformation among the occupied HF orbitals leaves the entire single-determinant HF wavefunction invariant up to a physically irrelevant phase factor. This freedom can be used to define different sets of "convenient" orbitals. Only the "canonical" HF orbitals have well-defined orbital energies as displayed in (5.7). There is no guarantee, of course, that the iterative HF process will converge and, furthermore, it is not even known how many solutions the HF equations may have.

In practice, unfortunately, not even the HF equations can be solved precisely due to the complicated shapes that the orbitals assume for general low-symmetry molecular environments. Hence, one introduces a fixed basis set $\{\varphi\}$ that is used to expand the HF orbitals:

$$\psi_i(\mathbf{x}) = \sum_{\mu} c_{\mu i} \varphi_{\mu}(\mathbf{x}). \tag{5.12}$$

If the basis set is mathematically complete, then the equation holds precisely. In practice, one has to work with an incomplete finite basis set and hence the equality is only approximate. Results close to the basis set limit (the exact HF solutions) can nowadays be found, but for all practical intents and purposes, one needs to live with a basis set incompleteness error that must be investigated numerically for specific applications.

The benefit is now that the HF equations are turned from complicated integro-differential equations into pseudo-eigenvalue equations for the unknown expansion coefficients $\boldsymbol{c}$.

$$\mathbf{F}\mathbf{c}_i = \varepsilon_i \mathbf{S}\mathbf{c}_i. \tag{5.13}$$

Here $\mathbf{c}_i$ is a vector that represents the ith column of the matrix $\boldsymbol{c}$. The matrix elements of the Fock (**F**) and overlap (**S**) matrices are given by:

$$F_{\mu\nu} = h_{\mu\nu} + \sum_{\kappa\tau} P_{\kappa\tau} \langle \mu\kappa \| \nu\tau \rangle, \tag{5.14}$$

$$S_{\mu\nu} = \langle \varphi_\mu | \varphi_\nu \rangle, \tag{5.15}$$

$$h_{\mu\nu} = \langle \varphi_\mu | h | \varphi_\nu \rangle, \tag{5.16}$$

$$P_{\mu\nu} = \sum_i c_{\mu i} c_{\nu i}. \tag{5.17}$$

5.2.3 Spin-Polarization and Total Spin

The basis functions are most commonly chosen such that the spin-function is either a pure spin-up function $\alpha(\sigma)$ or a pure spin-down function $\beta(\sigma)$. They are defined such that $\alpha(\frac{1}{2}) = \beta(-\frac{1}{2}) = 1$ and zero for any other argument σ. Since the BO and, consequently, the Fock operator do not contain any spin-dependent terms, the HF equations divide into spin-up and spin-down equations:

$$\mathbf{F}^\alpha \mathbf{c}_i^\alpha = \varepsilon_i^\alpha \mathbf{S} \mathbf{c}_i^\alpha \tag{5.18}$$

and

$$\mathbf{F}^\beta \mathbf{c}_i^\beta = \varepsilon_i^\beta \mathbf{S} \mathbf{c}_i^\beta, \tag{5.19}$$

where

$$F_{\mu\nu}^\alpha = h_{\mu\nu} + \sum_{\kappa\tau} (P_{\kappa\tau}^\alpha + P_{\kappa\tau}^\beta)(\mu\nu|\kappa\tau) - P_{\kappa\tau}^\alpha(\mu\kappa|\nu\tau). \tag{5.20}$$

An analogous equation holds for the spin-down Fock matrix. The two-electron integrals in round brackets are defined by "chemist's" (11|22) rather than the usual "physicist's" $\langle 12|12 \rangle$ notation as:

$$(\mu\nu|\kappa\tau) = \int \int \varphi_\mu(\mathbf{r}_1)\varphi_\nu(\mathbf{r}_1) r_{12}^{-1} \varphi_\kappa(\mathbf{r}_2)\varphi_\tau(\mathbf{r}_2) d\mathbf{r}_1 d\mathbf{r}_2. \tag{5.21}$$

The density matrices are defined as:

$$P_{\mu\nu}^\sigma = \sum_{i=1}^{\mathrm{N}_\sigma} c_{\mu i}^\sigma c_{\nu i}^\sigma, \tag{5.22}$$

where N_σ is the number of electrons with spin σ.The spin-up and spin-down Fock matrices differ only in the final term that represents the exchange contribution. This exchange contribution solely arises from the antisymmetry requirement of the total wavefunction and is entirely electrostatic in origin. This must be emphasized, since there is a widespread misconception about the existence of a "magnetic exchange interaction." There is no such force in nature: the "exchange splitting" only arises from antisymmetry in combination with the electron–electron repulsion. It is obvious that the exchange term exists only between electrons of the same spin. As a consequence, as soon as $N_\alpha \neq N_\beta$, the eigenfunctions of F^α and F^β will be different. In general, it can be shown that the exchange contribution to the orbital energy is always negative. Since there are, by convention, more spin-up than spin-down electrons (calculations are typically done for $M_S > 0$), the orbital energies of the spin-up electrons will be lower than those of the spin-down electrons – a phenomenon that is referred to as "spin-polarization." This can have fairly dramatic consequences. For example, in high-spin ferric compounds, the strong exchange stabilization drives the orbital energies of the spin-up metal *d*-based orbitals so far down that they usually fall in between or even below the ligand valence orbitals [12].

The method described previously is known as the "unrestricted" HF (UHF) method. It appears to be a logical choice. However, it does have one significant drawback: the total single-determinantal wavefunction:

$$\Psi_{\mathrm{UHF}} = \left| \psi_1^\alpha ... \psi_{N_\alpha}^\alpha \psi_1^\beta ... \psi_{N_\beta}^\beta \right| \tag{5.23}$$

violates one of the fundamental symmetries of the many-electron Schrödinger equation. Since the BO Hamiltonian does not contain spin-operators, it commutes with the total spin squared (a two-electron operator):

$$\hat{S}^2 = \hat{S}\hat{S} = \sum_{i,j} \hat{s}_i \hat{s}_j \tag{5.24}$$

as well as with its *z*-component (a one-electron operator):

$$\hat{S}_z = \sum_i \hat{s}_{iz}. \tag{5.25}$$

Since commuting operators have simultaneous eigenfunctions, it follows that correct eigenfunctions of the BO Hamiltonian must also be eigenfunctions of $\hat{S}^2$ and $\hat{S}_z$ with eigenvalues $S(S+1)$ and $M_S = S, S-1, ..., -S$. All $2S+1$ members of a given total spin-multiplet with spin-*S* with wavefunctions Ψ^{SM_S} are exactly degenerate at the level of the BO Hamiltonian. They are called the "magnetic sublevels" of the given state. This is not true, unfortunately, for Ψ_{UHF} , which is an eigenfunction of $\hat{S}_z$ with eigenvalue $M_S = \frac{1}{2}(N_\alpha - N_\beta)$ but *not* – in general – of $\hat{S}^2$. While this fact tends to be largely ignored by the solid state physics community, it may have some significant consequences and implications for calculations on open-shell *molecules* as will be described here.

It is possible to construct a HF method for open-shell molecules that does maintain the proper spin symmetry. It is known as the "restricted open-shell HF" (ROHF) method. Rather than dividing the electrons into spin-up and spin-down classes, the ROHF method partitions the electrons into closed- and open-shell. In the easiest case of the high-spin wavefunction ($n_{\mathrm{op}} = N_\alpha - N_\beta$ electrons in n_{op} orbitals coupled to a total spin of $S = \frac{1}{2}n_{\mathrm{op}}$), the ROHF wavefunction is a single-determinant of the form:

$$\Psi_{\mathrm{ROHF}} = \left|\psi_1\bar{\psi}_1 ... \psi_{n_{cl}}\bar{\psi}_{n_{cl}}\psi_{o1} ... \psi_{o_{nop}}\right|, \tag{5.26}$$

where the overbar denotes occupation with a spin-down electron and n_{cl} is the number of closed-shell orbitals. The energy of the ROHF wavefunction is always higher than that of the UHF, since it contains fewer variational parameters. This is one manifestation of the "symmetry dilemma": in variational calculations, the maintenance of the correct spin- and space-symmetries may come at the price of a higher energy. The ROHF method is quite flexible and also allows for the correct calculation of open-shell singlet states, multiplet states, and other wavefunctions that can be represented by a single configuration but not by a single determinant.

5.2.4 Electron Density and Spin-Density

Particular physical meaning is attached to the "density matrix" **P**. The electron density at any given point in space can be represented by:

$$\rho(\mathbf{r}) = \rho^\alpha(\mathbf{r}) + \rho^\beta(\mathbf{r}) = \sum_{\mu\nu} (P^\alpha_{\mu\nu} + P^\beta_{\mu\nu})\varphi_\mu(\mathbf{r})\varphi_\nu(\mathbf{r}) \tag{5.27}$$

and analogously, there exist a spin-density that is given by:

$$\rho^{\alpha-\beta}(\mathbf{r}) = \rho^\alpha(\mathbf{r}) - \rho^\beta(\mathbf{r}) = \sum_{\mu\nu} (P^\alpha_{\mu\nu} - P^\beta_{\mu\nu})\varphi_\mu(\mathbf{r})\varphi_\nu(\mathbf{r}). \tag{5.28}$$

Both objects are much less complicated than the total N-particle wavefunction itself, since they only depend on three spatial variables. The electron density is manifestly positive (or zero) everywhere in space while the spin-density can be positive or negative. If, by convention, there are more spin-up than spin-down electrons, the positive part of the spin-density will prevail and there will usually be only small regions of negative spin-density that arise from "spin-polarization." This spin-polarization is physically important and is already included in the UHF method but not in the ROHF method that, by construction, can only describe the positive part of the spin-density. This is one of the reasons for preferring the UHF over the ROHF method in calculations on open-shell systems. However, one should also notice that the "bare" UHF method describes spin-polarization poorly and grossly over-exaggerates its importance.

The electron- and spin-densities are the only building blocks of a much more powerful theory: the theory of reduced density matrices. Such one-particle, two-particle, ... electron- and spin-density matrices can be defined for any type of wavefunction, no matter whether it is of the HF type, another approximation, or even the exact wavefunction. A detailed description here would be inappropriate [13].

5.2.5 Post-Hartree–Fock Theory

Taken at face value, the HF method is surprisingly successful – typically around 99.8% of the exact nonrelativistic energy is recovered. Unfortunately, the remaining 0.2% of the missing energy is still a large number in the chemical context. In order to predict chemical phenomena accurately, one needs to reach an accuracy of $\sim$1 kcal mol^{-1}, which is $\sim 10^{-3}$ Eh. Given that the total energy of an ordinary iron complex is around 2,000 Eh, one concludes that the HF energy is in error by about 4 Eh or $\sim$2,500 kcal mol^{-1}. Thus, the errors are tremendous and unless they cancel to a very high degree of precision – which they usually do not – they are unacceptably large. In particular, the HF method has a built-in imbalance that has significant consequences in practical applications: the interaction between electrons of like spin and the interaction of electrons with opposite spin is not treated to the same level of precision. In fact, if two electrons have the same spin, the single determinantal form of the HF wavefunction already guarantees that these two electrons are never found in immediate vicinity. This is called the "Fermi hole." To the contrary, two electrons with opposite spin are allowed to come arbitrarily close in HF theory. Thus, HF theory has a tremendous preference for electrons of like spin. As a consequence, spin-polarization effects are strongly over-exaggerated in the HF method, high-spin states are artificially over-stabilized compared to low-spin states, and coordinative (and all other bonds) are predicted to be too polar. As a consequence, the spin-densities at the metal are predicted too high and those on the ligands are predicted too low – this can be clearly seen, for example, in calculated hyperfine parameters [14]. There are many other qualitative and quantitative failures of the HF method such as the underestimation of bond energies.

In principle, the deficiencies of HF theory can be overcome by so-called correlated wavefunction or "post-HF" methods. In the majority of the available methods, the wavefunction is expanded in terms of many Slater-determinants instead of just one. One systematic recipe to choose such determinants is to perform single-, double-, triple-, etc. substitutions of occupied HF orbitals by virtual orbitals. Pictorially speaking, the electron correlation is implemented in this way by allowing the electrons to "jump out" of the HF "sea" into the virtual space in order to avoid each other. The number of possible determinants grows very quickly, and only for the smallest few-electron systems can all of them be taken into account. In practical cases of interest to mainstream chemistry, one is limited to single- and double-substitutions with at most a rough estimate of triple-substitution effects. There are various flavors of estimating the importance of excited determinants: (a) many body perturbation theory (MBPT), (b) coupled-cluster theory (CC), (c) configuration interaction (CI), or (d) coupled-electron pair approaches (CEPA). Experience indicates that one can indeed obtain pleasingly accurate results for all observables simultaneously. Hence, these methods are systematic in approaching the exact solution of the nonrelativistic Schrödinger equation. They are also known to be computationally expensive and are often rejected from consideration for this reason. However, there are good reasons to believe that excellent approximations to the rigorous wavefunction based methods will become available for large-scale chemical use in the foreseeable future.

One specific problem becomes very acute in wavefunction based methods: the basis set problem. The introduction of a finite basis set is not highly problematic in HF theory since the results converge quickly to the basis set limit. This is, unfortunately, not true in post-HF theory where the results converge very slowly with basis set size – which is another reason why the methods become computationally intractable for more than a few heavy atoms (heavy being defined as nonhydrogen in this context). These problems are now understood and appropriate approaches have been defined to overcome the basis set problem but a detailed description is not appropriate here.

5.2.6 Density Functional Theory

A completely different route to the N-electron problem is provided by DFT. On an operational level it can be thought of as an attempt to improve on the HF method by including correlation effects into the self-consistent field procedure.

Formally, DFT rests on the "Hohenberg–Kohn" (HK) theorems which provided an elaborate justification for some largely intuitive methods that had been suggested very early in the history of quantum mechanics. Loosely speaking, the Hohenberg–Kohn theorems state that: (1) the total energy of the ground state of an N-electron system in some external potential V is a unique functional of the electron density ρ and (2) inserting a trial density $\tilde{\rho}$ into this unique and universal functional yields an energy that is greater than or equal to the exact ground state energy.

The existence of the first HK theorem is quite surprising since electron–electron repulsion is a two-electron phenomenon and the electron density depends only on one set of electronic coordinates. Unfortunately, the universal functional is unknown and a plethora of different forms have been suggested that have been inspired by model systems such as the uniform or weakly inhomogeneous electron gas, the helium atom, or simply in an ad hoc way. A recent review describes the major classes of presently used density functionals [10].

The connection to HF theory has been accomplished in a rather ingenious way by Kohn and Sham (KS) by referring to a fictitious reference system of noninteracting electrons. Such a system is evidently exactly described by a single Slater determinant but, in the KS method, is constrained to share the same electron density with the real interacting system. It is then straightforward to show that the orbitals of the fictitious system fulfil equations that very much resemble the HF equations:

$$\left\{h + \int \frac{\rho(\mathbf{r}')}{|\mathbf{r}-\mathbf{r}'|}\mathrm{d}\mathbf{r}' + V_{\mathrm{XC}}(\mathbf{r})\right\}\psi_i(\mathbf{x}) = \varepsilon_i\psi_i(\mathbf{x}). \tag{5.29}$$

The first term is the familiar one-electron operator, the second term represents the Coulomb potential, and the third term is called "exchange–correlation potential." HF and DFT differ only in this last term. In HF theory there is only a nonlocal exchange term, while in DFT the term is local and supposed to cover both exchange and correlation. It arises as a functional derivative with respect to the density:

$$V_{\mathrm{XC}}(\mathbf{r}) = \frac{\delta E_{\mathrm{XC}}[\rho]}{\delta\rho(\mathbf{r})}. \tag{5.30}$$

Here $E_{XC}[\rho]$ is the "magic" exchange–correlation functional that provides the correct sum of the exchange and correlation energies given an input $\rho(\mathbf{r})$ and the functional derivative can be thought of as measure of how strongly the exchange and correlation energies change as a result of an infinitesimal fluctuation of the density ρ at point $\mathbf{r}$.

There are several things known about the exact behavior of $V_{\mathrm{XC}}(\mathbf{r})$ and it should be noted that the presently used functionals violate many, if not most, of these conditions. Two of the most dramatic failures are: (a) in HF theory, the exchange terms exactly cancel the self-interaction of electrons contained in the Coulomb term. In exact DFT, this must also be so, but in approximate DFT, there is a sizeable self-repulsion error; (b) the correct KS potential must decay as $-1/r$ for long distances but in approximate DFT it does not, and it decays much too quickly. As a consequence, weak interactions are not well described by DFT and orbital energies are much too high (5–6 eV) compared to the exact values.

Nevertheless, DFT has been shown over the past two decades to be a fairly robust theory that can be implemented with high efficiency which almost always surpasses HF theory in accuracy. Very many chemical and spectroscopic problems have been successfully investigated with DFT. Many trends in experimental data can be successfully explained in a qualitative and often also quantitative way and therefore much insight arises from analyzing DFT results. Due to its favorable price/performance ratio, it dominates present day computational chemistry and it has dominated theoretical solid state physics for a long time even before DFT conquered chemistry. However, there are also known failures of DFT and in particular in spectroscopic applications one should be careful with putting unlimited trust in the results of DFT calculations.

5.2.7 Relativistic Effects

The parameters that Mössbauer spectroscopy is sensitive to depend on the immediate vicinity of the absorbing nucleus. However, for heavier elements, the electrons close to the nucleus move at high velocities, close to the speed of light. In this situation, relativistic effects become significant and should be treated at some level of detail.

In a rigorous treatment, one replaces the one-electron operator h by the four-component Dirac-operator h_D and perhaps supplement the two-electron operator by the Breit interaction term [15]. Great progress has been made in such four-component ab initio and DFT methods over the past decade. However, they are not yet used (or are not yet usable) in a routine way for larger molecules.

As is well known from fundamental relativistic quantum mechanics, the four-component Dirac spinor that describes a single relativistic electron (in place of the one-component spin-orbital used in the previous paragraphs) consists of two "small components" (positron-like) and two "large components" (electron-like). It seems natural to try to eliminate the small components and pass on to a two-component formalism. Such a reduction of complexity from four- to two-components can be achieved in a number of ways. The earliest is the straightforward Pauli expansion of the Dirac equation in powers of the *fine structure constant* (that, in atomic units, is equal to the inverse of the speed of light; $c \approx 137$ in atomic units). This expansion has fallen into disfavor since it leads to diverging potentials that are unstable in a variational context. More successful approaches are the zero'th order regular approximation (ZORA), the Douglas–Kroll–Hess (DKH) transformation, the "direct-perturbation theory" (DPT), or the "normalized elimination of the small component" (NESC). All these approaches lead to two-component equations that can be solved at 4–8 times the computational cost of a nonrelativistic approach.

However, there also exists a third possibility. By using a famous relation due to Dirac, the relativistic effects can be (in a nonunique way) divided into spin-independent and spin-dependent terms. The former are collectively called "scalar relativistic" effects and the latter are subsumed under the name "spin–orbit coupling" (SOC). The scalar relativistic effects can be straightforwardly included in the one-electron Hamiltonian operator h. Unless the investigated elements are very heavy, this recovers the major part of the distortion of the orbitals due to relativity. The SOC terms may be treated in a second step by perturbation theory. This is the preferred way of approaching molecular properties and only breaks down in the presence of very heavy elements or near degeneracy of the investigated electronic state.

The orbital distortions due to relativity can be pictured in a qualitative way as follows: the relativistic potential appears as a short-range attractive term that, above all, stabilizes s-orbitals and makes them more compact as well as steep close to the nucleus. The same happens to a lesser extent for the p-orbitals. Since the low-lying s- and p-orbitals are now closer to the nucleus they do shield the nuclear charge better than in the nonrelativistic case. Consequently, d- and f-orbitals are

Table 5.1 Effect of relativity on Hartree–Fock orbital energies (in eV) for the neutral Hg and Fe atoms. Scalar relativistic effects were treated with the DKH2 approximation

	$Hg^0(^1S)$			$Fe^0(^5D)$		
Orbital	DKH2 (eV)	Non-Rel. (eV)	Change (eV)	DKH2 (eV)	Non-Rel. (eV)	Change (eV)
1*s*	−83,321.2	−75,615.5	−7,705.7	−7,174.6	−7,112.2	−62.4
2*s*	−14,939.1	−12,813.8	−2,125.2	−882.5	−868.9	−13.6
2*p*	−12,959.0	−12,308.7	−650.3	−748.3	−745.9	−2.4
3*s*	−3,618.7	−3,082.5	−536.1	−115.5	−113.4	−2.1
3*p*	−3,032.5	−2,842.9	−189.6	−74.9	−74.5	−0.4
3*d*	−2,381.0	−2,401.3	20.3	−17.1	−17.5	0.4
4*s*	−8,34,3	−697.5	−136.7	−7.1	−7.0	−0.1
4*p*	−637.0	−591.6	−45.4			
4*d*	−391.2	−398.0	6.7			
5*s*	−138.9	−113.9	−25.0			
4*f*	−119.7	−136.5	16.8			
5*p*	−83.0	−77.5	−5.5			
5*d*	−16.3	−19.1	2.8			
6*s*	−8.9	−7.0	−1.9			

destabilized, move to higher energies and also become more diffuse. Hence, one has to expect significant effects of these distortions on the electric- and magnetic-hyperfine parameters, as described here.

This effect is illustrated in Table 5.1 for the neutral Hg and Fe atoms. It becomes evident that for Hg the changes due to relativity are very large. The innermost orbital energies change by several thousand eV and even the valence orbitals are shifted relative to each other by several electron volts. Hence, structure, bonding, and spectroscopy of such heavy elements cannot be understood without recourse to relativity. The situation is somewhat different for the first transition row. Here, the relativistic effects are much more limited. The valence orbital energies change by no more than a few tenths of an electron volt and even in the core the only reasonably large effects are observed for the 1*s* and 2*s* orbitals. It is clear that for highly accurate calculations it will be necessary to treat the relativistic effects but their impact on the structure and bonding of first-row transition metal complexes is limited.

Nevertheless, Mössbauer spectroscopy is sensitive to the core region of the absorber and therefore, it is of interest to investigate how the radial expectation values are influenced by relativity. This will be further elaborated here.

5.2.8 Linear Response and Molecular Properties

Given the result of an HF or DFT calculation, one aims at calculating molecular properties such as spectroscopic observables. The presumably best way to think about such properties is to regard the spectroscopic experiment as a perturbation of the molecular electronic structure and then calculate the limited changes that occur in the orbitals and the total energy in the presence of the perturbation. Since the perturbations are small, a Taylor expansion is appropriate and it can be truncated after the linear term. In this way, the majority of properties can be represented by first- or mixed second-derivative with respect to one- or two-perturbating operators. For variational methods such as HF or DFT, the first derivative (linear response) coincides with the expectation value that is familiar from standard perturbation theory. However, two perturbations are treated in linear response theory by a mixed second derivative, while in perturbation theory they would require an infinite sum over excited many electron states that is impossible to perform in practice. For exact wavefunctions the two formulations are equivalent, but for approximate electronic structure methods they are not. Since only the linear response approach is computationally feasible, it is the preferred route to molecular properties in actual applications. Mathematical details are given in Appendix 1 (Part III, 3 of CD-ROM).

5.3 Mössbauer Properties from Density Functional Theory

5.3.1 Isomer Shifts

5.3.1.1 Calibration Approach

From the discussion in Sect. 4.2 in Chap. 4, it became evident that the isomer shift is linearly related to the electron density at the Mössbauer absorbing nucleus. Thus, one is tempted to write:

$$\delta_{MB} = a + b[\rho(0) - c], \tag{5.31}$$

where a and b are fit parameters to be determined by linear regression and c is a number merely introduced for convenience. This equation implies that the only quantity that must be known in order to predict the isomer shift is the electron density at the nucleus (assumed to be at the origin $\mathbf{0}$). In the sense of Appendix 1 (see CD-ROM, Part III), this can be regarded as a first-order property and is readily calculated by DFT as:

$$\rho(0) = \sum_{\mu\nu} P_{\mu\nu} \langle \mu | \delta(\mathbf{r} - \mathbf{0}) | \nu \rangle = \sum_{\mu\nu} P_{\mu\nu} \varphi_\mu(\mathbf{0}) \varphi_\nu(\mathbf{0}). \tag{5.32}$$

What is then missing are merely the numerical values of the constants a and b (c is of no physical relevance). These values are best obtained from a linear regression analysis of calculated electron densities versus observed isomer shifts. Important early work along these lines was reported by Nieuwport who has already demonstrated the potential of this approach [16]. More recently, several workers have

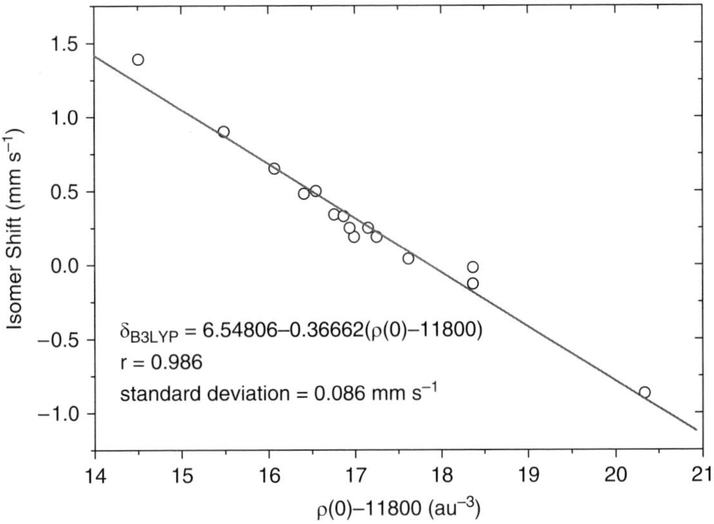

Fig. 5.1 Linear regression analysis between calculated electron densities at the iron nucleus and measured isomer shifts for a collection of iron complexes. (From [19])

provided plots of the calculated electron density at the iron nucleus versus the experimentally obtained isomer shift for a range of complexes [17–26].

Figure 5.1 shows the results that are usually obtained in such a regression analysis.

It becomes evident that the correlation between theory and experiment is very good and hence that isomer shift predictions on the basis of the simple regression analysis will be successful. These calibration curves have already seen dozens, if not hundreds, of successful applications (e.g. [8, 18–23, 25–65]). Importantly, the experience gained from these applications indicates that the quality of the calibration does not depend on the charge-state of the iron centers, not on their spin state, not on their coordination number or the nature of ligands or whether the iron is involved in spin-coupling or not. Thus, these calculations, despite their simplicity, are successful and robust. All of these plots show that very good linearity is obtained which allows the prediction of isomer shifts with an uncertainty that is smaller than 0.1 mm s^{-1}. Some workers have preferred to construct plots for a limited set of iron oxidation states and coordination environments while others have argued in favor of a single, unique calibration.

5.3.1.2 An Example

As an example of the successful application of an isomer shift calculation, the discovery of genuine Fe(V) may be quoted [47]. Upon irradiation of the crystallographically and spectroscopically well-characterized low-spin Fe(III)-azide precursor [Fe(cyclam-acetate)N_3]$^+$, a species is formed in nearly quantitative yield that displays an altered Mössbauer spectrum (Fig. 5.2). The isomer shift changes from 0.271 to −0.017 mm s^{-1} and the quadrupole splitting from 2.301 to 1.602 mm s^{-1}. Given the well-known photochemical reactivity of transition metal azides, it was hoped that the newly generated species actually represented the desired unique Fe(V)-nitrido species [Fe(cyclam-acetate)N]$^+$. The observed isomer shift is broadly consistent with this expectation, but certainly not proof of such a formulation. Consequently, quantum chemical calculations were carried out for [Fe(cyclam-acetate)N]$^+$. In analogy to the isoeletronic Cr(III) species as well as earlier observations [66], a $S = 3/2$ ground state was assumed. However, the calculated isomer shift of this species was found to be much higher than the experimentally observed one and, at the same time, the calculated bond distance (R(Fe–N) = 1.738 Å) did not match the one derived for the photoproduct from EXAFS studies (a scatterer at 1.61 Å was observed). Initially, these two findings appeared to contradict the assignment of the photoproduct to a genuine Fe(V) species. Consequently various decay products were examined theoretically, including hydrogen atom abstraction products such as [Fe(cyclam-acetate)NH]$^+$. However, none of the investigated species matched the experimental Mössbauer parameters. Finally, the alternative spin state of [Fe(cyclam-acetate)N]$^+$ – namely $S = 1/2$ – was investigated. The calculations on this species immediately yield a

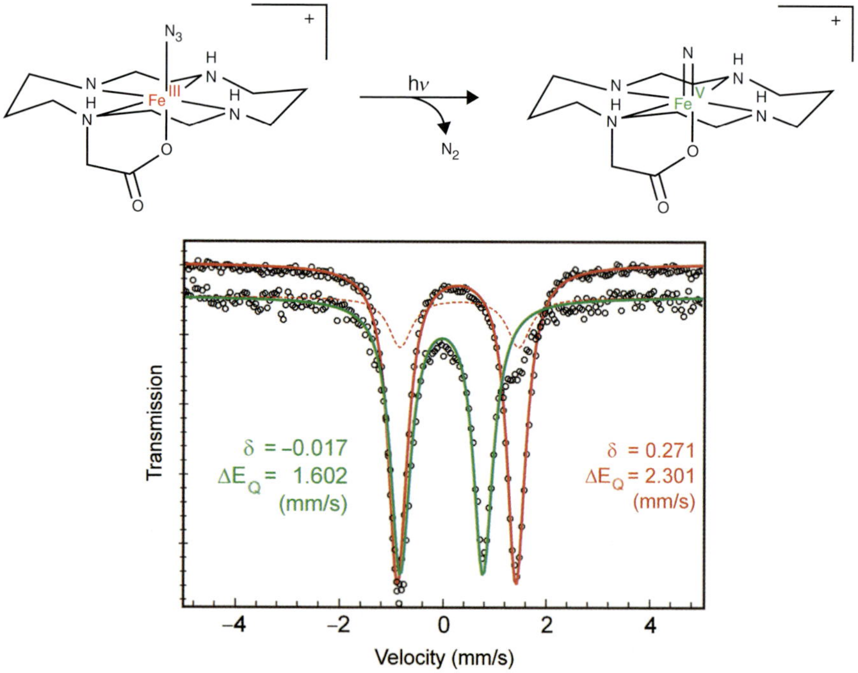

Fig. 5.2 *Top*: Photochemical production of Fe(V) from a ferric–azide precursor. *Bottom*: Mössbauer spectra observed for the precursor (*green*) and the photoproduct (*red*)

predicted bond distance of R(Fe–N) = 1.609 Å as well as a calculated isomer shift of −0.033 mm s^{-1}. Since these values were in excellent agreement with experiment, the magnetic susceptibility of the photoproduct was investigated. The analysis unambiguously confirmed that the photoproduct had a ground state spin of S = 1/2 and, consequently, it was assigned with confidence to [Fe(cyclam-acetate) N]$^+$, S = 1/2. This study therefore serves well as an example how quantum chemistry and experiment can go hand in hand in the characterization of short lived species with Mössbauer spectroscopy. A very similar approach was subsequently taken in the combined experimental and theoretical characterization of the first genuine Fe(VI) complex other than FeO_4^{2-} [56].

5.3.1.3 Advanced Considerations

Despite the simplicity and the outstanding success of the simple approach to the isomer shift, there are a number of limitations from the theoretical point of view that should be mentioned:

(a) Since the electron density shows a cusp at the nucleus, it appears to be necessary to carry out calculations with basis functions that show the correct behavior at the nuclei. This is not the case for the typically employed Gaussian basis sets that decay too slowly for small distances (and too fast for large distances). Consequently, the basis set limit, known from numerical HF calculations to be $\rho(0)$ ~11,903.987 a.u.$^{-3}$ [16] for the neutral iron atom, is difficult to reach with Gaussian basis sets. With good bases (say 17–19 uncontracted s-primitives [19]), one reaches electron densities at the nucleus that are around 11,820 a.u.$^{-3}$ (cf. Table 5.3). Obviously, the percentage error is small but the absolute error of several dozen a.u.$^{-3}$ is large compared to the limited variation of the electron density over the chemical range of Fe(VI) to Fe(I) compounds that amount to only ~10 a.u.$^{-3}$.

(b) Secondly, the relativistic effects on the electron density at the nucleus are very large for iron. Proper account for relativistic effects shifts $\rho(0)$ to around 15,704 a.u.$^{-3}$ [67]. Thus, if one pursues a nonrelativistic treatment the electron densities from DFT (or HF) calculations are off by thousands of a.u.$^{-3}$ while one interprets changes on the order of a fraction of 1 a.u.$^{-3}$. Simply including a relativistic potential in the BO Hamiltonian is not a satisfactory solution to the problem [25] because, with a point nucleus, the relativistic orbitals (and also the quasi-relativistic one- or two-component orbitals) are known to diverge in the basis limit [15, 68]. Thus, in order to obtain a systematically correct relativistic electron density at the iron nucleus one needs to resort to a finite-nucleus model.

Based on these two comments, one might be tempted to conclude that the calculation of Mössbauer isomer shifts is a very involved subject where accuracy is difficult to achieve. The reason why this is not the case is revealed in Fig. 5.3, which analyzes the contributions of the iron 1s, 2s, 3s, and 4s (valence shell) to the

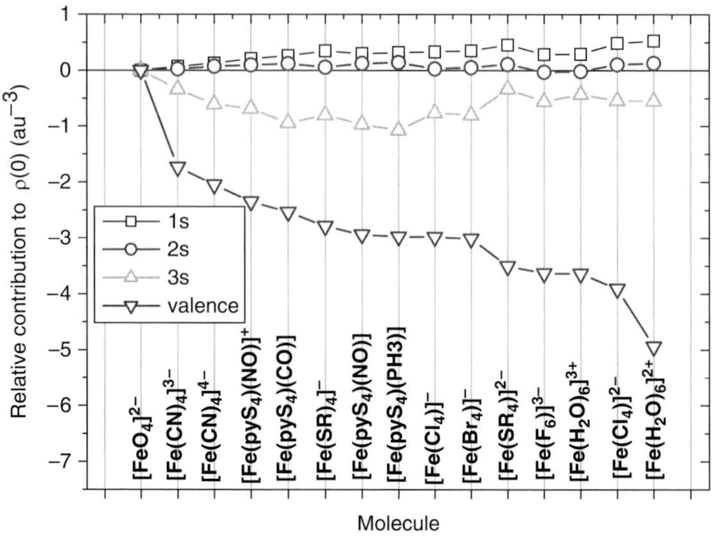

Fig. 5.3 Variation of the 1*s*, 2*s*, 3*s* and 4*s* contributions to the total electron density at the iron nucleus for a collection of iron complexes. Nonrelativistic B3LYP DFT calculations with the CP (PPP) basis set (taken from [19])

variation in the electron density at the nucleus from nonrelativistic calculations [19]. It is clear that by far the largest contributions to the electron density of ~11,800 come from the Fe-1*s* and Fe-2*s* orbitals. The 1*s* and 2*s* contributions are to an excellent approximation constant in molecules and practically all variation occurs in the semicore 3*s* and valence 4*s* shells. In particular, the 4*s* contributions are mainly influenced by the effects of bonding and these in turn are very well described by DFT. The changes in the 3*s* contributions to the electron density at the iron nucleus are strongly correlated with the corresponding changes in the 4*s* shell. This becomes understandable from the correlation of the shape of the 4*s* radial function with metal–ligand bond distance in conjunction with the orthogonality requirement as discussed further here. In conclusion, even the nonrelativistically calculated ρ(0) values accurately follow the chemical variations and provide a reliable tool for Mössbauer property predictions.

There is a technical point that deserves a closer discussion. This concerns the assignment of electron densities at the nucleus to the 1*s* through 4*s* shells. First, it is realized that for a single-determinantal wavefunction such as HF or KS DFT, the electron density can be written in two alternative forms:

$$\rho(\mathbf{0}) = \sum_{\mu\nu} P_{\mu\nu}\varphi_\mu(\mathbf{0})\varphi_\nu(\mathbf{0}), \tag{5.33}$$

$$= \sum_i n_i|\psi_i(\mathbf{0})|, \tag{5.34}$$

where $\mathbf{P}$ is the density matrix, $\{\varphi\}$ basis functions, n_i occupation numbers, and $\{\psi\}$ molecular orbitals. Equation (5.33) emphasizes the relationship between the electron density at the nucleus and the basis functions, while (5.34) emphasizes the relationship between the same quantity and the actual molecular orbitals. In the early days of molecular orbital calculations it was customary to employ so-called "minimal basis sets" that contained just one basis function for each orbital that is occupied in the free atom. In this case, the basis functions were required to resemble the shapes of the atomic orbitals. They were frequently represented by single Slater-type orbitals. In modern quantum chemistry, Gaussian orbitals are prevalent because of the ease with which the numerous molecular integrals can be calculated. Since individual Gaussian functions do not resemble atomic orbitals, it is necessary to take linear combinations of Gaussian functions to represent atomic or molecular orbitals. In general about six to eight Gaussian functions are necessary to represent the shape of an atomic orbital to high-precision. In most calculations performed today, the core orbitals are treated by fixed (e.g., preoptimized) linear combinations of Gaussian functions (so-called "contracted Gaussians") since the shape of the core orbitals does not change much upon bond formation. In this case, there is, again, a clean relationship between basis functions and atomic orbital contributions. However, in Mössbauer isomer shift calculations, it is the small changes of the radial functions of all s-orbitals that is of interest. Hence, it is not a good idea to employ contracted Gaussian basis functions. Rather, the most accurate approach is to use a sufficiently large number of uncontracted Gaussian basis functions and let the variational principle decide in which proportions they are combined to give the final molecular orbitals. Hence, in the present calculations where 17 primitive Gaussian of varying steepness have been used, the s-part of each and every molecular orbital that carries some iron s-character is of the form:

$$\psi_{i(s)}^{(\mathrm{Fe})}(\mathbf{0}) = \sum_{\mu=1}^{17} c_{\mu i}\varphi_{\mu}^{(\mathrm{Fe})}(\mathbf{0}) + \cdots. \tag{5.35}$$

In general, all 17 s-primitives contribute to each s-derived molecular orbital. Obviously, the tighter Gaussians will contribute more strongly to the inner-shell molecular orbitals and the more diffuse Gaussians to the valence s-orbitals. Nevertheless, it is impossible (and also not desired) to make a connection between basis functions and atomic orbitals.

However, the division of the electron density at the iron nucleus into contributions arising from $1s$ through $4s$ contributions can be done conveniently at the level of the canonical molecular orbitals. This arises because the iron $1s$, $2s$, and $3s$ orbitals fall into an orbital energy range where they are well isolated and hence do not mix with any ligand orbital. Hence, the $1s$, $2s$, and $3s$ contributions are well defined in this way. The $4s$ contribution then arises typically from several, if not many, molecular orbitals in the valence region that have contributions from the iron s-orbitals. Thus, the difference between the total electron density at the nucleus and the 1*s* through 3*s* contributions defines the valence part that is attributed to the iron 4*s* orbital.

To illustrate this point, the contributions of the occupied molecular orbitals to the total electron density at the nucleus are summarized in Table 5.2 for FeF_4^- ($S = 5/2$). It is evident from the table that the contributions coming from the orbitals at −6,966 eV must be assigned to the iron 1*s* orbital, those from orbitals at −816 eV to the iron 2*s* orbital, and those from orbitals at −95 eV to the iron 3*s* orbital. In this highly symmetric complex, only two valence orbitals contribute to $\rho(0)$, i.e. the −25 eV contribution from the totally symmetric ligand-group orbital that is derived from the *F* 2*s* orbitals and the −7 eV contribution from the totally symmetric

Table 5.2 Contributions to the electron density at the iron nucleus in FeF_4^- ($S = 5/2$) from spin-unrestricted B3LYP calculations with the CP(PPP) basis set. Orbitals that do contribute to the electron density at the iron nucleus are printed in boldface

Spin-up		Spin-down		Comment
Energy (eV)	$\|\psi_i(0)\|^2$	Energy (eV)	$\|\psi_i(0)\|^2$	
−6,966.2103	**5,349.614063**	**−6,966.1982**	**5,349.671994**	**Fe 1*s***
−816.4314	**486.021456**	**−813.8562**	**488.038191**	**Fe 2*s***
−703.5877	0.000000	−701.4629	0.000000	Fe 2*p*
−703.5877	0.000000	−701.4629	0.000000	Fe 2*p*
−703.5877	0.000000	−701.4629	0.000000	Fe 2*p*
−665.2459	0.000000	−665.0455	0.000000	F 1*s*
−665.2457	0.000000	−665.0453	0.000000	F 1*s*
−665.2456	0.000000	−665.0452	0.000000	F 1*s*
−665.2455	0.000000	−665.0451	0.000000	F 1*s*
−95.2432	**70.316125**	**−89.2464**	**68.995052**	**Fe 3*s***
−61.9102	0.000000	−55.5992	0.000000	Fe 3*p*
−61.9102	0.000000	−55.5992	0.000000	Fe 3*p*
−61.9102	0.000000	−55.5992	0.000000	Fe 3*p*
−25.6376	**0.316163**	**−25.1722**	**0.326837**	**F 2*s*+Fe 4*s***
−25.4722	0.000000	−24.9772	0.000000	F 2*s*
−25.4721	0.000000	−24.9771	0.000000	F 2*s*
−25.4721	0.000000	−24.9770	0.000000	F 2*s*
−8.9134	**0.000000**	**−6.6509**	**1.626746**	**F 2*p*+Fe 4*s***
−8.9134	0.000000	−6.3594	0.000000	F 2*p*
−8.9134	0.000000	−6.3594	0.000000	F 2*p*
−8.4408	0.000000	−6.3594	0.000000	F 2*p*
−8.4407	0.000000	−5.8465	0.000000	F 2*p*
−7.3264	**1.674431**	**−5.8464**	**0.000000**	**F 2*p*+Fe 4*s***
−6.1703	0.000000	−5.5892	0.000000	F 2*p*
−6.1702	−0.000000	−5.5891	0.000000	F 2*p*
−6.1701	0.000000	−5.5891	0.000000	F 2*p*
−5.3100	0.000000	−4.8180	0.000000	F 2*p*
−5.3099	0.000000	−4.8179	0.000000	F 2*p*
−5.3099	0.000000	−4.8179	0.000000	F 2*p*
−4.7665	−0.000000			e-shell Fe 3*d*+F 2*p*
−4.7664	0.000000			e-shell Fe 3*d*+F 2*p*
−3.9865	0.000000			t_2-shell Fe 3*d*+F 2*p*
−3.9864	−0.000000			t_2-shell Fe 3*d*+F 2*p*
−3.9864	0.000000			t_2-shell Fe 3*d*+F 2*p*

combination of *F* 2*p* orbitals. In larger iron complexes of low symmetry, however, the valence contribution is often "smeared" over many valence orbitals. Nevertheless, even then the 1*s*, 2*s*, and 3*s* contributions remain well defined and have been used to construct Fig. 5.3.

Numerically, the contributions to the electron density at the nucleus are further illustrated in Table 5.3, where the contributions of the 1*s*, 2*s*, 3*s*, and 4*s* orbitals to the total electron density at the iron nucleus from nonrelativistic HF calculations are listed. The early relativistic calculations by Trautwein and coworkers show the same trends [67]. It is clear from the table that the 1*s* and 2*s* contributions are numerically dominant but are not sensitive to the charge state or the electronic configuration at the iron center. Hence, essentially all variation comes from the 3s semicore and the 4*s* valence orbitals. Evidently, the 3*s* contribution reacts as sensitively as the 4*s* contribution. The traditional interpretation is that this is due to the increasing nuclear attraction with decreasing Fe 3d population (Table 5.2). However, the distortion of the radial functions upon bond formation is also certainly a major factor and the computational data does not allow for an unambiguous disentanglement. The total spread of the variation is about 15 a.u. which is small relative to the total electron density at the iron nucleus. The changes in the 4*s* orbital with respect to the oxidation state of the iron atom can be nicely visualized from HF calculations as seen in Fig. 5.4. It is obvious how the 4*s* orbital becomes more and more diffuse as the number of *d*-electrons increases. Hence, the nodes are "driven" outwards due to the increased shielding of the nuclear charge by the 3*d* electrons and consequently the electron density contribution of the 4*s* orbitals at the nucleus decreases. Analogous considerations apply to the 3*s* orbital. It should be noted, however, that the situation in atoms/ions and molecules is rather different. In the calculations on the free atoms, the changes of the 3*s* contribution to the density at the iron nucleus are as large as those arising from the 4*s* shell. In molecules, the variations of the 4*s* shell contributions dominate over the 3*s* shell variations. This once more shows the irrelevance of calculations on highly charged, spherically symmetric, and nonbonded ions for making quantitative arguments about iron in various oxidation states in molecules. This subject will be more carefully discussed in relation to the electric field gradient (EFG) tensor here.

If one pursues the calibration approach, one has to stick to a given combination of density functional and basis set, since the calibration will change for each such combination. Calibration curves have been reported for a number of widely used density functionals and basis sets. The results of a relatively comprehensive study are collected in Table 5.4. The standard deviation of the best fits is on the order of 0.08 mm s^{-1} which appears to be the intrinsic reliability of DFT for predicting Mössbauer isomer shifts.

The important points to appreciate are:

(a) The DFT potential provided by the standard functionals is not only wrong in the long range, but is also in error close to the nucleus. Since the different functionals differ in this respect, the absolute values of the electron density differ strongly.

Table 5.3 Contributions of *s*-orbitals to the total electron density at the iron nucleus (in a.u.$^{-3}$) as a function of oxidation state and configuration. Calculations were done with the spin-averaged Hartree–Fock method and a large uncontracted Gaussian basis set. (17*s* 11*p* 5*d* 1*f*)

	$Fe^{(0)}$			Fe^{1+}			Fe^{2+}			Fe^{3+}		
	s^2d^6	s^1d^7	s^0d^8	s^2d^5	s^1d^6	s^0d^7	s^2d^4	s^1d^5	s^0d^6	s^2d^3	s^1d^4	s^0d^5
1*s*	10,689.72	10,690.01	10,690.18	10,689.37	10,689.77	10,690.02	10,688.86	10,689.43	10,689.79	10,688.24	10,688.92	10,689.45
2*s*	981.99	982.04	982.19	981.99	981.92	982.00	982.19	981.86	981.85	982.63	982.06	981.80
3*s*	134.80	132.63	131.46	137.35	134.37	132.53	141.09	136.82	134.24	145.65	140.65	136.81
4*s*	6.12	2.05	–	9.55	4.09	–	12.44	5.51	–	15.68	6.89	–

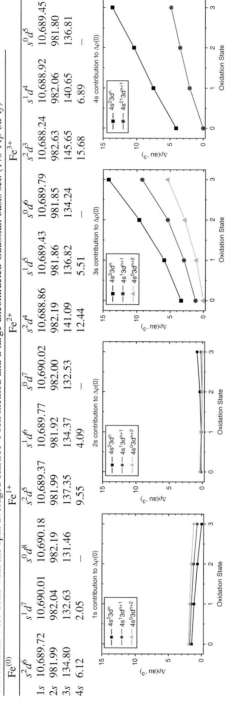

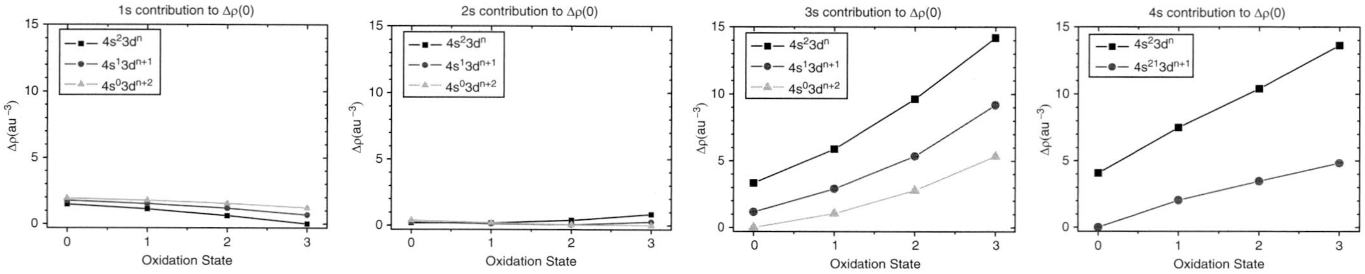

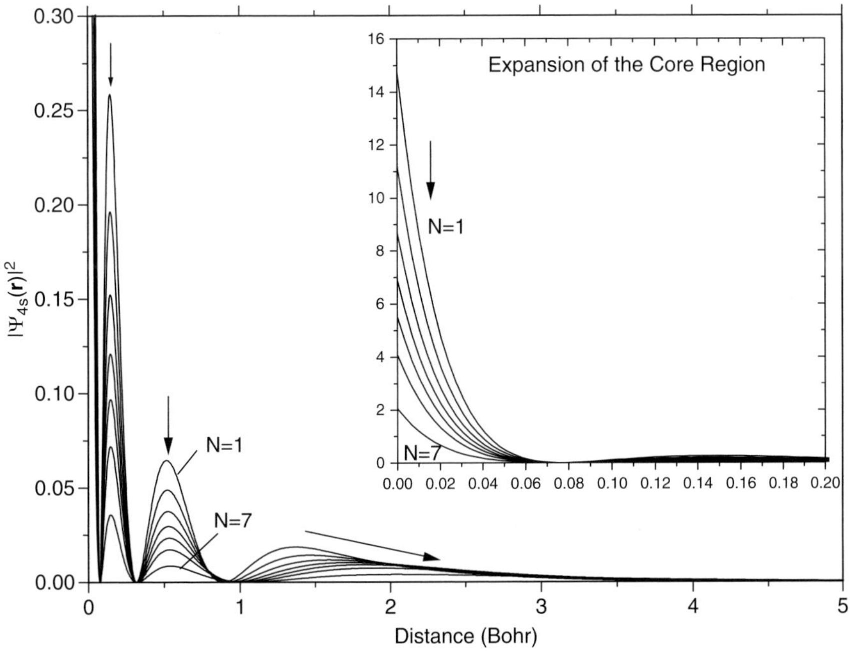

Fig. 5.4 Changes in the iron 4*s* orbital with the number of *d*-electrons as predicted by Hartree–Fock calculations for the configuration $3d^N4s^1$ ($N = 1-7$). (Taken from [19])

Table 5.4 Linear fit data for ^{57}Fe Mössbauer isomer shift predictions using the linear equation $\delta = b(\rho - c) + a$. A collection of 21 iron complexes with varying charge, oxidation- and spin-states have been studied (taken from [11])

Functional	Basis set	*a*	$b^{(1)}$	*C*	R^2	Std.dev (mm s^{-1})
BP86	CP(PPP)	7.916	−0.425	11,810	0.95	0.11
	TZVP	1.034	−0.340	11,580	0.93	0.12
	TZVP[a]	4.957	−0.362	13,800	0.92	0.12
B3LYP	CP(PPP)	2.852	−0.366	11,810	0.98	0.09
	TZVP	1.118	−0.298	11,580	0.94	0.11
	TZVP[a]	4.045	−0.307	13,770	0.97	0.08
TPSS	CP(PPP)	5.154	−0.421	11,810	0.96	0.10
	TZVP	1.327	−0.336	11,580	0.90	0.14
	TZVP[a]	1.385	−0.365	13,800	0.94	0.10
TPSSh	CP(PPP)	4.130	−0.376	11,810	0.97	0.08
	TZVP	1.466	−0.321	11,580	0.96	0.10
	TZVP[a]	1.830	−0.322	13,780	0.97	0.08
B2PLYP	CP(PPP)	2.642	−0.336	11,810	0.97	0.09
	TZVP	1.483	−0.261	11,580	0.84	0.12
	TZVP[a]	2.256	−0.311	13,790	0.97	0.08

[(1)]Based on the results of Trautwein and coworkers [67], the calibration constant carries a factor of 1.3 for compensating for the neglect of relativistic effects

(b) Different basis sets approach the basis set limit to a different extent and, consequently, the calibration only holds for the special basis set with which the calibration was performed. Despite claims to the contrary [21], one obtains better results if one provides additional flexibility in the *s*-part of the basis set in order to allow the core-orbitals to properly distort in the molecular environment [19].

(c) The slope one obtains from the linear regression varies with functional and basis set between −0.261 and −0.421. Relativity contributes a factor 1.3 to the nonrelativistic calibration constant. This brings the DFT-calculated calibrations to slopes in the range −0.200 and −0.323. Experiments were designed to measure directly the change of nuclear radius $\Delta\langle r^2\rangle$ during gamma resonance absorption, with $b = \beta\,\Delta\langle r^2\rangle$ ($\beta = 6.079\,Z / E_\gamma$ in fm^{-2} a.u.^{3} mm s^{-1}; $Z = 26$ and $E_\gamma = 14.4$ keV for ^{57}Fe). Such experiments have been performed on the basis of lifetime variations in the electron capture decay of ^{57}Fe yielding $-\Delta\langle r^2\rangle = (33 \pm 3)\ 10^{-3}\ \mathrm{fm}^2$ [69] and on the basis of conversion, electron spectroscopy with ^{57}Co(^{57}Fe) sources yielding $-\Delta\langle r^2\rangle < 9 \cdot 10^{-3}\ \mathrm{fm}^2$ [70]. These values correspond to an "experimentally" determined range of slopes from −0.100 to −0.395. Thus, the corresponding DFT-calculated calibrations listed previously with slopes varying between −0.200 and −0.323 are a very successful addition to the interpretation of isomer shifts, irrespective of their semiempirical nature.

5.3.1.4 Linear Response Treatment

More recently, Filatov has developed a linear response theory for the isomer shift and used it in conjunction with ab initio methods [71–73]. In many respects, this theory is more rigorous than the calibration approach described earlier. Hence, it will be briefly outlined here. Filatov considered a linear response treatment in which the perturbation was taken as the radial expansion of a finite sized nucleus. The resulting equation for the isomer shift is:

$$\delta = \frac{c}{E_\gamma}\left(\left.\frac{\partial E^{(absorber)}}{\partial R_N}\right|_{R=R_N} - \left.\frac{\partial E^{(standard)}}{\partial R_N}\right|_{R=R_N}\right)\Delta R, \tag{5.36}$$

where c is the speed of light and E_γ the energy of the γ-photon. ΔR is the expansion of the nucleus. The relative isomer shift then follows from:

$$\delta_{absober} - \delta_{standard} = -0.1573\left[\mathrm{mm\,s^{-1}\,a.u.^3}\right]\left\{\bar{\rho}_{\mathrm{e}}^{(absorber)} - \bar{\rho}_{\mathrm{e}}^{(standard)}\right\}, \tag{5.37}$$

where the "standard" is (presumably) a neutral iron atom thought to represent iron foil. The average density across the nuclear volume has been calculated according to:

$$\bar{\rho}_{\mathrm{e}} = \frac{5}{4\pi Z R_{\mathrm{N}}} \frac{\partial E}{\partial R_{\mathrm{N}}}\bigg|_{R=R_{\mathrm{N}}}. \tag{5.38}$$

The charge distribution of the nucleus was taken to be spherically symmetric and represented by a single Gaussian of the form:

$$\rho_{\mathrm{N}}(\mathbf{r}) = \frac{Z}{\pi^{3/2}}\left(\frac{3}{2\langle R^2\rangle}\right)\exp\left(-\frac{3}{2}\frac{r^2}{\langle R^2\rangle}\right), \tag{5.39}$$

where Z is the nuclear charge and $\langle R^2\rangle^{1/2}$ is the root-mean-square nuclear radius that defines the volume of the nucleus. This is the actual quantity with respect to which the derivative of the energy is taken. The normalized elimination of the small component has been taken in scalar relativistic one-component form as a relativistic method in which the scheme is employed.

This theory appears not to involve adjustable parameters (other than the nuclear radius parameters that were taken from the literature). In particular, it was criticized that the calibration approach involved a slope that is too high by about a factor of two. However, in actual calculations with the linear response approach, it was found that the slope of the correlation line between theory and experiment (dependent on the quantum chemical method) is close to 0.5. Thus, it also requires a scaling factor of about 2 in order to reach quantitative agreement with experiment. The standard deviations between the calibration and linear response approaches are comparable thus indicating that the major error in both approaches still stems from errors in the description of the bonding that is responsible for the actual valence shell electron distribution.

5.3.1.5 Solid State and Semiempirical Methods

In molecular DFT calculations, it is natural to include all electrons in the calculations and hence no further subtleties than the ones described arise in the calculation of the isomer shift. However, there are situations where other approaches are advantageous. The most prominent situation is met in the case of solids. Here, it is difficult to capture the effects of an infinite system with a finite size cluster model and one should resort to dedicated solid state techniques. It appears that very efficient solid state DFT implementations are possible on the basis of plane wave basis sets. However, it is difficult to describe the core region with plane wave basis sets. Hence, the core electrons need to be replaced by pseudopotentials, which precludes a direct calculation of the electron density at the Mössbauer absorber atom. However, there are workarounds and the subtleties involved in this subject are discussed in a complementary chapter by Blaha (see CD-ROM, Part III).

A second situation where core electrons may be avoided is in the field of semiempirical quantum chemistry. In these approaches, the HF or KS operators are replaced by simple parameterized expressions that are very fast to evaluate. Typically, semiempirical calculations only treat valence electrons in a minimal basis set. Consequently, such calculations are about three orders of magnitude faster than genuine HF or DFT calculations. This opens the way for calculations on very large molecules or extended molecular dynamics studies that would be unaffordable on the basis of genuine first-principles approaches. Much early work on the semiempirical calculation of Mössbauer parameters has been done by Trautwein and coworkers. The complementary chapter by Grodzicki describes the semiempirical approach to Mössbauer parameters (see CD-ROM, Part III).

5.3.1.6 Interpretation of the Isomer Shift

Given the success of DFT in the calculation of the isomer shift, it seems appropriate to return to the issue of interpretation: which factors are controlling the qualitative behavior of the isomer shift in iron compounds? Traditionally, one assumes that there is a correlation of the isomer shift and the charge at the iron center as is suggested from the well-known sensitivity of the isomer shift with respect to the oxidation state. However, things turn out to be more subtle than what is perhaps commonly perceived.

The traditional interpretation of the isomer shift in ^{57}Fe Mössbauer spectra is based on the following assumptions: (a) the influence of the 3*d* electron configuration on the IS occurs via the shielding effect of the 3*d* electrons on the 3*s* and 4*s* electrons, (b) the variations in the 3*s* shell are dominant, and (c) the influence of the 4*s* shell occurs via the 4*s* population.

The findings of [19] are summarized as follows:

(a) The trend of the ^{57}Fe isomer shifts along the series of molecules studied here is dominated to ~70% by the valence contribution with the remaining ~30% being mainly contributed by changes in the 3*s* shell (see Fig. 5.3)
(b) Within the valence contribution there is no discernible correlation with the 4*s* Löwdin population. This is a strong warning against any conclusion about the isomer shift that is derived from inspection of the charge distribution delivered by a quantum chemical calculation. A calculation of the electron density at the nucleus is unavoidable if useful predictions are to be achieved (Fig. 5.5)
(c) The valence contribution shows sizeable contributions from covalency effects, bondlength (overlap) changes, and changes in the shielding due to 3*d* electrons in a fairly subtle way. All of these effects make significant contributions to the changes in the electron density at the nucleus. Thus, a careful consideration of all these factors is necessary for a detailed understanding of isomer shifts. Explanations that only include parts of these considerations may work for a

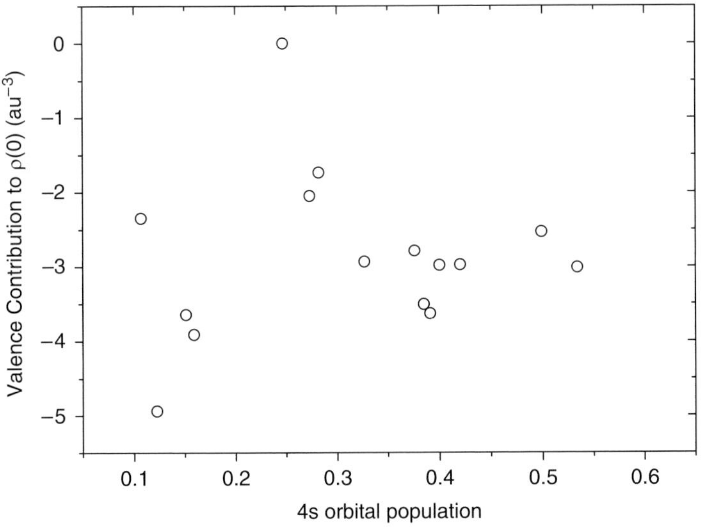

Fig. 5.5 Lack of correlation between the 4*s* orbital population and the valence contribution to the isomer shift (taken from [19])

series of closely related complexes but probably have a rather limited range of applicability

Perhaps the most decisive factor for the isomer shift is the metal ligand bond distance. It is this distance that governs the distortion of the radial wavefunctions of the absorbing atom. This distortion is propagated from the valence region into the core and determines the contribution of the valence orbital to the electron density at the nucleus. The underlying *s*-orbitals simply react to these changes due to orthogonality requirements. This can be seen in Fig. 5.6. Upon oxidation from the Fe(II) to the Fe(III) state, the Fe–F distance in $[FeF_4]^{n-}$decreases. Hence, the nodes of the 4*s* orbital are "driven inwards" with the consequence that the electron density contribution at the nucleus increases and the isomer shift decreases accordingly.

Finally, even if the importance of the 3*d* shielding is pronounced (see Fig. 5.6), it must be clearly recognized that between the valencies of Fe(VI) and Fe(II) the effective *d*-population does not change by more than ~1 electron. Arguments that are based solely on the ionic crystal field picture should therefore be viewed with utmost caution. In general, the metal–ligand bondlength is perhaps the most important factor. Thus, low-spin and high-valent complexes will tend to have short bonds and more negatively shifted ISs, while high-spin complexes will tend to show longer bondlengths and more positively shifted ISs. These correlations also help to understand the low isomer shifts commonly observed with strongly back-bonding ligands such as CO and NO^+ which also display low-spin states and short iron–ligand bondlengths.

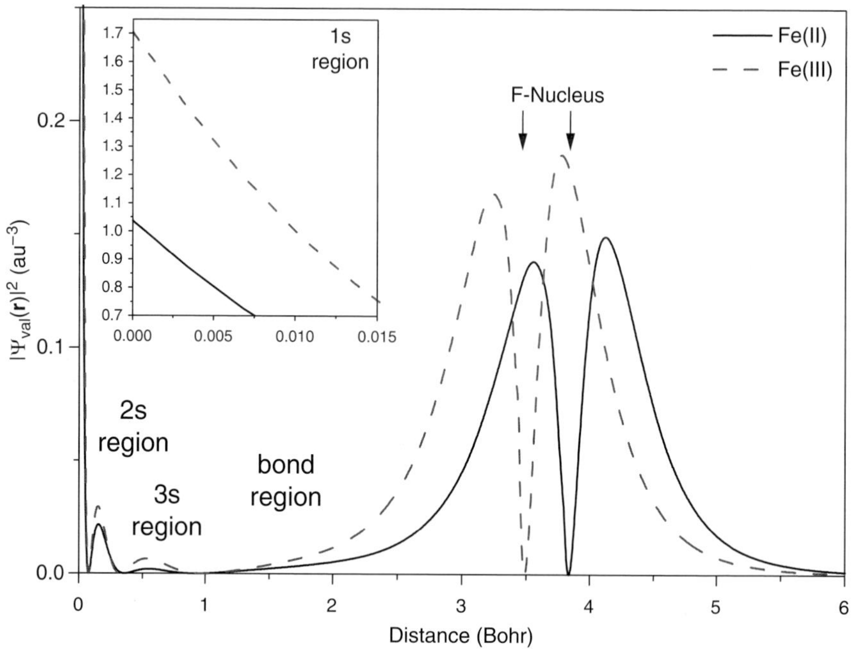

Fig. 5.6 Changes in the shape of the valence 4*s* contribution due to geometric and electronic relaxation in $[FeF_4]^{-,2-}$. *Full line*: $[FeF_4]^{2-}$ at its equilibrium geometry, *dashed line*: $[FeF_4]^{-}$ at its equilibrium geometry. The square of the valence orbital that mainly contributes to $\rho(0)$ along the Fe–F bond (distances are in units of the Bohr radius) is also drawn (from [19])

5.3.2 *Quadrupole Splittings*

From a theoretical point of view, the calculation of the quadrupole splitting is relatively straightforward since it can be calculated directly from the elements of the EFG tensor at the iron nucleus (nucleus "*A*") as:

$$V_{KL}^{(A)} = -\sum_{\mu\nu} P_{\mu\nu} \langle \mu | r_A^{-5} \left(r_A^2 \delta_{KL} - 3 r_{A;K} r_{A;L} \right) | \nu \rangle + \sum_{B \neq A} Z_B R_{AB}^{-5} \left(R_{AB}^2 \delta_{KL} - 3 R_{AB;K} R_{AB;L} \right). \tag{5.40}$$

Here $\mathbf{r}_A = \mathbf{r} - \mathbf{R}_A$ is the electronic position relative to nucleus "*A*" (magnitude $r_A = |\mathbf{r}_A|$, component $r_{A,K}$ K = x,y,z), and $\mathbf{R}_{AB} = \mathbf{R}_A - \mathbf{R}_B$ is the distance between nuclei "*A*" and "*B*."

Once available, and supplemented by the nuclear contribution, the EFG tensor can be diagonalized. The numerically largest element V_{max} (in atomic units) defines the value of q which is in turn used to calculate the quadrupole splitting parameter as $e^2qQ = 235.28\,V_{\max}Q$ where Q is the quadrupole moment of the nucleus in barn. Transformed to its eigensystem the quadrupole splitting enters the nuclear Hamiltonian in the following form [74]:

$$\hat{H}_Q = \hat{I}\mathbf{Q}\hat{I} = \frac{e^2qQ}{4I(2I-1)}\hat{I}\begin{pmatrix} -(1-\eta) & 0 & 0 \\ 0 & -(1+\eta) & 0 \\ 0 & 0 & 2 \end{pmatrix}\hat{I}. \tag{5.41}$$

The asymmetry parameter η is defined as:

$$\eta = \frac{\left|V_{yy} - V_{xx}\right|}{V_{zz}}. \tag{5.42}$$

The quadrupole splitting is then given by:

$$\Delta E_Q = \frac{1}{2}eQV_{zz}\sqrt{1+\frac{1}{3}\eta^2}. \tag{5.43}$$

V_{xx}, V_{yy}, and V_{zz} are the principal components of the EFG tensors in a coordinate system with $|V_{zz}| \geq |V_{yy}| \geq |V_{xx}|$, e is the positive elementary charge, and $Q(^{57}\text{Fe})$ is the nuclear quadrupole moment of the I = 3/2 excited state of ^{57}Fe (measured in barn).

5.3.2.1 Correlation with Experiment

A number of workers have reported calibration studies for iron quadrupole splittings in heme and nonheme iron complexes [23, 26, 31, 40, 41, 45, 75–77]. A typical study by Sinnecker et al. has investigated the quadrupole splitting of ten iron complexes of varying charge- and spin-state [25]. In addition, both nonrelativistic as well as scalar-relativistic calculations with the ZORA Hamiltonian were performed using the "gold standard" B3LYP functional. All the investigated model systems show noncubic ligand and valence electron distributions. In consequence, nonzero ^{57}Fe EFGs were obtained throughout. Linear fits of these data were used to redetermine the ^{57}Fe nuclear quadrupole coupling constant. From the nonrelativistic B3LYP and ZORA-B3LYP calculations a similar value of $Q(^{57}\text{Fe})$ of ~0.16 barn was obtained. This is in good agreement with the values that were reported and used in other studies (0.15–0.17 barn) [78]. A comparison of calculated and measured quadrupole splittings is given in Fig. 5.7. Correlation coefficients of 0.97 (B3LYP) and 0.95 (ZORA-B3LYP) were obtained, while the RMSD values amount to 0.57 mm s^{-1} (B3LYP) and 0.55 mm s^{-1} (ZORA-B3LYP). The high level of correlation for all model systems gives us confidence that the different oxidation states, charges, and ligand types have only a minor influence on the quality of the calculated data. Furthermore, the results clearly show that a

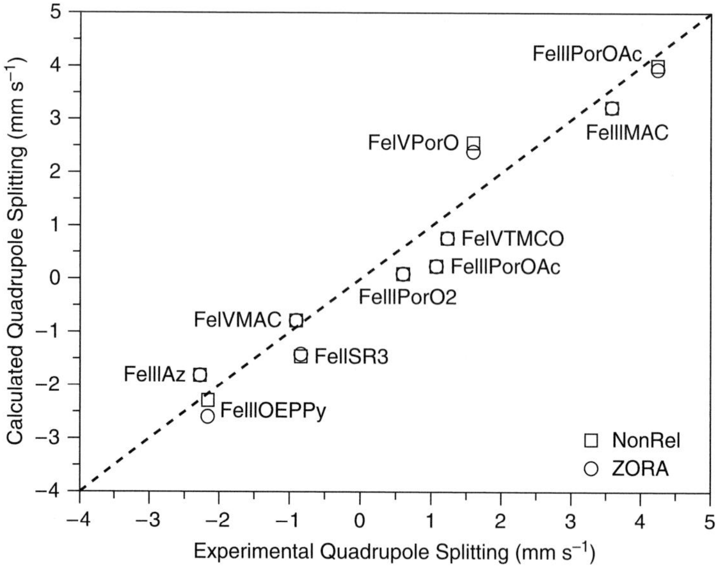

Fig. 5.7 Comparison of calculated and measured ^{57}Fe quadrupole splittings. Nuclear quadrupole moments of 0.158 barn (nonrelativistic DFT) and 0.156 barn (ZORA) were used in the calculations (taken from [25])

nonrelativistic treatment of the EFGs is satisfactory for ^{57}Fe despite the $\langle r^{-3} \rangle$ dependence of V_{zz} which would potentially seem to make the relativistic corrections (that are largest in the core region) mandatory. The significant RMSD values of more than 0.5 mm s^{-1} are also typical of DFT calculations of quadrupole splittings – even errors of up to 1 mm s^{-1} are not uncommon. Other workers have made similar observations although occasionally better correlation for more restricted sets of molecules (e.g. low-spin Fe(II)-porphyrins) have been reported. The achievable accuracy is significantly worse than that which is typically obtained for the isomer shift. Thus, the DFT calculations are apparently unable to detect the more subtle asymmetry in the charge distribution as accurately as the variation of the charge density at the iron nucleus. Consequently, there is considerable room for improvement of the theoretical predictions.

5.3.2.2 Physical Interpretation of the Electric Field Gradient Tensor

By recognizing that the basis functions are attached to parent atoms, the EFG tensor can be analyzed in a transparent way by breaking it down into components that represent multicenter interactions:

$$V_{KL}^{(A)} = V_{KL}^{(A);1c} + V_{KL}^{(A);1c-pc} + V_{KL}^{(A);1c-\text{bond}} + V_{KL}^{(A);3c} \tag{5.44}$$

with the individual components being given by:

$$V_{KL}^{(A);1c} = -\sum_{\mu\in A}\sum_{\nu\in A} P_{\mu_A\nu_A}\left\langle\mu_A\left|r_A^{-5}\left(r_A^2\delta_{KL} - 3r_{A;K}r_{A;L}\right)\right|\nu_A\right\rangle, \tag{5.45}$$

$$\begin{aligned} V_{KL}^{(A)2c-pc} = &-\sum_{\mu\in B\neq \grave{A}}\sum_{\nu\in B\neq \grave{A}} P_{\mu_B\nu_B}\left\langle\mu_B\left|r_A^{-5}\left(r_A^2\delta_{KL} - 3r_{A;K}r_{A;L}\right)\right|\nu_B\right\rangle \\ &+ \sum_{B\neq \grave{A}} Z_B R_{AB}^{-5}\left(R_{AB}^2\delta_{KL} - 3R_{AB;K}R_{AB;L}\right), \end{aligned} \tag{5.46}$$

$$V_{KL}^{(A);2c-\mathrm{bond}} = -2\sum_{\mu\in B\neq A}\sum_{\nu\in A} P_{\mu_B\nu_A}\left\langle\mu_B\left|r_A^{-5}\left(r_A^2\delta_{KL} - 3r_{A;K}r_{A;L}\right)\right|\nu_A\right\rangle, \tag{5.47}$$

$$V_{KL}^{(A);3c} = -\sum_{\mu\in B\neq A}\sum_{\nu\in C\neq A,B} P_{\mu_B\nu_C}\left\langle\mu_B\left|r_A^{-5}\left(r_A^2\delta_{KL} - 3r_{A;K}r_{A;L}\right)\right|\nu_C\right\rangle. \tag{5.48}$$

One Center Contributions

The one-center terms $V_{KL}^{(A);1c}$ represent the contributions of the iron electrons to the EFG tensor. These are the ones usually held responsible for the EFG tensor since the integrals should decay fairly quickly owing to the r_A^{-3} dependence. In HF or DFT theory, the one-center terms can be further analyzed by making use of the fact that the density is a simple function of the molecular orbitals $\hat{P} = \sum_i |i\rangle\langle i|$. By dividing the iron orbitals into core- and valence-orbitals, one obtains:

$$V_{KL}^{(A);1c-core} = -\sum_{i\in core}\left\langle i^{(A)}\left|r_A^{-5}\left(r_A^2\delta_{KL} - 3r_{A;K}r_{A;L}\right)\right|i^{(A)}\right\rangle, \tag{5.49}$$

$$V_{KL}^{(A);1c-val} = -\sum_{i\in valence}\left\langle i^{(A)}\left|r_A^{-5}\left(r_A^2\delta_{KL} - 3r_{A;K}r_{A;L}\right)\right|i^{(A)}\right\rangle, \tag{5.50}$$

where $\left|i^{(A)}\right\rangle$ is the part of the ith molecular orbital that is centred on atom "A." In this way one obtains insight into the part of the EFG that is created by the asymmetric electron distribution of the valence shell of the iron (the part covered by ligand field theory to some extent) and the part that arises from the small polarization of the inner shells (particularly the $3p$ shell).

One-Center Core Polarization

The core polarization may be considered to be not particularly large. However, the inner shell electrons are closer to the iron nucleus and, hence, the integrals are much larger than those in the valence shell. For example, the $\langle r^{-3}\rangle$ value is 461.7 a.u.$^{-3}$ for the 2p shell, 54.4 a.u.$^{-3}$ for the 3p shell, and 4.9 a.u.$^{-3}$ for the 3d shell of the neutral iron atom. Hence, a minimal polarization of the inner shells may introduce a sizeable EFG.

One Center Valence Contributions

For the one-center valence contribution, there are essentially three factors that control its value: (a) the radial wavefunction of the 3d orbitals, (b) the covalent "dilution" of the 3d orbitals with ligand orbitals, and (c) the occupation pattern of the 3d shell. An additional factor may be low-symmetry induced 3d/4p mixing. We will focus on the first three factors here.

A. Radial wavefunction. With respect to the radial wavefunction, a pronounced dependence of the $\langle r^{-3}\rangle_{3d}$ value on the oxidation state and configuration of the central iron may be expected. HF values for this matrix element for free ions are collected in Table 5.5. Roughly speaking, the $\langle r^{-3}\rangle_{3d}$ value increases by 0.5–0.6 a.u.$^{-3}$ for each unit increase in oxidation state and decreases by 0.5–0.6 a.u.$^{-3}$ for each additional d-electron.

However, one should view these matrix elements with great caution when it comes to their application within actual iron complexes. There are two main reasons for advising caution.

First, these extreme charges that are present in the free ions are not realistic for iron in any kind of molecular compound. In fact, the electrostatic forces that hold together molecules are so strong that all atoms in all molecules are reasonably close to neutral. Hence, even if the formal oxidation state is Fe(VI), the iron is very far from feeling a genuine Fe^{6+} potential. Perhaps, charges of around +1 or −1 are the most extreme kind of polarity that survives after bond formation. Although every quantum chemical program prints partial charges on individual atoms, it should be clearly recognized that these charges are not quantum mechanical observables. They therefore represent elements of "chemical poetry" that are very useful for investigating trends in bonding for related compounds but have no absolute and fundamental meaning in terms of many particle quantum mechanics. This does *not* imply that the concept of an oxidation state is not a highly useful one: it is a very powerful ordering principle for chemistry and gives sense and meaning to great many chemical and spectroscopic facts. One is well advised not to attach absolute physical meaning to the concept as such, in particular when it comes to actual numbers.

Table 5.5 $\langle r^{-3}\rangle_{3d}$ values for some iron ions in different electronic configurations (in a.u.$^{-3}$)

	$3d^n3s^2$	$3d^{n+1}4s^1$	$3d^{n+2}4s^0$
Fe^0	4.91	4.42	3.95
Fe^{1+}	5.51	4.97	4.44
Fe^{2+}	6.08	5.61	5.00
Fe^{3+}	6.68	6.20	5.63

The second reason for advising caution is that bond formation is associated with great changes in the radial wavefunctions of the constituent atoms. The pioneering works of Rüdenberg have demonstrated that, as a consequence of the virial theorem, atomic orbitals tend to contract in bonding molecular orbitals and expand in antibonding molecular orbitals. An example has been formulated for the case of $FeCl_4^-$ [79]. In Fig. 5.8, the radial wavefunctions of the *d*-based orbitals are compared to the corresponding radial functions obtained for the free ions. It is clear that the molecular radial wavefunctions for the ferric complex $FeCl_4^-$ are roughly intermediate between those for Fe^{1+} and Fe^{2+} but nowhere close to the radial wavefunction of Fe^{3+} as might have been expected. Furthermore, the expansion is anisotropic and is stronger for the more antibonding t_2 orbitals as compared to the essentially nonbonding *e*-orbital of the tetrahedral complex. Consequently, realistic values for $\langle r^{-3} \rangle_{3d}$ are never those of the free ion of the same charge as the formal oxidation state but will usually be somewhere between Fe^0 and, at most, Fe^{2+}. The anisotropy in the $\langle r^{-3} \rangle_{3d}$ value was found to be limited (~3%) in [79].

B. Covalent dilution. Given the caution with respect to the $\langle r^{-3} \rangle_{3d}$ values, we now turn to the subject of covalent dilution. In the simplest possible way, one can assume that only one *d*-orbital contributes to each *d*-based molecular orbital. If we further neglect overlap (which is a drastic oversimplification), then each of the metal *d*-based MOs is of the form:

$$\psi_i^{(\mathrm{Fe-3}d)} \approx \alpha_i |d_i\rangle - \sqrt{1-\alpha_i^2}\,|L_i\rangle, \tag{5.51}$$

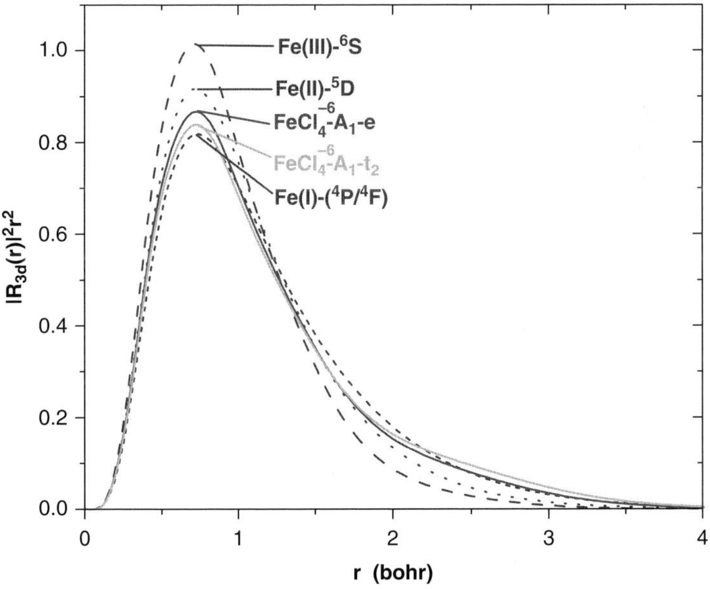

Fig. 5.8 Molecular radial wavefunctions for the ferric complex $FeCl_4^-$ compared to the radial wavefunctions of the free ions Fe^{1+}, Fe^{2+}, and Fe^{3+} (taken from [79])

where $|d_i\rangle$ is the appropriate d-orbital and $|L_i\rangle$ a ligand group orbital. If one neglects the multicenter integrals, then each of the matrix elements of the field gradient operator assumes the form:

$$\left\langle \psi_i^{(\mathrm{Fe}-3d)} \middle| \hat{f}_{KL} \middle| \psi_j^{(\mathrm{Fe}-3d)} \right\rangle \approx \left\langle r^{-3} \right\rangle_{3d} \alpha_i \alpha_j g_{ij}^{KL} \tag{5.52}$$

(g_{ij}^{KL} is the angular integral explained here). Thus, there is an anisotropic covalent reduction of the matrix elements that is proportional to the ligand admixture into the metal orbitals. This in turn depends on the electronegativity difference between the metal center and the ligands. Typically, soft anionic ligands like thiolates bind more covalently to the metal than hard neutral ligands like H_2O. Secondly, the bonding becomes more covalent for the higher oxidation states of the metal. As has been stressed in the works of Solomon and coworkers [80, 81], the covalency of metal–ligand bonds plays a decisive role not only for the spectroscopic properties of transition metal complexes but also for the redox properties and their reactivity. It should, however, be equally clear from the discussion above that the anisotropic covalency is only one of the many factors that contribute to quadrupole splitting (and isomer shift).

C. Occupation pattern. Finally, the third important factor – and the one that is often assumed to be the only important factor – is the occupation pattern of the d-orbitals. This concerns the angular integrals g_{ij}^{KL} in (5.52). Their values are well known and listed in many textbooks on Mössbauer or EPR spectroscopy and here in Table 5.6. Using these tables, one simply multiplies g_{ii}^{KL} with the occupation number of the orbital in question and the value $\left\langle r^{-3} \right\rangle_{3d} \alpha_i^2$ and sums over all d-based orbitals. The values α_i^2 could come from some kind of MO calculation or qualitative considerations. It should, however, be clear from the discussion above that the treatment is so strongly oversimplified that great accuracy cannot be expected. An example is provided here.

Two Center Point-Charge Contributions

The two-center "point-charge" contribution, $V_{KL}^{(A)2c-pc}$, is also intuitively appealing. It contains an electronic and a nuclear part. In the event that the basis functions on atom "B" can be compressed to delta functions (or if B is sufficiently far from "A"), the electronic integrals become:

$$V_{KL}^{(A)2c-pc} \approx \sum_{B \neq \grave{A}} \left(Z_B - \sum_{\mu \in B \neq \grave{A}} \sum_{\nu \in B \neq \grave{A}} P_{\mu_B \nu_B} \right) R_{AB}^{-5} \left(R_{AB}^2 \delta_{KL} - 3 R_{AB;K} R_{AB;L} \right). \tag{5.53}$$

By identifying the first term in round brackets with the charge of atom "B" – which is plausible given the assumption that the sum over density matrix

Table 5.6 Values for the angular integrals g_{ij}^{KL} used in the qualitative theory of the quadrupole interaction

	d_{z^2}	d_{xz}	d_{yz}	$d_{x^2-y^2}$	d_{xy}
xx					
d_{z^2}	$-2/7$			$-{}^2\!/_7\sqrt{3}$	
d_{xz}		$2/7$			
d_{yz}			$-4/7$		
$d_{x^2-y^2}$	$-{}^2\!/_7\sqrt{3}$			$2/7$	
d_{xy}					$2/7$
yy					
d_{z^2}	$-2/7$			${}^2\!/_7\sqrt{3}$	
d_{xz}		$-4/7$			
d_{yz}			$2/7$		
$d_{x^2-y^2}$	${}^2\!/_7\sqrt{3}$			$2/7$	
d_{xy}					$2/7$
zz					
d_{z^2}	$4/7$				
d_{xz}		$2/7$			
d_{yz}			$2/7$		
$d_{x^2-y^2}$				$-4/7$	
d_{xy}					$-4/7$
xy					
d_{z^2}					$-{}^2\!/_7\sqrt{3}$
d_{xz}			$3/7$		
d_{yz}		$3/7$			
$d_{x^2-y^2}$					
d_{xy}	$-{}^2\!/_7\sqrt{3}$				
xz					
d_{z^2}		${}^1\!/_7\sqrt{3}$			
d_{xz}	${}^1\!/_7\sqrt{3}$			$3/7$	
d_{yz}					$3/7$
$d_{x^2-y^2}$		$3/7$			
d_{xy}			$3/7$		
yz					
d_{z^2}			${}^1\!/_7\sqrt{3}$		
d_{xz}					$3/7$
d_{yz}	${}^1\!/_7\sqrt{3}$			$-3/7$	
$d_{x^2-y^2}$			$-3/7$		
d_{xy}		$3/7$			

elements roughly yields the number of electrons cantered on atom "B" – one obtains:

$$V_{KL}^{(A)2c-pc} \approx \sum_{B \neq A} Q_B R_{AB}^{-5} \left(R_{AB}^2 \delta_{KL} - 3R_{AB;K} R_{AB;L} \right). \tag{5.54}$$

This is known as the "lattice" contribution to the EFG.

Two-Center-Bond Contributions

Proceeding along the same lines, the other two terms can be transparently interpreted as well. The "2-center-bond" contribution, $V_{KL}^{(A);2c-\text{bond}}$, represents the contribution of the immediately bonding electrons (e.g. those cantered in the iron–ligand bonds) to the EFG tensor. It may be expected to become large when the metal–ligand bonds are particularly short and covalent and the ligand field is particularly anisotropic.

Three Center Contributions

Finally, the "three-center" term, $V_{KL}^{(A);3c}$, represents the contributions of the electrons in distant bonds to the EFG at the iron centers. Their contributions can always be expected to be small.

5.3.2.3 An Example

Quadrupole splittings are often interpreted from ligand field models with simple rules for the contributions from each occupied *d*-orbital (see discussion above). However, these models fail even qualitatively in the case of more covalent metal–ligand bonds. An example concerns the quadrupole splittings of Fe(IV)-oxo sites in their $S = 1$ or $S = 2$ spin states. Here, ligand field considerations do not even provide the correct sign of the quadrupole splitting [60].

Consider the simple model complex $[FeO(NH_3)_5]^{2+}$ in its lowest triplet and quintet states (Fig. 5.9).

The electronic structure of such systems is well understood and is shown in Fig. 5.10. The metal *d*-based orbitals are $1b_2$ (d_{xy}), $3e$ (d_{xz},d_{yz}), $2b_1$ ($d_{x^2-y^2}$), and

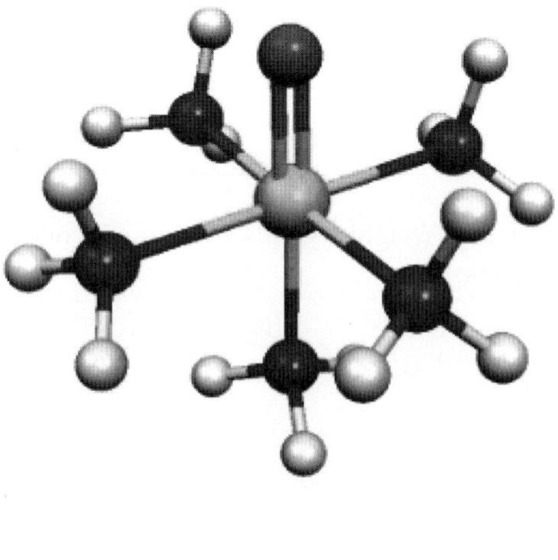

Fig. 5.9 A simple model for a high-valent Fe(IV) system: $[FeO(NH_3)_5]^{2+}$ (see [60])

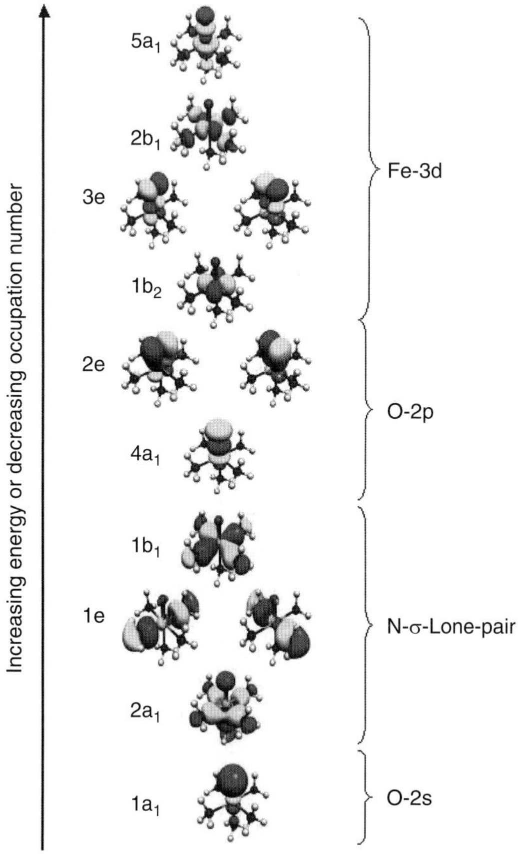

Fig. 5.10 Qualitative molecular orbital diagram for 1 and 2. In 1 the ground state electron configuration is ...$(1b_2)^2(3e)^2(2b_1)^0$ while for 2 it is ...$(1b_2)^1(3e)^2(2b_1)^1$ (taken from [60])

$5a_1$ (d_{z^2}). It is clear from inspection that the d_{xy} orbital is essentially nonbonding and a fairly pure metal d-orbital, while d_{xz} and d_{yz} are strongly π-antibonding and covalently diluted (roughly 40% oxygen character). The $d_{x^2-y^2}$ orbital is strongly antibonding with the equatorial ligands (~70% metal character) and the d_{z^2} orbital forms a strong σ-antibond with the oxo-ligand and the other axial ligand. This latter orbital is also strongly covalently diluted (perhaps ~70–80% metal character).

The shape of these orbitals does not change strongly between the lowest triplet state ($^3A_{2g}$) with the configuration ...$(1b_2)^2(3e)^2(2b_1)^0$ and the lowest quintet state ($^5A_{1g}$) with the configuration ...$(1b_2)^1(3e)^2(2b_1)^1$.

From the ligand field considerations outlined previously, and the values in Table 5.6, the EFG along the Fe–O bond should be dominated by the one-center valence contributions around the iron and given by:

$$V_{zz}\left(^3A_{2g}\right) \cong \tfrac{4}{7}\left\langle r^{-3}\right\rangle_{3d}\left[2\alpha_{xy}^2 - \alpha_{xz,yz}^2\right], \tag{5.55}$$

$$V_{zz}\left({}^5A_{1g}\right) \cong \tfrac{4}{7}\left\langle r^{-3}\right\rangle_{3d}\left[\alpha_{xy}^2 + \alpha_{x^2-y^2}^2 - \alpha_{xz,yz}^2\right]. \tag{5.56}$$

Thus, since $\alpha_{xy}^2 \approx 1$, $\alpha_{x^2-y^2}^2 \approx 0.8$, and $\alpha_{xz,yz}^2 \approx 0.6$ together with $\left\langle r^{-3}\right\rangle_{3d} \approx 5.0$ a.u.$^{-3}$ for the free iron ion, one expects similar and positive EFGs for the $S = 1$ and $S = 2$ species.

However, the calculated quadrupole splittings are relatively small and of *opposite* sign: $+0.54$ mm s^{-1} for $S = 1$ and -1.21 mm s^{-1} for $S = 2$. This is in qualitative agreement with the experimental results of Que's model complex with $S = 1$ ground state spin and the TauD ferryl intermediate with an $S = 2$ ground state spin which show positive and negative quadrupole splittings respectively [50, 82, 83]. Thus, the ligand field model breaks down in this case. It is therefore interesting to look in more detail at the individual contribution to the field gradient at the iron nuclei. The breakdown of the contributions to the calculated EFGs for **1** and **2** is shown in Table 5.7.

It is evident that the ligand field picture is not even qualitatively reasonable. First of all, there is a large positive core-polarization contribution to the field gradient which arises from partial $3p$ involvement in the Fe–O bond. Second, the local valence contributions are in both cases *negative* but differ considerably in magnitude. This can be readily traced back to large contributions to the EFG made by the iron character in the bonding valence orbitals which also includes the $4p$ orbitals. A more detailed analysis of these contributions has not been highly illuminating. However, it is clear that a property which depends on the total electron density such as the EFG at the iron nucleus cannot focus on the partially occupied d-shell if it aims at a realistic description. This is somewhat different for the spin-density for which the spin-polarization is much less important in the formally doubly occupied formally ligand-based valence orbitals. Thirdly, the calculations show, in agreement with previous results on other iron containing complexes, that the two-center bond contributions to the EFG are sizeable and cannot be neglected for a realistic modeling. By contrast, the point charge contributions, which are sometimes taken into account as the only nonlocal contribution in Mössbauer analysis, are an order of magnitude smaller for the present systems than the two-center bond contributions.

In summary, it appears that a positive EFG is typical for the ${}^3A_{2g}$ ground state, while a negative EFG appears to be characteristic for the ${}^5A_{1g}$ ground state. In either

Table 5.7 Breakdown of the contributions to the largest field gradient tensor element V_{zz} in a.u.$^{-3}$ from spin-polarized B3LYP DFT calculations (from [60])

	$S = 1$	$S = 2$
Core-polarization	+0.544	+0.638
One-center valence	−0.119	−1.221
Two-center Point charge	−0.014	+0.009
Two-center Bond	−0.107	−0.193
Three-center	+0.032	+0.022
Total	+0.336	−0.746

case, ligand field theory, even if corrected for anisotropic covalency in the metal d-shell, is unrealistic for the interpretation of the EFG.

5.3.2.4 Temperature-Dependent Quadrupole Splitting

An advanced subject in the theory of quadrupole splitting is the fact that the quadrupole splitting can become temperature dependent. At the heart of this effect is the change in Boltzmann populations of electronically nearly degenerate many-electron states with temperature.

While a rigorous analysis of this effect may become rather involved, it is possible to get a glimpse of the physics that is involved by studying the hexaquo ferrous ion in an idealized perfectly cubic coordination geometry (considering the possible proton orientations this amounts to T_h symmetry [6]). In this symmetry, the ground state of the system is $^5T_2$ that is threefold orbitally degenerate. Each of the three orbital components has fivefold-spin degeneracy in the absence of SOC and magnetic fields. Thus, the total degeneracy is – initially – 15. In terms of molecular orbitals, the five d-orbitals might be written as in the previous paragraph:

$$\psi_i^{(\mathrm{Fe}-3d)} \approx \alpha_i |d_i\rangle - \sqrt{1-\alpha_i^2}|L_i\rangle \tag{5.57}$$

and symmetry dictates $\alpha_{xy}^2 = \alpha_{xy}^2 = \alpha_{yz}^2 \equiv \alpha_{t_2}^2$ and $\alpha_{z^2}^2 = \alpha_{x^2-y^2}^2 \equiv \alpha_e^2$. The principle-states ($M_\mathrm{S} = S$) of the three orbital components of the $^5T_2$ state may then be conveniently represented as:

$$\left|T_{xy}^{22}\right\rangle = \left|(\mathrm{core})\psi_{xy}\bar{\psi}_{xy}\psi_{xz}\psi_{yz}\psi_{x^2-y^2}\psi_{z^2}\right|, \tag{5.58}$$

$$\left|T_{xz}^{22}\right\rangle = \left|(\mathrm{core})\psi_{xy}\psi_{xz}\bar{\psi}_{xz}\psi_{yz}\psi_{x^2-y^2}\psi_{z^2}\right|, \tag{5.59}$$

$$\left|T_{yz}^{22}\right\rangle = \left|(\mathrm{core})\psi_{xy}\psi_{xz}\psi_{yz}\bar{\psi}_{yz}\psi_{x^2-y^2}\psi_{z^2}\right|. \tag{5.60}$$

A particular feature of the $^5T_2$-state is that it is subject to in-state SOC ("in-state" refers to the SOC between the three spatial components of the $^5T_2$ state). This means that SOC is strongly mixing the 15 magnetic sublevels. Such mixing is maximal if the nonrelativistic states $\left|T_{xy}^{22}\right\rangle$, $\left|T_{xz}^{22}\right\rangle$, and $\left|T_{yz}^{22}\right\rangle$ are degenerate and rapidly reduces if the degeneracy is lifted under the influence of low-symmetry perturbations. The spatial degeneracy is, of course, lifted by the Jahn–Teller effect. However, if one considers that in the divalent ferrous hexaquo complex the pi-bonding effects are weak, the Jahn–Teller effect is small and may well be dynamic at which case the analysis becomes very complex. Hence, we restrict ourselves to the case of perfect cubic symmetry here noting that this is not fully realistic.

Diagonalization of the 15 × 15 matrix $\langle T_I^{2M}|\hat{H}_{BO}+\hat{H}_{SOC}|T_J^{2M'}\rangle$ ($I, J = xy$, xz, yz, $M = 0, \pm1, \pm2$) yields the spin–orbit corrected eigenstates:

$$|\Theta_K\rangle = \sum_{IM} d_{IM}^K |T_I^{2M}\rangle \tag{5.61}$$

(K=1–15) and the expansion coefficients d_{IM}^K are complex. In the present case, the T-state behaves as having orbital angular momentum L=1 and hence the diagonalization yields states that can be classified according to the total angular momentum J=1, J=2, and J=3 that are threefold, fivefold, and sevenfold degenerate. These states are at energies 0, $\frac{1}{2}\zeta_{\text{eff}}$, and $\frac{5}{4}\zeta_{\text{eff}}$ respectively, where ζ_{eff} is the effective and slightly covalently reduced one-electron SOC constant of the ferrous ion.

Each of these 15 eigenstates has its own EFG tensor:

$$V_{pq}^{(K)} = \sum_{IM,JM'} d_{IM}^{K*} d_{JM'}^{K} \left\langle T_I^{2M}|\hat{V}_{pq}|T_J^{2M'}\right\rangle. \tag{5.62}$$

We restrict ourselves to the local valence part of the EFG tensor to illustrate the principle. Since the EFG operator is spin-free, there are no off-diagonal elements $M \neq M'$ and an inspection of Table 5.6 reveals that there are also no off-diagonal components between different configurations $I \neq J$. Hence:

$$\left\langle T_I^{2M}|\hat{V}_{pq}|T_J^{2M'}\right\rangle = \delta_{IJ}\delta_{MM'}\langle T_I^{22}|\hat{V}_{pq}|T_I^{22}\rangle \tag{5.63}$$

and one obtains:

$$V_{pq}^{(K)} = \sum_I \langle T_I^{22}|\hat{V}_{pq}|T_I^{22}\rangle \sum_M \left|d_{IM}^K\right|^2. \tag{5.64}$$

Defining the total weight of configuration I in state K:

$$W_I^{(K)} = \sum_M \left|d_{IM}^K\right|^2 \tag{5.65}$$

one obtains more explicitly:

$$V_{xx}^{(K)} = \frac{2}{7}\alpha_{t_2}^2 \langle r^{-3}\rangle_{3d}\left(W_{xy}^{(K)} + W_{xz}^{(K)} - 2W_{yz}^{(K)}\right), \tag{5.66}$$

$$V_{yy}^{(K)} = \frac{2}{7}\alpha_{t_2}^2 \langle r^{-3}\rangle_{3d}\left(W_{xy}^{(K)} + W_{yz}^{(K)} - 2W_{xz}^{(K)}\right), \tag{5.67}$$

$$V_{zz}^{(K)} = \frac{2}{7}\alpha_{t_2}^2 \langle r^{-3}\rangle_{3d}\left(W_{xz}^{(K)} + W_{yz}^{(K)} - 2W_{xy}^{(K)}\right). \tag{5.68}$$

As long as the system is in thermal equilibrium the individual molecules in a sample are distributed among the 15 accessible states according to Boltzmann statistics:

$$N_K(T) = Z^{-1} \exp(-E_K/kT), \tag{5.69}$$

$$Z = \sum_K \exp(-E_K/kT). \tag{5.70}$$

Hence, the probability to meet a molecule in the K'th energy eigenstate is $N_K(T)$. During the time course of the Mössbauer measurement, thermal fluctuations will cause each individual molecule to visit all of the 15 available eigenstates with probability $N_K(T)$ and hence one obtains an averaged EFG tensor:

$$\langle\langle V_{pq}\rangle\rangle(T) = \sum_K N_K(T) V_{pq}^{(K)}. \tag{5.71}$$

In terms of quantum chemistry, one needs to employ a method that can properly represent the 15 magnetic sublevels of the $^5T_2$ manifold. This is, unfortunately, not the case for DFT since it is restricted to nondegenerate mono-determinantal states. Thus, the simplest method which does justice to the actual physics is the CASSCF method.

A CASSCF calculation of the cubic $[Fe(H_2O)_5]^{2+}$ complex in T_h symmetry (Fe–O distance 2.114 Å) and the def2-TZVP basis is shown here. Initially, the CASSCF procedure is converged on the average of the three degenerate $^5T_2$ components, thus yielding a set of molecular orbitals. The SOC is represented by the accurate spin–orbit mean-field operator [84, 85] that is then diagonalized over the 15 components of the $^5T_2$ manifold to yield the 15 spin–orbit eigenstates. The effective one-electron SOC constant is calculated to be 410 cm^{-1} which is realistic and hence the $J = 2$ levels are predicted at 205 cm^{-1} and the $J = 3$ levels at 515 cm^{-1} respectively. At room temperature, $kT \approx 200$ cm^{-1} and therefore 56% of the population is in the three $J = 1$ states, 34% of the population in the five $J = 2$ states, and 10% of the population in the seven $J = 3$ states. Owing to the artificially high symmetry used in the example, the EFG tensors summed over the three $J = 1$, the five $J = 2$, or the seven $J = 3$ components are individually zero. However, the EFG tensors corresponding to the individual M_J components of each J manifold are not. Hence, even slight geometric perturbations will split these levels apart which will then lead to temperature dependent quadrupole couplings that will tend towards zero in the high-temperature limit. Alternatively, an external magnetic field can split the M_J levels apart which leads, at low temperatures, to magnetically induced quadrupole splitting – also a well known phenomenon in Mössbauer spectroscopy.

Further experimental examples of temperature-dependent and magnetically-induced quadrupole splittings are provided in the first volume of this book (Chap. 6, see CD-ROM, Part VI) and also in Sect. 9.4 in Chap. 9 of the present volume.

5.3.3 *Magnetic Hyperfine Interaction*

The third prominent interaction in iron Mössbauer spectroscopy is the magnetic hyperfine interaction of the ^{57}Fe nucleus with a local magnetic field. As explained in detail in Chap. 4, it can be probed by performing the Mössbauer experiment in the presence of an applied external magnetic field.

The theory of the hyperfine interaction is, in some respects, similar to that of the quadrupole interaction but is more involved.

5.3.3.1 Theory

It is well-known that the hyperfine interaction for a given nucleus "A" consists of three contributions: (a) the isotropic Fermi contact term, (b) the spin–dipolar interaction, and (c) the spin–orbit correction. One finds for the three parts of the magnetic hyperfine coupling (HFC), the following expressions [3, 9]:

$$A_{kl}^{(A;c)} = \delta_{kl}\frac{8\pi}{3}\frac{P_A}{2S}\rho^{\alpha-\beta}(\mathbf{R}_A) \tag{5.72}$$

$$A_{kl}^{(A;d)} = \frac{P_A}{2S}\sum_{\mu\nu} P_{\mu\nu}^{\alpha-\beta}\left\langle\varphi_\kappa\middle|r_A^{-5}\left(r_A^2\delta_{\mu\nu} - 3r_{A;\mu}r_{A;\nu}\right)\middle|\varphi_\tau\right\rangle \tag{5.73}$$

$$A_{kl}^{(A;SO)} = -\frac{P_A}{S}\sum_{\mu\nu}\frac{\partial P_{\mu\nu}^{\alpha-\beta}}{\partial \hat{I}_k^{(A)}}\langle\varphi_\mu|z_l^{SOMF}|\varphi_\nu\rangle \tag{5.74}$$

with $P_A = g_e g_N \beta_e \beta_N$ and z_l^{SOMF} is a specific approximation to the l'th component of the spatial part of an effective one-electron SOC operator [85]. Thus, the first two terms are straightforward expectation values while the SOC contribution is a response property [86]. In this case, one has to solve a set of coupled-perturbed equations (Appendix 1, Part III, 3 of CD-ROM) with the three spatial components of the nucleus–orbit interaction operator $(P_A \sum_i (-i\nabla \times (\mathbf{r}_i - \mathbf{R}_A))r_{iA}^{-3})$ taken as the perturbation. As the solution of the coupled-perturbed equations becomes time consuming for larger molecules, it only matters for the metal nuclei of heavy ligand atoms (such as bromine or iodine). For light nuclei, the SOC correction is usually negligible [14]. Note that the dipolar HFC tensor is traceless, while the spin–orbit tensor is not. Thus, both the isotropic as well as the anisotropic part of the HFC do contain contributions from the spin–orbit interaction.

5.3.3.2 Correlation with Experiment

Isotropic Magnetic Hyperfine Couplings

A detailed study of the magnetic hyperfine structure in Mössbauer spectra and the performance of DFT methods is available [25]. It is known that DFT typically underestimates the values of the isotropic hyperfine interaction in magnitude and this is also found for ^{57}Fe. The reason concerns insufficient account of core level spin-polarization – a typical weakness of contemporary DFT. As explained earlier, the HF method greatly overestimates the spin-polarization phenomenon due to a lack of balance between the correlation effects of electrons of the same- and of opposite-spin. Hence, it is not surprising that hybrid DFT provides better numbers than either pure DFT or HF theory. However, an "optimal" and transferable fraction of HF exchange that works for all metals in all oxidation states has not been found. It has, nevertheless, been reported that the TPSSh functional provides better predictions than other functionals [87].

In the study of Sinnecker et al. [25], the magnetic HFCs of ten iron complexes were studied. For most model systems, the typical underestimation of A^{iso} was found [25]. Furthermore, the RMSD values decrease with increasing computational levels: the scalar-relativistic ZORA model, as well as the inclusion of spin–orbit effects, leads to an improvement of the calculated ^{57}Fe isotropic HFCs (recall that in the last case A^{iso} refers to the sum of the Fermi contact term and the isotropic part of the spin–orbit contribution). The best agreement between theory and experiment was found in the ZORA calculations including spin–orbit contributions (ZORA+SOC). The magnitudes of changes introduced by the ZORA approach are typically in the range of 1–2 MHz, while the spin–orbit contributions differ strongly depending on the system. It was argued that relatively good agreement was achieved for systems with large spin–orbit contributions (low-spin Fe(III) and high-spin Fe(II)), while those configurations that are dominated by the first-order terms (high-spin Fe (III) or Fe(IV)), show a significant underestimation of the HFCs. It was found that the underestimation of the HFCs is systematic and that a single scaling factor with value 1.8 for the nonrelativistic B3LYP approach and $f = 1.7$ for the ZORA-B3LYP calculations brings the isotropic HFCs into much better agreement with experiment (Fig. 5.11). This implies that the good results for systems with large spin–orbit interactions arise from a cancellation of errors in the Fermi contact term (not

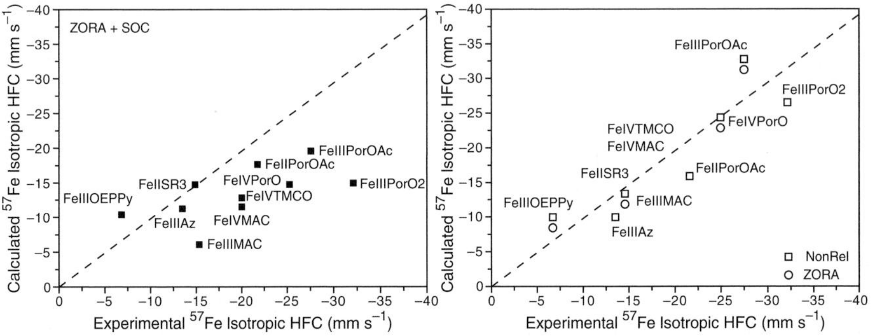

Fig. 5.11 Correlation between theory and experiment for the isotropic HFC of ^{57}Fe complexes. *Left*: before scaling, *right*: after scaling with 1.8 (from [25])

sufficiently negative) with correspondingly large errors in the spin–orbit part (not sufficiently positive).

Anisotropic Hyperfine Interaction

A comparison of anisotropic ^{57}Fe HFCs with the experimental results shows good agreement between theory and experiment for the ferryl complexes and reasonable agreement for ferrous and ferric complexes. Inspection reveals that the ZORA corrections are mostly small (~0.1 MHz) but can approach ~2 MHz and improve the agreement with the experiment. The SOC contributions are distinctly larger than the scalar-relativistic corrections for the majority of the investigated iron complexes. They can easily exceed 20%.

5.3.3.3 Problems with Density Functional Theory

The behavior described above is typical for magnetic hyperfine interaction calculations on transition metal nuclei. The three contributions can all be of the same order of magnitude but may be different in sign. The three contributions have very different physical origin and are all difficult to compute. The Fermi contact term arises from the spin-density at a single point in space (the iron nucleus) and depends strongly on core-level spin-polarization. This is usually underestimated by DFT. The dipolar hyperfine tensor is mostly sensitive to the spin-distribution in the valence shell. This is the easiest to calculate among the three contributions. DFT has a tendency to over-delocalize the spin from the metal to the ligands thus leading to dipolar couplings that are too small [14, 86]. The errors in this contribution may, however, not be very large compared to those of other sources. The spin–orbit part of the SOC, on the other hand, is a response property and depends on the "stiffness" of the system with respect to an external perturbation which is mainly governed by the excitation spectrum. Typically, DFT behaves "too stiff" and hence the response properties such as the closely related *g*-tensor and the SOC contribution to the HFC come out too low. One can certainly hope to improve the situation. This can partially be done by scaling, since the errors are fairly systematic. However, it would be more pleasing to have a more balanced theory.

5.3.3.4 Physical Interpretation

Isotropic Magnetic Hyperfine Interaction

The concept of spin-polarization has been found to be extremely useful for understanding the magnetic HFCs of organic radicals which are dominated by the Fermi contact contribution. The situation for transition metal complexes is rather different in several respects. The idea of spin-polarization is relatively simple and is best understood from the UHF method (5.20). According to this equation, the electron repulsion that a given electron feels is obtained by summing over all other electrons. Electrons of like spin contribute a positive Coulomb integral and a *positive* exchange integral but with a *negative* sign to the repulsion, while electrons of opposite spin only contribute positive Coulomb integrals. Consequently, the interelectronic repulsion is minimized within the set of spin-up electrons if the exchange interaction with the singly occupied MOs is maximized. This means that the optimum situation for spin-up electrons is different than for spin-down electrons because there is a larger number of spin-up than spin-down electrons in open shell molecules (by convention one tries to approximate the $M_s = S$ component of a given spin multiplet). Consequently, the shape of the spin-up orbitals will be slightly different than that for the spin-down electrons and upon taking the difference of the squares of the orbitals, a net spin-density arises even from the formally doubly occupied MOs. However, the spin-up orbitals must stay mutually orthogonal and overall the virial theorem (potential energy divided by kinetic energy equals two) must be obeyed. In general, orbitals that have little spatial overlap with the singly occupied orbitals are distorted towards the singly occupied orbitals and orbitals that occupy the same region in space are "repelled." In transition metal complexes, the singly occupied MOs are derived from the metal 3*d* orbitals. Thus the 3*s* shell is polarized to leave *positive* spin-density at the nucleus because the 3*s* and 3*d* shells occupy a similar region of space. By contrast, the 2*s* shell is polarized in the opposite direction and leaves a *negative* spin-density at the nucleus. The 1*s* shell is also polarized to give negative spin-density but here the spin-polarization is found to be very small.

In Fig. 5.12, the radial distribution functions for the neutral iron-atom are plotted. It is evident that the orbitals with the same main quantum number occupy similar regions in space and are relatively well separated from the next higher and next lower shell. In particular, the 4*s* orbital is rather diffuse and shows its maximum close to typical bonding distances while the 3*d* orbitals are much more compact.

At the resolution of the plot in Fig. 5.12, the spin-polarization of the spin-up and spin-down components is barely visible. Therefore the difference of the densities of each orbital is plotted in the inner atomic region in Fig. 5.13. This plot shows how the 3*s* shell is polarized to give a positive spin-density at the nucleus while the 2*s* shell is negatively polarized. The spin-polarization of the 1*s* shell is very small and contributes very little to the spin-density at the nucleus. The contribution of the 4*s* shell is also large and parallels that of the 3*s* shell. As long as the 4*s* shell is (formally!) not occupied, as is the case with most di- and tri-valent ions of the first transition row, the negative contribution of the 2*s* shell is larger than the contribution from the 3*s* shell and the net spin-density at the nucleus is negative which results in a large negative contribution to the metal magnetic HFC.

The analysis by Munzarova et al. [88] suggests the following interpretation of these findings: the polarization of the metal 2*s* shell is due to the enhanced exchange interaction with the singly occupied metal 3*d* orbitals and causes the spin-up 2*s* orbital to move closer to the metal 3*d* shell which leaves a net negative spin-density

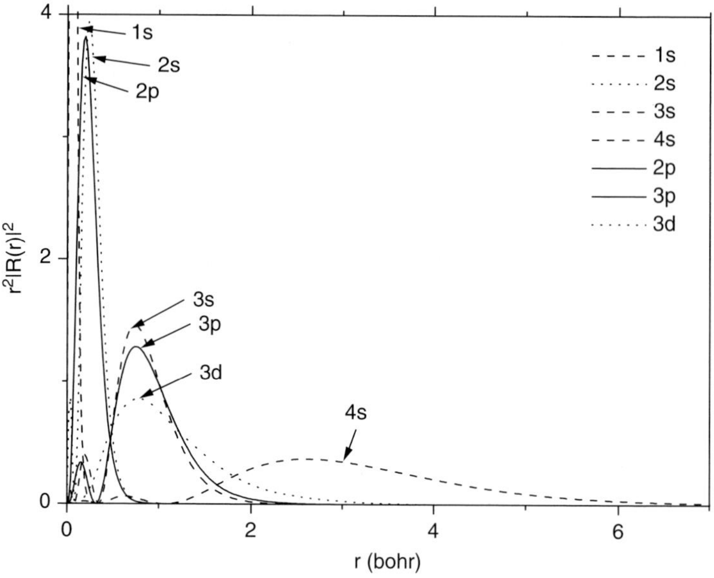

Fig. 5.12 Normalized radial distribution function for the neutral Mn atom ($3d^54s^2$; ^6S) as calculated by the UHF method

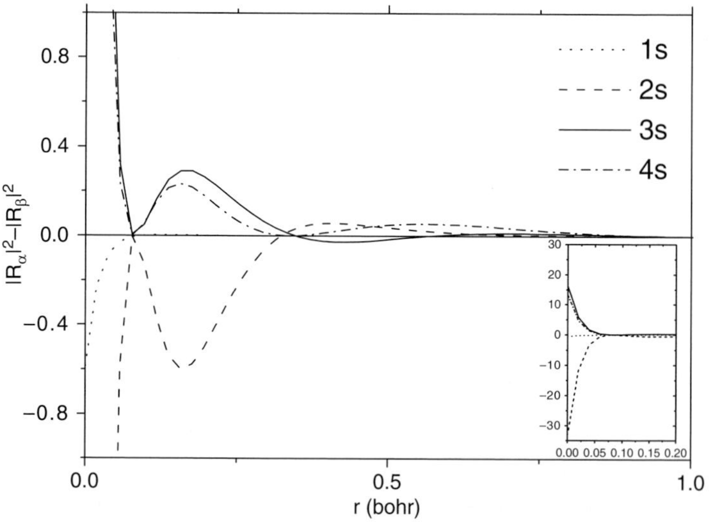

Fig. 5.13 Spin-polarization of the *s*-orbitals in the neutral Mn atom ($3d^54s^2$; ^6S) as calculated by the UHF method. The *inset* shows the region at the nucleus in more detail

at the nucleus. The behavior of the 3*s* shell is strictly opposite to that of the 2*s* shell. The reason for this is the orthogonality requirement between the 2*s* and the 3*s* orbital. Thus, the exchange interaction between the 3*d* and 3*s* shells is *not* optimized by the spin-polarization. On the contrary, this exchange interaction is increased. The reason for the 2*s* shell spin-polarization dominating over the 3*s* one is that the *gain* in 2*s*/3*d* exchange energy is larger than the possible 3*s*/3*d* exchange gain. In other words, there is little to gain by distorting the 3*s* shell but much to gain by distorting the 2*s* shell. Therefore, the 3*s* shell merely reacts to the distortion of the 2*s* shell rather than taking an "active" role in the spin-polarization and these changes are necessary in order to "re-equilibrate" the total energy and wavefunction according to the virial theorem and the orthogonality requirements. Similar considerations apply to the very small spin-polarization of the 1*s* shell.

Dipolar Magnetic Hyperfine Interaction

The physical interpretation of the dipolar hyperfine interaction is closely analogous to that pursued for the quadrupole interaction. In fact, the dipolar magnetic HFC and field gradient calculations use exactly the same integrals. Merely, for the EFG tensor one contracts these integrals with the total electron density, while for the dipolar magnetic HFC the same happens with the spin-density. Therefore, the latter also has no contributions from distant nuclei. Otherwise, the same partitioning applies as for the EFG tensor and a similar interpretation can be given for the one-center, two-center point-charge, two-center bond, as well as three-center terms. The most important term is the valence-shell one-center part of the dipolar HFC. The same ligand field type expression is also obtained for this contribution:

$$\left\langle \psi_i^{(\mathrm{Fe}-3d)} \middle| \hat{f}_{KL} \middle| \psi_j^{(Fe-3d)} \right\rangle \approx \left\langle r^{-3} \right\rangle_{3d} \alpha_i \alpha_j g_{ij}^{KL} \tag{5.75}$$

except that now the sum is taken over only the *singly* occupied orbitals, rather than *all* orbitals. Hence, the dipolar part of the HFC critically depends on which metal based orbitals are singly occupied and how much they are diluted by ligand orbitals or – in other words – how covalent the metal–ligand bonds are.

Spin–Orbit Coupling Contribution to the Magnetic HFC

The physical origin of the SOC contribution to the HFC is the orbital motion of the unpaired electrons. This is introduced into the ground state wavefunction through the SOC of the ground state with excited states. In the case of the HFC, only the excited states of the same multiplicity as the ground state play a role. This is similar for the *g*-tensor but different for the zero-field splitting (ZFS) (*vide infra*). The angular momentum in the ground state creates a magnetic dipole moment that can then interact with the magnetic dipole moment of the nuclei and hence contribute to the HFC arises. Rather than thinking in terms of excited states, it is better to calculate this contribution as a linear response to the nucleus–orbit interaction tensor as explained in detail in [86]. Since similar arguments apply for the interaction of the orbital angular momentum induced magnetic dipole moment with an external magnetic field, it is hardly surprising that the SOC contribution to the HFC and the g-tensor are closely related. In fact, in ligand field theory, the proportionality is direct and one obtains the SOC contribution to the HFC by multiplying the g-shift (deviation from the free-electron g-value) with $P_A = g_e g_N \beta_e \beta_N$ [3]. In a more rigorous framework this strict proportionality no longer holds, since the theory of the g-tensor involves the rather nonlocal global angular momentum operator while the theory of the HFC involves the rather local nucleus–orbit operator. Hence, one should not overemphasize the proportionality for numerical calculations.

5.3.3.5 An Example

Let us return to the example of the triplet and quintet state of the oxo-ferryl model complex studied previously. The magnetic hyperfine couplings (HFCs) for the two different spin states show interesting differences. The calculations predict a significantly larger HFC for the $S = 2$ compared to the $S = 1$ species. In both cases the contributions from the SOC are rather small (0.5–2 MHz in magnitude for $S = 1$ and ~0.5 MHz for $S = 2$). It is therefore sufficient to consider only the Fermi contact and dipolar contributions for the qualitative discussion. Ligand field theory provides the following expressions:

$$A_{\|}\left({}^3A_{2g}\right) \cong P_{Fe}\left\{\frac{4\pi}{3}\rho(0) + \frac{2}{7}\left\langle r^{-3}\right\rangle_{3d}\alpha^2_{xz,yz}\right\}, \tag{5.76}$$

$$A_{\perp}\left({}^3A_{2g}\right) \cong P_{Fe}\left\{\frac{4\pi}{3}\rho(0) - \frac{1}{7}\left\langle r^{-3}\right\rangle_{3d}\alpha^2_{xz,yz}\right\}, \tag{5.77}$$

$$A_{\|}\left({}^5A_{1g}\right) \cong P_{Fe}\left\{\frac{4\pi}{6}\rho(0) + \frac{1}{7}\left\langle r^{-3}\right\rangle_{3d}\left[\alpha^2_{xz,yz} - \alpha^2_{xy} - \alpha^2_{x^2-y^2}\right]\right\}, \tag{5.78}$$

$$A_{\perp}\left({}^5A_{1g}\right) \cong P_{Fe}\left\{\frac{4\pi}{6}\rho(0) - \frac{1}{14}\left\langle r^{-3}\right\rangle_{3d}\left[\alpha^2_{xz,yz} - \alpha^2_{xy} - \alpha^2_{x^2-y^2}\right]\right\} \tag{5.79}$$

with $P_{Fe} = g_e g_{^{57}Fe} \beta_e \beta_N \approx 17.25$ MHz a.u.$^{-3}$ and $\rho(0)$ is the spin density at the iron nucleus. The expressions suggest at first glance that the HFC of the $S = 2$ state should be about half of that of the $S = 1$ state owing to the $1/S$ dependence and that the sign of the dipolar HFC components should be opposite for the two spin states. The factor of two holds in the limit of no anisotropy in the covalency (all α's equal), but it is evident that anisotropic covalency will work to enhance the dipolar HFCs of

Table 5.8 Contributions to the ^{57}Fe hyperfine interaction in $[Fe(O)(NH_3)_5]^{2+}$ in the lowest $S = 1$ and $S = 2$ states from spin-polarized B3LYP calculations (in MHz) (from [60])

	$S = 1$		$S = 2$	
	$A_{\perp}$	$A_{\parallel}$	$A_{\perp}$	$A_{\parallel}$
Isotropic[a]	−20.9	−20.9	−36.7	−36.7
Dipolar[b]	−7.6	+15.2	+6.5	−13.1
Spin–Orbit[b]	−1.6	+0.8	−0.6	−0.5
Total	−30.1	−4.9	−29.6	−50.3

[a]According to the results of [60], the isotropic term was scaled by 1.8 in order to compensate for the intrinsic underestimation of the core polarization by DFT methods
[b]The interactions in the x, y plane were slightly averaged to axial symmetry

the $S = 2$ state since $\alpha^2_{xz,yz} < \alpha^2_{x^2-y^2} < \alpha^2_{xy}$. Since the core polarization will give a negative spin density at the iron nucleus, it is expected that in the $^3A_{2g}$ state the isotropic and dipolar contributions partially cancel each other along the Fe–O bond direction while they will reinforce each other along this direction in the $^5A_{1g}$ state.

In contrast to the EFG analyzed before, all of these expectations from ligand field theory are largely confirmed by the DFT calculations. Despite the fact that the $S = 2$ state has a smaller prefactor for the isotropic Fermi contact term, the core polarization in the presence of four unpaired electrons is much larger and, consequently, the isotropic ^{57}Fe-HFC is predicted to be roughly a factor of two larger in magnitude for $^5A_{1g}$ as compared to $^3A_{2g}$. Similarly, the dipolar HFCs are comparable for both spin states, which must be due to a considerable contribution from anisotropic covalency in the $S = 2$ species which partially compensates for the smaller prefactor.

The predicted signs of all HFC components match the expectations from ligand field theory and, consequently, the HFC component along the Fe–O bond (or alternatively along the largest EFG component) is diagnostic for the spin state of the $(FeO)^{2+}$ core. However, based on the previous discussion, the absolute numbers calculated for these hypothetical model systems should be viewed with caution. Nevertheless, the value of −29.6 MHz for $A_{\perp}$ of the S=2 species is in good agreement with the value of –27.8 MHz measured by Stoian et al. for the $[FeO(H_2O)_5]^{2+}$ system [54] (Table 5.8).

5.3.4 *Zero-Field Splitting and g-Tensors*

Since Mössbauer spectroscopy is sensitive to all terms in the SH, it is also sensitive to the ZFS and the g-tensor. The theory of both interactions can be approached along the same lines as explained in some detail in Appendix 1 (Part III, 3 of CD-ROM) [89, 90]. This becomes somewhat elaborate for the ZFS while the g-tensor is more readily approached. Both quantities have been previously treated in some detail and a protracted discussion would be inappropriate here [9, 79, 89–93]. In general, the accuracy with which both quantities can be calculated from DFT is rather moderate and a combination of ligand field theory and DFT or some moderate amount of multiconfigurational ab initio wavefunction theory (such as complete active space sel-consistent field theory, CASSCF, perhaps supplemented by a second order perturbation correction, CASPT2) can lead further than the direct application of DFT itself [93].

There is at least one issue that needs to be mentioned in this context. In the Mössbauer literature one frequently finds the use of a ligand field model by Oosterhuis and Lang [94]. In this model one obtains the g-tensor by exploiting a proportionality to the D-tensor. However, this correlation is, in general, not justified. As explained elaborately in [79], the ZFS tensor involves the SOC between the ground state with spin S and excited states with spin $S' = S, S + 1$ and $S - 1$. Quite frequently, the contributions with $S' = S - 1$ even dominate the ZFS – trivially so for high-spin Fe(III) ($S = 5/2$) where there is no other d–d excited state with $S = 5/2$ but also for many other configurations [93]. By contrast, only excited states with $S'=S$ enter the expression for the g-tensor. Hence, there is no proportionality between the g- and D-tensors and in many instances the predictions of the simple ligand field model are simply wrong. For example, the true g-shifts for high-valent Fe(IV) configuration with $S = 1$ are very small while the D-value is very large due to SOC with the low-lying $S = 2$ state [95].

5.4 Nuclear Inelastic Scattering

Conventional Mössbauer spectroscopy is based on the often appreciable probability that nuclei absorb X-rays without a change in the vibrational states of the molecule or lattice, i.e. the absorption is recoilless. NIS on the other hand is based on the simultaneous excitation of a nuclear resonance and molecular vibrations. Compared to conventional Mössbauer spectroscopy, the NIS signal is comprised of a number of resonances that are broadened by vibrational lifetimes, typically on the order of 1 meV. Thus, strong vibrational sidebands with a total integrated cross-section on the order of 1% of the recoilless resonance typically have a peak cross-section six to seven orders of magnitude weaker as compared to the recoilless band with a linewidth of a few nano electron volts. The NIS signal provides selective information on those vibrations that feature a high involvement of the Mössbauer absorber atom(s). Materials surrounding the sample that do not contain resonant nuclei produce no unwanted background. The most suitable nuclear resonance isotope is ^{57}Fe which is a fortunate circumstance given the importance of iron in chemistry and physics. NIS studies have been conducted on several model compounds for iron proteins [96–102] as well as on the proteins themselves [103–107].

Here the selectivity of NIS is particularly important since only the first coordination sphere of the iron atoms contribute appreciably to the NIS signal.

NIS provides an absolute measurement of the so-called "normal mode composition factors" that characterize the extent of involvement of the resonant nucleus in a given normal mode. On the basis of the analysis of experimental NIS data, one can therefore construct a partial vibrational density of states (PVDOS) that can be viewed as a vibrational density of states weighted with the normal mode composition factors.

The first-principles calculation of NIS spectra has several important aspects. First of all, they greatly assist the assignment of NIS spectra. Secondly, the elucidation of the vibrational frequencies and normal mode compositions by means of quantum chemical calculations allows for the interpretation of the observed NIS patterns in terms of geometric and electronic structure and consequently provide a means of critically testing proposals for species of unknown structure. The first-principles calculation also provides an unambiguous way to perform consistent quantitative parameterization of experimental NIS data. Finally, there is another methodological aspect concerning the accuracy of the quantum chemically calculated force fields. Such calculations typically use only the experimental frequencies as reference values. However, apart from the frequencies, NIS probes the shapes of the normal modes for which the iron composition factors are a direct quantitative measure. Thus, by comparison with experimental data, one can assess the quality of the calculated normal mode compositions.

5.4.1 *The NIS Intensity*

The full quantitative theory of NIS is somewhat involved and hence delegated to Appendix 2 (Part III, 3 of CD-Rom) that is based on [98, 7]. Here, we outline only the most important features.

From quantum chemistry one obtains the force field of the molecule in Cartesian coordinates by taking the second derivative of the quantum chemically calculated ground state energy:

$$V_{ij} = \frac{\partial^2 E}{\partial x_i \partial x_j}, \tag{5.80}$$

where x_i, x_j are Cartesian coordinates of nuclei. Following mass weighting and diagonalization, the harmonic vibrational frequencies $\{\omega_\alpha\}$ and normal modes of the molecule are obtained. Such calculations are implemented in almost all quantum chemical program packages. The eigenfunctions of the vibrational Hamiltonian in harmonic approximation are Hermite functions that can be completely specified by a set $|\{n_\alpha\}\rangle$ of occupation numbers for individual modes.

One useful way to write the normal modes is:

$$Q_\alpha = \sum_{j=1}^{N} \vec{e}_{j\alpha} \cdot \vec{r}_j \sqrt{m_j}, \tag{5.81}$$

where $\vec{r}_j$ is the displacement of atom j with mass m_j from its equilibrium position and $\vec{e}_{j\alpha}$ is the normal mode weighting factor that determines how much a given atom actually moves in mode α. Since the dimensionless vectors $\vec{e}_{j\alpha}$ describe the orthogonal transformation between the normal modes and mass weighted Cartesian coordinates, they obey the following normalization conditions.

$$\sum_{j=1}^{N} e_{j\alpha}^2 = 1 \quad \text{and} \quad \sum_{\alpha=1}^{3N} e_{j\alpha}^2 = 3. \tag{5.82}$$

One can hence think of (normal-mode composition factor) $e_{j\alpha}^2 = \vec{e}_{j\alpha}\vec{e}_{j\alpha}$ as the fractional involvement of atom j in normal mode α.The dimensionless vector $\vec{e}_{j\alpha}$ also specifies the direction of the motion of atom j in the α-th normal mode. Interestingly, the mode composition factors are also related to the magnitude of the atomic fluctuations. In a stationary state $|\{n_\alpha\}\rangle$ of a harmonic system, the mean square deviation (msd) of atom j from its equilibrium position may be expressed as a sum over modes of nonzero frequency:

$$\left\langle r_j^2 \right\rangle = \sum\nolimits_\alpha (2n_\alpha + 1)\left\langle r_{j\alpha}^2 \right\rangle_0, \tag{5.83}$$

where

$$\left\langle r_{j\alpha}^2 \right\rangle_0 = \frac{\hbar}{2m_j\omega_\alpha} e_{j\alpha}^2 \tag{5.84}$$

is the contribution of mode α to the zero-point fluctuations of atom j in the ground state.

In the context of the inelastic scattering process that is the basis for NIS, one observes inelastic transitions $|g\rangle|\{n_\alpha\}\rangle \rightarrow |e\rangle|\{n_\alpha + \Delta n_\alpha\}\rangle$, where $|g\rangle$ and $|e\rangle$ are the ground and the excited states of the nucleus. Thus, the transition energies shifted from the recoilless absorption (E_0) as $E = E_0 + \sum \Delta n_\alpha \omega_\alpha$. In terms of the normal-mode composition factors, the intensity of a vibrational fundamental transition $\Delta n_\alpha = 1$ is given by:

$$S(\bar{v}) = fL_0(\bar{v}) + \phi L_\upsilon(\bar{v} - v_\alpha). \tag{5.85}$$

Here, $L(v)$ is a lineshape function that integrates to unity, v is the frequency, f is the Lamb–Mössbauer factor, and the desired side bands have an area fraction ϕ that is proportional to $e_{j\alpha}^2$ which hence determines the relative peak heights in a NIS spectrum. More details are provided in Appendix 2 (Part III, 3 of CD-ROM). An equivalent and often more suggestive display of the NIS spectrum is the PVDOS approach, which describes the NIS signal in terms of the partial vibrational density of states:

$$D_j(\bar{v}) = \sum_\alpha e_{j\alpha}^2 L_f(\bar{v} - \bar{v}_\alpha). \tag{5.86}$$

The PVDOS directly characterizes the involvement of the probe nucleus in different normal modes and provides a graphical representation of the calculated normal mode composition factors.

5.4.2 Example 1: NIS Studies of an Fe(III)–azide(Cyclam-acetato) Complex

The Fe(III)–azide(cyclam-acetato) complex, as seen earlier, serves as a precursor for generation of the "superoxidized" Fe(V)–nitrido(cyclam-acetato) complex by means of photolysis with 420 nm light [63, 108] (Fig. 5).

The Fe(III)–azide-precursor and the photolysed product were characterized by NIS spectroscopy coupled to detailed DFT calculations [63]. The result of the study provides additional evidence in favor of a low-spin $S = 1/2$ ground state of the Fe(V)–nitrido complex. Here we show how first-principles calculations assist in quantitative analysis of experimental NIS data for the Fe(III)–azide complex.

In order to arrive at a detailed assignment of the spectrum (Fig. 5.14), a multi-step analysis has been performed that combines elements of least-squares fitting and electronic structure calculations. The fit was done in the spectral range 130–1,000 cm^{-1}. The spectral region below 130 cm^{-1} had almost unresolved structure due to the presence of rather intense low-frequency and acoustic bands, and was therefore excluded from further consideration. The subtraction of a broad featureless background was also necessary. Such a background may arise from two major factors: (a) secondary electron scattering processes, whereby the nuclear scattering observed in the NIS experiment is not completely separated from the

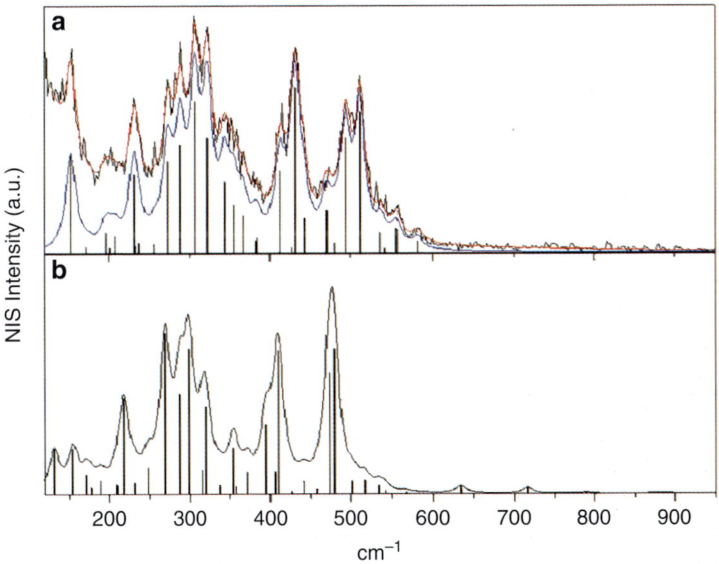

Fig. 5.14 Experimental (**a**, *black curve*), fitted (**a**, *red*) and simulated (**b**) NIS spectrum of the Fe (III)–azide complex obtained at the BP86/TZVP level ($T = 20$ K). *Bar graphs* represent the corresponding intensities of the individual vibrational transitions. The *blue curve* represents the fitted spectrum with a *background line* removed (taken from [63])

electronic one and (b) acoustic phonons which are coexcited with the molecular vibrations under consideration [98].

The DFT frequency calculations performed at the BP86/TZVP level of theory provided a number of vibrational modes in the investigated frequency range, as well as initial estimates of the normal mode composition factors $e^2_{\mathrm{Fe}\,\alpha}$ and vibrational frequencies. It was also found that for the given experimental temperature (T= 20 K), the NIS signal was mainly represented by fundamental contributions in the spectral range 130–1,000 cm^{-1}. Multiquanta contributions contribute no more than ~0.3% of the total integrated excitation probability. Thus, in the present case, only fractional contributions for fundamental transitions ϕ_1 defined in Appendix 2 (Part III, 3 of CD-ROM) were included into the expression for $S(\overline{v})$. Importantly, the number of vibrational bands considered in the fit was fixed from the DFT frequency calculations. In order to avoid ambiguities and to minimize the number of free parameters, all linewidths (FWHM, full width at half height) were kept identical. The lineshape was best represented by Lorentzians with a FWHM of 15.1 cm^{-1}.

According to the DFT calculations there are 33 vibrations in the range 130–600 cm^{-1} for the Fe(III)–azide complex. Nearly all the experimental and calculated NIS intensity is in this region. Indeed, as follows from the normal mode analysis, all vibrations with significant involvement of Fe motion actually fall into this range. Consistent with this finding, the sum of the Fe normal mode composition factors corresponding to the spectral region 130–600 cm^{-1} is 2.24 which is not much less than the value of 2.52 corresponding to the sum of all nontrivial normal modes. Since only relative NIS intensities were fitted, the variations of $e^2_{\mathrm{Fe}\,\alpha}$were performed such that the sum over all modes included in the fit was kept equal to the corresponding value obtained from the DFT calculations (2.24).

Figure 5.14 presents experimental, fitted, and purely quantum-chemically calculated NIS spectra of the ferric–azide complex. It is clear that the fitted trace perfectly describes the experimental spectra within the signal-to-noise ratio. Furthermore, the purely theoretical spectrum agrees well with the fitted spectrum. This indicates that the calculations provide highly realistic force field and normal mode composition factors for the molecule under study and are invaluable as a guide for least-square fittings.

The fitted and calculated vibrational frequencies and normal mode composition factors corresponding to the 17 most important NIS bands are presented in Table 5.9. It is evident that the vibrational peaks in the calculated NIS spectrum are typically 0–30 cm^{-1} lower than to the experimental values. In the calculated NIS spectra, there are two small peaks at 635 and 716 cm^{-1} (Fig. 5.14b) that are not visible in the experimental spectrum. According to the normal mode calculations these are Fe–N–N and Fe–O–C deformation vibrations. Small admixtures of Fe–N and Fe–O stretching modes account for the calculated nonzero normal mode composition factors. Although the calculated relative intensities are slightly above detection limit dictated by the signal-to-noise ratio, they are determined by values of $e^2_{\mathrm{Fe}\,\alpha}$, which are very small (0.028 and 0.026 for the peaks at 635 and 716 cm^{-1}). They must be considered to be within the uncertainties of the theoretical treatment. According to the comparison between fitted and calculated normal mode composition factors, the calculations have an accuracy of about 0.02 for $e^2_{\mathrm{Fe}\,\alpha}$ compared to the most intense features which have $e^2_{\mathrm{Fe}\,\alpha}\sim0.3$.

Table 5.9 Experimental and calculated at the BP86/TZVP level frequencies and corresponding values of the iron normal mode composition factors of the most important vibrations that appear in the NIS signal of the Fe(III)–azide complex (taken from [101])

Vibration	Frequency in cm^{-1}		$e^2_{\mathrm{Fe}\,\alpha}$	
	Calc.	Exp.	Calc.	Exp.
ν_1 N–Fe–N bending	131.7	153	0.029	0.06
ν_2 N–Fe–N bending	171.1	196	0.016	0.02
ν_3 N–Fe–N bending	189.4	207	0.014	0.02
ν_4 N–Fe–N bending	217.7	231	0.098	0.07
ν_5 N–Fe–N bending	231.5	256	0.013	0.01
ν_6 N–Fe–N bending	269.2	273	0.205	0.10
ν_7 N–Fe–N bending N–Fe stretching	287.4	289	0.135	0.12
ν_8 N–Fe–N bending N–Fe stretching	298.9	307	0.205	0.19
ν_9 N–Fe–N bending Fe–N stretching	315.2	323	0.037	0.04
ν_{10} N–Fe–N bending	319.9	323	0.132	0.15
ν_{11} Fe–N stretching	354.0	356	0.078	0.07
ν_{12} N–Fe–N bending	357.7	367	0.014	0.05
ν_{13} Fe–N stretching	394.8	413	0.129	0.13
ν_{14} Fe–N stretching	410.5	432	0.278	0.28
ν_{15} Fe–N stretching	426.6	443	0.007	0.06
ν_{16} Fe–N stretching	473.4	494	0.272	0.23
ν_{17} Fe–N stretching Fe–O stretching	479.5	512	0.328	0.29
ν_{18} Fe–N stretching	634.5	–	0.028	–

5.4.2.1 Normal Mode Compositions

The normal-mode analysis has shown that there are ~17 vibrational modes that are characterized by significant involvement of the Fe nucleus (i.e. large values of $e^2_{\mathrm{Fe}\,\alpha}$). The frequencies and normal mode composition factors corresponding to these vibrations are described in Table 5.9.

The shapes of the most important normal modes in the vicinity of the core region around the Fe atom are shown in Fig. 5.15. The entire set of vibrations that are active in the NIS spectrum may be roughly partitioned into three groups: (1) N–Fe–N bending vibrations (spanning the spectral region 100–300 cm^{-1}); (2) Mixed modes involving N–Fe–N bending and Fe–N stretching (300–400 cm^{-1}); (3) Fe–N stretching vibrations (>400 cm^{-1}). The in-plane Fe–N stretching modes ν_{13}–ν_{17} appear in the range ~400–500 cm^{-1}, while the out-of-plane mode involving the Fe–N_5 bond stretching has frequency 634 cm^{-1} according to the calculations. The latter mode was not detected in NIS due to the low value of the mode composition factor.

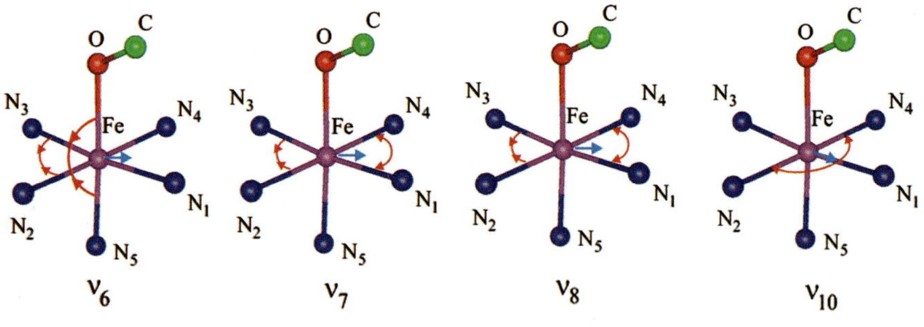

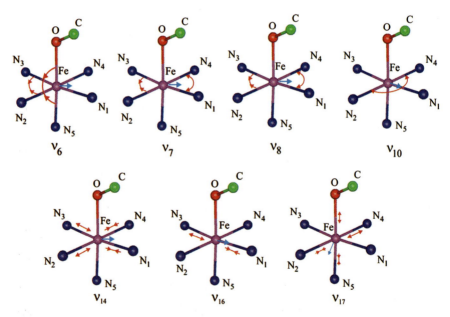

Fig. 5.15 Schematic representation of the normal modes of the Fe(III)–azide complex with the largest iron composition factors. The individual displacements of the Fe nucleus are depicted by a *blue arrow*. All vibrations except for ν_4 are characterized by a significant involvement of bond stretching and bending coordinates (*red arrows* and *archlines*). In such a case, the length of the *arrows* and *archlines* roughly indicate the relative amplitude of bond stretching and bending, respectively. Internal coordinates vibrating in antiphase are denoted by *inward* and *outward arrows* respectively (taken from [63])

5.4.3 *Example 2: Quantitative Vibrational Dynamics of Iron Ferrous Nitrosyl Tetraphenylporphyrin*

The current example illustrates PVDOS formulation as an effective basis for comparison of experimental and theoretical NIS data for ferrous nitrosyl tetraphenylporphyrin Fe(TPP)(NO), which was done [101] along with other ferrous nitrosyl porphyrins. Such compounds are designed to model heme protein active sites. In particular, the elucidation of the vibrational dynamics of the Fe atom provides a unique opportunity to specifically probe the contribution of Fe to the reaction dynamics. The geometrical structure of Fe(TPP)(NO) is shown in Fig. 5.16.

Experimental and theoretical PVDOS for Fe(TPP)(NO) single-crystal and powder samples are shown in Fig. 5.17. The DFT calculations were performed, using the B3LYP and BP86 functionals.

The upper panel in Fig. 5.17 compares the experimentally determined PVDOS for the oriented array of Fe(TPP)(NO) crystals with that for Fe(TPP)(NO) powder. The crystal data is recorded with the incident X-ray beam 6° above the mean porphyrin plane. Since the NIS measurement is only sensitive to Fe motion along a probe direction, the spectral contribution of modes with predominantly in-plane Fe motion are enhanced, and the contribution of out-of-plane modes is suppressed in the oriented crystal data.

The comparison of crystal and powder results in Fig. 5.17 identifies three dominant modes at 74, 128, and 540 cm^{-1} with Fe motion perpendicular to the porphyrin plane. In contrast, the Fe motion parallel to the porphyrin plane makes the dominant contribution to modes in the 200–500 cm^{-1} range. These observations are consistent with the assignment of the 540 cm^{-1} mode to the out-of-plane Fe–NO stretch (because its relative amplitude is reduced in the "in-plane" crystal spectrum). The corresponding experimental value $e^2_{Fe} = 0.30$ was based on fitting a

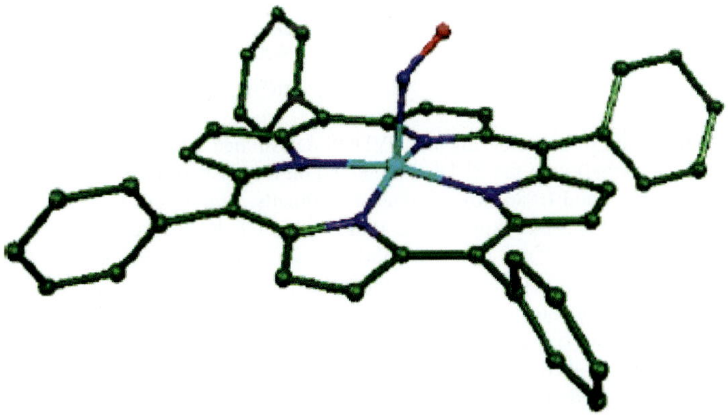

Fig. 5.16 Calculated structure of ferrous nitrosyl tetraphenylporphyrin Fe(TPP)(NO), resulting from geometric optimization with the B3LYP functional and 6-31G* + VTZ basis set [101]. Color scheme: *cyan* = iron, *green* = carbon, *blue* = nitrogen, *red* = oxygen (taken from [101])

Voigt function to the 540 cm^{-1} peak. It is very close to the expected value $e_{\mathrm{Fe}}^2 = 0.34$ for a two-body ^{57}Fe–NO oscillator. The calculated value $e_{\mathrm{Fe}}^2 = 0.39$ (B3LYP functional) is in agreement with the experimental one.

The dominant features in the 300–400 cm^{-1} range (Fig. 5.17) have been identified with stretching of the in-plane Fe–N$_{\mathrm{pyr}}$ bonds because their amplitudes are enhanced in the crystal spectrum. This is in agreement with the DFT calculations that show large iron normal mode composition factors for in-plane modes in this spectral range. The most important in-plane vibrations are shown in Fig. 5.18. The agreement between observed and predicted vibrational frequencies is good in all spectral ranges except for the Fe–NO bending modes at 540 cm^{-1} and below 150 cm^{-1}. In the latter case, it is conceivable that intermolecular interactions not included in the calculations on isolated molecules contribute to differences in frequencies and relative Fe amplitudes. The dominant peaks observed at 74 and 128 cm^{-1} were assigned to the calculated out-of-plane Fe modes at 77 and

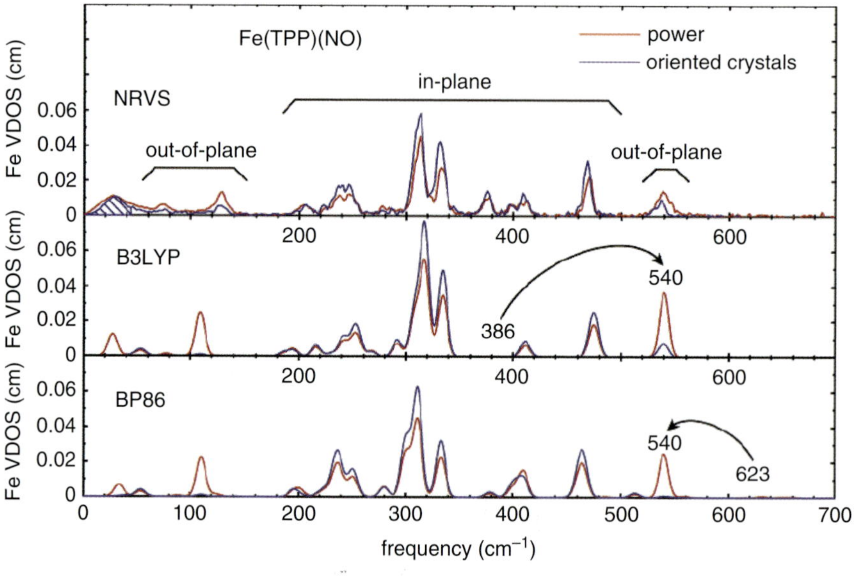

Fig. 5.17 Comparison of the experimental PVDOS determined from NIS measurements on Fe (TPP)(NO) (*upper panel*) with the PVDOS predicted on the basis of DFT calculations using the B3LYP (*center panel*) and BP86 (*lower panel*) functionals. *Blue traces* represent the PPVDOS $D_{\mathrm{Fe}\hat{k}}(\bar{\nu})$for oriented crystals (see Appendix 2, Part III, 3 of CD-ROM), scaled by a factor of 3 for comparison with the total PVDOS $D_{\mathrm{Fe}}(\bar{\nu})$of unoriented polycrystalline powder (*red traces*). Since the X-ray beam direction $\hat{k}$ lies 6° from the porphyrin plane, modes involving Fe motion in the plane of the porphyrin are enhanced, and modes with Fe motion primarily normal to the plane are suppressed, in the scaled oriented crystal PVDOS relative to the powder PVDOS. In-plane Fe modes dominate the 200–500 cm^{-1} range of the data, while Fe motion in modes observed at 74, 128, and 539 cm^{-1} is predominantly out-of-plane. Crosshatching in the upper panel indicates the area attributable to acoustic modes. In the lower two panels, the Fe–NO bend/stretch modes predicted at 386 and 623 cm^{-1}, have been artificially shifted to the observed 539 cm^{-1} frequency to facilitate comparison with the experimental results. Predicted PVDOS are convolved with a 10 cm^{-1} Gaussian (taken from [101])

109 cm^{-1} (Fig. 5.19). This is supported by the diminished amplitudes of both modes in oriented crystals. The out-of-plane mode at 109 cm^{-1} resembles the heme "doming" mode proposed to influence heme protein reactivity [109, 110].

In summary, the quantitative information on the frequencies, amplitudes, and directions of Fe motion from NIS measurements provides a definitive test of the detailed normal-mode predictions provided by modern quantum chemical calculations. However, first-principles calculations greatly assist in the analysis and interpretation of experimental NIS data, thus revealing a consistent picture of the vibrational dynamics of iron in molecules.

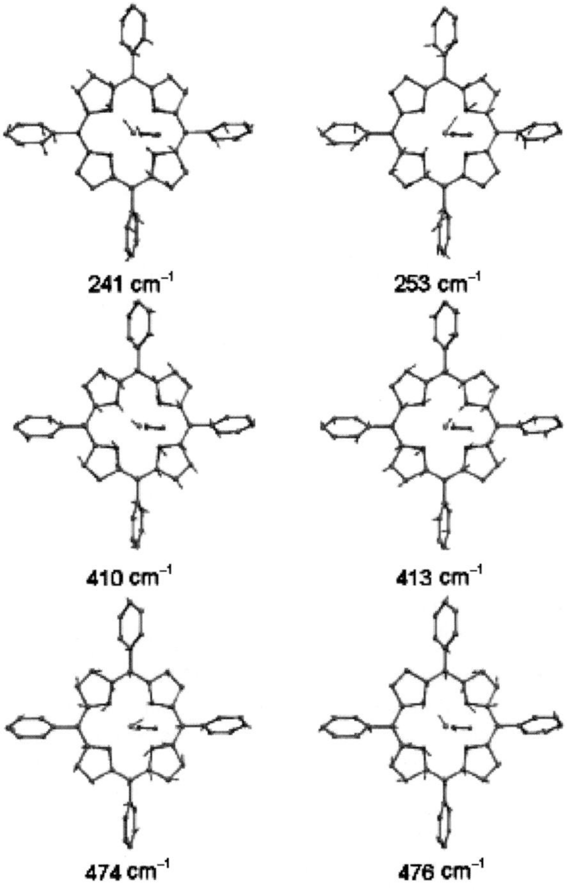

Fig. 5.18 Four in-plane Fe modes, predicted on the basis of B3LYP calculations, contributing to the pair of experimental features at 312 and 333 cm^{-1} in Fe(TPP)(NO). For ease of visualization, each *arrow* is $100(m_j/m_{\mathrm{Fe}})^{1/2}$ longer than the zero-point vibrational amplitude of atom *j*. Color scheme as in Fig. 5.15 (taken from [101])

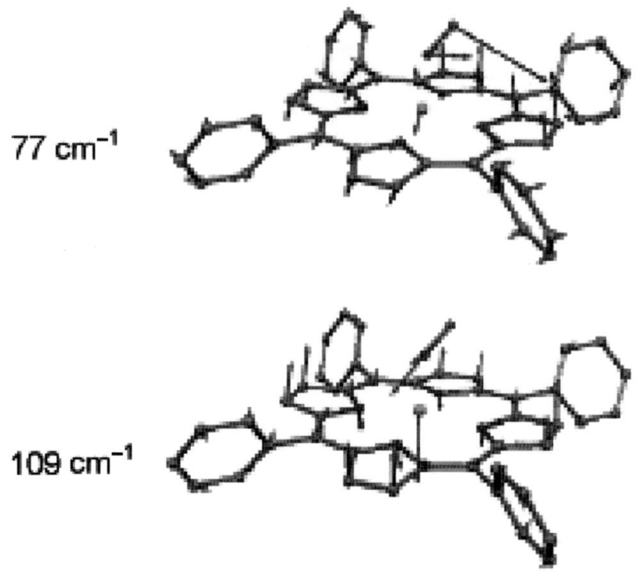

Fig. 5.19 Low-frequency Fe modes of Fe(TPP)(NO) predicted on the basis of B3LYP calculations. The modes mainly involve porphyrin core translation, Fe–NO torsion, Fe–N–O bending, and Fe out-of-plane motion coupled to doming of the porphyrin core. *Arrows* representing mass-weighted atomic displacements are $100(m_j/m_{\mathrm{Fe}})^{1/2}$ longer than the zero-point vibrational amplitude of atom *j*. Color scheme as in Fig. 5.15 (taken from [101])

References

1. Parr, R.G., Yang, W.: Density Functional Theory of Atoms and Molecules. Oxford University Press, Oxford (1989)
2. Koch, W., Holthausen, M.C.: A Chemist's Guide to Density Functional Theory. Wiley, Weinheim (2000)
3. Neese, F., Solomon, E.I.: Calculation and interpretation of spin-Hamiltonian parameters in transition metal complexes. In: Miller, J.S., Drillon, M. (eds.) Magnetoscience – From Molecules to Materials, vol. 4, p. 345. Wiley, Weinheim (2003)
4. Neese, F.: Zero-field splitting. In: Kaupp, M., Bühl, M., Malkin, V. (eds.) The Quantum Chemical Calculation of NMR and EPR Properties, p. 541. Wiley, Heidelberg (2004)
5. Neese, F.: J. Biol. Inorg. Chem. **11**, 702 (2006)
6. Neese, F., Petrenko, T., Ganyushin, D., Olbrich, G.: Coord. Chem. Rev. **251**, 288 (2007)
7. Petrenko, T., Sturhahn, W., Neese, F.: Hyperfine Interact. **175**, 165 (2007)
8. Sinnecker, S., Neese, F.: Theoretical bioinorganic spectroscopy. In: Reiher, M. (ed.) Atomistic Approaches in Modern Biology: from Quantum Chemistry to Molecular Simulations, p. 47. Springer, Berlin (2007)
9. Neese, F.: Biol. Mag. Res. **28**, 175 (2008)
10. Neese, F.: Coord. Chem. Rev. **253**, 526 (2009)
11. Römelt, M., Ye, S., Neese, F.: Inorg. Chem. **48**, 784 (2009)
12. Li, J., Noodleman, L., Case, D.A.: Electronic structure calculations: Density functional methods with applications to transition metal complexes. In: Solomon, E.I., Lever, A.B.P. (eds.) Inorganic Electronic Structure and Spectroscopy, vol. 1, p. 661. Wiley, New York (1999)
13. McWeeny, R.: Methods of Molecular Quantum Mechanics. Academic, London (1992)

14. Neese, F.: J. Phys. Chem. **A 105**, 4290 (2001)
15. Strange, P.: Relativistic Quantum Mechanics. Cambridge University Press, Cambridge (1998)
16. Nieuwport, W.C., Post, D., van Duijnen, P.T.: Phys. Rev. **B 17**, 91 (1978)
17. Lovell, T., Han, W.G., Liu, T., Noodleman, L.: J. Am. Chem. Soc. **124**, 5890 (2002)
18. Lovell, T., Li, J., Case, D.A., Noodleman, L.: J. Am. Chem. Soc. **124**, 4546 (2002)
19. Neese, F.: Inorg. Chim. Acta **337**, 181 (2002)
20. Vramasu, V.V., Bominaar, E.L., Meyer, J., Münck, E.: Inorg. Chem. **41**, 6358 (2002)
21. Zhang, Y., Mao, J., Oldfield, E.: J. Am. Chem. Soc. **124**, 7829 (2002)
22. Zhang, Y., Mao, J.H., Oldfield, E.: J. Am. Chem. Soc. **124**, 7829 (2002)
23. Liu, T.Q., Lovell, T., Han, W.G., Noodleman, L.: Inorg. Chem. **42**, 5244 (2003)
24. Neese, F.: Curr. Opin. Chem. Biol. **7**, 125 (2003)
25. Sinnecker, S., Slep, L.D., Bill, E., Neese, F.: Inorg. Chem. **44**, 2245 (2005)
26. Han, W.G., Liu, T.Q., Lovell, T., Noodleman, L.: J. Comp. Chem. **27**, 1292 (2006)
27. Lovell, T., Li, J., Liu, T.Q., Case, D.A., Noodleman, L.: J. Am. Chem. Soc. **123**, 12392 (2001)
28. Lovell, T., Li, J., Noodleman, L.: Inorg. Chem. **40**, 5251 (2001)
29. Lovell, T., Han, W.G., Liu, T.Q., Noodleman, L.: J. Am. Chem. Soc. **124**, 5890 (2002)
30. Lovell, T., Li, J., Case, D.A., Noodleman, L.: J. Biol. Inorg. Chem. **7**, 735 (2002)
31. Zhang, Y., Mao, J.H., Godbout, N., Oldfield, E.: J. Am. Chem. Soc. **124**, 13921 (2002)
32. Ghosh, P., Bill, E., Weyhermüller, T., Neese, F., Wieghardt, K.: J. Am. Chem. Soc. **125**, 1293 (2003)
33. Han, W.G., Lovell, T., Liu, T.Q., Noodleman, L.: Inorg. Chem. **42**, 2751 (2003)
34. Lim, M.H., Rohde, J.U., Stubna, A., Bukowski, M.R., Costas, M., Ho, R.Y.N., Münck, E., Nam, W., Que, L.: Proc. Natl. Acad. Sci. USA. **100**, 3665 (2003)
35. Lovell, T., Himo, F., Han, W.G., Noodleman, L.: Coord. Chem. Rev. **238**, 211 (2003)
36. Neese, F.: Curr. Opin. Chem. Biol. **7**, 125 (2003)
37. Slep, L.D., Mijovilovich, A., Meyer-Klaucke, W., Weyhermüller, T., Bill, E., Bothe, E., Neese, F., Wieghardt, K.: J. Am. Chem. Soc. **125**, 15554 (2003)
38. Slep, L.D., Neese, F.: Angew. Chem. Int. Ed. **42**, 2942 (2003)
39. Zhang, Y., Gossman, W., Oldfield, E.: J. Am. Chem. Soc. **125**, 16387 (2003)
40. Zhang, Y., Oldfield, E.: J. Phys. Chem. **A 107**, 4147 (2003)
41. Zhang, Y., Oldfield, E.: J. Phys. Chem. **B 107**, 7180 (2003)
42. Garcia Serres, R., Grapperhaus, C.A., Bothe, E., Bill, E., Weyhermüller, T., Neese, F., Wieghardt, K.: J. Am. Chem. Soc. **126**, 5138 (2004)
43. Han, W.G., Lovell, T., Liu, T.Q., Noodleman, L.: Inorg. Chem. **43**, 613 (2004)
44. Kaizer, J., Klinker, E.J., Oh, N.Y., Rohde, J.-U., Song, W.J., Stubna, A., Kim, J., Münck, E., Nam, W., Que Jr., L.: J. Am. Chem. Soc. **126**, 472 (2004)
45. Liu, T.Q., Lovell, T., Han, W.G., Noodleman, L.: Inorg. Chem. **43**, 6858 (2004)
46. Zhang, Y., Oldfield, E.: J. Am. Chem. Soc. **126**, 9494 (2004)
47. Aliaga-Alcade, N., DeBeer George, S., Bill, E., Wieghardt, K., Neese, F.: Angew. Chem. Int. Ed. **44**, 2908 (2005)
48. Han, W.G., Liu, T.Q., Lovell, T., Noodleman, L.: J. Am. Chem. Soc. **127**, 15778 (2005)
49. Jensen, M.P., Costas, M., Ho, R.Y.N., Kaizer, J., Mairata i Payeras, A., Münck, E., Que Jr., L., Rohde, J.-U., Stubna, A.: J. Am. Chem. Soc. **127**, 10512 (2005)
50. Pestovsky, O., Stoian, S., Bominaar, E.L., Shan, X.P., Münck, E., Que Jr., L., Bakac, A.: Angew. Chem. Int. Ed. **44**, 6871 (2005)
51. Praneeth, V.K.K., Neese, F., Lehnert, N.: Inorg. Chem. **44**, 2570 (2005)
52. Ray, K., Begum, A., Weyhermüller, T., Piligkos, S., van Slageren, J., Neese, F., Wieghardt, K.: J. Am. Chem. Soc. **127**, 4403 (2005)
53. Sastri, C.V., Park, M.J., Ohta, T., Jackson, T.A., Stubna, A., Seo, M.S., Lee, J., Kim, J., Kitagawa, T., Münck, E., Que Jr., L., Nam, W.: J. Am. Chem. Soc. **127**, 12494 (2005)

54. Stoian, S., Pestovsky, O., Bominaar, E. L., Münck, E., Que, L., Jr., Bakac, A.: Mössbauer Evidence for an Fe(IV)=O Species in Acidic Aqueous Solution. In: ICBIC-12 Proceedings, Minneapolis 2005
55. Bart, S.C., Chlopek, K., Bill, E., Bouwkamp, M.W., Lobkovsky, E., Neese, F., Wieghardt, K., Chirik, P.J.: J. Am. Chem. Soc. **128**, 13901 (2006)
56. Berry, J.F., Bill, E., Bothe, E., George, S.D., Mienert, B., Neese, F., Wieghardt, K.: Science **312**, 1937 (2006)
57. Berry, J.F., Bill, E., Bothe, E., Neese, F., Wieghardt, K.: J. Am. Chem. Soc. **128**, 13515 (2006)
58. Berry, J.F., Bill, E., Garcia-Serres, R., Neese, F., Weyhermüller, T., Wieghardt, K.: Inorg. Chem. **45**, 2027 (2006)
59. Han, W.G., Liu, T.Q., Lovell, T., Noodleman, L.: J. Inorg. Biochem. **100**, 771 (2006)
60. Neese, F.: J. Inorg. Biochem. **100**, 716 (2006)
61. Patra, A.K., Bill, E., Bothe, E., Chlopek, K., Neese, F., Weyhermüller, T., Stobie, K., Ward, M.D., McCleverty, J.A., Wieghardt, K.: Inorg. Chem. **45**, 7877 (2006)
62. Kirchner, B., Wennmohs, F., Ye, S.F., Neese, F.: Curr. Opin. Chem. Biol. **11**, 134 (2007)
63. Petrenko, T., DeBeer George, S., Aliaga-Alcalde, N., Bill, E., Mienert, B., Xiao, Y., Guo, Y., Sturhahn, W., Cramer, S.P., Wieghardt, K., Neese, F.: J. Am. Chem. Soc. **129**, 11053 (2007)
64. Ray, K., Petrenko, T., Wieghardt, K., Neese, F.: Dalton Trans. 1552 (2007)
65. DeBeer George, S., Berry, J.F., Neese, F.: Phys. Chem. Chem. Phys. **10**, 4361 (2008)
66. Meyer, K., Bill, E., Mienert, B., Weyhermüller, T., Wieghardt, K.: J. Am. Chem. Soc. **121**, 4859 (1999)
67. Trautwein, A.X., Harris, F.E., Freeman, A.J., Desclaux, J.P.: Phys. Rev. **B 11**, 4101 (1975)
68. Mastalerz, R., Lindh, R., Reiher, M.: Chem. Phys. Lett. **456**, 157 (2008)
69. Meykens, A., Coussement, R., Ladrière, J., Cogneau, M., Bogè, M., Auric, P., Bouchez, R., Benabed, D., Godard, J.: Phys. Rev. **B 21**, 3816 (1980)
70. Daniel, H., Hartmann, F., Pitesa, B.: Z. Naturforsch. **A 40**, 539 (1985)
71. Filatov, M.: J. Chem. Phys. **127**, 084101 (2007)
72. Kurian, R., Filatov, M.: J. Chem. Theo. Comp. **4**, 278 (2008)
73. Filatov, M.: Coord. Chem. Rev. **253**, 594 (2009)
74. Schweiger, A., Jeschke, G.: Principles of Pulse Electron Paramagnetic Resonance. Oxford University Press, Oxford (2001)
75. Godbout, N., Havlin, R., Salzmann, R., Debrunner, P.G., Oldfield, E.: J. Phys. Chem. **A 102**, 2342 (1998)
76. Havlin, R.H., Godbout, N., Salzmann, R., Wojdelski, M., Arnold, W., Schulz, C.E., Oldfield, E.: J. Am. Chem. Soc. **120**, 3114 (1998)
77. Wolny, J.A., Paulsen, H., Winkler, H., Trautwein, A.X., Tuchagues, J.P.: Hyperfine Interact. **166**, 495 (2006)
78. Dufek, P., Blaha, P., Schwarz, K.: Phys. Rev. Lett. **75**, 3545 (1995)
79. Neese, F., Solomon, E.I.: Inorg. Chem. **37**, 6568 (1998)
80. Deaton, J.C., Gebhard, M.S., Solomon, E.I.: Inorg. Chem. **28**, 877 (1989)
81. Gebhard, M.S., Deaton, J.C., Koch, S.A., Millar, M., Solomon, E.I.: J. Am. Chem. Soc. **112**, 2217 (1990)
82. Price, J.C., Barr, E.W., Tirupati, B., Bollinger, J.M., Krebs, C.: Biochemistry **42**, 7497 (2003)
83. Rohde, J.-U., In, J.-H., Lim, M.H., Brennessel, W.W., Bukowski, M.R., Stubna, A., Münck, E., Nam, W., Que Jr., L.: Science **299**, 1037 (2003)
84. Hess, B.A., Marian, C.M., Wahlgren, U., Gropen, O.: Chem. Phys. Lett. **251**, 365 (1996)
85. Neese, F.: J. Chem. Phys. **122**, 034107 (2005)
86. Neese, F.: J. Chem. Phys. **118**, 3939 (2003)
87. Kossmann, S., Kirchner, B., Neese, F.: Mol. Phys. **105**, 2049 (2007)
88. Munzarova, M., Kaupp, M.: J. Phys. Chem. **A 103**, 9966 (1999)
89. Neese, F.: J. Chem. Phys. **115**, 11080 (2001)
90. Neese, F.: J. Chem. Phys. **127**, 164112 (2007)

91. Neese, F.: Int. J. Quant. Chem. **83**, 104 (2001)
92. Ganyushin, D., Neese, F.: J. Chem. Phys. **125**, 024103 (2006)
93. Neese, F.: J. Am. Chem. Soc. **128**, 10213 (2006)
94. Oosterhuis, W.T., Lang, G.: J. Chem. Phys. **58**, 4757 (1973)
95. Schöneboom, J.C., Neese, F., Thiel, W.: J. Am. Chem. Soc. **127**, 5840 (2005)
96. Paulsen, H., Winkler, H., Trautwein, A.X., Grünsteudel, H., Rusanov, V., Toftlund, H.: Phys. Rev. **B 59**, 975 (1999)
97. Paulsen, H., Benda, R., Herta, C., Schünemann, V., Chumakov, A.I., Duelund, L., Winkler, H., Toftlund, H., Trautwein, A.X.: Phys. Rev. Lett. **86**, 1351 (2001)
98. Sage, J.T., Paxson, C., Wyllie, G.R.A., Sturhahn, W., Durbin, S.M., Champion, P.M., Alp, E. E., Scheidt, W.R.: J. Phys. Condens. Matt. **13**, 7707 (2001)
99. Rai, B.K., Durbin, S.M., Prohofsky, E.W., Sage, J.T., Wyllie, G.R.A., Ellison, M.K., Scheidt, W.R., Sturhahn, W., Alp, E.E.: Phys. Rev. **E 66**, 051904 (2002)
100. Rai, B.K., Durbin, S.M., Prohofsky, E.W., Sage, J.T., Wyllie, G.R.A., Scheidt, W.R., Sturhahn, W., Alp, E.E.: Biophys. J. **82**, 2951 (2002)
101. Leu, B.M., Zgiersky, M.Z., Graeme, R.A.W., Scheidt, W.R., Sturhahn, W., Alp, E.E., Durbin, S.M., Sage, J.T.: J. Am. Chem. Soc. **126**, 4211 (2004)
102. Leu, B.M., Silvernail, N.J., Zgiersky, M.Z., Graeme, R.A.W., Ellison, M.K., Scheidt, W.R., Zhao, J., Sturhahn, W., Alp, E.E., Sage, J.T.: Biophys. J. **92**, 3764 (2007)
103. Achterhold, K., Ostermann, A., van Bürck, U., Potzel, W., Chumakov, A. I., Baron, A. Q. R., Rüffer, R., Parak, F.: Eur. Biophys. J. 221 (1997)
104. Parak, F., Achterhold, K.: Hyperfine Interact. **123–124**, 825 (1999)
105. Sage, J.T., Durbin, S.M., Sturhahn, W., Wharton, D.C., Champion, P.M., Hession, P., Sutter, J., Alp, E.E.: Phys. Rev. Lett. **86**, 4966 (2001)
106. Achterhold, K., Keppler, C., Ostermann, A., van Bürck, U., Sturhahn, W., Alp, E.E., Parak, F.: Phys. Rev. **E 65**, 051916 (2002)
107. Xiao, Y., Fisher, K., Smith, M.C., Newton, W.E., Case, D.A., George, S.J., Wang, H., Sturhahn, W., Alp, E.E., Zhao, J., Yoda, J., Cramer, S.P.: J. Am. Chem. Soc. **128**, 7608 (2006)
108. Aliaga-Alcalde, M., George, S.D., Mienert, B., Bill, E., Wieghardt, K., Neese, F.: Angew. Chem. Int. Ed. **44**, 2908 (2005)
109. Sÿrajer, V., Reinisch, L., Champion, P.M.: J. Am. Chem. Soc. **110**, 6656 (1988)
110. Sage, J.T., Champion, P.M.: Small substrate recognition in heme proteins. In: Suslick, K.S. (ed.) Comprehensive Supramolecular Chemistry. Pergamon, Oxford (1996)

Chapter 6
Magnetic Relaxation Phenomena

Steen Mørup*

6.1 Introduction

Studies of magnetic relaxation phenomena are important for understanding the properties of magnetic materials, and Mössbauer spectroscopy has contributed much to the elucidation of the magnetic dynamics in solids. In experimental studies of relaxation phenomena, the time scale of the experimental technique is a crucial parameter, because it determines the range of relaxation times that can be measured. DC magnetization measurements have a time scale on the order of seconds, and in AC magnetization measurements the time scale can be varied from $\sim$1 to $\sim 10^{-7}$ s by varying the frequency. Neutron scattering is sensitive to very fast processes with relaxation times as short as $\sim 10^{-16}$ s. In Mössbauer spectroscopy, the time scale, τ_M, is on the order of nanoseconds and therefore Mössbauer spectroscopy covers a range of relaxation times that are not easy to measure by other conventionally used techniques. Mössbauer spectroscopy has been extensively used for the study of relaxation in paramagnetic materials [1–4], including biological molecules [5], and has also been used in numerous studies of relaxation phenomena in magnetic nanoparticles [6–8]. In this chapter, a brief overview of the application of Mössbauer spectroscopy for the examination of relaxation phenomena is given, with an emphasis on the use of ^{57}Fe Mössbauer spectroscopy. First, a discussion of the spin Hamiltonian of paramagnetic ions and how Mössbauer spectra of samples with very long paramagnetic relaxation times can be influenced by the various terms in the spin Hamiltonian is presented. In Sect. 3, the influence on the Mössbauer line shape of relaxation is discussed and, in Sect. 4, how spin–spin and spin–lattice relaxation processes in paramagnetic materials can be studied by Mössbauer spectroscopy is described. Section 6.5 deals with relaxation phenomena in magnetic nanoparticles, and in Sect. 6.6, the use of Mössbauer

*Department of Physics, Technical University of Denmark, Building 307, 2800 Kongens Lyngby, Denmark

P. Gütlich et al., *Mössbauer Spectroscopy and Transition Metal Chemistry*,
DOI 10.1007/978-3-540-88428-6_6,

spectroscopy to give information on transverse relaxation in canted spin structures in ferrimagnetic materials is discussed.

6.2 Mössbauer Spectra of Samples with Slow Paramagnetic Relaxation

Although the unpaired electrons of a paramagnetic ion usually give rise to a large magnetic field at the nucleus, the Mössbauer spectra of most paramagnetic materials consist of doublets or singlets, because the magnetic relaxation time is so short that the nucleus only experiences the average value of the magnetic hyperfine field. In some cases, the relaxation time is on the same order as the time scale of Mössbauer spectroscopy, τ_M, and, as discussed in Sect. 6.3, this results in broadened spectra. In extreme cases, i.e. in low-temperature studies of magnetically dilute samples, the paramagnetic relaxation time may be long compared to τ_M. The magnetic hyperfine interaction then gives rise to magnetically split spectra with a spectral shape that crucially depends on the detailed form of the electronic wave functions which are determined by the spin Hamiltonian [9–12]

$$\hat{H}_S = \hat{H}_{cf} + \hat{H}_Z + \hat{H}_{dd} + \hat{H}_{ex} + \hat{H}_{hf} + \hat{H}_Q. \tag{6.1}$$

Here, $\hat{H}_{cf}$ is the crystal field spin Hamiltonian, which can be expressed by (neglecting the fourth order axial term)

$$\hat{H}_{cf} = D\left[\hat{S}_z^2 - \tfrac{1}{3}S(S+1) + \lambda(\hat{S}_x^2 - \hat{S}_y^2)\right] + \tfrac{1}{6}a\left[\hat{S}_\xi^4 + \hat{S}_\eta^4 + \hat{S}_\varsigma^4 - \tfrac{1}{5}S(S+1)(3S^2 + 3S - 1)\right]. \tag{6.2}$$

S is the ionic spin and D, λD, and a are crystal field parameters describing the strength of the axial, the rhombohedral, and the cubic crystal field terms, respectively. The coordinate system of the cubic crystal field (ξ, η, ς) may differ from that used to describe the axial and the rhombohedral crystal field interactions (x, y, z).

If an external magnetic field, $\vec{B}$, is applied, one has to include the Zeeman Hamiltonian

$$\hat{H}_Z = g\mu_B \vec{B} \cdot \hat{\vec{S}}, \tag{6.3}$$

where g is the Landé factor and μ_B is the Bohr magneton.

The dipole and exchange interactions of an ion, i, with the neighboring magnetic ions, j, can be described by the Hamiltonians $\hat{H}_{dd}$ and $\hat{H}_{ex}$

$$\hat{H}_{dd} = \frac{\mu_0}{4\pi} \sum_{j \neq i} g_i g_j \mu_B^2 r_{ij}^{-5} \left[\hat{\vec{S}}_j r_{ij}^2 - 3\vec{r}_{ij}(\hat{\vec{S}}_j \cdot \vec{r}_{ij})\right] \cdot \hat{\vec{S}}_i \tag{6.4}$$

and

$$\hat{H}_{ex} = -\widehat{\vec{S}_i} \cdot \sum_{j \neq i} 2J_{ij}\widehat{\vec{S}_j}, \tag{6.5}$$

where μ_0 is the vacuum permeability, $\vec{r}_{ij}$ is the vector connecting the ions i and j, and J_{ij} is the exchange coupling constant.

Finally, the spin Hamiltonian also contains contributions from the magnetic and quadrupole hyperfine interactions, $\hat{H}_{hf}$ and $\hat{H}_Q$ where

$$\hat{H}_{hf} = \widehat{\vec{S}} \cdot \tilde{A} \cdot \widehat{\vec{I}} \tag{6.6}$$

and

$$\hat{H}_Q = \frac{eQV_{zz}}{4I(2I-1)}\left[\hat{I}_z^2 - \tfrac{1}{3}I(I+1) + \tfrac{1}{3}\eta(\hat{I}_x^2 - \hat{I}_y^2)\right] \tag{6.7}$$

where I is the nuclear spin, $\tilde{A}$ is a hyperfine interaction tensor (which for Fe^{3+} in the high-spin state often can be approximated by a scalar), Q is the nuclear quadrupole moment, V_{zz} is the z component of the electric field gradient (EFG), and η is the asymmetry parameter.

The crystal field interaction gives rise to an energy splitting into a number of Kramers doublets. In the case of high-spin Fe^{3+} with spin $S = 5/2$, there are three Kramers doublets, each of which give rise to separate contributions in the Mössbauer spectra of samples with slow paramagnetic relaxation. For $\lambda = 0$ and $a = 0$, they can be labeled $|\pm 1/2\rangle$, $|\pm 3/2\rangle$ and $|\pm 5/2\rangle$.

In the interpretation of the Mössbauer spectra of samples with slow paramagnetic relaxation, it is important to realize that the eigenstates of the electronic system depend critically on the perturbations that lift the degeneracy of the Kramers doublets and on the relative size of the crystal field splitting and the Zeeman splitting. This can be illustrated by comparing spectra obtained in different applied magnetic fields. Figure 6.1 shows the magnetic field dependence of Mössbauer spectra at 4.5 K of Fe^{3+} ions with very slow relaxation [12]. The sample was a frozen aqueous solution with 0.03 M $Fe(NO_3)_3$ and an appropriate amount of ionic glass-formers, such that it was possible for the sample to be frozen to an amorphous state with a homogeneous distribution of $[Fe(H_2O)_6]^{3+}$ complexes [13, 14]. This is essential to ensure slow spin–spin relaxation (see Sect. 4.2), because the solubility of Fe^{3+} in crystalline ice is very low, and if crystallization of ice takes place during freezing, the iron ions will be concentrated in the regions between the ice crystals and this gives rise to fast spin–spin relaxation [13]. The magnetic fields indicated in Fig. 6.1 were applied perpendicular to the γ-ray direction. At the smallest fields, $\hat{H}_Z$, $\hat{H}_{dd}$ and $\hat{H}_{ex}$ are negligible, and the hyperfine interaction alone lifts the degeneracy of the Kramers doublets. The spectra, obtained in applied fields of 0.0001 and 0.0020 T, essentially consist of an asymmetric component superimposed on two almost identical symmetric sextets. The asymmetric component is due to one of the Kramers doublets for which the

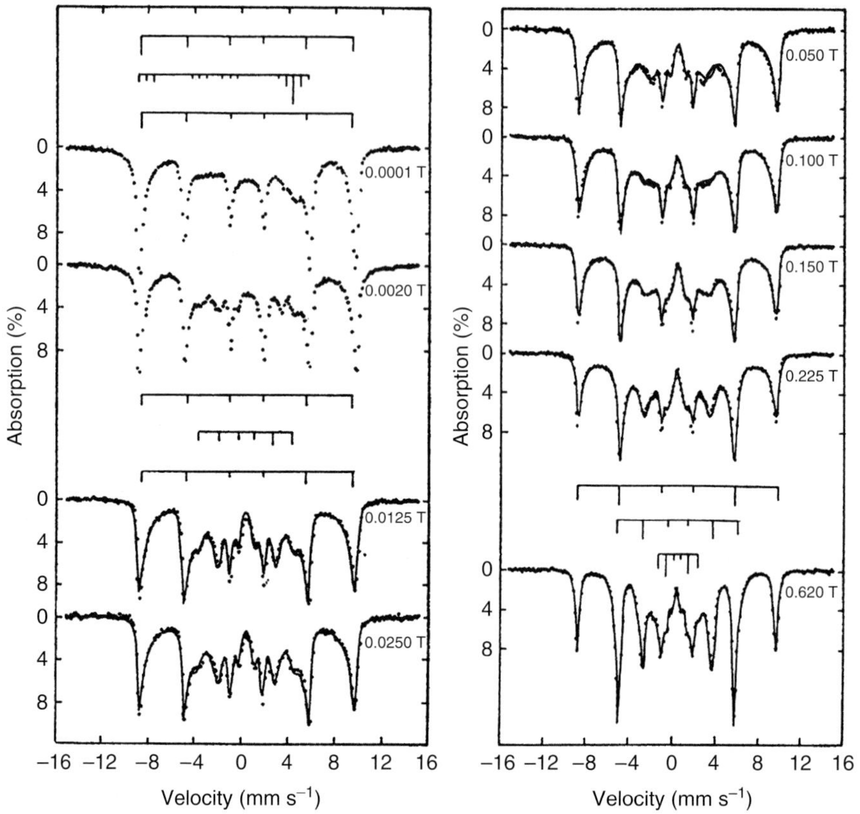

Fig. 6.1 Mössbauer spectra of an amorphous frozen aqueous solution of 0.03 M $Fe(NO_3)_3$, obtained at 4.5 K with various applied transverse magnetic fields. The bar diagrams indicate theoretical line positions of the spectral components. The lines are fits to the experimental data. (Reprinted with permission from [12]; copyright 1977 by Elsevier)

hyperfine interaction cannot be described in terms of an effective magnetic hyperfine field, because the eigenstates are combined electronic and nuclear states [1, 9, 12]. At applied magnetic fields of 0.0125 and 0.0250 T, the spectra still show the presence of the two nearly identical sextets, but the asymmetric component has transformed to a symmetric sextet with a relatively small splitting. This is because the Zeeman splitting with these applied fields is large compared to the hyperfine interaction such that the degeneracy of the Kramers doublets is lifted by the Zeeman interaction. With increasing applied fields, one can follow the evolution of the spectra when the Zeeman energy becomes comparable to, or larger than, the crystal field splitting. From fits of the whole series of spectra in Fig. 6.1, it was possible to estimate the crystal field parameters $D \cong 0.20\ \mathrm{cm}^{-1}$, $\lambda \cong 0.2$, and $a \cong 0.017\ \mathrm{cm}^{-1}$ [12].

When the Zeeman energy is large compared to the crystal field interaction, the electronic wave functions are approximately the $|S_z\rangle$ states with energy

$E \cong g\mu_B BS_z$. In this case, the magnetic hyperfine field is given by $B_{hf} \cong a_0 S_z$, where a_0 is a constant, which for Fe^{3+} in the high-spin state is on the order of 20 T. The magnetic splitting of the Mössbauer spectrum is then proportional to the vector sum of the magnetic hyperfine field and the applied magnetic field. At low temperatures, the populations of the electronic states differ, especially if large magnetic fields are applied, and this is reflected in the relative areas of the sextets [15].

Studies of the field dependence of Mössbauer spectra from other dilute frozen solutions with more than one type of complex have demonstrated that Mössbauer spectroscopy not only allows the determination of the crystal field parameters but can also give additional information about the relative amounts of different complexes such as $[Fe(H_2O)_{6-n}Cl_n]^{3-n}$ in frozen aqueous solutions containing Fe^{3+} and different concentrations of Cl^- [14]. It has been found that the lower symmetry of complexes with different ligands leads to increased crystal field splittings. The technique has also been used to study crystal field interaction in biological samples [5], and in a number of inorganic compounds with low iron concentration. Recently, it has been demonstrated that nuclear forward scattering of synchrotron radiation can also be used for studies of crystal field interactions in samples with slow paramagnetic relaxation [16].

6.3 Mössbauer Relaxation Spectra

The Mössbauer line shape in the presence of magnetic relaxation has been the subject of many theoretical studies [9, 17–26]. Mössbauer spectra are very sensitive to relaxation effects when the relaxation time is of the same order of magnitude as the nuclear Larmor precession time in the magnetic hyperfine field. In ^{57}Fe Mössbauer spectroscopy studies of Fe^{3+} with magnetic hyperfine fields on the order of 45–60 T, the Larmor precession time is on the order of nanoseconds. The detailed spectral shape depends on the way in which the hyperfine field fluctuates. In the simple case of longitudinal relaxation with a magnetic hyperfine field that can assume only the two values $+\vec{B}_{hf}$ and $-\vec{B}_{hf}$, the spectral shape can be calculated with relatively simple models [9, 22]. Theoretical spectra calculated by Wickman et al. [9] are shown in Fig. 6.2. For relaxation times $\tau \gg \tau_M$, the spectra consist of sextets with narrow lines. The lines start to broaden for relaxation times on the order of the mean life time of the excited nuclear state (~140 ns for ^{57}Fe). With decreasing relaxation time, the lines become further broadened and later they collapse pair-wise, first lines 3 and 4, then lines 2 and 5, and finally lines 1 and 6. The magnetic splitting of a pair of lines collapses at a critical relaxation time $\tau_{cr}(m_e, m_g)$ [22]

$$\tau_{cr}(m_e, m_g) = \frac{\hbar}{\left|(g_e m_e - g_g m_g)\mu_N B_{hf}\right|}, \tag{6.8}$$

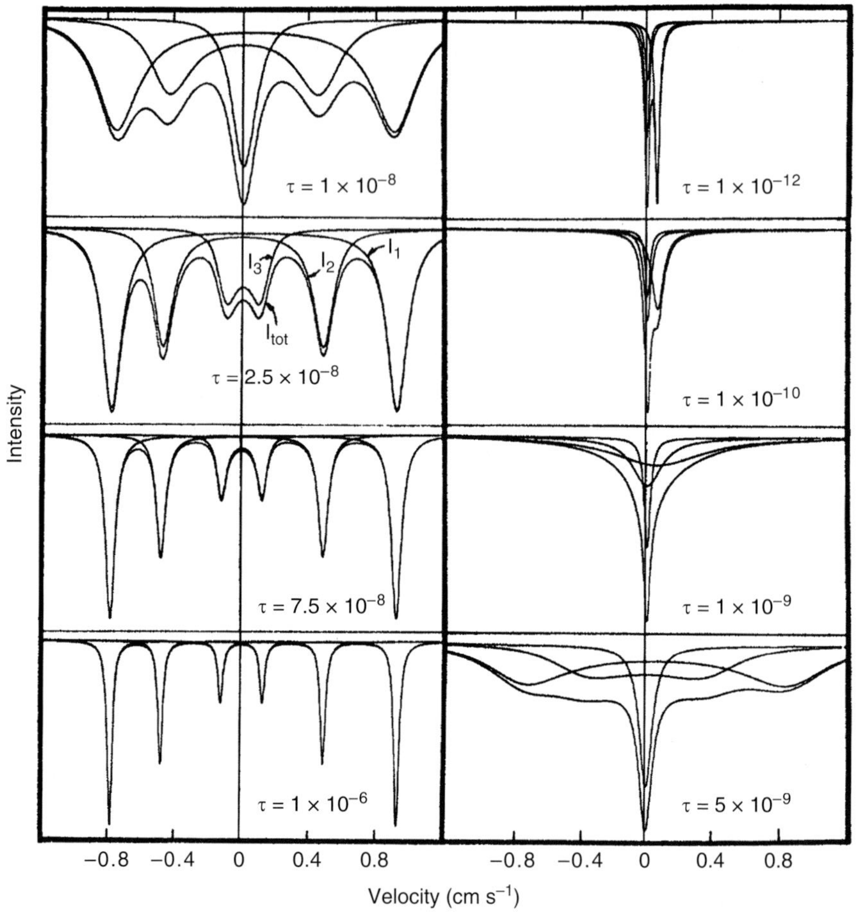

Fig. 6.2 Theoretical ^{57}Fe Mössbauer relaxation spectra for longitudinal relaxation with the indicated relaxation times and with a hyperfine field that can assume the values ± 55 T. The symmetry direction of the axially symmetric EFG is assumed parallel to the magnetic hyperfine field. (Reprinted with permission from [9]; copyright 1966 by the American Physical Society)

where g and m are the nuclear g-factors and the z components of the nuclear spins, respectively, and the subscripts e and g refer to the excited and the ground states. μ_{N} is the nuclear magneton. For $B_{\mathrm{hf}} = 55\,\mathrm{T}$, lines 1 and 6, lines 2 and 5, and lines 3 and 4 collapse for $\tau_{cr} = 1.6\,\mathrm{ns}$, $\tau_{cr} = 2.7\,\mathrm{ns}$ and $\tau_{cr} = 9.8\,\mathrm{ns}$, respectively. Thus, there is not a uniquely defined critical relaxation time for which the magnetic splitting of the whole spectrum collapses.

For relaxation times $\tau \lesssim 1$ ns, the spectra can be described as three Lorentzian lines with different line width, and for relaxation times around 10^{-10} s, the spectra appear as asymmetric doublets with line widths that decrease with decreasing relaxation time. In the theoretical spectra in Fig. 6.2, the EFG was assumed uniaxial with the symmetry direction parallel to the magnetic hyperfine field. In this case, the line corresponding to the $\pm 1/2 \rightarrow \pm 3/2$ transition is broader than the line corresponding to the $\pm 1/2 \rightarrow \pm 1/2$ and the $\pm 1/2 \rightarrow \mp 1/2$ transitions. If the symmetry direction of the EFG is perpendicular to the fluctuating magnetic hyperfine field, the asymmetry of the doublet spectra is opposite [22].

The simple model with a hyperfine field that fluctuates between only two values $+\vec{B}_{hf}$ and $-\vec{B}_{hf}$ may be a fair approximation in some studies of superparamagnetic relaxation in magnetic nanoparticles (see Sect. 6.5) but in the case of paramagnetic relaxation, the magnetic hyperfine field can usually assume several different values. As discussed in Sect. 6.2, Fe^{3+} in the high-spin state with spin $S = 5/2$ has six electronic states with different magnetic hyperfine interactions. If a large magnetic field is applied, such that $\hat{H}_Z$ is predominant in the spin Hamiltonian, the electronic states are the $|S_z\rangle$ states, and the relaxation is still longitudinal, but the magnetic hyperfine field can fluctuate between the six different values with $B_{hf} = a_0 S_z$. In other cases, the hyperfine field may fluctuate in different directions and this makes calculations of relaxation spectra more complex [1, 23–26]. The evolution of the spectra as a function of relaxation time is qualitatively similar to that of the longitudinal case, but the detailed line shape of the spectra depends on the way in which the magnetic hyperfine field fluctuates.

In many paramagnetic samples, such as ferric compounds with moderate dipole and exchange interaction between the Fe^{3+} ions, the paramagnetic relaxation results in spectra consisting of singlets or doublets with broadened lines [2, 3] due to relaxation times in the range 10^{-11}–10^{-9} s. Doublet spectra are usually asymmetrically broadened, and the asymmetry depends on the way in which the hyperfine field fluctuates relative to the EFG (e.g. longitudinal, isotropic, or isotropic transverse [26]). If the sample is exposed to a magnetic field, the directions of the fluctuating hyperfine fields are changed. Therefore, magnetic fields can have a significant influence on the asymmetry of the spectra [2, 3]. In $FeCl_3{\cdot}6H_2O$, the Fe^{3+} ions are surrounded by four water molecules and two Cl^- ions in the *trans* positions. This leads to a nearly axial symmetry around the Fe^{3+} ions. The Mössbauer spectrum of polycrystalline $FeCl_3{\cdot}6H_2O$ at 78 K in a small field (Fig. 6.3, upper spectrum) is an asymmetric doublet, because the hyperfine field fluctuates longitudinally along the symmetry direction of the EFG. When a field of 1.3 T is applied, the hyperfine field fluctuates in directions that are determined by the combined effect of the crystal field and the applied magnetic field. The angles between the hyperfine fields and the symmetry direction of the EFG are then more random and this gives rise to a more symmetric spectrum (Fig. 6.3, lower spectrum).

In samples with negligible quadrupole interaction, isotropic relaxation with relaxation times around 10^{-11}–10^{-9} s results in spectra consisting of a single, broad Lorentzian line [2, 3, 21, 26], but if the relaxation is longitudinal, the spectrum consists of a superposition of three singlets with line widths (FWHM) [2, 3]

$$\Gamma = \Gamma_0 + (g_e m_e - g_g m_g)^2 \mu_N^2 \hbar^{-1} \langle B_{hf}^2 \rangle \tau_1, \tag{6.9}$$

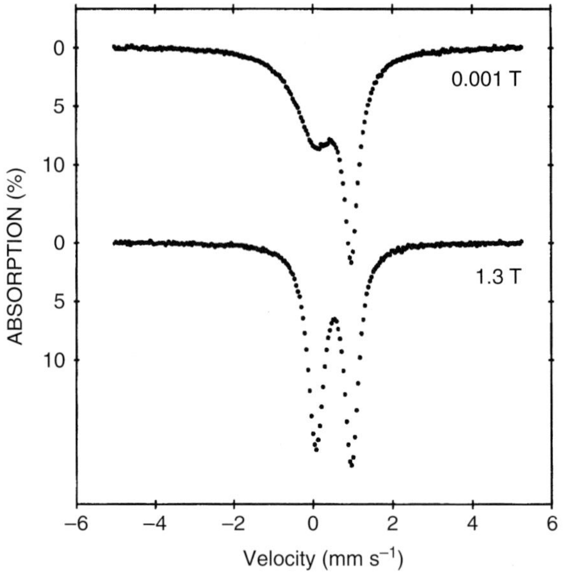

Fig. 6.3 Mössbauer spectra of $FeCl_3{\cdot}6H_2O$ at 78 K in applied magnetic fields of 0.001 and 1.3 T

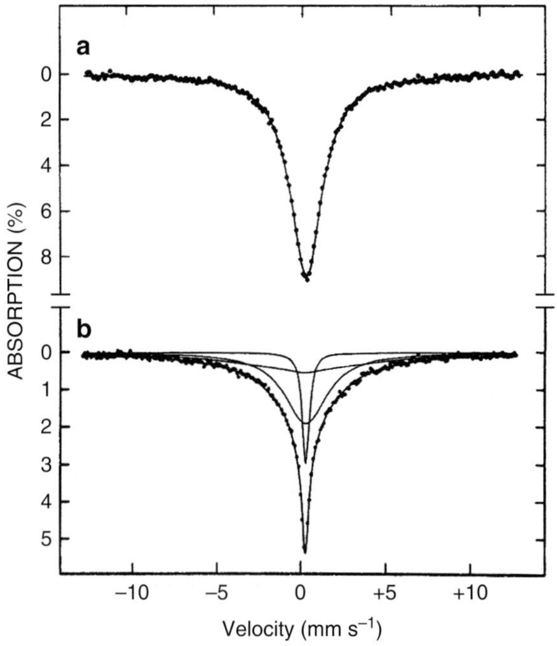

Fig. 6.4 Mössbauer spectra of $NH_4Fe(SO_4)_2{\cdot}12H_2O$ at 113 K. (**a**) In zero applied magnetic field; (**b**) In a transverse magnetic field of 0.8 T

where Γ_0 is the line width when the relaxation broadening is negligible and τ_1 is the spin correlation time for longitudinal relaxation. Figure 6.4 shows the experimental Mössbauer spectra of ferric alum ($NH_4Fe(SO_4)_2 \cdot 12H_2O$), which illustrate this [2, 3]. The spectrum (a) was obtained in zero applied magnetic field and is well fitted with a single Lorentzian line indicating isotropic relaxation. This is in accordance with the very small crystal field splitting in ferric alum resulting in a spin Hamiltonian in which the dipole interaction is predominant [2, 3]. When a magnetic field of 0.8 T is applied (Fig. 6.4b), the predominant term in the spin Hamiltonian is the Zeeman term (6.3), and the hyperfine field then fluctuates parallel and antiparallel to the applied field. As shown in Fig. 6.4, the spectrum can be fitted with broad and narrow components with line widths given by (6.9). By use of a polarized source of ^{57}Co in metallic iron, it is possible to separate the three components with different line widths and such measurements have confirmed the interpretation [27]. Similar data have been obtained for $Fe(ClO_4)_3{\cdot}6H_2O$ and cubic $(NH_4)_3FeF_6$ [28].

The aforementioned examples illustrate that singlet and doublet spectra with line broadening due to relaxation effects may be very sensitive to applied magnetic fields. Furthermore, as discussed in Sect. 6.4, the relaxation time may also be influenced by magnetic fields. Therefore, application of magnetic fields can be very useful to distinguish between line broadening due to relaxation effects and line broadening due to, for example, distributions of isomer shifts and quadrupole splittings.

If large magnetic fields are applied to a paramagnetic sample at low temperatures, the induced magnetization is proportional to the appropriate Brillouin function. The magnetic field at the nucleus can then be described as a sum of the average hyperfine field $\langle \vec{B}_{hf} \rangle$ (which is proportional to the Brillouin function), the applied field $\vec{B}$, and a fluctuating part of the hyperfine field, $\vec{B}_f(t)$. If the paramagnetic relaxation time is on the order of 10^{-11}–10^{-9} s, the spectra will show a line broadening that is proportional to the relaxation time and to $\langle B_f^2(t) \rangle$, which decreases with increasing applied fields [19, 29]. The broadening differs for the nuclear transitions, such that lines 1 and 6 will be most broadened and lines 3 and 4 least broadened. Furthermore, the line positions will be slightly shifted relative to the line positions in spectra of samples with infinitely fast relaxation because of relaxation effects [19, 29]. For example, Fig. 6.5 shows Mössbauer spectra of ferric alum obtained at 4.2 K in the indicated magnetic fields applied parallel to the γ-ray direction [29]. Lines 2 and 5 have negligible intensity because the magnetic field at the nuclei is parallel to the γ-ray direction. The lines are fits to a relaxation model and the bar diagrams indicate the expected line position in the case of infinitely fast relaxation [29].

For very short magnetic relaxation times, $\tau \lesssim 10^{-12}$ s, the line broadening due to relaxation is negligible, and the magnetic splitting of the spectra is proportional to the average value of the magnetic field at the nucleus, i.e. it vanishes when the average magnetic field at the nucleus is zero.

Relaxation phenomena can also be studied by nuclear forward scattering of synchrotron radiation [16, 30]. This is discussed in Chap. 9.

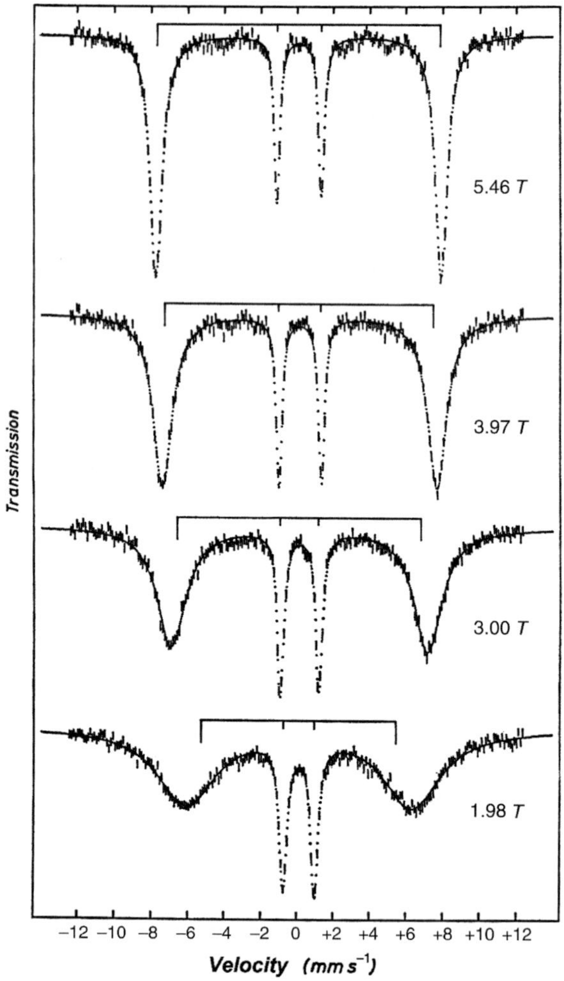

Fig. 6.5 Mössbauer spectra of $NH_4Fe(SO_4)_2{\cdot}12H_2O$ at 4.2 K and with the indicated magnetic fields applied parallel to the γ-ray direction. The lines indicate fits in accordance with a theoretical relaxation model [19, 29]. The bar diagrams indicate the theoretical line positions in the case of infinitely fast relaxation. (Adapted from [29]; copyright 1973 by Springer-Verlag)

6.4 Paramagnetic Relaxation Processes

In paramagnetic materials, the relaxation frequency is in general determined by contributions from both spin–lattice relaxation and spin–spin relaxation. Spin–lattice relaxation processes can conveniently be studied in samples with low concentrations of paramagnetic ions because this results in slow spin–spin relaxation. Spin–spin relaxation processes can be investigated at low temperatures where the spin–lattice relaxation is negligible. Paramagnetic relaxation processes have conventionally been studied by AC susceptibility measurements with time scale typically in the range from $\sim$1 to $\sim 10^{-7}$ s. Because Mössbauer spectroscopy has a time scale on the order of nanoseconds, this technique makes it possible to study spin–spin and spin–lattice relaxation processes that cannot be studied by AC susceptibility measurements. The following illustrates how Mössbauer spectroscopy can give information about spin–lattice and spin–spin relaxation processes.

6.4.1 Spin–Lattice Relaxation

As discussed in Sect. 6.2, the electronic states of a paramagnetic ion are determined by the spin Hamiltonian, (6.1). At finite temperatures, the crystal field is modulated because of thermal oscillations of the ligands. This results in spin–lattice relaxation, i.e. transitions between the electronic eigenstates induced by interactions between the ionic spin and the phonons [10, 11, 31, 32]. The spin–lattice relaxation frequency increases with increasing temperature because of the temperature dependence of the population of the phonon states. For high-spin Fe^{3+}, the coupling between the spin and the lattice is weak because of the spherical symmetry of the ^6S ground state. This leads to small crystal field splittings and relatively long spin–lattice relaxation times, τ_{sl}, which are often in the range where they can be studied by Mössbauer spectroscopy, even at room temperature. For nonspherical ions, like Fe^{2+}, the coupling to the lattice is stronger resulting in short spin–lattice relaxation times and the Mössbauer spectra may be influenced by relaxation effects only at low temperatures.

Several types of spin–lattice relaxation processes have been described in the literature [31]. Here a brief overview of some of the most important ones is given. The simplest spin–lattice process is the direct process in which a spin transition is accompanied by the creation or annihilation of a single phonon such that the electronic spin transition energy, Δ, is exchanged by the phonon energy, $\hbar\omega_q$. Using the Debye model for the phonon spectrum, one finds for $k_B T \gg \Delta$ that

$$\tau_{\mathrm{sl}}^{-1} \propto \Delta^2 T. \tag{6.10}$$

If the Zeeman splitting is large compared to the crystal field splitting, this leads to $\tau_{sl}^{-1} \propto B^2 T$. Usually, the direct process is important only compared to other spin–lattice processes at low temperatures, because only low-energy phonons with $\hbar\omega_q = \Delta$ contribute to the direct process.

At higher temperatures, the two-phonon (Raman) processes may be predominant. In such a process, a phonon with energy $\hbar\omega_q$ is annihilated and a phonon with energy $\hbar\omega_r$ is created. The energy difference $\hbar\omega_q - \hbar\omega_r$ is taken up in a transition of the electronic spin. In the Debye approximation for the phonon spectrum, this gives rise to a relaxation rate given by

$$\tau_{\mathrm{sl}}^{-1} \propto \int_0^{\omega_\mathrm{D}} \frac{\omega^6 \exp(\hbar\omega/k_\mathrm{B}T)\mathrm{d}\omega}{[\exp(\hbar\omega/k_\mathrm{B}T) - 1]^2}, \tag{6.11}$$

resulting in $\tau_{sl}^{-1} \propto T^7$ for $T \ll \theta_D$ and $\tau_{sl}^{-1} \propto T^2$ for $T \gg \theta_D$, where θ_D is the Debye temperature and $\omega_D = (k_B/\hbar)\theta_D$.

If optical phonons are responsible for the Raman processes, the Einstein model for the phonon spectrum is more appropriate. In this case, one finds

$$\tau_{sl}^{-1} = \frac{\exp(-\theta_E/T)}{[1 - \exp(-\theta_E/T)]^2}, \tag{6.12}$$

where θ_E is the Einstein temperature. In large applied magnetic fields, no field dependence of the Raman processes is expected. For $T \ll \theta_E$, one finds that $\tau_{sl}^{-1} \propto \exp(-\theta_E/T)$ and for $T \gtrsim \theta_E$, $\tau_{sl}^{-1} \propto T^2$.

The Orbach process is a two-phonon process that takes place via population of an excited electronic state with energy E_0. The temperature dependence of the relaxation rate is given by

$$\tau_{sl}^{-1} \propto \frac{1}{\exp(E_0/k_BT) - 1}. \tag{6.13}$$

As an example of a Mössbauer study of spin–lattice relaxation, Fig. 6.6 shows the spectra of Fe^{3+} in the ammonium alum, $NH_4(^{57}Fe_{0.02}Al_{0.98})(SO_4)_2 \cdot 12H_2O$ [32]. The crystal field splitting of Fe^{3+} in this compound is small ($D \approx 0.024\ \text{cm}^{-1}$). The spectra were obtained at the indicated temperatures in a magnetic field of 1.23 T applied perpendicular to the γ-ray direction. This field is sufficiently large to ensure that the Zeeman energy is much larger than the crystal field energy such that the electronic states are the $|S_z\rangle$ states. Because of the low concentration of Fe^{3+} ions, spin–spin relaxation is almost negligible and the relaxation is therefore dominated by spin–lattice relaxation. At the lowest temperatures, the spectra are magnetically split because of slow relaxation but, as the temperature increases, the spectra gradually collapse to a broadened singlet. The solid lines are fits to a model for longitudinal relaxation and the spin–lattice relaxation frequencies, shown in Fig. 6.7, were estimated from the fits. The temperature dependence of the relaxation frequency was in accordance neither with the Debye model for direct processes nor with the Debye model for Raman processes [32]. A fit based on the Einstein model for Raman processes (6.12) with $\theta_E = 450\,\text{K}$ is shown by the broken curve in Fig. 6.7. The model fits the data well at low and high temperatures but, at intermediate temperatures, there are significant deviations from the model. Moreover, it was found that in this temperature range, the relaxation time depends on the strength of the applied magnetic field in contrast to the expectation for a Raman process.

Ammonium alums undergo phase transitions at $T_c \approx 80\,\text{K}$. The phase transitions result in critical lattice fluctuations which are very slow close to T_c. The contribution to the relaxation frequency, shown by the dotted line in Fig. 6.7, was calculated using a model for direct spin–lattice relaxation processes due to interaction between the low-energy critical phonon modes and electronic spins.

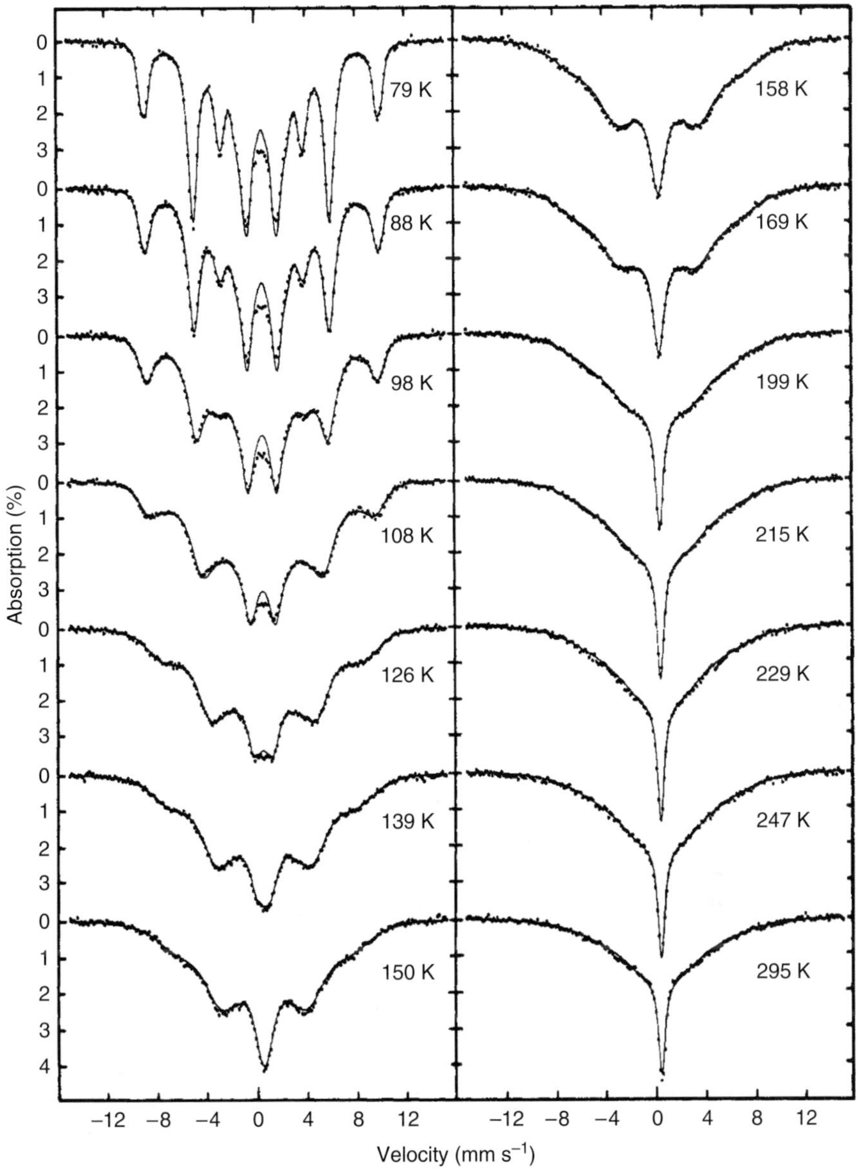

Fig. 6.6 Mössbauer spectra of $NH_4(^{57}Fe_{0.02}Al_{0.98})(SO_4)_2{\cdot}12H_2O$ obtained at the indicated temperatures in a transverse magnetic field of 1.23 T. The full lines are fits to a model for longitudinal relaxation. (Reprinted with permission from [32]; copyright 1979 by the Institute of Physics)

This model also explains the unexpected magnetic field dependence of the relaxation time in a temperature range in which Raman processes are normally expected to be predominant [32].

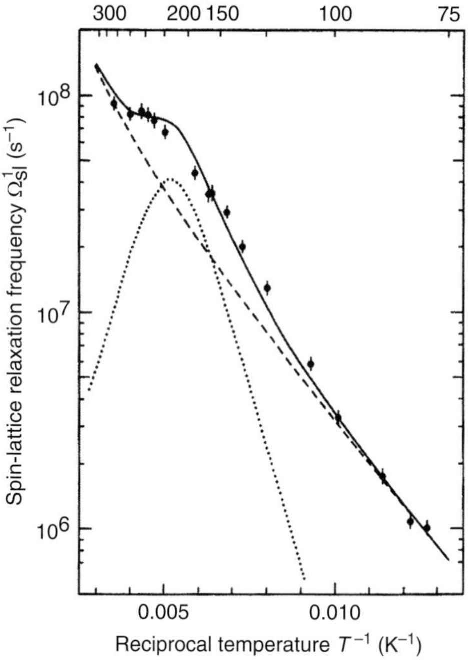

Fig. 6.7 Spin–lattice relaxation frequency of Fe^{3+} in $NH_4(^{57}Fe_{0.02}Al_{0.98})(SO_4)_2{\cdot}12H_2O$ as a function of the reciprocal temperature. The points are data obtained from fits of the Mössbauer spectra (Fig. 6.6). The broken curve is a fit to the Einstein model for a Raman process. The dotted curve corresponds to a contribution from a direct process due to interactions between the electronic spins and low-energy phonons associated with critical fluctuations near the phase transition temperature. (Reprinted with permission from [32]; copyright 1979 by the Institute of Physics)

Because Fe^{2+} ions in the high-spin state are more strongly coupled to the lattice than Fe^{3+}, relaxation effects may be seen only in low-temperature Mössbauer spectra of ferrous compounds [33–36]. Figure 6.8 shows Mössbauer spectra of Fe^{2+} in deoxy-myoglobin at temperatures from 4.2 to 50 K in applied fields of 2 and 6.2 T [34]. The temperature dependence of the spectra reflects both changes in the populations of the electronic states and changes in relaxation time. The solid lines are theoretical spectra, calculated by use of a relaxation model. In this study, it was found that the crystal field splitting was considerably larger than in typical Fe^{3+} high-spin compounds ($D = -10\ \mathrm{cm}^{-1}$) and that the temperature dependence of the relaxation time was in accordance with a direct process. Similar results have been obtained in Mössbauer studies of, for example, ferrous rubredoxin [33] and in other Fe^{2+} high-spin complexes [35, 36].

6.4.2 Spin–Spin Relaxation

Spin–spin relaxation is primarily induced by magnetic dipole interactions between paramagnetic ions. Usually, the most important spin–spin relaxation process is the so-called cross-relaxation process in which a transition of an ion i from the state $|M_a^i\rangle$ to the state $|M_{a+\alpha}^i\rangle$ is accompanied by a transition of another ion j from the

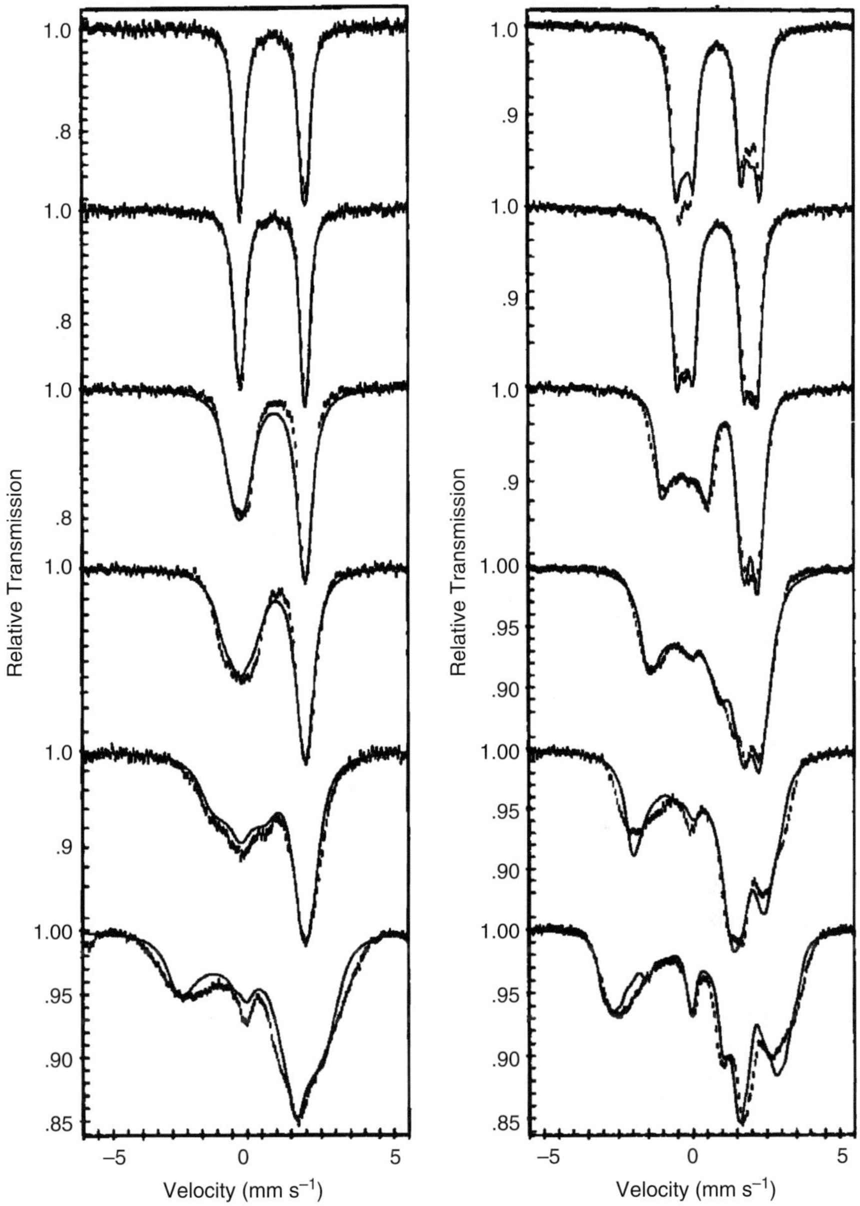

Fig. 6.8 Mössbauer spectra of deoxy-myoglobin, obtained in applied fields of 2 T (*left*) and 6.2 T (*right*) at temperatures of 4.2, 10, 15, 20, 30 and 50 K (*from bottom to top*). The solid lines were calculated using a relaxation model. (Reprinted from [34]; copyright 1994 by Springer-Verlag)

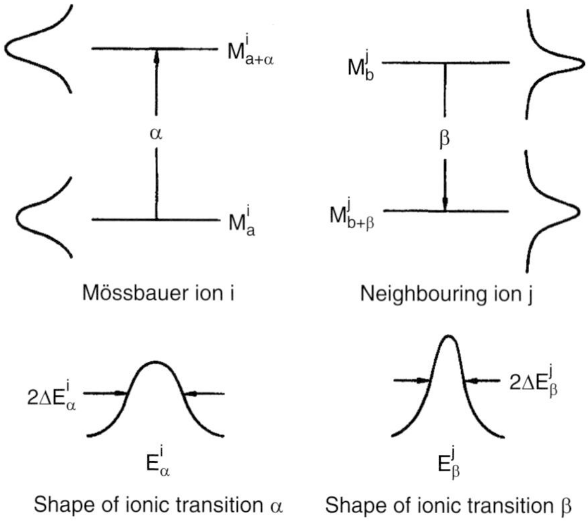

Fig. 6.9 Schematic illustration of a cross-relaxation process by which the ion i performs a transition α, accompanied by a transition β of a neighboring ion, j. (Reprinted with permission from [38]; copyright 1979 by Elsevier)

state $|M^j_b\rangle$ to the state $|M^j_{b+\beta}\rangle$. The process results in an increase of the energy of the ion i by an amount E^i_α and a decrease of the energy of the ion j by the amount E^j_β. The energy difference is taken up by the dipolar energy of the whole spin system [37]. Figure 6.9 shows a schematic illustration of a cross-relaxation process [38]. The energy levels are broadened, mainly because of random dipolar fields from the neighboring ions, and therefore the transition energies also show an energy distribution. If the shape functions of the transitions α and β are assumed Gaussian with second order moments $\left(\Delta E^i_\alpha\right)^2$ and $\left(\Delta E^j_\beta\right)^2$, the expression for the transition probability may be written as [37]

$$\Omega^{ij}_{ss} \cong \frac{(2\pi)^{-1/2}}{\hbar}\frac{\left|\left\langle M^i_{a+\alpha}, M^j_{b+\beta}\left|\hat{H}^{ij}_{dd}\right|M^i_a, M^j_b\right\rangle\right|^2}{\left[(\Delta E^i_\alpha)^2 + (\Delta E^j_\beta)^2\right]^{1/2}} \exp\left[-\frac{(E^i_\alpha - E^j_\beta)^2}{2\left[(\Delta E^i_\alpha)^2 + (\Delta E^j_\beta)^2\right]}\right] p_j(M^j_b). \tag{6.14}$$

$\hat{H}^{ij}_{dd}$ is the Hamiltonian describing the dipole interaction between the ions i and j (6.4), and $p_j(M^j_b)$ is the probability that the ion j is in the initial state $|M^j_b\rangle$. If the exchange interaction between the ions i and j is not negligible, $\hat{H}_{\mathrm{ex}}$ may also contribute to the spin–spin relaxation.

The dipole interaction depends on the distance between the ions (6.4). Therefore, the transition probability increases with increasing concentration of magnetic ions. Studies of the concentration dependence of the relaxation can be conveniently performed on samples of amorphous frozen solutions with a uniform distribution

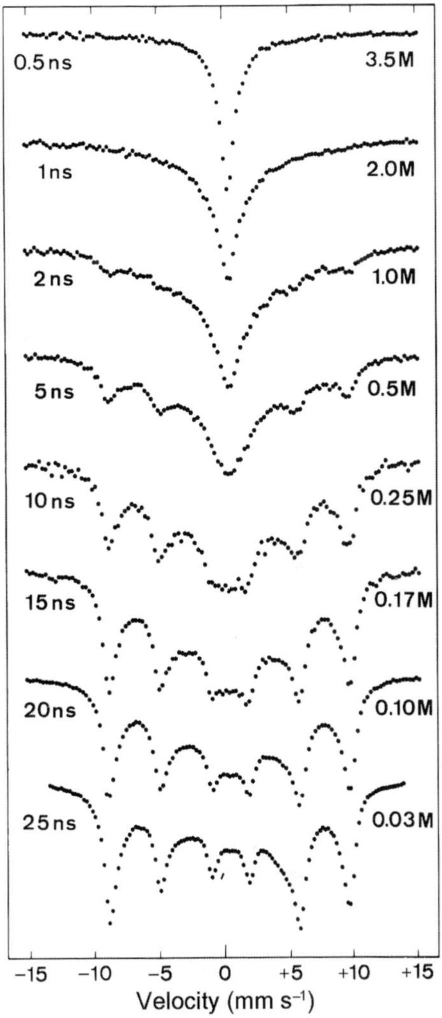

Fig. 6.10 Mössbauer spectra of amorphous frozen aqueous solutions with the indicated concentrations of $[Fe(H_2O)_6]^{3+}$. The spectra were obtained at 80 K. Rough estimates of the relaxation times are given

of the paramagnetic ions. Figure 6.10 shows Mössbauer spectra of such frozen aqueous solutions of $Fe(NO_3)_3$ at 80 K with the indicated concentrations of Fe^{3+} ions [14]. Approximate values of the relaxation times are also indicated. The data show that the spin–spin relaxation frequency is approximately proportional to the concentration of paramagnetic ions.

According to (6.14), spin–spin relaxation may be fast only if the transition energies E^i_α and E^j_β are not too different. In some cases, the difference between the transition energies can be varied by the application of a magnetic field and one can then investigate at least qualitatively the validity of (6.14). One example is $Fe(NO_3)_3 \cdot 9H_2O$ [39–41], which has a monoclinic crystal structure with space

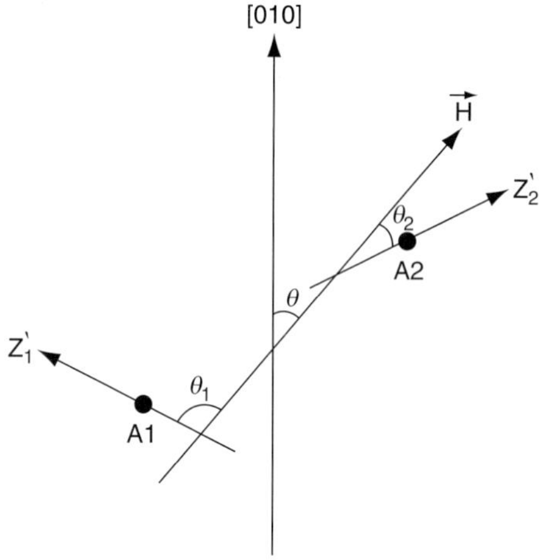

Fig. 6.11 Schematic illustration of two Fe^{3+} ions, A1 and A2, in $Fe(NO_3)_3 \cdot 9H_2O$, which are connected by a screw dyad axis parallel to the [010] direction. The crystal field axes z_1' and z_2' form the angles θ_1 and θ_2 with the applied magnetic field. (Reprinted with permission from [41]; copyright 1978 by Elsevier)

group $P2_1/c$. The four Fe^{3+} ions in the unit cell are connected in pairs by screw dyad axes parallel to the [010] direction. Therefore, the two Fe^{3+} ions in each pair have identical crystal field splittings, but the directions of the crystal field axes differ. When a magnetic field is applied at an angle θ to the [010] direction, the angles θ_1 and θ_2 between the crystal field axes and the magnetic field at the two sites in general differ, as shown schematically in Fig. 6.11 [41]. This gives rise to different energy splitting of the electronic states of the two ions, resulting in different transition energies. Consequently, the spin–spin relaxation time should in general increase when a magnetic field is applied [39–41]. However, as it can be seen from Fig. 6.11, the angles θ_1 and θ_2 will be identical if the magnetic field is applied parallel or perpendicular to the [010] direction, and the relaxation should therefore be faster in these cases. Figure 6.12 shows the Mössbauer spectra recorded from single crystals of $Fe(NO_3)_3 \cdot 9H_2O$ at 78 K with a magnetic field of 1.3 T applied at different angles to the [010] direction [40]. For $\theta = 0°$ and $\theta = 90°$, the spectra consist of broad components, but without indication of sextets. At other angles, there are magnetically split components in the spectra, indicating slower relaxation. This shows that the relaxation in fact is faster for $\theta = 0°$ and $\theta = 90°$ than for other orientations of the magnetic field.

In large applied magnetic fields at low temperatures, the ionic spins in a paramagnetic sample are to a large extent aligned and this leads to a field-dependent narrowing of the distribution of dipolar fields, i.e. $\left(\Delta E_\alpha^i\right)^2$ and $\left(\Delta E_\beta^j\right)^2$ decrease with increasing applied fields. For cross-relaxation processes involving two ions with identical transition energies ($E_\alpha^i = E_\beta^j$), this leads, according to (6.14), to an increase of the relaxation frequency with increasing applied field. This effect has been demonstrated in a study of $Fe(NO_3)_3 \cdot 9H_2O$ at 4.2 K in magnetic fields of up to 8 T [41].

In amorphous frozen solutions with only one type of species (e.g. $[Fe(H_2O)_6]^{3+}$) the crystal field interaction of the Fe^{3+} ions may be similar, but the orientations of the crystal field axes in general differ. When magnetic fields are applied, this

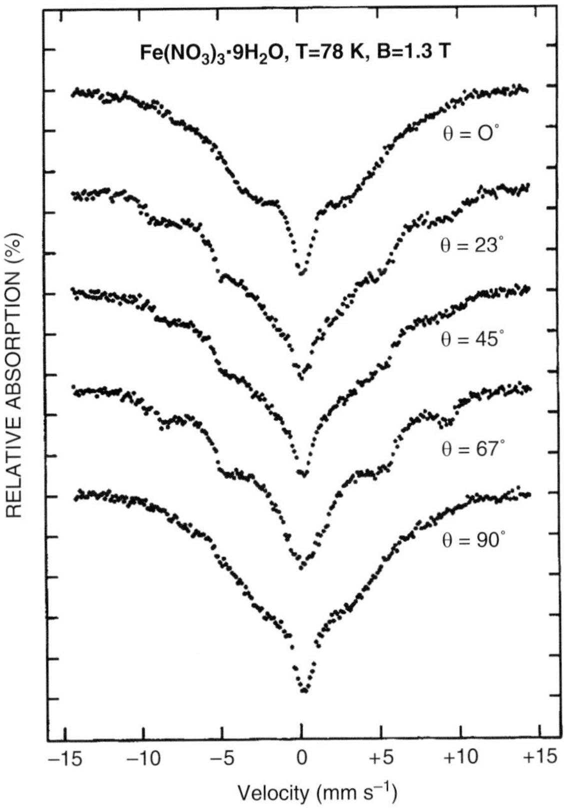

Fig. 6.12 Mössbauer spectra of a single crystal absorber of $Fe(NO_3)_3 \cdot 9H_2O$ with the absorber plane parallel to the (001) plane. The spectra were obtained at 78 K with a magnetic field of 1.3 T applied in various directions in the absorber plane. θ is the angle between the applied field and the [010] direction. (Reprinted with permission from [40]; copyright 1974 by Elsevier)

leads to different energy splitting of the ionic eigenstates, resulting in an increase in the spin–spin relaxation time as in crystalline $Fe(NO_3)_3 \cdot 9H_2O$ [2, 13, 39]. If more than one type of Fe^{3+} complex with different crystal field splitting is present, e.g. $[Fe(H_2O)_6]^{3+}$ and $[Fe(H_2O)_5Cl]^{2+}$, the relaxation time in zero applied field will be longer than if all Fe^{3+} ions were in identical local environments [42].

The spin–spin relaxation time is expected to be essentially temperature independent, because the dipole interaction does not depend on temperature. However, at low temperatures, a temperature dependence may be found due to a temperature dependence of the population of the electronic states. Consider an Fe^{3+} ion with spin Hamiltonian given by (6.2) with $\lambda = 0$ and $a = 0$. In this case the energies of the $|\pm 5/2\rangle$, the $|\pm 3/2\rangle$ and the $|\pm 1/2\rangle$ Kramers doublets are $(10/3)D$, $-(2/3)D$ and $-(8/3)D$, respectively. For $D > 0$, the $|\pm 1/2\rangle$ doublet has the lowest energy, and at low temperatures it therefore has the highest population. The matrix elements in (6.14) for the $|\pm 1/2\rangle \Leftrightarrow |\mp 1/2\rangle$ transitions are larger than those for other transitions [43]. Furthermore, at low temperatures, there is a large probability for an ion in the $|+1/2\rangle$ or the $|-1/2\rangle$ state to have neighbor ions in the appropriate state for such a cross-relaxation process. Therefore, the relaxation time decreases with decreasing temperature. This effect has been observed in ferric hemin [43] and in $FeCl_3 \cdot 6H_2O$ [44]. If $D < 0$, the $|\pm 5/2\rangle$ states have the lowest energy. Because the value of S_z only can be changed by ± 1 during a cross-relaxation process [37], the relaxation becomes very slow at low temperatures where the population of the other states is small. This has been seen in Mössbauer studies of iron(III) pyrrolidyl-dithiocarbamate [45].

6.5 Relaxation in Magnetic Nanoparticles

Mössbauer spectroscopy has been extensively used for studies of nanostructured materials and several reviews on magnetic nanoparticles have been published, see e.g. [6–8, 46–48]. The magnetic properties of nanoparticles may differ from those of bulk materials for several reasons. The most dramatic effect of a small particle size is that the magnetization direction is not stable at finite temperatures, but fluctuates.

6.5.1 Superparamagnetic Relaxation

Usually, it is assumed that the magnetic anisotropy in nanoparticles is uniaxial with the magnetic anisotropy energy given by the simple expression

$$E_a(\theta) = KV \sin^2\theta, \tag{6.15}$$

where K is the magnetic anisotropy constant which may have contributions from magnetocrystalline anisotropy, shape anisotropy, stress anisotropy and surface anisotropy [6]. V is the particle volume, and θ is the angle between the magnetization direction and the easy direction of magnetization. According to (6.15), there are energy minima at $\theta = 0°$ and $\theta = 180°$ separated by an energy barrier, KV. If this energy barrier is comparable to or smaller than the thermal energy, the magnetization direction of a magnetic nanoparticle may fluctuate spontaneously between the easy directions corresponding to the two minima. This phenomenon is known as superparamagnetic relaxation. For noninteracting particles in zero applied field, the relaxation time, i.e. the average time between two magnetization reversals, is for $KV/k_BT \gtrsim 1$ approximately given by the Néel–Brown expression [49, 50]

$$\tau = \tau_0 \exp\left(\frac{KV}{k_BT}\right). \tag{6.16}$$

τ_0 is typically in the range of 10^{-13}–10^{-9} s and depends weakly on temperature [7, 50]. Therefore, the superparamagnetic relaxation time is close to the time scale of Mössbauer spectroscopy for values of the parameter KV/k_BT on the order of 2–9, depending on the value of τ_0. In practice, a sample of magnetic nanoparticles will normally show a distribution of particle sizes and there is usually also a distribution of magnetic anisotropy constants. Because of the exponential dependence of the relaxation time on the energy barrier, KV, this results in a very broad distribution of relaxation times and Mössbauer spectra of samples of magnetic nanoparticles may therefore be dominated by sextets due to particles with relaxation time much longer than τ_M and singlets or doublets due to particles with much shorter relaxation times. Often, the broad components due to particles with relaxation time on the order of τ_M have a low relative area such that they are barely visible in the spectra. The area ratio of the sextets and doublets or singlets varies with temperature because of the temperature dependence of the relaxation time.

In studies of superparamagnetic relaxation the blocking temperature is defined as the temperature at which the relaxation time equals the time scale of the experimental technique. Thus, the blocking temperature is not uniquely defined, but depends on the experimental technique that is used for the study of superparamagnetic relaxation. In Mössbauer spectroscopy studies of samples with a broad distribution of relaxation times, the average blocking temperature is commonly defined as the temperature where half of the spectral area is in a sextet and half of it is in a singlet or a doublet form.

Figure 6.13 shows the Mössbauer spectra of ferritin [51], which is an iron-storage protein consisting of an iron-rich core with a diameter around 8 nm with a structure similar to that of ferrihydrite and which is surrounded by a shell of organic material. At 4.2 K essentially all particles contribute to a magnetically split component, but at higher temperatures the spectra show the typical superposition of a doublet and a sextet with a temperature dependent area ratio. At 70 K the sextet has disappeared since all particles have fast superparamagnetic relaxation at this temperature.

Hematite (α-Fe_2O_3) nanoparticles have been extensively studied by Mössbauer spectroscopy [52–54]. Temperature series of spectra, like the ferritin spectra in Fig. 6.13, can be fitted simultaneously taking into account the particle size distribution and assuming that the relaxation time is given by the Néel–Brown expression (6.16). From such fits it is possible to estimate both the anisotropy energy, KV, and the value of τ_0 [53, 54]. In studies of hematite nanoparticles it has been found that the anisotropy energy constant increases with decreasing particle size [54]. Similar results have been found in studies of nanoparticles of maghemite (γ-Fe_2O_3) [48] and α-Fe [55]. In nanoparticles, the low symmetry around surface atoms is expected to result in a large surface contribution to the magnetic anisotropy [56]. Therefore, the size dependence of the magnetic anisotropy constant may be explained by the increasing influence of surface anisotropy when the particle size decreases.

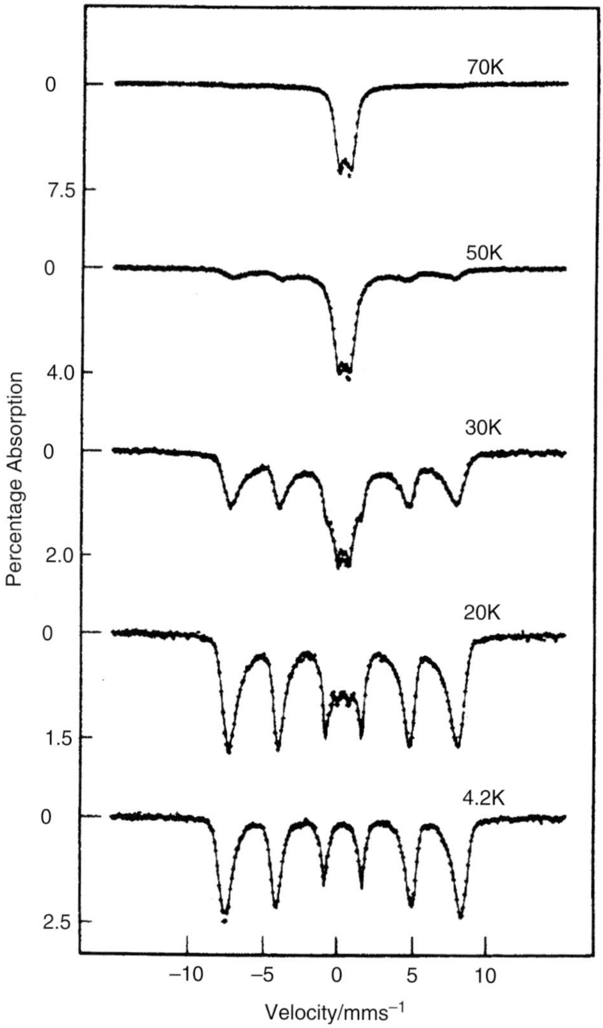

Fig. 6.13 Mössbauer spectra of horse spleen ferritin at the indicated temperatures. (Reprinted from [51]; copyright 1998 by Springer-Verlag)

When a magnetic field, $\vec{B}$, is applied to a ferromagnetic or ferrimagnetic particle with magnetic moment, $\vec{\mu}$, the magnetic energy has a contribution from the Zeeman term

$$E_Z = -\vec{\mu} \cdot \vec{B}. \tag{6.17}$$

Even for applied magnetic fields below 1 T, the Zeeman energy may be larger than the anisotropy energy. Above the blocking temperature application of a magnetic field results in an induced magnetization, which is approximately proportional to the Langevin function

$$L\left(\frac{\mu B}{k_{\mathrm{B}}T}\right) = \coth\left(\frac{\mu B}{k_{\mathrm{B}}T}\right) - \frac{k_{\mathrm{B}}T}{\mu B}. \tag{6.18}$$

For $\mu B \gtrsim 2KV$, there is only one energy minimum, i.e. there is no energy barrier between different magnetization directions [57], and the fluctuations of the magnetic hyperfine field can therefore be considered fast compared to τ_M. The induced magnetic hyperfine field is then proportional to the induced magnetization. Using the high field approximation of the Langevin function, $L(x) \approx 1 - x^{-1}$, one finds that the total average magnetic field at the nucleus is given by [57, 58]

$$\vec{B}_{\mathrm{obs}} \cong \vec{B}_0\left[1 - \frac{k_{\mathrm{B}}T}{\mu B}\right] + \vec{B}. \tag{6.19}$$

The magnetic hyperfine field of iron atoms is usually opposite to the magnetization, and therefore

$$B_{\mathrm{obs}} + B \cong B_0\left[1 - \frac{k_{\mathrm{B}}T}{\mu B}\right] \tag{6.20}$$

Thus, a plot of $B_{\mathrm{obs}} + B$ as a function of B^{-1} should give a straight line from which the magnetic moment μ can be determined.

Figure 6.14 shows the magnetic field dependence of Mössbauer spectra of carbon-supported ferromagnetic α-Fe nanoparticles at 80 and 300 K [58]. At both temperatures the spectrum in a small magnetic field consists of a singlet, indicating fast superparamagnetic relaxation. With increasing values of the applied field a magnetic splitting is induced, most prominent at 80 K. Figure 6.15 shows plots of $B_{\mathrm{obs}} + B$ as a function of B^{-1} at both temperatures. The data are in accordance with a linear behavior (6.20) and from the slope of the linear fits a magnetic moment of $\sim 1.5 \cdot 10^{-20}$ J T^{-1} was found, corresponding to a particle diameter of 2.5 nm assuming that the magnetization is equal to the bulk value.

6.5.2 Collective Magnetic Excitations

Below the blocking temperature, relaxation across the energy barrier is negligible, but the magnetization vector of a nanoparticle may still fluctuate in directions close to an easy magnetization direction. For a particle with magnetic energy given by (6.15), the probability that the magnetization vector forms an angle between θ and $\theta + \mathrm{d}\theta$ with the easy direction is given by [59–61]

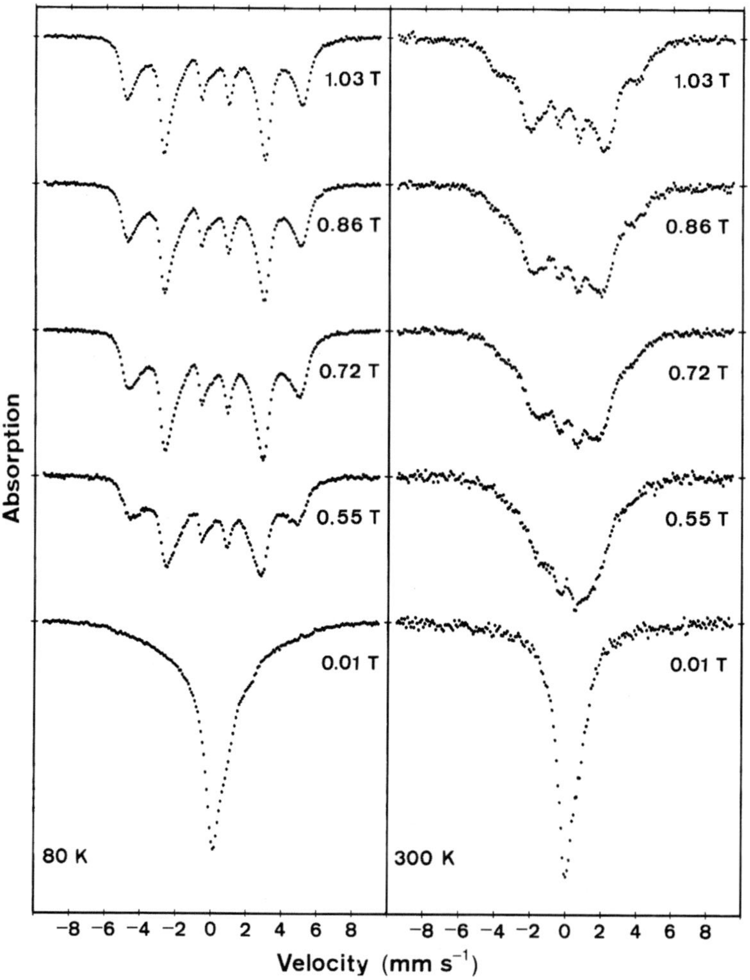

Fig. 6.14 Mössbauer spectra of α-Fe nanoparticles on a carbon support. The spectra were obtained at 80 and 300 K with the indicated magnetic fields applied perpendicular to the γ-ray direction. The asymmetry in the spectra is due to the presence of a small amount of iron carbide particles. (Reprinted with permission from [58]; copyright 1985 by the American Chemical Society)

$$p(\theta)\mathrm{d}\theta = \frac{\exp[-E(\theta)/k_{\mathrm{B}}T]\sin\theta\,\mathrm{d}\theta}{\int_0^{\pi/2}\exp[-E(\theta)/k_{\mathrm{B}}T]\sin\theta\,\mathrm{d}\theta} \tag{6.21}$$

The fluctuations of the magnetization direction around an easy axis, known as collective magnetic excitations, can be considered fast compared to the time scale of Mössbauer spectroscopy because there are no energy barriers between magnetization directions close to an easy direction, and the magnetic splitting in the

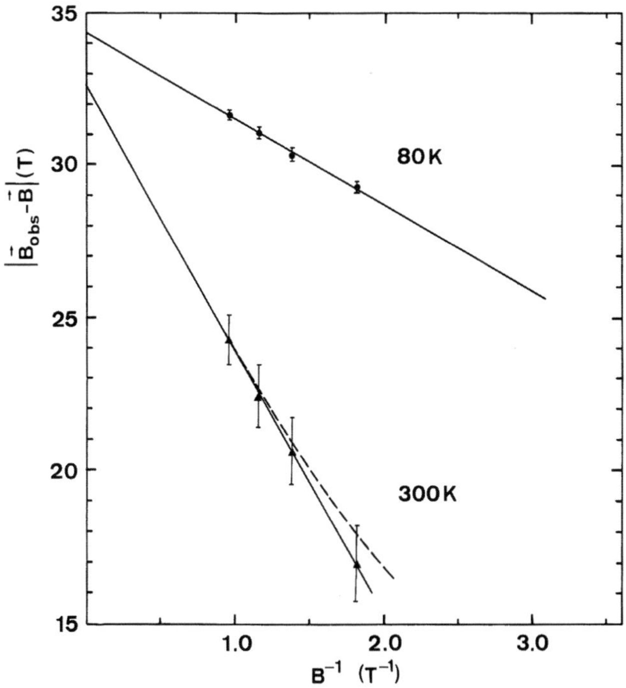

Fig. 6.15 Induced magnetic hyperfine fields, estimated from the spectra in Fig. 6.14, as a function of the reciprocal applied magnetic field. The full lines are linear fits in accordance with (6.20). The dotted line is a fit to the Langevin function. (Reprinted with permission from [58]; copyright 1985 by the American Chemical Society)

Mössbauer spectrum is then proportional to the average value of the hyperfine field which is given by

$$B_{obs} = B_{hf} \int_0^{\pi/2} p(\theta) \cos\, \theta \, d\theta \tag{6.22}$$

In the low-temperature limit $(k_B T \ll KV)$, one can use the approximation [59–61]

$$B_{obs} \cong B_{hf}\left[1 - \frac{k_B T}{2KV}\right] \tag{6.23}$$

Numerous experimental studies have shown that the magnetic hyperfine field of magnetic nanoparticles varies linearly with temperature at low temperatures, in accordance with (6.23). This is in contrast to bulk materials for which the decrease in the hyperfine field with increasing temperature in accordance with spin wave models is proportional to $T^{3/2}$ in ferromagnets and ferrimagnets and proportional to

T^2 in antiferromagnets. Using spin wave models to describe the magnetic dynamics in nanoparticles, collective magnetic excitations can be described as excitation of the so-called uniform mode (a spin wave with wave vector, $\vec{q} = 0$), which is the predominant excitation in nanoparticles [61].

6.5.3 Interparticle Interactions

If magnetic nanoparticles are not well separated, the magnetic interparticle interactions may influence the magnetic dynamics. In samples of ferromagnetic or ferrimagnetic nanoparticles, the long-range dipole interactions can be significant. The influence of the dipole interactions can conveniently be studied in solid suspensions of the particles in, for example, PVA [48, 62, 63] or oil [64], because the interparticle interactions can be easily varied by varying the particle concentration. Mössbauer studies of interacting nanoparticles of ferrimagnetic maghemite (γ-Fe_2O_3) have shown that weak dipole interactions can result in a decrease in the relaxation time [62–64], and this can be explained by a decrease in the average energy barriers [63–69]. In more concentrated samples the interactions lead to an increase of the relaxation time [66, 70, 71]. AC susceptibility measurements [72, 73] have shown that the relaxation time of nanoparticles with relatively strong interactions diverges at a finite temperature, T_p, which is on the order of [65, 66]

$$\mathrm{T}_p \cong \frac{\mu_0}{4\pi k_{\mathrm{B}}} \frac{\mu^2}{d^3}, \tag{6.24}$$

where d is the average distance between the particles. Below T_p, the interactions may result in a collective state, which has similarities to a spin-glass [66, 70–73].

Antiferromagnetic nanoparticles usually have magnetic moments due to uncompensated spins [46, 74], but the moments are so small that the influence of dipole interactions on the superparamagnetic relaxation time can be considered negligible [46, 75–77]. Nevertheless, interactions between antiferromagnetic nanoparticles can have a significant influence on the magnetic dynamics [75–79]. Figure 6.16 shows Mössbauer spectra of two samples of antiferromagnetic 8 nm hematite nanoparticles from the same batch [77]. In one of the samples, the particles were coated with phosphate in order to minimize interactions. The other sample was prepared by slowly drying an aqueous suspension of uncoated particles. The spectra of the coated particles (Fig. 6.16a) show the typical behavior of weakly interacting or noninteracting magnetic nanoparticles, i.e. a superposition of sextet and doublet components with relatively narrow lines and with a temperature dependent area ratio. At 80 K only a doublet is seen, indicating that all particles have a short relaxation time at this temperature. The spectra of the uncoated sample (Fig. 6.16b) show a completely different behavior. There is no doublet in the spectra, even at 295 K, but the lines gradually broaden with increasing temperature and the average

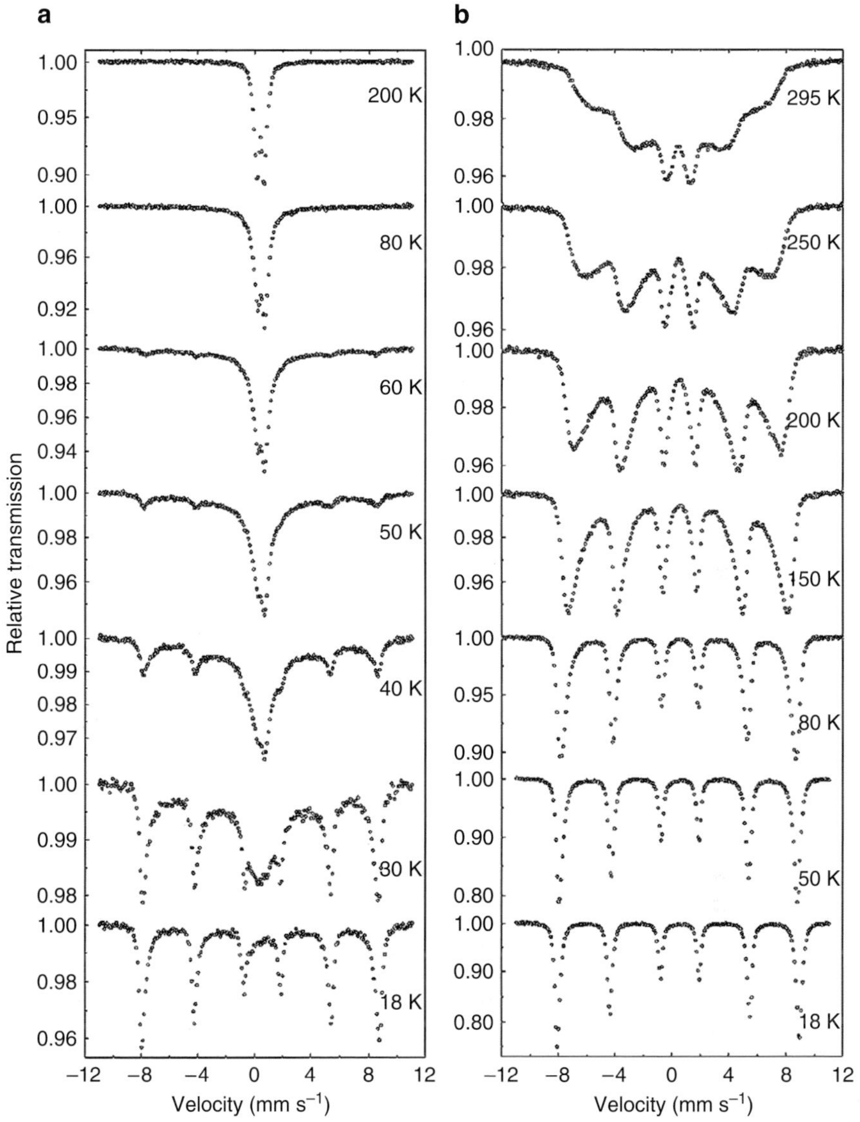

Fig. 6.16 Mössbauer spectra of 8 nm α-Fe_2O_3 particles at the indicated temperatures. (**a**) Data for phosphate coated (noninteracting) particles, and (**b**) data for the uncoated (interacting) particles. (Reprinted with permission from [77]; copyright 2006 by the American Physical Society)

hyperfine field decreases with temperature much faster than the bulk hyperfine field. A similar behavior has been seen in several other studies of uncoated antiferromagnetic nanoparticles [46, 75, 76, 78, 79], and it has been explained by a strong exchange coupling between surface atoms of neighboring particles.

The temperature dependence of the magnetic hyperfine splitting in spectra of interacting nanoparticles may be described by a mean field model [75–77]. In this model it is assumed that the magnetic energy of a particle, p, with volume V and magnetic anisotropy constant K, and which interacts with its neighbor particles, q, can be written

$$E_p = KV\sin^2\theta - \sum_{i,j} J_{ij}\vec{S}_i \cdot \vec{S}_j \tag{6.25}$$

where $\vec{S}_i$ and $\vec{S}_j$ represent the surface spins belonging to the particle, p, and the surface spins of the neighbor particles, q, respectively. J_{ij} is the exchange coupling constant. Using the mean field approximation, the summation in the last term is replaced by an effective interaction field acting on the sublattice magnetization of the particle p [75–77]:

$$E_p = KV\sin^2\theta - \vec{M}_p \cdot \left\langle \sum_q J_{pq}\vec{M}_q \right\rangle, \tag{6.26}$$

where $\vec{M}_p$ and $\vec{M}_q$ represent sublattice magnetization vectors of the particles p and q and J_{pq} is an effective exchange coupling constant. The effective interaction field may suppress the superparamagnetic relaxation, in a way that is similar to the influence of an applied magnetic field on a ferromagnetic or ferrimagnetic particle, and therefore give rise to magnetic splitting well above the blocking temperature of noninteracting particles [77].

The interactions also suppress the collective magnetic excitations at low temperatures. Figure 6.17 shows the temperature dependence of the magnetic hyperfine

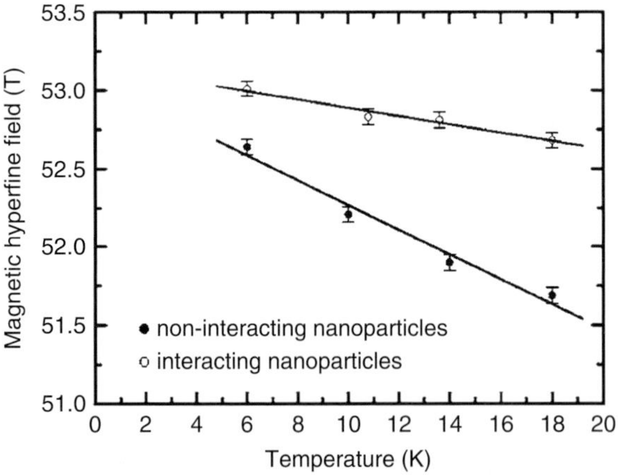

Fig. 6.17 The average magnetic hyperfine field at low temperatures, obtained from Mössbauer spectra of coated and uncoated 8 nm particles of α-Fe_2O_3 (Fig. 6.16). The lines are linear fits to the data in accordance with (6.23) and (6.25). (Reprinted with permission from [77]; copyright 2006 by the American Physical Society)

splitting of the coated and the uncoated 8 nm α-Fe_2O_3 particles at low temperatures. In both cases the data show a linear temperature dependence. For the coated sample the magnetic anisotropy can be estimated from (6.23). For interacting particles, (6.23) may be replaced by [58, 77]

$$B_{\text{obs}} \cong B_{\text{hf}}\left(1 - \frac{k_{\text{B}}T}{2KV + E_{int}}\right), \tag{6.27}$$

where E_{int} is related to the interaction energy. The data thus make it possible to estimate the strength of the interparticle interactions [77].

6.6 Transverse Relaxation in Canted Spin Structures

Defects in ferrimagnetic structures often lead to noncollinear (canted) spin structures. For example, a diamagnetic substitution or a cation vacancy can result in magnetic frustration which leads to spin-canting such that a spin may form an angle θ_{c} with the collinear spins in the sample [80, 81]. Similarly, the reduced number of neighbor ions at the surface can also lead to spin-canting [80–83].

Spin-canting is conveniently studied by Mössbauer spectroscopy with large magnetic fields applied parallel to the γ-ray direction. The relative areas of the six lines are given by 3:p:1:1:p:3, where

$$p = \frac{4\sin^2\theta_0}{1 + \cos^2\theta_0}, \tag{6.28}$$

and θ_0 is the angle between the total magnetic field at the nucleus (the vector sum of the hyperfine field and the applied field) and the γ-ray direction. Thus, in a perfect ferrimagnetic material, lines 2 and 5 vanish when the material is magnetically saturated in a field parallel to the γ-ray direction, but a finite intensity of lines 2 and 5 indicates that some spins are not aligned with the applied magnetic field.

It is a common feature in canted spin structures that a canted spin with canting angle θ_{c} may also be found in an equivalent canted state with canting angle $-\theta_{\text{c}}$ [81]. These two canted states are separated by an energy barrier that depends on the exchange coupling constants and the local magnetic anisotropy [81]. In many cases the energy barriers may be very low, and at finite temperatures the spin directions may therefore fluctuate between the two states. If these fluctuations are fast compared to τ_{M} the nucleus will only experience the average value of the hyperfine field.

In ferrites, the magnetization of the B-sublattice with octahedrally coordinated cations is usually parallel to the applied field whereas the magnetization of the A-sublattice with tetrahedrally coordinated cations is in the opposite direction.

Thus, in the case of fast relaxation between the two canted states the total average magnetic field, which is experienced by the nucleus, is given by [84, 85]

$$B_{\text{tot}} \approx |B_{\text{hf}}| \cos \theta_{\text{c}} \pm B, \tag{6.29}$$

where the plus is valid for A-site ions and the minus is valid for B-site ions [82–88].

Figure 6.18 shows spectra of the diamagnetically substituted ferrite, $Mn_{0.25}Zn_{0.75}Fe_2O_4$ [84]. The spectra were obtained in a magnetic field of 6 T

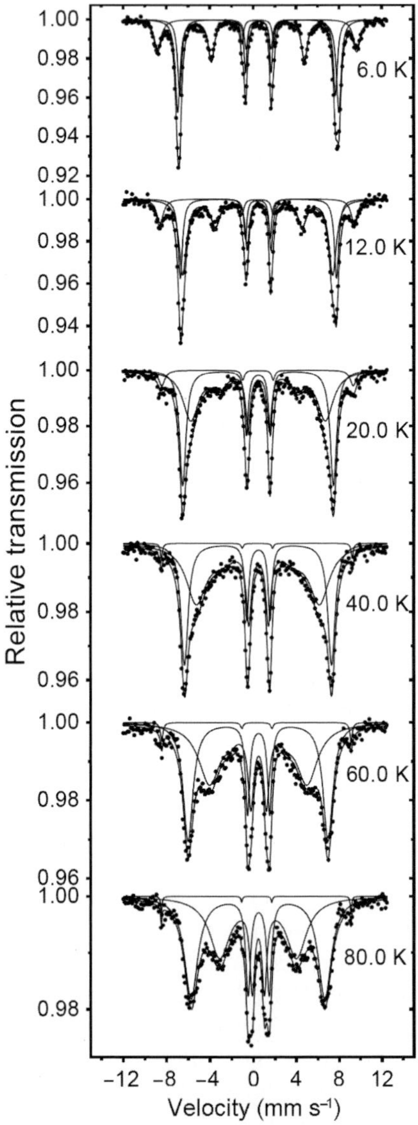

Fig. 6.18 Mössbauer spectra of $Mn_{0.25}Zn_{0.75}Fe_2O_4$, obtained at the indicated temperatures in a magnetic field of 6 T, applied parallel to the γ-ray direction. The lines show fits to three sextets as discussed in the text. (Reprinted with permission from [84]; copyright 2003 by Elsevier)

Fig. 6.19 Temperature dependence of the total magnetic field, B_{tot}, for the three sextets in the Mössbauer spectra of $Mn_{0.25}Zn_{0.75}Fe_2O_4$, shown in Fig. 6.18. (Adapted with permission from [84]; copyright 2003 by Elsevier)

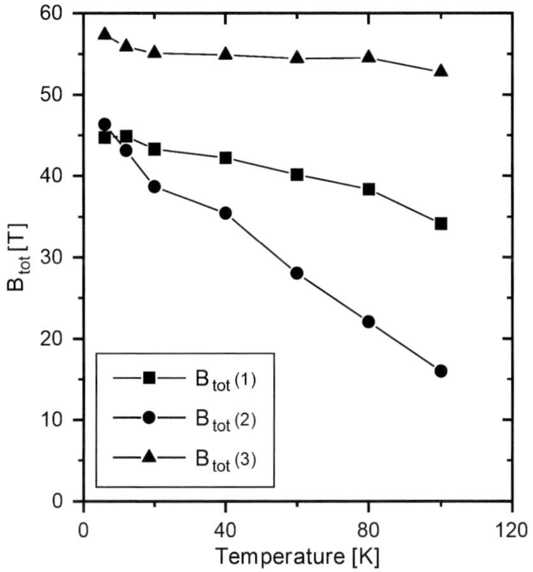

applied parallel to the γ-ray direction. At 6 K, lines 2 and 5 are relatively intense. However, with increasing temperature the relative area of these lines decrease, and above 60 K it is negligible. The reduction of the area of lines 2 and 5 with increasing temperature is accompanied by the appearance of a sextet with rapidly decreasing hyperfine field. The best fits to the spectra, shown by the solid lines, were obtained using three sextets, the first one due to Fe^{3+} in octahedral sites without spin-canting, the second due to Fe^{3+} in octahedral sites, which are influenced by canting and the third (weak) component mainly due to a small amount of Fe^{3+} without canting in tetrahedral sites in the spinel lattice [84]. Thus, the first and the third sextets were fitted with the area constraints 3:0:1:1:0:3, corresponding to perfect collinear spin structures ($\theta_c = 0°$). In the second sextet, the relative area of lines 2 and 5 was a free parameter. The temperature dependence of the total magnetic fields, estimated from the three sextets, is shown in Fig. 6.19. The sextets 1 and 3 have similar temperature dependence, but the magnetic field of the second sextet decreases more rapidly with increasing temperature and lines 1 and 6 of this sextet become considerably broadened. The disappearance of lines 2 and 5 and the decrease of the magnetic hyperfine field with increasing temperature can be explained by fast transverse relaxation in the canted spin structures, such that the nuclei only experience the total average hyperfine field, B_{tot} given by (6.29). The line broadening is presumably due to relaxation times for the transverse relaxation that are not very short compared to the time scale of Mössbauer spectroscopy. At temperatures above 100 K, the lines become narrower indicating faster transverse relaxation. Similar effects have been found in other diamagnetically substituted ferrimagnetic samples [86, 87] and ferrimagnetic maghemite nanoparticles [85, 88].

References

1. Wickman, H.H., Wertheim, G.K.: In: Goldanskii, V.I., Herber, R.H. (eds.) Chemical Applications of Mössbauer Spectroscopy, p. 548. Academic, New York (1968)
2. Mørup, S.: In: Gruverman, I.J., Seidel, C.W., Dieterly, D.K. (eds.) Mössbauer Effect Methodology, vol. 9, p. 127. Plenum, New York (1974)
3. Mørup, S.: In: International Conference on the Applications of the Mössbauer Effect Proceedings, Science Academy, New Delhi, 1982, p. 91
4. Cianchi, L., Moretti, P., Mancini, M., Spina, G.: Rep. Prog. Phys. **49**, 1243 (1986)
5. Schünemann, V., Winkler, H.: Rep. Prog. Phys. **63**, 263 (2000)
6. Mørup, S., Dumesic, J.A., Topsøe, H.: In: Cohen, R.L. (ed.) Applications of Mössbauer Spectroscopy, vol. 2, p. 1. Academic, New York (1980)
7. Dormann, J.L., Fiorani, D., Tronc, E.: Adv. Chem. Phys. **98**, 283 (1997)
8. Mørup, S., Hansen, M.F.: In: Kronmüller, H., Parkin, S. (eds.) Handbook of Magnetism and Advanced Magnetic Materials, vol. 4, p. 2179. Wiley, New York (2007)
9. Wickman, H.H., Klein, M.P., Shirley, D.A.: Phys. Rev. **152**, 345 (1966)
10. Pake, G.E.: Paramagnetic Resonance. W. A. Benjamin, New York (1962)
11. Abragam, A., Bleaney, B.: Electron Paramagnetic Resonance of Transition Ions. Clarendon, Oxford (1970)
12. Knudsen, J.E.: J. Phys. Chem. Solids **38**, 883 (1977)
13. Mørup, S., Knudsen, J.E., Nielsen, M.K., Trumpy, G.: J. Chem. Phys. **65**, 536 (1976)
14. Mørup, S., Knudsen, J.E.: Acta Chim. Hung. **121**, 147 (1986)
15. Sontheimer, E., Nagy, D.L., Dézsi, I., Lohner, T., Ritter, G., Seyboth, D., Wegener, H.: J. Phys. Colloq. **35**, C6–C443 (1974)
16. Leupold, O., Winkler, H.: Hyperfine Interact **123/124**, 571 (1999)
17. Afanas'ev, A.M., Kagan, Y.: Zh. Exper. Teor. Fiz **45**, 1660 (1963). Sov. Phys. JETP 18, 1139 (1964)
18. Kagan, Y., Afanas'ev, A.M.: Zh. Exper. Teor. Fiz. **47**, 1108 (1964). Sov. Phys. JETP 20, 743 (1965)
19. Wegener, H.: Z. Phys. **186**, 498 (1965)
20. Blume, M.: Phys. Rev. Lett. **14**, 96 (1965)
21. Bradford, E., Marshall, W.: Proc. Phys. Soc. **87**, 731 (1966)
22. Blume, M., Tjon, J.A.: Phys. Rev. **165**, 446 (1968)
23. Blume, M.: Phys. Rev. **174**, 351 (1968)
24. Clauser, M.J., Blume, M.: Phys. Rev. **B 3**, 583 (1971)
25. Clauser, M.J.: Phys. Rev. B **3**, 3748 (1971)
26. Afanas'ev, A.M., Gorobschenko, V.D.: Zh. Exper. Teor. Fiz. **66**, 1406 (1974). Sov. Phys. JETP 39, 690 (1974)
27. Mørup, S.: J. Phys. Colloq. **35**, C6–C683 (1974)
28. Mørup, S.: Hyperfine Interact **1**, 533 (1976)
29. Wegener, H.H.F., Braunecker, B., Ritter, G., Seyboth, D.: Z. Phys. **262**, 149 (1973)
30. Winkler, H., Meyer-Klaucke, W., Schwendy, S., Trautwein A.X., Matzanke B.F., Leupold, O., Rüter, H.D., Haas, M., Realo, E., Madon, D., Weiss, R.: Hyperfine Interact **113**, 443 (1998)
31. Shrivastava, K.N.: Phys. Stat. Sol. **B 117**, 437 (1983)
32. Bhargava, S.C., Knudsen, J.E., Mørup, S.: J. Phys. **C 12**, 2879 (1979)
33. Winkler, H., Schulz, C., Debrunner, P.G.: Phys. Lett. **A 69**, 360 (1979)
34. Winkler, H., Ding, X.-Q., Burkardt, M., Trautwein, A.X., Parak, F.: Hyperfine Interact **91**, 875 (1994)
35. Winkler, H., Kostikas, A., Petrouleas, V., Simopoulos, A., Trautwein, A.X.: Hyperfine Interact **29**, 1347 (1986)
36. Schulz, C.E., Nyman, P., Debrunner, P.G.: J. Chem. Phys. **87**, 5077 (1987)
37. Bloembergen, N., Shapiro, S., Pershan, P.S., Artman, J.O.: Phys. Rev. **114**, 445 (1959)

38. Bhargava, S.C., Knudsen, J.E., Mørup, S.: J. Phys. Chem. Solids **40**, 45 (1979)
39. Mørup, S., Thrane, N.: Chem. Phys. Lett. **21**, 363 (1973)
40. Mørup, S.: J. Phys. Chem. Solids **35**, 1159 (1974)
41. Mørup, S., Sontheimer, F., Ritter, G., Zimmermann, R.: J. Phys. Chem. Solids **39**, 123 (1978)
42. Mørup, S., Knudsen, J.E.: Chem. Phys. Lett. **40**, 292 (1976)
43. Blume, M.: Phys. Rev. Lett. **18**, 305 (1967)
44. Thrane, N., Trumpy, G.: Phys. Rev. **B 1**, 153 (1970)
45. Richards, R., Johnson, C.E., Hill, H.A.O.: J. Chem. Phys. **51**, 846 (1969)
46. Mørup, S., Madsen, D.E., Frandsen, C., Bahl, C.R.H., Hansen, M.F.: J. Phys. Condens. Matter **19**, 213202 (2007)
47. Mørup S.: In: Long, G.J. (ed.) Mössbauer Spectroscopy Applied to Inorganic Chemistry, vol. 2, p. 89. Plenum, New York (1987)
48. Tronc, E.: Nuovo Cimento **D 18**, 163 (1996)
49. Néel, L.: Ann. Geophys. **5**, 99 (1949)
50. Brown, W.F.: Phys. Rev. **130**, 1677 (1963)
51. Dickson, D.P.E.: Hyperfine Interact **111**, 171 (1998)
52. Kündig, W., Ando, K.J., Constabaris, G., Lindquist, R.H.: Phys. Rev. **142**, 327 (1966)
53. Bødker, F., Hansen, M.F., Koch, C.B., Lefmann, K., Mørup, S.: Phys. Rev. **B 61**, 6826 (2000)
54. Bødker, F., Mørup, S.: Europhys. Lett. **52**, 217 (2000)
55. Bødker, F., Mørup, S., Linderoth, S.: Phys. Rev. Lett. **72**, 282 (1994)
56. Néel, L.: J. Phys. Radium **15**, 225 (1954)
57. Mørup, S., Clausen, B.S., Christensen, P.H.: J. Magn. Magn. Mater. **68**, 160 (1987)
58. Christensen, P.H., Mørup, S., Niemantsverdriet, J.W.: J. Phys. Chem. **89**, 4898 (1985)
59. Mørup, S., Topsøe, H.: Appl. Phys. **11**, 63 (1976)
60. Mørup, S.: J. Magn. Magn. Mater. **37**, 39 (1983)
61. Mørup, S., Hansen, B.R.: Phys. Rev. **B 72**, 024418 (2005)
62. Prené, P., Tronc, E., Jolivet, J.P., Livage, J., Cherkaoui, R., Nogués, M., Dormann, J.L., Fiorani, D.: IEEE Trans. Magn. **29**, 2658 (1993)
63. Mørup, S., Tronc, E.: Phys. Rev. Lett. **72**, 3278 (1994)
64. Jiang, J.Z., Mørup, S., Jonsson, T., Svedlindh, P.: In: Ortalli, I. (ed.) ICAME'95 Proceedings, p. 529. SIF, Bologna (1996)
65. Hansen, M.F., Mørup, S.: J. Magn. Magn. Mater. **184**, 262 (1998)
66. Mørup, S.: Europhys. Lett. **28**, 671 (1994)
67. Berkov, D.V.: J. Magn. Magn. Mater. **186**, 199 (1998)
68. Jönsson, P.E., García-Palacios, J.L.: Europhys. Lett. **55**, 418 (2001)
69. Iglesias, O., Labarta, A.: Phys. Rev. **B 70**, 144401 (2004)
70. Mørup, S., Bødker, F., Hendriksen, P.V., Linderoth, S.: Phys. Rev. **B 52**, 287 (1995)
71. Fiorani, D., Dormann, J.L., Cherkaoui, R., Tronc, E., Lucari, F., D'Orazio, F., Spinu, L., Nogués, M., Garcia, A., Testa, A.M.: J. Magn. Magn. Mater. **196–197**, 143 (1999)
72. Zhang, J., Boyd, C., Luo, W.: Phys. Rev. Lett. **77**, 390 (1996)
73. Djurberg, C., Svedlindh, P., Nordblad, P., Hansen, M.F., Bødker, F., Mørup, S.: Phys. Rev. Lett. **79**, 5154 (1997)
74. Néel, L.: C. R. Acad. Sci. Paris **252**, 4075 (1961)
75. Hansen, M.F., Koch, C.B., Mørup, S.: Phys. Rev. **B 62**, 1124 (2000)
76. Frandsen, C., Bahl, C.R.H., Lebech, B., Lefmann, K., Kuhn, L.T., Keller, L., Andersen, N.H., Zimmermann, M., Johnson, E., Klausen, S.N., Mørup, S.: Phys. Rev. **B 72**, 214406 (2005)
77. Kuhn, L.T., Lefmann, K., Bahl, C.R.H., Ancona, S.N., Lindgård, P.-A., Frandsen, C., Madsen, D.E., Mørup, S.: Phys. Rev. **B 74**, 184406 (2006)
78. Frandsen, C., Mørup, S.: J. Magn. Magn. Mater. **266**, 36 (2003)
79. Bahl, C.R.H., Mørup, S.: Nanotechnology **17**, 2835 (2006)
80. Coey, J.M.D.: Can. J. Phys. **65**, 1210 (1987)
81. Mørup, S.: J. Magn. Magn. Mater. **266**, 110 (2003)
82. Coey, J.M.D.: Phys. Rev. Lett. **27**, 1140 (1971)

83. Morrish, A.H., Haneda, K.: J. Magn. Magn. Mater. **35**, 105 (1983)
84. Anhøj, T.A., Bilenberg, B., Thomsen, B., Damsgaard, C.D., Rasmussen, H.K., Jacobsen, C.S., Mygind, J., Mørup, S.: J. Magn. Magn. Mater. **260**, 115 (2003)
85. Helgason, Ö., Rasmussen, H.K., Mørup, S.: J. Magn. Magn. Mater. **302**, 413 (2006)
86. Brand, R.A., Georges-Gibert, H., Hubsch, J., Heller, J.A.: J. Phys **F 15**, 1987 (1985)
87. Dormann, J.L., El Harfaoui, M., Nogués, M., Jove, J.: J. Phys **C 20**, L161 (1987)
88. Tronc, E., Ezzir, A., Cherkaoui, R., Chaneac, C., Nogués, M., Kachkachi, H., Fiorani, D., Testa, A.M., Greneche, J.M., Jolivet, J.P.: J. Magn. Magn. Mater. **221**, 63 (2000)

Chapter 7
Mössbauer-Active Transition Metals Other than Iron

The previous chapters are exclusively devoted to the measurements and interpretation of ^{57}Fe spectra of various iron-containing systems. Iron is, by far, the most extensively explored element in the field of chemistry compared with all other Mössbauer-active elements because the Mössbauer effect of ^{57}Fe is very easy to observe and the spectra are, in general, well resolved and they reflect important information about bonding and structural properties. Besides iron, there are a good number of other transition metals suitable for Mössbauer spectroscopy which is, however, less extensively studied because of technical and/or spectral resolution problems. In recent years, many of these difficulties have been overcome, and we shall see in the following sections a good deal of successful Mössbauer spectroscopy that has been performed on compounds of nickel (^{61}Ni), zinc (^{67}Zn), ruthenium (mainly ^{99}Ru), tantalum (^{181}Ta), tungsten (mainly ^{182}W, ^{183}W), osmium (mainly ^{189}Os), iridium (^{191}Ir, ^{193}Ir), platinum (^{195}Pt), and gold (^{197}Au). The nuclear γ-resonance effect in the case of technetium (^{99}Tc), silver (^{107}Ag), hafnium (^{176}Hf, ^{177}Hf, ^{178}Hf, ^{180}Hf), rhenium (^{187}Re), and mercury (^{199}Hg, ^{201}Hg) has been of relatively little use to the chemists, so far. There are various reasons responsible for this, viz., (1) extraordinary difficulties in measuring the resonance effect because of the long lifetime of the excited Mössbauer level and hence the extremely small transition line width (e.g., in ^{67}Zn), (2) poor resolution of the resonance lines due to either very small nuclear moments or the very short lifetime of the excited Mössbauer level resulting in very broad resonance lines, (3) insufficient resonance effects due to unusually high transition energies between the excited and the ground nuclear levels, which in turn increase the recoil energy and thus reduces the recoilless fraction of emitted and observed γ-rays.

P. Gütlich et al., *Mössbauer Spectroscopy and Transition Metal Chemistry*,
DOI 10.1007/978-3-540-88428-6_7, © Springer-Verlag Berlin Heidelberg 2011

In Table 7.1 (*at the end of the book*), nuclear data are collected for those Mössbauer transitions of transition metal nuclides that are used in Mössbauer spectroscopy. The symbols used in this table have the following meaning:

a	Natural abundance of resonant nuclide (in %)
E_γ	Energy of γ-ray transition (in keV)
$t_{1/2}$	Half-life of excited nuclear state (in ns)
Γ	Natural line width (in mm s^{-1})
I_g, I_e	Nuclear spin quantum number of ground (g) and excited (e) state (the sign refers to the parity)
α_T	Total internal conversion coefficient
E_R	Recoil energy (in 10^{-3} eV)
σ_0	Resonant absorption cross section (in 10^{-20} cm^2)
MP	Multipolarity of γ-radiation
μ_g, μ_e	Nuclear magnetic dipole moment of ground (g) and excited (e) state (in nuclear magnetons, n.m.)
Q_g, Q_e	Nuclear electric quadrupole moment of ground (g) and excited (e) state (in barn $= 10^{-28}$ m^2)
$\Delta\langle r^2\rangle/\langle r^2\rangle$	$=(\langle r^2\rangle_e - \langle r^2\rangle_g/\langle r^2\rangle)$, relative change of mean-square nuclear radius in going from excited (e) to ground (g) state (in 10^{-4})

In the last column of Table 7.1, the most popular radioactive precursor nuclide is given together with the nuclear decay process (EC = electron capture, β^- = beta decay) feeding the Mössbauer excited nuclear level.

In the following sections, we discuss the decay schemes for all Mössbauer-active transition metal nuclides other than iron. For the sake of completeness, the decay scheme for ^{57}Fe (see Fig. 7.1) is inserted here. The relevant nuclear data,

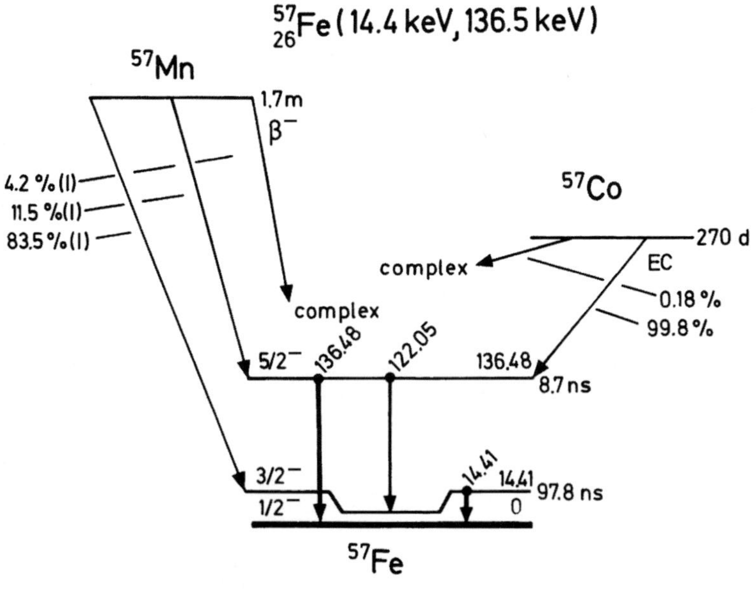

Fig. 7.1 Decay scheme of ^{57}Fe and parent nuclei (from [1])

which are of interest in ^{57}Fe Mössbauer spectroscopy, may be taken from Table 7.1 (end of the book) or the table of Mössbauer isotopes provided by the Mössbauer Effect Data Center (MEDC) in Part IV of the attached CD-ROM.

7.1 Nickel (^{61}Ni)

The nuclear γ-resonance effect in ^{61}Ni was first observed in 1960 by Obenshain and Wegener [2]. However, the practical application to the study of nickel compounds was hampered for several years by (1) the lack of a suitable single-line source, (2) the poor resolution of the overlapping broad hyperfine lines due to the short excited state lifetime, and (3) the difficulties in producing and handling the short-lived Mössbauer sources containing the ^{61}Co and ^{61}Cu parent nuclides, respectively.

Single-line sources are now available which cut down the number of resonance lines in a spectrum and thereby reduce the resolution problems considerably. Since many laboratories have access to electron and ion accelerators to produce the parent nuclides ^{61}Co and ^{61}Cu, the major experimental obstacles to ^{61}Ni spectroscopy have been overcome and a good deal of successful work has been performed in recent years. Moreover, the development of synchrotron radiation instead of conventional Mössbauer sources is of additional advantage for future Mössbauer applications (see below).

7.1.1 Some Practical Aspects

The sources used in ^{61}Ni Mössbauer work mainly contain ^{61}Co as the parent nuclide of ^{61}Ni; in a few cases, ^{61}Cu sources have also been used. Although the half-life of ^{61}Co is relatively short (99 m), this nuclide is much superior to ^{61}Cu because it decays via β^- emission directly to the 67.4 keV Mössbauer level (Fig. 7.2) whereas ^{61}Cu ($t_{1/2} = 3.32$ h) decays in a complex way with only about 2.4% populating the 67.4 keV level. There are a number of nuclear reactions leading to ^{61}Co [4]; the most popular ones are ^{62}Ni(γ, p)^{61}Co with the bremsstrahlung (about 100 MeV) from an electron accelerator, or ^{64}Ni(p, α)^{61}Co via proton irradiation of ^{64}Ni in a cyclotron.

In nickel metal and in many metallic systems, the nuclear spin of ^{61}Ni undergoes magnetic hyperfine interaction with relatively strong magnetic fields causing a splitting of the 67.4 keV emission line into 12 partially resolved lines (Fig. 7.3). The use of such a source for the study of magnetically ordered nickel compounds as absorbers would produce a large number of resonance lines which mostly overlap and would make the evaluation of the spectra rather cumbersome [2]. Much effort has therefore been invested into the development of nonmagnetic single-line sources. The alloy $Ni_{0.85}Cr_{0.15}$ has been found to provide a single-line source at temperatures down to 4 K [3, 5–7], yielding an experimental line width of 0.97 mm s^{-1} and a resonance effect of about 10% at 77 K [3]. Obenshain et al.

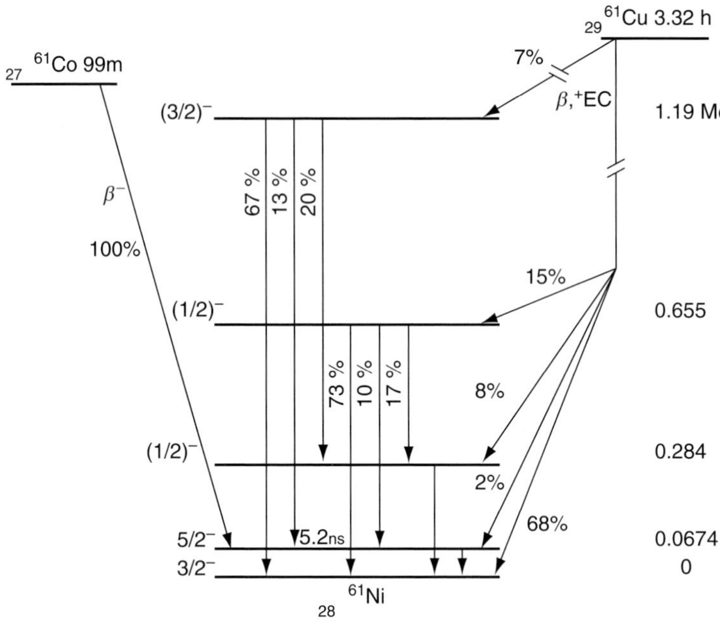

Fig. 7.2 Decay schemes of the parent nuclides of ^{61}Ni (from [3])

preferred a $Ni_{0.86}V_{0.14}$ alloy, which is also nonmagnetic at 4.2 K, and activated it by proton irradiation to give ^{64}Ni(p, α)^{61}Co [8, 9]; the line width is very near to the natural width. A Japanese group has also prepared a single-line source containing ^{61}Cu [10].

The relatively low Lamb–Mössbauer factors encountered in ^{61}Ni Mössbauer spectroscopy (a diagram showing the Lamb–Mössbauer factor as a function of Debye temperature is given in [3]) require the cooling of both source and absorber, preferentially to temperatures $\lesssim$80 K. The cryogenic systems have been described, e.g., in [2, 4].

7.1.2 Hyperfine Interactions in ^{61}Ni

7.1.2.1 Isomer Shifts

The essential nuclear data of ^{61}Ni may be taken from Table 7.1 (end of the book) or Part IV of the CD-ROM. More about nuclear properties has been summarized in [4].

Isomer shifts have been measured in a variety of nickel compounds. In most cases, however, the information concerning chemical bond properties was not very impressive. The reason is that the second-order Doppler (SOD) shift is, in many systems, of comparable magnitude as the real chemical isomer shift, which causes

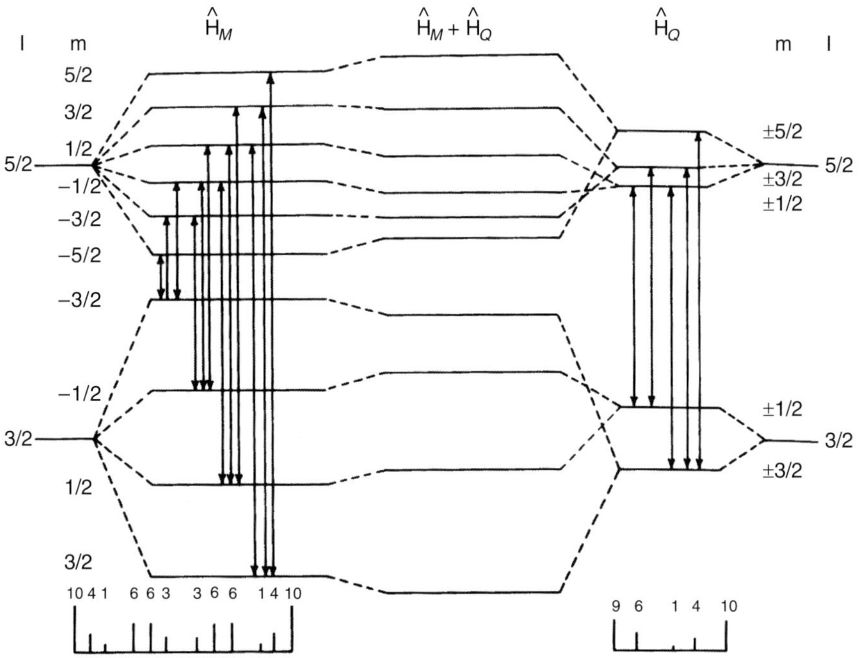

Fig. 7.3 Effect of magnetic dipole interaction ($\widehat{H}_M$), electric quadrupole interaction ($\widehat{H}_Q$), and combined interaction $\widehat{H} = \widehat{H}_M + \widehat{H}'_Q$, $E_M > E'_Q$ on the Mössbauer nuclear levels of ^{61}Ni. The larger spacings between the sublevels of the ground state are due to the somewhat larger magnetic dipole moment of the nuclear ground state as compared to the excited state. The relative transition probabilities for a powder sample as well as the relative positions of the transition lines are indicated by the stick spectra below

serious difficulties in correlating the electron density at the nickel nucleus with chemically important concepts such as electronegativity of the coordinated ligand and the ionicity of the nickel–ligand bond. If, however, the lattice dynamical properties are sufficiently known from infrared and optical spectroscopy, one may calculate the SOD shift and subtract it from the experimentally observed energy shift to find the real chemical isomer shift δ_{IS} according to [8, 9]

$$\delta_{exp} = \delta_{IS}^{(A)} - \delta_{IS}^{(S)} + \delta_{SOD}^{(A)} - \delta_{SOD}^{(S)}, \tag{7.1}$$

where the superscripts stand for the absorber (A) and the source (S). For cases where information about the lattice dynamics is not available, Love et al. [8] have suggested that δ_{SOD} may be estimated according to the relation $\delta_{SOD} = 1.23\delta_{SOD}^{(Debye)}$, where $\delta_{SOD}^{(Debye)} = +178/\ln f$ (in μm s^{-1}) and may be calculated in the framework of a Debye model. Following this approach, Obenshain et al. [8, 9] have determined the isomer shifts for a number of inorganic compounds as listed in the table below and have attempted to correlate the corrected isomer shift

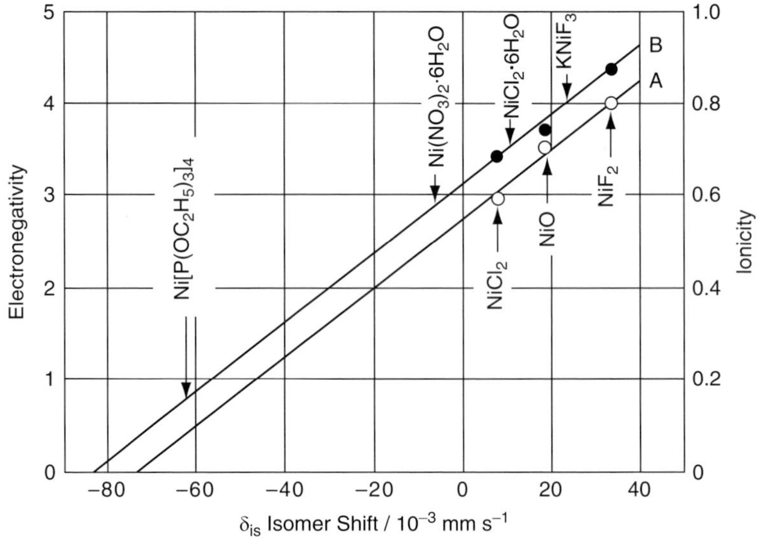

Fig. 7.4 Isomer shift (relative to Ni metal and corrected for SOD) as a function of ligand electronegativity (plot A) and as a function of nickel–ligand ionicity (plot B) (from [9])

δ_{IS} with the electronegativity (Pauling) of the ligand as well as with the ionicity of the nickel–ligand bond (Fig. 7.4 taken from [9]). As $\Delta\langle r^2\rangle/\langle r^2\rangle$ is negative for ^{61}Ni, as is the case for ^{57}Fe (Table 7.1 (at the end of the book)), the isomer shifts observed in nickel compounds follow the same trends as in ^{57}Fe Mössbauer spectroscopy. It is clear from Fig. 7.4 that the nickel fluorides with the highest values of ligand electronegativity and bond ionicity show the most positive isomer shifts and thus the smallest electron densities at the ^{61}Ni nucleus, due to minimal population in the 4*s* valence shell. Following the plots A and B to the left, we find nickel compounds of more complex bond properties with electron densities resulting from the combined σ- and π-interaction between nickel and the ligands. At any rate, a quantitative analysis of the isomer shift in relation to bond properties should be performed in connection with molecular orbital calculations or rather with modern DFT calculations as described in Chap. 5.

^{61}Ni isomer shifts have also been measured from other nickel compounds such as K_2NiF_4 [7], $Ni(PCl_3)_4$ [6], $(NH_4)_6[NiMo_9O_{32}]$ [7], $K_2[Ni(CN)_4]$ [6], $K_4Ni_2(CN)_6$ [8], $NiSO_4{\cdot}7H_2O$ [8], $(Et_4N)[Ph_3PNiBr_3]$ [11], $(Ph_3MeAs)_2$ $[NiCl_4]$ [11], $(diap)_2Ni$ (diap = *N*,*N'*-diphenyl-1-amino-3-iminopropene) [11], $Ni(en)_2Cl_2$ [6], $[Ni(en)_3]$ $Cl_2{\cdot}6H_2O$ [11], $Ni(CO)_4$ [6], $(Et_4N)_2[NiBr_4]$ [11], and from a number of alloys such as CuNi(2%) [7], CuNi(20%) [7], FeNi (1.5%) [7] and Ni_2R (R = Gd, Ho, Er, Tm, Yb) [6]. Unfortunately, the isomer shift data reported for most of these compounds have not been corrected for SOD (except those given in [8, 9]) and therefore cannot be included in the correlation diagram in Fig. 7.4. For meaningful discussions, ^{61}Ni isomer shift studies should always include SOD corrections. SOD corrected δ-values should also be used in setting up an isomer shift versus electron configuration diagram for ^{61}Ni, as has been attempted by Travis and Spijkerman [3], in analogy to the Walker–Wertheim–Jaccarino diagram for ^{57}Fe. Such diagrams, however, are much inferior to a quantitative MO or DFT evaluation of the contributions to the electron density from nickel 4*s* and all other atomic orbitals of nickel and the ligands contributing to the molecular orbitals.

7.1.2.2 Magnetic Interactions

Most of the nickel-containing substances are paramagnetic, ferromagnetic, or antiferromagnetic because of their relatively large spin resulting from unpaired Ni 3*d* electrons. For instance, $S = 1$ is common in Ni^{2+} compounds with octahedral, tetrahedral, or square planar symmetry around the nickel atom. The relatively large electronic spin gives rise to a correspondingly large internal magnetic field at the nickel nucleus via core-polarization. Other contributions to the local magnetic field may come from the orbital motion of Ni 3*d* electrons and from conduction electron polarization, which is in most nickel alloys the dominating contribution.

The local magnetic field interacts with the relatively large magnetic dipole moments μ of the ^{61}Ni nucleus in both the 67.4 keV excited (μ_e) and the ground state (μ_g) (μ_e is about four times larger than μ_e of ^{57}Fe, μ_g is about eight times larger than μ_g of ^{57}Fe; the signs are opposite, cf. Table 7.1). However, as we have already pointed out in Sect. 6.7, the magnetic interaction is observable in the Mössbauer spectrum only if the internal field fluctuations and the electronic spin fluctuations (in the absence of an external field) are sufficiently slow compared with the lifetime τ_N of the 67.4 keV nuclear state of ^{61}Ni and the time τ_L corresponding to the Larmor precession frequency ω_L of the nuclear spin.

^{61}Ni Mössbauer spectra with pure magnetic dipole interaction (in addition to electric monopole interaction) generally show a more or less resolved four-line pattern; each of the four broad and asymmetric lines represents an unresolved triplet arising from the transitions shown in Fig. 7.3. Despite the relatively large nuclear magnetic moments of the ^{61}Ni Mössbauer states, the resolution of the magnetically split spectra is often quite poor. There are two reasons behind this: (a) relatively weak magnetic fields and (b) the broad line width, which is about 14 times broader than that of ^{57}Fe. Two examples for magnetically split ^{61}Ni spectra are shown in Fig. 7.5a, b. The spectrum of the $Fe_{0.95}Ni_{0.05}$ alloy (Fig. 7.5a, reproduced from [4]) is reasonably well resolved due to a relatively large internal magnetic field of 23.4 T at the ^{61}Ni nuclei. The theoretical spectrum [8] matches the experimental data reasonably well; it consists of 12 individual Lorentzian lines (also shown in the figure) of different intensities referring to the allowed transitions between substates of the 67.4 keV level and of the ground state (cf. Fig. 7.3, case of H_M). The magnitude of the internal magnetic field has been found to vary with the composition of the Fe_xNi_{1-x} alloys [8, 12]. In NiO, the internal magnetic field is much smaller (1.4 T) leading to a less pronounced magnetic splitting which in turn results in a poorly resolved spectrum (cf. Fig. 7.5b, reproduced from [3]).

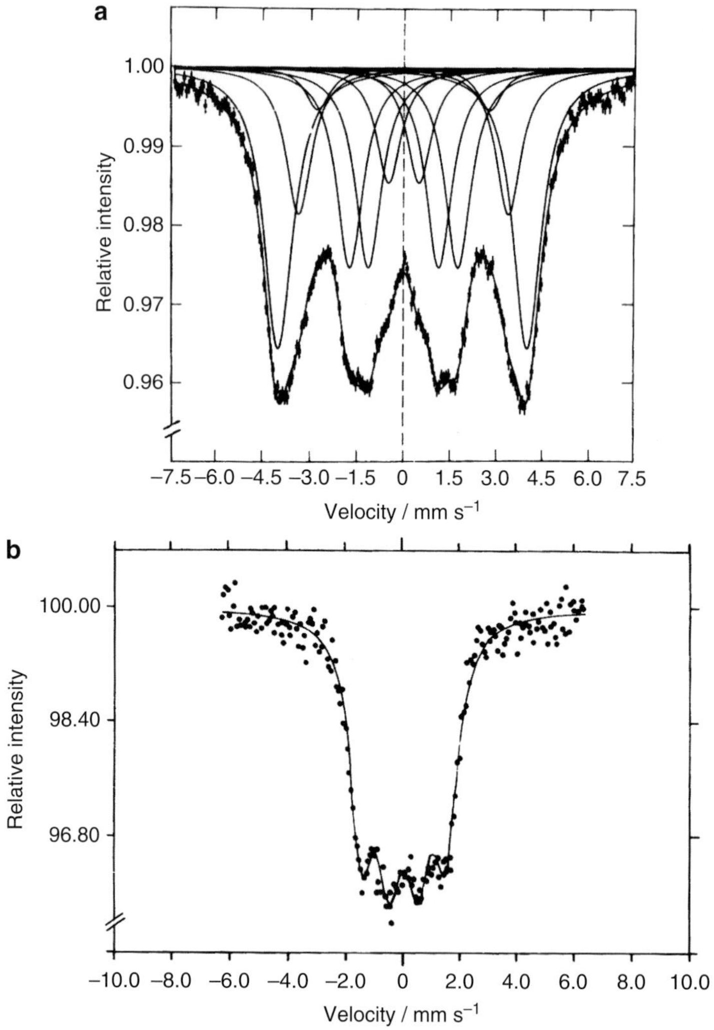

Fig. 7.5 ^{61}Ni Mössbauer spectra of (**a**) $Fe_{0.95}Ni_{0.05}$ obtained at 4.2 K with a $Ni_{0.86}V_{0.14}$ source (reproduced from [4]), (**b**) NiO obtained at 80 K with a $Ni_{0.85}Cr_{0.15}$ source (reproduced from [3])

7.1.2.3 Electric Quadrupole Interactions

Both the ground state and the 67.4 keV nuclear excited state of ^{61}Ni possess a nonzero electric quadrupole moment. If placed in an inhomogeneous electric field (electric field gradient, EFG $\neq 0$) the ^{61}Ni nucleus undergoes electric quadrupole interaction with the EFG at the nucleus, as a result of which the 67.4 keV level will split into three substates $|I, \pm m_{\mathrm{I}}\rangle = |5/2, \pm 5/2\rangle$, $|5/2, \pm 3/2\rangle$, and $|5/2, \pm 1/2\rangle$ and the ground level will split into two substates $|3/2, \pm 3/2\rangle$ and $|3/2, \pm 1/2\rangle$.

In practical ^{61}Ni spectroscopy, electric quadrupole interactions have been observed only in a few examples. The reason is that the contributions to the EFG from noncubic valence electron distributions frequently vanish (we shall consider some examples below), and the contributions from noncubic lattice surroundings are relatively small because of the $1/R^3$ effect (see Sect. 6.4). Furthermore, the quadrupole moment of the ground state, known from other spectroscopic measurements ($+0.16b$ [13]), turns out to be rather small. In the few examples of quadrupole interaction observed so far, e.g., in $(NH_4)_6NiMo_9O_{32}$ studied by Travis et al. [3], and in the spectrum of $NiCr_2O_4$ (Fig. 7.6a) measured at 77 and 4.2 K by Göring [14], the interaction manifested itself only in the asymmetry of the Mössbauer spectrum and additional broadening of the resonance lines. Göring [14] determined the quadrupole moment of the 67.4 keV excited nuclear state to be $Q = -0.20b$. The theoretical spectra were evaluated by considering five transition lines between the quadrupole split ground and excited states (see case of $\widehat{H}_Q$ on the far right side of Fig. 7.3).

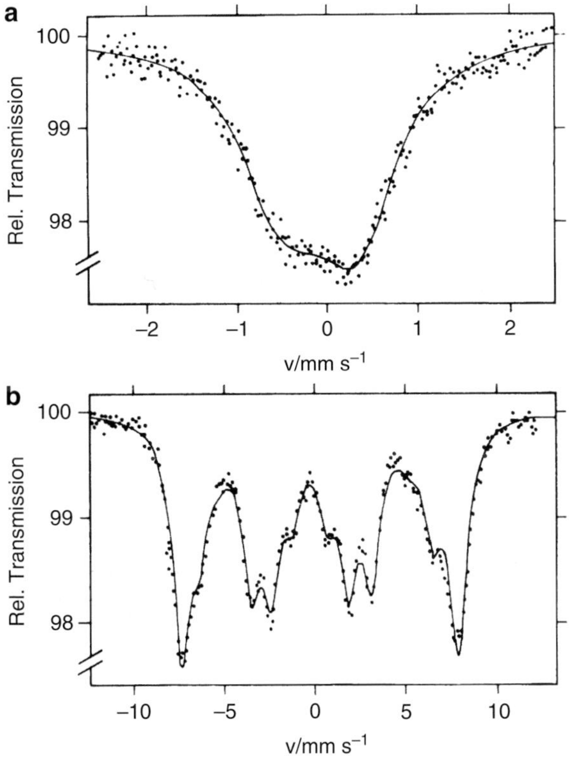

Fig. 7.6 ^{61}Ni Mössbauer spectra of $NiCr_2O_4$ using a $Ni_{0.85}Cr_{0.15}$ source. (**a**) Source and absorber at 77 K (pure quadrupole interaction). (**b**) Source and absorber at 4 K (combined magnetic and quadrupole interaction) (from [14])

An instructive description of the first-order perturbation treatment of the quadrupole interaction in ^{61}Ni has been given by Travis and Spijkerman [3]. These authors also show in graphical form the quadrupole-spectrum line positions and the quadrupole-spectrum as a function of the asymmetry parameter η; they give eigenvector coefficients and show the orientation dependence of the quadrupole-spectrum line intensities for a single crystal of a ^{61}Ni compound. The reader is also referred to the article by Dunlap [15] about electric quadrupole interaction, in general.

A quantitative consideration on the origin of the EFG should be based on reliable results from molecular orbital or DFT calculations, as pointed out in detail in Chap. 5. For a qualitative discussion, however, it will suffice to use the easy-to-handle one-electron approximation of the crystal field model. In this framework, it is easy to realize that in nickel(II) complexes of O_h and T_d symmetry and in tetragonally distorted octahedral nickel(II) complexes, no valence electron contribution to the EFG should be expected (cf. Fig. 7.7 and Table 4.2). A temperature-dependent valence electron contribution is to be expected in distorted tetrahedral nickel(II) complexes; for tetragonal distortion, e.g., $V_{zz} = -(4/7)e\langle r^{-3}\rangle_{3d}$ for compression along the z-axis, and $V_{zz} = +(2/7)e\langle r^{-3}\rangle_{3d}$ for elongation along the z-axis;

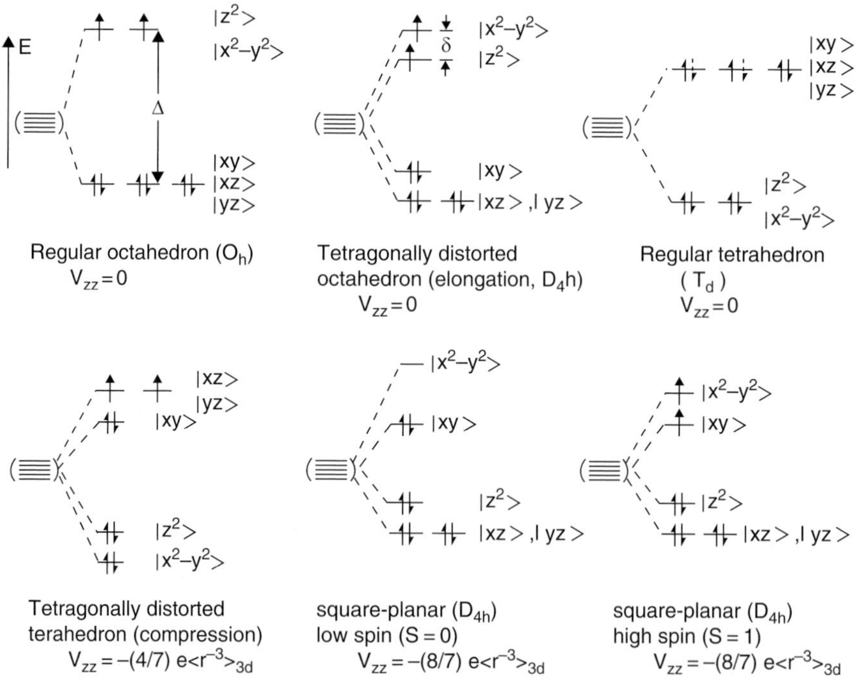

Fig. 7.7 Schematic diagrams for common electron configurations of Ni^{2+} complexes in the one-electron approximation. The resulting valence electron contributions V_{zz} are obtained from Table 4.2

e is the positive elementary charge. Valence electron contributions to the EFG are also expected in square-planar nickel(II) complexes; $V_{zz} = -(8/7)e\langle r^{-3}\rangle_{3d}$ both for high-spin ($S = 1$) and low-spin ($S = 0$) configurations. The lattice contributions to the EFG, of course, have to be treated separately, e.g., using the simple Townes–Dailey method [16, 17]. However, it should be emphasized again that much more reliable results are arrived at by finding the wave functions of the molecular orbitals involved and taking the Boltzmann weighted sum over all orbitals involved for each EFG tensor element. Since the spacings in the quadrupole spectrum of ^{61}Ni are unequal, the sign of the EFG (V_{zz}) can be extracted directly from the quadrupole-perturbed Mössbauer spectrum of a polycrystalline sample, with the known quadrupole moments of the excited and the ground nuclear state of ^{61}Ni (cf. Fig. 7.3).

7.1.2.4 Combined Magnetic and Quadrupole Interactions

The low-temperature Mössbauer spectra of the spinel type oxides, $NiCr_2O_4$ [14, 18] (Fig. 7.6b) and $NiFe_2O_4$ [3, 18], have been found to exhibit combined magnetic dipole and electric quadrupole interaction (Fig. 7.7). For the evaluation of these spectra, the authors have assumed a small quadrupolar perturbation and a large magnetic interaction, as depicted in Fig. 7.3 and represented by the Hamiltonian [3]

$$\widehat{H}_{\mathrm{I}} = \widehat{H}_{\mathrm{M}} + \widehat{H}'_{\mathrm{Q}}, \quad \text{with } \widehat{H}_{\mathrm{M}} = -g\beta_{\mathrm{N}}\widehat{H}_z\widehat{I}_z, \tag{7.2}$$

$$\widehat{H}'_{\mathrm{Q}} = [eQV_{zz}/4I(2I-1)]\left[3\left(-\sin\theta\widehat{I}_x + \cos\theta\widehat{I}_z\right)^2 - \widehat{I}^2\right].$$

First-order perturbation theory yields the eigenvalues (diagonal elements of the perturbation matrix added to the eigenvalues of the major interaction)

$$\begin{aligned} E^1_{\mathrm{m}} &= -g\beta_{\mathrm{N}}H_z m_i + [eQV_{zz}/4I(2I-1)]\left[3m_{\mathrm{I}}^2 - I(I+1)\right](3\cos^2\theta - 1)/2, \\ m_{\mathrm{I}} &= -I, -I+1, \ldots, I. \end{aligned} \tag{7.3}$$

θ is the angle between the quadrupolar quantization axis and that of the magnetic interaction in systems of axial symmetry (asymmetry parameter $\eta = 0$).

A rigorous quantum mechanical treatment of the combined interaction is laborious and computer-time consuming, because each interaction tends to have its own set of quantization axes, whereas the total perturbation does not quantize along either of the two sets of axes.

The alternative case of approximation analogous to the one mentioned above in (7.2) assumes a small magnetic perturbation and a large quadrupole interaction. This case, which is very rare and has not yet been observed in nickel systems, is expressed by the Hamiltonian [3]

$$\widehat{H}_2 = \widehat{H}_{\mathrm{Q}} + \widehat{H}'_{\mathrm{M}}, \quad \text{with } \widehat{H}_{\mathrm{Q}} = [eQV_{zz}/4I(2I-1)]\left(3\widehat{I}_z^{\,2} - \widehat{I}^2\right), \tag{7.4}$$

$$\widehat{H}'_{\mathrm{M}} = g\beta_{\mathrm{N}}H_z(-\sin\theta\widehat{I}_x + \cos\theta\widehat{I}_z).$$

The first-order eigenvalues for axial systems ($\eta = 0$) are

$$E_{\mathrm{m}}^2 = [eQV_{zz}/4I(2I-1)]\left[3m_{\mathrm{I}}^2 - I(I+1)\right] - g\beta_{\mathrm{N}}H_z m_{\mathrm{I}}\cos\theta, \tag{7.5}$$

Here, the magnetic perturbation is projected onto the quantization axes of $\widehat{H}_{\mathrm{Q}}$ as the major interaction.

7.1.3 Selected ^{61}Ni Mössbauer Effect Studies

From the applications of ^{61}Ni Mössbauer spectroscopy in solid-state research, it is clear that (1) information from isomer shift studies is generally not very reliable because of the smallness of the observed isomer shifts and the necessity of SOD shift corrections which turn out to be difficult, and (2) useful information about magnetic properties and site symmetry is obtained from spectra reflecting magnetic and/or quadrupolar interactions.

The magnitude and sign of the magnetic hyperfine field in the nickel spinel compounds $NiFe_2O_4$ (Fig. 7.8), $NiCr_2O_4$ (Fig. 7.9), $NiFe_{0.3}Cr_{1.7}O_4$, $NiRh_2O_4$, and

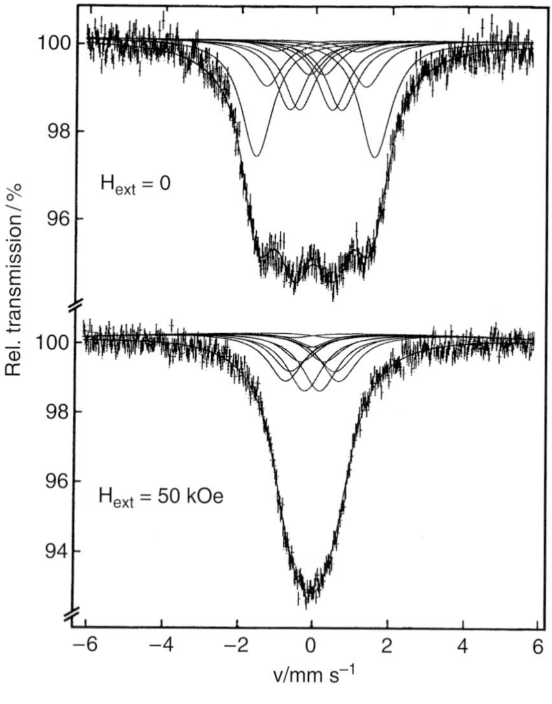

Fig. 7.8 ^{61}Ni Mössbauer spectra of $NiFe_2O_4$ without (*top*) and with (*bottom*) an external magnetic field of 5 T. The source was ^{61}Co in $^{61}Ni_{0.85}Cr_{0.15}$, enriched to 98% and activated in a 300 MeV electron accelerator at Mainz University making use of the reaction ^{62}Ni(γ, p)^{61}Co. Source and absorber were kept at 4.2 K (from [19])

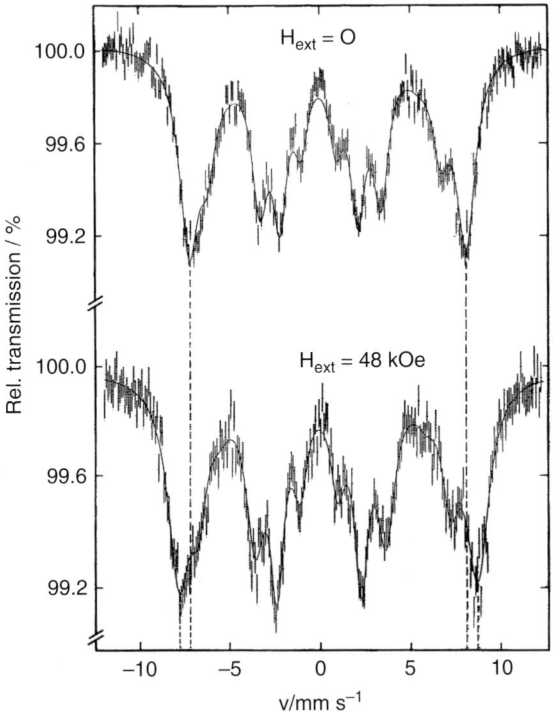

Fig. 7.9 ^{61}Ni Mössbauer spectra of $NiCr_2O_4$ without (*top*) and with (*bottom*) an external magnetic field of 5 T. Source and absorber were kept at liquid helium temperature (from [19])

$GeNi_2O_4$ were determined by means of ^{61}Ni Mössbauer spectroscopy in external magnetic fields [19]. The hyperfine fields of Ni^{2+} ions in octahedral lattice sites, e.g., $NiFe_2O_4$, were found to be negative and of the order of −10 T, and positive for Ni^{2+} at tetrahedral lattice sites ranging from +2.5 T in $NiRh_2O_4$ to +63 T in $NiFe_{0.3}Cr_{1.7}O_4$. The sign and the magnitude of the hyperfine fields have been explained in terms of ligand field theory, including the effect of 3*d*–4*p* orbital mixing.

The intermetallic compound, $LaNi_5$, which has been proposed as a hydrogen storage material because of its ability to absorb large amounts of hydrogen at room temperature and ambient pressure, has been shown to degrade slowly with each sorption–desorption cycle. This degradation was believed to lead to lanthanum hydride and metallic nickel. ^{61}Ni Mössbauer effect measurements on such materials after many cycles have indeed proven that nearly 50% of the original compound degrades to metallic nickel microprecipitates. The Mössabauer spectrum recorded after nearly 1,600 cycles (Fig. 7.10) could be deconvoluted into two subspectra, one representing the original material $LaNi_5$ and the other being the typical of metallic nickel [20].

An interesting study of oxidic spinel ferrites of the type $Co_xNi_{5/3-x}FeSb_{1/3}O_4$ was reported [21], where three different Mössbauer-active probes ^{57}Fe, ^{61}Ni and ^{121}Sb were employed on the same material. The results have been interpreted in terms of the cation distributions over spinel A- and B-lattice sites, magnetic moments and spin structure, and the magnitude of the supertransferred hyperfine

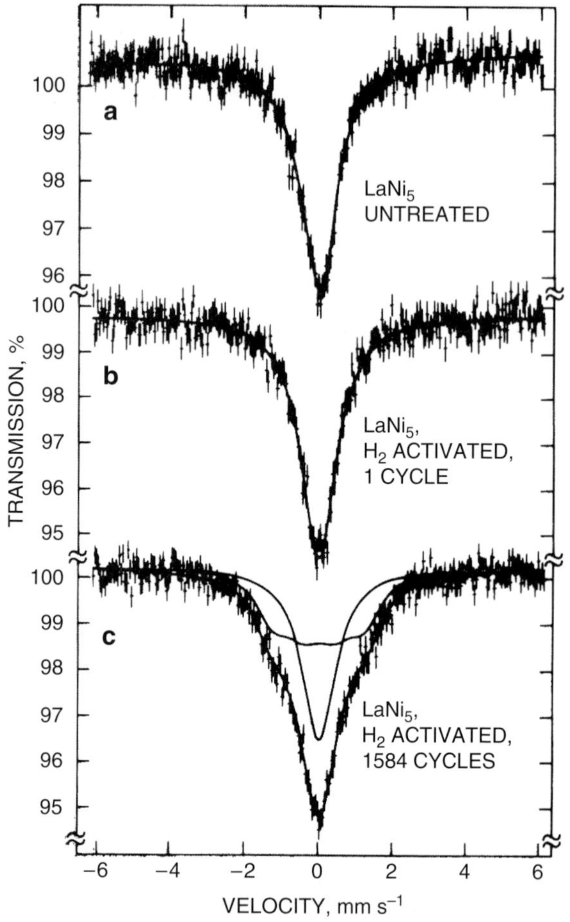

Fig. 7.10 ^{61}Ni Mössbauer spectra of $LaNi_5$ samples after various treatments: (**a**) no hydrogen exposure, (**b**) activated in hydrogen, (**c**) after 1,584 thermally induced absorption–desorption cycles, as described in the text. The *solid lines* are least-squares fits to a single Lorentzian in (**a**) and (**b**). (**c**) was fitted with a single Lorentzian (representing nonmagnetic nickel atoms) and a 12-line hyperfine spectrum (from [20])

field at Sb sites set up by covalent spin charge transfer from A-sites, occupied by Fe^{3+}, Ni^{2+} and Co^{2+} ions, into the 5*s* orbital of Sb^{5+} ions [21].

McCammon et al. have studied fine nickel particles using ^{61}Ni Mössbauer spectroscopy [22]. The measured average hyperfine field of 10 nm particles at 4.2 K was 7.7 T; for nickel foil, it was found to be 7.5 T. Application of an external magnetic field of 6 T caused a reduction of the hyperfine splitting to 1.5 T as a consequence of the negative hyperfine field at Ni nuclei.

A similar study using ^{61}Ni was carried out by Stadnik et al. [23]. Measurements were performed at 4.2 K on spherical nickel particles covered with a protective layer of SiO, with average diameter of 500 and 50 Å, respectively. The hyperfine magnetic field at ^{61}Ni nuclei for 500 Å particles was found to be 7.8 T, compared with the field for nickel foil of 7.5 T. The small difference is due to the demagnetization and dipolar fields in 500 Å particles. The spectrum of 50 Å particles had a surface component with the corresponding value of the hyperfine magnetic field of 4.0 T. This indicates that, in accordance with theoretical studies, there is a decrease of the hyperfine magnetic field in the surface layer of nickel. A further ^{61}Ni Mössbauer study, again at 4.2 K, on similarly prepared nickel particles with an averaged diameter of 100 and 30 Å, also covered with a protective layer of SiO, yielded spectra containing a surface component with a significantly reduced hyperfine magnetic field of 3.3 T for both particle sizes, as compared to the field of 7.5 T in the bulk [24].

The electronic structure of the spinel type compound $NiCo_2O_4$ has been investigated by XANES, EXAFS, and ^{61}Ni Mössbauer studies. On the basis of the derived cation valencies, the octahedral and tetrahedral site occupancies as well as the formula in standard notation for spinel compounds could be delineated [25].

Nasredinov et al. were the first to report on ^{61}Cu(^{61}Ni) emission Mössbauer spectroscopy to study hyperfine interactions in copper-based oxides (CuO_2, CuO) and HT_C superconductors ($YBa_2Cu_3O_{7-x}$, $La_{2-x}Sr_xCuO_4$) [26]. In this case, the $^{61}Ni^{2+}$ Mössbauer source nuclei produced by decay of ^{61}Cu also occupy the copper sites. The emission Mössbauer spectra were recorded at 80 K with $Ni_{0.86}V_{0.14}$ used as an absorber. Selected emission spectra are shown in Figs. 7.11 and 7.12. A linear relationship between the quadrupole coupling constant for ^{61}Ni and the calculated lattice EFG was established and the contribution of the Ni^{2+} valence electrons to the coupling constant was determined. Local magnetic fields were observed at ^{61}Ni

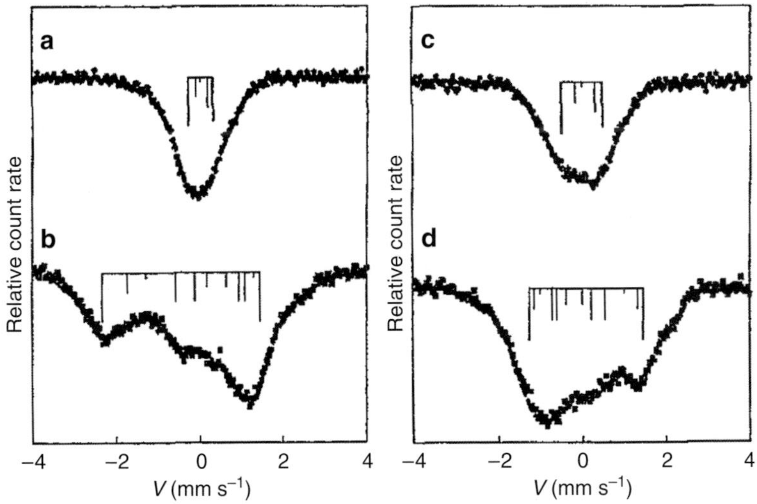

Fig. 7.11 ^{61}Cu(^{61}Ni) emission Mössbauer spectra of (**a**) Cu_2O, (**b**) CuO, (**c**) $La_{1.85}Sr_{0.15}CuO_4$ and (**d**) La_2CuO_4 ceramic samples. The positions of the components of (**a**) and (**c**), the quadrupole multiplets, and (**b**) and (**d**), Zeeman multiplets, are shown (from [26])

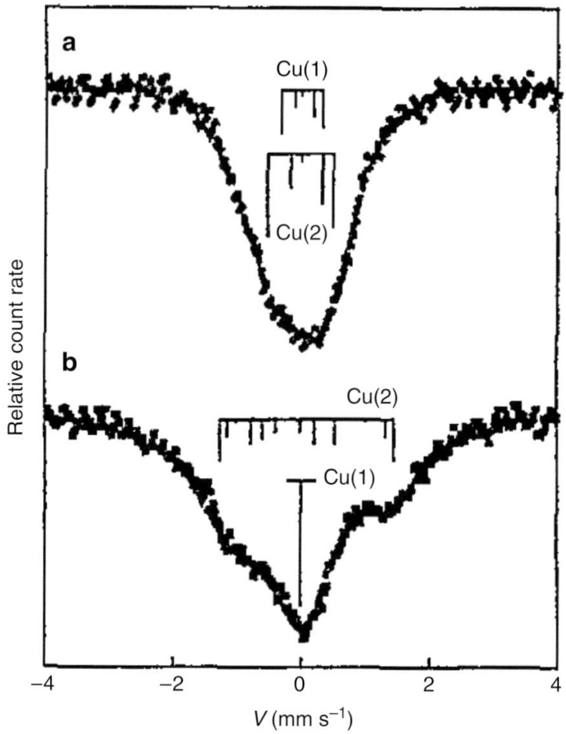

Fig. 7.12 ^{61}Cu(^{61}Ni) emission Mössbauer spectra of (**a**) $YBa_2Cu_3O_7$ and (**b**) $YBa_2Cu_3O_6$. Two quadrupole multiplets corresponding to $^{61}Ni^{2+}$ at the Cu(l) and Cu(2) sites are shown in spectrum (**a**). The Zeeman multiplet from the Cu(2) sites and the singlet from the Cu(1) sites are shown in spectrum (**b**) (from [26])

nuclei in those cases where the copper sublattices showed magnetic ordering. Further studies by the same research group using the technique of ^{61}Ni Mössbauer emission spectroscopy to determine the charges of atoms and the parameters of the EFG tensor at ^{61}Cu(^{61}Ni) lattice sites were reported for other HT_c superconductors [27–29].

Huge hyperfine magnetic fields have been observed in spinel chromites $Cu_{0.9}Ni_{0.1}Cr_2O_4$ and $Co_{0.9}Ni_{0.1}Cr_2O_4$ arising from $^{61}Ni^{2+}$ ions in compressed tetrahedral sites. The largest hyperfine field ever reported for ^{61}Ni was found to be 80 T in the spinel chromite $Cu_{0.9}Ni_{0.1}Cr_2O_4$ [30]. The large hyperfine field was attributed to the orbital angular momentum in the degenerate ground state of Ni^{2+} ions. ^{61}Ni Mössbauer spectra of the two spinel chromites measured at 5 K are shown in Fig. 7.13. The single line source of ^{61}Cu ($\rightarrow ^{61}$Ni) was produced by the nuclear reactions ^{58}Ni (α, p) ^{61}Cu and ^{58}Ni (α, n) ^{61}Zn by irradiation with 25 MeV α-particles in a cyclotron (RIKEN).

For the three nickel compounds CaNiN, $BaNiO_2$, and $BaNiO_3$, which contain nickel in three different oxidation states (I, II, and IV, respectively), the EFG at

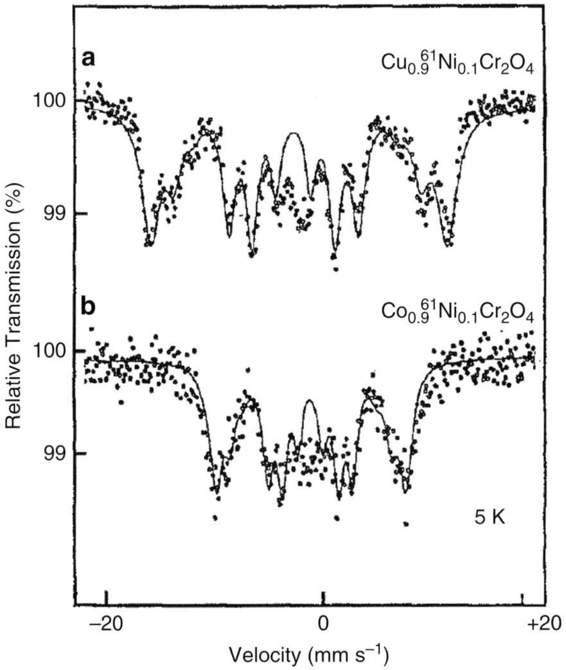

Fig. 7.13 ^{61}Ni Mössbauer spectra of spinel chromites (**a**) $Cu_{0.9}Ni_{0.1}Cr_2O_4$ and (**b**) $Co_{0.9}Ni_{0.1}Cr_2O_4$ at 5 K

the nickel site was investigated by ^{61}Ni spectroscopy and by band structure calculations using the FP-LMTO (Full Potential Linear Muffin-Tin Orbital) method [31]. The quantitative agreement between both methods was very satisfactory, and has explained the quadrupole interaction from the analysis of the bonding mechanism. It was found that besides the crystal field splitting in the ionic model, the covalent nickel–ligand bonds play an important role in the quadrupole interaction. The ^{61}Ni Mössbauer spectrum of $BaNiO_2$ recorded at 4.2 K reflects rather strong nuclear quadrupolar interaction and yields one of the best resolved quadrupole doublets observed so far with ^{61}Ni Mössbauer spectroscopy (Fig. 7.14). Calculated electron densities plotted versus measured isomer shifts are depicted in Fig. 7.15.

The first ^{61}Ni Mössbauer spectrum of nickel in a bioinorganic compound with determinable EFG and isomer shift was reported for a nickel complex compound with planar [NiS_4] core and considered as a model compound for hydrogenase. This Mössbauer spectrum from the formal Ni^{IV} compound is presented in Fig. 7.16. The observed quadrupolar interaction can be understood in terms of ligand field theory. In this approach, the b_{1g} and b_{2g} levels ($d_{x^2-y^2}$ and d_{xy}) are not occupied which is expected to cause a large negative EFG contribution [32].

Leupold et al. were the first to report on coherent nuclear resonant scattering of synchrotron radiation from the 67.41 keV level of ^{61}Ni. The time evolution of the forward scattering was recorded by employing the so-called nuclear lighthouse

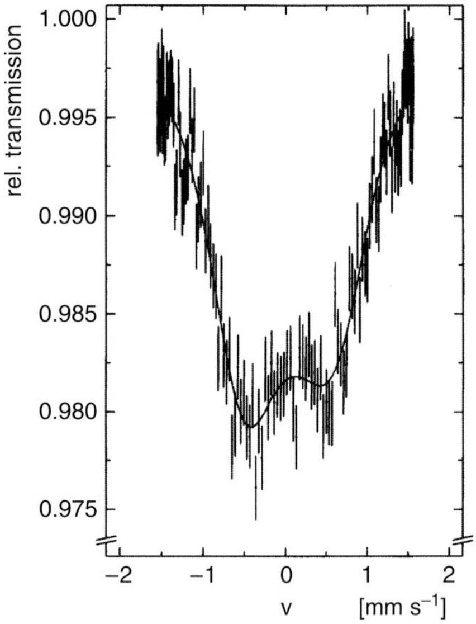

Fig. 7.14 ^{61}Ni Mössbauer spectrum of $BaNiO_2$ at 4.2 K. The source, also kept at 4.2 K, of the parent ^{61}Co was $^{62}Ni_{0.85}Cr_{0.15}$ (97% enriched) activated at Mainz Microtron via the nuclear reaction $^{62}Ni(\gamma, p)^{61}Co$. The spectrum shows resolved quadrupole splitting (from [31])

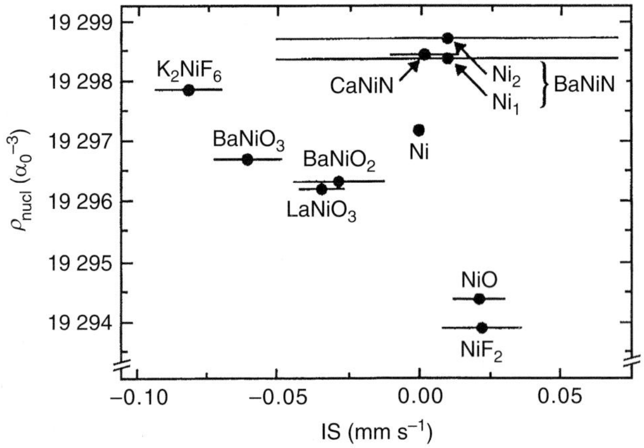

Fig. 7.15 Calculated (with FP-LMTO method) electron densities plotted versus measured isomer shifts (not corrected for SOD) (from [32])

effect. The authors have shown that this method can be used to investigate Mössbauer isotopes in a coherent scattering process with synchrotron radiation at high transition energies. The decay of the excited ensemble of nuclei in Ni metal showed quantum beats that allowed the determination of the magnetic hyperfine field at the ^{61}Ni nuclei [33]. Shortly thereafter, Sergueev et al. demonstrated the feasibility of nuclear forward scattering (NFS) for high-energy Mössbauer transitions by recording the time evolution of the NFS of synchrotron radiation of metallic nickel foil as a function of temperature (cf. Fig. 7.17) and nickel oxide powder at 3.2 K with and

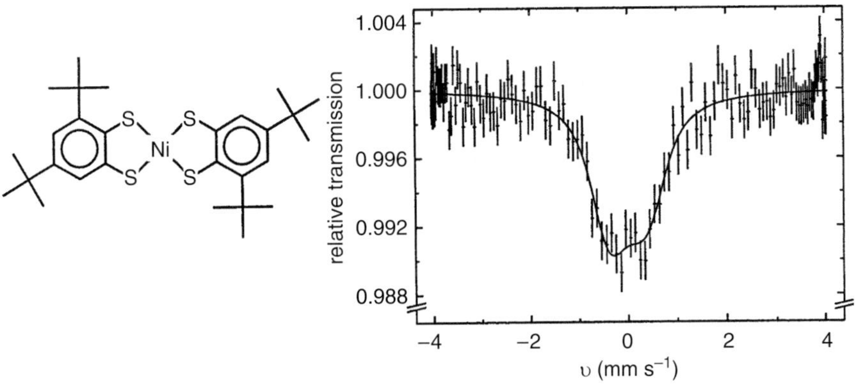

Fig. 7.16 ^{61}Ni Mössbauer spectrum at 4.2 K of a nickel complex compound with planar $[NiS_4]$ core known as a model compound for hydrogenase (source $^{62}Ni_{0.85}Cr_{0.15}$ (97% enriched) activated at Mainz Microtron) (from [32])

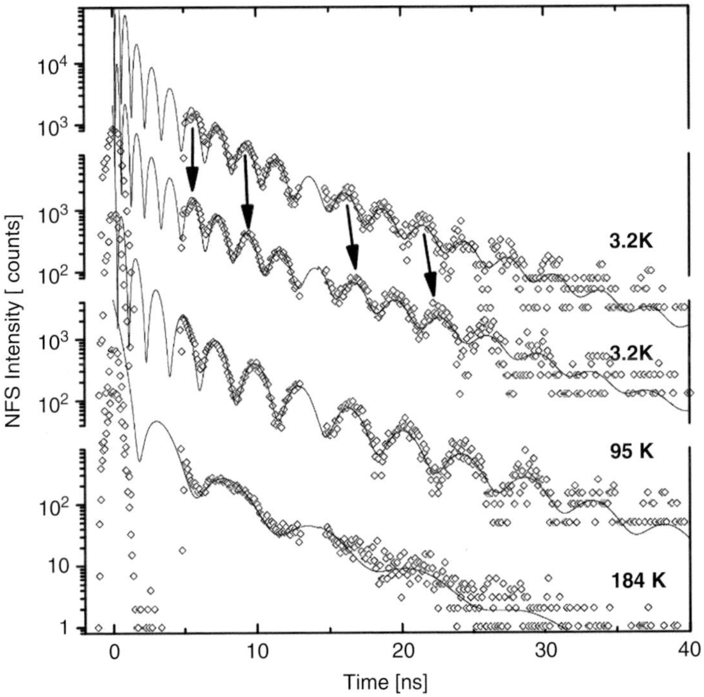

Fig. 7.17 Time evolution of the nuclear forward scattering for metallic Ni foil. All measurements except for the *upper* curve were performed with external magnetic field $B = 4$ T. The *solid lines* show the fit. The *arrows* emphasize stretching of the dynamical beat structure by the applied magnetic field. The data at times below 14.6 ns had to be rescaled (from [34])

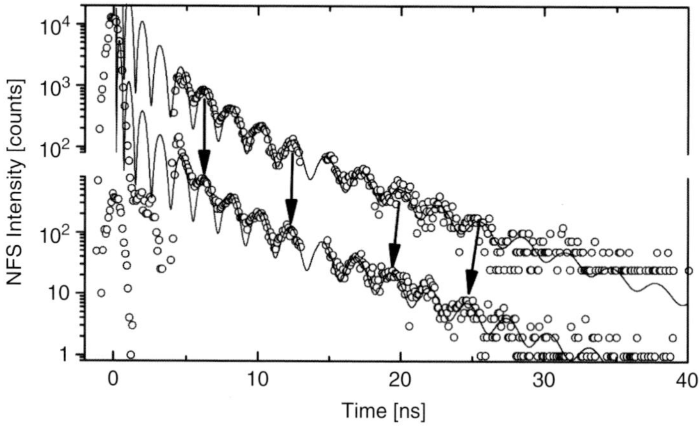

Fig. 7.18 Time evolution of the nuclear forward scattering for NiO powder at 3.2 K measured without external magnetic field (*top*) and with magnetic field of 4 T (*bottom*). The *solid lines* show the fit. The *arrows* emphasize compression of the dynamical beat structure by the applied magnetic field. The data at times below 14.6 ns had to be rescaled (from [34])

without a magnetic field (cf. Fig. 7.18). These results open new pathways (1) in using the method for studying nickel compounds in biology, geophysics, earth and materials sciences, etc., and (2) in extending the method to other isotopes with even higher nuclear transition energies.

In addition to the examples discussed above, we add the following other reports on ^{61}Ni Mössbauer studies where only the reference, the material under study, and some remarks about the study are quoted.

References	System	Remarks
[12]	$ErNi_2$	$\langle H^{int}\rangle$ at Ni sites
	$GdNi_2$	Dependence of $\langle H^{int}\rangle$ on x
	Ni_xFe_{1-x}	
[35]	Ni_xPd_{1-x}	Dependence of $\langle H^{int}\rangle$ on x
[36]	Ni–Pt	
	Ni–Pd	
[37]	$Ni_{1-x}Pd_x$(x: 0–0.995)	Absorber recoilless fraction, energy shift, and $\|H^{int}\|$ as function of x; δ values from temperature dependence of SOD; calculation of $\langle H^{int}\rangle$ and comparison with measurements in applied fields
[38]	$Ni_{1-x}Mn_x$ (x: 0.05–0.25)	$\langle H^{int}\rangle$ at Ni sites as function of x
[39]	$Ni_{1-x}Pt$ (x: 0.1–0.5)	$\langle H^{int}\rangle$ at Ni sites as function of x
	$Ni_{1-x}Pd_x$ (x: 0.05–0.9)	
[18]	$NiFe_2O_4$ (a)	$\|H^{int}\| = 7$–10 T for Ni in octahedral sites (a–d), about 45 T for Ni in tetrahedral sites (e);
	$NiCo_2O_4$ (b)	
	$NiMn_2O_4$ (c)	
	$GeNi_2O_4$ (d)	
	$NiCr_2O_4$ (e)	quadrupolar interaction in (e) and (f)
	$NiRh_2O_4$ (f)	

(*continued*)

References	System	Remarks
[8]	$K_4Ni(CN)_6$ NiF_2 $KNiF_3$ NiO $NiSO_4·7H_2O$ $(NH_4)_{12}[NiMo_9O_{32}]·13H_2O$	δ from measured energy shifts after SOD correction, comparison with calculated (relativistic) values;
	NiAl Fe_xNi_{1-x} (x: 0.2–0.99) Co_xNi_{1-x} (x: 0.1–0.5) Cu_xNi_{1-x} (x: 0.15–0.7)	H^{int} (^{61}Ni) in antiferromagnetic d^8 compounds; $\langle H^{int}\rangle$ (^{61}Ni) as function of x
[40]	NiS	Vibrational, magnetic, and electronic properties in semimetallic antiferromagnetic and in metallic phase; phase transition study
[41]	NiS_2	Study of magnetic dipole and electric quadrupole interactions, two Ni sites differing in angle between H and EFG axis; phase transitions in $NiS_{2.00}$ and $NiS_{1.96}$
[42]	NiS_2 $NiS_x(1.91 \leq x \leq 2.1)$ $Ni_{1-y}Cu_yS_{1.93}$ $(0.03 \leq y \leq 0.1)$	Investigation of structural, electronic, and magnetic properties by means of X-ray diffraction, densitometry, resistivity, susceptibility, and ^{61}Ni Mössbauer spectroscopy as function of x; temperature of phase transition from semimetallic to metallic state as function of x; different Ni sites with different $\langle \lvert H^{int}\rvert \rangle$ and different angle between H and EFG axis; effect of Cu impurities
[43]	$NiCr_2O_4$	Study of magnetic hyperfine and quadrupole splitting, $H(T)/H(T_0) = f(T/T_N)$, fit to Brillouin function with $S = 1$
[44]	$[Ni(niox)_2]$	Quadrupole splitting of a square-planar nickel complex, the sign of which demonstrates that the EFG of the nonbonding electrons outweighs that of the bonding electrons donated by the ligands
[45]	Heusler alloys $Ru_xY_{3-x}Z$ (Y = Fe, Ni; Z = Si, Sn)	Mössbauer studies with four probes: ^{99}Ru, ^{61}Ni, ^{57}Fe, ^{119}Sn; magnetic fields at Ru, Ni, Fe, and Sn sites
[46]	$Ni_3Sn_2S_2$	Valence states of nickel, tin, and sulfur in the ternary chalcogenide, ^{61}Ni and ^{119}Sn Mössbauer investigations, XPS and band structure calculations
[47]	$LiNiO_2$	Stoichiometry, cationic site assignment in $Li_{1-x}Ni_{1+x}O_2$
[48, 49]	CrNiAs, CrNiP	Magnetic properties, Debye temperature

7.2 Zinc (^{67}Zn)

7.2.1 Experimental Aspects

Craig and coworkers [50, 51] were the first to demonstrate the existence of recoilless nuclear resonance absorption for the 93.31 keV transition in ^{67}Zn. In an on–off type of experiment, they clamped the source of ^{67}Ga/ZnO rigidly to a ZnO absorber, both at 2 K. The absorber was placed in a magnetic field varying between 0 T and 0.07 T (700 G). As the magnetic field was increased, a variation in the transmission of the 99.3 keV γ-rays (maximum relative absorption was 0.3%) due to increasing magnetic hyperfine splitting in the absorber was detected. In a similar experiment, a Russian group also succeeded in observing nuclear resonance absorption in ^{67}Zn [52]. Attempts to record conventional velocity spectra were first undertaken by Alfimenkov et al. [53] using a piezoelectric quartz drive system but unambiguous results were not obtained. Later, de Waard and Perlow [54] succeeded in recording well-resolved ^{67}Zn Mössbauer hyperfine spectra using a piezoelectric quartz drive system.

The 93 keV resonance in ^{67}Zn has the highest relative energy resolution of the established Mössbauer isotopes. The comparatively long half-life, $t_{1/2} = 9.1$ μs, of the 93 keV level yields an extremely small natural line width of $2\Gamma = 0.32$ μm s^{-1}. The relative natural width $\Gamma/E_\gamma = 5.2 \cdot 10^{-16}$ is about 600 times smaller than that of the 14.4 keV level of ^{57}Fe. The noise vibration velocities in most spectrometers are generally two orders of magnitude larger than the natural line width of the 93 keV transition in ^{67}Zn and render Mössbauer spectroscopy of zinc compounds a difficult task to perform. Good spectra can only be obtained with a specially designed spectrometer with an unconventional drive system, e.g., the piezoelectric quartz drive system as described in [54, 55], or a transducer containing commercially available cylinder of PZT-4 (lead zirconate–titanate) fed with a sinusoidal voltage to produce periodical elongation and contraction of the cylinder [56]. Both source and absorber are rigidly clamped to the piezoelectric crystal and the transducer is kept at 4.2 K. Calibration was based on the known piezomodulus at that temperature.

The 93 keV Mössbauer level is populated either by β^- decay of ^{67}Cu or by EC of ^{67}Ga. The decay scheme, reproduced from [1], is shown in Fig. 7.19. The nuclear data of interest for ^{67}Zn Mössbauer spectroscopy may be taken from Table 7.1 (end of the book).

The sources most commonly used so far consisted of sintered disks containing about 100 mg ZnO enriched with 90% ^{66}Zn. The disks were irradiated with 12 MeV deuterons or 30 MeV ^{3}He particles, to yield the 78 h activity of ^{67}Ga, and then annealed by heating in oxygen to 700–1,000 K for about 12 h and cooling down slowly (about 50 K h^{-1}) to room temperature. A NaI scintillation counter, 2–3 mm thick, is suitable for the detection of the 93 keV γ-rays. Because of the relatively high transition energy, both source and absorber are generally kept at liquid helium temperature.

Most of the ^{67}Zn Mössbauer experiments so far have been carried out with ZnO as absorber. De Waard and Perlow [54] used polycrystalline ZnO enriched to 90% in ^{67}Zn with various pretreatments. They intended to determine (1) the quadrupole splitting in ZnO, (2) the influence of source and absorber preparation on the width and depth of a resonance, (3) the SOD shift, and (4) the influence of pressure on the source.

Figure 7.20 shows the ^{67}Zn Mössbauer spectrum of hexagonal ^{67}ZnO obtained with a ^{67}Ga/^{66}ZnO source by de Waard and Perlow [54, 55]. The full spectrum

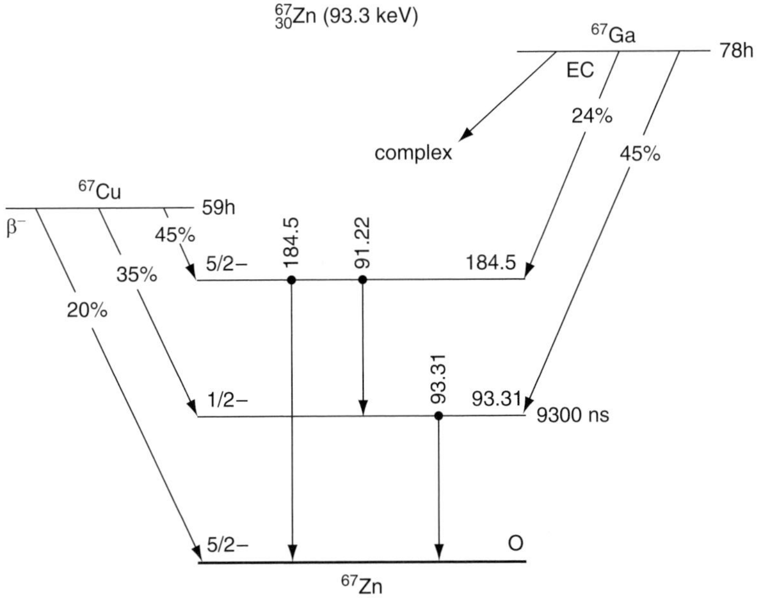

Fig. 7.19 Decay scheme of ^{67}Zn (from [1])

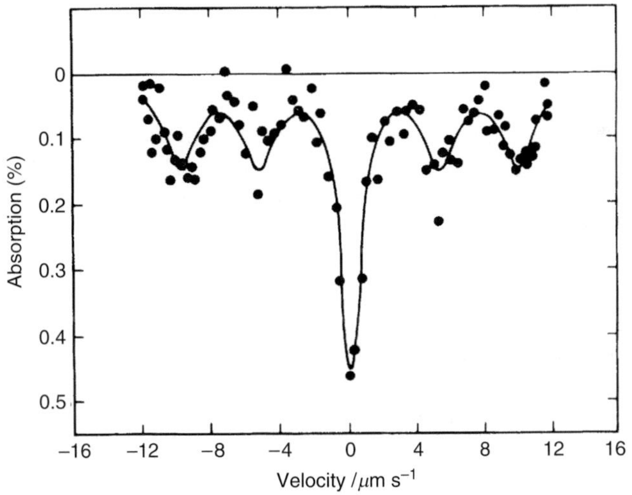

Fig. 7.20 ^{67}Zn Mössbauer spectrum of ^{67}ZnO at 4.2 K. *Source*: ^{67}Ga in ^{66}ZnO at 4.2 K. Absorber: 2.11 g cm^{-2} ZnO enriched to 90% in ^{67}Zn (from [54])

expected for electric quadrupole splitting in both source and absorber should contain seven resonance lines with intensity ratios of 1:1:1:3:1:1:1 according to the nine equally probable quadrupole resonance components for the pure $E2$

transition between the $5/2^-$ ground state and the $1/2^-$ excited state. Three of the nine components are degenerate and form the intense central line. The ^{67}ZnO spectrum of de Waard and Perlow did not show the outer two resonance lines because of the limited velocity range from -12 to $+12$ μm s^{-1} that they were able to scan. From the positions of the lines, these authors derived a quadrupole coupling constant $e^2qQ = 2.47 \pm 0.03$ MHz corresponding to 32.8 μm s^{-1}. From a small inequality of the spacings, they determined the asymmetry parameter η to be 0.23, which they later corrected to 0.19. They explained the nonzero asymmetry parameter by an exceedingly small (~0.001 Å) deviation of oxygen atoms from the ideal tetrahedron. With the known ground state quadrupole moment of ^{67}Zn, $Q = 0.17b$, they found an absolute value of $eq = V_{zz} = 6.0 \cdot 10^{16}$ V cm^{-2} for the z component of the EFG at the ^{67}Zn nucleus, which agrees well with their calculated value (lattice sum of point charges) of $5.4 \cdot 10^{16}$ V cm^{-2}.

De Waard and Perlow [54] observed a marked influence of the method of preparation of source and absorber on both the line width and the depth of resonance. For instance, the line width was found to be 2.7 μm s^{-1} for a powdered ZnO absorber as compared to 0.8 μm s^{-1} for a sintered one.

Because of the exceedingly small hyperfine splittings and shifts of resonance lines in ^{67}Zn Mössbauer spectroscopy, one should, in principle, consider the SOD shift as arising from different isotopic compositions or even more so from different chemical compositions of source and absorber. Lipkin [57] has derived a general expression for this shift, which the authors of [54] used to estimate a SOD shift for their experiments of $\Delta E_{\mathrm{D}} \approx -0.006$ nm s^{-1}.

The influence of pressure on the ^{67}Ga/^{66}ZnO source was found to be remarkable [54]. Applying a pressure of about 40 kbar, the authors observed (a) a shift of about -0.11 μm s^{-1} for the intense central line of Fig. 7.20, (b) a $(4 \pm 2)\%$ reduction of the splitting between the two outer lines, (c) about 25% broadening of the central line, and (d) a reduction of the ratio of center-line to outer-line intensity from 3.6 ± 0.5 at zero pressure to 2.4 ± 0.3 at high pressure. All these changes were compatible with an 8% reduction of the source quadrupole splitting as a result of compression.

In later experiments, Perlow et al. [58] used sources of ^{67}Ga in ZnO single crystals (natural isotopic abundance, cyclotron bombardment) and a compressed and sintered pellet of ZnO (2.1 g cm^{-2}, enriched to 90% in ^{67}Zn) as absorber. The source crystals were disks of 1 cm in diameter and 0.5 mm thick with the hexagonal symmetry axis (c axis) perpendicular to the faces. The velocity spectrum obtained in [58] at 4.2 K in the absence of a magnetic field is shown in Fig. 7.21. With the symmetry axis of the single crystal oriented along the $E2$ γ-emission, the ratio of line intensities of the full quadrupole split spectrum should be 1:2:1:3:2:0:0 (no transitions for $\Delta m = \pm 2$). The reduced spectrum of Fig. 7.21 ranging from -12 to $+12$ μm s^{-1} shows four lines with relative intensities of ~2:1:3:2 and is thus in very good agreement with prediction. The observed sequence of intensities shows that the quadrupole coupling is positive. Since the quadrupole moment is known to be positive, the EFG is also positive.

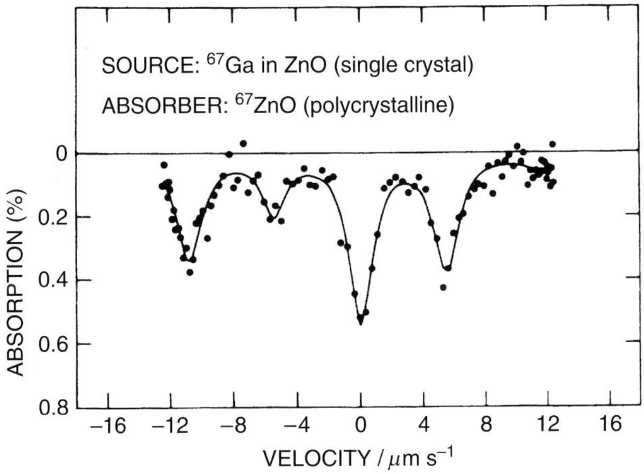

Fig. 7.21 ^{67}Zn Mössbauer spectrum of ^{67}ZnO using a source of ^{67}Ga in ZnO single crystal (from [58])

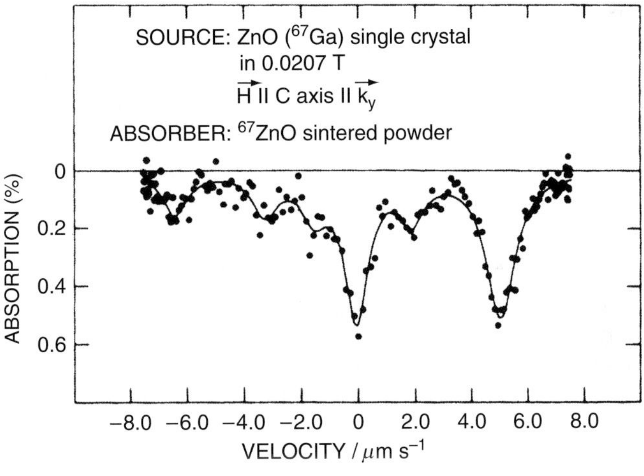

Fig. 7.22 ^{67}Zn Mössbauer spectrum of ^{67}ZnO as absorber with a single crystal ^{67}Ga/ZnO source in applied magnetic field (from [58])

In experiments with magnetic fields between 13 and 55 mT (130–550 G) applied to the single crystal source, the authors of [58] observed magnetic dipole interaction in addition to electric quadrupole splitting in a reduced spectrum (Fig. 7.22) [58]. They determined the magnetic moment of the excited $1/2^-$ state to be $\mu(1/2^-) = +(0.58 \pm 0.03)\mu_N$.

For the purpose of precise determination of the electric quadrupole interaction of ^{67}ZnO, Perlow et al. [59] applied the method of frequency modulation Mössbauer spectroscopy to a ^{67}Ga/ZnO single crystal source versus polycrystalline ^{67}ZnO absorber at 4.2 K. They determined the electric quadrupole coupling constant $e^2qQ = 2.408 \pm 0.006$ MHz, and the asymmetry parameter $\eta = 0.00\,\binom{+\,0.065}{-\,0}$.

Potzel et al. [60] used a ^{67}Ga/ZnO single crystal source in combination with a single crystal absorber of natural ZnO and observed a resonance line width of 0.36 ± 0.04 μm s^{-1} for the 93.3 keV transition in ^{67}Zn (at 4.2 K). This, after correction for finite absorber thickness, equals, within the limit of error, the minimum observable line width as deduced from the lifetime of 13.4 μs for the 93.3 keV state. The spectra observed by these authors are shown in Fig. 7.23.

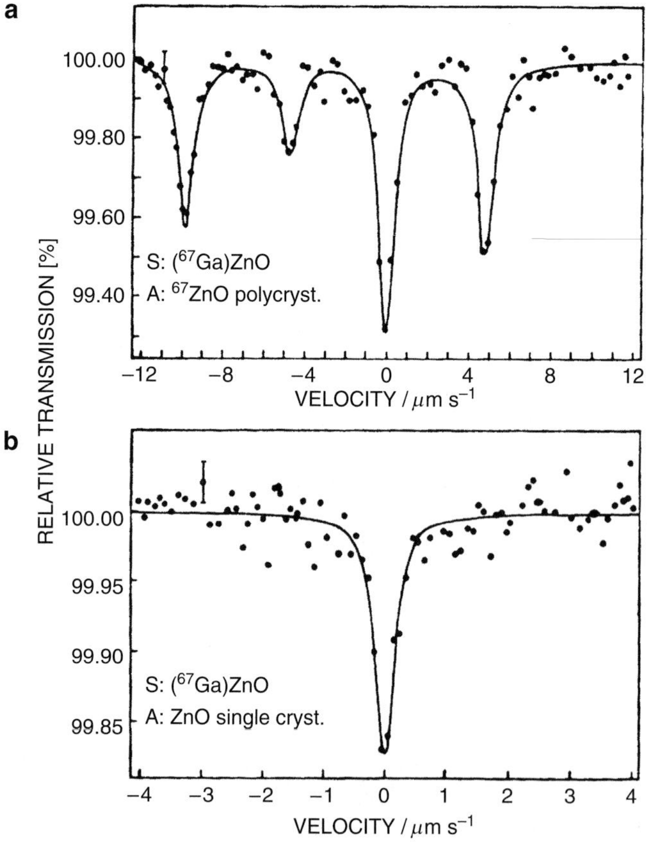

Fig. 7.23 ^{67}Zn Mössbauer spectra obtained with a single crystal source and (**a**) an enriched polycrystalline ZnO absorber (82.5% ^{67}Zn, 963 mg ^{67}Zn cm^{-2}, sintered in oxygen atmosphere at 1,000 K for 24 h), (**b**) a single crystal absorber of natural ZnO. Both source and absorber were at 4.2 K (from [60])

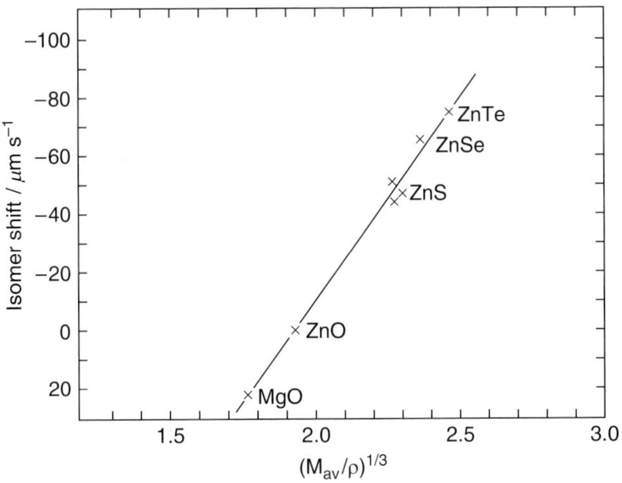

Fig. 7.24 Experimental isomer shifts of ^{67}Ga/XY sources (XY = ZnO, ZnS, ZnSe, ZnTe, MgO) relative to ZnO as absorber at 4.2 K are plotted against the lattice spacing parameter $(M_{av}/\rho)^{1/3}$ (from [55]). The isomer shift for MgO was taken from [61]

Griesinger et al. [56] recorded ^{67}Zn Mössbauer spectra with sources of ^{67}Zn diffused into ZnO, ZnS (both wurtzite and sphalerite), ZnSe, ZnTe, and Cu, and an enriched ^{67}ZnO absorber. The isomer shifts extracted from their spectra cover a velocity range of 112 μm s^{-1} and were found to follow linearly the lattice spacing parameter $(M_{av}/\rho)^{1/3}$, where ρ and M_{av} are the host density and average atomic weight, respectively (Fig. 7.24). The experimental data of Fig. 7.24 have been corrected for zero-point motion using Lipkin's formula [57]. The zero-point vibration calculations yielded much smaller shifts than are observed experimentally, from which the authors concluded that the shifts observed for the zinc chalcogenides were largely chemical in nature. From the observed sign of the variation in the isomer shift with $(M_{av}/\rho)^{1/3}$ and from the assumption that larger lattice spacings go along with less ionic bonds and thus higher 4*s* electron density at the zinc nucleus, the authors tentatively deduced the charge radius of the excited state as being greater than that of the ground state, $\Delta\langle r^2\rangle > 0$. In agreement with this suggestion are results reported by the Munich group, which performed ^{67}Zn Mössbauer experiments with ^{67}Ga/ZnO as source and single crystals of zinc chalcogenides as absorbers [62, 63]. For instance, for ZnS as absorber, an isomer shift of 53.9 ± 0.5 μm s^{-1} with respect to a ^{67}Ga/ZnO source (both source and absorber at 4.2 K) was measured.

The first comprehensive review article on ^{67}Zn Mössbauer spectroscopy, covering extensively the theory of hyperfine interactions, describing the spectrometer and cryogenic systems and reviewing the ^{67}Zn Mössbauer effect studies of the early stage appeared in 1983 [64].

7.2.2 Selected ^{67}Zn Mössbauer Effect Studies

7.2.2.1 Gravitational Red Shift Experiments

Katila et al. have carried out an unusual Mössbauer experiment of fundamental interest [65]. They have made use of the ultrahigh resolution of the 93.3 keV Mössbauer resonance of ^{67}Zn to study the influence of the gravitational field on electromagnetic radiation. A ^{67}Ga:ZnO source (4.2 K) was used at a distance of 1 m from an enriched ZnO absorber (4.2 K). A red shift of the photons by about 5% of the width of the resonance line was observed. The corresponding shift with ^{57}Fe as Mössbauer isotope would be only 0.01%. The result is in accordance with Einstein's equivalence principle. Further gravitational red shift experiments using the 93.3 keV Mössbauer resonance of ^{67}Zn were performed later employing a superconducting quantum interference device-based displacement sensor to detect the tiny Doppler motion of the source [66, 67].

^{67}Zn Mössbauer effect measurements are technically difficult to perform, and so far only few laboratories, mainly those in Munich [68–71] and St. Petersburg (see Sect. 7.2.2.4) have published significant contributions on the characterisation of zinc containing compounds and metallic materials with ^{67}Zn Mössbauer spectroscopy. Selected work is briefly discussed in the following sections.

7.2.2.2 Zinc Metal and Alloys

A theoretical method of determining the dynamic properties of atoms in three-dimensional lattices of arbitrary crystal structure was published by Vetterling et al. and applied to the case of metallic zinc, a highly anisotropic hexagonal crystal [72]. The authors calculated the second-order Doppler shift, the recoil-free fraction of γ-radiation, and the Goldanskii–Karyagin effect. The recoil-free fraction of γ-ray absorption of ^{67}Zn was measured between oriented single crystals of metallic zinc of natural constitution. The anisotropy of the recoil-free fraction was consistent with previous estimates of zero-point amplitudes [68, 73]. Shortly later, detailed measurements of the temperature dependence of hyperfine interactions and of the anisotropy of the recoil-free fraction (Lamb–Mössbauer factor, f) of ^{67}Zn in single crystals of metallic zinc in the temperature range between 4.2 and 47 K were reported by Potzel et al. [68]. The anisotropy of f was very much pronounced, and the f values perpendicular and parallel to the c-axis shown to vary by a factor of 2,100 at the highest temperature under study. This enormous anisotropy, as well as the second-order Doppler shift, could be interpreted on the basis of a Debye distribution characterized by two Debye temperatures Θ (perpendicular) and Θ (parallel). These findings demonstrate that the 93.3 keV resonance of ^{67}Zn is extremely sensitive to lattice dynamical effects. Similar measurements were extended by the Munich group to investigate the changes of s-electron densities at the Zn nucleus in Cu–Zn alloys; they observed the presence of short-range order in α-brass with only four configurations instead of the expected binomial distribution [74]. In Zn metal, no change of center shift was observed when crossing the superconducting phase transition.

The Munich group were the first to perform ^{67}Zn Mössbauer measurements under high pressure up to about 58 kbar [75]. A high pressure, low temperature Mössbauer spectrometer for the high resolution 93.3 keV resonance in ^{67}Zn is described in [76, 77]. The pressure is generated by applying the anvil (B_4C) technique. The piezo Doppler drive was mounted on top of the pressure clamp. The whole system can be cooled to liquid helium temperatures. ^{67}Zn Mössbauer absorption spectra of zinc metal and Cu–Zn alloys (brass) at ambient pressure and at 58 kbar, respectively, are shown in Figs. 7.25 and 7.26. In the case of zinc metal, the *c*/*a* ratio was found to play an important role in describing the change of the Mössbauer parameters with pressure. The reduction of the *c*/*a* ratio under pressure leads to a decrease of the EFG at the ^{67}Zn nucleus which is partially compensated for by an increase caused by a reduction in volume. Most striking is the observed increase of the recoil-free fraction by a factor of 3. The center shift also changes

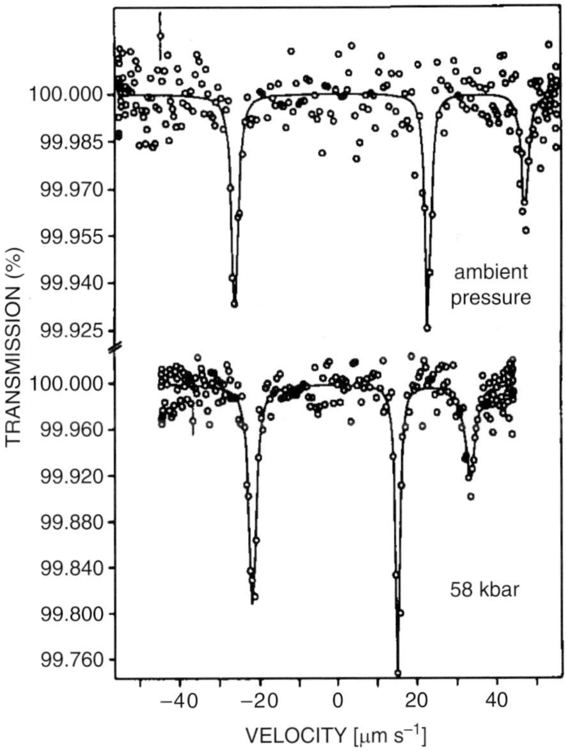

Fig. 7.25 Mössbauer absorption spectra of ^{67}Zn metal obtained at 4.2 K and at two different pressures. The source is $^{67}Ga/Cu$ (from [75])

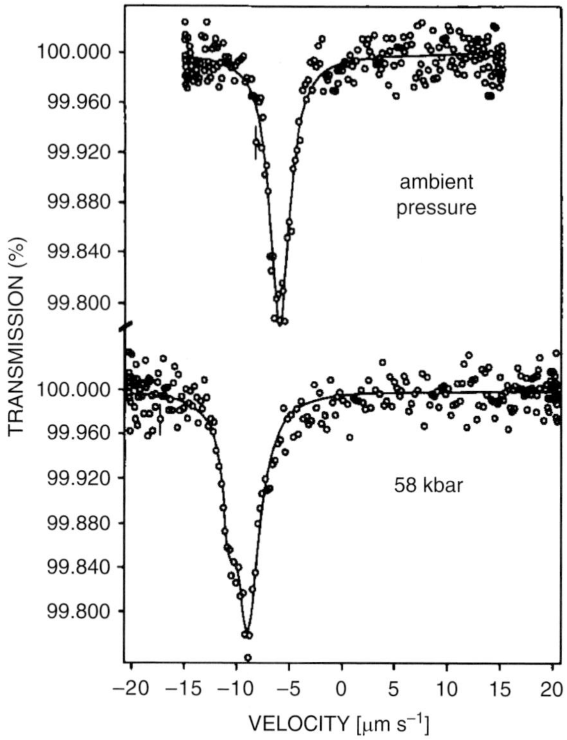

Fig. 7.26 Mössbauer absorption spectra of β-brass (48 at% Zn) recorded at 4.2 K and at two different pressures. The source is ^{67}Ga/Cu (from [75])

drastically under applied pressure; it decreases linearly with reduced volume [78]. In the case of brass, a new phase was observed at 4.2 K and high pressure which was tentatively identified as martensitic α-phase. The Munich group also reported on ^{67}Zn Mössbauer effect measurements on metal zinc under quasihydrostatic pressures up to 160 kbar [79]; they observed an electronic transition at 4.2 K and about 65 kbar where conduction electron states apparently move under the Fermi level and become occupied. The field gradient and the electron density at the zinc nucleus remain thereby unchanged, but the lattice dynamics were drastically affected as evidenced by a sharp drop in the Lamb–Mössbauer factor upon increase of pressure.

7.2.2.3 Inorganic Zinc Compounds

Potzel and Kalvius were the first to investigate zinc compounds with ^{67}Zn Mössbauer spectroscopy. The first measurements were carried out with ZnF_2 powder at 4.2 K [80]. The observed quadrupole splitting could not be explained by a simple lattice sum calculation. More detailed measurements were carried out at

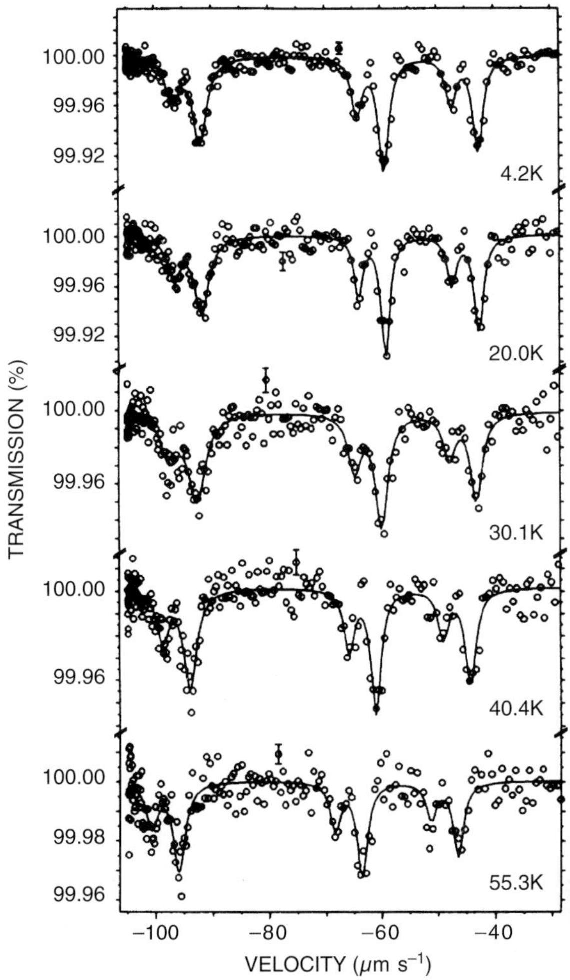

Fig. 7.27 Mössbauer absorption spectra of a polycrystalline $^{67}ZnF_2$ absorber recorded between 4.2 and 55.3 K. The source, ^{67}Ga in a ZnO single crystal, was kept at 4.2 K (from [81])

variable temperatures between 4.2 and 55.3 K [81]; the spectra are shown in Fig. 7.27. The Lamb–Mössbauer factor was found to be strongly temperature dependent; it decreased within this temperature range sharply from 1.23% at 4.2 K to 0.46% at 55.3 K. The quadrupole splitting was found to be temperature independent. The results regarding the isomer shift (relative to ZnO) and the quadrupole splitting were in very good agreement with theoretical cluster calculations. All-electron self-consistent Hartree–Fock cluster calculations have been carried out to derive electron densities at the zinc nucleus in various zinc compounds such as ZnF_2, ZnO, and the chalcogenides ZnS, ZnSe, and ZnTe [82]. The derived density differences show very good linear correlation with the experimental isomer shifts. A value of $13.9 \pm 1.4 \times 10^{-3}$ fm^2 for the change of the mean-square nuclear charge radius between the excited and the ground state of ^{67}Zn was calculated. These calculations point to the importance of the covalency of the Zn–ligand bond for the origin of the isomer shift and fully support the observed linear correlation between decreasing isomer shift values and electronegativity of the ligands. The most important contribution to the electron-density differences at the zinc nucleus was found to come from the 4*s* electrons of Zn with a smaller but significant contribution from the 3*s* electrons.

The semiconductors ZnO and ZnSe were investigated by X-ray and ^{67}Zn Mössbauer spectroscopy under high external pressures [69]. In ZnSe, the recoil free fraction increases from 0.50% at ambient pressure to 1.19% at about 6 GPa (60 kbar), but decreases again upon further increase of pressure to about 8 GPa due to softening of phonon modes which occurs below the crystallographic phase transition near 13.5 GPa. In the high-pressure phase of ZnO (NaCl structure), low-frequency acoustic modes become harder and high-frequency optical modes become softer as compared to ZnO (wurtzite structure). According to the results of Hartree–Fock cluster and full potential relativistic LAPW calculations, covalent contributions to the chemical bond are responsible for the change of *s*-electron density at the Zn nucleus as well as the EFG tensor in ZnO (wurtzite).

Using ^{67}Zn Mössbauer absorption and emission spectroscopy, the Munich group of Potzel and Kalvius et al. have investigated the electronic structure of Zn at the A (tetrahedral) and B (octahedral) sites in the spinels Zn[Al_2O_4], Zn[Fe_2O_4], and Zn [Ga_2O_4] [70]. The isomer shift was found to be more positive at the A sites, whereas the EFG was found to be negative at the B sites. The values of the isomer shift and of the field gradient correlate with the oxygen nearest-neighbor distance to Zn. In the Fe spinel, a transferred magnetic hyperfine field was observed at the zinc site below the antiferromagnetic ordering temperature $T_N = 10$ K. Hartree–Fock cluster calculations for the Al and Fe spinels have been performed, the results of which show that all hyperfine parameters are essentially determined by covalency effects, in accordance with similar work on the Zn chalcogenides.

Nanocrystalline particles (between 4 nm and 25 nm) of ZnO and the zinc ferrite spinel compound $ZnFe_2O_4$ were investigated with ^{67}Zn Mössbauer spectroscopy in conjunction with neutron diffraction and magnetic susceptibility measurements [71]. For ZnO, it was found that the asymmetry parameter changes drastically with nanocrystallinity, which was interpreted in terms of an off-axis displacement of either O or Zn atoms in finite size grains. The observed dependence of the Lamb–Mössbauer factor on particle size and temperature was described by a two-component model in which small crystallites are surrounded by a network of grain boundaries. In the case of Zn ferrite, the magnetic properties and their dependence on nanocrystallinity were of particular interest. A sample with a particle size of 9 nm was studied and compared with differently prepared crystalline samples. It was seen that the cation-site occupation, which changes significantly with nanocrystallinity, has a decisive influence on the magnetic properties.

7.2.2.4 ^{67}Zn Mössbauer Emission Spectroscopy

Nasredinov and Seregin et al. were the first to use the technique of ^{67}Zn Mössbauer emission spectroscopy to investigate the electronic structure of nucleogenic $^{67}Zn^{2+}$ ions generated by nuclear decay of ^{67}Cu in Cu(1) and Cu(2) positions in the lattices of the HT_c superconducting materials $YBa_2Cu_3O_7$ and $YBa_2Cu_3O_6$ [83]. These samples were prepared in the presence of radioactive ^{67}Cu and employed as Mössbauer sources at 4.2 K against a ^{67}ZnS absorber (4.2 K). The emission spectra yielded information about the charge state and the local symmetry of the precursor ^{67}Cu ions in different lattice positions of these compounds. For this purpose, the same research group from St. Petersburg has published a whole series of such ^{67}Zn Mössbauer emission spectroscopy studies on various HT_c superconducting materials. Table 7.2 gives an overview of these reports.

Table 7.2 Overview of ^{67}Zn Mössbauer emission spectroscopy studies

References	System/material under study	Remarks
[83]	$YBa_2Cu_3O_7$, $YBa_2Cu_3O_6$	EFG tensor, atomic charge at ^{67}Zn
[84]	$YBa_2Cu_3O_7$, $YBa_2Cu_3O_6$	EFG tensor, point charge calculation, charge states, hole on O positions
[85]	$La_{2-x}Sr_xCuO_4$	EFG tensor, disagrees with point charge model, charge state, holes from substitution of La^{3+} by Sr^{2+} are localized mainly at the oxygen sites in the Cu–O_2 plane
[86]	$Nd_{1.85}Ce_{0.15}CuO_4$	Substitution of Ce^{4+} for Nd^{3+} ions involves the formation of charge compensating electrons distributed among the copper sites
[87]	$RBa_2Cu_3O_7$ ceramics (R is a rare-earth metal or yttrium)	EFG tensor, comparison with point charge calculation, spatial distribution of electron defects in the lattice
[88, 89]	$RBa_2Cu_3O_7$ (R = rare earth element or Y), $La_{2-x}Sr_xCuO_4$ ($0 < x < 0.3$)	Eu-155(Gd-155) emission Mössbauer spectroscopy, EFG tensor at R sites, in good agreement with point charge model when holes are supposed to be mainly in sublattices of the chain and at oxygen in Cu–O plane
[90]	$Tl_2Ba_2CuO_6$, $Tl_2Ba_2CaCu_2O_8$	EFG tensor, comparison with point charge model and ^{63}Cu-NQR data, holes appearing after part of the thallium atoms lower their valency become localized primarily at the oxygen sites lying in the plane of the copper ions
[91]	About 20 HT_c superconducting compounds and copper oxidic systems	Correlations of the ^{63}Cu NQR/NMR data with the $^{67}Cu(^{67}Zn)$ emission Mössbauer data for HTSC lattices as a tool for the determination of atomic charges
[27]	$Tl_2Ba_2Ca_{n-1}Cu_nO_{2n+4}$ and $Bi_2Sr_2Ca_{n-1}Cu_nO_{2n+4}$ (n = 1,2,3)	Charge states, EFG tensor at the copper sites, ^{61}Cu (^{61}Ni) and $^{67}Cu(^{67}Zn)$ Mössbauer emission spectroscopy, location of electron holes

(*continued*)

Table 7.2 (continued)

References	System/material under study	Remarks
[28]	$Tl_2Ba_2Ca_{n-1}Cu_nO_{2n+4}$	EFG tensors for the Cu and Ba sites by $^{61}Cu(^{61}Ni)$, $^{67}Cu(^{67}Zn)$ and $^{133}Ba(^{133}Cs)$ Mössbauer emission spectroscopy, comparison with point charge approximation and literature data on the ^{63}Cu-NQR. Agreement of the measured and calculated parameters could be reached assuming the holes resulting from lowering the valence of some Tl atoms to be located mainly at oxygen sites of the Cu–O planes
[92, 93]	$RBa_2Cu_4O_8$ (R = Sm, Y, Er)	Nuclear–quadrupole coupling parameters at the rare-earth metal and copper sites from ^{67}Cu (^{67}Zn) and $^{67}Ga(^{67}Zn)$ Mössbauer emission spectroscopy, EFG tensor in comparison with point charge model, shows that holes in lattices are localized primarily at chain-oxygen sites
[94–96]	$HgBa_2Ca_{n-1}Cu_nO_{2n+2}$ ($n = 1, 2, 3$)	EFG tensor at the copper, barium, and mercury sites, by $^{67}Cu(^{67}Zn)$, $^{133}Ba(^{133}Cs)$, and ^{197}Hg (^{197}Au) Mössbauer emission spectroscopy. Comparison with point-charge approximation and ^{63}Cu NMR data showed that the holes originating from defects are localized primarily in the sublattice of the oxygen lying in the copper plane (for $HgBa_2Ca_2Cu_3O_8$, in the plane of the Cu(2) atoms)
[97]	$Nd_{1.85}Ce_{0.15}CuO_4$	Localization–delocalization of copper pairs on ^{67}Zn impurity centers in the copper sublattice of the HT_c superconductor $Nd_{1.85}Ce_{0.15}CuO_4$ was observed by ^{67}Zn Mössbauer emission spectroscopy
[98]	$Y_2Ba_4Cu_7O_{15}$	Nuclear quadrupole interaction at copper sites, EFG tensor at all sites is calculated using the point charge model, conclusion that holes in the $Y_2Ba_4Cu_7O_{15}$ lattice are localized predominantly at positions of chain oxygen
[99]	$^{67}Ga(^{67}Zn)$silicon	The two-electron acceptor impurity of Zn is present in silicon only in the form of neutral (ZnO) or doubly ionized Zn^{2-} centers depending on the Fermi-level position. Broadening of the spectra corresponding to the above centers indicates that the local symmetry of these centers is not cubic
[100–102]	$Nd_{1.85}Ce_{0.15}CuO_4$, $La_{1.85}Sr_{0.15}CuO_4$, $Tl_2Ba_2CaCuO_8$	$^{67}Cu(^{67}Zn)$ and $^{67}Ga(^{67}Zn)$ Mössbauer emission spectroscopy, in the range $T > T_c$ the temperature dependence of the center shift δ of the Mössbauer spectrum is determined by the SOD shift, while in the range $T < T_c$ the Bose condensation of Cooper pairs influences δ
[103, 104]	$Nd_{1.85}Ce_{0.15}CuO_4$, $La_{0.18}Sr_{0.15}CuO_4$,	Possible observation of Bose condensation by Mössbauer emission spectroscopy on ^{67}Cu

(*continued*)

Table 7.2 (continued)

References	System/material under study	Remarks
	$YBa_2Cu_3O_{6.6}$, $YBa_2Cu_3O_{6.9}$, $YBa_2Cu_4O_8$, $Bi_2Sr_2CaCu_2O_8$, $Tl_2Ba_2CaCu_2O_8$, $HgBa_2CuO_4$, and $HgBa_2CaCu_2O_4$	(^{67}Zn) and $^{67}Ga(^{67}Zn)$ isotopes, the transition to the superconducting state leads to a change of the electron density on the metal sites, above the transition the center shift is determined by SOD, below by the influence of Bose condensation, a correlation between the change in electron density and the temperature of the transition to the superconducting state is found
[105]	$^{67}Ga(^{67}Zn)$ and $^{67}Cu(^{67}Zn)$ in GaP, GaAs, and GaSb	$^{67}Ga(^{67}Zn)$ and $^{67}Cu(^{67}Zn)$ Mössbauer emission spectroscopy on bulk GaP, GaAs and GaSb semiconductors point at isolated zinc metal centers at Ga sites. The observed center shift to higher positive velocities at the transition from p- to n-type samples corresponds to the recharging of zinc impurity centers
[106]	$Pb_{1-x}Sn_xTe$	$^{67}Ga(^{67}Zn)$, ^{119}Sn, $^{129}Te(^{129}I)$ Mössbauer spectroscopy, no modifications of the local symmetry of lattice sites, electronic structure of atoms and intensity of electron–phonon interaction are revealed for $Pb_{1-x}Sn_xTe$ solid solutions in the gapless state at 80 and 295 K
[107]	Nb_3Al	Variation of electron density in the superconducting phase transition in the classical superconductor Nb_3Al with critical temperature $T_c = 18.6$ K was studied using ^{73}Ge emission Mössbauer spectroscopy, comparison of the results with data from ^{67}Zn Mössbauer emission spectroscopy on HT_c superconductors revealed a correlation between the electron density variation and the value of T_c
[108]	$La_{2-x}Sr_xCuO_4$, $Nd_{2-x}Ce_xCuO_4$	$^{67}Cu(^{67}Zn)$ and $^{67}Ga(^{67}Zn)$ Mössbauer emission spectroscopy: holes appearing as a result of the Sr^{2+} substitution for La^{3+} in the $La_{2-x}Sr_xCuO_4$ crystal lattice are localized predominantly at oxygen atoms occurring in the same atomic plane as the copper atoms, whereas electrons appearing as a result of the Ce^{4+} substitution for Nd^{3+} in the $Nd_{2-x}Ce_xCuO_4$ crystal lattice are localized in the copper sublattice. The results are consistent with a model assuming that a mechanism responsible for the HT_c superconductivity in $La_{2-x}Sr_xCuO_4$ and $Nd_{2x}Ce_xCuO_4$ crystal lattices is based on the interaction of electrons with two-site two-electron centers possessing negative correlation energies (negative-U centers)

7.3 Ruthenium (^{99}Ru, ^{101}Ru)

Kistner et al. [109] were the first to observe the Mössbauer effect in ^{99}Ru. Kistner also reported the ^{99}Ru Mössbauer spectroscopy study of ruthenium compounds and alloys [110].

Potzel et al. [111] have established recoil-free nuclear resonance in another ruthenium nuclide, ^{101}Ru. This isotope, however, is much less profitable than ^{99}Ru for ruthenium chemistry because of the very small resonance effect as a consequence of the high transition energy (127.2 keV) and the much broader line width (about 30 times broader than the ^{99}Ru line). The relevant nuclear properties of both ruthenium isotopes are listed in Table 7.1 (end of the book). The decay schemes of Fig. 7.28 show the feeding of the Mössbauer nuclear levels of ^{99}Ru (89.36 keV) and ^{101}Ru (127.2 keV), respectively.

7.3.1 *Experimental Aspects*

The source preparation imposes no particular difficulties except the accessibility of cyclotron radiation. The precursor of ^{99}Ru is ^{99}Rh ($t_{1/2}$ = 16 days), which is prepared by bombarding natural ruthenium metal with 10 MeV protons, ^{99}Ru (p, n)^{99}Rh, or 20 MeV deuterons, ^{99}Ru (d, 2n)^{99}Rh. Ru metal generally serves as the host lattice for ruthenium Mössbauer sources because it produces a single transition line close to the natural line width despite its hexagonal crystal structure. Spectra may be taken at liquid nitrogen temperature [109]; significantly larger effects, however, are obtained at lower temperatures using helium cryostats. All results reported so far were obtained using transmission geometry and counting the 89.36 keV γ-rays with a NaI(Tl) scintillation detector (preferably 3 mm NaI(Tl) crystal). The absorber thickness should vary between 50 and 100 mg of natural ruthenium per cm^2 to yield sizeable absorption effects. Isomer shifts are generally given with respect to metallic ruthenium.

The precursor of ^{101}Ru is ^{101}Rh ($t_{1/2}$ = 3 years). It is prepared by irradiating natural ruthenium metal with 20 MeV deuterons, ^{100}Ru (d, n)^{101}Rh. The target is then allowed to decay for several months to diminish the accompanying ^{99}Rh activity. In a report on ^{101}Ru Mössbauer spectroscopy [111], the authors reported on spectra of Ru metal, RuO_2, and $[Ru(NH_3)_4(HSO_3)_2]$ at liquid helium temperature in standard transmission geometry using a Ge(Li) diode to detect the 127 keV γ-rays. The absorber samples contained ~1 g of ruthenium per cm^2.

7.3.2 *Chemical Information from ^{99}Ru Mössbauer Parameters*

Kistner's early measurements of the ^{99}Ru Mössbauer effect in Ru metal, RuO_2, Ru $(C_5H_5)_2$, and $Ru_{0.023}Fe_{0.977}$ [110] demonstrated that

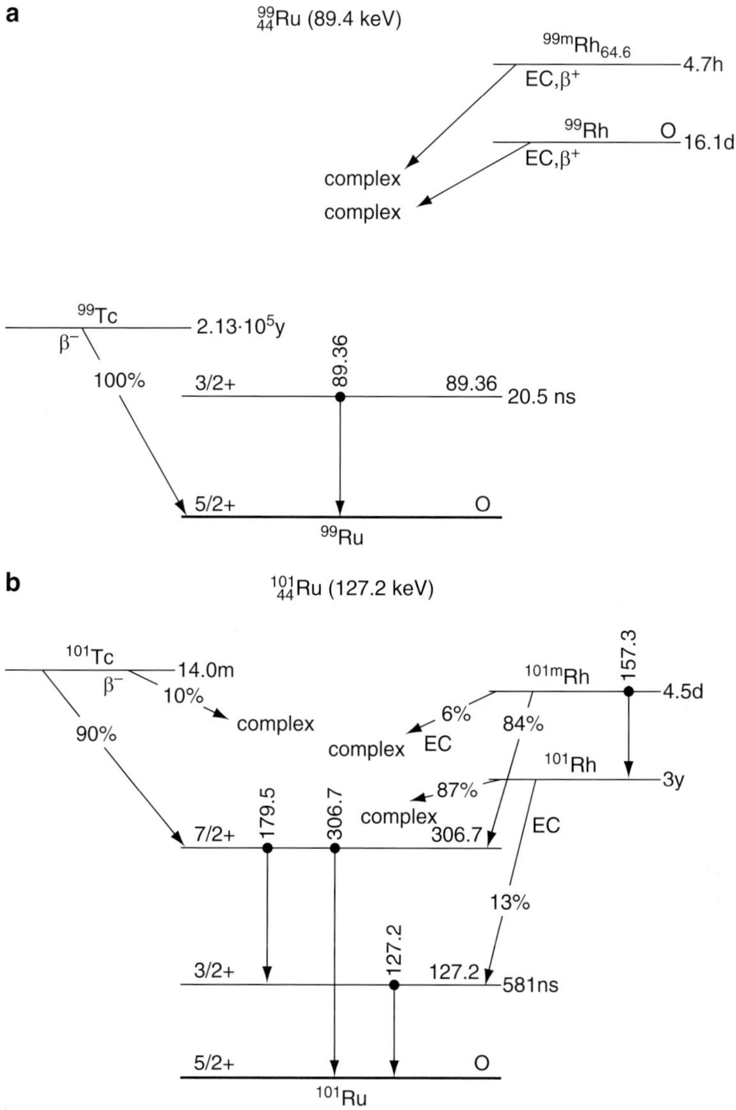

Fig. 7.28 Simplified decay schemes of (**a**) ^{99}Ru and (**b**) ^{101}Ru (reproduced from [112])

1. Isomer shift changes are sufficiently large for Ru atoms in different systems to distinguish between different oxidation states and different bond properties;
2. Quadrupolar perturbation reveals nonspherical charge distributions about Ru atoms;
3. Magnetic hyperfine splitting can be used for investigating local magnetic fields (sign and magnitude of $\vec{H}_{\text{int}}$) in systems exhibiting cooperative magnetism.

7.3.2.1 Isomer Shift

The relative change of the mean-square nuclear radius in going from the excited to the ground state, $\Delta\langle r^2\rangle/\langle r^2\rangle$, is positive for ^{99}Ru. An increase in observed isomer shifts δ therefore reflects an increase of the *s*-electron density at the Ru nucleus caused by either an increase in the number of *s*-valence electrons or a decrease in the number of shielding electrons, preferentially of *d*-character.

One of the interesting features in ruthenium chemistry is the large variety of oxidation states (0 to +8) of ruthenium, which may well be distinguished by isomer shift measurements. As an example, Fig. 7.29 shows a few representative single-line spectra [113] reflecting the significant isomer shift changes for different

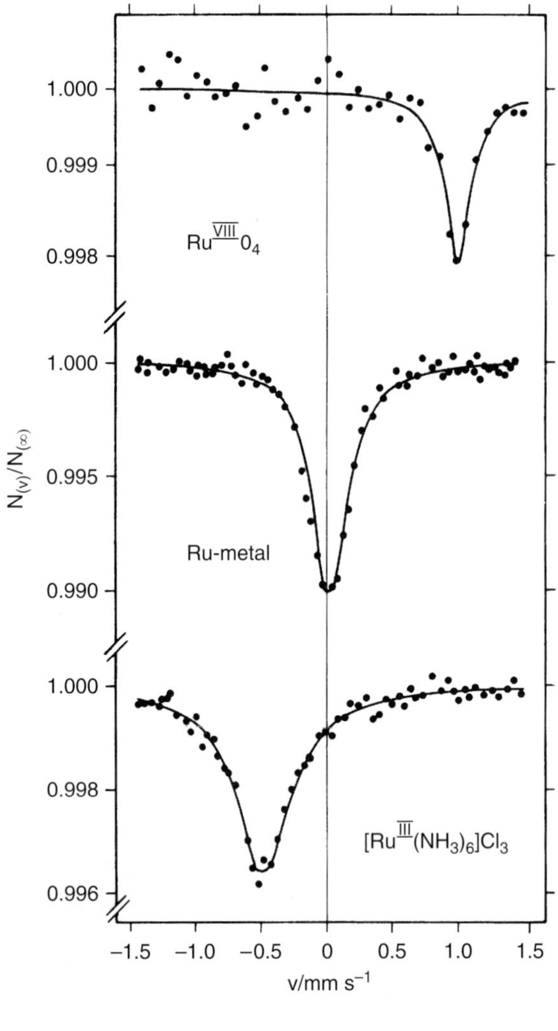

Fig. 7.29 ^{99}Ru Mössbauer spectra of RuO_4, Ru metal, and $[Ru^{III}(NH_3)_6]Cl_3$. The source (^{99}Rh in Ru metal) and the absorbers were kept at 4.2 K (from [113])

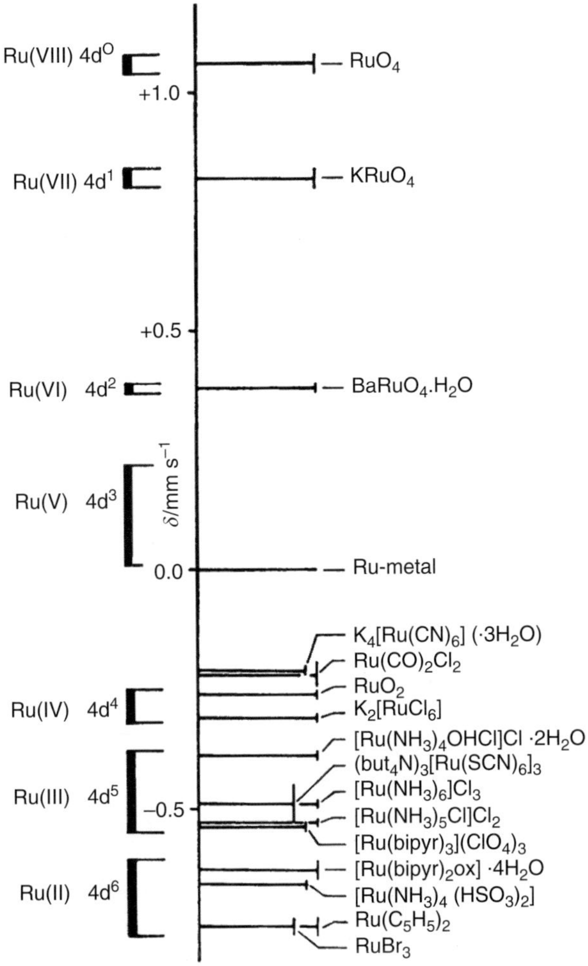

Fig. 7.30 Measured ^{99}Ru isomer shifts in ruthenium compounds with different oxidation states of Ru (from [113] and complemented by values obtained at 4.2 K for Ru(V) oxides of the form $Ba_3Ru_2MO_9$ (M = Mg; Ca, Sr; Co, Ni, Zn; and Cd) from [114])

oxidation states of ruthenium. For a variety of ruthenium compounds, the observed isomer shifts have been correlated with the oxidation state in a graphical representation as shown in Fig. 7.30 [113, 114]. From the diagram, it is clear that the isomer shift values increase monotonically with the oxidation state of ruthenium, and that each formal oxidation state embraces a certain range of isomer shifts, indicating different bond properties in terms of σ-donation ($\sigma_{L\to M}$; L = ligand, M = metal) and $d_\pi - p_\pi$ back donation ($\pi_{M\to L}$).

A similar situation with changing oxidation state of Ru within a series of compounds was observed for the ordered perovskites BaLaMRuO$_6$ (M = Mg, Fe, Co, Ni, Zn) [115]. The measured isomer shifts in the range +0.06 to +0.13 mm s^{-1}

are characteristic of Ru(V) (Fig. 7.30) with the exception of $BaLaFeRuO_6$ (−0.34 mm s^{-1}). This latter isomer shift is indicative of Ru(IV). The presence of Ru(IV) in this compound implies the coexistence of Fe(III) which was confirmed by ^{57}Fe Mössbauer spectroscopy [115].

The diagram shown in Fig. 7.30 was (and still is) of considerable use in characterizing the oxidation (valence) state of Ru in other Ru oxides: The isomer shifts (measured at 4.2 K) for $Na_4Ru_2O_4$ (0.0 mm s^{-1}) [117], $RuSr_2CdCu_2O_8$ (+0.03 mm s^{-1}) [114], Na_3RuO_4 (+0.11 mm s^{-1}) [117], and $MLnRuO_6$ (M = Ca, Ln = Y, La, Eu; M = Sr, Ln = Y; M = Ba, Ln = La, Eu) (in the range +0.13 to +0.18 mm s^{-1}) [118] confirm the Ru(V) oxidation state. Correspondingly, the values (measured at 4.2 K) for $SrRuO_3$ (−0.33 mm s^{-1}) [119], $Y_2Ru_2O_7$ (−0.25 mm s^{-1}) [120], and $Gd_2Ru_2O_7$ (−0.35 mm s^{-1}) [121] confirm the Ru(IV) oxidation state. Intermediate Ru oxidation states were observed in the solid solutions $La_xCa_{1-x}RuO_3$ ($0 \leq x \leq 1$), where the isomer shifts (4.2 K) move progressively from the value characteristic of Ru(IV) (−0.300 mm s^{-1} for $CaRuO_3$) to Ru (III) (−0.557 mm s^{-1} for $LaRuO_3$) [122].

It is further noteworthy that $K_4[Ru(CN)_6] \cdot 3H_2O$ shows an isomer shift close to the shifts of Ru(IV) compounds, from which it is inferred that practically two t_{2g} electrons have been delocalized from the Ru(II) cluster to the strong π-bonding CN^- ligands [123].

In another investigation of bond properties in ruthenium compounds, it was found [123] that the isomer shift increases in the series

$$\left[Ru^{II}(CN)_5NO_2\right]^{4-} < \left[Ru^{II}(CN)_6\right]^{4-} < \left[Ru^{II}(CN)_5NO\right]^{2-} \quad \text{and}$$

$$\left[Ru^{II}(NH_3)_6\right]^{2+} < \left[Ru^{II}(NH_3)_5N_2\right]^{2-} < \left[Ru^{II}(NH_3)_5NO\right]^{3+}$$

from which the authors derived the following order for the backdonation power of the ligands:

$$NO_2^- < CN^- < NO^+ \quad \text{and}$$

$$NH_3 < N_2 \ll NO^+,$$

i.e., the nitrosyl group exhibits the strongest backbonding capability, which parallels earlier observations made in analogous iron complexes. This has also been supported by the observed isomer shift tendency in the series [123]

$$\left[Ru^{II}(NH_3)_6\right]^{2+} < \left[Ru^{II}(NH_3)_4pyr_2\right]^{2+} < \left[Ru^{II}(NH_3)_4(SO_3H)_2\right]$$
$$< \left[Ru^{II}(NH_3)_4Cl(SO_2)\right]^{+} \ll \left[Ru(NH_3)_4(OH)(NO)\right]^{2+}.$$

The isomer shift changes in the series [123]

$$\left[\mathrm{Ru^{II}Cl_5NO}\right]^{2-} < \left[\mathrm{Ru^{II}(NH_3)_5NO}\right]^{3+} < \left[\mathrm{Ru^{II}(CN)_5NO}\right]^{2-} \quad \text{and}$$

$$\left[\mathrm{Ru^{II}(NH_3)_6}\right]^{2+} \ll \left[\mathrm{Ru^{II}(CN)_6}\right]^{4-}$$

reflect strong backdonation properties of the CN^- ligand as compared to NH_3. These δ-changes also reflect that a NO^+ ligand provokes less change in electron density in going from the pentaammine to the pentacyano complex as compared to the transition from $[Ru^{II}(NH_3)_6]^{2+}$ to $[Ru^{II}(CN)_6]^{4-}$. This suggests that some kind of a saturation value is reached when an isomer shift change formally corresponding to about 2.5 oxidation states has been induced by π-backbonding effects. In analogy to $[Fe^{II}(CN)_5NO]^{2-}$, the $[Ru(CN)_5NO]^{2-}$ complex shows the most positive isomer shift among divalent ruthenium compounds. From the isomer shift tendencies in the series

$$\left[\mathrm{Ru^{II}Cl_5NO}\right]^{2-} < \left[\mathrm{Ru^{II}(NH_3)_5NO}\right]^{3+} < \left[\mathrm{Ru^{II}(CN)_5NO}\right]^{2-},$$

$$\left[\mathrm{Ru^{II}(NH_3)_6}\right]^{2+} \ll \left[\mathrm{Ru^{II}(CN)_6}\right]^{4-} \quad \text{and}$$

$$\left[\mathrm{Ru^{III}F_6}\right]^{3-} < \left[\mathrm{Ru^{III}(NCS)_6}\right]^{3-} \approx \left[\mathrm{Ru^{III}(NH_3)_6}\right]^{3+},$$

it is clear that the ordering of the ligands parallels the spectrochemical series with the NCS^- ligand being N bonded to ruthenium.

Greatrex et al. [124] also studied bond properties in nitrosyl ruthenium(II) compounds, $[RuL_5(NO)]^{n+}$, by ^{99}Ru spectroscopy and found that the isomer shift δ decreases steadily as the ligand-field strength of L decreases in the order $L = CN^- > NH_3 > NCS^- > Cl^- > Br^-$, as represented graphically in Fig. 7.31. From these findings the authors derived a positive sign for $\Delta\langle r^2\rangle/\langle r^2\rangle$ in ^{99}Ru. A further example for the strong ligand-field dependence of the Ru isomer shift is the series of Ru nitrosyl complexes $[Ru(NO_2)_4(OH)NO]^{2-}$ (**1**), $[RuCl_5NO]^{2-}$ (**2**), [Ru $(NH_3)_5NO]^{3+}$ (**3**), and $[Ru(py)_5XNO]^{2+}$ with X = Cl, Br (**4**,**5**) [125]. Their measured isomer shifts (4.2 K) are -0.22 mm s^{-1} for **1**, -0.43 mm s^{-1} for **2**, and -0.20 mm s^{-1} for **3**–**5**. The important consideration for the deviation of these Ru (II) isomer shifts from the pattern shown in Fig. 7.30 are σ-donor and π-acceptor characteristics of the ligands. Both of these bonding interactions increase the *s*-electron density at the Ru nucleus. Backdonation to the π^* orbitals of the NO^+ ligand is very strong, dominates the ligand field and effectively shifts the isomer shift to more positive values than those indicated in Fig. 7.30 for Ru(II) complexes. Ru(II) in **2** experiences a much weaker ligand-field strength than in **1**, **3**–**5**.

A large number of ruthenium complexes have been investigated by ^{99}Ru Mössbauer spectroscopy, mainly by the Munich group and the groups of M.L. Good (New Orleans, USA) and N.N. Greenwood (England). It is now possible to set up much more complete isomer shift diagrams of the type shown in Fig. 7.30, as has

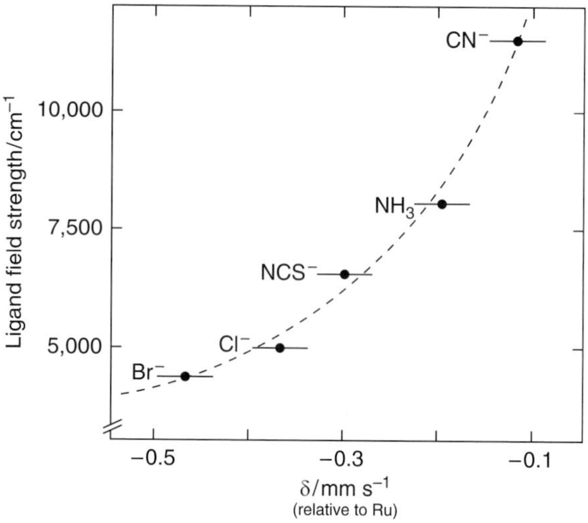

Fig. 7.31 Partial ligand-field strength for the ligand L in correlation with the isomer shift δ (relative to Ru metal at 4.2 K) of the nitrosylruthenium (II) compounds $[RuL_5(NO)]^{n\pm}$ (L = Br^-, Cl^-, NCS^-, NH_3, CN^-) (from [124])

been done, e.g., for ammine and halide complexes of Ru(II) and Ru(III) in an instructive review article by Good [126] and for Ru(V) by Fernandez et al. [114].

A correlation of isomer shift, electronic configuration, and calculated *s*-electron densities for a number of ruthenium complexes in analogy to the Walker–Wertheim–Jaccarino diagram for iron compounds has been reported by Clausen et al. [127]. Also useful is the correlation between isomer shift and electronegativity as communicated by Clausen et al. [128] for ruthenium trihalides where the isomer shift appears to increase with increasing Mulliken electronegativity.

^{99}Ru isomer shift studies have been demonstrated to be particularly powerful in characterizing the oxidation state of ruthenium in mixed-valence polynuclear ruthenium complexes. Clausen et al. [129] and later Wagner and Wordel [116] were able to distinguish clearly between different oxidation states of the three Ru sites in "ruthenium red," $[Ru_3O_2(NH_3)_{14}]^{6+}$, and in "ruthenium brown," $[Ru_3O_2(NH_3)_{14}]^{7+}$. The systematics of isomer shifts (Fig. 7.30) and quadrupole splittings [130] in Ru compounds allow to define noninteger oxidation states (delocalized mixed valencies). The Mössbauer results of "ruthenium red" (Fig. 7.32a and b top spectra) and its derivative are in general agreement with the notation that this compound is based on a Ru(III)–O–Ru(IV)–O–Ru(III) backbone. The formulation of "ruthenium brown" (Fig. 7.32 bottom) and its derivative as Ru(IV)–O–Ru(III)–O–Ru(IV) is, however, incompatible with the Mössbauer data. For such a complex, one would expect the Ru(III) and Ru(IV) sites to exhibit practically the same isomer shifts as in "ruthenium red," but the quadrupole pattern of Ru(IV) to be about twice as intense as that of Ru(III). This was not observed. Instead, in "ruthenium brown" (as well as its derivative) both quadrupole patterns had somewhat larger isomer shifts than the

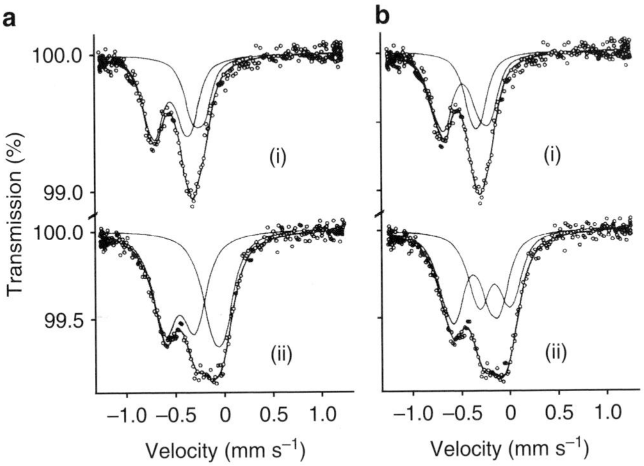

Fig. 7.32 Mössbauer spectra of (i) ruthenium red, $[Ru_3O_2(NH_3)_{14}]Cl_6$, and (ii) ruthenium brown, $[Ru_3O_2(NH_3)_{14}]Cl_7$, fitted with a side-by-side (**a**) and an overlapping (**b**) arrangement of two quadrupole patterns (from [116])

corresponding patterns in "ruthenium red" (and its derivative), while the pattern with the smaller shift always remains more intense. From this, the authors [116] concluded that oxidation of the hexavalent to the heptavalent cations increases the effective valencies of all three Ru atoms, i.e., that the heptavalent cation $[Ru_3O_2(NH_3)_{14}]^{7+}$ should be described by delocalized valencies, Ru(III + x)–O–Ru(IV + y)–O–Ru(III + x), where $2x + y = 1$. This conclusion does not give preference to either the overlapping or the side-by-side arrangement of the two quadrupole patterns in the individual complexes (Fig. 7.32a, b). The valencies in "ruthenium brown" are reasonably well approximated by $x = 0.25$ and $y = 0.5$.

A similar delocalization has been proposed for the mixed-valence complex

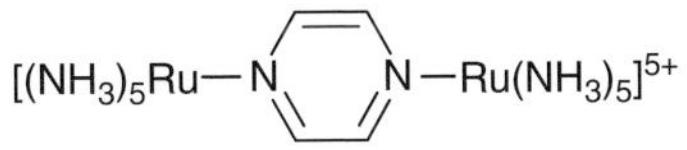

by various physical techniques including Mössbauer spectroscopy [131, 132].

7.3.2.2 Quadrupole Splitting

^{99}Ru has a nuclear ground state with spin $I_g = 5/2^+$ and a first excited state with $I_e = 3/2^+$. Electric quadrupole perturbation of ^{99}Ru was first reported by Kistner [110] for RuO_2 and ruthenocene; this author has also evaluated the ratio of the nuclear quadrupole moments to be $|Q_e/Q_g| \geq 3$. The sign and magnitude of the individual quadrupole moments are given in Table 7.1 (end of the book).

The effect of a positive quadrupole interaction on the ground and excited nuclear states of ^{99}Ru is shown schematically in Fig. 7.33 as adapted from a publication by

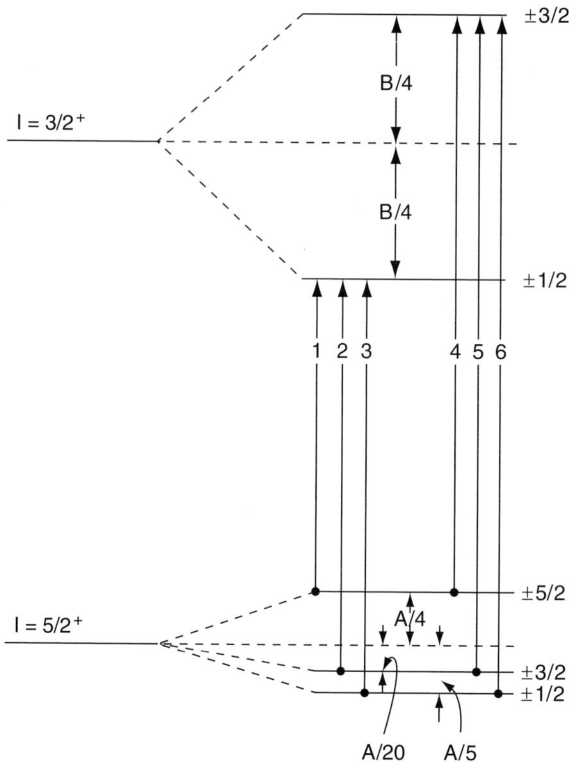

Fig. 7.33 Effect of a positive quadrupole interaction on the ground and first excited states of ^{99}Ru. The asymmetry parameter η is assumed to be zero. The ratio of the quadrupole moments is taken to be $Q_{3/2}/Q_{5/2} = 3$. $A = e^2qQ_{5/2}$ and $B = e^2qQ_{3/2}$ (from [124])

Greatrex et al. [124]. As the transition between the $I_g = 5/2^+$ ground state and the $I_e = 3/2^+$ excited state of ^{99}Ru occurs via combined $M1$ and $E2$ radiation with a mixing ratio of $\delta^2 = 2.7 \pm 0.6$ [110], the allowed transitions are those with $\Delta m_I = 0, \pm1, \pm2$. Following these selection rules, six transition lines should appear in a quadrupole-split ^{99}Ru spectrum using a single-line source; their relative intensities may be determined from the squares of the Clebsch–Gordan coefficients as given in [124]. An actual quadrupole-split ^{99}Ru spectrum, however, generally shows only two resonance lines, each of which consists of three unresolved lines due to the relatively small quadrupole moment of the ground state. Thus, the measurable quadrupole splitting reflects in most cases that of the $I_e = 3/2^+$ excited state of ^{99}Ru. As an example, the ^{99}Ru spectrum of Co_2RuO_4, measured at 4.2 K by Gibb et al. [133] is reproduced in Fig. 7.34.

As in the case of iron chemistry, most valuable information concerning bond properties (anisotropic electron population of molecular orbitals) and local structure may be extracted from quadrupole-split ^{99}Ru spectra. This has been

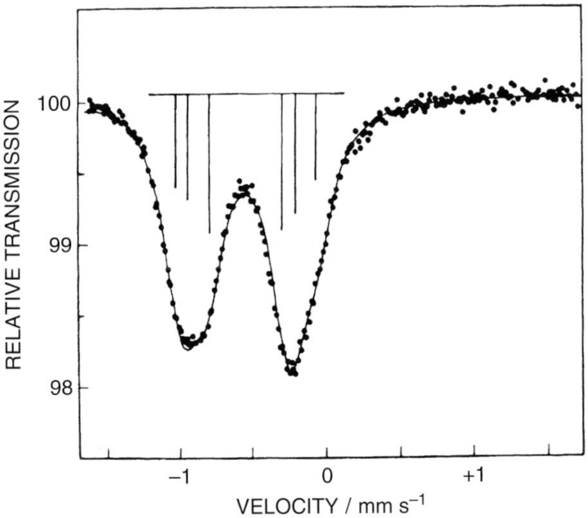

Fig. 7.34 Mössbauer spectrum of Co_2RuO_4 at 4.2 K with respect to the Ru metal source (from [133])

demonstrated in a number of communications from various groups. Clausen et al. [134], for example, have studied the cyano complexes of ruthenium(II),

1 $K_2[Ru(CN)_5NO]\cdot 2H_2O$
2 $K_4[Ru(CN)_6]$
3 $K_4[Ru(CN)_5NO_2]\cdot 2H_2O$

which are characterized by $4d^6$, $S = 0$ electronic configuration. The observed quadrupole splitting in **1** and **3** was interpreted as arising predominantly from an asymmetric expansion of the ruthenium t_{2g}^6 electrons towards the ligands. The sign of the EFG is expected to be positive for complex **1** because NO^+ is known [135] to form very strong π-bonds (stronger than π-bonds of CN^-), and negative for complex **3** because $NO_2{}^-$ is incapable of forming π-bonds. No quadrupole splitting has been observed in **2** because of O_h point symmetry of Ru which yields zero EFG.

Greatrex et al. [124] have observed quadrupole splitting in the octahedral nitrosylruthenium(II) complexes $[RuL_5(NO)]^{n\pm}$ with decreasing magnitudes in the order of the ligands L = $CN^- > NO^+ > NCS^- \approx Cl^- > Br^-$ and discussed these qualitatively in terms of the relative σ-donor and π-acceptor abilities of the ligands. They have found that, although σ-donation and π-backdonation reinforce one another in their effect on the isomer shift, they oppose one another in their effect on the quadrupole splitting.

Bancroft et al. [136] have used partial quadrupole-splitting values from low-spin iron(II) compounds to obtain the sign of the quadrupole coupling constant e^2qQ_e for a number of ruthenium(II) compounds. Their correlation between $1/2e^2qQ$ of iron (II) compounds and $|1/2e^2qQ_e|$ of analogous ruthenium(II) compounds strongly suggests that the bonding in corresponding Fe and Ru complexes is reasonably similar. In the $M(NH_3)_5L$ series (L = N_2, MeCN, CO) they found that $(\sigma\text{–}\pi)$ (σ-donor ability, π-acceptor ability) increases in the order N_2 < MeCN < CO. CO is most likely a better σ-donor and a better π-acceptor than NH_3; N_2 is probably a better π-acceptor, but poorer σ-donor than NH_3. They also established that e^2qQ_e becomes more negative as $(\sigma - \pi)$ of L increases, and that the isomer shift becomes more positive as $(\sigma + \pi)$ of L increases.

Foyt et al. [137] interpreted the quadrupole-splitting parameters of low-spin ruthenium(II) complexes in terms of a crystal field model in the strong-field approximation with the t_{2g}^5 configuration treated as an equivalent one-electron problem. They have shown that, starting from pure octahedral symmetry with zero quadrupole splitting, ΔE_Q increases as the ratio of the axial distortion to the spin–orbit coupling increases.

The lack of substantial quadrupole splitting in a number of Ru oxides implies that the RuO_6 octahedra are almost regular, irrespective of the oxidation state of ruthenium:

Compound	δ (4.2 K), mm s^{-1}	e^2qQ_e, mm s^{-1}	Ref.
$LaRu(III)O_3$	−0.56	0	[122]
$SrRu(IV)O_3$	−0.37	0	[138]
$La_4Ru(IV)_2O_{10}$	−0.30	~0.3	[139]
$CaRu(IV)O_3$	−0.30	0.12	[122]
$BaRu(IV)O_3$	−0.23	0	[122]
$M_2LnRu(V)O_6$ (M = Ca, Ln = Y, La; M = Sr, Ln = Y; M = Ba, Ln = La, Eu)	+0.11 to +0.18	0	[118]
$Sr_2Fe(III)Ru(V)O_6$	+0.11	0	[122]
$Na_3Ru(V)O_4$	+0.11	0	[140]
$Ru(V)Sr_2GdCu_2O_8$	+0.03	0.37	[114]

The largest quadrupole splittings ever found in ^{99}Ru Mössbauer spectra have been reported by Gibb et al. [141] for some nitrido complexes of ruthenium, $(Bu_4^nN)[RuNCl_4]$ and $(Ph_4As)[RuNBr_4]$. From the almost resolved six-component pattern, they derived new estimates for the ratio of the quadrupole moments of $Q_e/Q_g = +2.82 \pm 0.09$, with both Q_e and Q_g being positive, and for the $E2/M1$ mixing ratio of $\delta^2 = 2.64 \pm 0.17$. A simple interpretation of the extremely large excited-state quadrupole splitting in $[RuNX_4]^-$ (X = Cl, Br), $e^2qQ = 3.19$ mm s^{-1} is to attribute this value to an electron pair in the $4d_{xy}$ nonbonding orbital, implying a highly asymmetric electronic ground-state configuration $t_{2g}{}^2$ for the formally hexavalent ruthenium site in $[Ru^{6+}N^{3-}X_4{}^-]$. However, the isomer shifts of $[RuNCl_4]^-$ and $[RuNBr_4]^-$ of +0.083 and +0.038 mm s^{-1} [141] are significantly less than expected for Ru(VI) (see Fig. 7.30), but they are much closer to the values characteristic for Ru(V) complexes. The small difference in isomer shifts between the chloride and bromide complexes are consistent with the dominant effect of the nitrido ligand on the bonding, and the shift of about −0.30 mm s^{-1} away from a more ionic Ru^{6+} configuration is good evidence for a substantial augmentation of the effective $4d$ occupancy compared to $4d^2$. In this context, it is interesting to consider the Mössbauer parameters of $[RuNCl_5]^{2-}$ ($\delta = -0.36$ mm s^{-1} and $e^2qQ_e < 0.40$ mm s^{-1}) [141]. If, as seems likely, the addition of an axial chlorine in $[RuNCl_5]^{2-}$ does not weaken the N–Ru π-bond substantially, the isomer shift of -0.36 mm s^{-1} is indicative of a further increase in $4d$-occupation. The virtual elimination of the quadrupole splitting implies an increased occupation of the $4d_{z^2}$ orbital so that the increased electron density along the Cl–Ru–N axis compensates for the EFG produced by the "nonbonding" d_{xy} electron pair. Thus, attaching Cl^- to the vacant axial site in $[RuNCl_4]^-$ leads to a reduction of both isomer shift and quadrupole splitting.

In an account on the reduction of Mössbauer data, Foyt et al. [142] have described a general analysis of quadrupole split ^{99}Ru spectra.

7.3.2.3 Magnetic Splitting

Magnetic hyperfine splitting in ^{99}Ru Mössbauer spectra was observed in the 1960s by Kistner [110, 143] for an absorber of 2.3 at% ruthenium dissolved in metallic iron. The spectra obtained with an unpolarized absorber (a) and with polarized absorbers, i.e., magnetization parallel to incident γ-rays (b), and magnetization perpendicular to incident γ-rays (c) are shown in Fig. 7.35. The stick spectra on top

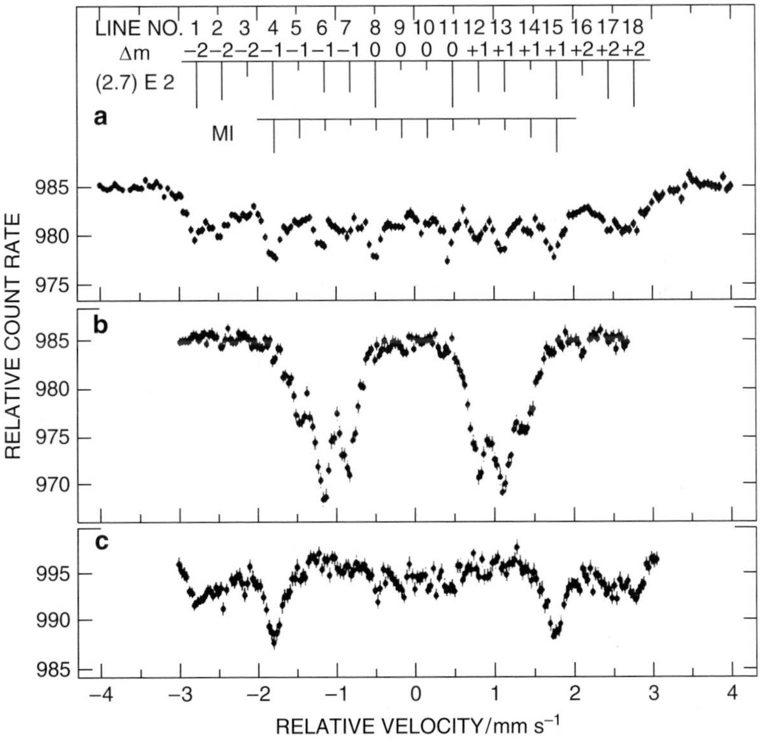

Fig. 7.35 ^{99}Ru Mössbauer spectra of $Ru_{0.023}Fe_{0.977}$ at 4.2 K (source ^{99}Rh in Ru metal), (**a**) unpolarized absorber; (**b**) absorber magnetized parallel and (**c**) perpendicular to incident γ-rays (from [110])

of Fig. 7.35a indicate the calculated [110] line positions and the relative intensities for the 18 allowed electric quadrupole ($E2$) transitions with the change in magnetic spin quantum number of $\Delta m = 0, \pm 1, \pm 2$ as well as for the 12 allowed magnetic dipole ($M1$) transitions with $\Delta m = 0, \pm 1$. The best-fit value for the mixing ratio $E2/M1$ in this case was found, by Kistner, to be $\delta^2 = 2.7$.

Extensive use of the magnetic hyperfine splitting of the ^{99}Ru Mössbauer resonance in ruthenium chemistry has been initiated by Greenwood's group with their study on $SrRuO_3$ [144], which orders ferromagnetically below 160 K and shows metallic properties. The magnetic field strength was found to be 35.2 ± 1.5 T at 4.2 K and 31.5 ± 1.5 T at 77 K. The sign of the field was not determined, but was assumed to be negative.

In a subsequent publication, the same research group reported on ^{99}Ru Mössbauer studies of the magnetic properties of the ternary oxides $CaRuO_3$, $SrRuO_3$, $BaRuO_3$, and $Y_2Ru_2O_7$, and the quaternary oxides $Sr(Ru_{1-x}Ir_x)O_3$ ($x = 0.1$ and 0.2) and $Sr(Ru_{0.7}Mn_{0.3})O_3$ [145]. The internal field of 35.2 T (4.2 K) in $SrRuO_3$ was found to be compatible with the ferromagnetic moment derived from neutron diffraction and susceptibility data and confirmed the collective electron magnetism model for this perovskite. The $E2/M1$ mixing ratio for the ^{99}Ru Mössbauer γ-transition was reevaluated to be $\delta^2 = 2.72 \pm 0.17$. The antiferromagnetic ordering in $CaRuO_3$ reported earlier by other authors could not be confirmed. Both $CaRuO_3$ and $BaRuO_3$ gave single-line spectra even at 4.2 K. Substitution of ruthenium in $SrRuO_3$ by iridium was found to reduce the magnetic field at Ru atoms by $\sim$6 T for each Ir nearest neighbor, and in the case of manganese substitution by only 2.2 T for each Mn neighbor.Y_2RuO_7 has revealed magnetic ordering at 4.2 K with a moderate hyperfine field of 12.6 T and zero quadrupole splitting [145] or 8 T and nonzero quadrupole splitting [120], depending on the fit procedure of the not very well resolved magnetic Mössbauer pattern. Similar fit values for the hyperfine field and the quadrupole splitting at the ruthenium site were observed for $Gd_2Ru_2O_7$; separate ^{99}Ru and ^{155}Gd Mössbauer measurements at 4.2 K confirmed that both ruthenium and gadolinium sublattices are magnetically ordered, with much weaker exchange interaction between Ru and Gd ions in comparison with the exchange interaction within the Ru sublattice [121].

In further studies along this line, Greenwood et al. have examined the magnetic superexchange interactions in the solid solution series $La_xCa_{1-x}RuO_3$ ($0 < x \leq 0.6$) and $La_xSr_{1-x}RuO_3$ ($0.4 \leq x \leq 0.75$) [122] and $Ca_xSr_{1-x}RuO_3$ ($0.1 \leq x \leq 0.5$) [146], which exhibit a distorted perovskite structure, and found that the proportion of ruthenium atoms experiencing magnetic fields close to 35 T decreases as x increases. In $Ca_xSr_{1-x}RuO_3$ samples with $x \geq 0.3$, the central "paramagnetic" component, imposed on the magnetic hyperfine split spectra, increases rapidly with increasing x, which was interpreted as being indicative of a rapid relaxation of the total electron spin and hence the magnetic field due to weakening of the coupling between Ru atoms by Ca substitution. This result has led to the conclusion that the greater electron-pair acceptor strength (Lewis acidity) of Ca(II) compared to Sr(II) causes a more effective competition with Ru atoms for the oxygen anion orbitals involved in the superexchange mechanism.

The magnetic properties of the new solid solution series $SrFe_xRu_{1-x}O_{3-y}$ ($0 \leq x \leq 0.5$) with distorted perovskite structure, where iron substitutes exclusively as Fe(III) thereby causing oxygen deficiency, has also been studied by Greenwood's group [147] using both ^{99}Ru and ^{57}Fe Mössbauer spectroscopy. Iron substitution was found to have little effect on the magnetic behavior of Ru(IV) provided that x remains small ($x \leq 0.2$).

Greenwood et al. published ^{99}Ru Mössbauer studies of the ternary sodium ruthenium oxides Na_3RuO_4, $Na_4Ru_2O_7$, and $NaRu_2O_4$. Na_3RuO_4 [117, 140] exhibits a magnetic hyperfine pattern, from which a magnetic field at the nucleus of 58.7 T was derived, being the largest yet observed for any ruthenium system. The field arises from antiferromagnetic coupling of the Ru^{5+} ions. Other examples for ruthenium oxides, which contain the pentavalent Ru(V) ion and which exhibits a similarly high magnetic hyperfine field, as observed in Na_3RuO_4, are $Na_4Ru_2O_7$ [117], $RuSr_2GdCu_2O_8$ [148–150], the perovskite phases Ca_2LnRuO_6 (Ln = Y, La, Eu), and the analogous strontium and barium compounds M_2LnRuO_6 (M = Sr, Ln = Y; M = Ba, Ln = La, Eu) [139]. The results obtained for the latter demonstrate that the Ru(V)–O–Ln(III)–O–Ru(V) superexchange interaction is stronger than the Ru(IV)–O–Ru(IV) superexchange in e.g., $CaRuO_3$, indicating that exchange interaction between Ru(V) ions is very much stronger than that involving Ru(IV) ions.

For an estimate of the magnetic hyperfine field H_{hf} in Ru(III), Ru(IV), and Ru(V) compounds with low-spin groundstate, $S = 1/2$ for Ru(III), $S = 1$ for Ru(IV), and $S = 3/2$ for Ru(V), one may use as a rule of thumb the finding that the contribution to H_{hf} at the Ru nucleus is about 20 T per unpaired 4*d* electron [151]; this value may be compared with 11 T for 3*d* transition metal elements in oxides.

A specific feature of the above-mentioned Ru(V) oxide $RuSr_2GdCu_2O_8$ [148–150] is that it orders magnetically at around 135 K and has a full transition to superconductivity at very low temperature (8.7 K) [148, 149]. The most interesting property of this oxide is the coexistence of magnetism and superconductivity, especially for the ferromagnetic compound. A magnetic ordering temperature of ~135 K requires a magnetic ordering between the RuO layers. The coupling strength is ~275 K for Ru(V)–Ru(V) coupling and ~30 K for Ru(V)–Gd(III) coupling.

The ^{99}Ru Mössbauer study of anhydrous ruthenium trichlorides, i.e., "black" α-$RuCl_3$ and "brown" β-$RuCl_3$, was performed at 5 K; this study was complemented by temperature-dependent magnetic susceptibility measurements and by crystal structure investigations [151]. In α-$RuCl_3$, the Ru(III) ions form a honeycomb-lattice layer whereas in β-$RuCl_3$, they are positioned in chains along the ***c***-axis perpendicular to the layers. These structural differences are responsible for the different magnetic and electric hyperfine interactions: α-$RuCl_3$ exhibits antiferromagnetic ordering ($T_N = 15.6$ K) and yields a magnetic hyperfine field at 5 K of 20.9 T, whereas β-$RuCl_3$ is paramagnetic at 5 K. For α-$RuCl_3$, the quadrupole splitting is zero and for β-$RuCl_3$ it takes the value -0.75 mm s^{-1} at 5 K. The delocalisation of the 4*d* electrons one-dimensionally along the ***c***-axis in β-$RuCl_3$ explains qualitatively the large quadrupole splitting and also the relatively small paramagnetic moment of $0.9\mu_B$ as compared to other Ru(III) compounds. This one-dimensional deformation is absent in α-$RuCl_3$ resulting in zero quadrupole splitting and an enhanced magnetic moment which makes magnetic interaction between Ru(III) sites detectable.

^{99}Ru and ^{57}Fe Mössbauer spectroscopy was performed on ternary intermetallic compounds, $Fe_{3-x}Ru_xSi$ ($0.1 \leq x \leq 1.5$) [152]. The alloys which contain ruthenium as the main substance are nonmagnetic. For $x = 0$ and $x = 1.0$ the magnetic hyperfine pattern of ^{99}Ru yields the values $H_{hf} = 34$ and 37 T, respectively. In the region $0.1 \leq x \leq 1.0$, the value for H_{hf} of Fe at the [B] site diminished with Ru substitution because of the decreasing number of Fe atoms at the [A, C] sites.

^{99}Ru, ^{61}Ni, ^{57}Fe, and ^{119}Sn Mössbauer measurements were performed on Heusler alloys, $Ru_xY_{3-x}Z$ (Y = Fe, Ni; Z = Si, Sn) [45, 153]. Conclusive evidence was obtained that Ru atoms are substituted preferentially on Fe[B] sites. The magnitude of H_{hf}(Ru) was found to decrease with an increase in the ruthenium concentration.

7.3.3 Further ^{99}Ru Studies

An overview of the work published after 1978 is collected in Table 7.3. Earlier reports are covered in the first edition of this book (Part V of CD-ROM).

Table 7.3 ^{99}Ru Mössbauer effect studies on various materials

References	System/material under study	Remarks
[154]	*T*-violation experiment	The effect of time-reversal (*T*) violation on the transition of photons through a system of magnetized foils that have a Mössbauer transition with comparable *M*1 and *E*2 strength (such as in ^{99}Ru or ^{197}Au)
[155]	Supported (alumina, silica) Ru catalysts	The Mössbauer data show that $RuCl_3{\cdot}(1\text{–}3)H_2O$ reacts chemically when supported onto alumina, but does not when impregnated on a silica support. The study further shows that a supported ruthenium catalyst converts quantitatively into RuO_2 upon calcination, and that the reduction of a supported ruthenium catalyst converts all of the ruthenium into the metallic state
[156]	Carbon-supported Ru–Sn catalyst	^{99}Ru and ^{119}Sn Mössbauer measurements were performed to investigate catalysts of ruthenium and tin supported on activated carbon (Ru–Sn/C). The samples were subjected to different reducing and oxidizing treatments. The presence of tin leads to a substantial increase of the Lamb–Mössbauer factor of the metallic Ru-particles showing that tin strengthens the attachment of the particles to the support. The close contact between the two metals appears to be decisive for the formation of catalytically active sites (Ru–Sn and Ru–SnO_x)
[157]	Pt–Ru anodes for methanol electrooxidation	The catalytically most active samples are highly dispersed and contain, as indicated by the Mössbauer data, a mixture of two Ru(IV) species

Table 7.3 (continued)

References	System/material under study	Remarks
[158]	Salts of ruthenocene with halides	Isomer shift and quadrupole splitting of salts, $[Ru(C_5H_5)X]$ Y (X = Cl, Br; Y = PF_6 and X = I, Y = I_3) are larger compared to those of ruthenocene. This indicates direct chemical bonding between Ru and Cl, Br and I and that the Ru ion in each salt is in an oxidation state higher than Ru(II) in ruthenocene
[159–161]	TDPAC and ^{99}Ru emission Mössbauer spectroscopy of various Ru-oxides	Various Ru-oxides, $YBa_2Cu_3O_{7-x}$ (I), Ba $Ru_{2/3}Gd_{1/3}O_3$ (II) as well as Ru-doped α-Fe_2O_3 (III), to probe the local chemical structure around the Ru atoms. Compound (I) has interesting properties: with $x \leq 0.2$ it is a superconductor and with $x \sim 1$ a semiconductor. Ru oxidation state and coordination are discussed on the basis of measured isomer shifts and quadrupole splittings: Ru(IV) ions exclusively occupy Cu-1 sites which form one-dimensional chains
[160, 161]	TDPAC and ^{99}Ru emission Mössbauer spectroscopy of Fe_3O_4	Radioactive ^{99}Ru is dilutely doped in Fe_3O_4. The value of the magnetic hyperfine field (at the Ru nucleus) obtained by TDPAC at 10 K (−16 T) is essentially in agreement with that derived from the Mössbauer spectrum at 5 K (14.5 T). The negative sign of H_{hf} means that the Ru ions are located at the octahedral (B) sites of Fe_3O_4. The hyperfine field detected by TDPAC consists of two parts: the major part (−13.5 T) is produced by the magnetic moments of the nearest-neighbor tetrahedral Fe(III) (A) sites and the minor part (∼−2.5 T) by that of the unpaired 4*d* electron of the Ru ion itself. The isomer shift indicates that the oxidation state of Ru is between II and III

7.4 Hafnium (^{176,177,178,180}Hf)

The Mössbauer effect has been observed in the following four hafnium isotopes ^{176}Hf, ^{177}Hf, ^{178}Hf, and ^{180}Hf, with the nuclear transitions of 88.36, 112.97, 93.2, and 93.33 keV. Apart from the ^{177}Hf resonance, which involves the $9/2^-$ excited and the $7/2^-$ ground spin states, the other three isotopes have $2^+ \leftrightarrow 0^+$ electric quadrupole (*E*2) transitions. The decay schemes of the hafnium isotopes are given in Fig. 7.36. The relevant nuclear parameters are collected in Table 7.1 (end of the book). The high energetic transitions as well as the relatively short half-lives of the parent nuclei work against hafnium it being a "good" Mössbauer element, i.e., not favorable for practical applications. Only some ten papers have been published on this subject, so far.

The first report on hafnium Mössbauer measurements by Wiedemann et al., in 1963 [162], deals with the 88.36 keV transition in ^{177}Hf. Observations of the ^{176}Hf and ^{180}Hf nuclear resonances were described in 1966, and that of the ^{178}Hf resonance in 1968 by Gerdau and coworkers [163, 164]. In the majority of hafnium

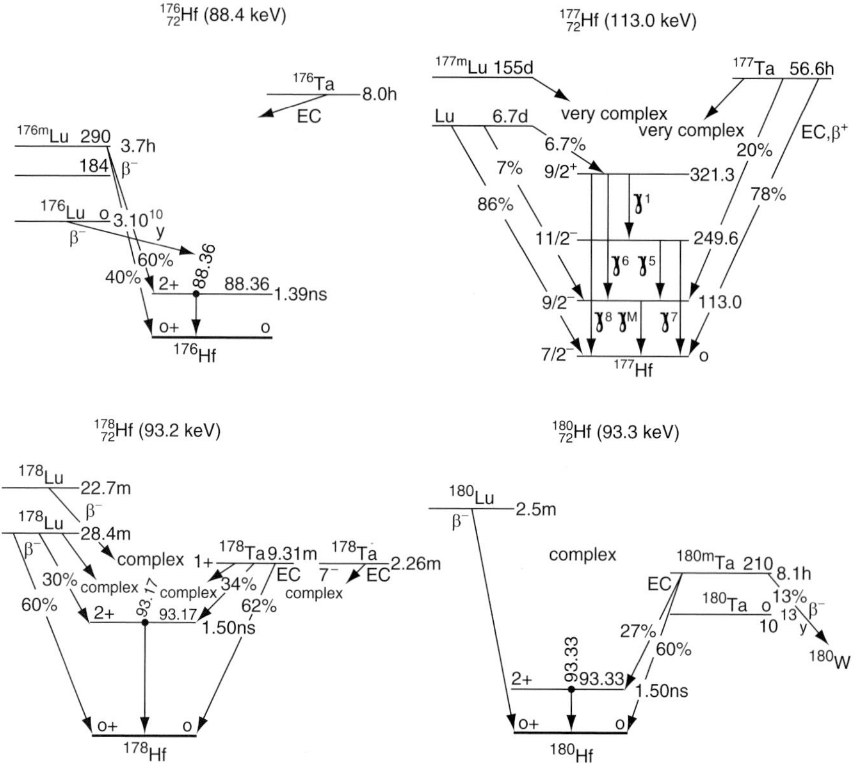

Fig. 7.36 Nuclear decay schemes for the hafnium isotopes ^{176,177,178,180}Hf (from [162])

Mössbauer experiments, the 93.2 keV transition in ^{178}Hf was used because of the reasonable half-life (21.5 days) of its parent nuclide ^{178}W.

7.4.1 Practical Aspects of Hafnium Mössbauer Spectroscopy

The parent nuclei of the hafnium Mössbauer isotopes can be produced by the following reactions:

$$^{175}\mathrm{Lu/Lu_2Rh_2(n,}\gamma)^{176\mathrm{m}}\mathrm{Lu} \underset{\beta^-}{\overset{3.7\mathrm{h}}{\rightarrow}} {}^{176}\mathrm{Hf} \tag{7.6}$$

$$^{175}\mathrm{Lu/Lu-met.(2n,}\gamma)^{177}\mathrm{Lu} \underset{\beta^-}{\overset{6.7\mathrm{d}}{\rightarrow}} {}^{177}\mathrm{Hf} \tag{7.7}$$

$$^{181}\mathrm{Ta/Ta.met.(P;S)}^{178}\mathrm{W} \underset{\mathrm{EC}}{\overset{21.5\mathrm{d}}{\rightarrow}} {}^{178}\mathrm{Ta}\frac{9.4^{\mathrm{m}}}{\mathrm{EC},\beta^+}{}^{178}\mathrm{Hf} \tag{7.8}$$

$$^{179}\mathrm{Hf}/\mathrm{HfO_2}(\mathrm{n},\gamma)^{180\mathrm{m}}\mathrm{Hf} \overset{5.5\mathrm{h}}{\rightarrow} {}^{180}\mathrm{Hf} \quad (7.9)$$

In addition, Coulomb excitation can be used to populate the Mössbauer levels of ^{177,178,180}Hf [165–167]. The experimental line width using these sources is only slightly larger than the natural line width (e.g., the thickness corrected line width of ^{178}Hf in a tantalum foil is in good agreement with the natural line width: $\Gamma_{\mathrm{exp}} = 1.90 \pm 0.07$ mm s^{-1}, $\Gamma_{\mathrm{nat}} = 1.99 \pm 0.04$ mm s^{-1} [168]).

The ratio of the quadrupole moments of ^{176,178,180}Hf has been determined by Snyder et al. [169], using the quadrupole splitting of HfB_2, HfO_2, and Hf-metal, and was found to be $Q_{2^+}(176)/Q_{2^+}(178)/Q_{2^+}(180) = (1.055 \pm 0.008)/(1.014 \pm 0.013)/1$. From the known half-life of the 93 keV level in ^{178}Hf, Boolchand et al. [170] estimated the quadrupole moment to be $Q_{2^+}(180) = -1.93 \pm 0.05b$ using the theory of deformed nuclei.

The isomer shifts in hafnium Mössbauer isotopes usually are of the order of some percent of the line width. Boolchand et al. [168] observed a relatively large isomer shift of +0.19 $\pm$ 0.06 mm s^{-1} between cyclopentadienyl hafnium dichloride ($Hf(Cp)_2Cl_2$) and Hf metal. From a comparison with $Os(Cp)_2$ and Os-metal, a value of $\delta\langle r^2\rangle$ (^{178}Hf) $= -0.37 \cdot 10^{-3}$ fm^2 has been derived, which implies a shrinking of the nuclear radius in the excited 2^+ state. Figure 7.37 shows some typical spectra for ^{178}Hf in various hafnium compounds (from [168]).

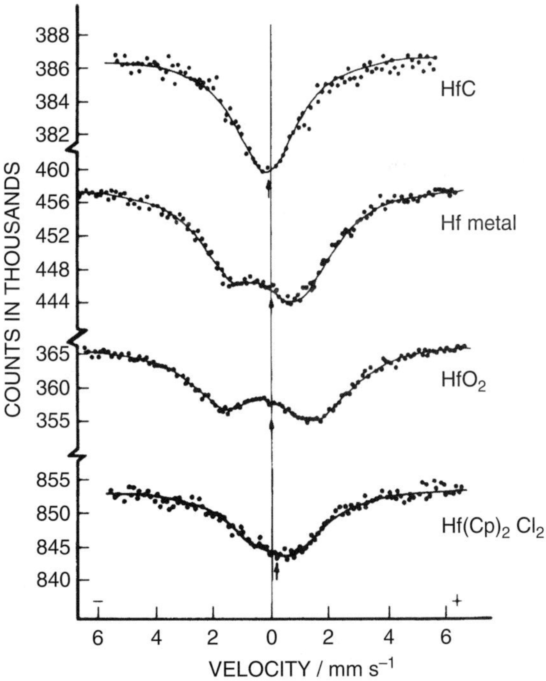

Fig. 7.37 Mössbauer spectra taken with a source of ^{178}W in tantalum metal using absorbers of: HfC, Hf metal, HfO_2, and $Hf(Cp)_2Cl_2$. The centers of gravity are indicated by the *arrows* (from [168])

7.4.2 Magnetic Dipole and Electric Quadrupole Interaction

Steiner et al. have measured the magnetic hyperfine field of ^{178}Hf and ^{180}Hf in iron [171] using sources of 1 at.% W in Fe and 1 at.% Hf in Fe, respectively. Taking the known magnetic moments (from γ–γ angular correlation measurements) of $\mu_{2+}(^{178}\text{Hf}) = 0.704(70)$ and $\mu_{2+}(^{180}\text{Hf}) = 0.741(64)$ [171], a magnetic hyperfine field of $H_{\text{eff}}(4.2\ \text{K}) = 60.6(7.0)$ T and $H_{\text{eff}}(4.2\ \text{K}) = 33.4(4.0)$ T could be deduced for ^{178}Hf in a W–Fe alloy and for ^{180}Hf in a Hf–Fe alloy, respectively. The difference in the hyperfine fields is attributed to the chemical constitution of the sources used. In the case of the ^{178}W source with 1 at% W in Fe, a homogenous solution of ^{178}Hf in Fe is expected, whereas the field in the ^{180}Hf experiment may be due to clustering and formation of the intermetallic compound $HfFe_2$. In addition, the temperature dependence of the hyperfine field in ^{178}Hf, $H_{\text{eff}}(77\ \text{K})/H_{\text{eff}}(4.2\ \text{K}) = 0.90(3)$, is much stronger than the temperature dependence of the magnetization of the iron host lattice $M(77\ \text{K})/M(4.2\ \text{K}) = 0.997$. This discrepancy has not been explained.

Chemical information from hafnium Mössbauer spectroscopy can primarily be deduced from the quadrupole-splitting parameter. In Table 7.4, we have listed the quadrupole coupling constants eQV_{zz} for some hafnium compounds. Schäfer et al. [172] have investigated the electric quadrupole interaction of ^{178}Hf in $PbTiO_3$ and compared their results with calculations in the framework of the ligand-field model, including a contribution to the EFG from the permanent electric dipoles of the perovskite structure. Covalency effects have been found to play a considerable role, even in these relatively ionic compounds.

Jacobs and Hershkowitz [166] observed an increase in the ^{178,180}Hf quadrupole coupling constants due to recoil radiation damage following Coulomb excitation in Hf-metal, HfB_2, HfC, HfN, and HfO_2. The anomalous interactions were discussed in terms of lattice distortions resulting from vacancies in the vicinity of the recoiling Mössbauer nucleus caused by Coulomb excitation. The extent to which local vacancies can affect the EFG at the Hf nuclei was related to the bonding properties of the above compounds.

Table 7.4 Quadrupole coupling constant eQV_{zz} and asymmetry parameter η for ^{178}Hf in some hafnium compounds

Compound	eQV_{zz} (mm s^{-1})	η	References
HfB_2	−7.18(3)	0.42 (5)	[169]
HfO_2	−8.00 (3)	0.48 (4)	[169]
Hf-metal	−5.94 (4)	0	[169]
$Hf(NO_3)_4$	8.18 (31)	0.57 (15)	[169]
$HfOCl_2 \cdot 8H_2O$	5.96 (20)	0.86 (12)	[169]
$HfCl_4$	6.26 (57)	0.71 (36)	[169]
$(NH_4)_2HfF_6$	8.50 (15)	0.90 (5)	[163]
$Hf(Cp)_2Cl_2$	−3.54 (32)	0	[168]
$^{178}Hf/Pb(Ti_{0.9}Hf_{0.1})O_3$	−6.74 (7)	0	[172]

7.5 Tantalum (^{181}Ta)

^{181}Ta provides two γ-transitions, the 136.25 keV transition ($E2/M1 = 0.19$) from the $9/2^+$ ($t_{1/2} = 40$ ps) state to the $7/2^+$ ground state and the 6.23 keV transition ($E1$) from the $9/2^-$ first excited state ($t_{1/2} = 6.8$ ns) to the ground state, both of which are suitable for recoilless nuclear resonance; cf. Fig. 7.38 and Table 7.1 (end of the book).

Nuclear resonance absorption for the 136 keV transition has been established by Steiner et al. [174]. The authors used a ^{181}W metal source and an absorber of metallic tantalum to determine the mean lifetime of the 136 keV level from the experimental line width (≈52.5 mm s^{-1} for zero effective absorber thickness) and found a value of 55 ps. This has been the only report so far on the use of the 136 keV excited state of ^{181}Ta for Mössbauer experiments.

The 6.2 keV excited state of ^{181}Ta appears to be most favorable for Mössbauer spectroscopy from the point of view of its nuclear properties:

1. The small natural width of $2\Gamma = 0.0065$ mm s^{-1} implies an extremely narrow relative line width of $\Gamma/E_\gamma = 1.1 \cdot 10^{-14}$, which is about 30 times narrower than that for the 14.4 keV state of ^{57}Fe.
2. The high atomic mass yields small recoil energies and thus a large recoil-free fraction even at room temperature ($f_{300} \approx 0.95$ for Ta metal [175]); this allows measurements of the resonance effect to be made over a wide temperature range up to about 2,300 K.
3. On one hand, the large magnitude of the change of the mean-squared nuclear charge radius, $\Delta\langle r^2\rangle = \langle r^2\rangle_e - \langle r^2\rangle_g$, between the excited state and the ground

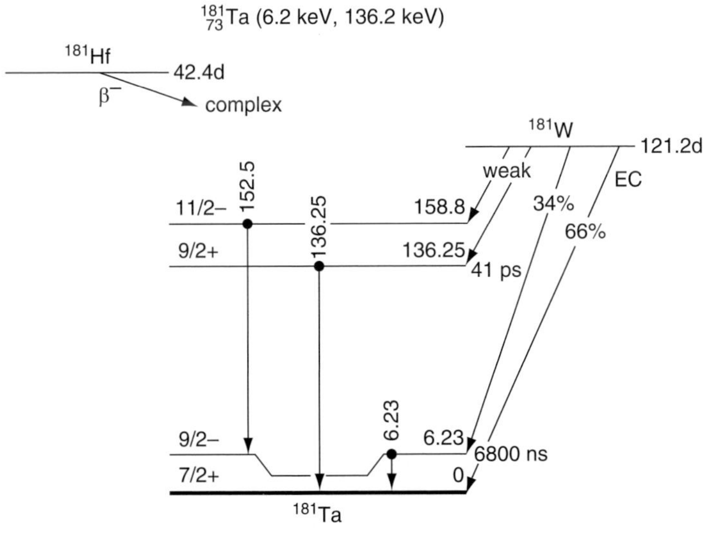

Fig. 7.38 Decay scheme of ^{181}Ta (from [173])

state of ^{181}Ta and the large electric quadrupole moments both in the ground state and in the excited state results in an unusually high sensitivity of the isomer shift and the quadrupole interaction to small changes in the local environment of a ^{181}Ta lattice site. On the other hand, the extremely narrow line width and the high spin of the transition render the resonance effect very susceptible even to minor perturbations. In fact, only strongly broadened resonance lines with experimental line widths of 30–40 times the natural width could be found by the early investigators [176–178], presumably because of the unusually high sensitivity of the 6.2 keV transition to electric quadrupole interaction.

7.5.1 Experimental Aspects

The EC decay of ^{181}W ($t_{1/2}$ = 140 days), which is produced from ^{180}W by thermal neutron activation in a nuclear reactor, populates the two Mössbauer levels of ^{181}Ta at excitation energies of 6.2 and 136.2 keV.

The high sensitivity against hyperfine interactions has severe consequences with respect to source and absorber preparations. Extreme care must be taken for the purity of the samples because of the large distortions caused even by small concentrations of impurities. It has even been shown that crystal defects could influence markedly the line width. Annealing at ≈2,000°C under ultrahigh vacuum (≈10^{-8}–10^{-10} Torr) is therefore necessary in source preparation after diffusion of ^{181}W into the host matrix under H_2 atmosphere at ≈1,000°C. For further details about source and absorber preparation, the readers are referred to the work of Sauer [179]. The best results concerning narrowest possible line width are obtained with single crystals, thereby avoiding crystal defects and preferred diffusion along grain boundaries. The experimental line widths observed to date remain more than one order of magnitude larger than the minimum theoretical width of 2Γ. The smallest value, Γ_{exp} = 0.069 mm s^{-1}, was observed by Wortmann [180] using a source of ^{181}W in a W single crystal and a 4.6 mg cm^{-2} thick tantalum foil as absorber which was treated as described by Sauer [179]. Thin foils of tantalum are required because of the large photoelectric absorption of the 6.2 keV γ-rays; for example, a tantalum metal foil of 2–3 μm thickness (3–5 mg cm^{-2}) absorbs 65–80% of the γ-ray intensity.

The layout of the spectrometer is determined by the narrow resonance width of the ^{181}Ta (6.2 keV) transition and the wide range of line positions. A high-stability drive system employing either a piezoelectric crystal transducer [176] or a conventional high-quality electromechanical transducer [181] in connection with a large number of channels (≈2,000) for recording the spectra is generally used. Special attention must be paid to rigid connections of source and absorber to avoid uncontrolled vibrations. The relatively large recoil-free fraction for the ^{181}Ta (6.2 keV) transition requires no cooling of source and absorber unless temperature effects of the hyperfine interaction are of particular interest.

A commonly used detector for the 6.2 keV γ-rays is an Ar-filled proportional counter, where the Mössbauer quanta are detected in the slope of the 8.15 keV L_α X-rays. The resolution becomes worse when working with higher counting rates, and a relatively large background of 2–3 times the counts in the 6.2 keV line could be expected. Pfeiffer [182] reported on the use of Si(Li) and intrinsic germanium (IG) detectors. The 6.2 keV γ-rays and the L_α X-rays can be resolved with both types of counters, but the IG is the more favorable detector because of the lower Compton background arising from the strong K_α and K_β X-rays (56.3 and 67.0 keV).

As a general feature in evaluating the Mössbauer spectra of ^{181}Ta, one has to take into account in the computer fitting procedure a modification of the Lorentzian lines by the interference effect between photoelectric absorption and Mössbauer absorption followed by internal conversion [183, 184]. This effect was first observed by Sauer et al. and has been found to be particularly large for the 6.2 keV γ-rays of ^{181}Ta due to their E1 multipolarity, low transition energy, and high internal conversion coefficient. It gives rise to an additional dispersion term besides the normal Lorentzian line shape leading to an asymmetry of the observed resonance line; cf. Fig. 7.39. The Mössbauer spectra are best fitted with dispersion-modified Lorentzian lines of the form [183]

$$N(v) = N(\infty)[l - \varepsilon(1 - 2\xi X)/(1 + X)^2],$$

with $X = 2(\upsilon - S)/W$. $N(\upsilon)$ is the intensity transmitted at relative velocity υ, S is the position of the line, W is the full line width at half-maximum, ε is the magnitude of the resonance effect, and ξ is a parameter determining the relative magnitude of the dispersion term ($2\xi = -0.31 \pm 0.01$ has been found to be a good value for ^{181}Ta (6.2 keV) Mössbauer spectra).

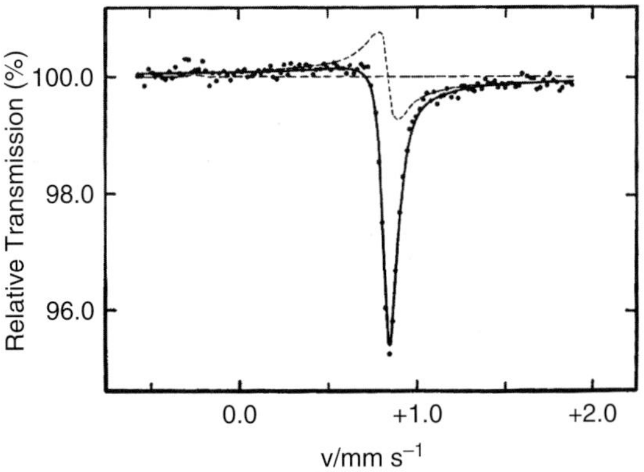

Fig. 7.39 ^{181}Ta (62 keV) Mössbauer spectrum of ^{181}W in W metal versus Ta metal absorber at room temperature. The *solid line* represents the fit of a dispersion-modified Lorentzian line to the experimental data; the *dashed line* shows the dispersion contribution (from [179, 185])

7.5.2 Isomer Shift Studies

The first isomer shifts in ^{181}Ta (6.2 keV) Mössbauer spectroscopy were reported in [177] for the combination of ^{181}W in W metal source and Ta metal absorber and ^{181}W in Ta source and Ta metal absorber, respectively. Sauer et al. [185] found large positive isomer shifts between 0.83 and 0.94 mm s^{-1} for various combinations of W sources and Ta absorbers at room temperature and suggested a considerable change of $\Delta\langle r^2\rangle$ in the nuclear charge radii associated with the γ-transition. Sauer et al. [179] studied the influence of plastic deformation and oxygen doping of a Ta metal absorber and observed remarkable influences on both isomer shift and line width. Kaindl et al. [186] have summarized the results from various isomer shift studies [180, 185–190]; the data are collected in Table 7.5. Some typical room temperature spectra showing the enormously large changes in isomer shift from one sample to another are reproduced in Figs. 7.40–7.42. A graphical representation of isomer shifts of the 6.2 keV γ-transition in ^{181}Ta for tantalum compounds versus ^{181}W/W (single crystal) source and for dilute impurities of ^{181}Ta(^{181}W) in transition metal hosts (vs. Ta metal as absorber) published by Kaindl et al. [186] is shown in Fig. 7.43. The isomer shift of TaC is, together with that of ^{237}Np in Li_5NpO_6 ($\delta = 68.9 \pm 2.6$ mm s^{-1} versus ^{241}Am/Th source at 4.2 K) [192], the largest one observed so far. The total range of isomer shifts (110 mm s^{-1}) in ^{181}Ta (6.2 keV)

Table 7.5 ^{181}Ta isomer shift, experimental line width and resonance effect, observed for sources of ^{181}W diffused into various transition metal hosts against tantalum metal as absorber, and for various tantalum compounds as absorber versus a source of ^{181}W/single crystal W. Source and absorber in all cases at room temperature (from [191])

	δ (mm s^{-1})	Γ_{exp} (FWHM) (mm s^{-1})	Effect (%)
Source lattice			
V	−33.2 (5)	5.0 (10)	0.1
Ni	−39.5 (2)	0.50 (8)	1.6
Nb	−15.26 (10)	0.19 (6)	1.5
Mo	−22.60 (10)	0.13 (4)	3.0
Ru	−27.50 (30)	1.3 (2)	0.7
Rh	−28.80 (25)	3.4 (5)	0.3
Pd	−27.80 (25)	1.3 (3)	0.3
Hf	−0.60 (30)	1.6 (4)	0.2
Ta	−0.075 (4)	0.184 (6)	2.4
W	−0.860 (8)	0.069 (1)	20
Re	−14.00 (10)	0.60 (4)	1.3
Os	−2.35 (4)	1.8 (2)	0.8
Ir	−1.84 (4)	1.60 (14)	0.5
Pt	+2.66 (4)	0.30 (8)	1.5
Absorber lattice			
$LiTaO_3$	−24.04 (30)	1.6 (2)	0.9
$NaTaO_3$	−13.26 (30)	1.0 (2)	0.9
$KTaO_3$	−8.11 (15)	1.5 (2)	0.3
TaC	+70.8 (5)	2.4 (4)	0.2

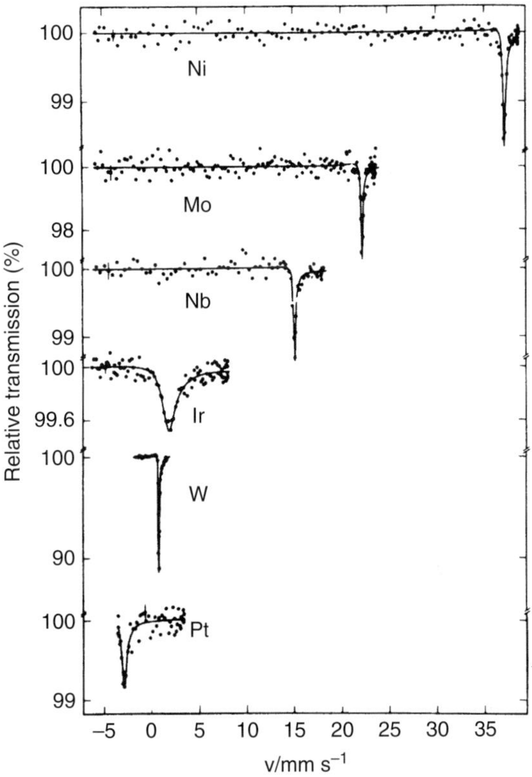

Fig. 7.40 ^{181}Ta (6.2 keV) Mössbauer spectra obtained with a tantalum metal absorber and sources of ^{181}W diffused into various cubic transition metal hosts. The *solid lines* represent the results of least-squares fits of dispersion-modified Lorentzian lines to the experimental spectra (from [186])

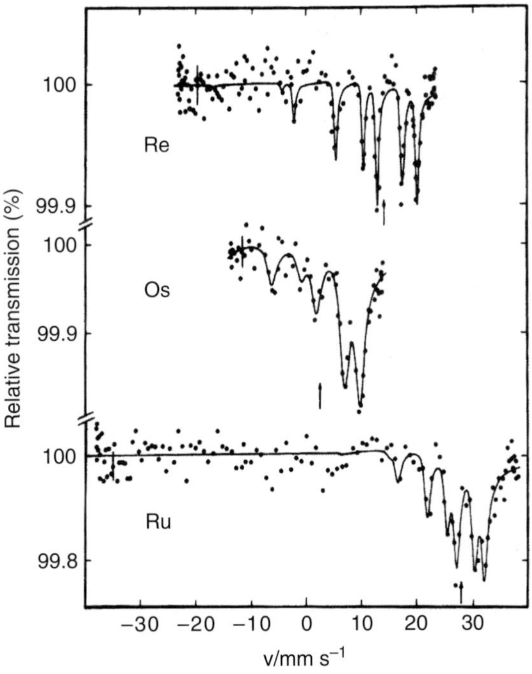

Fig. 7.41 Electric quadrupole split ^{181}Ta (6.2 keV) Mössbauer spectra for sources of ^{181}W diffused into hexagonal transition metals (Re, single crystal; Os, polycrystalline; Ru, single crystal). The isomer shifts relative to a tantalum metal absorber are indicated by the *arrows* (from [186, 189])

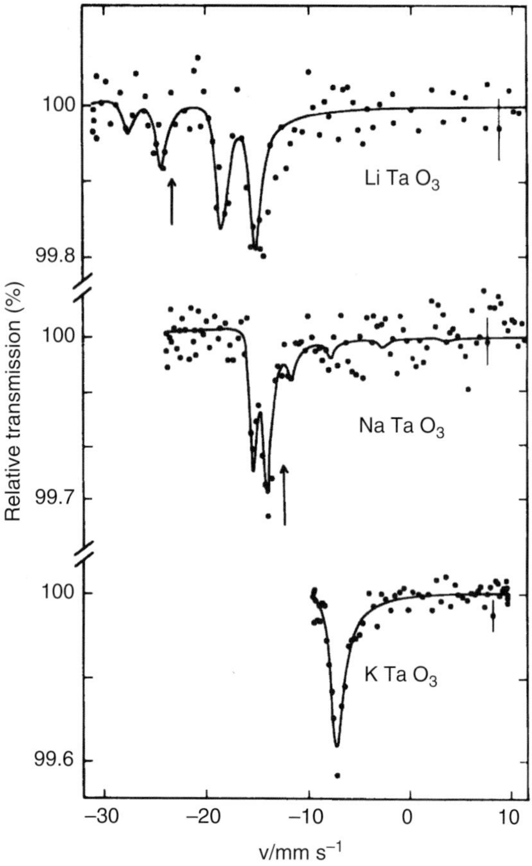

Fig. 7.42 ^{181}Ta (6.2 keV) Mössbauer spectra of alkali tantalates obtained at room temperature with a source of ^{181}W in W (single crystal). The centers (isomer shifts) of the quadrupole split spectra are indicated by *arrows* (from [186])

corresponds to about 30,000 times the natural width or 1,600 times the best experimental line width obtained so far.

Kaindl et al. [186] have plotted the isomer shift results for metallic hosts versus the number of outer electrons of the 3*d*, 4*d*, and 5*d* metals and found the transition energy to decrease when proceeding from a 5*d* to a 4*d* and further to a 3*d* host metal in the same column of the periodic table. This systematic behavior is similar to that observed for isomer shifts of γ-rays of ^{57}Fe(14.4 keV) [193], ^{99}Ru(90 keV), ^{197}Au (77 keV), and ^{193}Ir(73 keV) [194]. The changes of $\Delta\langle r^2\rangle = \langle r^2\rangle_e - \langle r^2\rangle_g$ for these Mössbauer isotopes are all reasonably well established. Kaindl et al. [186] have used these numbers to estimate, with certain assumptions, the $\Delta\langle r^2\rangle$ value for ^{181}Ta (6.2 keV) and found a mean value of $\Delta\langle r^2\rangle = -5 \cdot 10^{-2}$ fm^2 with some 50% as an upper limit of error. The negative sign of $\Delta\langle r^2\rangle$ is in agreement with the observed variation of the isomer shift of $LiTaO_3$, $NaTaO_3$, and $KTaO_3$, as well as with the isomer shift found for TaC [186].

The temperature dependence of the transition energy of the 6.2 keV γ-rays in ^{181}Ta requires particular attention, especially in studies over a wide temperature

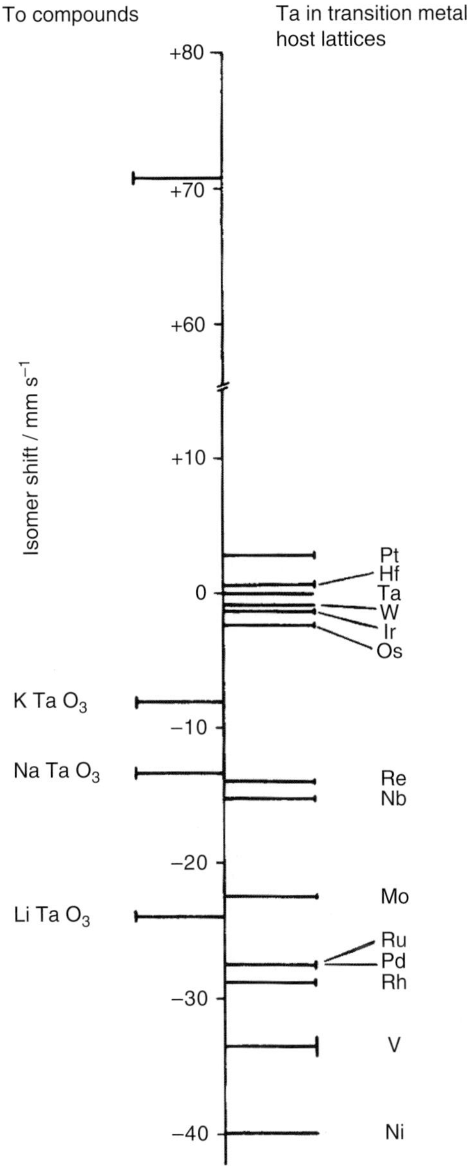

Fig. 7.43 Graphical representation of isomer shifts of the 6.2 keV γ-transition in ^{181}Ta for tantalum compounds and for dilute impurities of ^{181}Ta (^{181}W) in transition metal hosts (from [186])

range. In contrast to ^{57}Fe and ^{119}Sn, where the observed temperature shifts were found to be predominantly caused by the SOD shift [195], Kaindl and Salomon [196] have found that the 6.2 keV transition energy emitted from ^{181}Ta as a dilute impurity in transition metal hosts exhibits a strong temperature dependence far beyond the SOD shift. In the nickel hosts, for instance, the 6.2 keV transition energy increases with temperature with a slope which is 32 times larger and of opposite sign than the one expected from the SOD shift alone (cf. Fig. 1 of [196]). The line shift corresponds to 2.3 natural widths per degree. The slopes of the temperature shifts for W, Ta, and Pt have the same sign as the SOD shift, but are up to eight times larger. After [196], the experimentally observed isobaric temperature variation of the line position S may be written as

$$\left(\frac{\partial S}{\partial T}\right)_P = \left(\frac{\partial S_{SOD}}{\partial T}\right)_P + \left(\frac{\partial S_{IS}}{\partial T}\right)_V + \left(\frac{\partial S_{IS}}{\partial \ln V}\right)_T \left(\frac{\partial \ln V}{\partial T}\right)_P.$$

With the observed temperature shift data for $(\partial S/\partial T)_P$ and calculated (within the framework of the Debye model) numbers for the temperature shift of SOD and with the known thermal expansion coefficient as well as results from ^{181}Ta Mössbauer experiments under pressure, the authors [191] were able to evaluate the true temperature dependence of the isomer shift, $(\partial S_{\mathrm{IS}}/\partial T)$ as $-33 \cdot 10^{-4}$ and $-26 \cdot 10^{-4}$ mm s^{-1} degree^{-1} for Ta and W host metal, respectively.

Another study of the temperature dependence of the 6.2 keV Mössbauer resonance of ^{181}Ta has been carried out by Salomon et al. [197] for sources of ^{181}W/W metal and ^{181}W/Ta metal in the temperature range from 15 to 457 K. In more recent investigations, Salomon et al. [198] have extended such studies of the temperature behavior of the 6.2 keV Mössbauer transition of ^{181}Ta in tantalum metal to temperatures up to 2,300 K which has been the highest temperature range for any Mössbauer study so far.

Enormously large isomer shift changes have been observed by Heidemann et al. [199] in tantalum metal loaded with hydrogen (α-phase of Ta–H) as a function of hydrogen concentration. Some typical spectra of their work demonstrating this striking effect are shown in Fig. 7.44. A substantial line broadening with increasing hydrogen concentration was explained by the authors quantitatively in terms of isomer shift fluctuations arising from varying hydrogen configurations. With a motional narrowing approach, they evaluated the jump frequency and the activation energy of the hydrogen interstitials from the concentration and temperature dependence, respectively, of the line broadening.

7.5.3 *Hyperfine Splitting in ^{181}Ta (6.2 keV) Spectra*

The ground state as well as the 6.2 keV excited state of ^{181}Ta possesses sizeable electric quadrupole and magnetic dipole moments (cf. Table 7.1 at the end of the book) which, in cooperation with the extremely narrow line width, generally yield well-resolved Mössbauer spectra even in cases with relatively weak interactions.

7.5.3.1 Quadrupole Splitting

Electric quadrupole interaction splits the $I_e = 9/2^-$ excited state into five sublevels (with $m_I = \pm 9/2, \pm 7/2, \pm 5/2, \pm 3/2, \pm 1/2$) and the $I_g = 7/2^+$ ground state into

Fig. 7.44 ^{181}Ta (6.2 keV) Mössbauer spectra of Ta–H absorbers at 300 K with various hydrogen concentration (from [199])

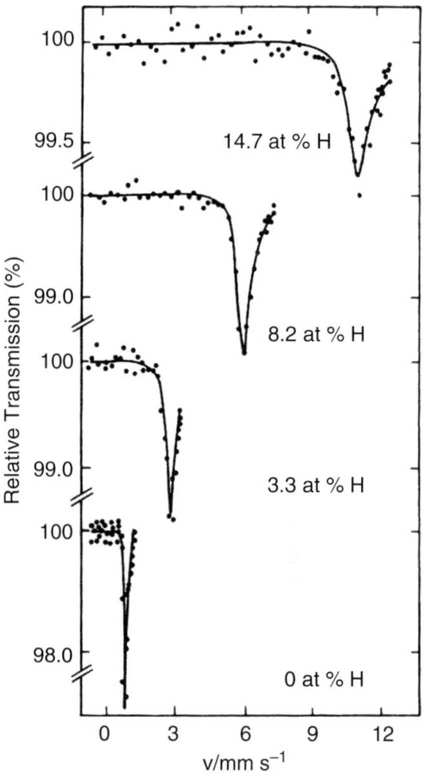

four sublevels (with $m_I = \pm 7/2, \pm 5/2, \pm 3/2, \pm 1/2$) and there are a total of 11 possible $E1$ transitions whose relative intensities are fixed by the relevant Clebsch–Gordan coefficients.

The broadened single-line spectrum of the early measurements by Cohen et al. [176], obtained with a TaC absorber and an annealed ^{181}W/W metal source at room temperature, was assigned by the authors to the $\pm 3/2 \leftrightarrow \pm 3/2$ transition with an estimated Q_e/Q_g ratio between 0.8 and 0.95.

Steyert et al. [177] observed quadrupole effects in their ^{181}Ta Mössbauer spectra of source/absorber combinations with "nominally" cubic environments of the ^{181}Ta atoms in both source and absorber. They ascribed this unexpected effect to interstitial impurities of O_2, N_2, and C.

Sauer et al. [185] derived a weak quadrupole interaction from the asymmetry of a poorly resolved Zeeman split spectrum of ^{181}W in W metal versus a Ta metal absorber. They also ascribed the unexpected weak quadrupole effect to deviations from cubic symmetry at the source or absorber atom arising from either interstitial impurities or crystal defects.

Well-resolved quadrupole split ^{181}Ta Mössbauer spectra were observed by Kaindl and coworkers with a direction-oriented source of ^{181}W in a Re single crystal with hexagonal symmetry versus a specially treated Ta metal absorber [187], from which they derived the sign and the magnitude of the quadrupole interaction and the quadrupole moment ratio $Q_e/Q_g = 1.333 \pm 0.010$. In further investigations, they used sources of ^{181}W in the hexagonal transition-metals Hf (single crystal), Os, and Ru (both polycrystalline) versus a single-line absorber of Ta metal foil (4 mg cm^{-2}) [189] (cf. Fig. 7.41), and a ^{181}W/W single crystal source versus hexagonal $LiTaO_3$ and orthorhombic $NaTaO_3$ absorbers [186] (cf. Fig. 7.42). All these spectra were recorded with the source and the absorber kept at room temperature. In all cases, dispersion-modified Lorentzians were fitted to the experimental spectra by a least-squares computer fit.

7.5.3.2 Magnetic Dipole Splitting

The nuclear Zeeman effect splits the $9/2^-$ excited state into ten sublevels and the $7/2^+$ ground state into eight sublevels and a large number of allowed transitions can occur in ^{181}Ta Mössbauer experiments on systems with magnetic dipole interaction. The number of transitions can be reduced by applying a small magnetic field (~1–2 kOe). In fact, in all studies of magnetic dipole splitting by ^{181}Ta Mössbauer spectroscopy published so far, the authors have applied magnetic fields of 0.15–0.17 T (1,500–1,700 Oe) so as to magnetize the source parallel to the γ-ray direction. All the $\Delta m = \pm 0$ transitions do not show up this way, and a ^{181}Ta Zeeman spectrum consists of only 16 hyperfine components with $\Delta m = \pm 1$ and relative intensities given by the Clebsch–Gordan coefficients.

Sauer et al. [185] determined the gyromagnetic ratio $g(9/2)/g(7/2)$ and the magnetic moment of the 6.2 keV level in ^{181}Ta in two ways, (1) from the Zeeman split velocity spectrum of a ^{181}W/W metal source in a longitudinal field versus a Ta metal absorber and (2) in a "Zeeman drive" experiment, in which the hyperfine components 9/2 ↔ 7/2 and 7/2 ↔ 5/2 were observed at zero velocity as a function of the applied magnetic field ranging up to 0.25 T. A typical spectrum obtained in this kind of experiment is shown in Fig. 7.45. The Zeeman split velocity spectrum

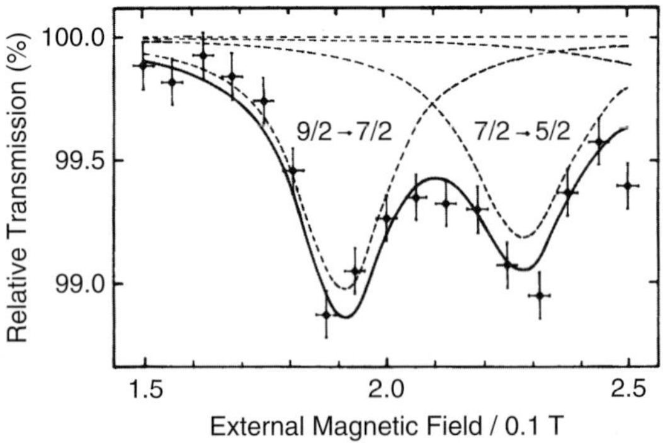

Fig. 7.45 Outer two components of the magnetic hyperfine spectrum of a ^{181}W/W source versus a Ta absorber, measured at zero velocity as a function of a longitudinal magnetic field ("Zeeman drive" experiment) (from [185])

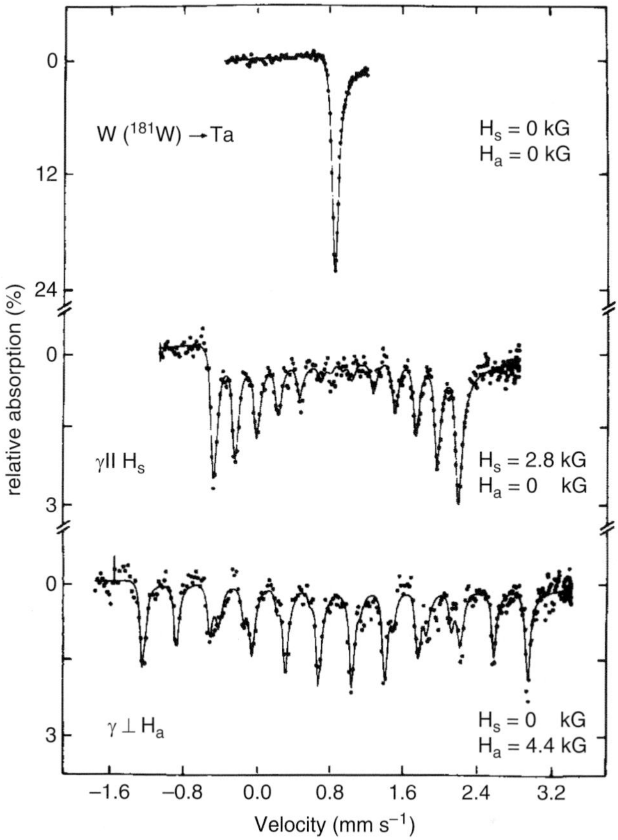

Fig. 7.46 ^{181}Ta Mössbauer spectra of a Ta metal absorber without magnetic field (*top*), with magnetic field applied to the ^{181}W/W single crystal source only, $H_s = 2.8$ kG (*middle*), and magnetic field applied to the absorber only, $H_a = 4.4$ kG (*bottom*). The magnetically split absorber spectrum shows the 16 transition lines with $\Delta m = \pm 1$ expected for magnetization parallel to the γ-beam (from [204])

yielded $g(9/2)/g(7/2) = +1.72 \pm 0.05$ and, with the known ground state moment, the excited state moment has been found to be $\mu_e = (+5.20 \pm 0.15)\mu_N$. The "Zeeman drive" experiment gave $g(9/2)/g(7/2 = 1.62 \pm 0.09$ and $\mu_e = (4.9 \pm 0.3)\mu_N$, somewhat less accurate and less reliable because of the weak quadrupole perturbation neglected in this case. Similar experiments have been performed by Kaindl and Salomon [200]. They found $g(9/2)/g(7/2) = 1.74 \pm 0.03$ and 1.76 ± 0.04 from the magnetically split velocity spectrum and a "Zeeman drive" experiment, respectively, in good agreement within experimental error with the results of [185].

Using a single-line Ta metal absorber (4 mg cm^{-2}) and a ^{181}W/Ni (single-crystal) source in a polarizing external field, Kaindl and Salomon [190] measured the sign and the magnitude of the supertransferred hyperfine field at the ^{181}Ta nucleus and found a value of $H = -9 \pm 1$ T at room temperature.

7.5.4 Methodological Advances and Selected Applications

Mössbauer sources for the 6.2 keV resonance of ^{181}Ta have mostly been produced by diffusing, under ultrahigh vacuum conditions, neutron-activated ^{181}W into highly pure matrices of transition metals [186, 191]. The narrowest line width was obtained with lowest-possible concentration of ^{181}W in tungsten metal. It is, therefore, important to produce ^{181}W with highest-possible specific activity. Dornow et al. [201] have reported on the use of cyclotron-produced carrier-free ^{181}W, and they obtained a narrower line width (53 μm s^{-1}) as compared to the procedure using neutron-activated ^{181}W activity. However, both procedures are suited for practical use, and the experimentalist will choose the one with easiest access to the production of the radioactive precursor.

^{181}Ta Mössbauer effect measurements have, up to now, not been carried out with a great variety of materials. An overview of the reports, that have appeared since the first edition of this book came out, is given in Table 7.6.

Table 7.6 ^{181}Ta Mössbauer effect studies on various materials and systems

References	System/material under study	Remarks
[202]	H_2/Ta	Study of the dynamic behavior of hydrogen/deuterium in Ta metal, observation of fluctuation in electron density at Ta atoms caused by diffusing hydrogen interstitials. Diffusion rate of deuterium changes by three orders of magnitude between 400 and 100 K, correspondingly the relaxation behavior of ^{181}Ta spectra change from "motional narrowing" to "slow relaxation"
[203]	1T-$TaSe_2$	Observation of charge density waves. At 300 K, two Ta sites are observed with isomer shifts of +51 and 87 mm s^{-1}, respectively, relative to Ta metal, with nearly equal population and the same EFG at the Ta nucleus
[204]	Ta metal in appl. magnetic field	Completely resolved spectra of 6.2 keV transition at room temperature in applied magnetic field (2.8 and 4.4 kOe) with ^{181}W/W single crystal source (s. Fig. 7.46). Precise determination of g-factor ratio $g_{9/2}/g_{7/2}$ to be 1.797 (3) and magnetic moment of 6.2 keV state to be 5.47 (2) n.m
[205]	2H-TaS_2, 2H-$TaSe_2$	The nuclear transition $\pm 3/2 \rightarrow \pm 1/2$ is completely resolved from the transition $\pm 9/2 \rightarrow \pm 7/2$, and a new high accuracy value for the quadrupole moment ratio $Q(9/2)/Q(7/2) = 1.1315 \pm 0.0002$ has been derived
[206]	Ta metal under proton irradiation	Study of radiation effects of the ^{181}Ta(p, n)^{181}W reaction in Ta foil, recording of emission spectra before and after annealing with metallic Ta absorber, and of absorption spectra before and after annealing ^{181}W/W source

Table 7.6 (continued)

References	System/material under study	Remarks
[207]	2H-TaS_2 intercalated with hydrated sodium	The quadrupole interaction increases upon Na intercalation, and the isomer shift increases to a higher (more positive) value reflecting a lower valence at the Ta site.
[208]	2H-Li_xTaS_2 ($0 < x < 0.95$)	Observation of charge transfer from Na to Ta ($Ta^{4+} \rightarrow Ta^{3+}$)
[209]	2H-$TaSe_2$ intercalated with lithium	The isomer shift increases upon intercalation with Li from 82.7 mm s^{-1} for 2H-$TaSe_2$ to 107 mm s^{-1} for 2H-$Li_{0.9}TaSe_2$. This reflects a change from $Ta^{4+}(5d^1)$ toward $Ta^{3+}(5d^2)$
[210]	2H-TaS_2, $LiTaS_2$, $SnTaS_2$	Electronic properties of the layered compound 2H-TaS_2 and its Li and Sn intercalates determined from first principles using the full potential LAPW method. Energy band structures, total and partial density of states, partial charges and electron densities are presented. EFGs at all nuclear positions are determined and the origin of EFG is given
[211]	1T-$TaSe_2$	Sign and amplitude of commensurate charge-density waves. The three inequivalent Ta sites predicted by the 1:6:6 population ratio models have been observed at 4.2 K in a single crystal of 1T-$TaSe_2$
[212]	1T-$LiTaS_2$ single crystal	The isomer shift of 105 mm s^{-1} is 35 mm s^{-1} larger than that of 2H-TaS_2 and comparable to that of 2H-$LiTaS_2$. This reflects a change in electron configuration from Ta^{4+} ($5d^1$) towards Ta^{3+} ($5d^2$) due to charge transfer from Li to Ta
[213]	Ta foil	First successful observation of nuclear resonance scattering of synchrotron radiation (NRS) with ^{181}Ta
[214]	Ta foil	Nuclear forward scattering of synchrotron radiation (NFS) at ^{181}Ta resonance in Ta foil without and with applied magnetic field, point out advantages over conventional ^{181}Ta Mössbauer spectroscopy
[215]	Computer program MOTIF	Evaluation of time spectra for NFS of synchrotron radiation, allows fully automatic fits of experimental data. Has been tested on several Unix platforms by fitting NFS time spectra of Fe-57, Sn-119, Eu-151, Dy-161, Ta-181 in various compounds with different time-independent and time-dependent hyperfine interactions
[216]	Ta metal	Stroboscopic detection of nuclear forward scattered synchrotron radiation is used to detect the high-resolution 6.2 keV Mössbauer resonance of ^{181}Ta in Ta metal

7.6 Tungsten (^{180,182,183,184,186}W)

There are five isotopes of tungsten for which the Mössbauer effect has been observed. In the even *A* isotopes (*A*, mass number), the γ-resonances from the 0^+ ground state to the 2^+ excited states at 103.65 keV ($t_{1/2} = 1.27$ ns), 100.10 keV (1.37 ns), 111.19 keV (1.28 ns), and 122.3 keV (1.01 ns) have been observed for ^{180}W, ^{182}W, ^{184}W, and ^{186}W, respectively. In ^{183}W, two nuclear levels are suitable for Mössbauer measurements, viz., the 46.48 keV ($t_{1/2} = 0.183$ ns) and the 99.08 keV ($t_{1/2} = 0.69$ ns) levels. The nuclear data for the Mössbauer transitions are given in Table 7.1 at the end of the book. In Fig. 7.47, the nuclear decay schemes and parent isotopes are depicted. The recoilless nuclear resonance absorption with the 100 keV transition in ^{182}W was observed shortly after Mössbauer's

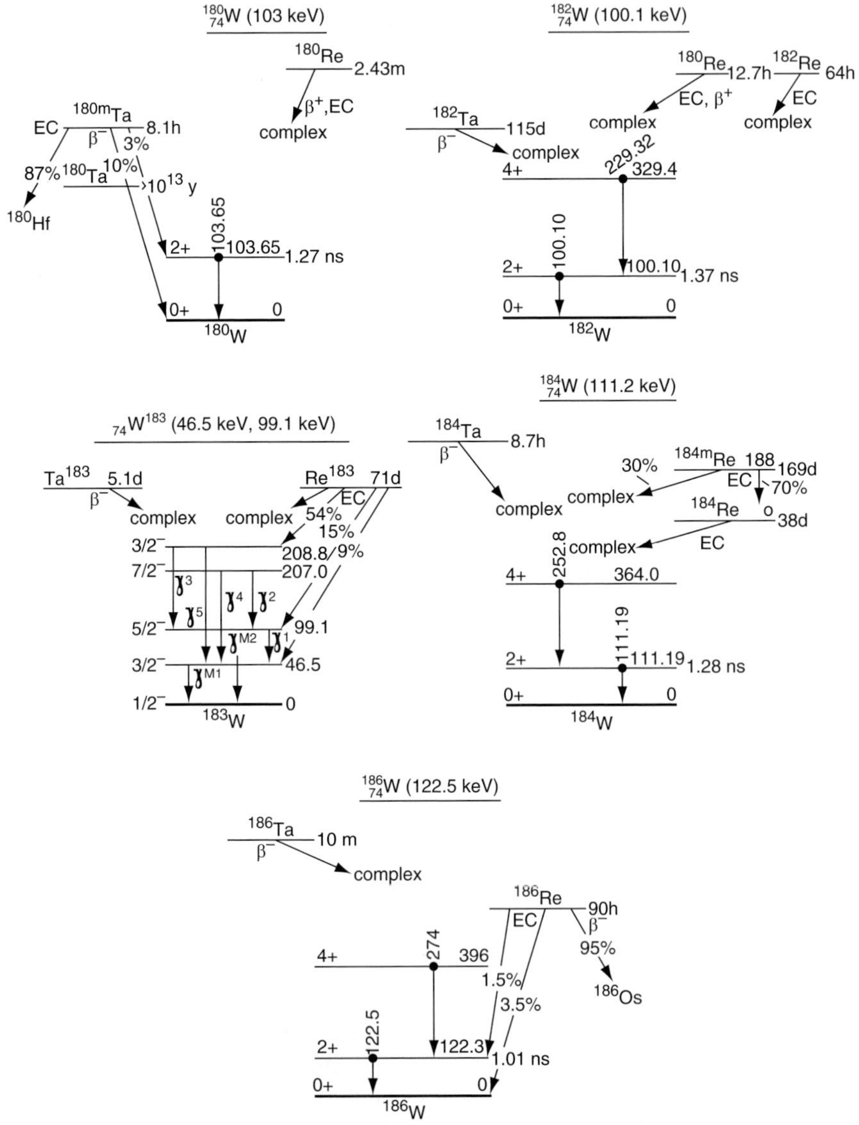

Fig. 7.47 Nuclear decay schemes for the tungsten isotopes ^{180,182,183,184,186}W (from [217])

discovery on iridium [218–220]; physicists have been mainly concerned with measurements of the nuclear γ-resonance in tungsten to determine nuclear properties like magnetic dipole and electric quadrupole moments. The reason why there are only a few articles devoted to applications in chemistry or solid-state physics are the small Debye–Waller factors caused by the relatively high transition energies, the short half-lives of the excited Mössbauer levels leading to broad resonance lines and therefore to poorly resolved spectra, and the small changes of the nuclear radii $\delta\langle r^2\rangle$ in the excited 2^+ (even W isotopes) or $3/2^-$ (^{183}W) rotational states causing only changes in the isomer shift of a few percent of the line width.

Nevertheless there has been some work, mainly from research groups in Cambridge, Darmstadt, and Munich, on measurements of isomer shift and quadrupole hyperfine interaction in ^{180,182}W under chemical aspects.

7.6.1 *Practical Aspects of Mössbauer Spectroscopy with Tungsten*

The source preparation for tungsten Mössbauer spectroscopy is in general cumbersome, apart from the production of ^{182}Ta, the parent nuclide of ^{182}W. The first excited rotational levels (2^+ in even W isotopes or $3/2^-$ in ^{183}W) can be achieved by Coulomb excitation and in beam Mössbauer measurements [221–224]. Other nuclear reactions leading to Mössbauer parent nuclei are:

$$^{181}\mathrm{Ta(d,p2n)}^{180}\mathrm{Ta}\xrightarrow[8.1\,\mathrm{h}]{\beta^-}{}^{180}\mathrm{W} \tag{7.10}$$

$$^{181}\mathrm{Ta}(\gamma,\mathrm{n})^{180}\mathrm{Ta}\xrightarrow[8.1\,\mathrm{h}]{\beta^-}{}^{180}\mathrm{W} \tag{7.11}$$

$$^{181}\mathrm{Ta}(2\mathrm{n},\gamma)^{183}\mathrm{Ta}\xrightarrow[5.1\,\mathrm{d}]{\beta^-}{}^{183}\mathrm{W} \tag{7.12}$$

$$^{181}\mathrm{Ta}(\mathrm{n},\gamma)^{182}\mathrm{Ta}\xrightarrow[115\,\mathrm{h}]{\beta^-}{}^{182}\mathrm{W} \tag{7.13}$$

The isotope ^{182}W is almost exclusively used for the investigation of chemical compounds. The resolving power of ^{182}W Mössbauer spectroscopy is demonstrated by the spectra from WS_2 powder and single-crystal measurements as shown in Fig. 7.48 (from [222]). On the right-hand side of the picture, the angular dependence of $E2$ transitions for $\Delta m = \pm 2, \pm 1, 0$ hyperfine components is indicated.

The ratio of the quadrupole moments in tungsten isotopes as measured by Mössbauer spectroscopy – collected from [222–228] – is

$$Q_{2^+}(180)/Q_{2^+}(182)/Q_{2^+}(184)/Q_{2^+}(186)/Q_{5/2^-}(183)/Q_{3/2^-}(183)$$
$$=(0.988\pm 0.012)/1/(0.955\pm 0.03)/(0.907\pm 0.014)/(0.86\pm 0.06)/(0.88\pm 0.14).$$

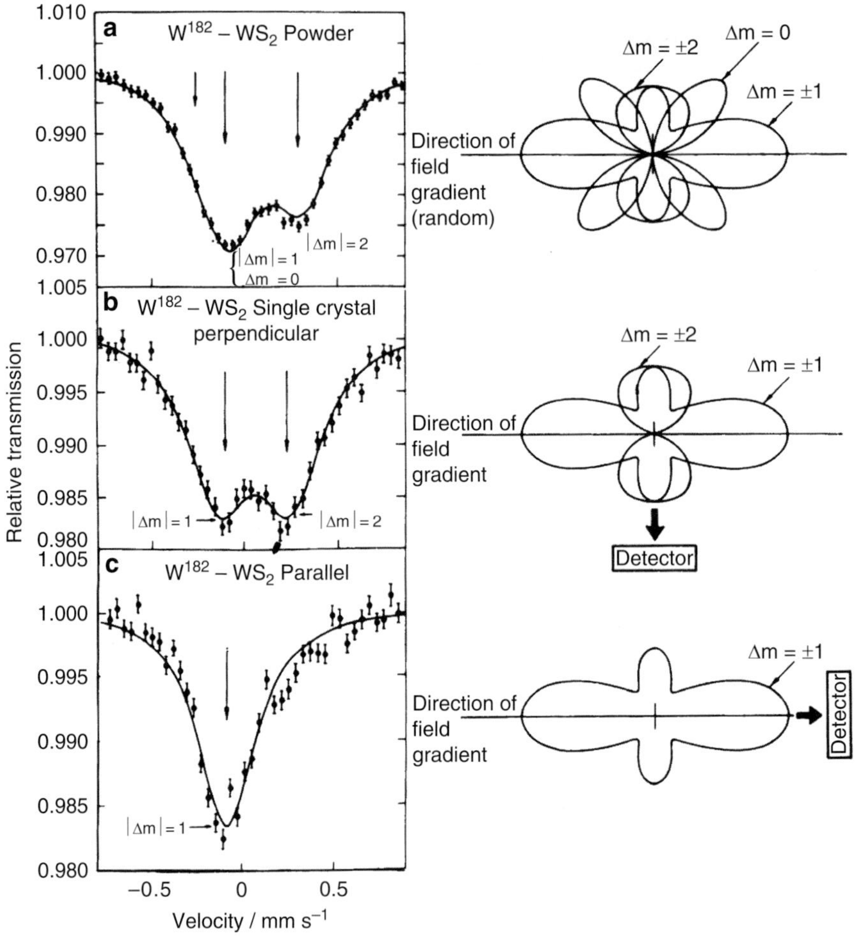

Fig. 7.48 Mössbauer spectrum of the $2^+ \leftrightarrow 0^+$ (100 keV) transition of ^{182}W in WS_2 powder and WS_2 single-crystals (from [222])

The nuclear g_{2^+} factors for the first excited 2^+ states in ^{180,182}W have been determined by Zioutas et al. [225] as $g_{2^+}(^{180}\mathrm{W}) = 0.260 \pm 0.017$ and by Frankel et al. [229] as $g_{2^+}(^{182}\mathrm{W}) = 0.22 \pm 0.02$ by applying external magnetic fields up to 12.8 T on ^{182}W diluted in iron. The internal magnetic field at the tungsten site in alloys of 0.5–5% tungsten in iron turned out to be -70.8 ± 2.5 T. This value, together with another value (63 ± 1.3 T [230]) obtained from NMR data, was used to derive $g_{5/2^-}(^{183}\mathrm{W}) = 0.307 \pm 0.015$, $g_{2^+}(^{184}\mathrm{W}) = 0.243 \pm 0.008$, and $g_{2^+}(^{186}\mathrm{W}) = 0.258 \pm 0.008$. An estimate of the $\delta\langle r^2\rangle$ value in ^{182}W(2^+–0^+) has been derived on the basis of a relativistic Hartree–Fock–Slater calculation [226] comparing the isomer shift of Rb_2WS_4 versus W-metal and WCl_6 versus K_2WCl_6: $-0.05 \cdot 10^{-4} > \delta\langle r^2\rangle/\langle r^2\rangle > -0.24 \cdot 10^{-4}$, which is in agreement with a value given by Wagner et al. [230] but in disagreement with $\delta\langle r^2\rangle = 0.65 \cdot 10^{-4}$ derived by Cohen et al. [231] from the isomer shift between tungsten metal and WCl_6.

The $E2/M1$ mixing ratio of the 46.48 keV transition in ^{183}W has been determined by Shikazono et al. [227] as $\delta = E2/M1 \simeq 0.5\%$.

7.6.2 Chemical Information from Debye–Waller Factor Measurements

The tungsten isotopes are very well suited to the investigation of bond properties and crystal dynamics obtainable from the Debye–Waller factor because of their high transition energies. Small changes in the mean square displacements of the vibrational amplitude will, therefore, lead to large changes in the Debye–Waller factor ($f = \exp(-k^2\langle x^2\rangle)$). Raj and Puri [232] have calculated the recoilless fraction and thermal shift (SOD shift) δ_{SOD} for ^{182}W and ^{183}W in tungsten metal by the relations

$$f = \exp\left\{-\frac{E_\gamma^2}{2Mc^2\hbar}\frac{1}{3N}\int_0^{\omega_{max}}\frac{g(\omega)}{\omega}\coth\left(\frac{\hbar\omega}{2k_BT}\right)d\omega\right\} \tag{7.14}$$

and

$$\delta_{SOD} = \frac{3\hbar}{4Mc}\frac{1}{3N}\int_0^{\omega_{max}} g(\omega)\omega\coth\left(\frac{\hbar\omega}{2k_BT}\right)d\omega, \tag{7.15}$$

which hold only for monatomic cubic crystals. M is the mass of the emitting nucleus and $g(\omega)$ is the phonon frequency distribution function. An experimentally determined phonon spectrum was used in the calculations of Raj and Puri. Ruth and Hershkowitz [233] have measured the temperature dependence of f in tungsten. They found very good agreement with the calculated values of Raj and Puri [232] (Fig. 7.49). The experimental results of [233] could not be interpreted in terms of one Debye temperature value θ_D; applying a Debye model, one could find θ_D to vary from 339 ± 3 K at 29 K to 265 ± 5 K at 100 K.

Kaltseis et al. [234] have investigated the recoil-free fraction of the 46.5 keV transition of ^{183}W in anhydrous lithium tungstate. Their results can be expressed by an effective Debye temperature of 172 ± 9 K which is in good agreement with a value of 205 ± 40 K derived from X-ray diffraction measurements of Li_2WO_4 powder.

Conroy and Perlow [235] have measured the Debye–Waller factor for ^{182}W in the sodium tungsten bronze $Na_{0.8}WO_3$. They derived a value of $f = 0.18 \pm 0.01$ which corresponds to a zero-point vibrational amplitude of $R = 0.044$ Å. This amplitude is small as compared to that of beryllium atoms in metallic beryllium (0.098 Å) or to that of carbon atoms in diamond (0.064 Å). The authors conclude that atoms substituting tungsten in bronze may well be expected to have a high recoilless fraction.

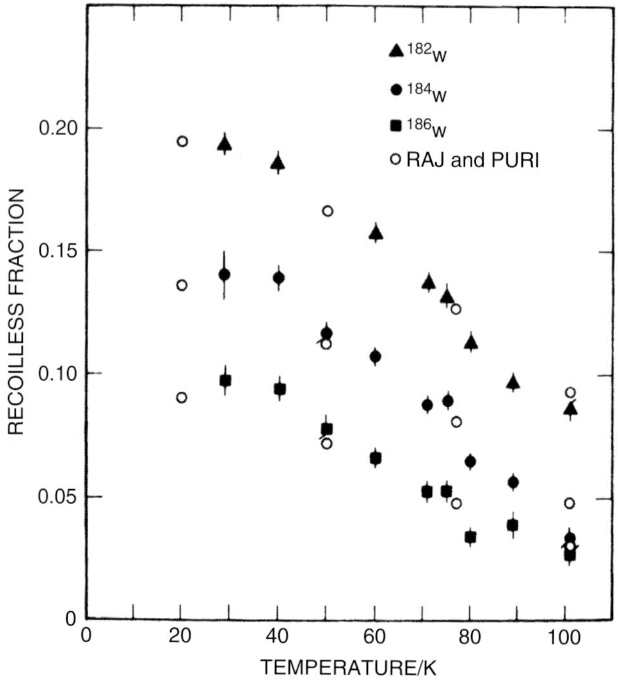

Fig. 7.49 Recoilless fraction as a function of temperature for ^{182}W, ^{184}W, and ^{186}W in tungsten metal (from [233])

As is well known, the recoil-free fraction of very small crystals differs markedly from that of bulk material. Roth and Hörl [236] observed a decrease of the *f*-factor from 0.61 to 0.57 in going from 1 μm crystals to microcrystals with a diameter of about 60 Å. Two effects will contribute to this decrease: (1) the low frequency cut-off, because the longest wavelength must not exceed the dimensions of the crystal, and (2) high frequency cut-off caused by the weaker bonds between surface atoms.

Wender and Hershkowitz [237] used the sensitivity of the recoil-free fraction in tungsten Mössbauer spectroscopy to deduce the effect of irradiation of tungsten compounds by Coulomb excitation of the resonance levels (2^+ states of $^{182,184,186}W$) with 6 MeV α-particles. While no effect of irradiation on the *f*-factors could be observed for tungsten metal in agreement with [233], a decrease of *f* was measured for WC, W_2B, W_2B_5, and WO_3 after irradiation.

7.6.3 *Chemical Information from Hyperfine Interaction*

Apart from the already mentioned (Sect. 7.6.1) determination of the nuclear *g*-factors of $^{180,182}W$ through Mössbauer measurements with tungsten diluted in an iron foil [225, 229] where a hyperfine field at the W site of -70.8 ± 2.5 T was

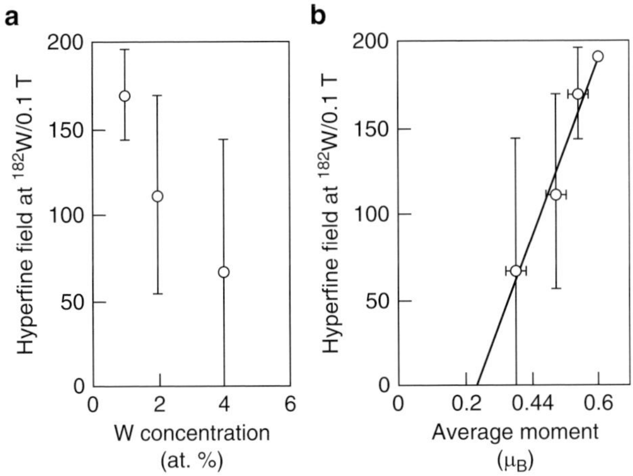

Fig. 7.50 Magnetic hyperfine field H_{hf} in Ni–W-alloys at the ^{182}W-site as a function (**a**) of the tungsten concentration and (**b**) of the average magnetic moment (from [238])

observed, there is only one further study of the magnetic hyperfine interaction in a tungsten alloy. Schibuya et al. [238] have investigated the hyperfine field H_{hf} at ^{182}W in Ni–W-alloys. They found a linear dependence of H_{hf} on the tungsten concentration and on the average magnetic moment (Fig. 7.50), which can be expressed by $H_{hf} = 540\langle\mu\rangle - 14$ T. The simple assumption leading to this relation is that the first term is due to the conduction electron polarization induced by the magnetic moment of the Ni host and that of W, and that the second term originates from the core electron polarization induced by the magnetic moment of W.

The chemical shift observed in tungsten Mössbauer spectroscopy is very small, about 1/20 of the line width for ^{182}W, due to the small changes of $\delta\langle r^2\rangle$ in going from the first excited rotational 2^+ state to the ground state. In spite of this difficulty, there are two articles by the Munich [230] and Darmstadt [226] groups in which the isomer shift has been determined in a number of tungsten compounds. The results are presented in Fig. 7.51. It is obvious that the assignment of the Mössbauer isomer shift to a certain oxidation state is not unambiguous in all cases.

Maddock and coworkers [239–242] and Kankeleit et al. [243] have published a series of articles reporting on the measurements and interpretation of the electric quadrupole interaction in a number of inorganic and metal-organic tungsten compounds. The results are collected in Table 7.7. The EFG is composed of the superposition of an electronic and lattice contribution. While the latter can often be neglected, the former strongly depends on the electronic configuration, the ionicity or covalency of the chemical bond, and on the geometrical arrangement of the surrounding ligands. The interplay of all these effects can be accounted, to a reasonable extent, only on the basis of electronic structure calculations.

Savvateev et al. [244] have reported on tungsten oxides with regular octahedral WO_6^{6-} structure (Ba_2CoWO_6, Ba_2CaWO_6, and Cr_2WO_6) and with deformed tetrahedral WO_4^{2-} structure (XWO_4 with X = Ca, Sr, Ba, Zn, Mg, Co). In these

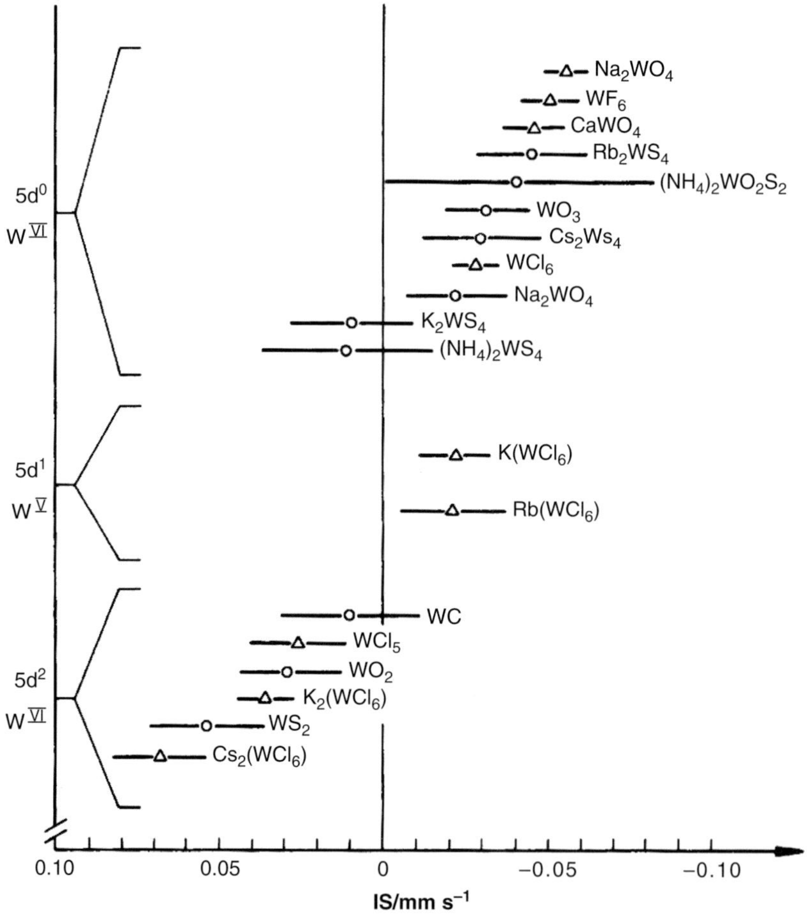

Fig. 7.51 Isomer shift versus formal electron configuration of tungsten compounds (from [226]). The *triangles* are data taken from [230], the *circles* from [226]

W(VI) compounds with $5d^0$ electronic configuration, the EFG is reasonably well described by the lattice contribution only. The regular octahedral WO_6^{6-} structure yields zero EFG in agreement with experiment, while the deformed tetrahedral WO_4^{2-} structure results in significant quadrupole coupling constants of either sign, $e^2qQ > 0$ and also <0, depending on the distortion of the O_4 tetrahedron.

In a ^{182}W Mössbauer study of various polycrystalline W(IV) and W(V) octocyanides and in frozen solutions, Clark et al. [242] were able to determine the quadrupole coupling constant. They concluded that the EFG in the W(IV) compounds with D_{2d} and D_{4d} symmetry arises from partial cancelation of the dominant d^2 contribution by populating the nominally empty d-orbitals through ligand-to-metal σ-bonding. The large values of the asymmetry parameter η observed for $Cd_2[W(CN)_8]{\cdot}8H_2O$ and $(EtNC)_4[W(CN)_4]$ (Table 7.7) are indicative of distortions from D_{2d} and D_{4d} geometry, respectively.

Table 7.7 Quadrupole coupling constants e^2qQ (mm s^{-1}), asymmetry parameter $\eta = V_{xx} - V_{yy}/V_{zz}$, and experimental line width Γ_{exp} (mm s^{-1}) of inorganic and metal-organic tungsten compounds

Compound	e^2qQ (mm s^{-1})	η	Γ_{exp} (mm s^{-1})	References
WO_3	−8.30 ± 0.20	0.62 ± 0.03	2.50 ± 0.20	[239]
	−8.34 ± 0.05	0.63 ± 0.20	2.42 + 0.10	[243]
$FeWO_4$	−8.80 ± 0.20	0.54 ± 0.05	2.76 ± 0.20	[239]
Na_2WO_4	–	–	2.76 ± 0.20	[239]
$W(CO)_6$	–	–	2.60 ± 0.20	[239]
2H-WSe_2	9.38 ± 0.20	–	2.34 ± 0.20	[240]
$W_{0.92}Ta_{0.08}Se_2$	9.09 ± 0.20	–	2.32 ± 0.20	[240]
$W_{0.5}Ta_{0.5}Se_2$	9.61 ± 0.20	–	2.57 + 0.20	[240]
WS_3	−7.72	–	3.43	[241]
W_2N	±2.25	–	3.39	[241]
W_2C	7.38	–	3.20	[241]
SiO_2, $12WO_3$, $26H_2O$	−18.13	0.42	3.60	[241]
$Na_2W_2O_7$ (2 sites)	10.29	0.95	2.31	[241]
	−3.42	0.95	2.24	
$Li_2W_2O_7$ (2 sites)	−11.66	0.83	2.78	[241]
	+5.95	0.71	2.78	
$WOCl_4$	0	0	3.52	[241]
Me_2SnWO_4	−9.32	0.71	2.81	[241]
$(NH_4)_2WS_4$	−8.31	0.76	2.88	[241]
$K_3[W_2Cl_9]$	8.89	0	2.24	[241]
$Et_4N[WCl_6]$	−12.58	0	3.95	[241]
$Et_4N[WCl_6]$	−12.3	0.4	4.0	[241]
$Na_{0.53}WO_3$	0	0	4.28	[241]
$Li_{0.36}WO_3$	0	0	5.25	[241]
$K_4[W(CN)_8]\cdot 2H_2O$	−16.02[a]	–	2.49	[242]
$K_4[W(CN)_8]$ Frozen soln.	−16.48[b]	–	2.13	[242]
$Li_4[W(CN)_8]\cdot nH_2O$	−15.90[a]	–	2.59	[242]
$H_4[W(CN)_8]\cdot 6H_2O$	+12.59[a]	–	2.81	[242]
$H_4[W(CN)_8]$ Frozen soln.	+14.79[b]	–	2.63	[242]
$Cd_2[W(CN)_8]\cdot 8H_2O$	+12.80[a]	0.82	3.00	[242]
$K_3[W(CN)_8]\cdot H_2O$	0	–	2.93	[242]
$Na_3[W(CN)_8]\cdot 4H_2O$	0	–	3.20	[242]
$Ag_3[W(CN)_8]$	0	–	2.42	[242]
$[Co(NH_3)_6]$ $[W(CN)_8]$	0	–	2.44	[242]
$(EtNC)_4[W(CN)_4]$	−8.59[a]	0.74	1.83	[242]

[a]Estimated errors 0.2 mm s^{-1}
[b]Estimated errors 0.5 mm s^{-1}

7.6.4 Further ^{183}W Studies

Quasielastic (Rayleigh) scattering of the ^{183}W 46.5 keV Mössbauer radiation was used to examine the liquid dynamics of glycerol [245, 246] and the harmonic vibrations of the nonhydrogen atoms in polycrystalline myoglobin [247] as a function of temperature. The γ-quanta emitted by the Mössbauer source are

Rayleigh scattered by the electrons of the sample under investigation and are detected at a scattering angle 2ϑ. The energy of the scattered radiation is analyzed using a Mössbauer absorber. It should be noted that the sample itself does not contain the Mössbauer isotope. If the Mössbauer absorber is placed between the source and the sample (position 0), the area of the resulting Mössbauer spectrum is proportional to the number of the recoil-free emitted γ-quanta. With the absorber placed between the sample and the detector (position 2ϑ), the area of the Mössbauer spectrum is proportional to the fraction of the recoil-free emitted γ-quanta which have been scattered elastically by the sample. From these values, a weighted elastic fraction of the scattered radiation is derived, which is a measure of the dynamic properties of the sample; this elastic fraction can be analyzed in terms of the mean-square displacements of the atoms or in terms of the diffusion of the molecules in the sample under study. The time resolution of these dynamic properties depends on the Mössbauer isotope used, i.e., $\tau = 0.265$ ns for ^{183}W.

The short wavelength ($\lambda = 0.267$ Å) of the 46.5 keV Mössbauer transition of ^{183}W limits the lowest accessible angle to $\sin(\vartheta)/\lambda = 0.12$ Å^{-1}. The measurements for glycerol were performed in the direction ϑ corresponding to the first liquid structure peak at 1.3 Å^{-1}, and those for polycrystalline myoglobin from 0.12 to 0.44 Å^{-1}.

Double absorber Mössbauer spectroscopy with ^{183}W [248] has been used for accurate line-shape measurements. This type of spectroscopy is like conventional Mössbauer transmission spectroscopy except that a stationary absorber is placed between the source and a moving absorber. The effect of the stationary absorber is to filter the emission spectrum about the resonance energy E_0, making the transmission signal more sensitive to effects such as interference between Rayleigh and nuclear resonance scattering, and between photoelectric and internal conversion electron production. These combined interference effects lead to an interference parameter β and introduce a slight asymmetry to the Mössbauer line shape [248–250]. The role of β is of importance to time reversal invariance (T) experiments of electromagnetic decay of mixed polarity (see also Table 7.3 and [154]).

7.7 Osmium (^{186,188,189,190}Os)

Among the Os isotopes, there are four (^{186,188,189,190}Os) for which the Mössbauer effect has been observed. In Fig. 7.52, the nuclear decay schemes and parent nuclei of these isotopes are shown. While only one Mössbauer transition has been observed in each of the even Os isotopes, there are three for ^{189}Os with transition energies of 95.2, 69.6, and 36.2 keV. The relevant nuclear data for all six Mössbauer transitions are collected in Table 7.1 at the end of the book. The transition energies of the even osmium isotopes are relatively high, 137.2, 155.0, and 187.0 keV for ^{186}Os, ^{188}Os, and ^{190}Os, respectively, resulting in recoil-free fractions of only some tenths of a percent at low temperatures (~70 K). In addition, their line widths are very large compared with expected isomer shifts, due to the small $\Delta\langle r^2\rangle$ values in going from the 0^+ ground to 2^+ excited nuclear rotational

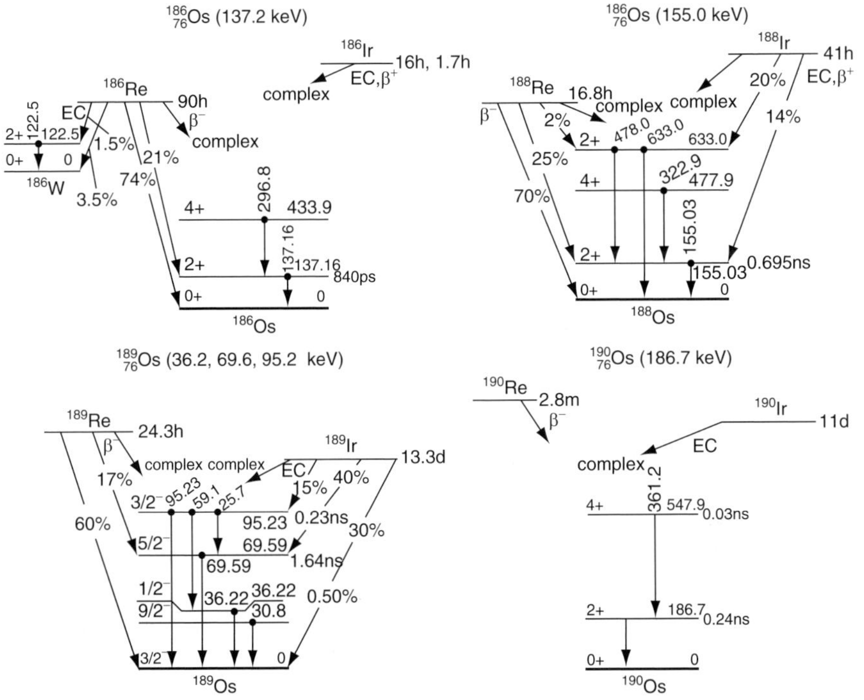

Fig. 7.52 Simplified nuclear decay schemes for the four osmium Mössbauer isotopes ^{186,188,189,190}Os (from [217])

states. Therefore, Mössbauer measurements of ^{186,188,190}Os have only been devoted so far to the determination of the nuclear quadrupole moment and nuclear g-factors ($\mu_{2^+} = 2g_{2^+}\mu_N$) in the 2^+-states.

Apart from the determination of nuclear parameters, the Mössbauer transition in ^{189}Os, especially the 36.2 and 69.6 keV transitions, are suited for chemical applications. As shown below, the 36.2 keV level, in spite of its large half-width, can be well used for the measurement of isomer shifts, whereas the 69.2 keV state is favorable for the characterization of electric quadrupole or magnetic dipole interactions. Both Mössbauer levels are populated equally well by the parent isotope ^{189}Ir, and simultaneous measurement is possible by appropriate geometrical arrangement.

7.7.1 *Practical Aspects of Mössbauer Spectroscopy with Osmium*

The sources for the even osmium isotopes ^{186,188}Os have been the same since the first observations of the Mössbauer effect in the scattering experiments by Morrison et al. [251] and Barret and Grodzins [252] for ^{186}Os and ^{188}Os, respectively. They are produced by neutron irradiation of ^{185}Re and ^{187}Re in Re metal foils making use of the nuclear reactions ^{185}Re(n, γ) ^{186}Re $\xrightarrow[90\text{h}]{\beta^-}$ ^{186}Os and ^{187}Re(n, γ) ^{188}Re $\xrightarrow[16.8\text{h}]{\beta^-}$ ^{188}Os.

Due to unresolved electric quadrupole interaction in the hexagonal Re metal lattice, the emission lines are slightly broadened by a factor of about 1.1 and 1.24 for ^{186}Os and ^{188}Os, respectively ($2\Gamma_{\text{nat}}(^{186}\text{Os}) = 2.42 \pm 0.08$ mm s^{-1}, $2\Gamma_{\text{nat}}(^{188}\text{Os}) = 2.50 \pm 0.08$ mm s^{-1}) [253]. ^{190}Os, for which the Mössbauer effect was first observed by Wagner et al. in 1972 [254], is populated via decay of ^{190}Ir as parent nucleus.

This source is produced by the reaction ^{190}Os(d, 2n) ^{190}Ir $\xrightarrow[11\text{d}]{\text{EC}}$ EC^{190}Os with a 30 μA beam of 13 MeV deuterons. The target, which consisted in Wagner's experiment [254] of 70 wt% Cu and 30 wt% Os (enriched to 95.5% ^{190}Os), was used as the source without further physical or chemical treatment and showed no line broadening ($\Gamma_{\text{exp}} = 4.0 \pm 0.4$ mm s^{-1}, $2\Gamma_{\text{nat}} = 3.67 \pm 0.08$ mm s^{-1}). In reality, because of the insolubility of osmium in copper, the target consisted of Os metal clusters in the host matrix. To get an impression of the resolving power, some Mössbauer spectra for the even Os isotopes are given in Fig. 7.53a, b for OsO_2- and OsP_2-absorbers, respectively (from [254]). The derived EFG components are $|V_{zz}| \simeq 2 \cdot 10^{18}$ V cm^{-2}, $\eta \simeq 0$ (OsO_2) and $|V_{zz}| \simeq 4 \cdot 10^{18}$ V cm^{-2},

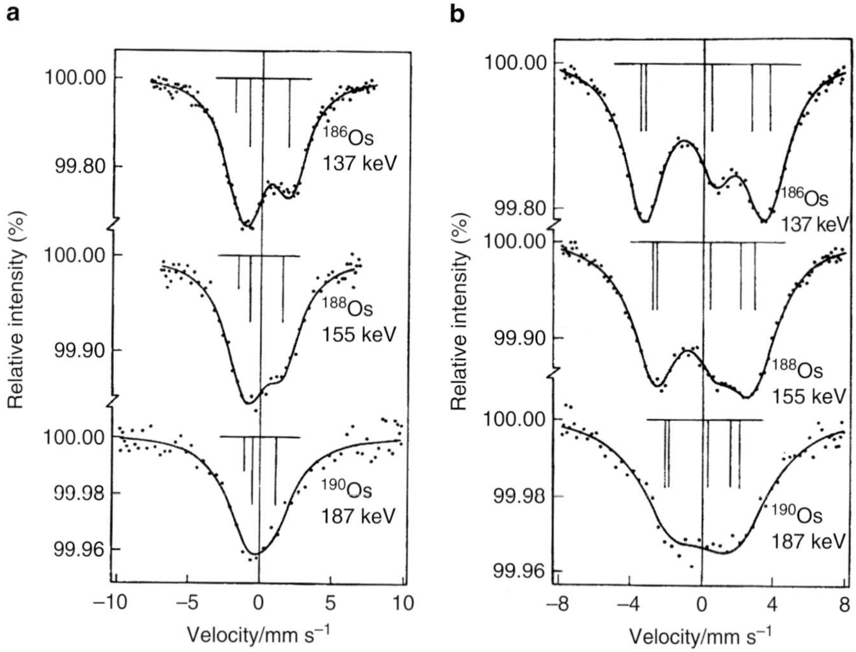

Fig. 7.53 Transmission Mössbauer spectra of the 137, 155, and 187 keV nuclear transitions of ^{186,188,190}Os taken with sources emitting an unsplit line and (**a**) OsO_2-absorber ($\eta \simeq 0$), (**b**) OsP_2-absorber ($\eta \simeq 0.74$). The *curves* are the results of least-squares fits. The *vertical bars* indicate the positions and relative intensities of the individual hyperfine components (from [254])

$\eta \simeq 0.74$ (OsP_2). It can be seen that the asymmetric hyperfine pattern – in the case of $\eta \simeq 0$ (OsO_2), which allows the determination of the sign of V_{zz} – becomes more symmetric with increasing asymmetry parameter η. For $\eta = 1$, the spectra become completely symmetric and the sign of V_{zz} no longer influences the shape of the spectra.

The odd mass isotope ^{189}Os possesses three nuclear states at 95.2, 69.6, and 36.2 keV, which are all suitable for Mössbauer measurements. The first observations of the resonance absorption of these three transitions were made by Gregory et al. [255] (95.2 keV), Iha et al. [256], Persson et al. [257] (69.6 keV), and by Wagner et al. [254] (36.2 keV). Single-line sources of ^{189}Ir in Ir metal can be obtained by use of the reactions ^{189}Os(p, n) ^{189}Ir [257] or ^{189}Os(d, 2n) ^{189}Ir [253, 254, 258] and subsequent chemical separation of osmium and ^{189}Ir. Wagner et al. [253, 254, 258] have used noncorrosive targets with good heat conductivity containing 70 wt% Cu and 30 wt% Os enriched to 87.4% ^{189}Os which were irradiated by a 30 μA deuterium beam of 13 MeV. After irradiation, the ^{189}Ir activity was chemically separated from the target and incorporated in Ir metal, giving nearly unbroadened emission lines since Ir metal is cubic and nonmagnetic. The ^{189}Ir source populates nearly equally well the three Mössbauer levels (cf. Fig. 7.52). However, the 69.6 keV transition is heavily masked by K_β X-rays ($E_{K\beta} \approx 70.8$ keV) which are difficult to separate energetically, even with high-resolution Ge (Li) detectors because of the high source activities. The 95.2 keV level up to now has only been used to estimate the lifetime of this state. A value of $\tau > 0.2$ ns has been reported for the mean life of this state [255]. In Fig. 7.54, the quadrupole spectra for a number of compounds are given for comparison of the 36.2 keV and 69.6 keV Mössbauer transitions [258]. Although the line width of the former is a factor of about 7 larger, the ratio of isomer shift to line width (δ/Γ_{exp}) is a factor of ~4 greater because of the larger quantity of $\Delta\langle r^2\rangle$ for the 36 keV level (Sect. 7.7.2). It is just opposite with respect to the electric quadrupole and magnetic dipole interactions. Here, the 69.6 keV transition is more favorable than the 36 keV transition.

7.7.2 Determination of Nuclear Parameters of Osmium Mössbauer Isotopes

The application of Mössbauer spectroscopy in chemistry requires a prior knowledge of the nuclear states and transitions involved. In this section, we shall describe the determination of nuclear parameters by means of Mössbauer experiments with Os nuclei.

7.7.2.1 Magnetic Moments and *E*2/*M*1 Mixing Parameter

The magnetic moment and the *E*2/*M*1 mixing parameter of the 69.6 keV level ($5/2^-$) of ^{189}Os have been determined by several authors [254, 255, 257, 259].

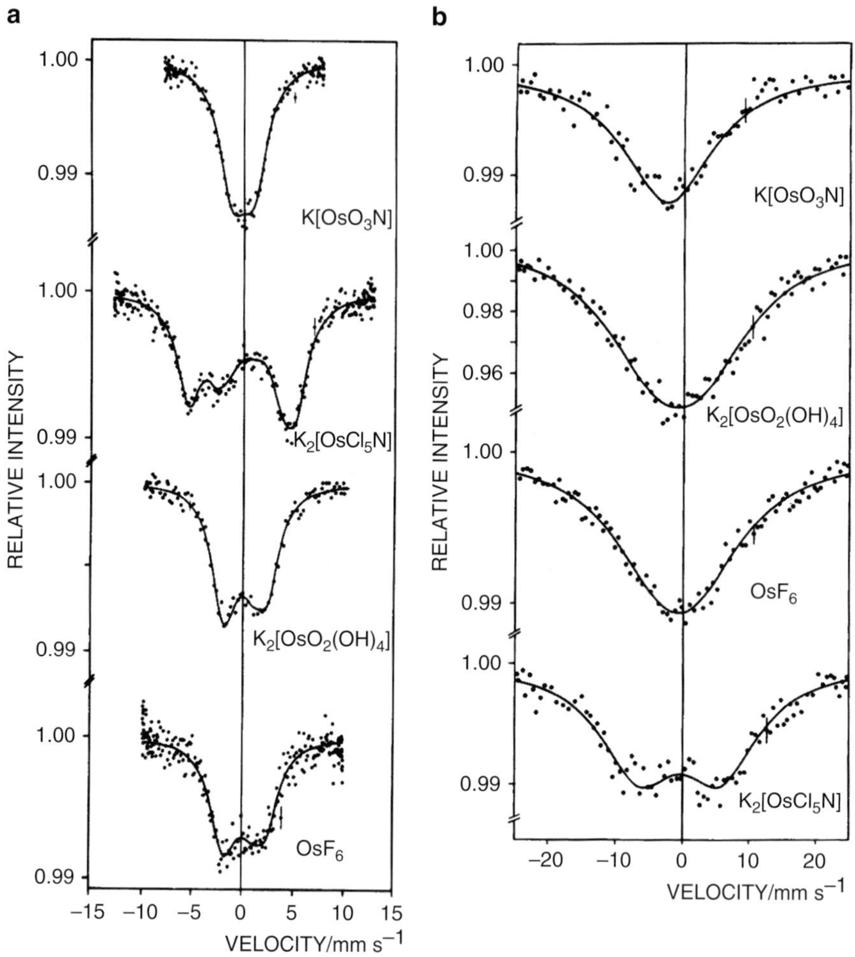

Fig. 7.54 Mössbauer spectra of osmium compounds obtained at 4.2 K with 69.6 keV (**a**) and 36.2 keV (**b**) transitions of ^{189}Os with a source of ^{189}Ir in iridium metal (from [258])

These measurements have been done using $^{189}Ir/Fe$ sources under applied external fields and single-line absorbers as, for example, $K_2[OsCl_6]$. The weighted average of three most reliable values leads to $E2/M1 = +0.69 \pm 0.04$ and μ (69.6 keV) $= 0.980 \pm 0.008\mu_N$ using the known magnetic moment of the ground state $\mu_{3/2}- = 0.6565 \pm 0.0003\mu_N$ [260].

The magnetic moment of the 36.2 keV level ($1/2^-$) in ^{189}Os was measured by Wagner et al. [261]. The source consisted of ^{189}Ir in an iron matrix ($Ir_{0.01}Fe_{0.99}$) which was produced by melting the ^{189}Ir with iron in an induction furnace. Applying an external magnetic field of 2.97 T, the hyperfine spectrum became partly resolved leading to a ratio of $\mu_{1/2^-}/\mu_{3/2^-} = 0.302 \pm 0.003$. The weighted average of the internal magnetic hyperfine field of ^{189}Os in the ^{189}Ir/Fe source was found to be -111.5 ± 2.0 T.

Wagner et al. [253] have also determined the magnetic moments of the first excited $I_e = 2^+$ states of ^{186}Os and ^{188}Os using ^{186}Re and ^{188}Re in $Re_{0.01}$ $Fe_{0.99}$ sources and Os metal absorbers. The sources were polarized by an external magnetic field of 2.97 T parallel to the direction of the γ-beam so that only $\Delta m = m_e - m_g = \pm 1$ hyperfine components appeared in the Mössbauer spectrum. From their measurements, they obtained for the ratio of the magnetic moments of the 2^+ states $\mu_{2^-}(^{188}\text{Os})/\mu_{2^+}(^{186}\text{Os}) = 1.08 \pm 0.05$. Using the value of 113 ± 2.5 T for the internal hyperfine field obtained from NMR measurements, which have to be corrected for the external and demagnetization field, the following magnetic moments could be derived:

$$\mu_{2^-}(^{186}\text{Os}) = 0.562 \pm 0.016\,\mu_N \text{ and } \mu_{2^+}(^{188}\text{Os}) = 0.610 \pm 0.030\,\mu_N.$$

7.7.2.2 Nuclear Quadrupole Moments

Mössbauer measurements with determination of the electric quadrupole moments have been reported in [253, 254, 259]. Wagner et al. [254] measured the quadrupole hyperfine interaction in OsO_2 and OsP_2 of the Mössbauer isotopes ^{186,188,189,190}Os. The ratios of the quadrupole moments of the $I_e = 2^+$ states in the even osmium isotopes and of the $I_e = 5/2^-$ (69.6 keV) and $I_g = 3/2^-$ states in ^{189}Os were deduced very accurately. In Table 7.8, the experimental results [254] are given, from which the following ratios can be calculated:

$$Q_{2^+}(^{186}\text{Os})/Q_{2^+}(^{188}\text{Os})/Q_{2^+}(^{190}\text{Os})/Q_{3/2^-}(^{189}\text{Os, g.s.})$$
$$= (+1.100 \pm 0.020)/1.0/(+0.863 \pm 0.051)/(-0.586 \pm 0.011)$$

$$\text{and} \quad Q_{5/2^-}(^{189}\text{Os, 69.6 keV})/Q_{3/2^-}(^{189}\text{Os, g.s.}) = -0.735 \pm 0.012.$$

From optical hyperfine measurements, a value of $Q_{3/2^-}(^{189}\text{Os, g.s.}) = 0.91 \pm 0.10b$ has been derived [262]. Theoretical values for the quadrupole moment of ^{186}Os and experimental $I_e = 2^+$ to $I_g = 0^+$ transition probabilities led the authors [254] to the conclusion that $Q_{2^+}(^{186}\text{Os})$ should be $-(1.50 \pm 0.10)b$, which in turn yielded $Q_{3/2^-}(^{189}\text{Os, g.s.}) = +(0.80 \pm 0.06)b$ which is somewhat smaller than the value from the optical measurements.

7.7.2.3 Change of Nuclear Charge Radii

An essential prerequisite for the determination of electron densities at the nucleus from isomer shift measurements is the knowledge of the change of the average

Table 7.8 Summary of results obtained for the four Os Mössbauer transitions studied. The absorber thickness d refers to the amount of the resonant isotope per unit area. The estimates of the effective absorber thickness t are based on Debye–Waller factors f for an assumed Debye temperature of $\theta = 400$ K. For comparison with the full experimental line widths at half maximum, Γ_{exp}, we give the minimum observable width $2\Gamma_{nat} = 2\,\hbar/\tau$ as calculated from lifetime data. The electric quadrupole interaction is described by $\Delta E_Q = (eQV_{zz}/4)(1+1/3\eta^2)^{1/2}$ and the asymmetry parameter η. The latter was assumed to be zero in the least-squares fits of the OsO_2 data. In the fit of the OsO_2 spectrum for ^{190}Os, the quadrupole splitting ΔE_Q was constrained to the value given in the table. For the 69.6 keV transition of ^{189}Os, the results for the ratio $Q_{5/2^-}/Q_{3/2^-}$ of the quadrupole moments and the $E2/M1$ mixing parameter are included (from [254])

Source	Absorber	d (mg cm^{-2})	t	Γ_{exp} (mm s^{-1})	ΔE_Q (mm s^{-1})	η
(a) ^{186}Os, 137.16 keV, 2^+–0^+; $2\Gamma_{nat} = 2.41 \pm 0.04$ mm s^{-1}; $f(\theta = 400\text{ K}) = 0.082$						
^{186}Re in Re	OsP_2	10	0.7	2.80 ± 0.05	-3.68 ± 0.05	0.74 ± 0.02
^{186}Re in Re	OsP_2	4	0.3	3.00 ± 0.15	-3.66 ± 0.08	0.78 ± 0.04
^{186}Re in Re	OsO_2	11	0.8	$2.86 + 0.12$	$+1.91 \pm 0.04$	<0.3
^{186}Re in Pt	OsO_2	8	0.6	2.75 ± 0.13	$+1.82 \pm 0.05$	<0.3
^{186}Re in Pt	OsO_2	5	0.4	$2.70 + 0.11$	$+1.87 \pm 0.04$	<0.3
(b) ^{188}Os, 155,02 keV, 2^+–0^+; $2\Gamma_{nat} = 2.50 \pm 0.05$ mm s^{-1}; $f(\theta = 400\text{ K}) = 0.043$						
^{188}Re in ^{187}Re	OsP_2	43	1.6	3.10 ± 0.10	-2.93 ± 0.05	0.74
^{188}Re in Pt	OsO_2	23	0.7	3.06 ± 0.20	$+1.51 \pm 0.06$	<0.3
^{188}Re in Pt	OsO_2	8	0.3	2.80 ± 0.47	$+1.61 + 0.13$	<0.3
(c) ^{190}Os, 186.7 keV, 2^+–0^+; $2\Gamma_{nat} = 3.67 + 0.08$ mm s^{-1}; $f(\theta = 400\text{ K}) = 0.011$						
^{190}Ir in ^{190}Os	OsP_2	150	1.3	4.32 ± 0.50 (4.04 ± 0.44)	-2.10 ± 0.10 ($+2.13 \pm 0.09$)	0.74
^{190}Ir in ^{190}Os	OsO_2	180	1.5	3.76 ± 0.34	+1.1	<0.3
(d) ^{189}Os, 69.6 keV, $5/2^-$–$3/2^-$; $2\Gamma_{nat} = 2.41 \pm 0.06$ mm s^{-1}; $f(\theta = 400\text{ K}) = 0.53$, $Q_{5/2^-}/Q_{3/2^-} = -0.735 \pm 0.012$; $\delta = +0.685 \pm 0.025$						
^{189}Ir in Ir	OsP_2	16	1.2	2.86 ± 0.05	$+3.83 \pm 0.05$	0.73 ± 0.03
^{189}Ir in Ir	OsO_2	11	0.8	3.30 ± 0.10	-1.99 ± 0.06	<0.3

squared nuclear charge radii in going from the ground ($\langle r^2\rangle_g$) to the excited ($\langle r^2\rangle_e$) nuclear state. As already mentioned, the $\Delta\langle r^2\rangle = \langle r^2\rangle_e - \langle r^2\rangle_g$ values of the 2^+ states in even Os isotopes are very small compared with the line width of the Mössbauer transition, so that there are only estimations from muonic atom spectroscopy available [263]:

$$\Delta\langle r^2\rangle/\langle r^2\rangle(^{188}\text{Os}) = -2.37 \cdot 10^{-5};$$

$$\Delta\langle r^2\rangle/\langle r^2\rangle(^{190}\text{Os}) = -6.05 \cdot 10^{-5};$$

$$\Delta\langle r^2\rangle/\langle r^2\rangle(^{192}\text{Os}) = -8.16 \cdot 10^{-5}.$$

Concerning the change of the nuclear charge radii in ^{189}Os, the situation is much better. Bohn et al. [264] have determined the ratio of $\Delta\langle r^2\rangle_{36.2\text{keV}}/$

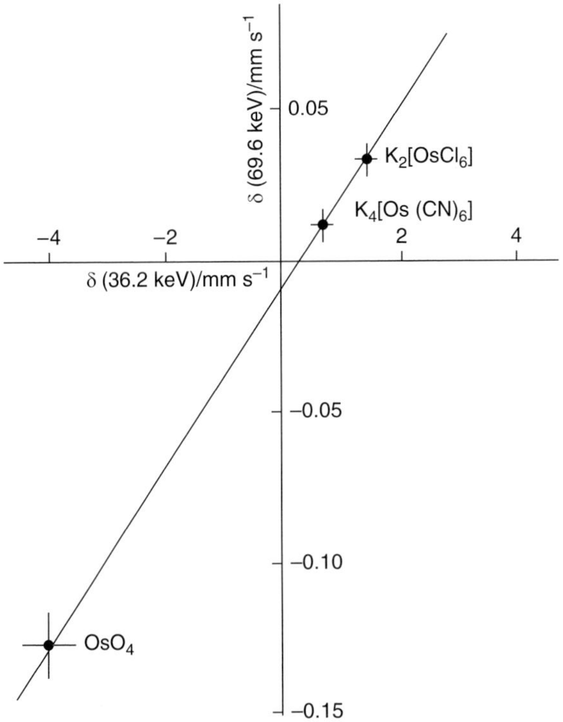

Fig. 7.55 Isomer shift $\delta = \delta_{exp} - \delta_{SOD}$ of the 69.6 keV Mössbauer transition versus the isomer shift of the 36.2 keV line in ^{189}Os (from [264])

$\langle r^2 \rangle_{69.6keV}$ by Mössbauer measurements with $K_2[OsCl_6]$, $K_4[Os(CN)_6]$ and OsO_4 as absorbers. Plotting isomer shifts of the 69.6 keV transition, corrected for the SOD shift, versus the isomer shift of the 36.2 keV transition, a straight line is obtained (Fig. 7.55). The slope of this line leads to the ratio of $\Delta\langle r^2 \rangle_{36.2}/\Delta\langle r^2 \rangle_{69.6} = 16.6 \pm 1.3$. Wagner et al. [258] (see Sect. 7.7.3) derived a value of $\Delta\langle r^2 \rangle_{36.2} = -2.0 \cdot 10^{-3}$ fm^2 from Mössbauer measurements on osmium compounds with Os in the oxidation states +8, +6, +4, +3, and +2, using the calculated electron densities from relativistic self-consistent field calculations for free Os ions. These calculations must be considered with care in the case of covalently bound Os atoms and the $\Delta\langle r^2 \rangle_{36.2}$ value should therefore be considered only as an estimate.

7.7.3 Inorganic Osmium Compounds

Wagner et al. [258] reported a systematic investigation of the isomer shift and quadrupole splitting in various osmium compounds. Of special interest is the comparison with similar or isoelectronic compounds of iridium and ruthenium.

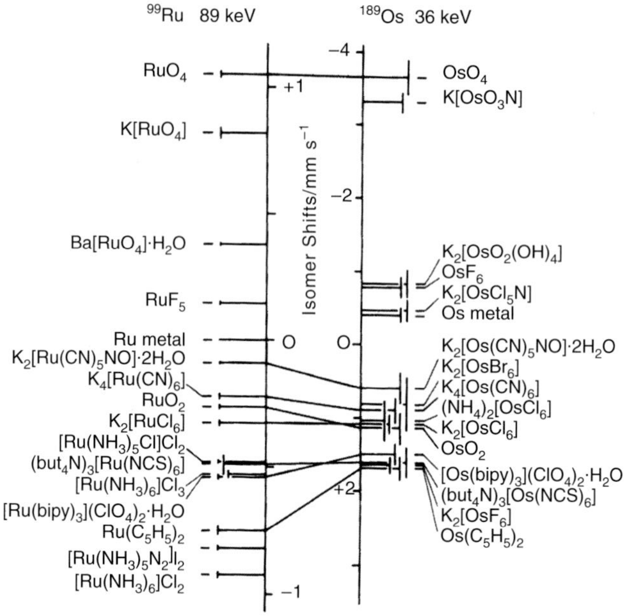

Fig. 7.56 Comparison of isomer shift results for compounds of Ru and Os. The velocity scales are chosen such that isoelectronic compounds of the two elements lie nearly on horizontal lines (from [258, 265])

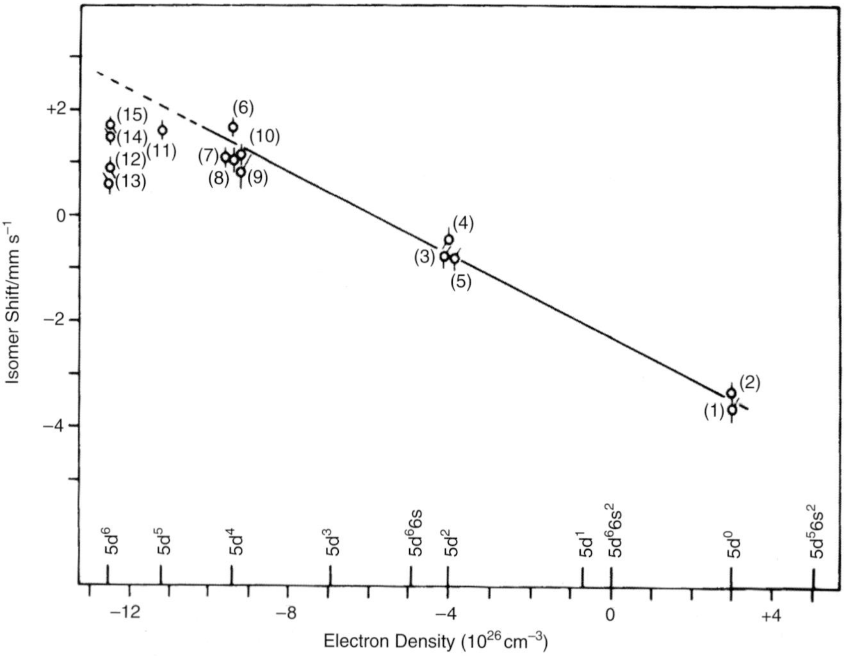

Fig. 7.57 Plot of the isomer shift δ of the 36.2 keV Mössbauer transition of ^{189}Os versus Dirac–Fock values for the electron density differences at the Os nuclei in free ion 5*d* configurations. The numbers of the data points refer to the numbering of the compounds in Table 7.9 (from [258])

Table 7.9 Summary of the results obtained from the Mössbauer spectra of the 36.2 and 69.6 keV γ-rays of ^{189}Os in absorbers of various osmium compounds containing d mg cm^{-2} of ^{189}Os. Γ_{exp} is the full experimental line width at half maximum, δ the isomer shift with respect to the source of ^{189}Ir metal. $\Delta E_Q = 1/4e^2qQ_{3/2}$ is the electric quadrupole coupling constant of the ^{189}Os ground state and V_{zz} the EFG at the Os nuclei as calculated from the ΔE_Q values with $Q_{3/2} = +(0.80 \pm 0.06)b$ (from [258])

	Compound and formal oxidation state		36.2 keV transition		δ (mm s^{-1})	69.6 keV transition		δ (mm s^{-1})	ΔE_Q (mm s^{-1})	V_{zz} (10^{18} V cm^{-2})
			d (mg cm^{-2})	Γ_{exp} (mm s^{-1})		d (mg cm^{-2})	Γ_{exp} (mm s^{-1})			
1.	OsO_4	+8	20	21.9 ± 1.7	−3.64 ± 0.24	43	3.07 ± 0.08	−0.109 ± 0.011	–	–
2.	$K[OsO_3N]$	+8	7	15.0 ± 0.7	−3.30 ± 0.15	24	2.84 ± 0.14	−0.154 ± 0.029	0.83 ± 0.03	0.96 ± 0.09
3.	OsF_6	+6	15	18.7 ± 0.9	−0.78 ± 0.16	40	2.82 ± 0.18	+0.054 ± 0.050	−1.37 ± 0.04	−1.58 ± 0.14
4.	$K_2[OsCl_5N]$	+6	5	14.2 ± 0.4	−0.47 ± 0.16	33	2.98 ± 0.10	+0.006 ± 0.021	+3.19 ± 0.04	+3.75 ± 0.31
5.	$K_2[OsO_2(OH)_4]$	+6	23	18.2 ± 1.3	−0.82 ± 0.19	23	2.62 ± 0.12	−0.009 ± 0.021	−1.39 ± 0.02	−1.61 ± 0.13
6.	$K_2[OsF_6]$	+4	8	16.9 ± 0.7	+1.64 ± 0.13	8	2.85 ± 0.14	+0.096 ± 0.042	–	–
7.	$K_2[OsCl_6]$	+4	21	19.0 ± 0.7	+1.08 ± 0.13	21	3.14 ± 0.18	+0.049 ± 0.005	–	–
8.	$(NH_4)_2[OsCl_6]$	+4	7	18.6 ± 1.5	+1.04 ± 0.27	16	3.15 ± 0.08	+0.043 ± 0.013	–	–
9.	$(NH_4)_2[OsBr_6]$	+4	5	15.2 ± 1.5	+0.83 ± 0.37	21	3.52 ± 0.08	+0.073 ± 0.020	–	–
10.	OsO_2	+4	11	16.4 ± 1.3	+1.13 ± 0.22	11	3.30 ± 0.10	+0.035 ± 0.020	−1.99 ± 0.04	−2.31 ± 0.19
11.	$(but_4N)_3[Os(NCS)_6]$	+3	11	19.9 ± 1.1	+1.59 ± 0.19	36	–	–	–	–
12.	$K_4[Os(CN)_6]$anhydr	+2	5	16.6 ± 0.9	+0.89 ± 0.17	27	2.86 ± 0.05	+0.026 ± 0.003	–	–
13.	$K_2[Os(CN)_5NO]\cdot H_2O$	+2	7	18.8 ± 1.5	+0.59 ± 0.21	20	6.20 ± 0.44	+0.066 ± 0.037	1.10 ± 0.10	1.27 ± 0.15
14.	$[Os(bipy)_3](ClO_4)_2\cdot H_2O$	+2	13	16.9 ± 0.7	+1.48 ± 0.14	21	3.36 ± 0.13	+0.077 ± 0.019	0.58 ± 0.04	0.67 ± 0.06
15.	$Os(C_5H_5)_2$	+2	9	15.4 ± 1.0	+1.67 ± 0.16	15	3.19 ± 0.06	0.184 ± 0.029	+1.53 ± 0.03	+1.77 ± 0.15
16.	Os metal		22	20.5 ± 0.6	−0.41 ± 0.10	65	4.30 ± 0.06	+0.008 ± 0.005	0.77 ± 0.03	0.89 ± 0.08

The appropriate Mössbauer transitions for the determination of the isomer shift and quadrupolar interaction are the 36.2 and 69.6 keV transitions, respectively, in ^{189}Os. A summary of the results from Wagner et al. [258] is given in Sect. 7.7.2. Magnetic hyperfine interaction has been neglected in the evaluation of the Mössbauer spectra, and only single Lorentzian lines or a pure quadrupole pattern has been adjusted to the experimental data points. In Fig. 7.56, the isomer shift of the 36.2 keV transition is compared with the shift found for the 89 keV Mössbauer resonance of ^{99}Ru. The velocity scale has been chosen such that isoelectronic compounds lie on nearly horizontal lines (e.g., RuO_4–OsO_4; $K_2[RuCl_6]$–$K_2[OsCl_6]$; $(but_4N)_3[Ru(NCS)_6]$–$(but_4N)_3[Os(NCS)_6]$). The agreement between both elements is good except for the compounds $A(C_5H_5)_2$ (A = Ru, Os). As in the cases of Fe, Ru, Ir, Pt, and Au, the authors found a monotonic dependence of the isomer shift for osmium compounds on the oxidation state (Fig. 7.57).

The change of the isomer shift of d^n configurations is mainly caused by the shielding of the *s*-electrons by the varying number of *d*-electrons. Only in those compounds where strong backbonding effects play a concomitant role strong deviations from this rule are expected and observed. In this case, d_π orbitals of the central ion and π-orbitals of the ligand build up the molecular orbital with a spatial spread of the *d*-electrons towards the ligands, thus decreasing the shielding capability of the *d*-electrons. This effect can be observed in the divalent Os compounds $K_4[Os(CN)_6]$ and $K_2[Os(CN)_5NO]\cdot 2H_2O$, where the Mössbauer isomer shift is close to the values observed for the tetravalent compounds OsO_2 or $K_2[OsCl_6]$.

7.8 Iridium (^{191,193}Ir)

There are two iridium isotopes, ^{191}Ir and ^{193}Ir, suitable for Mössbauer spectroscopy. Each of them possesses two nuclear transitions with which nuclear resonance absorption has been observed. Figure 7.58 (from [266]) shows the (simplified) nuclear decay schemes for both iridium Mössbauer isotopes; the Mössbauer transitions are marked therein with bold arrows. The relevant nuclear data known to date for the four Mössbauer transitions are collected in Table 7.1 at the end of the book.

It is a matter of historical interest that Mössbauer spectroscopy has its deepest root in the 129.4 keV transition line of ^{191}Ir, for which R.L. Mössbauer established recoilless nuclear resonance absorption for the first time while he was working on his thesis under Prof. Maier-Leibnitz at Heidelberg [267]. But this nuclear transition is, by far, not the easiest one among the four iridium Mössbauer transitions to use for solid-state applications; the 129 keV excited state is rather short-lived ($t_{1/2}$ = 90 ps) and consequently the line width is very broad. The 73 keV transition line of ^{193}Ir with the lowest transition energy and the narrowest natural line width (0.60 mm s^{-1}) fulfills best the practical requirements and therefore is, of all four iridium transitions, most often (in about 90% of all reports published on Ir Mössbauer spectroscopy) used in studying electronic structures, bond properties, and magnetism.

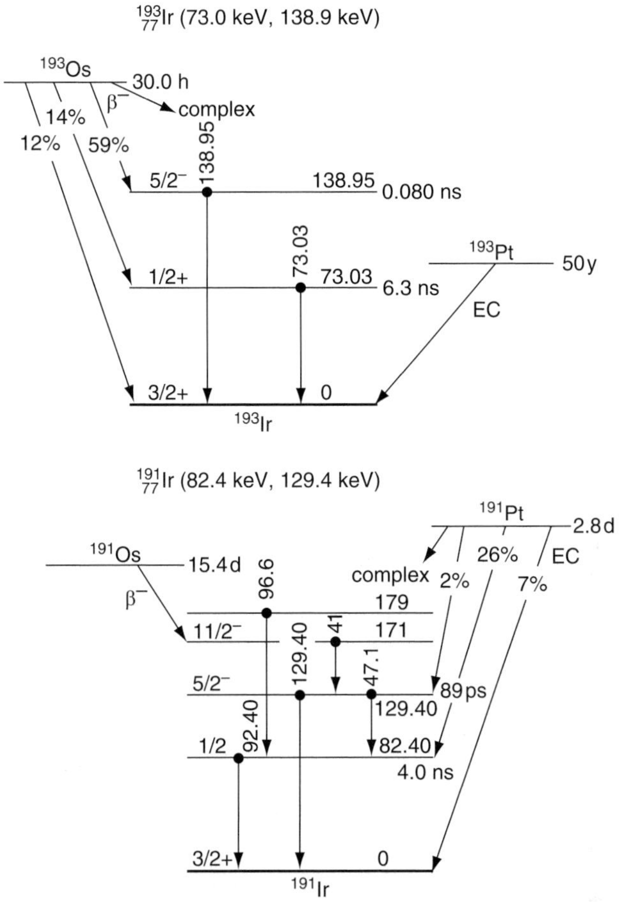

Fig. 7.58 Simplified decay scheme leading to the population of the four nuclear Mössbauer transitions of ^{191}Ir and ^{193}Ir (from [266])

7.8.1 *Practical Aspects of ^{193}Ir Mössbauer Spectroscopy*

The most popular precursor to feed the 73 keV level of ^{193}Ir is ^{193}Os ($t_{1/2} = 31$ h). This isotope is produced by neutron activation, ^{192}Os (n,γ)^{193}Os. Most laboratories use enriched (>95%) ^{192}Os metal as the target without further treatment after neutron irradiation. The only disadvantage in working with this source is the fact that osmium metal has hexagonal symmetry and therefore nonzero EFG at the Ir nucleus which causes an electric quadrupole splitting of the $I_g = 3/2$ ground state. As a consequence, the ^{193}Os/Os metal source emits an unresolved quadrupole doublet; the two overlapping emission lines have equal intensity and are separated by $\Delta E_Q = 0.48 \pm 0.02$ mm s^{-1} [268]. This splitting must be considered in the fitting procedures of measured spectra. Wagner and Zahn [268] have also tested an

$Os_{0.01}Pt_{0.99}$ alloy as source. Although it emits a single (unsplit) line, it is not as convenient to work with as the metallic osmium source, because the preparation involves induction melting of the activated ^{193}Os with platinum in argon atmosphere. Another single-line source for ^{193}Ir Mössbauer spectroscopy has been described by Davies et al. [269]. They used an $Os_{0.05}Nb_{0.95}$ alloy with 98.7% enriched ^{192}Os without any postirradiation annealing; the single emission line had a line width of 0.64 mm s^{-1} (4.2 K).

The ^{193}Ir Mössbauer experiments are usually carried out in transmission geometry with both source and absorber kept at liquid helium temperature and a Ge(Li) diode or a 3 mm NaI(Tl) crystal used to detect the 73 keV γ-rays. The absorbers typically contain 50–500 mg cm^{-2} of natural iridium, which contains 62.7% of the Mössbauer isotope ^{193}Ir. The isomer shifts are generally given with respect to iridium metal (the isomer shift between ^{193}Os/Os and Ir metal is $-(0.540 \pm 0.004)$ mm s^{-1} at 4.2 K ([268]).

7.8.2 *Coordination Compounds of Iridium*

After a first brief report by Thomson et al. [270] on isomer shifts of the 73 keV γ-rays of ^{193}Ir in Ir metal, IrO_2, $K_2[IrCl_6]$, and $IrCl_4$, it was only in 1967 that, at nearly the same period, the first extended articles on applications of ^{193}Ir Mössbauer spectroscopy to intermetallic and coordination compounds of iridium were published by an Israeli group [271] and the Munich laboratory [272, 273].

Atzmony et al. [271] have observed that the ^{193}Ir (73 keV) isomer shifts of the tetravalent Ir compounds IrO_2, $K_2[IrCl_6]$, $(NH_4)_2[IrCl_6]$, and $IrCl_4$ relative to the trivalent Ir compounds IrO_3, $K_3[IrCl_6]$, and $IrCl_3$ are positive. They assumed that in tetravalent Ir compounds ($[Xe]4f^{14}5d^5$ for Ir^{4+}) with one 5d electron less than in trivalent Ir compounds ($[Xe]4f^{14}5d^6$ for Ir^{3+}), the core-s electrons are less effectively shielded and that therefore the s-electron density $|\psi_s(o)|^2$ at the Ir nucleus in Ir^{4+} is larger than in the Ir^{3+} compounds. From this they concluded that the sign of $\Delta\langle r^2\rangle$, the difference of the mean-square radius of the charge distribution between the 73 keV excited state and the ground state of ^{193}Ir, is positive. An absolute value of $\Delta\langle r^2\rangle$ (^{193}Ir, 73 keV) was derived by Wagner et al. [273] from the isomer shifts measured with metallic sources of ^{191}Os(14.6 h), ^{191}Pt(3.0 d), and ^{193}Os(31 h), and Ir metal (with natual abundance of ^{193}Ir) as absorber.

Wagner and Zahn [268] have reported on an extensive study of a variety of hexacoordinated trivalent and tetravalent iridium compounds as well as on IrF_5 and IrF_6. Their observed isomer shifts (relative to Ir metal) are summarized in Fig. 7.59. The isomer shifts in halogen compounds of Ir slightly increase with the formal oxidation state of Ir, but deviations are observed for the complexes with CN^- and NO_2^- as ligands. The authors have interpreted their observations qualitatively on the basis of the shielding effect of the varying effective number of 5d electrons, paying attention to the fact that $5d^{n_{eff}}$ not only changes by altering the Ir oxidation state but also by backbonding (in complexes with CN^- and NO_2^- ligands). They have also pointed out the influence of varying σ-bond covalency (direct 6s contribution to $|\psi_s(o)|^2$ by σ-donation from the ligands) which accompanies both changes

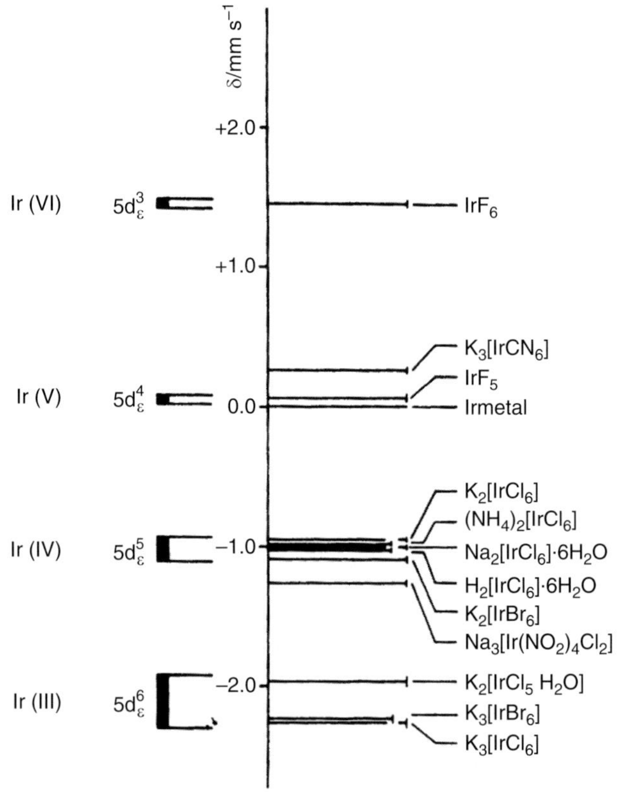

Fig. 7.59 ^{193}Ir (73 keV) isomer shifts as a function of oxidation state of iridium (from [268])

in the oxidation state of iridium and backdonation ($\pi_{M\to L}$) effects in order to fulfill the requirements of Pauling's electroneutrality principle. A quantitative interpretation of the isomer shifts in iridium compounds is only possible in connection with electronic structure calculations.

The compounds containing the hexahalogen complexes $[IrCl_6]^{2-}$, $[IrBr_6]^{2-}$ with Ir(IV) formal oxidation state ($5d^5$), and iridium hexafluoride with hexavalent iridium ($5d^3$) possess an odd number of electrons and have therefore Kramers degenerate electronic levels with an effective spin of $S = 1/2$ and $S = 3/2$ in the cases of $5d^5$ and $5d^3$, respectively. These compounds are, therefore, expected to order magnetically at sufficiently low temperatures, as indeed has been established by magnetic susceptibility measurements. The Néel temperatures occur around 2–3 K for the $5d^5$ compounds and around 8 K in IrF_6. Wagner and Zahn [268] have measured the Mössbauer hyperfine spectra of such compounds between 0.5 and 4.2 K. Some representative spectra of $K_2[IrCl_6]$, recorded with the single-line source $^{193}Os_{0.01}Pt_{0.99}$ for better resolution, are shown in Fig. 7.60. Using a single-line source, the magnetic hyperfine pattern is composed of eight lines – arising from the $M1/E2$ transitions between the sublevels of the excited and the ground nuclear states – as depicted in Fig. 7.61, for the general case of a combined magnetic dipole and electric quadrupole interaction. The spectra in Fig. 7.60 show a gradual decrease of the magnetic hyperfine interaction with increasing temperature, in favor of a single center line from the paramagnetic state of $K_2[IrCl_6]$, due to the gradual decrease of the effective magnetic field as a consequence of the thermally induced increase in electronic spin fluctuation rate. An example for a well-resolved ^{193}Ir (73 keV) magnetic hyperfine pattern is the spectrum of IrF_6 at 4.18 K, shown in Fig. 7.62, from the work of Wagner and Zahn [268].

Each of the eight hyperfine resonances is an unresolved quadrupole doublet, due to the quadrupole interaction of ^{193}Os in the hexagonal Os metal source. The authors have interpreted the hyperfine fields in terms of core polarization, orbital and spin-dipolar contributions.

In another paper from the Munich laboratory, Rother et al. [265] have reported on studies of chemical consequences of the β^- decay of ^{193}Os in osmium compounds by the recoilless resonance absorption of the 73 keV γ-rays of ^{193}Ir. They have investigated a number of ^{193}Os-labeled osmium compounds, $K_2[OsO_4]\cdot 2H_2O$, OsO_4, $K_2[OsCl_6]$, $K_2[OsBr_6]$, $(NH_4)_2[OsCl_6]$, $(NH_4)_2[OsBr_6]$, and $Os(C_5H_5)_2$, which they synthesized by standard methods either with active ^{193}Os or with inactive compounds with subsequent neutron activation. These Mössbauer sources were then measured at 4 K versus Ir metal as absorber. The prominent resonance lines of the emission spectra were assigned to different charge states on the basis of the measured isomer shift data and in comparison with the isomer shift/oxidation state scale for ^{193}Ir (73 keV) [268]. (Note that in the case of Mössbauer emission spectroscopy, for technical reasons, the isomer shift data have opposite sign as compared to those taken from absorption spectra.) The authors found that in the case of OsO_4, $K_2[OsO_4]\cdot 2H_2O$ and $Os(C_5H_5)_2$, the ^{193}Ir daughter species were isoelectronic with the parent compounds, which implies that species with iridium in the valence states +9 and +7 with isomer shifts of -3.78 and -2.14 mm s^{-1}, respectively (compare with Fig. 7.59), and $Ir(C_5H_5)_2$ might have been produced as a consequence of the β^- decay and have existed with lifetimes at least in the order of the lifetime of the ^{193}Ir (73 keV) excited state. In the case of the Os(IV) hexahalides, however, Ir(IV), Ir(III), and possibly Ir(II) species were observed, which the authors have explained in terms of fast electronic recombination. No major influence of the ^{192}Os(n, γ)^{193}Os reaction and of postirradiation annealing on the pattern of the emission spectra could be observed.

An interesting effect, called hyperfine anomaly, [1] has been observed by Wagner and Zahn [268] and by Perlow et al. [275] for the 73 keV transition of ^{193} Ir, which apparently has favorable characteristics to establish this phenomenon.

[1] The observed ratio of the nuclear magnetic moments μ_e/μ_g is generally taken to be constant for the nuclear states of a given Mössbauer transition. This, however, is not necessarily so, because there may be slight differences in the interactions between the nuclear magnetic moment (with radial distribution due to orbital and spin motions of unpaired nucleons) and an applied magnetic field (which is uniform over the nuclear magnetization distribution) on the one hand, and with an internal hyperfine field (which may also have radial distribution) on the other hand. *Such differences will be reflected in differences of the ratio* μ_e/μ_g observed for the two cases of interaction. This is called "hyperfine anomaly".

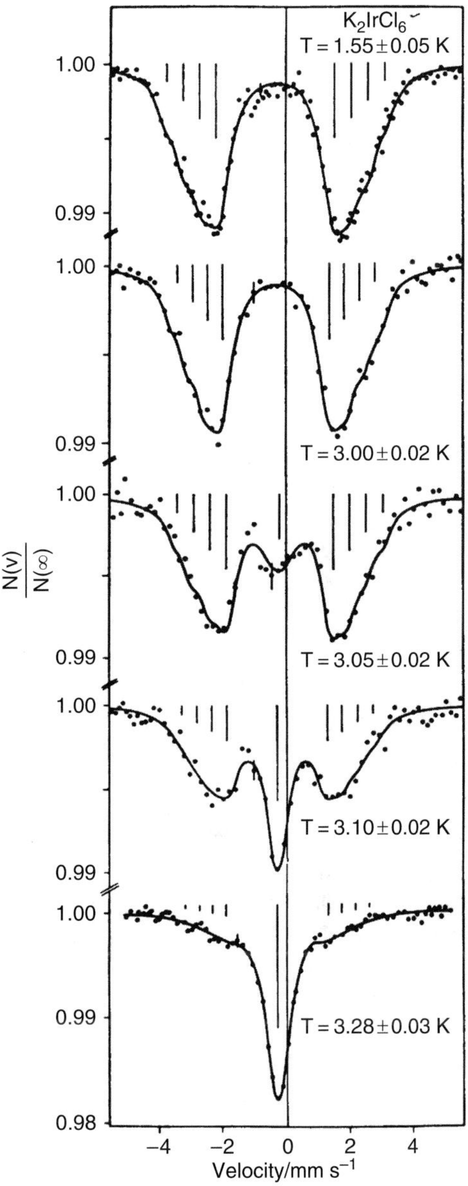

Fig. 7.60 Mössbauer spectra of $K_2[IrCl_6]$ at various temperatures, taken with the single-line source $^{193}Os_{0.01}Pt_{0.99}$. The positions and relative intensities of the resonance lines are indicated by *vertical bars* (from [268])

Wickman and Silverthorn [276] have investigated bond properties in molecular adducts of the planar Vaska type compound *trans-bis*(triphenyl-phosphine)iridium carbonyl chloride, $IrCl(CO)((C_6H_5)_3P)_2$, with small molecules such as H_2, O_2, Cl_2, I_2, CH_3I, and HCl. They essentially observed a decrease of the isomer shift in the following series of adduct molecules XY: $H_2 > HCl > CH_3J > O_2 > I_2 > Cl_2$,

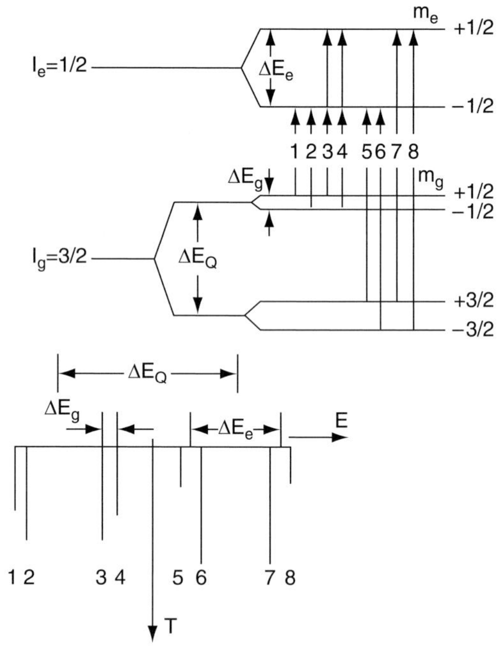

Fig. 7.61 Effect of combined magnetic dipole and electric quadrupole interaction on the nuclear excited (e) and ground (g) states of the 73 keV transition in ^{193}Ir. The relative intensities, given in the bar diagram, correspond to a *E2*/*M*1 mixing ratio of $\delta^2 = 0.340$ for random orientation (powder sample). ΔE_e = magnetic splitting of the excited state ($I_e = 1/2$), ΔE_g = magnetic splitting of the ground state ($I_g = 3/2$), ΔE_Q = electric quadrupole splitting. The relatively large magnetic splitting of the excited state reflects the fact that the *g*-factor of the 73 keV state is about nine times larger than that of the $3/2^+$ ground state (from [272])

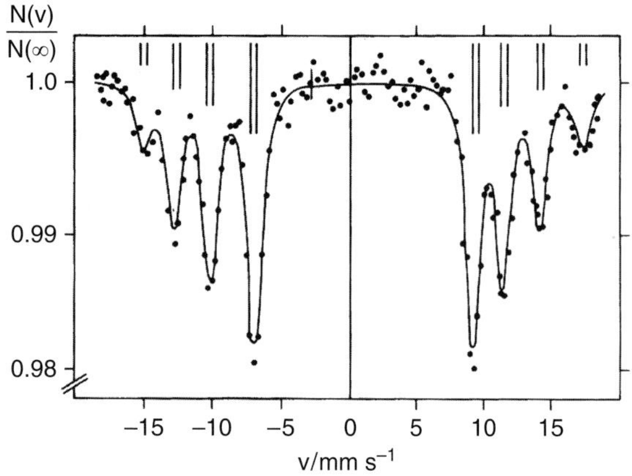

Fig. 7.62 Mössbauer spectrum of IrF_6 at 4.18 K, recorded with a ^{193}Os/Os source. The positions and relative intensities (in accordance with an *E2*/*M*1 mixing ratio of $\delta^2 = 0.311$ [274]) are indicated by the *vertical bars* (from [268])

which they explained by variations of the electron population in the d^2sp^3 hybrid orbitals owing to decreasing relative σ-donation power on going from H_2 to Cl_2 as the predominant effect. The π-backdonation from the filled t_{2g} orbitals (the adducts in question are formally Ir(III) systems) to these ligand molecules does not seem to contribute in this case. However, a π-effect is indicated by the observed trend in the Ir carbonyl frequencies. With the exception of the O_2 adduct, which differs from the others in that the O_2 bond is partially retained in forming the adduct, in the other ligand molecules XY of the above series, the bond is ruptured and the constituents form bonds only to the iridium atom [277], the carbonyl frequency increases on going from H_2 to Cl_2. This apparently implies that the tendency of charge delocalization from the filled t_{2g} orbitals to the π carbonyl orbital decreases in the same order of XY ligands, which in turn influences the electron density $|\psi_s(o)|^2$ at the iridium nucleus via shielding and thus alters the isomer shift in the same direction as the σ-donation effect. Both effects are superimposed; the isomer shift varies as a function of $\sigma(XY \rightarrow Ir)$–$\pi(Ir \rightarrow CO)$. It is assumed that the two effects are concomitant (synergic) to fulfill the requirements of the electroneutrality principle, e.g., H_2 with the strongest relative σ-donation power is expected to cause the largest loss of t_{2g} charge by backdonation into the π (CO) orbital.

In a similar study of a series of $((C_6H_5)_3P)_2Ir(CO)X$ complexes (X = Cl, Br, N_3, SCN; NO^+) by X-ray photoelectron and ^{193}Ir (73 keV) Mössbauer spectroscopy, Holsboer et al. [278] found that substitution of Cl^- by other halides and pseudohalides in the planar *trans*-$(PPh_3)_2Ir(CO)X$ complexes affects only slightly the total electron density and the EFG at the Ir nucleus. There is, however, a distinct influence of the variation of the anionic X^- ligands on the IR stretching frequency of the CO bond [279]. The changes in the electron density at the Ir nucleus caused by the variation of X are nearly compensated by the concomitant $\pi(Ir \rightarrow CO)$ effect. The quadrupole splittings for the $(PPh_3)_2Ir(CO)X$ complexes (X = Cl, Br, N_3, SCN) are among the largest ones observed in ^{193}Ir (73 keV) Mössbauer spectroscopy.

In another publication, Holsboer et al. [280] have investigated bond properties in the five-coordinated Ir(I) complexes $(PPh_3)_3Ir(CO)H$, $(PPh_3)_3Ir$ (CO)CN, and the tetracyanoethylene, fumaronitrile, and acrylonitrile adducts of $(PPh_3)_2Ir(CO)Cl$ by X-ray photoelectron and ^{193}Ir Mössbauer spectroscopy.

Ginsberg et al. [281] have applied ^{193}Ir (73 keV) Mössbauer spectroscopy and magnetic susceptibility measurements to the one-dimensional conductor iridium carbonyl chloride, $Ir(CO)_3Cl$. By carbon and chlorine analysis, they confirmed the earlier report by Krogmann et al. [282] that this compound is partially oxidized. However, the ^{193}Ir Mössbauer spectra measured in the temperature range from 1.8 to 35 K are not in accordance with Krogmann's suggested formulation of $Ir(CO)_{3-x}Cl_{1+x}$ ($x = 0.07$). They rather suggested the formulation of $Ir(CO)_3Cl_{1+x}$ ($x = 0.10 \pm 0.03$), because the ^{193}Ir (73 keV) Mössbauer spectra exhibit no line attributable to a species other than partially oxidized $Ir(CO)_3Cl$ units. Ginsberg et al. presumed that the "extra" chloride is localized in disordered interchain positions. The results from both susceptibility and Mössbauer measurements imply that the charges arising from partial oxidation of $Ir(CO)_3Cl$ chains are not localized on individual atoms.

Another paper dealing with ^{193}Ir Mössbauer studies of planar Ir(I) compounds with metal–metal bonds was published by Aderjan et al. [283]. They investigated chelato-dicarbonyl-iridium(I) compounds of composition $Ir(CO)_2L$ and $Ir(CO)_2L'Cl$ (L = singly charged bidentate organic ligand; L′ = neutral monodentate organic base) and found that the formation of columnar stacks (one-dimensional chains) with metal–metal bonds results in a reduction of both the magnitude of the quadrupole splitting and the isomer shift. They interpreted these findings in terms of a model which assumes that in the columnar stacks a wide

d_{z^2} band overlaps with a narrow energetically higher $d_{x^2-y^2}$ band, and that the partial filling of the antibonding $d_{x^2-y^2}$ band would then destabilize the bonds with σ-donating and π-accepting ligands.

Ginsberg et al. [284] have explored the dependence of Ir (73 keV) isomer shifts on the oxidation state in a set of closely related low-valent iridium carbonyl complexes, $(C_6H_5)_4As[Ir(CO)_2Cl_2]$, $Ir(CO)_3Cl_{1.1}$, $K_{0.6}[Ir(CO)_2Cl_2] \cdot 0.5H_2O$, and $[(C_6H_5)_4As]_2[Ir(CO)_4Cl_2]$ with formal iridium oxidation states + 1.0, + 1.1, + 1.4, and + 2, respectively. In contradiction to the generally observed increase in isomer shift with increasing oxidation state (cf. Fig. 7.59), the measured isomer shifts decrease with increasing formal oxidation state and are therefore opposite in direction to what is expected for decreased shielding by removal of 5*d* electrons. The authors assumed that a combination of two effects may account for the phenomenon: (1) the increase in electron density $|\psi_s(o)|^2$ expected from 5*d* electron removal will largely be canceled by an accompanying decrease in $\pi(5d, t_{2g}) \rightarrow \pi^*(CO)$ backbonding; (2) the electronic charge removed in going down the series from + 1.0 to + 2.0 will contain a significant 6*s* component. A more satisfactory answer to this problem could be obtained from electronic structure calculations.

A study of six-coordinate ammine complexes of iridium, $[Ir(NH_3)_5X]Y_n$ (X = I^-, Cl^-, $NO_2{}^-$, $HCOO^-$, CH_3COO^-, NCS^-, and NH_3; Y = Cl^-, $ClO_4{}^-$, of $(but_4N)_3$ $[Ir(SCN)_6]$, and of the *cis* and *trans* isomers of $[IrCl_4pyr_2]$, C_5H_5NH $[IrCl_4pyr_2]$, and $[IrCl_2pyr_4]Cl$ (pyr = pyridine) was carried out by Wagner et al. [285]. Some of their measured spectra are shown in Fig. 7.63. The spectra are more or less resolved quadrupole doublets with each line being an unresolved doublet by itself due to the quadrupole splitting in the source. The ratio of the quadrupole splittings for the *trans* and *cis* isomers of the pyridine complexes was found to be about 2:1 (cf. upper two spectra of Fig. 7.63), as has been predicted on the basis of a point charge model [286, 287] and also observed, e.g., in low-spin iron(II) complexes [286]. The isomer shift tendencies are understood in terms of varying powers of σ-donation and π-backbonding. The authors have attempted to interpret the isomer shifts in the framework of the angular overlap model [288–290], which enables one to make separate estimates of the strength of σ and π bonds.

Two other publications on ^{193}Ir (73 keV) Mössbauer spectroscopy of complex compounds of iridium have been reported by Williams et al. [291, 292]. In their first article [291], they have shown that the “additive model” suggested by Bancroft [293] does not account satisfactorily for the partial isomer shift and partial quadrupole splitting in Ir(III) complexes. Their second article [292] deals with four-coordinate formally Ir(I) complexes. They observed, like other authors on similar low-valent iridium compounds [284], only small differences in the isomer shifts, which they attributed to the interaction between the metal–ligand bonds leading to compensation effects. Their interpretation is supported by changes in the ^{31}P NMR data of the phosphine ligands and in the frequency of the carbonyl stretching vibration.

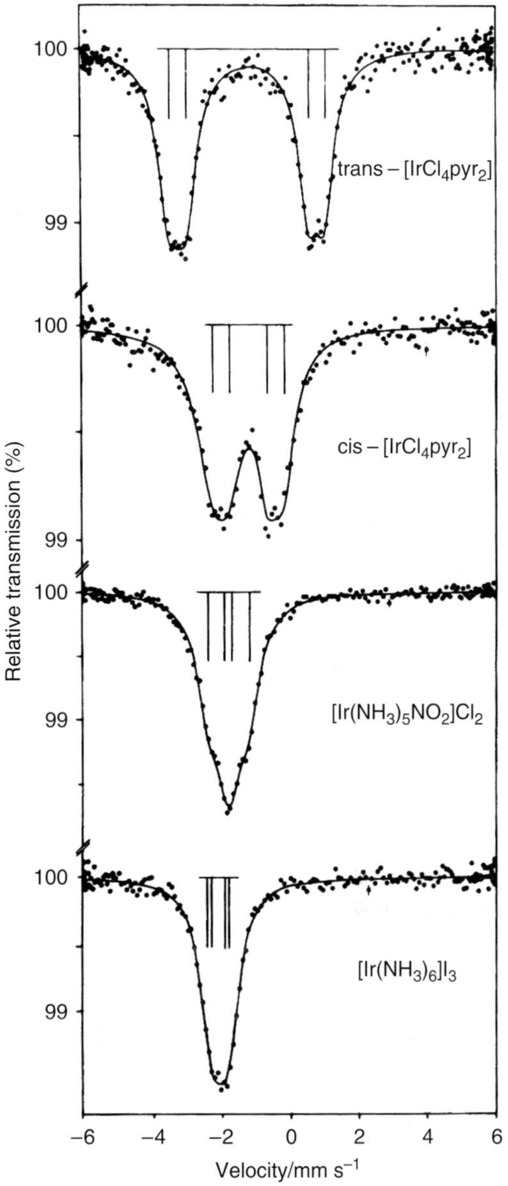

Fig. 7.63 Mössbauer spectra of some hexacoordinated ammine and pyridine complexes of trivalent iridium taken at 4.2 K with a source of ^{193}Os in Os metal. The stick spectra indicate the positions and relative intensities of the individual resonance lines (from [285])

7.8.3 Intermetallic Compounds and Alloys of Iridium

Heuberger et al. [272] and an Israeli group [271] have, independently from each other and at nearly the same period, measured the hyperfine splitting and the isomer shift of the ^{193}Ir (73 keV) Mössbauer transition in intermetallic compounds of iridium with rare earths, RIr_2. The systems studied by the former group included R = Sm, Gd, Tb, Dy, Ho, and Er; those of the latter research group contained R = Nd, Pr, Sm, Gd, Tb, Dy, and Ho. The intermetallic compounds RIr_2 belong to the family of fcc Laves phases. They are magnetically ordered with varying Curie temperatures below 100 K. Above the Curie temperature, T_C, these phases exhibit a quadrupole doublet resulting from a quadrupole interaction due to the nonzero EFG at the noncubic Ir site. Below T_C, the resonances are generally split into four lines by both electric quadrupole and magnetic hyperfine interaction, and into two broad lines only, if the electric quadrupole interaction is practically zero; the magnetic splitting of the $I = 3/2$ ground state has never been resolved in these systems. The hyperfine structure of the ^{193}Ir (73 keV) transition of mixed $E2/M1$ multipolarity is depicted in Fig. 7.61, together with a bar spectrum indicating the relative positions and intensities of the eight ($E2 + M1$) allowed transitions. Typical spectra of $TbIr_2$ with pure quadrupole interaction (above $T_C = 45$ K) and combined quadrupole and magnetic hyperfine interaction (at $T < T_C$) are shown in Fig. 7.64.

The main object of these studies was to investigate the dependence of the effective magnetic field acting on the nucleus of the nonmagnetic Ir atom on the properties of the rare earth constituents, and to learn the mechanism of the formation of the hyperfine fields at the Ir nuclei induced by the magnetic rare earth neighbor atoms. Essentially, it has been found that the saturation field at the ^{193}Ir nucleus in RIr_2 is a linear function of $(g - 1)J$, i.e., of the projection of the spin of the rare earth atoms onto their total angular momentum. This, in turn, implies that the induced fields predominantly originate from a spin polarization of the s conduction electrons by their interactions with the 4f electrons of the R atoms (s–f interaction). The polarized conduction electrons are believed to contribute to the effective field H_{eff} at the Ir nucleus mainly through two mechanisms: (1) direct contribution through the Fermi contact interaction and (2) indirect contribution of the non s conduction electrons polarized by the s conduction electrons and of the core electrons of the iridium atom polarized by overlap with the polarized conduction electrons. These effects lead to polarized core s-electrons, which build up the field contributions at the nucleus by Fermi contact interaction. Both mechanisms (1) and (2) depend on the spin of the rare earth atoms and on the number of s conduction electrons, provided the Kasuya–Yosida theory [288] on conduction electron polarization is applied. This was confirmed in the studies of the RIr_2 systems by both research groups.

The same Israeli group [271] has also measured the magnetic hyperfine interaction in the alloy $Ir_{0.01}Fe_{0.99}$. The spectrum taken at 4.2 K with a ^{193}Os/Os metal source looks very similar to the spectrum of IrF_6 shown in Fig. 7.62. The supertransferred field at the nuclear site of the "nonmagnetic" iridium induced by the magnetic iron atoms of the host lattice is so large that all eight ($E2$ + M1) allowed transitions (cf. Fig. 7.61) are observed and the magnetic splitting of the ground state is well resolved. Slight broadening of the eight resonance lines is due to the quadrupole interaction intrinsic to the Os metal source.

Mössbauer et al. [294] studied Ir–Fe and Ir–Ni alloy systems over the whole composition range by means of ^{193}Ir (73 keV) and ^{57}Fe nuclear resonance

Fig. 7.64 ^{193}Ir (73 keV) Mössbauer spectra of $TbIr_2$ at various temperatures, taken with a ^{193}Os/Os source kept at the absorber temperature (from [272])

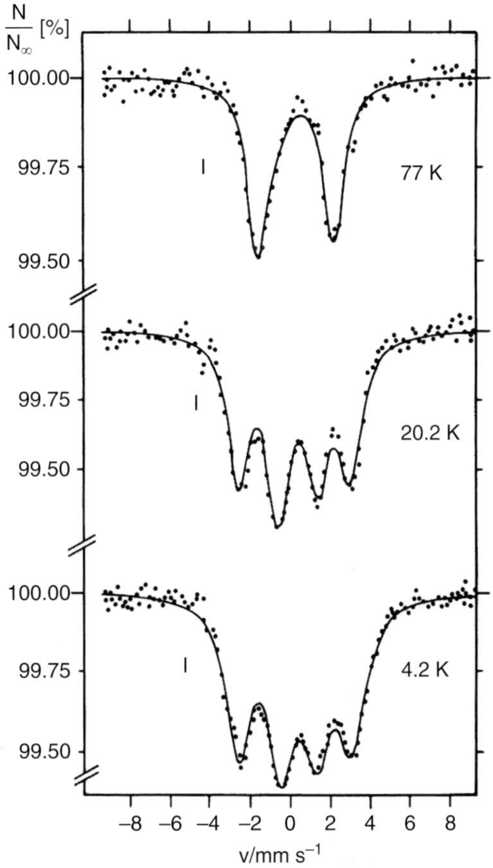

absorption. They found that the magnetic hyperfine field at the Ir nuclei in Ir–Ni alloys decreases approximately linearly with the Ir concentration from −46.0 T at 4.2 K in very dilute alloys to zero at about 20 at% Ir. Some representative ^{193}Ir (73 keV) spectra from their work are shown in Fig. 7.65; they demonstrate the gradual disappearance of the magnetic hyperfine pattern with increasing Ir concentration in favor of a central absorption line characteristic of nonmagnetic Ir metal. The concentration dependence of the hyperfine fields has been discussed in terms of a rigid 3*d* band model combined with local shielding. It is suggested that the conduction electron polarization is the primary source of the magnetic hyperfine fields at dilute Ir impurities in the ferromagnetic host lattices; this view is supported by the observation that the hyperfine fields are approximately proportional to the host magnetic moment.

Vogl et al. [295] have employed the Mössbauer effect in ^{193}Ir (73 keV) and ^{197}Au to study radiation damage in α-iron which was doped with 0.6 at% Os and 4 at% Pt,

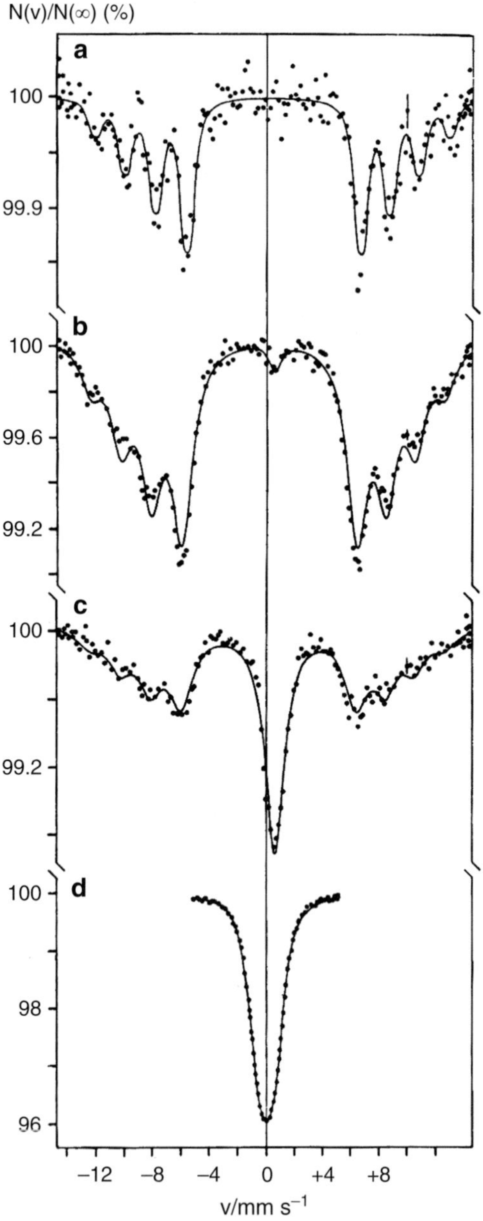

Fig. 7.65 ^{193}Ir (73 keV) Mössbauer spectra of Ir_xFe_{1-x} alloys containing $x = 0.7$ at% (**a**), 13.5 at% (**b**), 16.2 at% (**c**), and 53.6 at% (**d**) iridum. The spectra were recorded at 4.2 K using a ^{193}Ir/Os metal source (from [294])

respectively, and irradiated in a nuclear reactor at 4.6 K. Thermal neutron capture produces the precursor nuclides ^{193}Os and ^{197}Pt, respectively, for the Mössbauer isotopes ^{193}Ir and ^{197}Au and could lead to radiation damage. The influence of annealing on the average magnetic field and on line broadening was studied.

Salomon and Shirley [296] have performed Mössbauer-source experiments on Fe(^{193}Os) alloys with less than 0.1 at% osmium against a metallic Ir absorber. Their spectra were well-resolved magnetic eight-line patterns, which the authors could satisfactorily analyze using the Hamiltonian

$$\widehat{H} = -g\mu_N B_z \widehat{I}_z + \frac{eV_{zz}Q}{4I(2I-1)}[3\widehat{I}_z^2 - I(I+1)].$$

Their results show that quadrupole coupling persists in domains at high dilution, for which a reasonably accurate magnitude and the sign may be determined, even in more concentrated systems and even in the presence of substantial

Table 7.10 ^{193}Ir Mössbauer effect studies on various iridium compounds and alloys

References	System/material under study	Remarks
[298]	Review article	Excellent source of information about Mössbauer spectroscopy of iridium compounds and alloys
[299]	^{193}Ir/transition metals	Description of a new model (Atomic cell model) for the interpretation of isomer shift values, with electronegativity and cell boundary electron density as parameters
[300]	^{193}Ir in Tb, Dy, Ho, Er	Magnetic hyperfine field and EFG acting on dilute ^{193}Ir in ferromagnetic RE metals (Gd, Tb, Dy, Ho and Er) were measured at 4.2 K. The EFG remains practically constant, while the hyperfine field decreases strongly on going from Gd to Er, more strongly than expected from the spin polarization of the host metals, which is possibly due to nonzero angular momentum of the host ions
[301]	Fe/Ir catalysts on silica and alumina	^{57}Fe and ^{193}Ir Mössbauer spectroscopy: silica- and alumina-supported Fe–Ir catalysts formed by calcination in air contain mixtures of small particles of Fe(III) oxide and Ir(IV) oxide. IrO_2 is reduced in hydrogen to metallic Ir. α-Fe_2O_3 on SiO_2 is reduced in hydrogen to an Fe–Ir alloy, whilst supported on alumina stabilizes in hydrogen as Fe(II). Possible use for methanol formation is discussed
[302]	$Pt_{1-x}Ir_x$ catalysts	The chemical form of Ir in bimetallic Pt–Ir catalysts supported on amorphous silica has been determined after coexchange, calcinations and reduction in hydrogen. The composition of the highly dispersed bimetallic $Pt_{1-x}Ir_x$ clusters as determined from the measured isomer shifts reveal a strong tendency for segregation of iridium and platinum
[303, 304]	Fe/Ir catalysts	In situ ^{57}Fe and ^{193}Ir Mössbauer spectroscopy of silica-supported Fe/Ir catalysts with different iron to iridium ratios following pretreatment in hydrogen show that the reduction of the Fe component is enhanced by the presence of Ir metal. The presence of Ir was found to increase the catalytic activity in hydrogenation of carbon monoxide and also to influence selectivity

(*continued*)

Table 7.10 (continued)

References	System/material under study	Remarks
[305]	Pt/Ir catalysts	The chemical state of Ir in Pt/Ir catalysts prepared by impregnation of amorphous silica with H_2IrCl_6 and H_2PtCl_6 was studied by ^{193}Ir Mössbauer spectroscopy after different steps of preparation treatment. Ir is adsorbed in its trivalent state, presumably as $[IrCl_6]^{3-}$. Calcinaion in air at 450°C converts this to IrO_2. Metallic clusters formed by subsequent reduction in hydrogen at 200°C show a strong tendency towards segregation of Ir and Pt and reoxidize partially when exposed to air at ambient temperature (see Fig. 7.66)
[306]	Hydrides of Nb_3M (M = Au, Ir, Sn)	The hydrides with A15 structure were prepared at hydrogen pressures up to 7 GPa and studied by ^{197}Au, ^{193}Ir, and ^{119}Sn Mössbauer spectroscopy. Hydrogenation was found to lead to a reduction of the electron density at the Mössbauer nuclei, the decrease for ^{119}Sn was much smaller than for ^{197}Au and ^{193}Ir. This suggests that in Nb_3Au and Nb_3Ir the hydrogen occupies interstitial sites which are much closer to the metal atoms than in Nb_3Sn
[307]	Iridium(I), iridium(III) and iridium(I)/gold (I) complexes	Comparison of the ^{193}Ir and ^{197}Au Mössbauer spectra of the series *trans*-$(Ph_3P)_2Ir(CO)X$, *trans,cis*-$(Ph_3P)_2(H)_2Ir(CO)(X)$ (X = Cl or pz-N; pzH = 3,5-dimethyl-, 3,5-dimethyl-4-nitro-, 3,5-*bis* (trifluoromethyl)pyrazole) and *trans*-$(Ph_3P)_2(CO)Ir[\mu$-(3,5-dimethylpyrazolato-*N,N′*)]AuX′ (X′ = Cl, Br) shows that the substituents on the heterocycle influence the electron density at the iridium nucleus and that the bridging pyrazolato ligand transmits electronic effects from gold to iridium through three bonds
[308]	^{193}Ir CEMS	Conversion electron Mossbauer spectra of 73.0 keV γ-rays from iridium ^{193}Ir were obtained at liquid helium temperature by detecting low-energy electrons emitted from the resonant iridium absorber with a channel electron multiplier. The CEMS technique analyzes iridium within a depth of about 1 μm from the surface of materials, with a sensitivity several times greater than in conventional Mössbauer transmission measurements. A resonant effect of up to 10% has been observed for thin metallic iridium and IrO_2 films having natural isotopic composition (see Fig. 7.67)
[309] [310]	$DyIr_2Si_2$ Er Ir_2Si_2	^{161}Dy, ^{166}Er, and ^{193}Ir Mössbauer spectroscopy, magnetic measurements: The intermetallic compounds were shown to order antiferromagnetically at 40 K (Dy) and 10 K (Er) to exhibit metamagnetic-like behavior. No transferred field was observed at the ^{193}Ir nuclei
[311]	Ba_2CaIrO_6	For the first time Ir(VI) was stabilized in an octahedral site of the perovskite lattice Ba_2CaIrO_6l as confirmed by chemical analysis, structural, magnetic and ^{193}Ir Mössbauer measurements (see Fig. 7.68)
[312–314]	$(IrCl(CO)(PPh_3)_2)$ and adducts with TCNE and C_{60}	^{193}Ir Mössbauer spectra were recorded on chloro (carbonyl)-*bis*(triphenyl-phosphine) iridium $(IrCl(CO)(PPh_3)_2)$, (CCTI) and on its adducts with

(*continued*)

Table 7.10 (continued)

References	System/material under study	Remarks
		tetracyanoethylene (TCNE) and buckminsterfullerene, C_{60}. The Mössbauer measurements confirm that C_{60} acts as a π-acceptor, but has a weaker acceptor capability than TCNE
[314, 315]	$Ir_2Cl_2(COD)_2$ and adduct with C_{60}	Comparison with adducts of [312]
[316]	$CuIr_2(S_{1-x}Se_x)$	The thiospinel $CuIr_2S_4$ exhibits a temperature-induced metal–insulator (M–I) transition around 226 K with hysteresis, that manifests itself as a gap in the electronic density of states with increasing electrical resistivity at low temperatures. Conversely, $CuIr_2Se_4$ remains metallic down to 0.5 K. The effect of substitution of S for Se on the structural, electrical, and magnetic properties of $CuIr_2(S_{1-x}Se_x)$ was studied. Mössbauer measurements of ^{193}Ir have been performed for $CuIr_2S_4$ and $CuIr_2Se_4$. The M–I transition of $CuIr_2(S_{1-x}Se_x)$ for $x \leq 0.15$ is accompanied by a structural transformation from tetragonal (low-temperature insulating phase) to cubic (high-temperature metallic phase) symmetry. With increasing Se concentration x, the sharp M–I transition shifts to lower temperature. There is a general trend toward increasing metallicity with increasing x, which is consistent with the magnetic susceptibility results
[317, 318]	Fe–Ir/MgO catalysts	Fe–Ir/MgO catalysts derived from the $[Et_4N]_2[Fe_2Ir_2(CO)_{12}]$ cluster precursor, which exhibit a high activity in the synthesis of methanol from CO and H_2, were studied by ^{193}Ir and ^{57}Fe Mössbauer spectroscopy. The study extends from the precursors via the fresh to the aged catalysts. The presence of iridium in the metallic state as well as the presence of trivalent, divalent and alloyed iron is detected. Representative ^{193}Ir and ^{57}Fe Mössbauer spectra are shown in Fig. 7.69. Information about the adsorption on the surface of MgO
[319]	Pt–Ir/silica catalysts, Fe–Ir/MgO catalysts review	Information on the chemical state of iridium on going from the molecular precursors, and its adsorption on the surface of the support can be obtained by ^{193}Ir Mössbauer spectroscopy. It allows to estimate the composition of the Ir-containing alloys that are possibly formed during the activation treatment of supported bimetallic systems. The main results obtained in the application of ^{193}Ir Mössbauer spectroscopy to characterize two Ir-containing bimetallic supported nanoparticles, i.e., Pt–Ir on amorphous silica and Fe–Ir on magnesia are presented and discussed
[320]	Ir(I) complexes with fullerene ligands	^{193}Ir Mössbauer, ^{31}P-NMR, and magnetic measurements, and DFT calculations. The Ir(I) complexes have strong covalent character due to direct donation of the ligands into the 6*s* orbital of Ir. Coordination of further

(*continued*)

Table 7.10 (continued)

References	System/material under study	Remarks
		π-acceptor ligands decreases the population of the 6s orbital. First attempt to obtain Mössbauer parameters of ^{193}Ir by DFT calculations. Magnetization measurements proved that Ir(I) has a diamagnetic, singlet electronic structure in all the studied compounds. This finding was in accordance with DFT calculations
[321]	$(PPh_4)[Fe_2Ir_2(CO)_{12}\{\mu_3\text{-}Au(PPh_3)\}]$	The mixed-metal cluster compound $(PPh_4)[Fe_2Ir_2(CO)_{12}\{\mu_3\text{-}Au(PPh_3)\}]$, which has a trigonal bipyramidal core consisting of five atoms of three different Mossbauer isotopes, was studied by ^{193}Ir, ^{197}Au and ^{57}Fe Mössbauer spectroscopy. The nature and the chemical character of the atoms located at the different sites are discussed with respect to their Mössbauer spectra

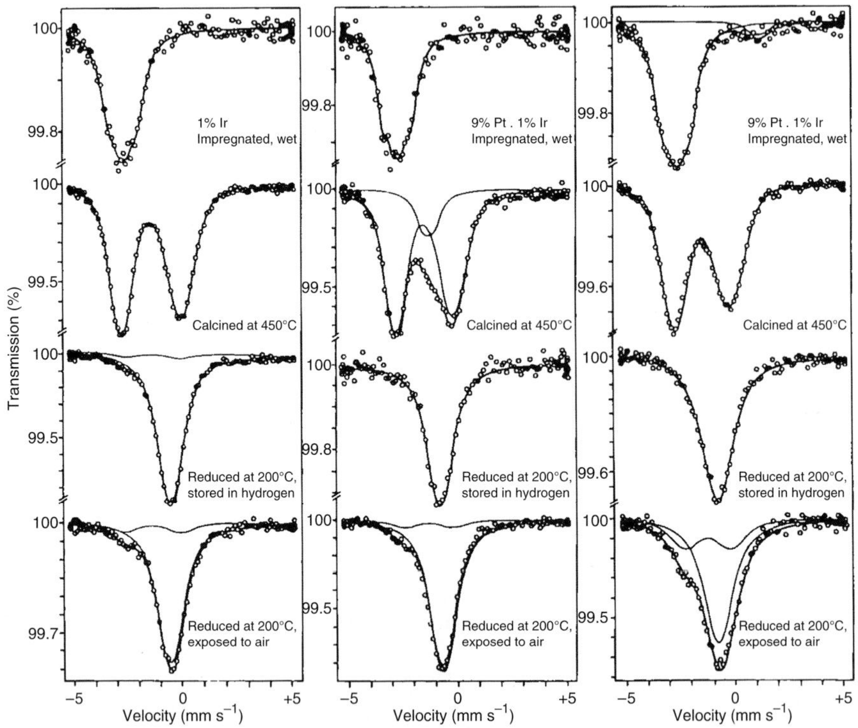

Fig. 7.66 ^{193}Ir Mössbauer spectra of amorphous silica impregnated with H_2IrCl_6 and H_2PtCl_6 and treated in different ways. The samples contain 1% Ir (*left*) and 9% Pt + 1% Ir (*middle and right*). The spectra on the *right* represent the specimen upon which the Pt was calcined before the Ir was deposited (from [304])

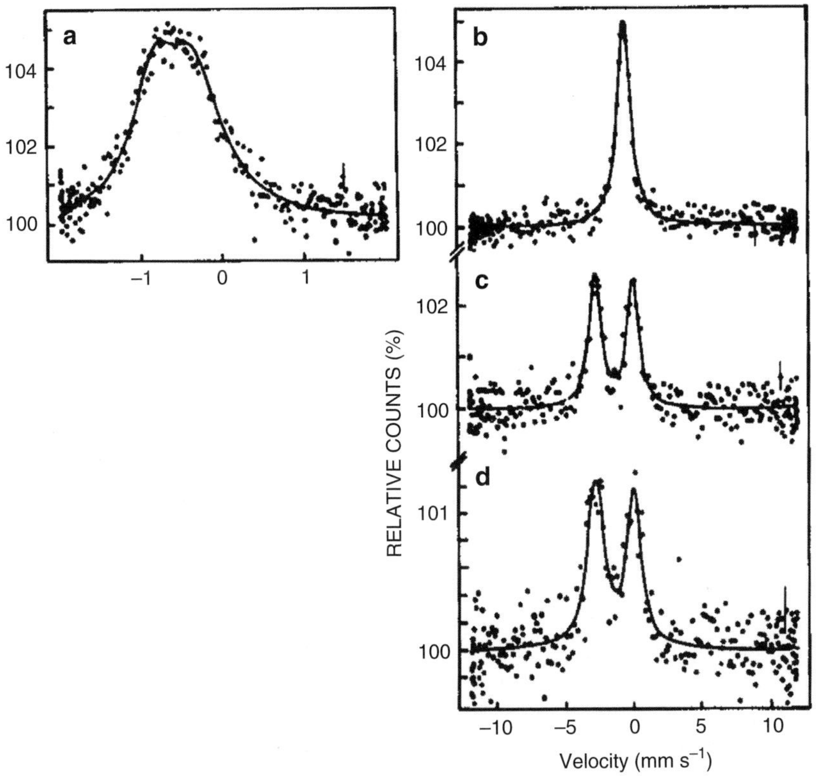

Fig. 7.67 Conversion electron Mössbauer spectra of the 73 keV γ-rays in ^{193}Ir recored at 4.2 K using a ^{193}Os metal source. (**a**, **b**) metallic iridium, 5 mg cm^{-2}, (**c**, **d**) iridium dioxide, 5 and 1 mg cm^{-2}, respectively. Measuring time about 10 h for spectra **a**–**c** and 20 h for spectrum **d** (from [308])

solute–solute-induced magnetic and quadrupole perturbations. The derived magnetic hyperfine field at Ir was 148.1 T.

A Japanese group [297] measured the hyperfine field at Ir nuclei in the alloys $Fe_{0.7}Pt_{0.3-x}Ir_x$ ($0.03 \le x \le 0.2$) using the ^{193}Ir (73 keV) Mössbauer transition, in order to understand the mechanism associated with the decrease of the magnetic moment (and negative hyperfine field) in these alloys. The isomer shift was found to increase with increasing x, which could be rationalized by the contraction of the lattice volume.

7.8.4 Recent ^{193}Ir Mössbauer Studies

The early Mössbauer studies using (mostly) the ^{193}Ir (73 keV) transition up to about 1983 has been reviewed in an article by Wagner [298]. The author discusses all aspects of practical Mössbauer spectroscopy with this iridium probe, such as

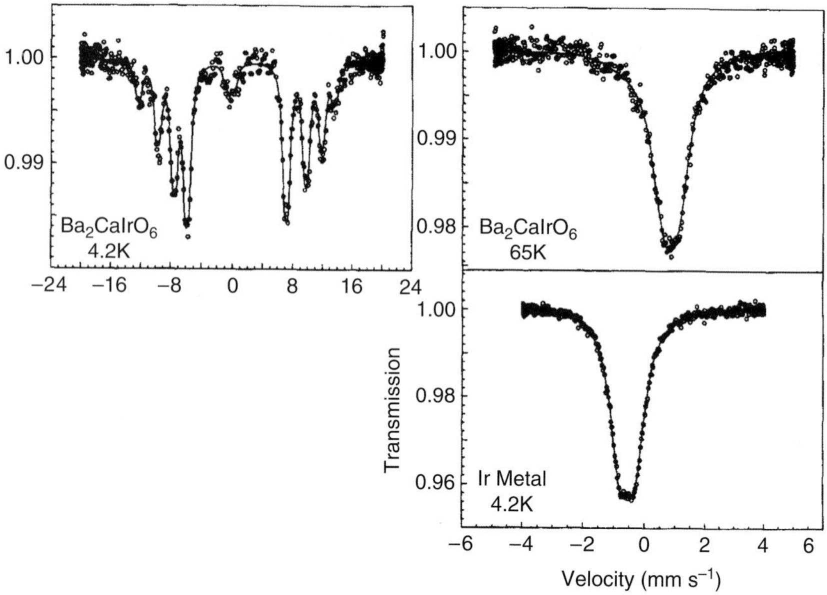

Fig. 7.68 ^{193}Ir Mössbauer spectra of Ba_2CaIrO_6 at 4.2 and 65 K (notice the different velocity scale) and of Ir metal (the isomer shift reference) at 4.2 K (from [310])

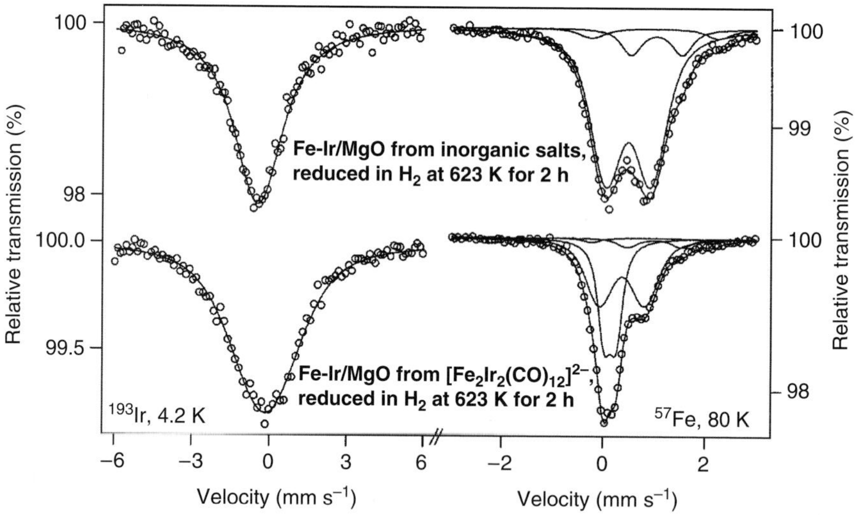

Fig. 7.69 Representative ^{193}Ir and ^{57}Fe Mössbauer spectra of Fe–Ir/MgO catalysts (from [317])

preparation of sources, spectrometer requirements, compilation of relevant data for all four Mössbauer transitions, nuclear properties, multipole mixing ratios and their implication for the line intensities, electric quadrupole moments, changes of nuclear charge radii, magnetic moments and magnetic anomaly, magnetic hyperfine data and isomer shifts of a large variety of iridium alloys and compounds, scattering experiments, to name a few.

Nearly all Mössbauer measurements on iridium compounds and alloys have been carried out with the ^{193}Ir (73 keV) and not with the ^{191}Ir (129 keV) transition, though R.L. Mössbauer originally discovered the nuclear resonance phenomenon with the latter nuclide. An overview of noteworthy publications, which were published after 1978, when the first edition of this book appeared, is collected in Table 7.10.

7.9 Platinum (^{195}Pt)

Mössbauer spectroscopy with ^{195}Pt started only in 1965, when Harris et al. [322] measured the Mössbauer absorption spectra of the 99 keV transition of ^{195}Pt in platinum metal as a function of temperature (between 20 and 100 K) and of absorber thickness and derived the temperature dependence of the Debye–Waller factor.

There are two γ-transitions in ^{195}Pt amenable to the Mössbauer effect – the 130 keV transition between the $5/2^-$ excited state and the $1/2^-$ ground state and the 99 keV transition between the first excited $3/2^-$ state and the ground state. Figure 7.70 shows the simplified decay scheme of ^{195}Pt. The relevant nuclear data may be taken from Table 7.1 (at the end of the book).

It is much more difficult to observe the Mössbauer effect with the 130 keV transition than with the 99 keV transition because of the relatively high transition energy and the low transition probability of 130 keV transition, and thus the small cross section for resonance absorption. Therefore, most of the Mössbauer work with ^{195}Pt, published so far, has been performed using the 99 keV transition. Unfortunately, its line width is about five times larger than that of the 130 keV transition, and hyperfine interactions in most cases are poorly resolved. However, isomer shifts in the order of one-tenth of the line width and magnetic dipole interaction, which manifests itself only in line broadening, may be extracted reliably from ^{195}Pt (99 keV) spectra.

7.9.1 Experimental Aspects

The two Mössbauer levels of ^{195}Pt, 99 keV and 130 keV, are populated by either EC of ^{195}Au($t_{1/2}$ = 183 days) or isomeric transition of ^{195m}Pt($t_{1/2}$ = 4.1 days). Only a few authors, e.g., [323, 324] reported on the use of ^{195m}Pt, which is produced by thermal neutron activation of ^{194}Pt via ^{194}Pt(n, γ)^{195m}Pt. The source used in the early measurements by Harris et al. [322, 325] was carrier-free ^{195}Au diffused into platinum metal. Walcher [326] irradiated natural platinum metal with deuterons to obtain the parent nuclide ^{195}Au by (d, xn) reactions. After the decay of short-lived isotopes, especially ^{196}Au($t_{1/2}$ = 6.18 days), ^{195}Au was extracted with ethyl acetate, and the ^{195}Au/Pt source prepared by induction melting. Buyrn and Grodzins [323] made use of (α, xn) reactions when bombarding natural iridium with

30 MeV α-particles and used the ^{195}Au/Ir source after annealing without any further chemical or physical treatment. Commercially available sources are produced via ^{195}Pt(p, n) ^{195}Au. The most popular source matrix into which ^{195}Au is diffused is platinum metal although it has the disadvantage of being a resonant matrix – natural platinum contains 33.6% of ^{195}Pt. Using copper and iridium foils as host matrices for the ^{195}Au parent nuclide, Buyrn et al. [327] observed natural line widths and reasonable resonance absorption of a few percent at 4.2 K.

^{195}Pt (99 keV) Mössbauer spectra have been recorded at temperatures ranging from 4 K to about 100 K for source and absorber. There is an increase in the percentage resonance effect of about one order of magnitude on going from 77 K to 20 K (cf. Fig. 3 in [322]).

Typical absorbers contain 50–700 mg cm^{-2} of natural platinum. The observed experimental line widths in ^{195}Pt (99 keV) spectra range from values close to the natural width ($2\Gamma_{nat}$ (99 keV) = 16.28 mm s^{-1}) to ~25 mm s^{-1}. With respect to the line width, the 130 keV transition with a natural width of $2\Gamma_{nat}$ (130 keV) = 3.40 mm s^{-1} seems to be more favorable for the study of hyperfine interaction in platinum compounds; in practice, experimental line widths of 3.4 ± 0.4 [328] and 3.5 ±0.7 mm s^{-1} [329] have been measured. The considerably higher energy resulting in a much smaller recoilless fraction and the lower probability for the population

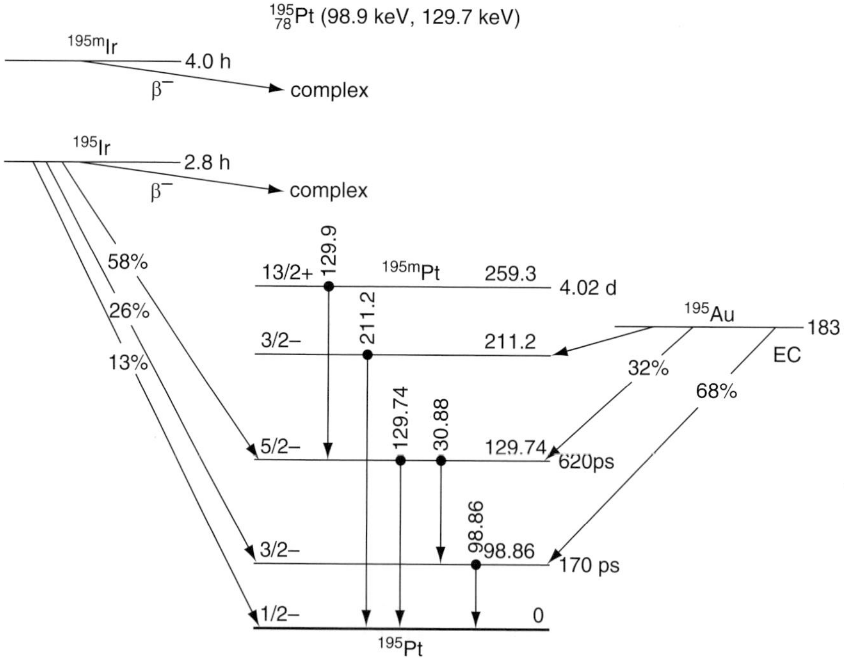

Fig. 7.70 Simplified decay scheme of ^{195}Pt. The two Mössbauer transitions have energies of 98.86 keV and 129.74 keV (from [1])

of the 130 keV excited state, however, make the use of the 130 keV transition less favorable for ^{195}Pt Mössbauer spectroscopy than the 99 keV transition.

The 99 keV γ-quanta are usually counted with NaI(Tl) scintillation counters or Ge(Li) diodes in transmission geometry. A Cd absorber should be used to reduce the background counting rate of the *K* X-rays and to avoid pile-up of the different X- and γ-rays (cf. Fig. 4 in [325]).

Atac et al. [330] reported on the use of the Mössbauer scattering technique. They recorded Mössbauer scattering spectra of platinum in iron, nickel, and cobalt at 29 K by counting the resonantly scattered 99 keV γ-quanta.

7.9.2 Platinum Compounds

First applications of ^{195}Pt (99 keV) Mössbauer spectroscopy to study platinum compounds have been attempted by Agresti et al. [331]. For Pt(II) and Pt(IV) oxides and chlorides at 4.2 K, they measured the following isomer shifts (with respect to the source of ^{195}Au diffused into Pt foil): PtO, −0.34(11); PtO_2, −0.40(8); $PtCl_2$, −0.1 (2); $PtCl_4$, −0.3(3) mm s^{-1}. Although the errors given here are relatively large, it seems as though the isomer shift tends to become less negative with increasing oxidation state. It was, however, too early to draw any definite conclusion from these results with respect to the sign of $\Delta\langle r^2\rangle$ of the 99 keV transition.

This became possible by the thorough ^{195}Pt Mössbauer study of a number of Pt (II) and Pt(IV) complex compounds by Walcher [326]. The isomer shifts observed with the 99 keV transition at 4.2 K are given (with respect to metallic Pt) in Table 7.11. Comparing these data with isomer shifts of isoelectronic compounds of iridium and gold, Walcher concluded that the change in the mean-square nuclear radius on going from the 99 keV state to the ground state is negative, $\Delta\langle r^2\rangle_{99} < 0$. From the observed difference in isomer shift of $[PtCl_4]^{2-}$ and $[PtCl_6]^{4-}$ and from the estimated difference in electron density at the Pt nucleus for $5d^8$ and $5d^6$ electron configurations, respectively, and using the results from relativistic free-ion Hartree–Fock–Slater calculations, Walcher derived the value of $\Delta\langle r^2\rangle_{99}/\langle r^2\rangle = -\left(1.6^{+4.4}_{-0.9}\right)\cdot 10^{-4}$.

Walcher's isomer shift data of Table 7.11 may be interpreted qualitatively in a similar way as described for gold (Sect. 7.10) or iridium (Sect. 7.8). The difference in isomer shift on going from Pt(II) to Pt(IV) in compounds with the same ligands is predominantly due to the changing number of 5*d* electrons, $5d^8$ for Pt(II) and $5d^6$ for Pt(IV). This causes less shielding of the 5*s* electrons from the nuclear charge and, along with it, lower electron density at the nucleus and, with $\Delta\langle r^2\rangle_{99} < 0$, a more positive isomer shift in Pt(II) than in Pt(IV) compounds. Moreover, within a given oxidation state, the differences in bond properties are extractable from the isomer shifts of Table 7.11. Substituting halide for cyanide as ligand in either Pt(II) or Pt(IV) complexes causes *d*-electron delocalization from the metal ion into the antibonding π-orbitals of the CN^- ligand and results in a decrease of shielding and thus an increase in electron density at the nucleus and therefore a decrease in isomer

Table 7.11 ^{195}Pt (99 keV) isomer shifts δ in platinum(II) and platinum(IV) complex compounds at 4.2 K (relative to platinum metal, from [326])

Compounds	δ (mm s^{-1})
$K_2[Pt(CN)_4]$	−1.7 (1)
$K_2[PtBr_4]$	+0.85 (10)
$K_2[PtCl_4]$	+0.85 (10)
$K_2[Pt(CN)_6]$	−2.1 (4)
$K_2[Pt(NO)_4Cl_2]$	−0.85 (10)
$K_2[PtBr_6]$	−0.02 (7)
$K_2[PtCl_6]$	−0.22 (7)

Table 7.12 Isomer shift δ (with respect to the source of ^{195}Au in Pt metal at 4.2 K), line width Γ, and percentage resonance effect f from ^{195}Pt (99 keV) Mössbauer measurements on platinum compounds at 4.2 K by Rüegg and coworkers [329, 332, 333]

Compound	δ^a (mm s^{-1})	Γ^b (mm s^{-1})	f (%)
PtO_2	−0.34 (2) [−0.40 (8)]	19.0 (8)	6.5 (21)
$PtCl_4$	0.03 (5) [−0.3 (3)]	16.6 (5)	1.7 (6)
$PtCl_2$	0.87 (6) [−0.1 (2)]	20.0 (11)	1.05 (45)
PtI_2	0.58 (6)	19.2 (8)	1.35 (50)
$K_2[Pt(CN)_4]Br_{0.30}\cdot 3H_2O$	−1.68 (4)	16.8 (3)	7.5 (20)
$K_2[Pt(CN)_4]\cdot 3H_2O$	−1.13 (9)	18.7 (15)	1.4 (5)
$K_2[Pt(CN)_4Br_2]$	−1.45 (6)	16.9 (3)	7.3 (20)

[a]Data given in square brackets from [331]
[b]Data from extrapolation to zero absorber thickness

shift. Even a gradual change in π-backbonding power of the ligand sphere is evident from the observed isomer shifts. Although NO^+ is known to be a stronger π-backbonding ligand than CN^-, the average π-backdonation power of the total ligand sphere is most probably larger in case of $K_2[(Pt(CN)_6]$ than in $K_2[Pt(NO)_4Cl_2]$ (which still contains two Cl^- ligands that are not capable of π-backbonding), and one therefore finds the latter complex placed between $K_2[Pt(CN)_6]$ and $K_2[PtCl_6]$ on the isomer shift scale [326]. Walcher did not observe any quadrupole interaction in compounds with nonzero EFG, which is not surprising in view of the very large line width. Buyrn et al. [327] have pointed out that electric quadrupole interactions of usual strengths hardly affect the ^{195}Pt (99 keV) spectrum.

Further studies of platinum compounds using the 99 keV transition in ^{195}Pt have been performed by Rüegg and coworkers. In addition to measurements on PtI_2, they remeasured PtO_2, $PtCl_4$, and $PtCl_2$ [329, 332]. Their isomer shift data, which are given in Table 7.12 together with data of the line width and percentage resonance effect, are considerably more accurate than, and for the latter two compounds quite different from, those observed earlier by Agresti et al. [331]. The more accurate isomer shift data by Rüegg et al. have allowed to distinguish between Pt(II) and Pt (IV) compounds similarly to the work of Walcher [326], viz., isomer shifts of Pt(II) compounds are clearly more positive than those of Pt(IV) compounds. The difference in isomer shift between $PtCl_2$ and PtI_2 most probably originates from a more effective σ-donation of s-like charge to the metal ion in case of PtI_2 due to the smaller electronegativity of iodine (2.5) as compared to chlorine (3.0). The experimental line width in the case of $PtCl_4$ is practically identical to, and in the other compounds only slightly more than twice, the natural line width ($2\Gamma_{nat} = 16.28$ mm s^{-1}).

The relatively small percentage resonance effects require measuring times on the order of 1 week. A representative ^{195}Pt (99 keV) Mössbauer spectrum, that of PtO_2 taken at 4.2 K [332], is shown in Fig. 7.71.

In another investigation using the ^{195}Pt (99 keV) Mössbauer effect, Rüegg et al. [332, 333] attempted to elucidate the problem about the charge localization in the one-dimensional conductor $K_2[Pt(CN)_4]Br_{0.30} \cdot 3H_2O$. Optical properties of this compound indicate metallic behavior. The d_c conductivity, along the Pt–Pt chains, is thermally activated below 200 K. At 4.2 K, however, the compound is an excellent insulator. All models which have been proposed to describe the one-dimensional conductors presume that the insulating state at 0 K is indicative of a certain localization of the conduction electrons. Rüegg et al. showed that the Mössbauer spectrum of $K_2[Pt(CN)_4]Br_{0.30} \cdot 3H_2O$, taken at 4.2 K (cf. Fig. 7.72), could be fitted ideally to a single Lorentzian absorption line. The fit assuming a sum of two Lorentzian lines with an intensity ratio of 85:15 according to the possible formation of a double salt $\{0.85K_2[Pt^{II}(CN)_4] + 0.15K_2[Pt^{IV}(CN)_4Br_2] \cdot 3H_2O\}$ (which is chemically equivalent to $K_2[Pt(CN)_4]Br_{0.30}{\cdot}3H_2O$) gave worse results. The authors, therefore, concluded that charge is not localized on an atomic scale. The relatively high electron density obtained from the isomer shift of the

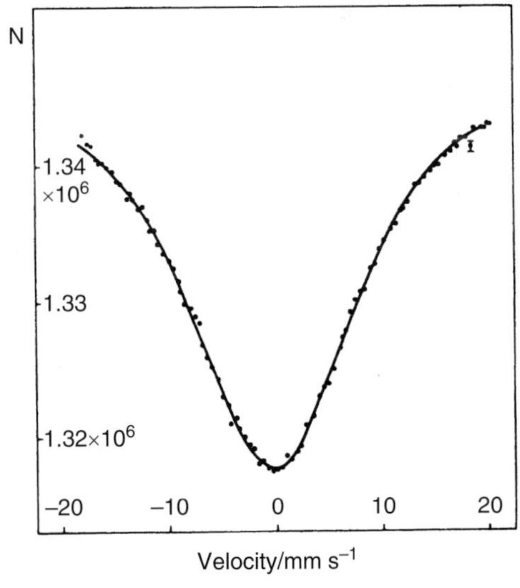

Fig. 7.71 ^{195}Pt (99 keV) Mössbauer spectrum of PtO_2 taken at 4.2 K with a source of ^{195}Au in platinum at 4.2 K (from [332])

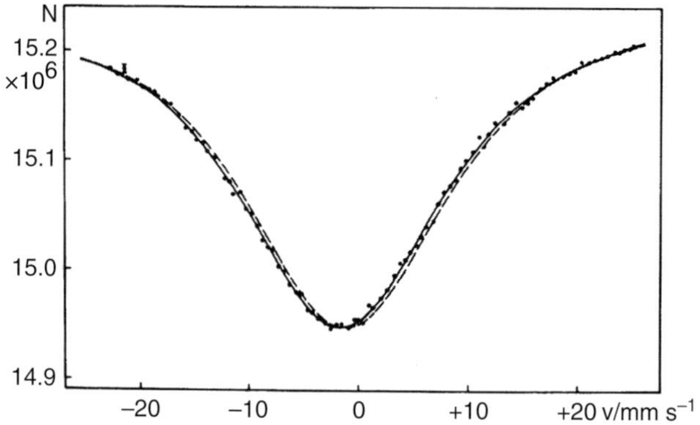

Fig. 7.72 ^{195}Pt (99 keV) Mössbauer spectrum of the one-dimensional conductor $K_2[Pt(CN)_4]Br_{0.30}\cdot 3H_2O$ at 4.2 K (source: ^{195}Au in platinum at 4.2 K). The *solid line* represents a single Lorentzian line fitted to the measured spectrum. The *dashed line* represents the best fit using a sum of two Lorentzian lines with an intensity ratio of 85:15 and with the isomer shifts of the spectra of $K_2[Pt(CN)_4]\cdot 3H_2O$ and $K_2[(Pt(CN)_4Br_2]$ (from [333])

one-dimensional conductor system as compared to the insulating compounds $K_2[(Pt(CN)_4] \cdot 3H_2O$ and $K_2[Pt(CN)_4Br_2]$ (cf. Table 7.11) appears to be a consequence of the strong metal–metal bond.

7.9.3 Metallic Systems

Much of the ^{195}Pt Mössbauer work performed so far has been devoted to studies of platinum metal and alloys in regard to nuclear properties (magnetic moments and lifetimes) of the excited Mössbauer states of ^{195}Pt, lattice dynamics, electron density, and internal magnetic field H^{int} at the nuclei of Pt atoms placed in various magnetic hosts. The observed changes in the latter two quantities, $|\psi(o)|^2$ and H^{int}, within a series of platinum alloys are particularly informative about the conduction electron delocalization and polarization.

In the only ^{195}Pt (99 keV) backscattering experiments communicated so far, Atac et al. [330] observed magnetic hyperfine interaction in ferromagnetic alloys of natural platinum with iron, cobalt, and nickel ($Pt_{0.3}Fe_{0.7}$, $Pt_{0.07}Co_{0.93}$, $Pt_{0.07}Ni_{0.93}$). Some typical spectra reproduced from their work are shown in Fig. 7.73. The spectra are unresolved due to the large natural line width of the 99 keV transition. Nevertheless, by fitting a Zeeman pattern consisting of six lines of equal width and with an intensity ratio appropriate for a pure dipole transition, reliable values for the magnitude and even the sign (from the angular correlation pattern observed with a polarized scatterer) of the internal field at the Pt nuclei could be evaluated (at 29 K) to be −124 T in $Pt_{0.3}Fe_{0.7}$, −77 T in $Pt_{0.07}Co_{0.93}$, and −23 T in $Pt_{0.07}Ni_{0.93}$ (error in H^{int} about 10%). From the approximate constancy of $|H^{int}|/\mu_B$, where μ_B is the

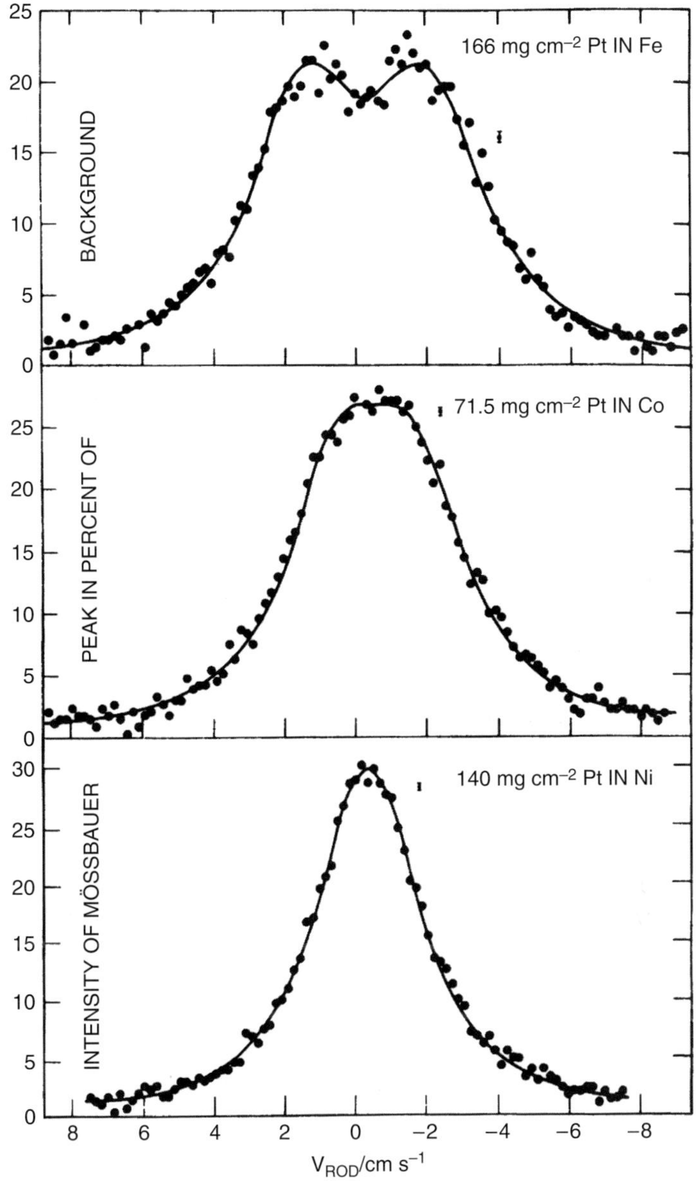

Fig. 7.73 Mössbauer spectra of Pt (99 keV) in ferromagnetic alloys measured in backscattering geometry with the scatterer kept at 29K. The count rate at infinite velocity is normalized to 1 (from [330])

magnetic moment in Bohr magnetons of the host atoms, the authors concluded that the internal field at the Pt nuclei originates predominantly from conduction electron polarization. Buyrn et al. [327] arrived at the same conclusion in their study of the

^{195}Pt (99 keV) Mössbauer effect in Pt_xFe_{1-x} alloys ($0.03 \leq x \leq 0.5$). By applying an external magnetic field, they also found the nuclear g-factor ratio to be $g(3/2)/g(1/2) < 0$ (longitudinal polarization eliminates the $\Delta m = 0$ transitions and the intensity pattern should be 3:1:1:3 for $g(3/2)/g(1/2) < 0$ but 1:3:3:1 for $g(3/2)/g(1/2) > 0$). H^{int} was found to be nearly constant, about 126 T (at 4.2 K), over a range of composition from 3 to 30 at% Pt.

Agresti et al. [331] investigated alloys of 3 at% Pt in iron, cobalt, and nickel at 4.2 K in transmission geometry. They applied a small magnetic field (0.25 T) parallel and perpendicular to the γ-ray direction modifying the intensities of the transition between the nuclear magnetic sublevels and thereby increasing the obtainable information as compared to measurements using unpolarized absorbers. Their results concerning the magnitude of H^{int} are in satisfactory agreement with those measured in scattering geometry by Atac et al. [330]. Agresti et al. could also extract isomer shifts from their recorded spectra. From the observed decrease of δ with decreasing electronegativity (see Fig. 7.74), they concluded that the mean-square radius of the ^{195}Pt nucleus in the 99 keV state is smaller than that in the ground state.

The first clear resonance effect using the 130 keV transition in ^{195}Pt was observed on a platinum metal absorber by Wilenzick et al. [328] (resonance effect 0.16%, line width 4.4 ± 0.4 mm s^{-1}) and by Wolbeck and Zioutas [324] (resonance effect 0.044%, line width 2.6 ± 1.5 mm s^{-1}). Considerably improved

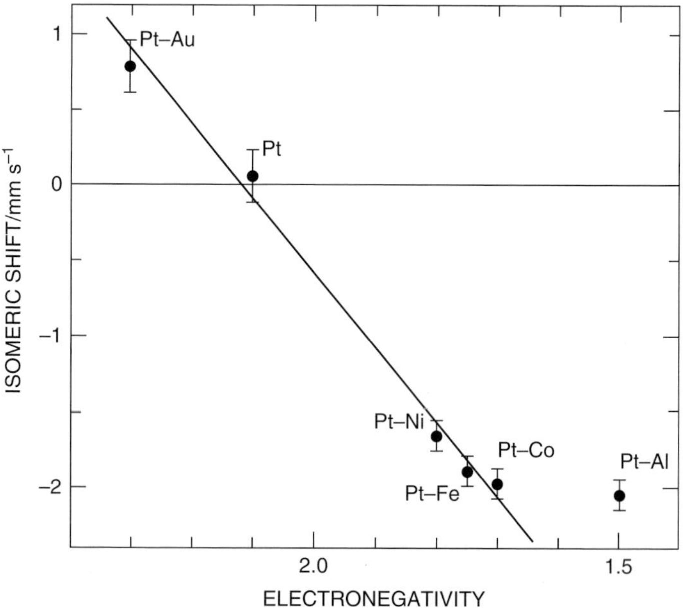

Fig. 7.74 ^{195}Pt (99 keV) isomer shift in platinum alloys (3 at% Pt) as a function of the electronegativity of the host element (from [331])

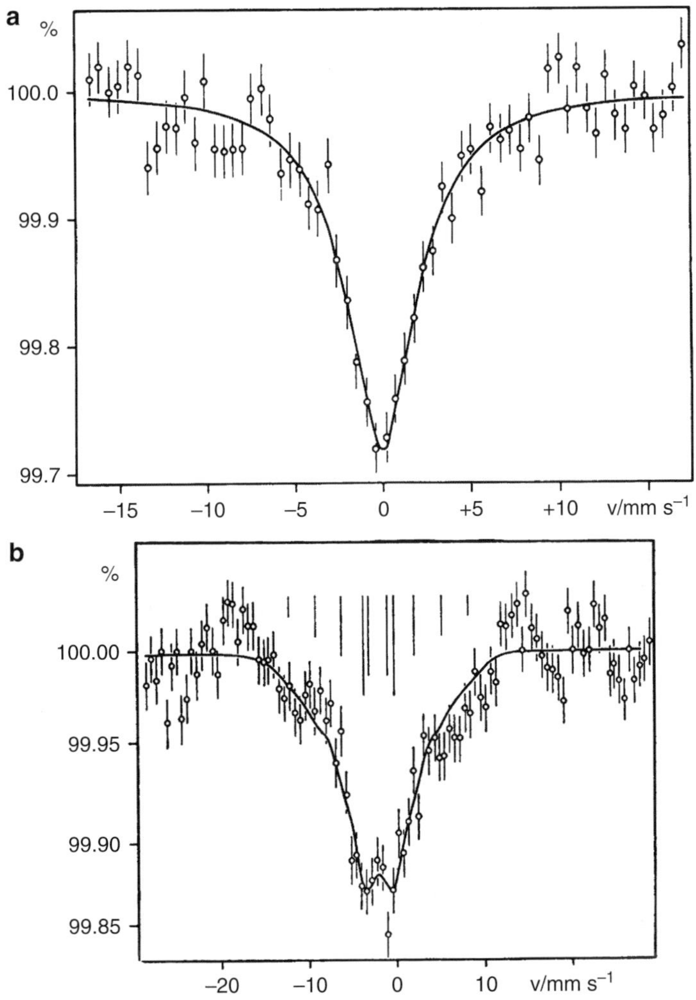

Fig. 7.75 ^{195}Pt (130 keV) Mössbauer spectra of (**a**) natural Pt metal (700 mg cm^{-2}) and (**b**) $Pt_{0.1}Fe_{0.9}$ taken at 4.2 K with a ^{195}Au/Pt metal source (from [326])

^{195}Pt (130 keV) spectra were recorded by Walcher [326]. Two representative spectra are shown in Fig. 7.75. The spectrum of $Pt_{0.1}Fe_{0.9}$ is broadened due to magnetic hyperfine interaction. The solid line in Fig. 7.75b was obtained by adjusting ten hyperfine components of the 130 keV transition to the measured spectrum, with relative intensities and positions as indicated by the bars. From this fit, Walcher was able to determine the magnetic moment of the 130 keV state to be $\mu = (0.81^{+0.13}_{-0.25})\mu_N$. From the isomer shift data observed for both the 99 keV and 130 keV transitions in ^{195}Pt in various alloys (cf. Table 7.13), Walcher found that $\Delta\langle r^2\rangle_{130}/\Delta\langle r^2\rangle_{99} = 1.5(2)$. The same value, within

Table 7.13 Isomer shift data observed with the 99 and the 130 keV transition in ^{195}Pt in various alloys at 4.2 K (relative to Pt metal)

Alloy	δ_{99} (mm s^{-1})	δ_{130} (mm s^{-1})
$Pt_{0.1}Fe_{0.9}$	−1.52 (7)	−1.70 (14)
$Pt_{30}Cu_{70}$	−0.72 (10)	−0.72 (13)
$CePt_2$	−1.34 (11)	−1.61 (15)

experimental error, was found by Rüegg [329, 332]. Two of the ^{195}Pt Mössbauer spectra obtained with a Fe–Pt alloy absorber (3 at% Pt) at 4.2 K by Rüegg are shown in Fig. 7.76; the upper spectrum (a) has been recorded with the 99 keV transition and the lower spectrum (b) with the 130 keV transition. The theoretical spectrum in Fig. 7.76a is composed of six hyperfine components (according to a pure $M1$ transition); in Fig. 7.76b, it is composed of ten hyperfine components (according to a pure $E2$ transition); the individual Lorentzian lines are indicated on top of the figures.

Ianarella et al. [334] used the ^{195}Pt (99 keV) Mössbauer effect to study the influence of hydrogen loading of palladium. As in similar experiments using the Mössbauer effect in ^{197}Au, ^{193}Ir, and ^{99}Ru, they observed a decrease of the electron density at the nuclei of the Mössbauer probe on hydrogenation, but they could not decide whether the effect is due to the volume expansion of the palladium lattice or due to the filling of the host conduction band with electrons donated by the hydrogen.

Van der Woude and Miedema [335] have proposed a model for the interpretation of the isomer shift of ^{99}Ru, ^{193}Ir, ^{195}Pt, and ^{197}Au in transition metal alloys. The proposed isomer shift is that derived from a change in boundary conditions for the atomic (Wigner–Seitz) cell and is correlated with the cell boundary electron density and with the electronegativity of the alloying partner element. It was also suggested that the electron density mismatch at the cell boundaries shared by dissimilar atoms is primarily compensated by $s \rightarrow d$ electron conversion, in agreement with results of self-consistent band structure calculations.

7.10 Gold (^{197}Au)

Recoilless nuclear resonance of the 77.34 keV transition in ^{197}Au was first reported in 1960 using gold foil absorbers and both $^{197}Pt/Pt$ and $^{197}Hg/Hg$ sources at 4.2 K [336]. During the first decade after its discovery, the ^{197}Au resonance was applied almost exclusively to metallic systems, mostly in studies of the band structure and magnetic hyperfine structure of gold as a function of various host metals. Chemical applications demonstrate the usefulness of the technique in distinguishing oxidation states and bond properties of gold in various surroundings.

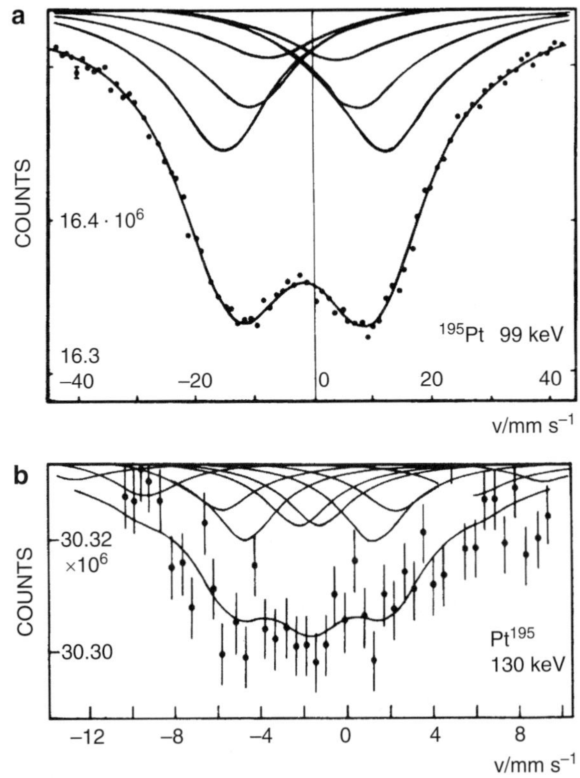

Fig. 7.76 ^{195}Mössbauer spectra obtained with a Fe–Pt absorber (3 at% Pt) at 4.2 K using (**a**) the 99 keV transition and (**b**) the 130 keV transition of a ^{195}Au/Pt metal source (from [329])

7.10.1 Practical Aspects

The 77.34 keV level may be populated by decaying ^{197}Pt and ^{197}Hg precursor nuclides, as shown in the decay scheme of Fig. 7.77. ^{197}Hg, however, is hardly used because of the relatively low recoil-free fraction in gold amalgam. ^{197}Pt in platinum metal foil, activated by thermal neutron irradiation of natural or enriched platinum metal in a nuclear reactor according to ^{196}Pt(n, γ)^{197}Pt, is the preferred source for ^{197}Au Mössbauer spectroscopy. It can be reirradiated repeatedly.

The relatively high transition energy of 77.34 keV requires cooling of both source and absorber; most experiments are therefore carried out at temperatures between 77 and 4 K in transmission geometry. Diodes like Ge(Li) are most suitable as detectors.

The data of the nuclear moments and other quantities relevant for ^{197}Au Mössbauer spectroscopy are collected in Table 7.1 (at the end of the book). The spectra taken of chemical compounds are generally very simple, singlets in the absence and quadrupole doublets in the presence of electric quadrupole interaction, with line widths only slightly (some 10–20%) above the natural line width of 1.882 mm s^{-1}. Measured isomer shift changes and quadrupole splittings are several times the experimental line widths and allow the conclusive distinction of chemical species to be made. Magnetic hyperfine splitting has been observed in metallic systems with magnetic fields up to ~140 T, presumably due to conduction electron polarization by the magnetic host lattice. Here, the *E2*/*M*1 mixing ratio of $\delta^2 = 0.11$ for the multipolarity of the γ-radiation determines the relative intensities of the eight $E2 + M1$ allowed transition lines.

7.10.2 Inorganic and Metal-Organic Compounds of Gold

Mössbauer studies on gold compounds were first reported by Roberts et al. [337] and Shirley et al. [338, 339]. They observed rather large isomer shift changes in some simple gold (I) and gold (III) halides and in halogeno-complexes such as AuX (X = Cl, Br, I), AuX_3 (X = F, Cl), K[AuX_4] (X = F, Cl, Br), and K[$Au(CN)_2$].

In 1970, Charlton and Nichols [340] published their results of a ^{197}Au Mössbauer study on 31 gold complexes, for which they describe a general correlation between Mössbauer parameters and chemical features. They investigated the series

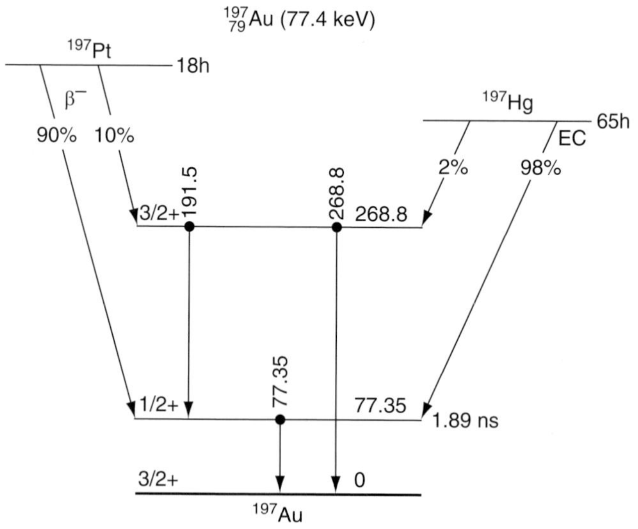

Fig. 7.77 Decay scheme of ^{197}Au [from 266]

$$LAuCl, LAuCl_3(L = Ph_3P, (C_6F_5)Ph_2P, Ph_3As, C_5H_5N, Me_2S, \text{ and } p\text{-tolylisocyanide})$$

$$Ph_3PAuX(X = I, Br, Cl, N_3, OCOMe, CN, \text{ and } ME)$$

$$M[AuX_4]\ (X = F, Cl, Br, \text{ and } I;\ M = H, Na, K, NH_4, Cs)$$

and discussed the variations in isomer shift and quadrupole splitting in terms of *sp* and *dsp*^{2} hybridization of the gold atom. Some general points could be made by these authors.

1. The range of isomer shifts and quadrupole splittings is larger than that for tin and iron.
2. The range of isomer shifts is larger for aurous, Au(I), than for auric, Au(III), compounds, most probably due to the larger amount of *s*-character in *sp* than in *dsp*^{2} hybrid orbitals, and also to the smaller variety of ligands in the auric compounds under study.
3. The quadrupole splittings of the aurous compounds are generally larger than those of the auric compounds.
4. A positive correlation between isomer shift and quadrupole splitting exists for both the aurous and auric series. Moreover, in the LAuCl series, they found an increasing isomer shift and thus increase in donation of charge by the ligand into the *sp* hybrid orbital in the following order of ligands: $Me_2S < C_5H_5N < Ph_3As < (C_6F_5)Ph_2P \simeq Ph_3P$. A similar order has been found for the isomer shift of the series $LAuCl_3$: *p*-tolylisocyanide $< Me_2S < C_5H_5N < Ph_3P$. In the series Ph_3PAuX, the order of increasing isomer shift is $I < Br < Cl < OCOMe \simeq N_3 < CN < Me$, which parallels the spectrochemical series. With the exception of fluoride, the isomer shift order also parallels the spectrochemical series in the AuX_4^- complexes (X = F, Cl, Br, I).

More systematic investigations of Au(I) and Au(III) compounds using the ^{197}Au Mössbauer effect were described by Faltens and Shirley [341], by the Mössbauer group in Munich [342], and in a review by Parish [343]. All groups found for each oxidation state a linear correlation between the isomer shift and the quadrupole splitting. From this finding, the oxidation state of gold could be determined straightforward. This is demonstrated by the δ–ΔE_Q diagrams shown in Fig. 7.78 [341] and in Fig. 7.79 [342]. Some representative ^{197}Au transmission spectra are shown in Fig. 7.80 [342].

Employing electronic structure calculations on Au(I) and Au(III) complexes provides a qualitative and quantitative mean for the understanding of this δ–ΔE_Q relation. For two sets, i.e., linear Au(I) and square-planar Au(III) complexes, such calculations have been carried out [344]. In the case of $[AuCl_2]^-$, the bonding, nonbonding, and antibonding MOs with Au 5*d* character are doubly occupied, which means that the Au 5*d* shell is practically filled with ten electrons. From this consideration, one might conclude that the EFG and thus, also ΔE_Q, for linear

Au(I) complexes is very small; this is not the case, however (Fig. 7.79). The electronic structure calculations reveal that, besides contributions from Au 5*d*, other contributions to the EFG play an important role (*vide infra*). Going from the linear Au(I) to the square-planar Au(III) complexes has the consequence that the nonbonding MOs with mainly $5d_{xy}$ and $5d_{x^2-y^2}$ character (in the case of Au(I)) will split into bonding and antibonding parts (in the case of Au(III)). Parallel to this, the bonding and antibonding MOs in $[AuCl_2]^-$ with mainly d_z^2 character become nonbonding in $[AuCl_4]^-$, because there are no ligands along the *z*-axis in the square-planar complex. Since the borderline between highest occupied and lowest unoccupied MOs in $[AuCl_4]^-$ is just below the antibonding MO with mainly $5d_{x^2-y^2}$ character, one might conclude that the deficiency in $5d_{y^2-y^2}$ occupation yields a negative EFG in square-planar Au(III) complexes. The contrary, however, is the case (*vide infra*) here.

Electronic structure calculations reveal that the actual EFG in linear Au(I) and square-planar Au(III) complexes is not merely a matter of 5*d* occupation. Instead, the EFG is markedly determined by AO populations, including Au core 5*p* as well as valence 6*p* orbitals and further by the charge in the overlap area ($AO_{Au}AO_{lig}$) and on the ligands (Table 7.14).

Any change in the populations of the MOs with mainly Au character and in overlap and ligand charges has its bearing on the EFG V_{zz} and thus on the quadrupole splitting

$$\Delta E_Q = \frac{1}{2} e Q_g V_{zz} (1 + \eta^2/3)^{1/2}.$$

The value of the electric quadrupole moment of the ground state of the ^{197}Au nucleus was reported as $Q_g = +0.547(16)b$ [345], $+0.64b$ [346], $+0.566(1)b$ [347].

In Au(I) complexes, the populations in the $6p_z$- and $5d_{z^2}$-like MOs both contribute negatively to the EFG. Thus, the increasing σ-donor capacity of the ligand will increase the magnitude of the presumably negative EFG. The concomitant synergic decrease in the population of the $6p_{x,y}$-like MOs will reduce the positive contribution of the $6p_{x,y}$ electrons and therefore further enhance the total (negative) EFG. The π-backbonding effects decrease the population of the $5d_{xz,yz}$-like MOs and thus cause positive contributions to the total EFG. This together with the negative contributions from (negative) overlap and ligand charge along the *z*-axis yields large calculated ΔE_Q values (and negative EFG) for linear Au(I) complexes, in agreement with experimental results (Table 7.14).

In Au(III) compounds, the "hole" in the antibonding $5d_{x^2-y^2}$-like MO produces a negative contribution to the EFG. The contributions from the various 6*p* populations have opposite signs: the $6p_{x,y}$-AOs contribute positively, the $6p_z$-AO contributes negatively. As the σ-overlap, and thus σ-bonding, is usually larger than π-overlap, it is expected that the contributions to the EFG are mainly determined by the $6p_{x,y}$ populations rather than the $6p_z$ population. With increasing σ-donor properties of the ligands also the 5*d*-"hole" will be

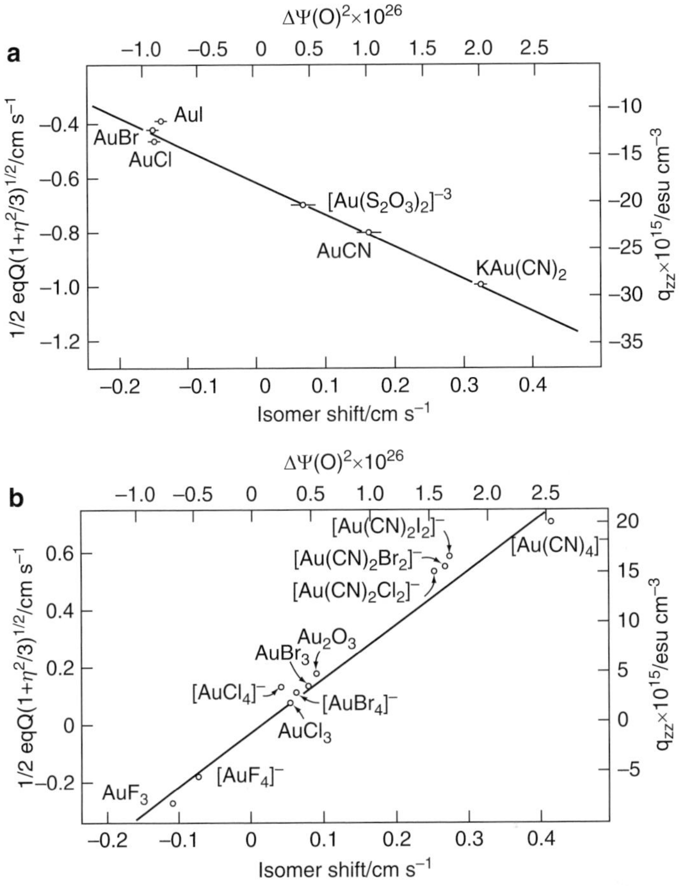

Fig. 7.78 Linear relation of the quadrupole splitting $\Delta E_Q = (1/2)eqQ\ (1 + \eta^2/3)1/2$ and the isomer shift δ for aurous (**a**) and auric (**b**) compounds. Also included is a correlation with the relative change in electron density at the gold nucleus, $\Delta|\psi(0)|^2$, as derived from Dirac–Fock atomic structure calculations for several electron configurations of gold. An approximate scale of the EFG q_{zz} (in the principal axes system) is given on the right-hand ordinate (from [341])

more and more filled; this way the EFG will become more and more positive. The effect of π-backdonation reduces the atomic population of the $5d_{xy}$- and the $5d_{xz,yz}$-like MOs. These MOs contribute to the EFG with opposite signs. In summary, the ΔE_Q of square-planar Au(III) complexes takes moderate positive or negative values in the case of halo-complexes but large and positive values in the case of cyano-complexes in agreement with experimental results (Table 7.14). Obviously, the σ-donor strength turns out to be stronger in cyano- than in halo-complexes.

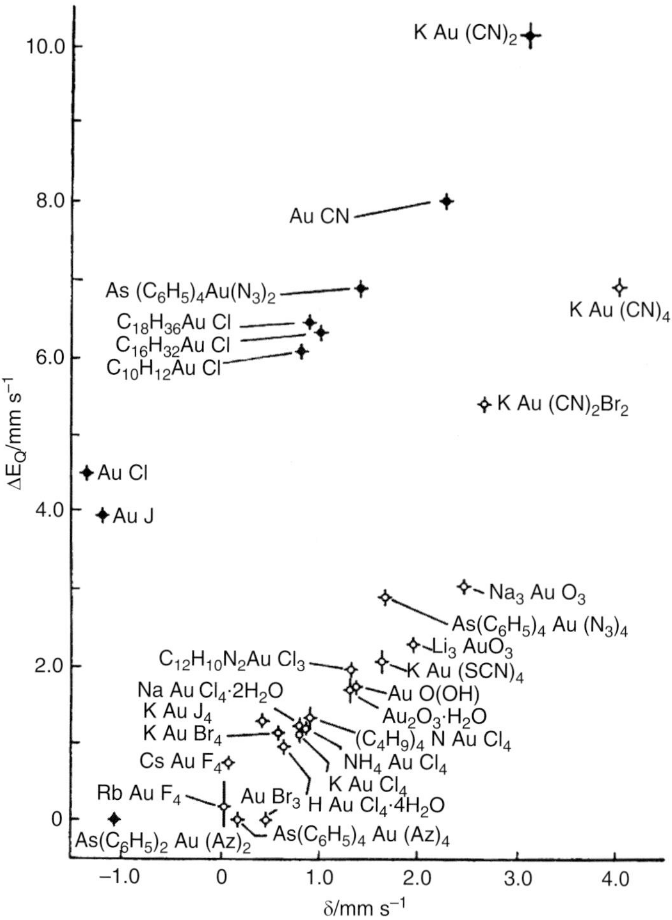

Fig. 7.79 Quadrupole splitting QS versus isomer shift δ in gold (I) and gold (III) compounds (from [342])

The isomer shift δ of a specific Au-containing complex is a measure of electronic density $\varrho(0)$ at the ^{197}Au nucleus of this complex with respect to a reference $\varrho_{\text{ref}}(0)$

$$\delta = \alpha[\varrho(0)_{\text{ref}} - \varrho(0)],$$

with $\alpha = (\Delta R/R)\,(3Ze^2cR^2)/(5\varepsilon_0 E_0)$. For ^{197}Au ($E_0 = 77.34$ keV, $R = 6.928$ fm), the isomer shift calibration constant was determined as $\alpha = +0.0665(4)$ mm s^{-1} a.u.^{3} [347], corresponding to $\Delta R/R = +6.65 \cdot 10^{-5}$. Elsewhere, $\Delta R/R$ was reported to take values in the range $1.5 \cdot 10^{-4}$, $2.5 \cdot 10^{-4}$ [1] or 1.9–$3.1 \cdot 10^{-4}$ [341]. It is unclear from where this discrepancy in $\Delta R/R$ values stems from,

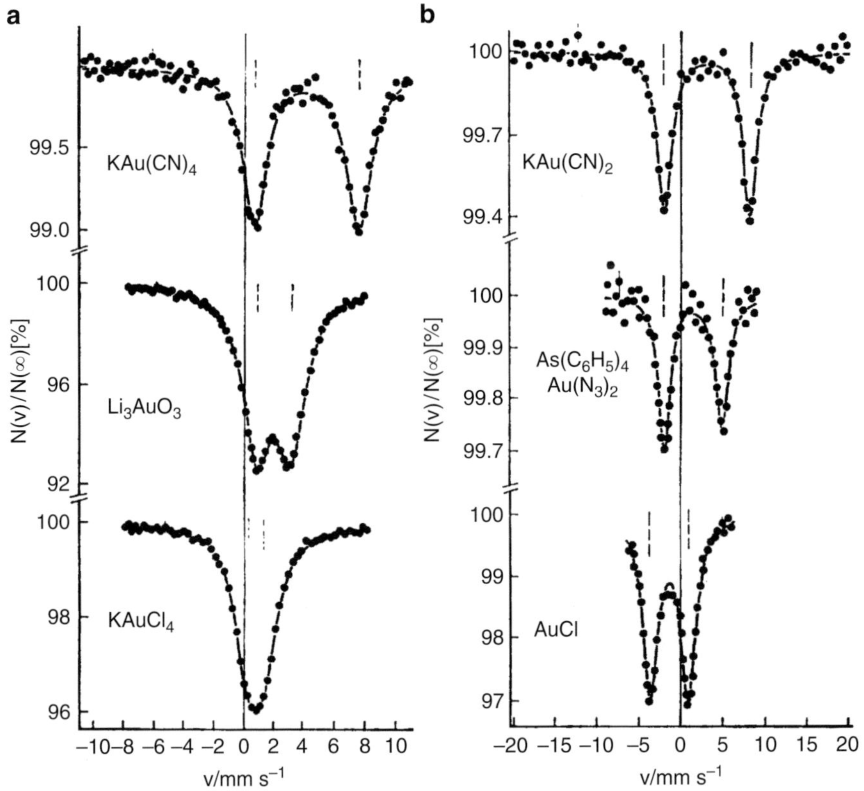

Fig. 7.80 ^{197}Au Mössbauer spectra of some gold (III) (**a**) and gold (I) (**b**) compounds. ^{197}Pt/Pt source and Au-containing absorbers were kept at 4.2 K (from [342])

Table 7.14 Results from electronic structure calculations for linear Au(I) and square-planar Au (III) complexes: $[AuCl_2]^-$ with $R_{Au-Cl} = 2.30$ Å and $[Au(CN)_2]^-$ with $R_{Au-C} = 2.12$ Å. $[AuF_4]^-$ with $R_{Au-F} = 1.95$ Å, $[AuCl_4]^-$ with $R_{Au-Cl} = 2.28$ Å, $[AuCl_2(CN)_2]^-$ with $R_{Au-Cl} = 2.28$ Å, $R_{Au-C} = 2.10$ Å (*trans*) and $[Au(CN)_4]^-$ with $R_{Au-C} = 2.10$ Å (from [344])

Complex	AO population			Contributions to ΔE_Q (in mm s^{-1})^a				ΔE_Q (in mm s^{-1})^a	
	5*d*	6*s*	6*p*	5*p*	5*d*	6*p*	Σ^b	Σ^c	exp
$[AuCl_2]^-$	9.88	0.46	0.33	−4.4	1.1	−0.1	−4.3	−7.7	−4.7
$[Au(CN)_2]^-$	9.85	0.48	0.33	−3.8	2.1	−1.6	−8.3	−11.6	−10.2
$[AuF_4]^-$	9.18	0.38	0.49	2.2	−5.8	−1.7	3.5	−1.8	−1.8
$[AuCl_4]^-$	9.11	0.33	0.53	4.3	−6.7	−0.5	4.2	1.3	1.3
$[AuCl_2(CN)_2]$	9.05	0.35	0.70	3.5	−6.0	1.0	6.1	5.1^d	5.3
$[Au(CN)_4]^-$	8.96	0.40	0.82	2.6	−5.5	2.5	7.4	7.0	7.0

^{a}Sign refers to EFG component V_{zz}
^{b}Sum of orbital contributions
^{c}Total calculated ΔE_Q, including orbital, overlap and ligand contribution
^{d}Calculated with $\eta = 0.8$

though Ruebenbauer et al. [347] as well as Faltens et al. [341] had evaluated their $\rho(0)$ values relativistically. In most publications, δ is given either relative to the reference "^{197}Au in metallic Pt" or relative to "^{197}Au in metallic Au." The two values are related by

$$\delta_{\mathrm{Au}} = \delta_{\mathrm{Pt}} + 1.21\ \mathrm{mm\,s^{-1}} \tag{7.16}$$

The isomer shift in gold compounds mainly depends on the Au 6*s* and 5*d* populations. Positive contributions to the electron density at the gold nucleus arise from the 6*s* populations, whereas a decrease of the electron density is caused by the increase of the atomic 5*d* populations due to shielding effect. The 6*p* populations are less significant here. Kopfermann has reported [348] that the addition of one 5*d* (6*p*) electron corresponds to the screening of 25% (10%) of one 6*s* electron. The direct contribution to the electron density by one 6*p* electron through relativistic effects is only about 5% of $\varrho(0)$ of one 6*s* electron. In this view, it is expected that in gold complexes with appreciable π-donation (e.g., halides as ligands), π-charge can only be donated into the $6p_{x,y}(D_{4h})$, $6p_z(D_{4h})$, and $6p_{x,y}(D_{\infty h})$ metal orbitals, respectively, because the $5d_{xz,yz}(D_{4h})$ and $5d_{xz,yz}(D_{\infty h})$ metal orbitals are fully occupied; therefore, the isomer shift is hardly influenced by π-donation. The interpretation of the isomer shift of gold compounds can therefore be based mainly on σ-donation and π-backdonation properties of the ligands.

In square-planar Au(III) complexes, the metal orbitals $5d_{x^2-y^2}$, $5d_{z^2}$, 6*s*, and $6p_{x,y}$ are suitable for σ-bonding. The experimental data available show that the isomer shift increases with increasing σ-covalency, and thus reflects the dominating role of the 6*s* population. In the presence of π-backbonding (e.g., CN^- as ligand), charge will be delocalized from the $5d_{xz,yz}$ and $5d_{xy}$ metal orbitals into π^*-antibonding orbitals of the ligand. This leads to the further increase of the electron density at the nucleus owing to decreasing shielding effect. Similar considerations hold for linear Au(I) complexes, where the metal orbitals $5d_{z^2}$, 6*s*, and $6p_z$ are suitable for σ-bonding, whereas the orbitals $5d_{xz,yz}$ are engaged in π-backbonding, and the $6p_{x,y}$ orbitals are employed in π-donor bonds.

An extension of ^{197}Au isomer shift studies has been contributed by investigating unusual Au(V) fluorides $A^+[AuF_6]^-$ with $A^+ = Xe_2F_{11}^+$, XeF_5^+, and Cs^+ [349, 350]. Structural data indicate an octahedral $[AuF_6]^-$ anion; accordingly, the Mössbauer spectra consist of a single line with approximately twice the natural line width (after correction for line broadening caused by finite absorber thickness). The spectrum of $Xe_2F_{11}^+$ $[AuF_6]^-$ is shown, as an example, in Fig. 7.81. The isomer shift data are almost the same for all complexes supporting the existence of similar $[AuF_6]^-$ anions in all Au(V) fluorides. The magnitude of the shifts is well above the range of isomer shifts of Au(III) compounds, as shown in Fig. 7.82, where the isomer shifts of Au(I), Au(III), and Au(V) compounds with halogen ligands are plotted. A decrease of the atomic 5*d* population is assumed on going from Au(I) to Au(III) and to Au(V) with the concomitant increase of the isomer shift due to reduced shielding with reduced 5*d* occupancy.

Schmidbaur et al. [351] have reported results from ESCA and ^{197}Au Mössbauer studies of a series of ylide complexes of Au(I), Au(II), and Au(III). The ESCA data reveal a pronounced influence of the Au oxidation state on the Au ($4f_{7/2}$) binding energies, the differences of which are large enough to distinguish between the various oxidation states. The ^{197}Au Mössbauer spectra of these complexes show a systematic decrease of the isomer shift with increasing oxidation state of gold. This behavior (parallel to that of the $KAu(CN)_2$ and $KAu(CN)_2Br_2$ or Ph_3PAuCl and Ph_3PAuCl_3 pairs) is opposite to the usually observed increase of the isomer shift with increasing oxidation state of gold (see Fig. 7.82).

In his thesis, Viegers [352] studied some 50 gold compounds, AuX_4^- with X = Br, Cl, and different cations, *bis*-dithiolato Au(III) complexes, gold(I) complexes with bidentate sulfur-donor ligands, triphenylphosphine Au(I) complexes, polynuclear organo-gold compounds, and some others. The data are collected in Table 7.15; the relevant abbreviations used in the chemical formulae are given in Table 7.16. The general features observed by Viegers follow those found in earlier ^{197}Au Mossbauer effect studies by other authors [340–343]: the linear relationship between the quadrupole splitting (ΔE_Q) and the isomer shift (δ) still exists, i.e., the relation is on the average given by

$$\Delta E_Q(\text{Au(III)}) = 1.74\delta - 0.22 \text{ mm s}^{-1},$$

$$\Delta E_Q(\text{Au(I)}) = 1.06\delta + 5.05 \text{ mm s}^{-1}.$$

Comparing an Au(I) complex with an Au(III) complex with the same ligands, the isomer shift of the latter is more positive and its quadrupole splitting is smaller.

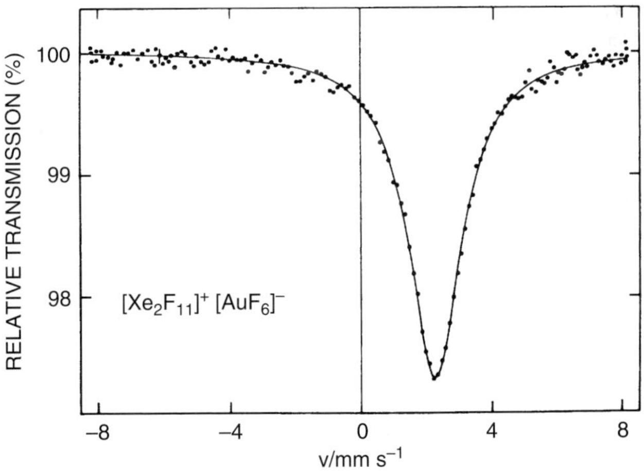

Fig. 7.81 Mössbauer spectrum of $Xe_2F_{11}{}^+[AuF_6]^-$ taken with a ^{197}Pt/Pt source at 4.2 K (from [350])

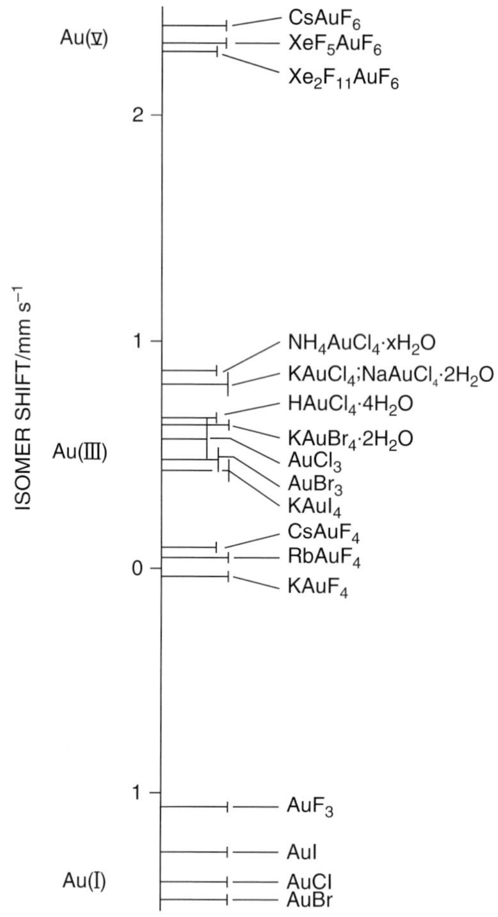

Fig. 7.82 ^{197}Au isomer shifts (with respect to a ^{197}Pt/Pt source, both source and absorbers at 4.2 K) of Au(I), Au(III), and Au(V) compounds with halogen ligands (from [350]). The isomer shifts δ_{Au} and δ_{Pt} are related by $\delta_{Au} = \delta_{Pt} + 1.21$ mm s^{-1}

The Mössbauer parameters are largest, if the coordinating atoms belong to Group IV, and smallest if they belong to Group VII. The following rule seems to apply: the Mössbauer parameters of compounds with mixed ligands are averages of those of the pure complexes. Although the δ–ΔE_Q correlation lines approach each other for large δ and ΔE_Q values, it still appears possible to determine the oxidation state of gold unambiguously using this plot.

The δ–ΔE_Q relation in Au(I) and Au(III) compounds was the matter of numerous publications. An excellent review covering the work up to the year 1984 was provided by Parish [343]. Later, Wagner et al. [353–359], Schmidbaur et al. [360], Hadjiliadis et al. [361], Reiff et al. [355], Stanek et al. [362], and Sakai et al. [363, 364] have described a number of mononuclear gold complexes making use of the above-described $\delta - \Delta E_Q$ characteristics to identify Au(I) or Au(III) oxidation states. These characteristics were also employed when investigating binuclear gold complexes to distinguish between Au(I) and Au(III) sites as described by Katada et al. [365, 366], Wagner et al. [367], Kojima et al.

Table 7.15 ^{197}Au Mössbauer parameters observed for various gold compounds at 4.2 K. The isomer shift data refer to the used source of ^{197}Au in Pt (from [352])

Compound	Isomer shift ($mm\ s^{-1}$)	Quadrupole splitting ($mm\ s^{-1}$)
(N-Et. pip)$AuCl_4$	1.02 ± 0.02	1.4 ± 0.1
Bu_4NAuBr_4	0.92 ± 0.02	1.5 ± 0.1
$KAuBr_4{\cdot}2H_2O$	0.67 ± 0.02	1.3 ± 0.1
$Bu_4NAu(tdt)_2$	2.99 ± 0.02	2.73 ± 0.02
Au(mnt)(dtc)	2.64 ± 0.02	2.57 ± 0.02
$Bu_4NAu(mnt)_2$	2.92 ± 0.01	2.33 ± 0.01
$Au(H_2dtc)_2Br$	2.26 ± 0.02	3.01 ± 0.04
$Au(Me_2dtc)_2Br$	2.04 ± 0.04	2.60 ± 0.05
$Au(Et_2dtc)_2Br$	2.24 ± 0.03	3.06 ± 0.05
$Au(Pr_2dtc)_2Br$	2.22 ± 0.02	3.02 ± 0.04
$Au(Bu_2dtc)_2Br$	2.20 ± 0.08	2.8 ± 0.2
$Au(Ph_2dtc)_2Br$	2.06 ± 0.03	2.89 ± 0.05
$Bu_4NAu(cdc)$	2.15 ± 0.05	2.71 ± 0.07
$(CH_3)_2Au(Me_2dtc)$	3.97 ± 0.01	5.07 ± 0.02
$(CH3)_2Au(Pr_2dtc)$	3.95 ± 0.01	4.98 ± 0.01
$(CH3)_2Au(Bu_2dtc)$	3.99 ± 0.01	5.11 ± 0.02
$I_2Au(Bu_2dtc)$	1.26 ± 0.03	2.10 ± 0.05
$Br_2Au(Bu_2dtc)$	1.47 ± 0.01	2.20 ± 0.01
BrI $Au(Bu_2dtc)$	1.28 ± 0.01	1.77 ± 0.02
$Cl_2Au(Hep_2dtc)$	1.35 ± 0.01	1.95 ± 0.02
$Au(Et_2dtc)$	1.70 ± 0.01	5.98 ± 0.02
$Au(Pr_2dtc)$	1.80 ± 0.01	6.39 ± 0.02
$Au(Bu_2dtc)$	1.62 ± 0.01	5.94 ± 0.01
Au(Oct·xan)	1.42 ± 0.01	6.16 ± 0.02
Au(dtp)	0.96 ± 0.01	6.09 ± 0.02
Ph_4PAuWS_4	0.86 ± 0.01	5.58 ± 0.02
$Ph_4AsAuWS_4$	1.10 ± 0.01	5.71 ± 0.02
Ph_4PAuCS_3	1.91 ± 0.01	6.43 ± 0.01
Ph_3PAuBr	2.77 ± 0.01	7.35 ± 0.01
Ph_3PAuCl	2.93 ± 0.01	7.50 ± 0.01
$Ph_3PAuSCN$	2.85 ± 0.01	7.65 ± 0.01
$Ph_3PAuSC(S)pip$	2.79 ± 0.02	7.60 ± 0.03
$Ph_3PAu(Ph_2dtc)$	2.95 ± 0.02	7.79 ± 0.04
$Ph_3PAuSeC(O)pip$	3.16 ± 0.02	7.93 ± 0.02
$Ph_3PAuSC(O)pip$	3.24 ± 0.02	8.35 ± 0.03
$Ph_3PAu(nds)$	3.74 ± 0.02	8.42 ± 0.02
Ph_3PAuCN	3.82 ± 0.01	10.11 ± 0.01
$(dmap)_4Au_2Cu_4I_2$	4.00 ± 0.01	9.37 ± 0.01
$(dmap)_4Au_2Cu_4(otf)_2$	3.88 ± 0.01	9.14 ± 0.01
$(dmap)_4Au_2Cu_2$	4.42 ± 0.01	9.86 ± 0.01
$(dmap)_4Au_2Li_2$	5.27 ± 0.01	11.29 ± 0.01
$(dmap)_4Au_2Li_2$	5.65 ± 0.01	12.01 ± 0.01
(dmap)Au	3.69 ± 0.01	6.73 ± 0.01
(dmap)Au	4.44 ± 0.01	7.32 ± 0.01
$AuCl(PPh_3)_2$	1.15 ± 0.01	8.03 ± 0.01
$(dmap)AuPPh_3$	4.84 ± 0.01	10.11 ± 0.01
$(dmap)AuCN(C_6H_{11})$	4.62 ± 0.01	10.35 ± 0.01
(dmap)SnAuBr	2.34 ± 0.02	6.51 ± 0.03
AuCN	2.35 ± 0.01	7.97 ± 0.01

Table 7.16 Abbreviations used in the chemical formulae of Table 7.15 (from [352])

Me = Methyl, Et = Ethyl, Pr = *n*-Propyl, Bu = *n*-Butyl, Hep = Heptyl, Oct = Octyl, Ph = Phenyl
Bu_4N^+ = Tetra-butylammonium
R_2dtc = *N,N*-Di-alkyl (aryl) dithiocarbamate
Rxan = *O*-Alkylxanthate
cdc = *N*-Cyanodithiocarbimate
pip = Piperidyl
OC(O)pip = *N*-Piperidyl-carbamate
ac = Acetate
dtp = Dialkyldithiophosphate
tdt = Toluene-3,4-dithiolate
mnt = Maleonitriledithiolate
nds = Naphthalene-1,8-dithiolate
dmamp = 2-(Dimethylamino)methyl phenyl
dmap = 2-Dimethylaminophenyl
otf = O_3SCF_3

[368–373], and Burini et al. [374] or when studying polynuclear metal complexes including three or more gold atoms or even metal atoms other than gold: trinuclear complexes with Au/Au/Au [375–377], dinuclear complexes with Au/Ag [378, 379], Au/Fe [380], Au/Pt [381], Au/Ir [307, 382], and Au/Ru [383, 384].

An interesting aspect of Au oxidation states is provided by the investigation of the pressure-induced transition from the mixed-valence state of Au(I)/Au(III) to the single valence state of Au(II) as described for $M_2[Au(I)X_2][Au(III)X_4]$ (M = Rb, Cs; X = Cl, Br, I) [385, 386]. The valence states of Au(I) and Au(III) at ambient pressure were clearly distinguishable. With increasing pressure, the doublets gradually increase their overlap. Finally, the ^{197}Au Mössbauer spectrum of $Cs_2Au_2I_6$ shows, at 12.5 GPa of applied pressure, only one doublet which was associated with Au(II).

The Au(II) oxidation state was also observed in cyclometalated gold dimers [386, 387] and in gold complexes of *bis*(diphenylphosphino)amine [388]. The specific feature of these Au(II) complexes is that their δ – ΔE_Q relation is located midway between that of Au(I) and Au(III) complexes.

The Au(0) oxidation state was observed after UV irradiation of adsorbed Au(III) ions on MnO_2 [389] and after the coprecipitation of Au(III) ions with $Mn(OH)_2$ which causes stoichiometric reduction to Au(0) [390]. For both cases, δ values (relative to bulk Au) and ΔE_Q values were reported to be ~0 mm s^{-1}. The isomer shift of Au(0) ($\delta_{Au} \sim 0$, $\delta_{Pt} \sim -1.2$ mm s^{-1}) is comparable to that of Au(I) (Fig. 7.82) because of the considerable screening of 6*s* electron density at the Au nucleus by the fully occupied 5*d* shell in Au(0) (with formal $5d^{10}6s$ configuration).

For the rarely observed oxidation state Au(−I) (with formal $5d^{10}6s^2$ configuration) in systems like CsAu, RbAu, and KAu [391], isomer shifts δ_{Au} as large as 7–8 mm s^{-1} have been detected. Obviously, the extra 6*s* electron with repect to Au (0) causes this sizable increase of $\rho(0)$.

Deviations from linear two-coordinate Au(I) geometry cause changes both in isomer shift and quadrupole splitting. Three-coordinate Au(I) complexes show considerably lower δ and greater ΔE_Q values than two-coordinate complexes as a result of the change in hybridization of the Au atoms [392–394]. In two- and three-coordinate Au(I) complexes, the ΔE_Q values range from 7 to 10 mm s^{-1}. Four-coordinate Au(I) complexes with identical ligands and ideal tetrahedral geometry exhibit zero quadrupole splitting [395, 396].

7.10.3 *Specific Applications*

7.10.3.1 Gold in Medicine

Chrysotherapy, the gold-based treatment of rheumatoid arthritis, is an important and established mode of therapy for this crippling desease. Au(I) sodium thiomalate (AuSTm) and Au(I) thioglucose (AuSTg) are widely used in chrysotherapy. Bovine serum albumin (BSA) is the principle binding site in blood for intramuscularly administered AuSTm and AuSTg. ^{197}Au Mössbauer spectroscopy was used to study these gold–protein complexes and to derive information about the environment of gold bound to macromolecules. The spectrum of BSA–AuSTm, for example, yields two well-resolved quadrupole doublets with $\delta_1 = 1.88$ mm s^{-1}, $\Delta E_{Q1} = 6.68$ mm s^{-1} and $\delta_2 = 1.70$ mm s^{-1}, $\Delta E_{Q2} = 6.50$ mm s^{-1}, consistent with two slightly different Au(I)S_2 environments [397]. Comparable parameters were reported for a number of related Au(I) thiol derivatives [398, 399]. Common for these drugs is that thiol-exchange reactions play a key role in the biology of gold: metabolic transformations of gold drugs begin with the displacement of their thiolate ligands, via ligand-exchange reactions with serum proteins, and culminate at the inflammatory sites where dicyanoaureate, $[Au(CN)_2]^-$, is formed. Mössbauer spectroscopy is a suitable technique to characterize the Au(I)S_2 environment of the drug and then to distinguish among thiol-exchange reactions [400] such as

1. Direct adduct formation of $[Au(CN)_2]^-$,
2. Cyanide displacement at cysteine 34 to form [–S–Au–(CN)], and
3. Formation of three-coordinate cysteine 34 or several of the 17 histidine side chains $[–S–Au(CN)_2]$ or $[–N–Au(CN)_2]^-$.

Despite the importance of Au(I) thiolates for the treatment of rheumatoid arthritis and inflammatory disorders, little is known of how the structures of these drugs are affected while being biologically active.

7.10.3.2 Gold Substitution in High-T_c Superconductors

Gold is the only known dopant to $YBa_2Cu_3O_{7-\delta}$, which increases the critical temperature T_c of transition to the superconducting state. In this way, T_c is enhanced from 97 K by 1.5 K. Eibschütz et al. [401] observed by Mössbauer studies that the valence state of gold changes as a function of oxygen stoichiometry from Au(III) at full oxygenation to Au(I) for the deoxygenated compound. Tomkowicz et al. [402] found that Au substitutes Cu(1) in chain positions while excess Cu atoms enter probably the Y sites. For $YBa_2Cu_{2.9}Au_{0.1}O_{7-\delta}$, an isomer shift of 3.29 mm s^{-1} (relative to metallic gold) and a quadrupole splitting of 3.61 mm s^{-1} was reported [403]; these values can be attributed to Au(III) in a square-planar oxygen coordination, as is expected for Au replacing Cu(1). Au(III) is the only known trivalent substituent for Cu(1) that is not detrimental to superconductivity, a fact that may be important for the outstandingly good properties of gold contacts for 1:2:3 superconductors [403].

7.10.3.3 Gold Minerals and Ores

In the field of gold mining, Mössbauer spectroscopy has potential advantages in analyzing gold minerals and ores [404], in studying treated gold concentrates and extracted gold species and, thus, contributing to the optimization of the extraction process of gold from ores. The main problem has been that of "hidden" gold, that is gold which is contained in the ore but does not show up under any of the normal observational techniques such as an optical or scanning electron microscope. Unless the chemical bonding of gold is known, the extraction scheme to recover it is by trial and error. Several articles are devoted to the characterization of the chemical state of "hidden" gold: "The extractive metallurgy of gold," [405] "The gold mineralogy and extraction," [406] "The Mössbauer study of gold in a copper refinery," [407] "the ^{197}Au and ^{57}Fe Mössbauer study of the roasting of refractory gold ores," [408] and "the ^{197}Au Mössbauer study of gold ores, mattes, roaster products and gold minerals" [409].

Gold is usually recovered from its ores by cyanide leaching, which converts the metallic gold in the ore into soluble $[Au(CN)_2]^-$. From some ores, however, the gold cannot, or at least can only partly be recovered in this way. The gold in such so-called refractory ores is often made amenable to cyanidation by roasting of floating concentrates, which consist mainly of pyrite (FeS_2) and arsenopyrite (FeAsS), the two minerals with which gold is usually associated [408].

A technology for treating low-grade gold-containing pyritic ores is the bacterial oxidation of these ores [406]. The principal bacterial culture used is *Thiobacillus ferrooxidans*. The oxidation occurs most rapidly along grain boundaries increasing the porosity of the material. This procedure operates efficiently to make gold accessible to the leachant.

The characterization of the "hidden" gold species and of the extraction products is possible by comparing their Mössbauer parameters with reference data obtained from systematic studies of gold alloys, natural and synthetic gold minerals, and of gold in ore concentrates [404]. In this context, the Mössbauer investigation of gold-bearing pyrite-rich materials plays an important role [410–412]: it seems that chemically bound gold substitutes for S in $[AsS]^{3-}$ in arsenopyrite or arsenian pyrite. This chemical state of gold is one of the existing stable states of gold in nature. In a publication dealing with the nature of gold in sulfides from a Chinese Au–Sb–W deposit, the authors speculate on the basis of the Lamb–Mössbauer factor that 81% of the gold in pyrite crystals occurs in the metallic state and that only 11% is chemically bound [413]. This finding indicates that the composition of gold ores could vary significantly from one deposit to another. Gold in ores, actually, could be present as submicroscopic metallic gold clusters, as submicroscopic particles of any of the known gold minerals, or it could form a dilute solid solution as an impurity in the lattices of minerals like pyrite or arsenopyrite [409]. In fact, any of these explanations may be correct for some ore and wrong for others.

7.10.3.4 Gold-Containing Catalysts

Bulk gold is regarded to be catalytically inactive. However, small well-dispersed supported gold particles were found to have high catalytic activity for different reactions [413]. It was argued that the catalytic activity is due to either (1) the highly abundant metal/support interface, (2) the presence of ionic gold or (3) to coordinatively unsaturated atoms at the surface of small gold clusters. Out of these suggestions, which one is relevant for a specific gold-containing catalyst has to be investigated in each case separately.

Extensive knowledge of the behavior of gold molecular cluster compounds and colloids, with well-defined particle sizes, using ^{197}Au Mössbauer absorption and emission spectroscopy has been gathered [414, 415]. The local sensitivity of these techniques makes it possible to resolve different sites within the particles.

Catalysts with gold particles ranging from 2 to 25 nm supported on mixed oxides consisting of TiO_2, SiO_2, Al_2O_3, MgO, and MnO_x were characterized in various Mössbauer studies [416–419]. The measured spectra were explained in terms of an inner-core (bulk-like) component plus a surface contribution with δ- and ΔE_Q-values determined by the local atomic coordination. The presence of ionic gold in active catalysts, as suggested by other work [420], however, could not be confirmed in these studies.

Wagner et al. have investigated a series of catalysts, their reaction pathways, and reaction products:

1. Three series of Au nanoparticles on oxidic iron catalysts were prepared by coprecipitation, characterized by ^{197}Au Mössbauer spectroscopy, and tested for their catalytic activity in the room-temperature oxidation of CO. Evidence was found that the most active catalyst comprises a combination of a noncrystalline and possibly hydrated gold oxyhydroxide, $AuOOH{\cdot}xH_2O$, and poorly crystallized ferrihydrate, $FeHO_8{\cdot}4H_2O$ [421]. This work represents the first study to positively identify gold oxyhydroxide as an active phase for CO oxidation. Later, it was confirmed that the activity in CO_2 production is related with the presence of –OH species on the support [422].

2. In another study, it was shown that the activity of hydrochlorination catalysts made by impregnation of activated carbon with tetrachloroaureate is due to the adsorption of $[AuCl_4]^-$ anions on the activated carbon [423]. Similar studies, dealing with the adsorption of aurocyanide, $[Au(CN)_2]^-$, onto activated carbon, have revealed that aurocyanide does not convert to metallic gold up to temperature of about 240°C [424, 425]
3. Colloids embedded in a silica sol–gel matrix were prepared by using fully alloyed Pd–Au particles. The Mössbauer data have yielded evidence that alloying Pd with Au in bimetallic colloids leads to enhanced catalytic hydrogenation and also to improved selectivity [426].

7.10.3.5 Gold Clusters

Small gold clusters (<100 atoms) have become the subject of interest because of their use as building blocks of nanoscale devices and because of their quantum-size effects and novel properties such as photoluminescence, magnetism, and optical activity [427].

Molecular clusters are monodisperse identical molecules, which are usually protected by organic ligands. An example is the ^{197}Au Mössbauer investigation of a series of gluthathionate-protected Au_x clusters (Au:SG) with $x = 11, 13, 20, 39$, and 55 probing the local environments of the Au sites. These studies provide evidence for the formation of Au:SG rings around a metallic Au core [427]. Cluster stabilization and subsequent Mössbauer investigation were also reported for thiol-capped Au nanoparticles, with only one sulfur atom being linked to three metal atoms [428], and for $Au_{55}(PPh_3)_{12}Cl_6$ molecules, exhibiting specific core and surface properties [429, 430].

Colloidal systems of particles are different from molecular clusters in having a small distribution in sizes. Also, colloidal systems may be stabilized by organic molecules, resulting in chemical bonds between these and the outer monolayer of metal atoms. The Mössbauer study of such systems includes rather diverse fields:

1. Nano-sized Au particles (>1.5 nm) obtained by evaporation of metallic gold on Mylar [431] exhibit differences in δ and ΔE_Q for surface and core atoms when the particle sizes are less than 6 nm.
2. Gold ruby glass is a silicate glass containing nanometer-sized bimetallic particles of Au (about 200 ppm) and up to 2 wt% of Sn. These particles impart the glass its distinctive color, which changes from purple red for virtually tin-free to brown for tin-rich glasses. On the basis of the ^{197}Au and ^{119}Sn Mössbauer investigation of these glasses, the action of tin as condensation nucleus for particle formation could be explained: tin induces a decrease of the average particle size compared with the size of particles with no or very little tin [432].
3. Au and Pt nanoparticles in the size range 1–17 nm exhibit different core and surface effects, detected via δ and ΔE_Q. The Au particles were measured in the

usual ^{197}Au transmission geometry, whereas for the Pt particles the emission method ($^{197}Pt \rightarrow {}^{197}Au^{+} + e^{-} + \nu_e + 0.6$ MeV) was applied [433, 434].

4. Bimetallic Au/Pd nanoparticles were prepared by ultrasound irradiation of a mixture solution of $NaAuCl_4 \cdot H_2O/PdCl_2$ $2NaCl \cdot 3H_2O$ by which the Au and Pd ions were reduced to the metallic state. The Mössbauer spectra of AuPd–SDS particles, with SDS (sodium dodecyl sulfate) representing the surfactant of the system, consist of two components, one for the pure Au core and the other for the alloy layer at the interface of Au core and Pd shell [435].

7.10.3.6 Gold Multilayers

In Fe/X multilayers, because of antiferromagnetic interlayer coupling through a layer of nonmagnetic atoms (i.e., X = Cr, Sn, Au), the magnetizations of adjacent Fe layers are oriented antiparallel. By applying an external field, the magnetizations of Fe layers are oriented parallel and then the resistivity decreases greatly. This effect is known as giant magnetoresistance (GMR) effect [436]. It is supposed that the GMR effect is a result of spin-dependent scattering of electrons, which occurs at interface sites. An issue raised in this context is the magnetism in nonmagnetic layers. For the study of such nonmagnetic spacer layers, hyperfine-field measurements by nonmagnetic Mössbauer isotopes (i.e., ^{197}Au and ^{119}Sn) are chosen as probing sensors [437].

Since natural Au consists solely of ^{197}Au, the interface-selective enrichment technique cannot be applied in ^{197}Au studies. The absorber thickness for ^{197}Au is required to be large and therefore multilayered samples of ^{197}Au layers/3*d* metal layers have to be prepared. The spectra for Au/Fe with varying Au-layer thickness are shown in Fig. 7.83 [437]. The results were interpreted as follows: large magnetic hyperfine fields at Au sites exist only within two monolayers at the interface region, which are supposed to be induced by direct coupling with antiferromagnetically oriented Fe 3*d* atoms.

Shinjo et al. have investigated a number of multilayers and interfaces, also by ^{119}Sn Mössbauer spectroscopy [437–439], as well as different kinds of multilayers of Au/TM with transition metals TM = Fe, Co, Ni [440–442].

Conversion electron Mössbauer spectroscopy (CEMS) measurements with back scattering geometry have the merit that spectra can be obtained from a sample with much less isotope content compared with transmission measurements. Another merit is that a sample, deposited on a thick substrate, could be measured, and that because of the limited escape depth of the conversion electrons, depth-selective surface studies are possible. The CEMS technique was found to be best applicable to specimens of 10–100 μg Au cm^{-2}, i.e., about two orders of magnitudes thinner than required for measurements in transmission mode [443]. This way (1) very thin films of gold alloys, as well as laser- and in beam-modified surfaces in the submicrometers range of depth [443], and (2) metallic gold precipitates in implanted MgO crystals [444] were investigated.

A specific feature of the CEMS technique is the possibility for nondestructive testing of the surface composition of paintings, ancient coins, and pottery or other valuable objects. Wagner et al. [445] have investigated Celtic gold coins (from the time period 480–15 B.C.) and have shown that the surface of the coins consist of two phases, one of which is strongly enriched in gold compared to the bulk composition.

7.10.3.7 Intermetallic Compounds

Dunlap et al. [446] measured the temperature dependence of the ^{197}Au Mössbauer resonance in ferromagnetic Au_4V and found a line broadening below the Curie temperature due to magnetic hyperfine interaction. The magnitude of the effective field indicates that the magnetic moment in Au_4V is localized on the V atoms and little moment, if any, resides on the Au atoms. The temperature dependence of the hyperfine field approximately follows the bulk magnetization. The internal field at

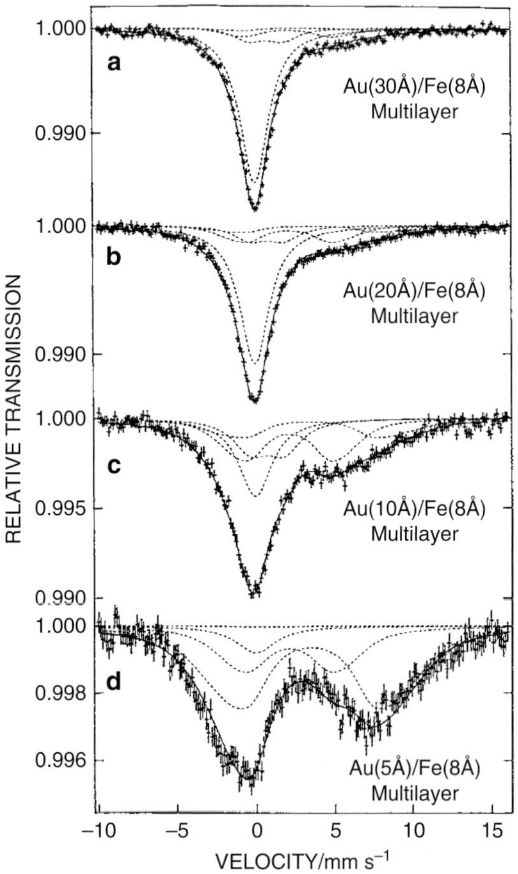

Fig. 7.83 ^{197}Au Mössbauer transmission spectra of Au/Fe multilayer systems with varying Au-layer thickness, measured at 16 K and fitted by a four-component model, including magnetic hyperfine interaction at the Au layer atoms (from [437])

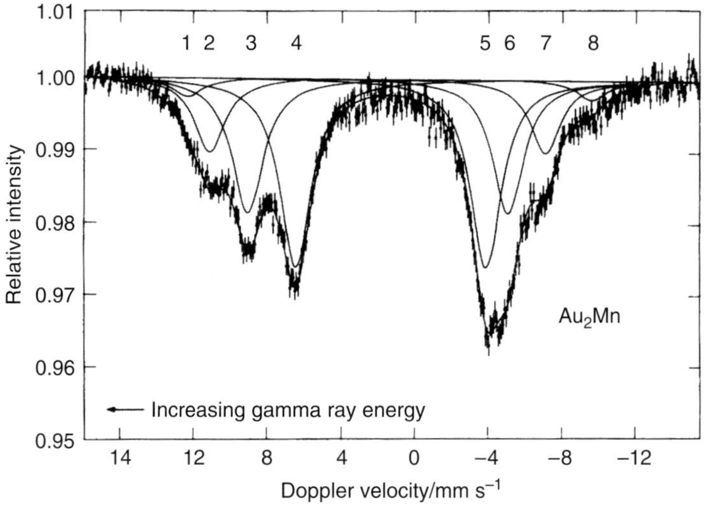

Fig. 7.84 ^{197}Au Mössbauer spectrum of Au_2Mn at 4.2 K (from [448])

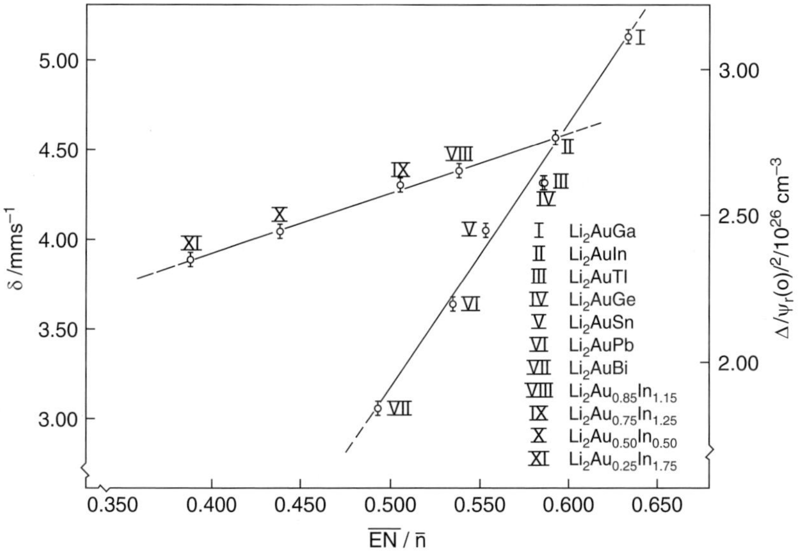

Fig. 7.85 ^{197}Au isomer shift and difference in electron density at the gold nucleus, $\Delta|\psi_r(o)|^2$ (relative to metallic gold), in Li_2AuX and $Li_2Au_{2-x}In_x$ as a function of the average electronegativity (Allred–Rochow) normalized to the average number of outer electrons, $\overline{EN}/\overline{n}$, of the atoms in the first three coordination spheres around a gold atom (from [451])

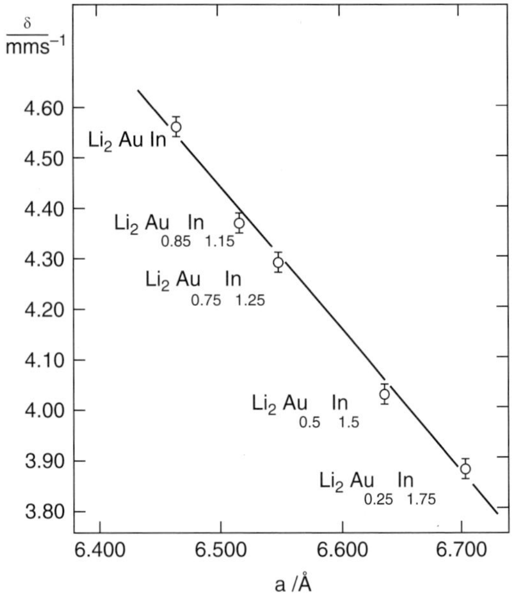

Fig. 7.86 ^{197}Au isomer shift in $Li_2Au_{2-x}In_x$ as a function of the lattice constant (from [451])

the gold sites (measured at 4.2 K) in Au_4V takes the value 18.5 ± 2.5 T and in Au_4Mn the value 84.7 ± 2.5 T [447].

Extensive investigations of the intermetallic compounds Au_4Mn, Au_3Mn, Au_2Mn, AuMn, $AuMn_3$ and of dilute alloys of gold in manganese have been carried out by Patterson et al. [448]. One of their measured spectra of Au_2Mn is shown in Fig. 7.84. It is composed of eight ($M1 + E2$)-allowed resonance lines (with $E2/(E2 + M1) \sim 0.1$) arising from magnetic hyperfine interaction, somewhat shifted towards each other by electric quadrupole interaction. The isomer shift, magnetic dipole, and electric quadrupole splitting observed in these alloys were discussed in terms of the magnetic and crystalline structure. For Au_2Mn, the authors found an effective field at the ^{197}Au nucleus of 159.0 T. The spectra shown in Fig. 7.84 are consistent with the helical magnetic structure assumed for this alloy. Mössbauer measurements under applied pressure revealed a 7% increase in the magnitude of the effective field between 0 and 30 MPa, which is believed to be due to (partial) uncoiling of the Mn spin helix. The observed large quadrupole splitting in Au_2Mn has been attributed to about 0.17 of one 5d hole in the $5d_{x^2-y^2}$ AO of gold.

Erickson et al. [449] have carefully investigated the ^{197}Au Mössbauer resonance line shape and recoilless fraction as a function of temperature, absorber thickness, and lattice order in Cu_3Au. The spectra have been interpreted in terms of a pseudoquadrupole splitting, which has been introduced to describe line broadening due to noncubic local atomic environments of gold atoms, and of the temperature-dependent recoilless fraction for gold. The pseudoquadrupole splitting has been found to be smaller and the recoilless fraction to be larger in the more highly ordered state.

Other ^{197}Au Mössbauer measurements have been carried out on $AuAl_2$, $AuGa_2$, $AuIn_2$, and $AuSb_2$ by Thomson et al. [450]. The observed isomer shifts with respect to the source of ^{197}Au in Pt are −1.22 (Au metal), +2.29 ($AuSb_2$), +3.49 ($AuIn_2$), +4.45 ($AuGa_2$), and +5.97 ($AuAl_2$) mm s^{-1}, and reflect increasing electron density at the Au nucleus in this order of alloys as compared to that of Au metal. The authors concluded that, when volume changes for the alloys with respect to the pure metal are taken into account, approximately one 6*s*-like electron is transferred to the Au in each of the alloys. From the temperature dependence of the recoil-free fraction, the authors evaluated Debye temperatures for the various intermetallic compounds.

The series of ternary intermetallic gold compounds Li_2AuX (X = Ga, In, Tl; Ge, Sn, Pb, Bi) and $Li_2Au_{2-x}In_x$ ($1.0 \leq x \leq 1.75$), all of which crystallize in the cubic NaTl structure, have been studied at 4.2 K by Gütlich et al. [451]. The isomer shifts derived from the single-line spectra have been correlated with the average Allred–Rochow electronegativity [452], $\overline{\mathrm{EN}}$, of the first three coordination spheres around the gold atoms, normalized to the average number of outer electrons $\overline{n}$ and corrected for the distance from the gold atom according to the expression

$$\overline{\mathrm{EN}}/\overline{n} = \frac{\mathrm{EN(Li1)} + \mathrm{EN(Li2)}(r_1^2/r_2^2) + 2\mathrm{EN(Au)}(r_1^2/r_3^2) + \mathrm{ENX}}{2n(\mathrm{Li}) + 3n(\mathrm{Au}) + n(\mathrm{X})}$$

(Li1 and Li2 are crystallographically different) for the Li_2AuX series and a similar expression for the series $Li_2Au_{2-x}In_x$ [451]. Figure 7.85 shows this correlation $\delta = f(\overline{\mathrm{EN}}/\overline{n})$ for the two series, and also the correlation with changes in electron density at the nucleus, $\Delta|\psi_r(o)|^2$, evaluated from results given in [341]. The isomer shift in these ternary gold alloys also correlates well with structural data from X-ray diffraction studies, as shown in Fig. 7.86. From these investigations, it is suggested that substantial 5*d* – 6*s* mixing occurs with nearly matching charge compensation and additional depletion of 5*d* charge through an interaction of the 5*d* band of gold with host orbitals of proper symmetry. The experimental findings are consistent with a net charge-flow off the gold sites in these gold compounds.

7.10.3.8 Alloys

A number of publications deal with ^{197}Au Mössbauer studies of various gold alloys. As the primary goal of this review is directed towards applications of Mössbauer spectroscopy to chemical compounds rather than metallic systems, we present here a brief enumeration of most of the work by giving reference to relevant publications and providing a note concerning the essential points of interest in each case:

References	Studied system	Points of interest
[336]	Au (metal)	First observation of ^{197}Au Mossbauer effect
[453]	^{197}Au embedded in Au, Pt, stainless steel, Fe, Co, Ni	Recoil-free emission fraction, isomer shift, $\|H^{int}\|$
[454]	Dilute alloys of Au with Fe, Co, Ni	^{197}Au nuclear data, $\|H^{int}\|$, line width as function of absorber thickness
[455]	Au (metal)	Resonance effect as function of absorber thickness, isomer shift, Debye temperature of source and absorber
[456]	^{197}Au in 19 metals and semiconductors versus gold metal absorber	Electron transfer from isomer shifts, correlation between isomer shift and host electronegativities
[457]	^{197}Au dissolved in ferromagnetic hosts of Fe, Co, Ni as sources versus Au metal absorber	Nuclear Zeeman effect in Au atoms, super-transferred hf fields, $\|H^{int}\|$ at Au sites
[458]	Dilute gold alloys with Cu, Ag, Ni, Pd, and Pt as absorbers	Correlation of isomer shift with residual electrical resistivity, wave function at Fermi level, *s*-band population of gold
[459]	Au microcrystals	Effect of crystal size on recoil-free fraction
[460]	Au microcrystals	Debye temperature for microcrystals
[461]	^{197}Au in iron metal	Magnetic hyperfine splitting of ^{197}Au, nuclear moment of 77.3 keV state, $\|H^{int}\|$ of supertransferred field at ^{197}Au
[462]	^{197}Au and ^{119}Sn in various metallic matrices as absorber	Interpretation of isomer shift
[463]	Gold alloys with transition metals (Au_4V, Au_4Mn, Au_4Cr), crystallographically ordered and disordered phases	Magnetic behavior as function of disorder, model for magnetic behavior of Au–V system
[464]	Alloys of Pd–Au–Fe (2 at%)	Mössbauer effect in ^{57}Fe and ^{197}Au, study of band filling, hyperfine fields, isomer shifts
[465]	Au microcrystals supported in gelatin	Correlation of isomer shift with lattice contraction
[466]	Neutron-irradiated platinum	Debye–Waller factor of Au in Pt after low-temperature neutron irradiation, lattice defect
[467]	$Au_{0.8}Cr_{0.2}$	Hyperfine field and isomer shift in crystallographically ordered and disordered phases
[468]	Au–Sn alloys	Isomer shift study
[469]	Au–Hg alloys	Isomer shift as function of Hg concentration
[470]	Au–Cu and Au–Ag alloys	Isomer shift and electrical resistivity as function of alloy composition and, in Cu_3AU, of pressure; model to describe δ in terms of average atomic volume, of short-range parameter and alloy composition; average charge density on Au
[471]	Au metal	Thick absorber line-shape analysis and interference effects

(*continued*)

References	Studied system	Points of interest
[472]	Au–Ni and Cu–Ni–Au alloys	Magnetic hyperfine spitting at ^{197}Au, $\|H^{int}\|$ and isomer shift as function of composition, model to describe charge density distribution
[473]	Various gold alloys	Review of ^{197}Au Mossbauer work on metallic systems of gold
[474]	Alloys of Au with Ag, Sn, Ni, Pd, Pt, Al	Charge between host and Au atoms from trends of isomer shift and work function data (ESCA) correlated with electronegativity of host elements
[475]	Dilute Au in Zn and Cd single crystals	Study of the microscopic vibrational properties of the Au impurity in the host lattice
[476]	Au–Cu and Au–Ag alloys	Correlation of isomer shift with Au 5*d* population
[477]	Au-substituted Al–Cu–Fe quasi crystalline alloys, a ^{197}Au and ^{57}Fe study	Replacement of Cu by Au on BC1 sites and strong hybridization of Au with Al neighbors
[478]	Au–Al alloys	Comparison of measured δ and ΔE_Q values with calculated $\varrho(0)$ and EFG data yields $\Delta R/R = 9.2 \cdot 10^{-5}$ (comparable to $\Delta R/R = 6.65 \cdot 10^{-5}$ from [347]) and $Q = 0.56b$ for ^{197}Au
[479]	Half-Heusler ferromagnetic MnAuSn in the antiferromagnetic CdAuSn matrix, a ^{119}Sn, ^{155}Cd, and ^{197}Au study	Investigation of the atomic order, local surroundings, and magnetic hyperfine fields
[480]	CdAuSn/MnAuSn, a ^{119}Sn,^{155}Cd and ^{197}Au study	Mössbauer, X-ray diffraction and band structure analysis reveals that GdAuSn can be described as $[Gd]^{3+}$ stuffing the $[AuSn]^{3-}$ sublattice

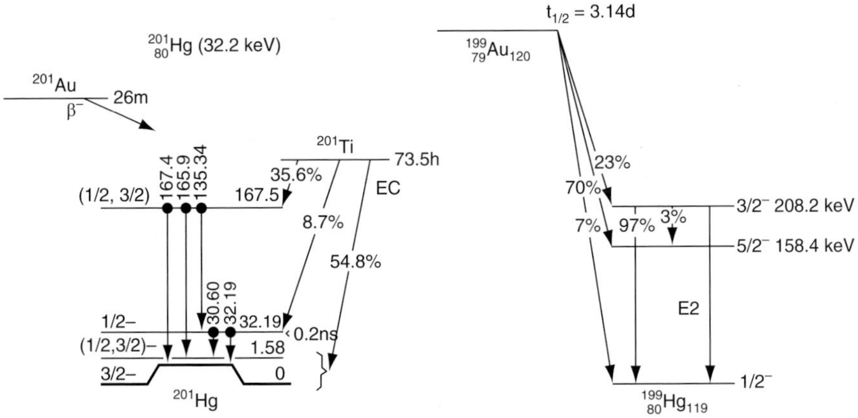

Fig. 7.87 Nuclear decay schemes for the ^{201}Hg ([217]) and ^{199}Hg ([482]) Mössbauer isotopes

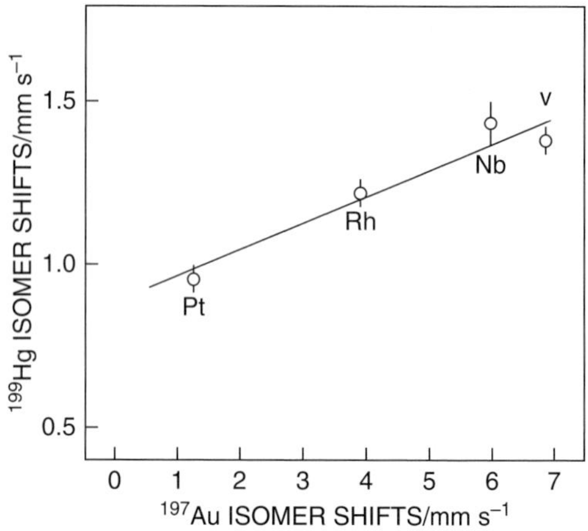

Fig. 7.88 Isomer shifts observed for ^{199}Hg in Pt, Rh, Nb, and V matrices plotted versus the corresponding shifts for the 77 keV Mössbauer resonance in ^{197}Au. The sign of the ^{199}Hg shifts has been reversed with respect to the experimental results in order to be consistent with the absorber convention. The shifts are given relative to HgF_2 and metallic gold for ^{199}Hg and ^{197}Au, respectively (from [483])

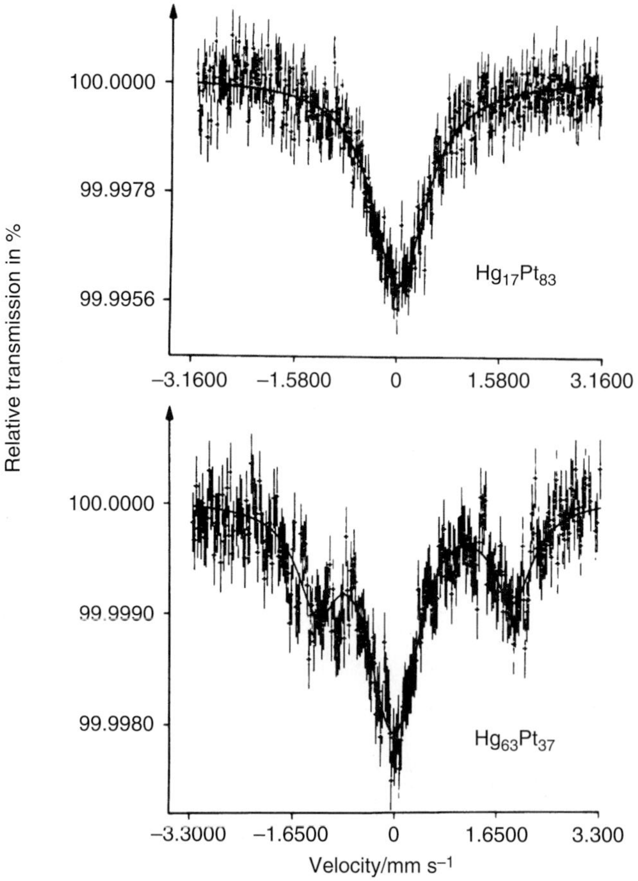

Fig. 7.89 Mössbauer spectra of $^{199}Hg_{17}Pt_{83}$ and $^{199}Hg_{63}Pt_{37}$ taken at 4.2 K with a ^{199}Au/Pt source (from [482])

7.11 Mercury (^{199,201}Hg)

The first Mössbauer measurements involving mercury isotopes were reported by Carlson and Temperley [481], in 1969. They observed the resonance absorption of the 32.2 keV γ-transition in ^{201}Hg (Fig. 7.87). The experiment was performed with zero velocity by comparing the detector counts at 70 K with those registered at 300 K. The short half-life of the excited state (0.2 ns) leads to a natural line width of 43 mm s^{-1}. Furthermore, the internal conversion coefficient is very large ($\alpha = 39$) and the ^{201}Tl precursor populates the 32 keV Mössbauer level very inefficiently ($\approx$10%).

In 1971, Walcher [326] succeeded in observing a resonance effect of about 0.6% in ^{201}Hg as a function of the Doppler velocity using a Tl_2O_3 source and an enriched (81% ^{201}Hg) HgO absorber at 4.2 K. The half-width turned out to be $\Gamma_{exp} = 76$ (10) mm s^{-1}, corresponding to a lower limit of the half-life of $t_{1/2} \geq 0.1$ ns. It is clear that the properties of the ^{201}Hg Mössbauer isotope do not render it an interesting isotope from a chemical point of view.

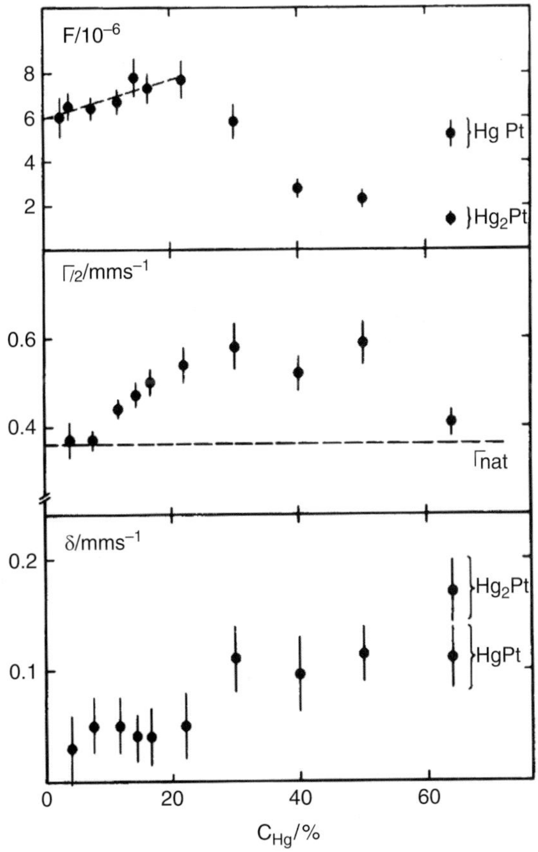

Fig. 7.90 $F = f_A f_S \delta_R$ (f_S, f_A Debye–Waller factors of the source and the absorber, respectively; δ_R, experimental correction factor, which is constant as function of the mercury concentration), experimental line width $\Gamma/2$ and isomer shift δ as a function of the Hg content of the PtHg alloy (taken from [482])

In 1976, Koch et al. [483] and Wurtinger [482] simultaneously succeeded in observing the resonance effect of the high energetic (158.4 keV) $5/2^{-}–1/2^{-}$ $E2$ transition in ^{199}Hg, which is of the order of around 10^{-5}, using the backscattering geometry [483] and the transmission geometry with the current integration technique [482, 483], respectively. A source material involving a Pt metal foil with ^{198}Pt enriched to about 95% was activated, making use of the reaction ^{198}Pt(n, γ) ^{199}Pt. ^{199}Pt decays with a half-life of 30 min to the radioactive ^{199}Au parent nucleus ($t_{1/2} = 3.14$ days) of ^{199}Hg (Fig. 7.87).

The line width of $\Gamma_{exp} = 0.74 \pm 0.06$ mm s^{-1} [482] with a ^{199}HgPt alloy absorber is in very good agreement with the value of $t_{1/2} = 2.37 \pm 0.07$ ns as measured by γ–γ coincidence technique. Koch et al. [483] have investigated sources of Pt metal and alloys of the composition $Pt_{0.05}Rh_{0.95}$, $Pt_{0.02}V_{0.98}$, and $Pt_{0.05}Nb_{0.95}$ with HgF_2 as an absorber. In Fig. 7.88, the isomer shift of ^{199}Hg is plotted versus the corresponding shifts of the 77 keV resonance in ^{197}Au. A linear relationship with a positive slope is obtained, which implies the sign of $\Delta\langle r^2\rangle$ to be the same for ^{199}Hg as for ^{197}Au, for which a positive value has been established. Comparison with ^{193}Ir, ^{195}Pt, and ^{197}Au isomer shifts led to an estimate of $\Delta\langle r^2\rangle_{^{199}Hg} \approx 1 \cdot 10^{-3}$ fm^2 [483]. Wurtinger [482] has studied HgPt alloys with a

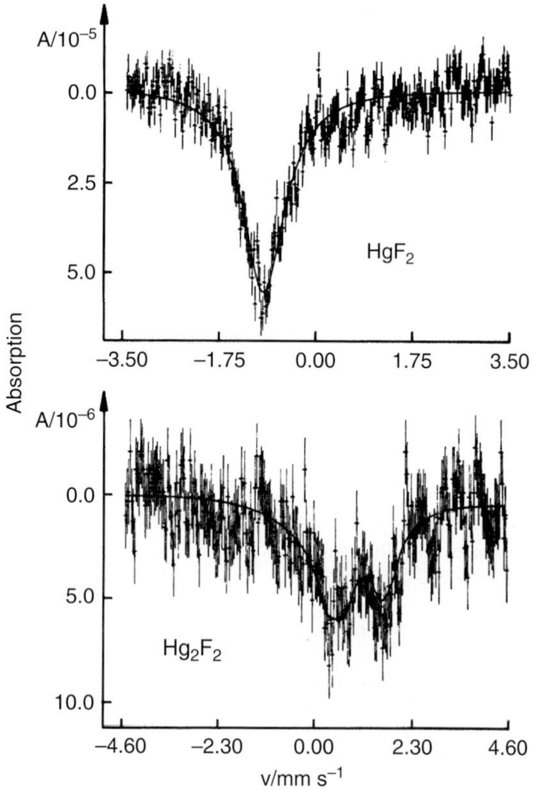

Fig. 7.91 Mössbauer spectra of $^{199}HgF_2$ and $^{199}Hg_2F_2$ taken at 4.2 K with a ^{199}Au/Pt source (from [484])

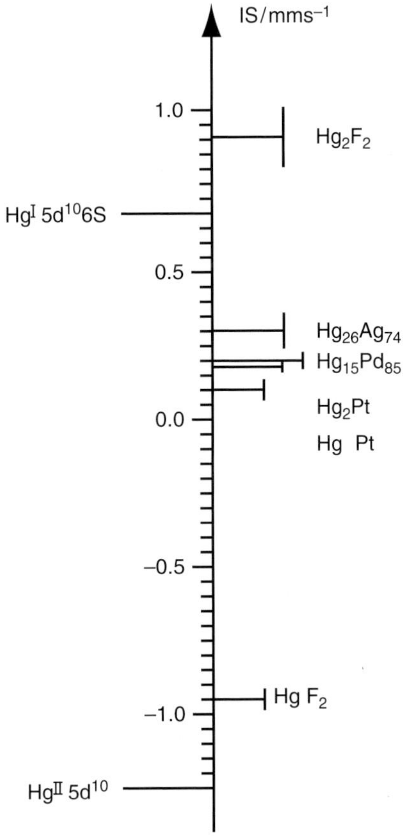

Fig. 7.92 Mercury compounds arranged according to increasing isomer shift values. Using the calculated electron densities, the isomer shifts expected for free Hg^{+}- and Hg^{2+}-ions are shown. The experimental error is given by the *bars* at the *right* side (from [485])

concentration of mercury up to 65%. Two representative spectra are depicted in Fig. 7.89. From analyzing the Mössbauer spectra, the recoil-free fraction, the half-width, and the isomer shift could be derived (Fig. 7.90). The alloy system with <18% mercury content consists of a solid solution with a slightly increasing Debye temperature θ, around 220 K, as a function of the Hg concentration; the Mössbauer spectrum shows a single resonance line. The spectrum of $Hg_{63}Pt_{37}$, however, consists of a quadrupole splitting superimposed by a single line. This has been explained as a mixture of HgPt and Hg_2Pt intermetallic compounds, where the tetragonal structure of Hg_2Pt leads to a quadrupole coupling constant $eQV_{zz}/4 = (1.83 \pm 0.03)$ mm s^{-1}. Comparing this large value with the EFG, as obtained from point-charge calculations on Hg_2Pt [484], yielded the nuclear quadrupole moment of the first excited $5/2^{-}$ state in ^{199}Hg of $Q = -0.8(4)b$. The large error of Q of 50% results from limitations of the applied point-charge model. In HgPt, no observable quadrupole splitting is expected because of its small tetragonal distortion from cubic symmetry. The corresponding Debye temperatures are $\theta_D = (167 \pm 4)$ K and $\theta_D = (206 \pm 7)$ K for Hg_2Pt and HgPt, respectively.

Wurtinger [484, 485] succeeded in measuring the Mössbauer effect of ^{199}Hg in Hg_2F_2 (Fig. 7.91), which shows a large quadrupole interaction with $eQV_{zz}/4 = 0.76$ (9) mm s^{-1}. The difference in the isomer shift between HgF_2 and Hg_2F_2 of $\delta = -1.77$ mm s^{-1} (Fig. 7.92) in connection with the electron density difference derived from relativistic Hartree–Fock and MO calculations, which were calibrated with isoelectronic gold compounds, led to a $\Delta\langle r^2\rangle$ value in ^{199}Hg of $\delta\langle r^2\rangle_{^{199}\mathrm{Hg}} = (3.1 \pm 1.1) \cdot 10^{-3}$. This is in excellent agreement with muonic data [263]. Although the ^{199}Hg isotope is not very easy to handle, improvements in Mössbauer technology (current integration technique) allows the observation of the very small resonance effect and the deduction of interesting information from ^{199}Hg hyperfine interaction and recoilless fraction.

References

References to Sect. 7.1

1. Stevens, J.G., Stevens, V.E.: Mössbauer Effect Data Index, Covering the 1975 Literature. IFI Plenum, New York (1976)
2. Obenshain, F.E., Wegener, H.H.F.: Phys. Rev. **121**, 1344 (1961)
3. Travis, J.C., Spijkerman, J.J.: In: Gruverman, I.J. (ed.) Mössbauer Effect Methodology, vol. 4, p. 237. Plenum, New York (1970)
4. Obenshain, F.E.: In: Stevens, J.G., Stevens, V.E. (eds.) Mössbauer Effect Data Index, p. 13. Adam Hilger, London (1972)
5. Erich, U., Quitmann, D.: In: Matthias, E., Shirley, D.A.: (eds.) Hyperfine Struct. Nucl. Radiat., Proceedings, p. 130. Amsterdam, North-Holland (1968)
6. Erich, U.: Z. Phys. **227**, 25 (1969)
7. Spijkerman, J.J.: The Mössbauer Effect. In: Symp. Faraday Soc, 1967, vol. 1, p. 134. Butterworth, London (1968)
8. Love, J.C., Obenshain, F.E., Czjzek, G.: Phys. Rev., 2827 (1971)
9. Obenshain, F.E., Williams, J.C., Houk, L.W.: J. Inorg. Nucl. Chem. **38**, 19 (1976)
10. Ambe, F., Ambe, S., Takeda, M., Wei, H.H., Ohki, K., Saito, N.: Radiochem. Radioanal. Lett. **1**, 341 (1969)
11. Erich, U., Fröhlich, K., Gütlich, P., Webb, G.A.: Inorg. Nucl. Chem. Lett. **5**, 855 (1969)
12. Erich, U., Kankeleit, E., Prange, H., Hüfner, S.: J. Appl. Phys. **40**, 1491 (1969)
13. Childs, W.J., Goodman, L.S.: Phys. Rev. **170**, 136 (1968)
14. Göring, J.: Z. Naturforsch. **A 26**, 1931 (1971)
15. Dunlap, B.D.: Mössbauer effect data index. In: Stevens, J.G., Stevens, V.E. (eds.) An Introduction to Electric Quadrupole Interactions in Mössbauer Spectroscopy. Adam Hilger, London (1972)
16. Townes, C.H., Dailey, B.P.: J. Chem. Phys. **17**, 782 (1949)
17. Dailey, B.P., Townes, C.H.: J. Chem. Phys. **23**, 118 (1955)
18. Sekizawa, H., Okada, T., Okamoto, S., Ambe, F.: J. Phys. (Paris) Colloq. **32**(Suppl.), C1–C326 (1971)
19. Gütlich, P., Hasselbach, K.M., Rummel, H., Spiering, H.: J. Chem. Phys. **81**, 1396 (1984)
20. Rummel, H., Cohen, R.L., Gütlich, P., West, K.W.: Appl. Phys. Lett. **40**, 477 (1982)
21. Dehe, G., Griesbach, P., Gütlich, P., Suwalski, J.: Appl. Phys. **A 43**, 213 (1987)
22. McCammon, C.A., Morrish, A.H., Pollard, R.J.: Hyperfine Interact. **28**, 483 (1986)
23. Stadnik, Z.M., Griesbach, P., Dehe, G., Gütlich, P., Kohara, T., Stroink, G.: Phys. Rev. **B 35**, 6588 (1987)

24. Stadnik, Z.M., Griesbach, P., Gütlich, P., Stroink, G., Kohara, T.: Hyperfine Interact. **41**, 705 (1988)
25. Lenglet, M., Guillamet, R., Durr, J., Gryffroy, D., Vandenberghe, R.E.: Solid State Commun. **74**, 1035 (1990)
26. Nasredinov, F.S., Seregin, P.P., Masterov, V.F., Seregin, N.P., Prikhodko, O.A., Sagatov, M.A.: J. Phys. Condens. Matter **7**, 2339 (1995)
27. Masterov, V.F., Nasredinov, F.S., Seregin, N.P., Seregin, P.P.: J. Exp. Theor. Phys. **87**, 588 (1998)
28. Masterov, V.F., Nasredinov, F.S., Seregin, N.P., Seregin, P.P.: Physica Status Solidi B Basic Res. **207**, 223 (1998)
29. Nemov, S.A., Seregin, P.P., Seregin, N.P., Davydov, A.V.: Phys. Solid State **46**, 997 (2004)
30. Okada, T., Noro, Y., Kobayashi, Y., Kitazawa, H., Ambe, F.: Phys. Lett. **A 209**, 241 (1995)
31. Hannebauer, B., Schmidt, P.C., Kniep, R., Jansen, N., Walcher, D., Gütlich, P., Gottschall, R., Schollhorn, R., Methfessel, M.: Z. Naturforsch. **A 51**, 515 (1996)
32. Jansen, N., Walcher, D., Gütlich, P., Haussinger, D., Hannebauer, B., Kniep, R., Lutar, K., Schmidt, P.C., Sellmann, D., Zemva, B.: Nuovo Cimento Soc. Ital. Fis. D – Condens. Matter At. Mol. Chem. Phys. Fluids Plasmas Biophys. **18**, 231 (1996)
33. Roth, T., Leupold, O., Wille, H.C., Ruffer, R., Quast, K.W., Rohlsberger, R., Burkel, E.: Phys. Rev. **B 71**, 140401 (2005)
34. Sergueev, I., Chumakov, A.I., Beaume-Dang, T.H.D., Rüffer, R., Strohm, C., van Bürck, U.: Phys. Rev. Lett. **99**, 097601 (2007)
35. Erich, U., Göring, J., Hüfner, S., Kankeleit, E.: Phys. Lett. **A 31**, 492 (1970)
36. Ferrando, W.A., Segnan, R., Schindler, A.I.: Phys. Rev. **B5**, 4657 (1972)
37. Tansil, J.E., Obenshain, F.E., Czjzek, G.: Phys. Rev. **B6**, 2796 (1972)
38. Göring, J.: Phys. Status Solidi **B 57**, K11 (1973)
39. Göring, J.: Phys. Status Solidi **B 57**, K7 (1973)
40. Fink, J., Czjzek, G., Schmidt, H., Rübenbauer, K., Coey, J.M.D., Brusetti, R.: J. Phys. (Paris) Colloq. **C 35**(Suppl. 6), C6–C657 (1974)
41. Czjzek, G., Fink, J., Schmidt, H., Krill, G., Gautier, F., Lapierre, M.F., Robert, C.: J. Phys. (Paris) Colloq. **35**(Suppl.), C6–C621 (1974)
42. Krill, G., Lapierre, M.F., Gautier, F., Robert, C., Czjzek, G., Fink, J., Schmidt, H.: J. Phys. C: Solid State Phys. **9**, 761 (1976)
43. Göring, J., Wurtinger, W., Link, R.: J. Appl. Phys. **49**, 269 (1978)
44. Dale, B.W., Dickinson, R.J., Parish, R.V.: Chem. Phys. Lett. **64**, 375 (1979)
45. Kobayashi, Y., Katada, M., Sano, H., Okada, T., Asai, K., Iwamoto, M., Ambe, F.: Hyperfine Interact. **54**, 585 (1990)
46. Gütlich, P., Range, K.J., Felser, C., Schultz-Munzenberg, C., Tremel, W., Walcher, D., Waldeck, M.: Angew. Chem. Int. Ed. **38**, 2381 (1999)
47. Ksenofontov, V., Reiman, S., Walcher, D., Garcia, Y., Doroshenko, N., Gütlich, P.: Hyperfine Interact. **139**, 107 (2002)
48. Stadnik, Z.M., Wang, P., Jansen, N., Walcher, D., Gütlich, P., Kanomata, T.: J. Phys. Condens. Matter **20**, 325230 (2008)
49. Stadnik, Z.M., Wang, P., Jansen, N., Walcher, D., Gütlich, P., Kanomata, T.: J. Phys. Condens. Matter **20**, 285227 (2008)

References to Sect. 7.2

50. Nagle, D.E., Craig, P.P., Keller, W.E.: Nature **186**, 707 (1960)
51. Craig, P.P., Nagle, D.E., Cochran, D.R.F.: Phys. Rev. Lett. **4**, 561 (1960)

52. Aksenov, S.I., Alfimenkov, V.P., Lushchikov, V.I., Ostanevich, Yu M., Shapiro, F.L., Yen, W.: Zh. Eksperim. i Teor. Fiz. **40**, 88 (1961). Soviet Phys. JETP **13**, 63 (1961)
53. Alfimenkov, V.P., Ostanevich, Yu M., Ruskov, T., Strelkov, A.V., Shapiro, F.L., Yen, W.K.: Zh. Eksperim. i Teor. Fiz. **42**, 1029 (1962). Soviet Phys. JETP **15**, 713 (1962)
54. de Waard, H., Perlow, G.J.: Phys. Rev. Lett. **24**, 566 (1970)
55. Perlow, G.J.: In: Cohen, S.G., Pasternak, M. (eds.) Perspectives in Mössbauer Spectroscopy, p. 221. Plenum, New York (1972)
56. Griesinger, D., Pound, R.V., Vetterling, W.: Phys. Rev. **B 15**, 3291 (1977)
57. Lipkin, H.J.: Ann. Phys. (Leipz.) **23**, 28 (1963)
58. Perlow, G.J., Campbell, L.E., Conroy, L.E., Potzel, W.: Phys. Rev. **B7**, 4044 (1973)
59. Perlow, G.J., Potzel, W., Kash, R.M., de Waard, H.: J. Phys. Colloq. **35**, C6–C197 (1974)
60. Potzel, W., Forster, A., Kalvius, G.M.: J. Phys. Colloq. **37**, C6–C691 (1976)
61. Breskrovny, A.I., Lebedev, N.A., Ostanevich, Y.M.L.: Joint Institute for nuclear research report, Dubna (1971) (private communication)
62. Forster, A.A.: Diplomarbeit, Physik-Department E15, Technische Universitat München (1976)
63. Forster, A., Potzel, W., Kalvius, G.M.: Z. Phys. **B 37**, 209 (1980)
64. Katila, T., Riski, K.: Hyperfine Interact. **13**, 119 (1983)
65. Katila, T., Riski, K.J.: Phys. Lett. **A 83**, 51 (1981)
66. Ikonen, E., Seppa, H., Potzel, W., Schäfer, C.: Rev. Sci. Instrum. **62**, 441 (1991)
67. Potzel, W., Schäfer, C., Steiner, M., Karzel, H., Schiessl, W., Peter, M., Kalvius, G.M., Katila, T., Ikonen, E., Helisto, P., Hietaniemi, J., Riski, K.: Hyperfine Interact. **72**, 197 (1992)
68. Potzel, W., Adlassnig, W., Narger, U., Obenhuber, T., Riski, K., Kalvius, G.M.: Phys. Rev. **B 30**, 4980 (1984)
69. Karzel, H., Potzel, W., Kofferlein, M., Schiessl, W., Steiner, M., Hiller, U., Kalvius, G.M., Mitchell, D.W., Das, T.P., Blaha, P., Schwarz, K., Pasternak, M.P.: Phys. Rev. **B 53**, 11425 (1996)
70. Schiessl, W., Potzel, W., Karzel, H., Steiner, M., Kofferlein, M., Kalvius, G.M., Melzer, K., Dietzmann, G., Martin, A., Halevy, I., Gal, J., Schäfer, W., Will, G., Mitchell, D.W., Das, T. P.: Hyperfine Interact. **90**, 359 (1994)
71. Potzel, W., Schäfer, W., Kalvius, G.M.: Hyperfine Interact. **130**, 241 (2000)
72. Vetterling, W.T., Candela, D.: Phys. Rev. **B 27**, 5394 (1983)
73. Pound, R.V., Niesen, L., Dewaard, H., Zhang, G.L.: Phys. Rev. **B 29**, 6086 (1984)
74. Obenhuber, T., Adlassnig, W., Zankert, J., Narger, U., Potzel, W., Kalvius, G.M.: Hyperfine Interact. **33**, 69 (1987)
75. Adlassnig, W., Potzel, W., Moser, J., Schiessl, W., Schäfer, C., Kalvius, G.M.: Hyperfine Interact. **41**, 531 (1988)
76. Adlassnig, W., Potzel, W., Moser, J., Schäfer, C., Steiner, M., Kalvius, G.M.: Nucl. Instrum. Methods Phys. Res. Sect. A Accel. Spectrom. Dect. Assoc. Equip. **277**, 485 (1989)
77. Adlassnig, W., Potzel, W., Moser, J., Schiessl, W., Potzel, U., Schäfer, C., Steiner, M., Karzel, H., Peter, M., Kalvius, G.M.: Phys. Rev. **B 40**, 7469 (1989)
78. Potzel, W., Adlassnig, W., Moser, J., Schäfer, C., Steiner, M., Kalvius, G.M.: Phys. Rev. **B 39**, 8236 (1989)
79. Steiner, M., Potzel, W., Karzel, H., Schiessl, W., Kofferlein, M., Kalvius, G.M., Blaha, P.: J. Phys. Condens. Matter **8**, 3581 (1996)
80. Potzel, W., Kalvius, G.M.: Phys. Lett. **A 110**, 165 (1985)
81. Steiner, M., Potzel, W., Schäfer, C., Adlassnig, W., Peter, M., Karzel, H., Kalvius, G.M.: Phys. Rev. **B 41**, 1750 (1990)
82. Mitchell, D.W., Das, T.P., Potzel, W., Kalvius, G.M., Karzel, H., Schiessl, W., Steiner, M., Kofferlein, M.: Phys. Rev. **B 48**, 16449 (1993)
83. Nasredinov, F.S., Masterov, V.F., Seregin, N.P., Seregin, P.P.: Z. Eksperim. Noi Teor. Fiz. **99**, 1027 (1991)

84. Seregin, N.P., Nasredinov, F.S., Masterov, V.F., Daribaeva, G.T.: Supercond. Sci. Technol. **4**, 283 (1991)
85. Seregin, N.P., Masterov, V.F., Nasredinov, F.S., Saidov, C.S., Seregin, P.P.: Supercond. Sci. Technol. **5**, 675 (1992)
86. Seregin, N.P., Nasredinov, F.S., Masterov, V.F., Seregin, P.P., Saidov, C.S.: Solid State Commun. **87**, 345 (1993)
87. Masterov, V.F., Nasredinov, F.S., Seregin, N.P., Seregin, P.P.: Phys. At. Nucl. **58**, 1467 (1995)
88. Masterov, V.F., Nasredinov, F.S., Seregin, N.P., Seregin, P.P., Sagatov, M.A.: J. Phys. Condens. Matter **7**, 2345 (1995)
89. Masterov, V.F., Seregin, P.P., Nasredinov, F.S., Seregin, N.P., Sagatov, M.A.: Phys. Status Solidi B Basic Res. **196**, 11 (1996)
90. Masterov, V.F., Nasredinov, F.S., Seregin, N.P., Seregin, P.P.: Phys. Solid State **39**, 1559 (1997)
91. Seregin, P.P., Masterov, V.F., Nasredinov, F.S., Seregin, N.P.: Physica Status Solidi B Basic Res. **201**, 269 (1997)
92. Masterov, V.F., Nasredinov, F.S., Seregin, N.P., Seregin, P.P.: Phys. Solid State **41**, 1580 (1999)
93. Nasredinov, F.S., Masterov, V.F., Seregin, N.P., Seregin, P.P.: J. Phys. Condens. Matter **11**, 8291 (1999)
94. Masterov, V.F., Nasredinov, F.S., Seregin, N.P., Seregin, P.P.: Phys. Solid State **41**, 1590 (1999)
95. Masterov, V.F., Nasredinov, F.S., Seregin, N.P., Seregin, P.P.: Phys. Solid State **41**, 890 (1999)
96. Nasredinov, F.S., Masterov, V.F., Seregin, N.P., Seregin, P.P.: J. Phys. Condens. Matter **12**, 7771 (2000)
97. Nasredinov, F.S., Seregin, N.P., Seregin, P.P.: JETP Lett. **70**, 641 (1999)
98. Nasredinov, F.S., Seregin, N.P., Seregin, P.P.: Phys. Solid State **42**, 621 (2000)
99. Nasredinov, F.S., Seregin, N.P., Seregin, P.P., Bondarevskii, S.I.: Semiconductors **34**, 269 (2000)
100. Seregin, N.P., Seregin, P.P.: J. Exp. Theor. Phys. **91**, 1230 (2000)
101. Seregin, N.P., Nasredinov, F.S., Seregin, P.P.: Phys. Solid State **43**, 609 (2001)
102. Seregin, N.P., Nasredinov, F.S., Seregin, P.P.: J. Phys. Condens. Matter **13**, 149 (2001)
103. Nemov, S.A., Seregin, N.P., Irkaev, S.M.: Semiconductors **36**, 1267 (2002)
104. Seregin, N.P.: Phys. Solid State **45**, 11 (2003)
105. Seregin, N.P., Nemov, S.A., Irkaev, S.M.: Semiconductors **36**, 975 (2002)
106. Seregin, N.P., Nemov, S.A., Stepanova, T.R., Kozhanova, Y.V., Seregin, P.P.: Semicond. Sci. Technol. **18**, 334 (2003)
107. Nemov, S.A., Seregin, P.P., Kozhanova, Y.V., Troitskaya, N.N., Volkov, V.P., Seregin, N.P., Shamrai, V.F.: Phys. Solid State **46**, 231 (2004)
108. Bordovskii, G.A., Marchenko, A.V., Seregin, P.P., Terukov, E.I.: Tech. Phys. Lett. **34**, 397 (2008)

References to Sect. 7.3

109. Kistner, O.C., Monaro, S., Segnan, R.: Phys. Lett. **5**, 299 (1963)
110. Kistner, O.C.: Phys. Rev. **144**, 1022 (1966)
111. Potzel, W., Wagner, F.E., Mössbauer, R.L., Kaindl, G., Seltzer, H.E.: Z. Phys. **241**, 179 (1971)

112. Stevens, J.G., Stevens, V.E. (eds.): Mössbauer Effect Data Index, vols. 1973, 1975. Plenum, New York (1975, 1976)
113. Kaindl, G., Potzel, W., Wagner, F., Zahn, U., Mössbauer, R.L.: Z. Phys. **226**, 103 (1969)
114. Fernandez, I., Greatrex, R., Greenwood, N.N.: J. Solid State Chem. **34**, 121 (1980)
115. Fernandez, I., Greatrex, R., Greenwood, N.N.: J. Solid State Chem. **32**, 97 (1980)
116. Wagner, F.E., Wordel, R., Griffith, W.P., McManus, N. T.: J. Chem. Soc., Dalton Trans. 1679 (1988)
117. Greenwood, N.N., de A da Costa, F.M., Greatrex, R.: Rev. Chim. Mineral. **13**, 133 (1976)
118. Greatrex, R., Greenwood, N.N., Lal, M., Fernandez, I.: J. Solid State Chem. **30**, 137 (1979)
119. de Marco, M., Cao, G., Crow, J.E., Coffey, D., Toorongian, S., Haka, M., Fridmann, J.: Phys. Rev. **B 62**, 14297 (2000)
120. Kmiec, R., Swiatkowska, Z., Gurgul, J., Rams, M., Zarzycki, A., Tomala, K.: Phys. Rev. **B 74**, 104425 (2006)
121. Gurgul, J., Rams, M., Swiatkowska, Z., Kmiec, R., Tomala, K.: Phys. Rev. **B 75**, 064426 (2007)
122. Dacosta, F.M., Greatrex, R., Greenwood, N.N.: J. Solid State Chem. **20**, 381 (1977)
123. Potzel, W., Wagner, F.E., Zahn, I.L., Mössbauer, R.L., Danon, J.: Z. Phys. **240**, 308 (1970)
124. Greatrex, R., Greenwood, N.N., Kaspi, P.: J. Chem. Soc. A, 1873 (1971)
125. Goodman, M.S., de Marco, M.J., Haka, M.S., Toorongian, S.A., Fridmann, J.: J. Chem. Soc., Dalton Trans. 117 (2002)
126. Good, M.L.: In: Stevens, J.G., Stevens, V.E. (eds.) Mössbauer Effect Data Index, vol. 1972, p. 61. Plenum, New York (1973)
127. Clausen, C.A., Prados, R.A., Good, M.L.: In: Gruverman, I.J. (ed.) Mössbauer Effect Methodology, vol. 6, p. 41. Plenum, New York (1971)
128. Clausen, C.A., Prados, R.A., Good, M.L.: Chem. Phys. Lett. **8**, 565 (1971)
129. Clausen, C.A., Prados, R.A., Good, M.L.: Inorg. Nucl. Chem. Lett. **7**, 485 (1971)
130. Wagner, F.E., Wagner, U.: Mössbauer Isomer Shifts, North Holland, Amsterdam (1978), p. 431
131. Creutz, C., Good, M.L., Chandra, S.: Inorg. Nucl. Chem. Lett. **9**, 171 (1973)
132. Führholz, U., Bürgi, H.-B., Wagner, F.E., Stebler, A., Ammeter, J.H., Krausz, E., Clarke, R.J.H., Stead, M.J., Ludi, A.: J. Am. Chem. Soc. **106**, 121 (1984)
133. Gibb, T.C., Greatrex, R., Greenwood, N.N., Puxley, D.C., Snowdon, K.G.: Chem. Phys. Lett. **20**, 130 (1973)
134. Clausen, C.A., Prados, R.A., Good, M.L.: J. Am. Chem. Soc. **92**, 7482 (1970)
135. Manoharan, P.T., Gray, H.B.: Inorg. Chem. **5**, 823 (1966)
136. Bancroft, G.M., Butler, K.D., Libbey, E.T.: J. Chem. Soc. Dalton 2643 (1972)
137. Foyt, D.C., Siddall, T.H., Alexander, C.J., Good, M.L.: Inorg. Chem. **13**, 1793 (1974)
138. De Marco, M., Coffey, D., Heary, R., Dabrowski, B., Klamut, P., Maxwell, M., Toorongian, S., Haka, M.: Phys. Rev. **B 71**, 104403 (2005)
139. Heary, R., Coffey, D., De Marco, M., Khalifah, P., Toorongian, S., Haka, M.: Physica **B 393**, 78 (2007)
140. Gibb, T.C., Greatrex, R., Greenwood, N.N.: J. Solid State Chem. **31**, 153 (1980)
141. Gibb, T.C., Greatrex, R., Greenwood, N.N., Meinhold, R.H.: Chem. Phys. Lett. **29**, 379 (1974)
142. Foyt, D.C., Cosgrove, J.G., Collins, R.L., Good, M.L.: J. Inorg. Nucl. Chem. **37**, 1913 (1975)
143. Kistner, O.C.: In: Gruverman, I.J. (ed.) Mössbauer Effect Methodology, vol. 3, p. 217. Plenum, New York (1967)
144. Gibb, T.C, Greatrex, R., Greenwood, N.N., Kaspi, P.: Chem. Commun. 319 (1971)
145. Gibb, T.C., Greatrex, R., Greenwood, N.N., Kaspi, P.: J. Chem. Soc. Dalton 1253 (1973)
146. Gibb, T.C., Greatrex, R., Greenwood, N.N., Puxley, D.C., Snowdon, K.G.: J. Solid State Chem. **11**, 17 (1974)
147. Gibb, T.C., Greatrex, R., Greenwood, N.N., Snowdon, K.G.: J. Solid State Chem. **14**, 193 (1975)

148. de Marco, M., Coffey, D., Tallon, J., Haka, M., Toorongian, S., Fridmann, J.: Phys. Rev. **B 65**, 212506 (2002)
149. Coffey, D., de Marco, M., Dabrowski, B., Kolesnik, S., Toorongian, S., Haka, M.: Phys. Rev. **B 77**, 214412 (2008)
150. Kruk, R., Kmiec, R., Klamut, P.W., Dabrowski, B., Brown, D.E., Maxwell, M., Kimball, C.W.: Physica **C 370**, 71 (2002)
151. Kobayashi, Y., Okada, T., Asai, K., Katada, M., Sano, H., Ambe, F.: Inorg. Chem. **31**, 4570 (1992)
152. Kobayashi, Y., Asai, K., Ohada, T., Ambe, F.: Hyperfine Interact. **84**, 131 (1994)
153. Kobayashi, Y., Katada, M., Sano, H., Okada, T., Asai, K., Ambe, F.: J. Radioanal. Nucl. Chem. Lett. **136**, 387 (1989)
154. Schäfer, A., Adelberger, E.G.: Z. Phys. **A 339**, 305 (1991)
155. Clausen, C.A., Good, M.L.: J. Catal. **38**, 92 (1975)
156. Stievano, L., Calogero, S., Wagner, F.E., Calvagno, S., Milone, C.: J. Phys. Chem. **B 103**, 9445 (1999)
157. Hamnet, A., Kennedy, B.J., Wagner, F.E.: J. Catal. **124**, 30 (1990)
158. Kobayashi, Y., Okada, T., Ambe, F., Watanabe, M., Katada, M., Sano, H.: Hyperfine Interact. **68**, 189 (1991)
159. Ohkubo, Y., Kobayashi, Y., Harasawa, K., Ambe, S., Okada, T., Asai, K., Shibata, S., Takeda, M., Ambe, F.: Hyperfine Interact. **84**, 83 (1994)
160. Ohkubo, Y., Kobayashi, Y., Harasawa, K., Ambe, S., Okada, T., Ambe, F., Asai, K., Shibata, S.: J. Phys. Chem. **99**, 10629 (1995)
161. Ambe, F., Ohkubo, Y., Ambe, S., Kobayashi, Y., Okada, T., Yanagida, Y., Nakamura, J., Asai, K., Kawase, Y., Uehara, S.: J. Radioanal. Nucl. Chem. **190**, 215 (1995)

References to Sect. 7.4

162. Wiedemann, W., Kienle, P., Stanek, F.: Z. Angew. Phys. **15**, 7 (1963)
163. Gerdau, E., Steiner, P., Steenken, D.: In: Matthias, E., Shirley, D.A. (eds.) Hyperfine Structure and Nuclear Radiations, p. 261. North Holland, Amsterdam (1968)
164. Gerdau, E., Körner, H.J., Lerch, J., Steiner, P.: Z. Naturforsch. **A 21**, 941 (1966)
165. Jacobs, C.G., Hershkowitz, N., Jeffries, J.B.: Phys. Lett. **A 29**, 498 (1969)
166. Jacobs, C.G., Hershkowitz, N.: Phys. Rev. **B 1**, 839 (1970)
167. Hershkowitz, N., Jacobs, C.G.: In: Gruverman, I.J. (ed.) Mössbauer Effect Methodology, vol. 6, p. 143. Plenum, New York (1971)
168. Boolchand, P., Langhammer, D., Ching-Lu, Lin, Jha, S., Peek, N.F.: Phys. Rev. **C 6**, 1093 (1972)
169. Snyder, R.E., Ross, J.W., Bunbury, D.P.: Proc. Phys. Soc. **10**, 1662 (1968)
170. Boolchand, P., Robinson, B.L., Jha, S.: Phys. Rev. **187**, 475 (1969)
171. Steiner, P., Gerdau, E., Steenken, D.: Proc. R. Soc. **A 311**, 177 (1969)
172. Schäfer, G., Herzog, P., Wolbeck, B.: Z. Phys. **257**, 336 (1972)

References to Sect. 7.5

173. Stevens, J.G., Stevens, V.E.: Mössbauer Effect Data Index, Covering the 1975 Literature, p. 166. IFI Plenum, New York (1976)
174. Steiner, P., Gerdau, E., Hautsch, W., Steenken, D.: Z. Phys. **221**, 281 (1969)

175. Boyle, A.J.F., Hall, H.E.: Rep. Prog. Phys. **25**, 441 (1962)
176. Cohen, S.G., Marinov, A., Budnick, J.I.: Phys. Lett. **12**, 38 (1964)
177. Steyert, W.A., Taylor, R.D., Storms, E.K.: Phys. Rev. Lett. **14**, 739 (1965)
178. Muir Jr., A.H., Nadler, H.: Bull. Am. Phys. Soc. **12**, 202 (1967)
179. Sauer, G.: Z. Phys. **222**, 439 (1969)
180. Wortmann, G.: Phys. Lett. **A 35**, 391 (1971)
181. Kaindl, G., Maier, M.R., Schaller, H., Wagner, F.: Nucl. Instrum. Methods **66**, 277 (1968)
182. Pfeiffer, L.: Nucl. Instrum. Methods **140**, 57 (1977)
183. Trammel, G.T., Hannon, J.T.: Phys. Rev. **180**, 337 (1969)
184. Kagan, Y.M., Afans'ev, A.M., Vojtovetskii, V.K.: Zh. Eksperim. Teor. Fiz. Pis'ma Red. **9**, 155 (1969). JETP Lett. **9**, 91 (1969)
185. Sauer, C., Matthias, E., Mössbauer, R.L.: Phys. Rev. Lett. **21**, 961 (1968)
186. Kaindl, G., Salomon, D., Wortmann, G.: Phys. Rev. **B 8**, 1912 (1973)
187. Kaindl, G., Salomon, D., Wortmann, G.: Phys. Rev. Lett. **28**, 952 (1972)
188. Salomon, D., Kaindl, G., Shirley, D.A.: Phys. Lett. **A 36**, 457 (1971)
189. Kaindl, G., Salomon, D.: Phys. Lett. **A 40**, 179 (1972)
190. Kaindl, G., Salomon, D.: Phys. Lett. **A 42**, 333 (1973)
191. Kaindl, G., Salomon, D., Wortmann, G.: In: Gruverman, I.J. (ed.) Mössbauer Effect Methodology, vol. 8, p. 211. Plenum, New York (1973)
192. Fröhlich, K., Gütlich, P., Keller, G.: J. Chem. Soc. Dalton 971 (1972)
193. Quaim, S.M.: Proc. Phys. Soc. London **90**, 1065 (1967)
194. Wagner, F.E., Wortmann, G., Kalvius, G.M.: Phys. Lett. **A 42**, 483 (1973)
195. Josephson, B.D.: Phys. Rev. Lett. **4**, 341 (1960)
196. Kaindl, G., Salomon, D.: Phys. Rev. Lett. **30**, 579 (1973)
197. Salomon, D., Triplett, B.B., Dixon, N.S., Boolchand, P., Hanna, S.S.: J. Phys. Colloq. **35**, C6–C285 (1974)
198. Salomon, D., Wallner, W., West, P. J.: In: Proceedings of International Conference on Mössb. Spec., Cracow, p. 105 (1975)
199. Heidemann, A., Kaindl, G., Salomon, D., Wipf, H., Wortmann, G.: Proceedings of International Conference on Mössb. Spec., Cracow, p. 411 (1975)
200. Kaindl, G., Salomon, D.: Phys. Lett. **B 32**, 364 (1970)
201. Dornow, V.A., Binder, J., Heidemann, A., Kalvius, G.M., Wortmann, G.: Nucl. Instrum. Methods **163**, 491 (1979)
202. Heidemann, A., Wipf, H., Wortmann, G.: Hyperfine Interact. **4**, 844 (1978)
203. Pfeiffer, L., Eibschütz, M., Salomon, D.: Hyperfine Interact. **4**, 803 (1978)
204. Salomon, D., West, P.J.: Z. Phys. A Hadrons Nuclei **288**, 291 (1978)
205. Eibschütz, M., Salomon, D., Disalvo, F.J.: Phys. Lett. **A 93**, 259 (1983)
206. Zhetbaev, A.K., Ozernoi, A.N.: Physica Status Solidi A Appl. Res. **77**, 63 (1983)
207. Salomon, D., Lerf, A., Biberacher, W., Butz, T., Saibene, S.: Chem. Phys. Lett. **119**, 238 (1985)
208. Eibschütz, M., Salomon, D., Murphy, D.W., Zahurak, S., Waszczak, J.V.: Chem. Phys. Lett. **135**, 591 (1987)
209. Eibschutz, M., Salomon, D., Zahurak, S., Murphy, D.W.: Phys. Rev. **B 37**, 3082 (1988)
210. Blaha, P.: J. Phys. Condens. Matter **3**, 9381 (1991)
211. Eibschütz, M.: Phys. Rev. **B 45**, 10914 (1992)
212. Eibschütz, M., Murphy, D.W., Zahurak, S., Waszczak, J.V.: Appl. Phys. Lett. **61**, 2976 (1992)
213. Chumakov, A.I., Baron, A.Q.R., Arthur, J., Ruby, S.L., Brown, G.S., Smirnov, G.V., van Bürck, U., Wortmann, G.: Phys. Rev. Lett. **75**, 549 (1995)
214. Leupold, O., Chumakov, A.I., Alp, E.E., Sturhahn, W., Baron, A.Q.R.: Hyperfine Interact. **123**, 611 (1999)
215. Shvyd'ko, Y.V.: Hyperfine Interact. **125**, 173 (2000)

216. Serdons, I., Callens, R., Coussement, R., Gheysen, S., Ladriere, J., Morimoto, S., Nasu, S., Odeurs, J., Yoda, Y., Wortmann, G.: Nucl. Instrum. Meth. Phys. Res. Sect. B Beam Interact. Mater. Atoms **251**, 297 (2006)

References to Sect. 7.6

217. Stevens, J.G., Stevens, V.E.: Mössbauer Effect Data Index. IFI Plenum, New York (1969–1975)
218. Lee, L.L., Meyer-Schützmeister, L., Schiffer, J.P., Vincent, D.: Phys. Rev. Lett. **3**, 223 (1959)
219. Bussiere de Nercy, A., Langevin, M., Spighel, M.: J. Phys. Radium **21**, 288 (1960)
220. Kankeleit, E.: Z. Phys. **164**, 442 (1961)
221. Hardy, K.A., Russell, D.G., Wilenzick, R.M.: Phys. Lett. **A 27**, 422 (1968)
222. Chow, Y.W., Greenbaum, E.S., Howes, R.H., Hsu, F.H.H.: Phys. Lett. **B 30**, 171 (1969)
223. Oberley, L.W., Hershkowitz, N., Wender, S.A., Carpenter, A.B.: Phys. Rev. **C 3**, 1585 (1971)
224. Hershkowitz, N., Wender, S.A., Carpenter, A.B.: Phys. Rev. **C 3**, 219 (1972)
225. Zioutas, K., Wolbeck, B., Perscheid, B.: Z. Phys. **262**, 413 (1973)
226. Bokemeyer, H., Wohlfahrt, K., Kankeleit, E., Eckardt, D.: Z. Phys. **A 274**, 305 (1975)
227. Shikazono, N., Takekoshi, H., Shoji, T.: J. Phys. Soc. Jpn **21**, 829 (1966)
228. Gedikli, A., Winkler, H., Gerdau, E.: Z. Phys. **267**, 61 (1974)
229. Frankel, R.B., Chow, Y., Grodzins, L., Wulff, J.: Phys. Rev. **186**, 381 (1969)
230. Wagner, F.E., Schaller, H., Felscher, R., Kaindl, G., Kienle, P.: In: Goldring, G., Kalish, R. (eds.) Hyperfine Interactions in Excited Nuclei, vol. 2, p. 603. Gordon and Breach, New York (1971)
231. Cohen, S.G., Blum, N.A., Chow, Y.W., Frankel, R.B., Grodzins, L.: Phys. Rev. Lett. **16**, 322 (1966)
232. Raj, D., Puri, S.P.: Phys. Lett. **A 29**, 510 (1969)
233. Ruth, R.D., Hershkowitz, N.: Phys. Lett. **A 34**, 203 (1973)
234. Kaltseis, J., Posch, H.A., Vogel, W.: J. Phys. C: Solid State Phys. **5**, 2523 (1972)
235. Conroy, L.E., Perlow, G.J.: Phys. Lett. **A 31**, 400 (1970)
236. Roth, S., Hörl, E.M.: Phys. Lett. **A 25**, 299 (1967)
237. Wender, S.A., Hershkowitz, A.: Phys. Rev. **B 8**, 4901 (1973)
238. Schibuya, N., Tsunoda, Y., Nishi, A., Kunitomi, N.: J. Phys. Soc. Jpn **33**, 564 (1972)
239. Bancroft, G.M., Garrod, R.E.B., Maddock, A.G.: Inorg. Nucl. Chem. Lett. **7**, 1157 (1971)
240. Clark, M.G., Gancedo, J.R., Maddock, A.G., Williams, A.F., Yoffe, A.D.: J. Phys. C: Solid State Phys. **6**, L474 (1973)
241. Maddock, A.G., Platt, R.H., Williams, A.F., Gancedo, R.: J. Chem. Soc. Dalton 1314 (1974)
242. Clark, M.G., Gancedo, J.R., Maddock, A.G., Williams, A.F.: J. Chem. Soc. Dalton 120 (1975)
243. Agresti, D., Kankeleit, E., Persson, B.: Phys. Rev. **155**, 1342 (1969)
244. Savvateev, N.N., Pokholok, K.V., Babechkin, A.M., Dzevitsky, B.E., Tviadadze, G.N.: Solid State Commun. **39**, 793 (1981)
245. Schupp, G., Yelon, W.B., Mullen, J.G., Wagoner, R.A.: Hyperfine Interact. **93**, 491 (1994)
246. Ruebenbauer, K., Mullen, J.G., Nienhaus, G.U., Schupp, G.: Phys. Rev. **B 49**, 15607 (1994)
247. Zach, C., Keppler, C., Huenges, E., Achterhold, K., Parak, F.: Hyperfine Ineract. **126**, 83 (2000)
248. Wagoner, R.A., Mullen, J.G., Schupp, G.: Phys. Lett. **B 279**, 25 (1992)
249. Bullard, B.R., Mullen, J.G., Schupp, G.: Phys. Rev. **B 43**, 7405 (1991)
250. Wagoner, R.A., Mullen, J.G., Schupp, G.: Phys. Rev. **C 47**, 1951 (1993)

References to Sect. 7.7

251. Morrison, R.J., Atac, M., Debrunner, P., Frauenfelder, H.: Phys. Lett. **12**, 35 (1964)
252. Barret, P.H., Grodzins, L.: Rev. Mod. Phys. **36**, 971 (1964)
253. Wagner, F., Kucheida, D., Kaindl, G., Kienzle, P.: Z. Phys. **230**, 80 (1970)
254. Wagner, F.E., Spieler, H., Kucheida, D., Kienle, P., Wäppling, R.: Z. Phys. **254**, 112 (1972)
255. Gregory, M.C., Robinson, B.L., Iha, S.: Phys. Rev. **180**, 1158 (1969)
256. Iha, S., Owens, W.R., Gregory, M.C., Robinson, B.L.: Phys. Lett. **B 25**, 115 (1967)
257. Persson, B., Blumberg, H., Bent, M.: Phys. Rev. **174**, 1509 (1968)
258. Wagner, F.E., Kucheida, D., Zahn, U., Kaindl, G.: Z. Phys. **266**, 223 (1974)
259. Kucheida, D., Wagner, F., Kaindl, G., Kienle, P.: Z. Phys. **216**, 346 (1968)
260. Schrenk, A., Zimmermann, G.: Phys. Lett. **A 16**, 258 (1968)
261. Wagner, F., Kaindl, G., Bohn, H., Biebl, U., Schaller, H., Kienle, P.: Phys. Lett. **B 28**, 548 (1969)
262. Himmel, G.: Z. Phys. **211**, 68 (1968)
263. Walter, H.K.: Nucl. Phys. **A 234**, 504 (1974)
264. Bohn, H., Kaindl, G., Kucheida, D., Wagner, F.E., Kienle, P.: Phys. Lett. **B 32**, 346 (1970)
265. Rother, P., Wagner, F., Zahn, U.: Radiochim. Acta **11**, 203 (1969)

References to Sect. 7.8

266. Stevens, J.G., Stevens, V.E. (eds.): Mössbauer Effect Data Index, vol. 1972, 1975. Plenum, New York (1973, 1976)
267. Mössbauer, R.L.: Z. Phys. **151**, 124 (1958). Naturwiss. **45**, 538 (1958); Z. Naturforsch. **A 14**, 211 (1959)
268. Wagner, F., Zahn, U.: Z. Phys. **233**, 1 (1970)
269. Davies, G.J., Maddock, A.G., Williams, A.F.: J. Chem. Soc. Chem. Commun., 264 (1975)
270. Thomson, J.O., Werkheiser, A.H., Lindauer, M.W.: Rev. Mod. Phys. **36**, 357 (1964)
271. Atzmony, U., Bauminger, E.R., Lebenbaum, D., Mustachi, A., Ofer, S., Wernick, J.J.: Phys. Rev. **163**, 314 (1967)
272. Heuberger, A., Pobell, F., Kienle, P.: Z. Phys. **205**, 503 (1967)
273. Wagner, F., Klöckner, J., Körner, H.J., Schaller, H., Kienle, P.: Phys. Lett. **B 25**, 253 (1967)
274. Wagner, F., Kaindl, G., Kienle, P., Körner, H.-J.: Z. Phys. **207**, 500 (1967)
275. Perlow, G.J., Henning, W., Olson, D., Goodman, G.L.: Phys. Rev. Lett. **23**, 680 (1969)
276. Wickman, H.H., Silverthorn, W.E.: Inorg. Chem. **10**, 2333 (1971)
277. Vaska, L.: Acc. Chem. Res. **1**, 335 (1968)
278. Holsboer, F., Beck, W., Bartunik, H.D.: J. Chem. Soc. Dalton 1829 (1973)
279. Vaska, L., Peone, J.: Chem. Commun. 418 (1971)
280. Holsboer, F., Beck, W., Bartunik, H.D.: Chem. Phys. Lett. **18**, 217 (1973)
281. Ginsberg, A.P., Cohen, R.L., DiSalvo, F.J., West, K.W.: J. Chem. Phys. **60**, 2657 (1974)
282. Krogmann, K., Binder, W., Hausen, H.D.: Angew. Chem. Int. Ed. **7**, 812 (1968)
283. Aderjan, R., Keller, H.J., Rupp, H.H., Wagner, F.E., Wagner, U.: J. Chem. Phys. **64**, 3748 (1976)
284. Ginsberg, A.P., Koepke, J.W., Cohen, R.L., West, K.W.: Chem. Phys. Lett. **38**, 310 (1976)
285. Wagner, F.E., Potzel, W., Wagner, U., Schmidtke, H.-H.: Chem. Phys. **4**, 284 (1974)
286. Berret, R.R., Fitzsimmons, B.W.: J. Chem. Soc. A., 525 (1967)
287. Clark, M.G.: Mol. Phys. **20**, 257 (1971)
288. Jørgensen, C.K., Pappalardo, R., Schmidtke, H.-H.: J. Chem. Phys. **39**, 1422 (1963)
289. Schäffer, C.E., Jørgensen, C.K.: Mol. Phys. **9**, 401 (1965)

290. Schmidtke, H.-H.: Theor. Chim. Acta **20**, 92 (1971)
291. Williams, A.F., Jones, G.C.H., Maddock, A.G.: J. Chem. Soc. Dalton 1952 (1975)
292. Williams, A.F., Bhaduri, S., Maddock, A.G.: J. Chem. Soc. Dalton 1958 (1975)
293. Bancroft, G.M.: Coord. Chem. Rev. **11**, 247 (1973)
294. Mössbauer, R.L., Lengsfeld, M., von Lieres, W., Potzel, W., Teschner, P., Wagner, F.E., Kaindl, G.: Z. Naturforsch. **A 26**, 343 (1971)
295. Vogl, G., Schaefer, A., Mansel, W., Prechtel, J., Vogl, W.: Physica Status Solidi **B 59**, 107 (1973)
296. Salomon, D., Shirley, D.A.: Phys. Rev. **B 9**, 29 (1974)
297. Kanashiro, M., Nishi, M., Kunitomi, N., Sakai, H.: J. Phys. Soc. Jpn **38**, 897 (1975)
298. Wagner, F.E.: Hyperfine Interact. **13**, 149 (1983)
299. Vanderwoude, F., Miedema, A.R.: Solid State Commun. **39**, 1097 (1981)
300. Forker, M., Krusch, K.: Hyperfine Interact. **9**, 399 (1981)
301. Berry, F.J., Jobson, S.: Hyperfine Interact. **41**, 613 (1988)
302. Wagner, F.E., Sawicki, J.A., Rolston, J.H.: Hyperfine Interact. **41**, 733 (1988)
303. Berry, F.J., Jobson, S.: Hyperfine Interact. **46**, 559 (1989)
304. Berry, F.J., Jobson, S., Zhang, T., Marco, J.F.: Catal. Today **9**, 137 (1991)
305. Vonbrandis, H., Wagner, F.E., Sawicki, J.A., Marcinkowska, K., Rolston, J.H.: Hyperfine Interact. **57**, 2127 (1990)
306. Baier, M., Wordel, R., Wagner, F.E., Antonova, T.E., Antonov, V.E.: J. Less-Common Met. **172**, 358 (1991)
307. Bandini, A.L., Banditelli, G., Bonati, F., Calogero, S., Wagner, F.E.: J. Organomet. Chem. **410**, 241 (1991)
308. Sawicki, J.A., Sawicka, B.D., Wagner, F.E.: Nucl. Instrum. Meth. Phys. Res. Sect. B-Beam Interact. Mater. Atoms **62**, 253 (1991)
309. Tomala, K., Blaise, A., Kmiec, R., Sanchez, J.P.: J. Magn. Magn. Mater. **117**, 275 (1992)
310. Sanchez, J.P., Blaise, A., Ressouche, E., Malaman, B., Venturini, G., Tomala, K., Kmiec, R.: J. Magn. Magn. Mater. **128**, 295 (1993)
311. Demazeau, G., Jung, D.Y., Sanchez, J.P., Colineau, E., Blaise, A., Fournes, L.: Solid State Commun. **85**, 479 (1993)
312. Vertes, A., Gal, M., Wagner, F., Tuczek, F., Gütlich, P.: Magy. Kem. Foly. **99**, 259 (1993)
313. Vertes, A., Gal, M., Wagner, F.E., Tuczek, F., Gütlich, P.: Inorg. Chem. **32**, 4478 (1993)
314. Tuczek, F., Gal, M., Wagner, F.E., Vertes, A.: Fullerene Sci. Technol. **5**, 443 (1997)
315. Vertes, A.: Magy. Kem. Foly. **105**, 99 (1999)
316. Nagata, S., Matsumoto, N., Kato, Y., Furubayashi, T., Matsumoto, T., Sanchez, J.P., Vulliet, P.: Phys. Rev. **B 58**, 6844 (1998)
317. Stievano, L.: Hyperfine Interact. **126**, 101 (2000)
318. Stievano, L., Calogero, S., Psaro, R., Guidotti, M., Della Pergola, R., Wagner, F.E.: J. Cluster. Sci. **12**, 123 (2001)
319. Stievano, L., Calogero, S., Psaro, R., Wagner, F.E.: Comm. Inorg. Chem. **22**, 275 (2001)
320. Gal, M., Kiss, L.F., Klencsar, Z., Papai, I., Schubert, G., Rohonczy, J., Vertes, A., Wagner, F.E.: J. Radioanal. Nucl. Chem. **260**, 133 (2004)
321. Stievano, L., Della Pergola, R., Wagner, F.E.: Hyperfine Interact. **156**, 365 (2004)

References to Sect. 7.9

322. Harris, J.R., Benczer-Koller, N., Rothberg, G.M.: Phys. Rev. **A 137**, 1101 (1965)
323. Buyrn, A., Grodzins, L.: Phys. Lett. **21**, 389 (1966)
324. Wolbeck, B., Zioutas, K.: Z. Naturforsch. **A 25**, 1779 (1970)
325. Harris, J.R., Rothberg, G.M., Benczer-Koller, N.: Phys. Rev. **B 138**, 554 (1965)

326. Walcher, D.: Z. Phys. **246**, 123 (1971)
327. Buyrn, A., Grodzins, L., Blum, N.A., Wulff, J.: Phys. Rev. **163**, 286 (1967)
328. Wilenzick, R.M., Hardy, K.A., Hicks, J.A., Owen, W.R.: Phys. Lett. **A 29**, 678 (1969)
329. Rüegg, W., Launaz, J.P.: Phys. Lett. **A 37**, 355 (1971)
330. Atac, M., Debrunner, P., Frauenfelder, H.: Phys. Lett. **21**, 699 (1966)
331. Agresti, D., Kankeleit, E., Persson, B.: Phys. Rev. **155**, 1339 (1967)
332. Rüegg, W.: Helv. Phys. Acta **46**, 735 (1974)
333. Rüegg, W., Kuse, D., Zeller, H.R.: Phys. Rev. **B 8**, 952 (1973)
334. Ianarella, L., Wagner, F.E., Wagner, U., Danon, J.: J. Phys. Colloq. **35**, C6–517 (1974)
335. Van der Woude, F., Miedema, A.R.: Solid State Commun. **39**, 1097 (1981)

References to Sect. 7.10

336. Nagle, D., Craig, P.P., Dash, J.G., Reiswig, R.D.: Phys. Rev. Lett. **4**, 237 (1960)
337. Roberts, L.D., Pomerance, H., Thomson, J.O., Dam, C.F.: Bull. Am. Phys. Soc. **7**, 565 (1962)
338. Shirley, D.A.: Rev. Mod. Phys. **36**, 339 (1964)
339. Shirley, D.A., Grant, R.W., Keller, D.A.: Rev. Mod. Phys. **36**, 352 (1964)
340. Charlton, J.S., Nichols, D.I.: J. Chem. Soc. A 1484 (1970)
341. Faltens, M.O., Shirley, D.A.: J. Chem. Phys. **53**, 4249 (1970)
342. Bartunik, H.D., Potzel, W., Mössbauer, R.L., Kaindl, G.: Z. Phys. **240**, 1 (1970)
343. Parish, R.V.: In: Lang, G.L. (ed.) Mössbauer Spectroscopy Applied in Inorganic Chemistry, vol. I, p. 577. Plenum, New York (1984)
344. Trautwein, A.X., Bläs, R., Lauer, R., Hasselbach, K.M.: In: Gütlich, P., Kalvius, G.M. (eds.) Trends in Mössbauer Spectroscopy, 2nd Seeheim Conference on Mössbauer Spectroscopy, Buchdruck Universität Mainz, Mainz, 1983, p. 68
345. Powers, J., Martin, P., Miller, G.H., Welsh, R.E., Jenkins, R.H.: Nucl. Phys. **A 230**, 413 (1974)
346. Schwerdtfeger, P., Bast, R., Gerry, M.C.L., Jacob, C.R., Jansen, M., Kello, V., Mudring, A.V., Sadlej, A.J., Saue, T., Sohnel, T., Wagner, F.E.: J. Chem. Phys. **122**, 124317 (2005)
347. Wdowik, U.D., Ruebenbauer, K.: J. Chem. Phys. **129**, 104504 (2008)
348. Kopfermann, H.: Kernmomente, 2nd edn. Akademische Verlagsgesellschaft, Frankfurt/Main (1956)
349. Leary, K., Bartlett, N.: Chem. Commun. **131** (1973)
350. Kaindl, G., Leary, K., Bartlett, N.: J. Chem. Phys. **59**, 5050 (1973)
351. Schmidbaur, H., Mandl, J.R., Wagner, F.E., van de Vondel, D.F., van der Kelen, G.P.: J. Chem. Soc. Chem. Commun., 170 (1976)
352. Viegers, M.P.A.: Ph.D. thesis, Catholic University of Nijmegen, Netherlands (1976)
353. Banditelli, G., Bonati, F., Calogero, S., Valle, G., Wagner, F.E., Wordel, R.: Organometallics **5**, 1346 (1986)
354. Bonati, F., Burini, A., Pietroni, B.R., Calogero, S., Wagner, F.E.: J. Organomet. Chem. **309**, 363 (1986)
355. Bonati, F., Cingolani, A., Calogero, S., Wagner, F.E.: Inorg. Chim. Acta **127**, 87 (1987)
356. Bonati, F., Burini, A., Pietroni, B.R., Torregiani, E., Calogero, S., Wagner, F.E.: J. Organomet. Chem. **408**, 125 (1991)
357. Adams, M.D., Friedl, J., Wagner, F.E.: Hydrometallurgy **31**, 265 (1992)
358. Calogero, S., Wagner, F.E., Ponticelli, G.: Polyhedron **12**, 1459 (1993)
359. Stievano, L., Calogero, S., Storaro, L., Lenarda, M., Wagner, F.E.: J. Chem. Soc. **94**, 2627 (1998)
360. Grohmann, A., Riede, J., Schmidbaur, H.: J. Chem. Soc. Dalton 783 (1991)
361. Calis, C.H.M., Hadjiliadis, N.: Inorg. Chim. Acta **91**, 203 (1984)

362. Stanek, J., Hafner, S.S., Miczko, B.: Phys. Rev. **57**, 6219 (1998)
363. Sakai, H., Ando, M., Maeda, Y.: Hyperfine Interact. **68**, 201 (1991)
364. Sakai, H., Ando, M., Ichiba, S., Maeda, Y.: Chem. Lett. 223 (1991)
365. Katada, M., Uchida, Y., Sato, K., Sano, H., Sakai, H., Maeda, Y.: Bull. Chem. Soc. Jpn **55**, 444 (1982)
366. Katada, M., Ssato, K., Uchida, Y., Iijima, S., Sano, H., Wie, H.H., Sakai, H., Maeda, Y.: Bull. Chem. Soc. Jpn **56**, 945 (1983)
367. Schmidbaur, H., Hartmann, C., Wagner, F.E.: Angew. Chem. Int. Ed. Engl. **26**, 1148 (1987)
368. Kitagawa, H., Kojima, N., Sakai, H.: J. Chem. Soc. Dalton Trans. 3211 (1991)
369. Kojima, N., Tanaka, A., Sakai, H., Maeda, Y.: Nucl. Instrum. Methods **B 76**, 366 (1993)
370. Kojima, N., Matsushita, N.: Coord. Chem. Rev. **198**, 251 (2000)
371. Ikeda, K., Kojima, N., Ono, Y., Kobayashi, Y., Seto, M., Liu, X.J., Moritomo, Y.: Hyperfine Interact. **156/157**, 311 (2004)
372. Kojima, N., Amita, F., Kitagawa, H., Sakai, H., Maeda, Y.: Nucl. Instrum. Methods **B 76**, 321 (1993)
373. Ikeda, K., Kojima, N., Kobayashi, Y., Seto, M.: Hyperfine Interact. **166**, 403 (2005)
374. Burini, A., Pietroni, B.R., Galassi, R., Valle, G., Calogero, S.: Inorg. Chim. Acta **229**, 299 (1995)
375. Bovio, B., Calogero, S., Wagner, F.E., Burini, A., Pietroni, B.R.: J. Organomet. Chem. **470**, 275 (1994)
376. Albinati, A., Eckert, J., Hofmann, P., Rüegger, H., Venanzi, L.M.: Inorg. Chem. **32**, 2377 (1993)
377. Parish, R.V., Moore, L.S., Dens, A.J.D., Mingos, D.M.P., Sherman, D.J.: J. Chem. Soc. Dalton Trans. 781 (1988)
378. Wagner, F.E., Sawicki, J.A., Friedl, J., Mandarino, J.A., Harris, D.C., Cabri, L.J.: Can. Mineral **32**, 189 (1994)
379. Wagner, F.E., Sawicki, J.A., Friedl, J., Mandarino, J.A., Harris, D.C.: Can. Mineral **30**, 327 (1992)
380. Greffié, C., Benedetti, M.F., Parron, C., Amouric, M.: Geochim. Cosmochim. Acta **60**, 1531 (1996)
381. Parish, R.V., Moore, L.S., Dens, A.J.D., Mingos, D.M.P., Sherman, D.J.: J. Chem. Soc. Dalton Trans. 539 (1989)
382. Stievano, L., Pegola, R.D., Wagner, F.E.: Hyperfine Interact. **156/167**, 365 (2004)
383. Moore, L.S., Parish, R.V., Brown, S.S.D., Salter, I.D.: J. Chem. Soc. Dalton Trans. 2333 (1987)
384. Brown, S.S.D., Salter, I.D., Dyson, D.B., Parish, R.V., Bates, P.A., Hursthouse, M.B.: J. Chem. Soc. Dalton Trans. 1795 (1988)
385. Kojima, N.: Bull. Chem. Soc. Jpn **73**, 1445 (2000)
386. Ikeda, K., Ono, Y., Enomoto, M., Kojima, N., Kobayashi, Y., Seto, M., Koyama, K., Uwatoko, Y.: Ceramics **48**, 159 (2004)
387. Bhargava, S.K., Mohr, F., Takahashi, M., Takeda, M.: Bull. Chem. Soc. Jpn **74**, 1051 (2001)
388. Kitadai, K., Takahashi, M., Takeda, M., Bhargava, S.K., Benett, M.A.: Bull. Chem. Soc. Jpn **79**, 886 (2006)
389. Ohashi, H., Ezoe, H., Okaue, Y., Kobayashi, Y., Matsuo, S., Kurisaki, T., Miyazaki, A., Wakita, H., Yokoyama, T.: Anal. Sci. **21**, 789 (2005)
390. Yamashita, M., Ohashi, H., Kobayashi, Y., Okaue, Y., Kurisaki, T., Wakita, H., Yokoyama, T.: J. Colloid. Interface Sci. **319**, 25 (2008)
391. Bachelor, R.J., Birchall, T., Burns, R.C.: Inorg. Chem. **25**, 2009 (1986)
392. Usón, R., Laguna, A., Navarro, A., Parish, R.V., Moore, L.S.: Inorg. Chim. Acta **112**, 205 (1986)
393. Houlton, A., Roberts, R.M.G., Silver, J., Parish, R.V.: J. Organomet. Chem. **418**, 269 (1991)
394. Viotte, M., Gautheron, B., Kubicki, M.M., Mugnier, Y., Parish, R.V.: Inorg. Chem. **34**, 3465 (1995)

395. Al-Sa´ady, A.K.H., McAuliffe, C.A., Moss, K., Parish, R.V., Fields, R.: J. Chem. Soc. Dalton Trans. 491 (1984)
396. Healy, P.C., Loughrey, B.T., Bowmaker, G.A., Hanna L.V.: Dalton Trans. 3723 (2008)
397. Shaw III, C.F., Schaeffer, N.A., Elder, R.C., Eidness, M.K., Trooster, J.M., Calis, G.H.M.: J. Am. Chem. Soc. **106**, 3511 (1984)
398. Brown, K., Parish, R.V., McAuliffe, C.A.: J. Am. Chem. Soc. **103**, 4943 (1981)
399. Hill, D.T., Sutton, B.M., Isab, A.A., Razi, T., Sadler, P.J., Trooster, J.M., Calis, G.H.M.: Inorg. Chem. **22**, 2936 (1983)
400. Canumalla, A., Shaw III, C.F., Wagner, F.E.: Inorg. Chem. **38**, 3268 (1999)
401. Eibschutz, M., Lines, M.E., Reiff, W.M., van Dover, B., Waszczak, J.V., Zahurak, S., Felder, R.J.: Appl. Phys. Lett. **62**, 1827 (1993)
402. Tomkowicz, Z., Stanek, J., Szytula, A., Bajorek, A., Balanda, M., Sciesinska, E., Sciesinski, J., Guillot, M.: J. Alloys Compd. **224**, 274 (1995)
403. Makarov, E.F., Stukan, R.A., Panfilov, A., Wagner, F.E., Friedl, J.: Hyperfine Interact. **93**, 161 (1994)
404. Friedl, J., Wagner, F.E., Sawicki, J.A., Harris, D.C., Mandarino, J.A., Marion, Ph: Hyperfine Interact. **70**, 945 (1992)
405. Kongolo, K., Mwema, M.D.: Hyperfine Interact. **111**, 281 (1998)
406. Cashion, J.D., Brown, L.J.: Hyperfine Interact. **111**, 271 (1998)
407. Sawicki, J.A., Dutrizac, J.E., Friedl, J., Wagner, F.E., Chen, T.T.: Nucl. Instrum. Methods **B 76**, 378 (1993)
408. Wagner, F.E., Swash, P.M., Marion, P.H.: Hyperfine Interact. **46**, 681 (1989)
409. Wagner, F.E., Marion, P.H., Regnard, J.-R.: Hyperfine Interact. **41**, 851 (1988)
410. Wagner, F.E., Friedl, J., Sawicki, J.A., Harris, D.C.: Hyperfine Interact. **91**, 619 (1994)
411. Wu, X., Delbove, F., Touray, J.C.: Miner. Deposita **25**(Suppl.), S8 (1990)
412. Jiuling, L., Feng, Q., Qingsheng, X.: N. J. Miner. Mh. **5**, 193 (2003)
413. Yang, S., Blum, N., Rahders, E., Zhang, Z.: Can. Mineral. **36**, 1361 (1998)
414. Schmid, G. (ed.): Clusters and Colloids. VCH, Weinheim (1994)
415. de Jogh, L.J. (ed.): Physics and Chemistry of Metal Cluster Compounds. Kluwer Academic, Dordrecht (1994)
416. Goossens, A., Craje, M.W.J., Van der Kraan, A.M., Zwijnenburg, A., Makkee, M., Moulijn, J.A., Grisel, R.J.H., Nieuwenhuys, B.E., de Jongh, L.J.: Hyperfine Interact. **139/140**, 59 (2002)
417. Goossens, A., Craje, M.W.J., Van der Kraan, A.M., Zwijnenburg, A., Makkee, M., Moulijn, M., de Jongh, L.J.: Catal. Today **72**, 95 (2002)
418. Zwijnenburg, A., Goossens, A., Sloof, W.G., Craje, M.W.J., van der Kraan, A.M., de Jongh, L.J., Makkee, M., Moulijn, J.A.: J. Phys. Chem. **B 106**, 9853 (2002)
419. Lee, S.-J., Gavriilidis, A., Pankhurst, Q.A., Kaek, A., Wagner, F.E., Wong, P.C.-L., Yeung, K.L.: J. Catal. **200**, 298 (2001)
420. Kobayashi, Y., Nasu, S., Tsubota, S., Haruta, M.: Hyperfine Interact. **126**, 95 (2000)
421. Finch, R.M., Hodge, N.A., Hutchings, G.J., Meagher, A., Pankhurst, Q.A., Siddiqui, M.R.H., Wagner, F.E., Whyman, R.: Phys. Chem. Chem. Phys. **1**, 485 (1999)
422. Daniells, S.T., Owerweg, A.R., Makkee, M., Moulijn, J.A.: J. Catal. **230**, 52 (2005)
423. Friedl, J., Wagner, F.E., Nkosi, B., Towert, M., Coville, N.J., Adams, M.D., Hutchings, G.J.: Hyperfine Interact. **69**, 767 (1991)
424. Kongolo, K., Bahr, A., Friedl, J., Wagner, F.E.: Hyperfine Interact. **57**, 1929 (1990)
425. Adams, M.D., Friedl, J., Wagner, F.E.: Hydrometallurgy **37**, 33 (1995)
426. Parvulescu, V.I., Parvulescu, V., Endruschat, U., Filoti, G., Wagner, F.E., Kübel, C., Richards, R.: Chem. Eur. J. **12**, 2343 (2006)
427. Ikeda, K., Kobayashi, Y., Negishi, Y., Seto, M., Iwasa, T., Nobusada, K., Tsukada, T., Kojima, N.: J. Am. Chem. Soc. **129**, 7230 (2007)
428. Garitaonandia, J.S., Insausti, M., Goikolea, E., Suzuki, M., Cashion, J.D., Kawamura, N., Ohsawa, H., de Muro, I.G., Suzuki, K., Plazaola, F., Rojo, T.: Nano Lett. **8**, 661 (2008)

429. Thiel, F.G., Benfield, R.E., Zanoni, R., Smit, H.H.A., Dirken, M.W.: Z. Phys. **D 26**, 162 (1993)
430. van de Straat, D.A., Thiel, R.C., de Jongh, L.J., Gubbens, P.C.M., Schmid, G., Cerotti, A., della Pergola, R.: Z. Phys. **D 40**, 574 (1997)
431. Stievano, L., Santucci, S., Lozzi, L., Calogero, S., Wagner, F.E.: J. Non-Cryst. Solids **232–234**, 644 (1998)
432. Haslbeck, S., Martinek, K.-P., Stievano, L.: Hyperfine Interact. **165**, 89 (2005)
433. Paulus, P.M., Goossens, A., Thiel, R.C., Schmid, G., van der Kraan, A.M., de Jongh, L.J.: Hyperfine Interact. **126**, 199 (2000)
434. Paulus, P.M., Goossensw, A., Thiel, R.C., van der Kraan, A.M., Schmid, G., de Jongh, L.J.: Phys. Rev. **B 64**, 205418 (2001)
435. Kobayashi, Y., Kiao, S., Seto, M., Takatani, H., Nakanishi, M., Oshima, R.: Hyperfine Interact. **156/157**, 75 (2004)
436. Grünberg, P., Schreiber, R., Pang, Y., Brodsky, M.B., Sower, H.: Phys. Rev. Lett. **57**, 2442 (1986)
437. Shinjo, T., Mibu, K.: Hyperfine Interact. **136/137**, 253 (2001)
438. Shinjo, T.: Surf. Sci. **438**, 329 (1999)
439. Shinjo, T., Keune, W.: J. Magn. Magn. Mater. **200**, 598 (1999)
440. Kobayashi, Y., Nasu, S., Emoto, T., Shinjo, T.: Hyperfine Interact. **111**, 129 (1998)
441. Kobayashi, Y., Nasu, S., Emoto, T., Shinjo, T.: Physica **B 237**, 249 (1997)
442. Kobayashi, Y., Nasu, S., Emoto, T., Shinjo, T.: J. Magn. Magn. Mater. **156**, 45 (1996)
443. Sawicki, J.A.: Nucl. Instrum. Methods **B 93**, 469 (1994)
444. Sawicki, J.A., Abouchacra, G., Serughetti, J., Perez, A.: Nucl. Instrum. Methods B **16**, 355 (1986)
445. Kyek, A., Wagner, F.E., Lehrberger, G., Pankhurst, Q.A., Ziegaus, B.: Hyperfine Interact. **126**, 235 (2000)
446. Dunlap, B.D., Darby Jr., J.B., Kimball, C.W.: Phys. Lett. **A 25**, 431 (1967)
447. Cohen, R.L., Sherwood, R.C., Wernick, J.H.: Phys. Lett. **A 26**, 462 (1968)
448. Patterson, D.O., Thomson, J.O., Huray, P.G., Roberts, L.D.: Phys. Rev. **B 2**, 2440 (1970).
449. Erickson, D.J., Roberts, L.D.: Phys. Rev. **B 9**, 3650 (1974)
450. Thomson, J.O., Obernshain, F.E., Huray, P.G., Love, J.C., Burton, J.: Phys. Rev. **B 11**, 1835 (1975)
451. Gütlich, P., Odar, S., Weiss, A.: J. Phys. Chem. Solids **37**, 1011 (1976)
452. Allred, A.L., Rochow, E.G.: J. Inorg. Nucl. Chem. **5**, 264 (1958)
453. Shirley, D.A., Kaplan, M., Axel, P.: Phys. Rev. **123**, 816 (1961)
454. Roberts, L.D., Thomson, J.O.: Phys. Rev. **129**, 664 (1963)
455. Andrä, H.J., Hashmi, C.M., Kienle, P., Stanek, F.W.: Z. Naturforsch. **A 18**, 687 (1963)
456. Barrett, P.H., Grant, R.W., Kaplan, M., Keller, D.A., Shirley, D.A.: J. Chem. Phys. **39**, 1035 (1963)
457. Grant, R.W., Kaplan, M., Keller, D.A., Shirley, D.A.: Phys. Rev. **133**, 1062 (1964)
458. Roberts, L.D., Becker, R.L., Obenshain, F.E., Thomson, J.O.: Phys. Rev. A **137**, 895 (1965)
459. Marshall, S.W., Wilenzick, R.M.: Phys. Rev. Lett. **16**, 219 (1966)
460. Schroeer, D.: Phys. Lett. **21**, 123 (1966)
461. Cohen, R.L.: Phys. Rev. **171**, 344 (1968)
462. Chekin, V.V.: Sov. Phys.-JETP **27**, 983 (1968)
463. Cohen, R.L., Wernick, J.H., West, K.W., Sherwood, R.C., Chin, G.Y.: Phys. Rev. **188**, 684 (1969)
464. Longworth, G.: J. Phys. C: Met. Phys., Suppl. 1, S81 (1970)
465. Schroeer, D., Marzke, R.F., Erickson, D.J., Marshall, S.W., Wilenzick, R.M.: Phys. Rev. **B 2**, 4414 (1970)
466. Mansel, W., Vogl, G., Vogl, W., Wenzl, H., Barb, D.: Physica Status Solidi **40**, 461 (1970)
467. Kohgi, M., Yamada, T., Kunitomi, N., Maeda, Y.: J. Phys. Soc. Jpn **28**, 793 (1970)
468. Charlton, J.S., Harris, I.R.: Physica Status Solidi **39**, Kl (1970)

469. Cohen, R.L., Yafet, Y., West, K.W.: Phys. Rev. **B 5**, 2872 (1971)
470. Huray, P.G., Roberts, L.D., Thomson, J.O.: Phys. Rev. **B 4**, 2147 (1971)
471. Erickson, D.J., Prince, J.F., Roberts, L.D.: Phys. Rev. **C 8**, 1916 (1973)
472. Burton, J.W., Thomson, J.O., Huray, P.G., Roberts, L.D.: Phys. Rev. **B 7**, 1773 (1973)
473. Roberts, L.D., Prince, J.F., Erickson, D.J.: In: Cohen, S.G., Pasternak, M. (eds.) Perspectives in Mössbauer Spectroscopy. Plenum, New York (1973)
474. Chou, T.S., Perlman, M.L., Watson, R.E.: Phys. Rev. **B 14**, 3248 (1976)
475. Perscheid, B., Haas, H., Isolde Collaboration: Hyperfine Interact. **15/16**, 227 (1983)
476. Kuhn, M., Bzowski, A., Sham, T.K.: Hyperfine Interact. **94**, 2267 (1994)
477. Kyek, A., Wagner, F.E., Palade, P., Jianu, A., Macovei, D., Popescu, R., Manaila, R., Filoti, G.: J. Alloys Compd. **313**, 13 (2000)
478. Palade, P., Wagner, F.E., Jianu, A.D., Filoti, G.: J. Alloys Compd. **353**, 23 (2003)
479. Ksenofontov, V., Kroth, K., Reimann, S., Casper, F., Jung, V., Takahashi, M., Takeda, M., Felser, C.: Hyperfine Interact. **168**, 1201 (2006)
480. Casper, F., Ksenofontov, V., Kandpal, H.C., Reiman, S., Shishido, T., Takahashi, M., Takeda, M., Felser, C.: Z. Anorg. Allg. Chem. **632**, 1273 (2006)

References to Sect. 7.11

481. Carlson, D.E., Temperley, A.A.: Phys. Lett. **B 30**, 322 (1969)
482. Wurtinger, W.: J. Phys. **C 6**, C6–C697 (1976)
483. Koch, W., Wagner, F.E., Flach, D., Kalvius, G.M.: J. Phys. **C 6**, C6–C693 (1976)
484. Wurtinger, W.: Dissertation, Technische Hochschule, Darmstadt (1977)
485. Wurtinger, W., Kankeleit, E.: Z. Phys. **A 293**, 219 (1979)

Chapter 8
Some Special Applications

We have learned from the preceding chapters that the chemical and physical state of a Mössbauer atom in any kind of solid material can be characterized by way of the hyperfine interactions which manifest themselves in the Mössbauer spectrum by the isomer shift and, where relevant, electric quadrupole and/or magnetic dipole splitting of the resonance lines. On the basis of all the parameters obtainable from a Mössbauer spectrum, it is, in most cases, possible to identify unambiguously one or more chemical species of a given Mössbauer atom occurring in the same material. This – usually called phase analysis by Mössbauer spectroscopy – is nondestructive and widely used in various kinds of physicochemical studies, for example, the studies of

- Phase transitions
- Spin and magnetic transitions
- Dynamic solid-state phenomena
- Solid-state reactions (e.g., thermolysis, radiolysis)
- Problems of metallurgy
- Surface phenomena
- Catalysis
- Corrosion
- After effects of nuclear transformations in solids
- Biological systems
- Minerals
- Archeological artifacts
- Industrial applications

Because of the limited scope of this monograph, it is impossible to give a rigorous account of the work that has been accomplished in these fields. By the end of 2008, about 60,000 publications dealing with the use of Mössbauer spectroscopy had been documented in the literature. Excellent review articles on "Mössbauer Spectroscopy Applied to Inorganic Chemistry" (3 volumes) are given in [30, 34–35] in Chap. 1 and on "Mössbauer Spectroscopy Applied to Magnetism and Materials Science" (2 volumes) in [38, 40] in Chap. 1.

P. Gütlich et al., *Mössbauer Spectroscopy and Transition Metal Chemistry*,
DOI 10.1007/978-3-540-88428-6_8, © Springer-Verlag Berlin Heidelberg 2011

The reader is also referred to the CD-ROM enclosed at the end of this book, where we have collected a large variety of examples of applications of Mössbauer spectroscopy in many disciplines. The first part of the CD-ROM presents a lecture series on Mössbauer spectroscopy (Principles and typical applications), which has been arranged particularly for teaching purposes. The second part comprises a large number of special applications contributed by various research groups specialized on Mössbauer spectroscopy.

The authors of this book consider it appropriate to include in this section two contributions from their own laboratories, one on Mössbauer spectroscopy of spin crossover (SCO) phenomena in iron(II) compounds and the other on applications to biological systems. Both chapters will demonstrate the effectiveness of Mössbauer spectroscopy in these particular fields.

8.1 Spin Crossover Phenomena in Fe(II) Complexes

8.1.1 Introduction

The behavior of SCO compounds is among the most striking and fascinating shown by relatively simple molecular species. As a consequence of the splitting of the energy of the *d*-orbitals into the t_{2g} and e_g sets in a ligand field, octahedral complexes of certain transition metal ions, nearly exclusively those of the first transition series with configurations d^4–d^7 may exist in either the high-spin (HS) or low-spin (LS) state, depending on the nature of the ligand field about the metal ion. In weak fields, the ground state is HS where the spin multiplicity is a maximum, the *d*-electrons being distributed over the t_{2g} and e_g sets, whereas strong fields stabilize the LS state with minimum multiplicity, the t_{2g} set being completely occupied before electrons are added to the e_g set. For the d^6 configuration of iron(II), for example, the two states are illustrated by $[Fe(H_2O)_6]^{2+}$, which, with the configuration $t_{2g}{}^4 e_g{}^2$, has four unpaired electrons and is thus strongly paramagnetic ($^5T_{2g}$ state in octahedral symmetry), and by $[Fe(CN)_6]^{4-}$ ($t_{2g}{}^6 e_g{}^0$), which has no unpaired electrons ($^1A_{1g}$ state) (Fig. 8.1). For intermediate fields, the energy difference (ΔE^0_{HL}) between the

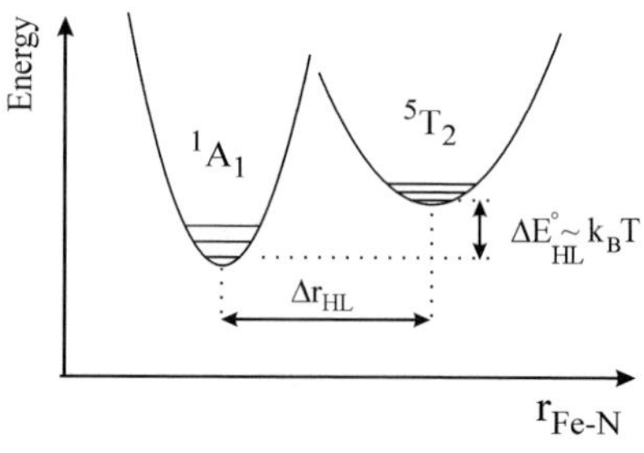

Fig. 8.1 Representation of the potential wells for the $^1A_1$ and $^5T_2$ states of an iron(II) SCO system, the nuclear coordinate being the metal–donor atom distance

lowest vibronic levels of the potential wells of the two states may be sufficiently small such that relatively minor external perturbation causes a change in the spin state. This phenomenon is known as spin transition (ST) or SCO and its origin is illustrated in Fig. 8.1. ST will be induced thermally when $\Delta E^0_{HL} \sim k_B T$, and when this criterion is met, pressure- and light-induced STs may also be observed (*vide infra*).

Although STs have been observed for the first (3*d*) transition series and also partially for the second transition series, by far the greatest number has been reported for iron(II). The earlier developments in iron(II) ST systems were reviewed by König [1], Goodwin [2] and Gütlich [3]. Since then, numerous reviews have appeared on specific aspects of the phenomenon [4, 5 and references therein].

From the measurement of the temperature-dependent properties of a ST system, it is possible to evaluate the relative concentrations of HS and LS states as a function of temperature and, thus, to plot the HS molar fraction versus T. Such curves are diagnostic of the nature of the ST. The steepness of the change, for example, is indicative of the extent of cooperativity involved in the propagation of the spin change throughout the solid lattice, that is, how electronic and structural changes in a molecule proceed to neighboring molecules. When this cooperativity is low, the transition will be a gradual process, but as cooperativity increases, the transition becomes more abrupt and may occur within a very narrow range of temperature. In this case, the transition can be associated with a phase change and/or a thermal hysteresis. The transition temperature $T_{1/2}$ is defined as that temperature at which the fractions of HS and LS isomers, which take part in the transition, are equal. For transitions displaying hysteresis, two transition temperatures $T_{1/2}\downarrow$ and $T_{1/2}\uparrow$ define the width of the hysteresis loop. The appearance of a thermally induced ST is manifested in strong temperature-dependence of properties, which are related to the electronic structure of the system, that is, color, magnetism, and molecular structure. Since the antibonding e_g orbitals of iron(II) are unoccupied in the LS state, while they are partially occupied in the HS state, the metal–donor atom distance is remarkably sensitive to changes of the spin-state. The structural change amounts to an increase of the Fe–N bond distances by about 0.2 Å as a consequence of the singlet ($^1A_1$) $\leftrightarrow$ quintet ($^5T_2$) transition. For systems in the solid state, this change in the molecular dimension may bring about fundamental changes in the crystal lattice.

It is possible to monitor an ST by a variety of techniques. The measurement of magnetic susceptibility is frequently used to characterize ST since there is a marked difference in the magnetic moments of HS (ca. 5 B.M.) and LS (ca. 0 B.M.) iron(II) and, thus, a strong temperature-dependence of the magnetic moment is expected. Magnetism alone may not be sufficient to establish the existence of ST, since magnetic properties refer to the bulk material rather than to a local atomic site. Techniques such as optical spectroscopy in the visible region, vibrational spectroscopy, and magnetic resonance methods, are frequently applied to identify and characterize ST. A survey of physical techniques and methods suitable for the investigation of SCO systems have been discussed in [6].

^{57}Fe Mössbauer spectroscopy is particularly suitable to study ST since (1) the spectral parameters associated with the HS and LS states of iron(II) clearly differ and (2) the time-scale of the technique ($\sim 10^{-7}$ s) allows the detection of the separate spin states in the course of the transition. Typically, Mössbauer spectra of HS iron(II) show relatively high quadrupole splitting ($\Delta E_Q \approx 2-3$ mm s^{-1}) and isomer shift ($\delta \approx 1$ mm s^{-1}), while for LS iron(II), these parameters are generally smaller ($\Delta E_Q \leq 1$ mm s^{-1}, $\delta \leq 0.5$ mm s^{-1}). Among the early applications of Mössbauer spectroscopy to study ST phenomena in iron(II) complexes is the work of Dézsi et al. [7] on [Fe^{II}(phen)$_2$(NCS)$_2$] (phen = 1,10-phenanthroline) as a function of temperature (Fig. 8.2). The transition from the HS ($^5T_2$) state (quadrupole doublet of outer two lines with $\Delta E_Q \approx 3$ mm s^{-1}) to the LS ($^1A_1$) state (quadrupole

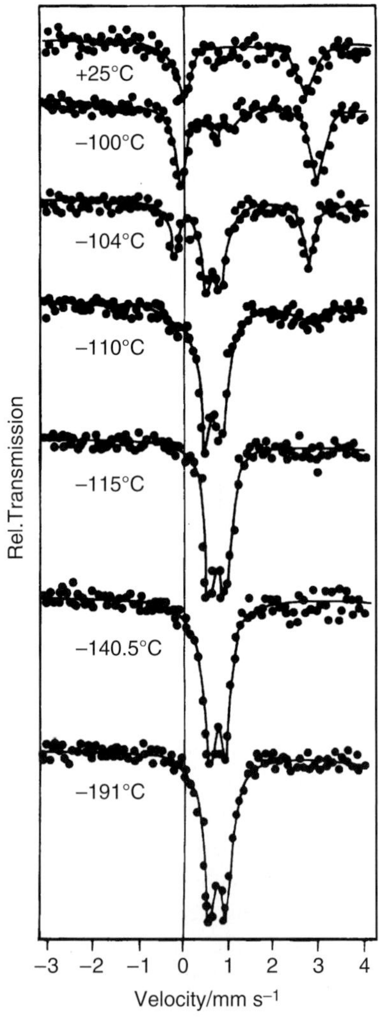

Fig. 8.2 ^{57}Fe Mössbauer spectra of [Fe^{II}(phen)$_2$(NCS)$_2$] as a function of temperature. At room temperature, the spectrum shows only a HS quadrupole doublet characteristic of the HS ($^5T_2$) state. On lowering the temperature, a new quadrupole doublet (inner two lines with $\Delta E_Q \sim 0.5$ mm s^{-1} characteristic of the LS state) develops. The thermally induced ST occurs abruptly near 178 K (from [7])

doublet of inner two lines with $\Delta E_Q \approx 0.5$ mm s^{-1}) with decreasing temperature occurs at $T_{1/2} \sim 172$ K. Jesson et al. [8, 9] studied the $^5T_2$–$^1A_1$ transition in an iron (II) complex with hydrotris(1-pyrazolyl)borate as ligands, and Long et al. [10] remeasured the temperature-dependent Mössbauer spectra of this SCO complex several decades later (Fig. 8.3).

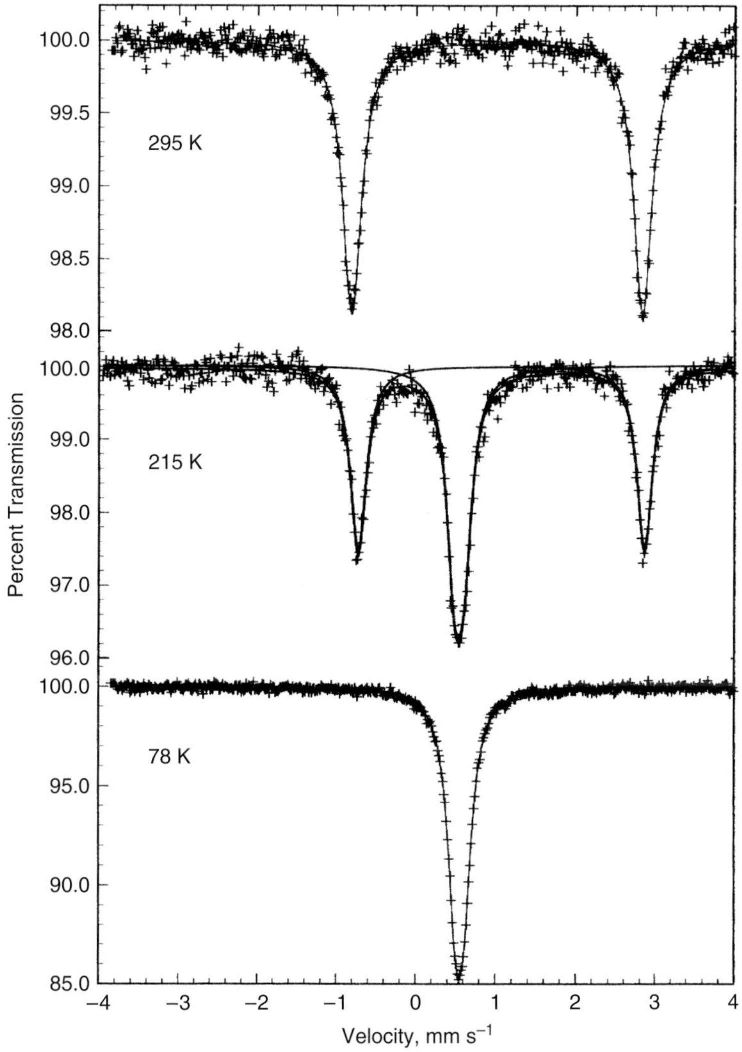

Fig. 8.3 ^{57}Fe Mössbauer spectra of [Fe(HB(3,5-(CH$_3$)$_2$pz)$_3$)$_2$] (pz = pyrazolyl) obtained at the indicated temperatures. The spectra were fitted with two symmetric quadrupole doublets; the outer two lines refer to the HS state and the inner (unresolved quadrupole doublet) to the LS state (from [10])

8.1.2 Spin Crossover in $[Fe(2\text{-}pic)_3]Cl_2{\cdot}Sol$

Gütlich et al. [4, 6] have studied SCO in solid $[Fe(2\text{-}pic)_3]Cl_2{\cdot}EtOH$ (2-pic = 2-picolylamine), particularly the influence of dilution with Zn and Co, the nature of noncoordinated anions and crystal solvent, the *H*/*D* and $^{14}N/^{15}N$ isotope effect, and the influence of pressure on ST properties. This SCO system has proven particularly suitable for mechanistic studies because of the excellent resolution of the resonance signals of the HS and LS states. As an example, Fig. 8.4 shows a series of temperature-dependent ^{57}Fe Mössbauer spectra of $[Fe(2\text{-}pic)_3]Cl_2{\cdot}EtOH$, and the spectra in Fig. 8.5 demonstrate the effect of metal dilution in the isomorphous mixed crystal series $[Fe_xZn_{1-x}(2\text{-}pic)_3]Cl_2{\cdot}EtOH$ [11, 12]. On decreasing the iron concentration, the intensity of the HS quadrupole doublet increases at a given temperature and the ST curves become less steep and, additionally, they are shifted toward lower temperatures, as seen in Fig. 8.6. The inset in Fig. 8.6 shows the strong concentration dependence of the ST temperature. Clearly, this effect originates from cooperative interactions, that is, the distances between the iron(II) sites become larger and the cooperative interactions weaker with decreasing iron concentration. These cooperative interactions have been interpreted in a thermodynamic approach in the context of elasticity theory [4, 13].

The SCO properties of the system $[Fe(2\text{-}pic)_3]Cl_2{\cdot}Sol$ (Sol = EtOH, MeOH, H_2O, and $2H_2O$) have been found to be strongly dependent on the nature of the solvent [14]. Temperature-dependent ^{57}Fe Mössbauer spectra of deuterated systems have proven that hydrogen bonding plays a significant role in communicating the ST from one iron site to neighboring sites via cooperative interaction [15]. ST curves for $[Fe(2\text{-}pic)_3]Cl_2{\cdot}EtOH$, deuterated in different positions of the hydrogen bonding network, are shown in Fig. 8.7; the changes are particularly dramatic when the deuteration positions are direct constituents of the communication chain as in the case of C_2H_5OH/ND_2, but they are less dramatic when deuteration occurs only as an attachment to the chain, as in the case of C_2D_5OH as crystal solvent. The high-resolution Mössbauer data obtained for $[Fe(2\text{-}pic)_3]Cl_2{\cdot}C_2H_5OH$ unraveled for the first time that ST in this system actually takes place in two steps, as is clearly seen in Fig. 8.7 [16].

The ST curve is sensitive to the external pressure because of the difference in volume of the HS and LS isomers involved in the transition. The primary effect of an increase in pressure is favoring the LS state in which the metal–donor atom distances are shorter (by ca. 10%) than those of the HS state. This generally results in a displacement of the transition temperature to higher values, and this also implies an increase in the zero-point energy separation ΔE^0_{HL} in Fig. 8.1. The study of the effect of pressure on ST systems has been pursued for many years, primarily to elucidate the subtle effects associated with the transitions and to gain insight into their mechanistic aspects. The pioneering work of Drickamer [17] on the effect of pressure on the spin state of $[Fe(phen)_2(NCS)_2]$ and related iron(II) ST systems provided unequivocal evidence for the HS $\rightarrow$ LS conversion with increasing pressure. The effect of pressure on the ST behavior of $[Fe(2\text{-}pic)_3]Cl_2{\cdot}C_2H_5OH$

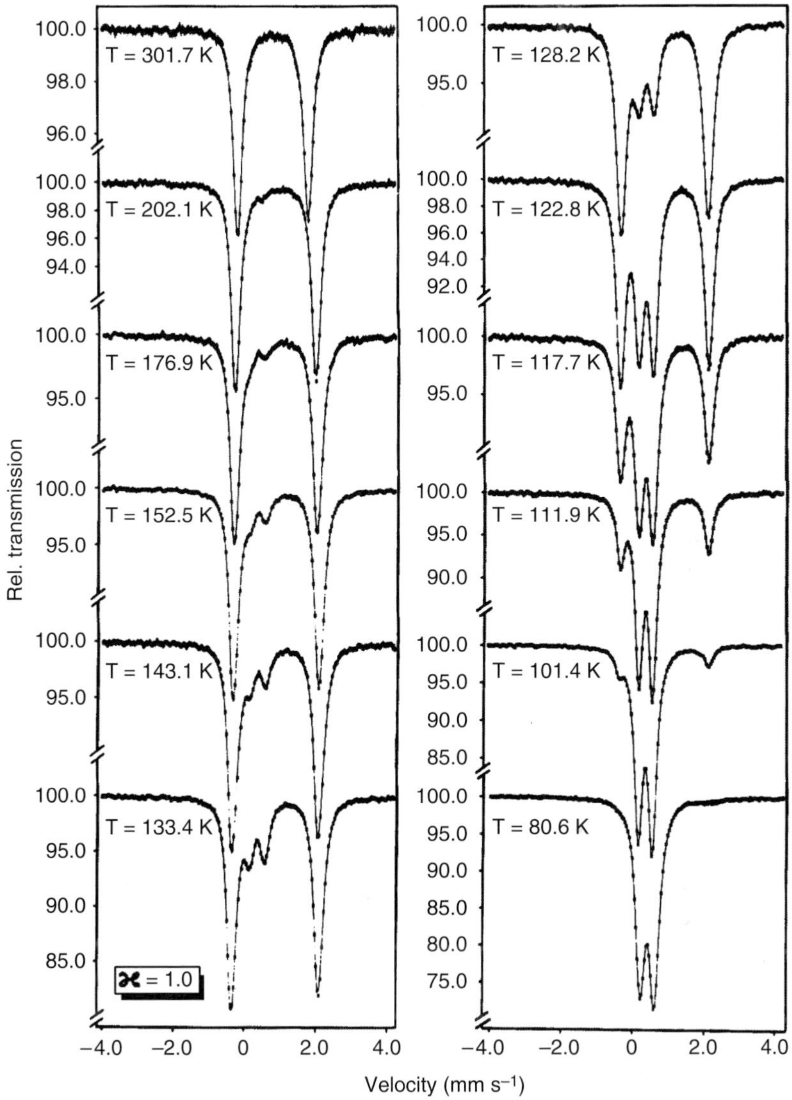

Fig. 8.4 ^{57}Fe Mössbauer spectra of [Fe(2-pic)$_3$]Cl$_2$·C$_2$H$_5$OH as a function of temperature (*source*: ^{57}Co/Cu at room temperature). The *outer two lines* represent the quadrupole doublet of the $^5T_2$ (HS) state, the *inner two lines* that of the $^1A_1$ (LS) state. The *solid lines* are obtained by a least-squares computer fit of Lorentzian lines to the experimental spectra (from [11])

was also investigated by placing the sample in a custom-made hydrostatic gas cell for pressures up to ca. 12 kbar [18] (Fig. 8.8). As expected, the intensity of the LS quadrupole doublet increased and the ST curve was shifted to higher temperatures with increase of applied pressure and, interestingly, the step in the ST curve disappeared, as is seen in Fig. 8.9 [19].

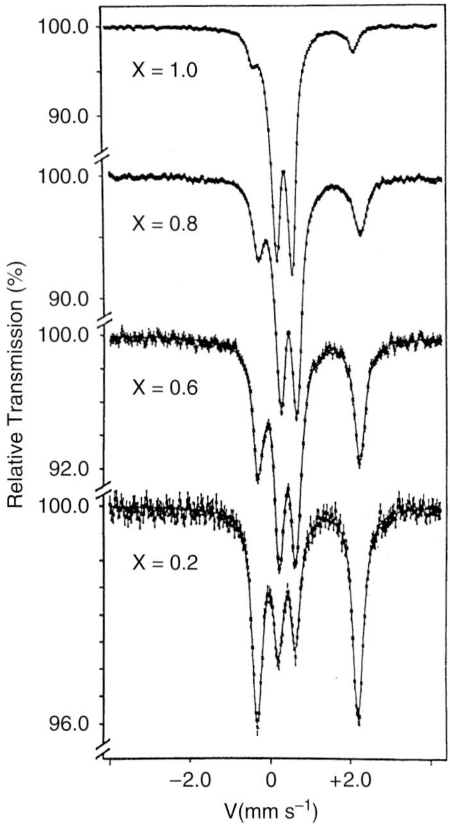

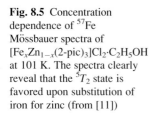

Fig. 8.5 Concentration dependence of ^{57}Fe Mössbauer spectra of $[Fe_xZn_{1-x}(2\text{-pic})_3]Cl_2 \cdot C_2H_5OH$ at 101 K. The spectra clearly reveal that the $^5T_2$ state is favored upon substitution of iron for zinc (from [11])

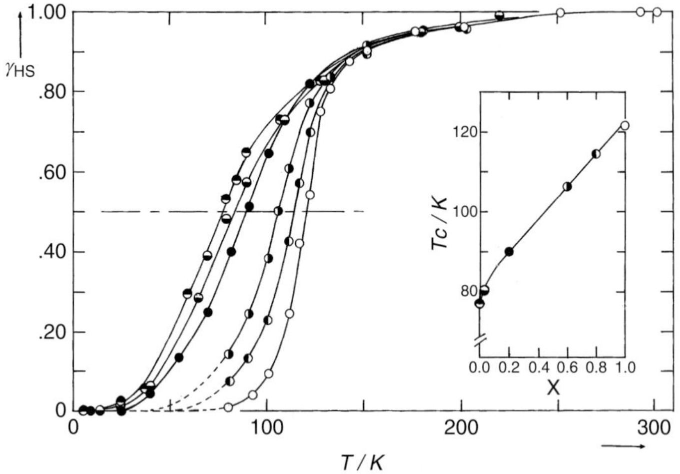

Fig. 8.6 Concentration dependence of the ST temperature for $[Fe_xZn_{1-x}(2\text{-pic})_3]Cl_2 \cdot C_2H_5OH$ at 101 K (from [11, 12])

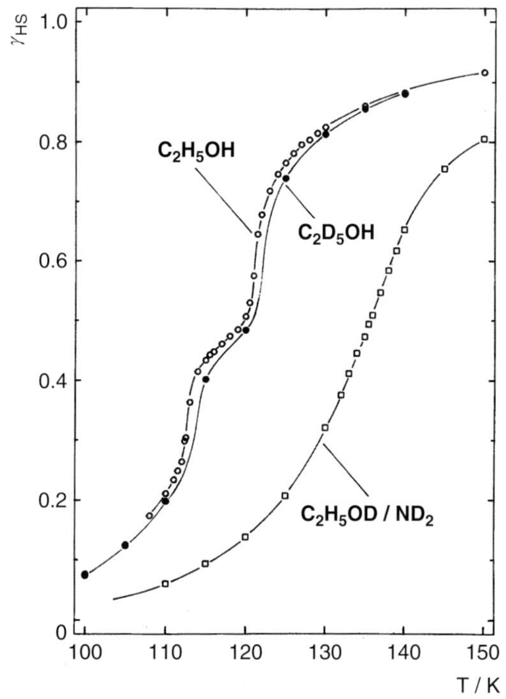

Fig. 8.7 ST curves (molar fraction γ_{HS} as a function of temperature) for $[Fe(2\text{-pic})_3]Cl_2{\cdot}C_2H_5OH$ deuterated in different positions of the solvent (from [15])

8.1.3 Effect of Light Irradiation (LIESST Effect)

In the mid-1980s, it was accidentally observed that the LS state of the iron(II) SCO compound $[Fe(ptz)_6](BF_4)_2$ (ptz = 1-H-propyltetrazole) was converted by green light into the metastable HS state which had a virtually infinite lifetime at sufficiently low temperatures [20, 21]. This phenomenon became known as the LIESST effect (Light-Induced-Excited-Spin-State-Trapping). Later, it was found that the reverse process is also possible using red light [22]. The mechanism for these photoswitching processes, which turned out to be a common feature of most Fe (II) SCO systems, is sketched in Fig. 8.10 [21, 22]. Green light (514 nm from an Ar laser) is used for the spin-allowed excitation $^1A_1 \rightarrow {}^1T_1$ with $^1T_1$ lifetimes typically of nanoseconds. A fast relaxation cascade over two successive intersystem crossing steps, $^1T_1 \rightarrow {}^3T_1 \rightarrow {}^5T_2$, populates the metastable $^5T_2$ state. Radiative relaxation $^5T_2 \rightarrow {}^1A_1$ is forbidden, and decay by thermal tunneling to the ground state $^1A_1$ is slow at low temperatures. Reverse LIESST is achieved by the application of red light (ca. 820 nm), whereby the transition from the $^5T_2$ state to the ^{5}E state is induced with two subsequent intersystem crossing processes, $^5E \rightarrow {}^3T_1 \rightarrow {}^1A_1$, leading back to the LS ground state. As demonstrated later by Hauser [22], photoswitching from LS to HS is also possible via $^1A_1 \rightarrow {}^3T_1 \rightarrow {}^5T_2$ transitions using red light of 980 nm. LIESST has also been observed for the conversion of a HS ground state to the metastable LS state by red light [23].

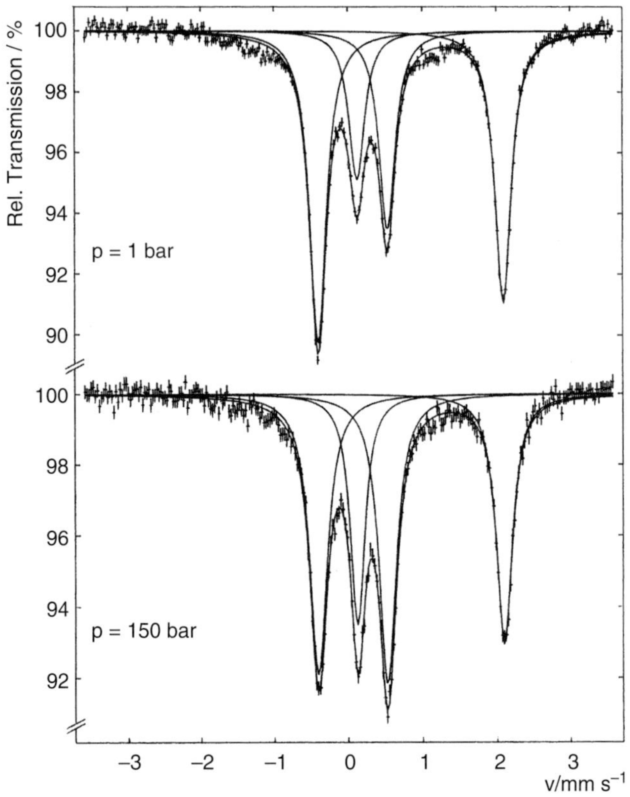

Fig. 8.8 Pressure dependence of the ^{57}Fe Mössbauer spectra of $[Fe(2\text{-}pic)_3]Cl_2{\cdot}C_2H_5OH$ at 121 K. The area fraction of the LS doublet (inner two lines) increases by application of pressure (from [18])

As the lifetimes of the light-induced excited spin states are relatively long in the low temperature region, Mössbauer spectroscopy can easily be employed to identify the spin states. As an example, Mössbauer spectra measured on $[Fe(mtz)_6](BF_4)_2$ (mtz = 1-H-methyltetrazole) under various experimental conditions are shown in Fig. 8.11 [23]. $[Fe(mtz)_6](BF_4)_2$ is an interesting SCO complex, which, according to single-crystal structure analysis, possesses two slightly inequivalent lattice sites for the iron(II) ions, that is, sites A and B. Both are in the HS state at room temperature. On lowering the temperature, only A sites undergo thermally induced ST, whereas B sites remain in the HS state at all temperatures. This is obvious from the spectrum (recorded at 50 K) in the center of Fig. 8.11. The outer lines refer to the quadrupole doublet of HS isomers at B sites; the intensity asymmetry is due to texture, a well-known effect for plate-like crystals. The poorly resolved quadrupole doublet in the middle refers to LS isomers at A sites which have undergone thermal ST from HS(A) to LS(A). Irradiating the sample with green light (514 nm) from an Ar laser at 20 K induces LS to HS conversion at A

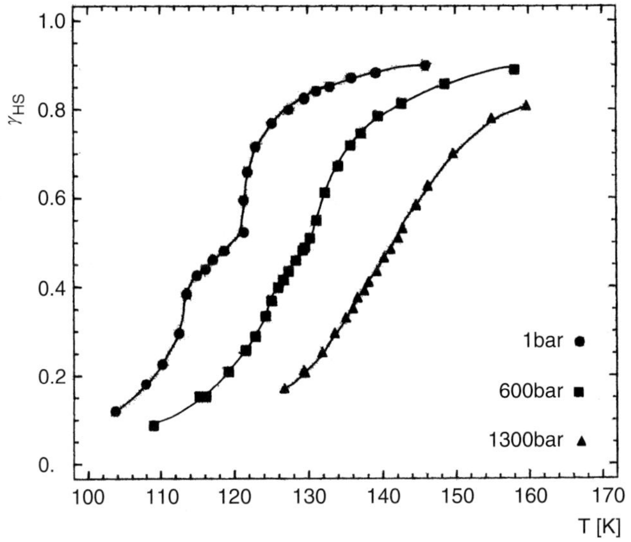

Fig. 8.9 ST curves $\gamma_{HS}(T)$ as obtained from Mössbauer measurements on $[Fe(2\text{-pic})_3]Cl_2{\cdot}C_2H_5OH$ at ambient and applied hydrostatic pressures of 600 and 1,300 bar (from [19])

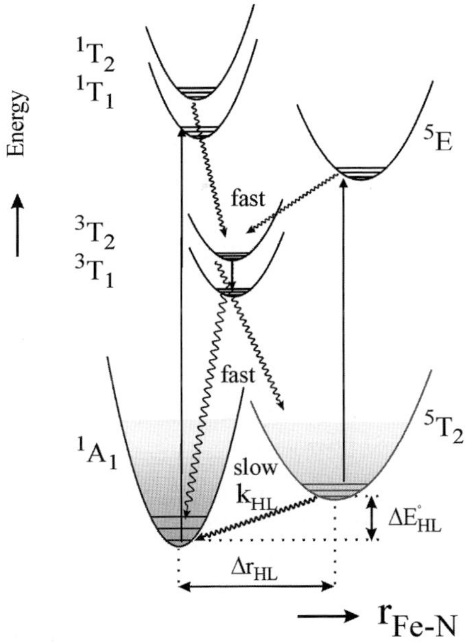

Fig. 8.10 Schematic illustration of LIESST and reverse LIESST of a d^6 complex. Spin allowed *d–d* transitions are denoted by *arrows* and the radiationless relaxation processes by *wavy lines* (from [21])

sites (see the Mössbauer spectrum at the top of Fig. 8.11); the LS(A) quadrupole doublet has disappeared in favor of a new HS(A) doublet, also showing line intensity asymmetry due to texture. The metastable HS(A) site relaxes on heating

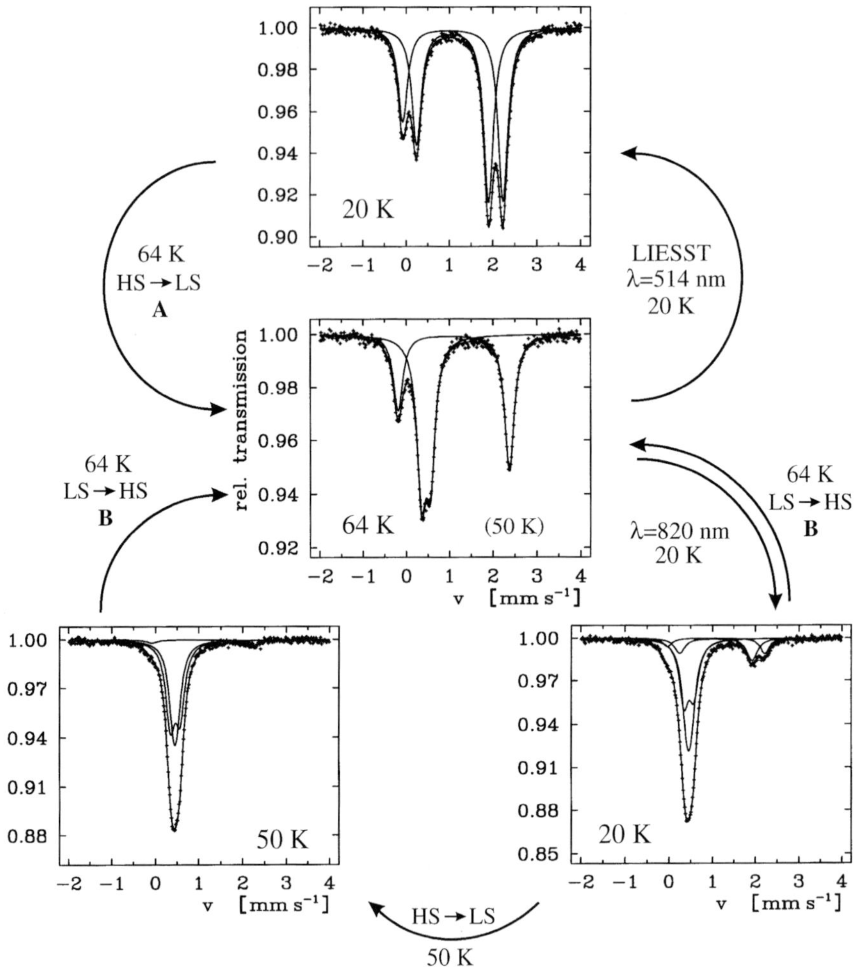

Fig. 8.11 ^{57}Fe Mössbauer spectra of $[Fe(mtz)_6](BF_4)_2$ recorded under various experimental conditions as described in the text. This Mössbauer study demonstrates the LIESST (LS $\leftrightarrow$ HS) phenomenon at A sites, where thermal ST occurs, and LIESST (HS $\leftrightarrow$ LS) at B sites, where no thermal ST takes place (from [23])

the sample; the spectrum measured at 64 K corresponds to the original spectrum displayed in the center of Fig. 8.11 with the resonance signals for HS(B) and LS(A). If the sample is irradiated with red light (820 nm) at 20 K, the spectrum on the lower right is obtained, where essentially, the HS(B) doublet disappeared in favor of a LS signal which overlaps with the LS(A) signal. There is also a small fraction of HS (A) doublet overlapping with the HS(B) doublet; this is formed via the above-mentioned $^1A_1 \rightarrow {}^3T_1 \rightarrow {}^5T_2$ LIESST pathway that is possible with red light, though with lower efficiency than with green light. The light-induced conversion of stable HS(B) to metastable LS(B) was the first LIESST (HS $\leftrightarrow$ LS) effect reported for an iron(II) SCO complex. After thermal relaxation at 64 K, the metastable LS(B) relaxes back to the stable HS(B) state yielding the original spectrum (center of Fig. 8.11). If, however, thermal relaxation is conducted at only 50 K, the Mössbauer spectrum shown at the lower left is obtained, where essentially, the small fraction of metastable HS(A) has relaxed. At this point, the sample contained ca. 50% each of metastable LS(B) and stable LS(A) ions. Further heating to 64 K caused thermal relaxation of the metastable LS(B) isomers and yielded again the Mössbauer spectrum displayed in the center of Fig. 8.11. These experiments have demonstrated the possible existence of both HS and LS states in both A and B lattice sites at one and the same temperature, a case of bistability [23].

8.1.4 Spin Crossover in Dinuclear Iron(II) Complexes

The interplay between magnetic coupling and ST in 2,2′-bipyrimidine (bpym)–bridged iron(II) dinuclear compounds has been studied by magnetic susceptibility and ^{57}Fe Mössbauer spectroscopy [24]. The molecular structures of {[Fe(bpym)$(NCS)_2]_2$(bpym)} and {[Fe(bt)$(NCS)_2]_2$(bpym)} (bt = 2,2′-bithiazoline) are shown in Fig. 8.12. The magnetic behavior in the form of $\chi_M T$ versus T plots, χ_M being the magnetic susceptibility, of four members of this series is drawn in Fig. 8.13. The complex (bpym, *S*) does not undergo thermal ST; instead, it exhibits intramolecular antiferromagnetic coupling between the two iron(II) ions through the bpym bridge ($J = -4.1\ \text{cm}^{-1}$, $g = 2.18$) [25]. When thiocyanate is replaced by selenocyanate, the resulting (bpym, Se) derivative shows an abrupt ST in the 125–115 K temperature region with a small hysteresis loop of 2.5 K width. Only 50% of the iron atoms undergo ST. The decrease of the $\chi_M T$ values at lower temperatures is due to the zero-field splitting (ZFS) of the $S = 2$ state and does not result from a further ST (Fig. 8.14e). The magnetic properties of the compounds abbreviated as (bt, S) and (bt, Se) are similar and show a complete ST with the remarkable feature that it takes

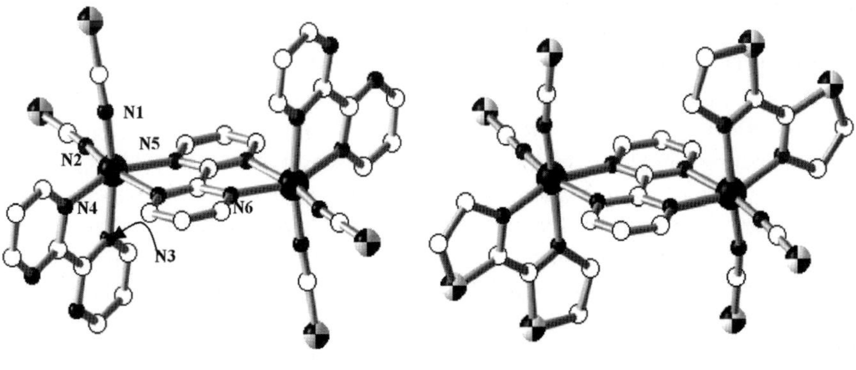

Fig. 8.12 Molecular structures of {[Fe(bpym)$(NCS)_2]_2$(bpym)} (*left*) and of {[Fe(bt)$(NCS)_2]_2$(bpym)} (*right*) (from [24, 25])

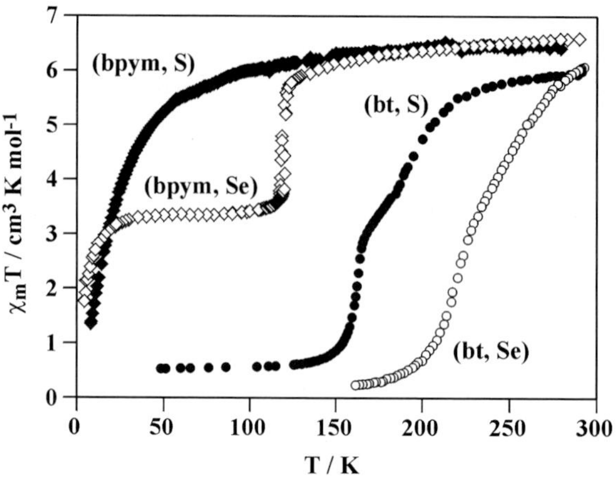

Fig. 8.13 Temperature dependence of $\chi_M T$ for {[Fe(bpym)(NCS)$_2$]$_2$(bpym)} (L = bpym and X = S (bypm, S) or Se (bypm, Se) and L = bt and X = S (bt, S) or Se (bt, Se)) (from [25])

place in two steps, though somewhat less pronounced than in the (bpym, Se) derivative. ST is centered at 197 and 163 K for (bt, S) and at 265 and 223 K for (bt, Se). In both cases, the region between the two steps corresponds approximately to 50% spin conversion. These steps, also detected by Mössbauer and calorimetric measurements, were interpreted in terms of a microscopic two-step transition between the three possible spin pairs of each individual dinuclear molecule [26]: [HS–HS] $\leftrightarrow$ [HS–LS] $\leftrightarrow$ [LS–LS]. Of particular interest has been the nature of the plateau (region between the two steps) in the ST curve of dinuclear iron(II) complexes. In the case of the (bpym, Se) derivative, for instance, it may either be a 1:1 mixture of [HS–HS] and [LS–LS] pairs, or it may consist only of the mixed spin-state pairs [HS–LS]. Both situations are not distinguishable in the $\chi_M T$ versus T plot. In this situation, conventional Mössbauer spectroscopy is not sensitive, since the spectra corresponding to the HS state of iron(II) in the [HS–LS] and [HS–HS] spin pairs are indistinguishable. Differentiation of three spin pairs [HS–HS], [HS–LS], [LS–LS], however, is possible by Mössbauer measurements in applied field [27]. This is demonstrated for the (bpym, S), (bpym, Se) and (bt, S) derivatives by the spectra shown in Fig. 8.14.

The effective hyperfine field H_{eff} at the iron nuclei of a paramagnetic nonconducting sample in an external field H_{ext} may be estimated as $H_{\mathrm{eff}} \approx H_{\mathrm{ext}} - (220 - 600(g - 2))\langle S\rangle$ where $\langle S\rangle$ is the expectation value of the iron spin and g is the Landé splitting factor [28, 29]. The difference between the expectation values of S for the LS and the HS states in [HS–LS] and [HS–HS] pairs permits unambiguous distinction between these dinuclear units. To do so, the strength of the external field should be sufficiently high and the temperature sufficiently low in order to avoid magnetic relaxation taking place within the characteristic time window of the Mössbauer experiment. Figure 8.14 displays the Mössbauer spectra of (bpym, S), (bpym, Se), and (bt, S) recorded at 4.2 K in zero-field and at 50 kOe,

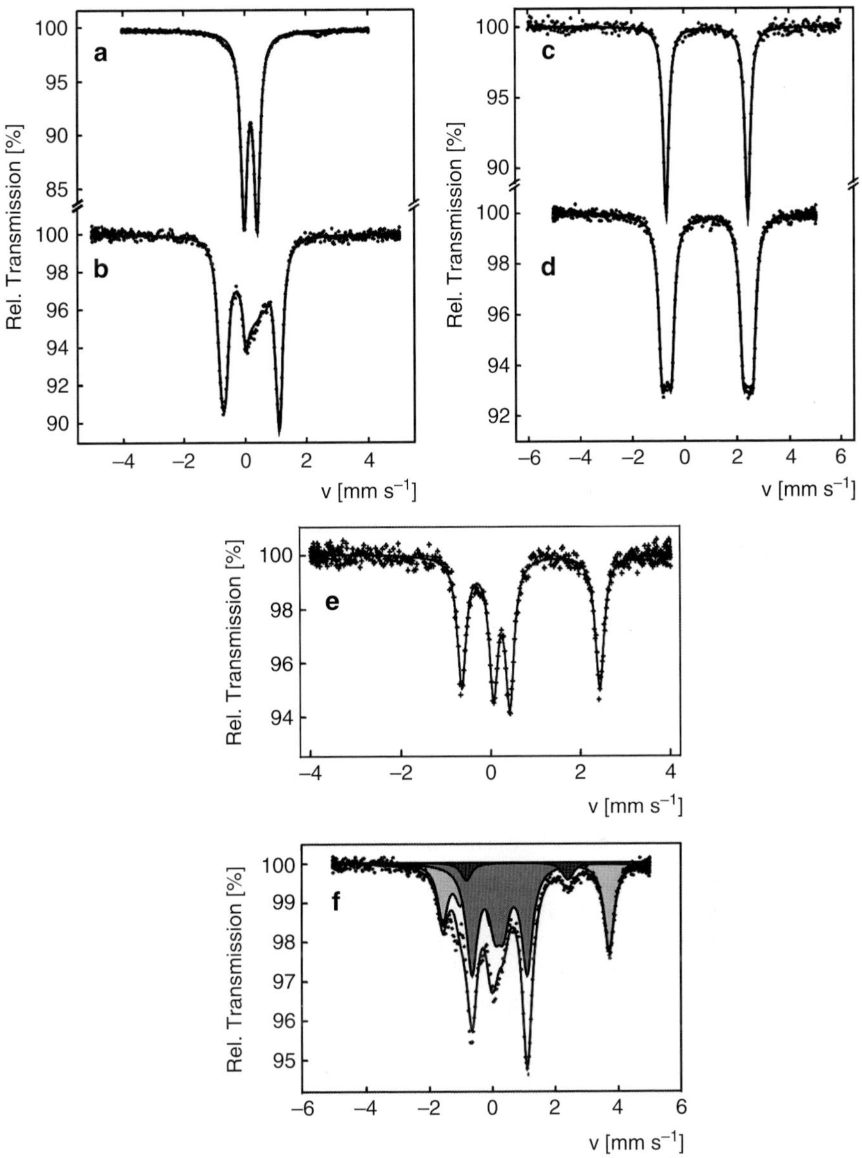

Fig. 8.14 Mössbauer spectra of (bt, S) recorded at 4.2 K in zero-field (**a**) and in a field of 50 kOe (**b**), of (bpym, S) recorded at 4.2 K in zero-field (**c**) and in a magnetic field of 50 kOe (**d**), of (bpym, Se) recorded at 4.2 K in zero-field (**e**) and in a magnetic field of 50 kOe (**f**). LS in [HS–LS] and [LS–LS] pairs (*gray*), HS in [HS–LS] pairs (*light gray*), HS in [HS–HS] pairs (*dark gray*) (from [27])

respectively. The zero-field spectrum of the (bt, S) complex with only [LS–LS] pairs present at low temperatures shows the expected typical iron(II)-LS quadrupole doublet with isomer shift δ_{LS}(bt, S) = 0.19(1) mm s^{-1} and quadrupole splitting

$\Delta E_{Q(LS)}$(bt, S) = 0.43(2) mm s^{-1} at 4.2 K (Fig. 8.14a). In an applied field of H_{eff} = 50 kOe, magnetic splitting is observed with a local effective field of $H_{eff} \approx 50$ kOe as is expected for $\langle S \rangle = 0$ of the [LS–LS] pair (Fig. 8.14b). The zero-field spectrum of (bpym, S), consisting of only [HS–HS] pairs, is characterized by a typical HS doublet (Fig. 8.14c). The presence of an external magnetic field causes a slight broadening of the doublet lines (Fig. 8.14d). The value of the effective field, H_{eff}, calculated from this spectrum is 15 kOe. The difference between H_{ext} = 50 kOe and the observed field at the iron nucleus, H_{eff}, arises from the antiferromagnetic nature of the [HS–HS] pairs. In fact, the $\langle S \rangle$ value deduced from the corresponding partition function is around 0.5, which is consistent with the parameters $J = -4.1$ cm^{-1} and $g = 2.2$ for (bpym, S) at 4.2 K. The zero-field spectrum of the (bpym, Se) complex recorded at 4.2 K reflects the nearly 50% ST by virtue of the nearly 1:1 intensity ratio of the HS:LS resonance signals with parameters δ_{HS} = 0.86(1) mm s^{-1}, ΔE_Q(HS) = 3.11(2) mm s^{-1}, and δ_{LS} = 0.22(1) mm s^{-1}, ΔE_Q(LS) = 0.36(1) mm s^{-1}, respectively (Fig. 8.14e). Measurements in a magnetic field of 50 kOe at 4.2 K yield a spectrum with three components (Fig. 8.14f). One of them, with a relative intensity of ca. 52.0% and with isomer shift and quadrupole splitting being equal to δ_{LS}(bt, S) and $\Delta E_{Q(LS)}$(bt, S), can be identified as the LS state ($H_{eff} \approx H_{ext}$). The second low-intensity contribution (ca. 4.0%) is a broadened doublet with parameters δ_{HS}(bpym, S) and $\Delta E_{Q(HS)}$ (bpym, S) and H_{eff} = 14 kOe corresponding to iron(II) sites in antiferromagnetically coupled [HS–HS] pairs. The third component (relative intensity ca. 44.0%) with parameters δ_{HS}(bpym, S) and $\Delta E_{Q(HS)}$ (bpym, S) can be unambiguously assigned to the HS state in [HS–LS] pairs because the measured effective field at the iron nuclei of 81 kOe originates from an $S = 2$ spin state. As a result, the complete distinction of dinuclear units becomes possible. It follows from the area fractions of the subspectra intensities that, at 4.2 K, the sample (bpym, Se) contains 88.0% [HS–LS], 4.0% [HS–HS] and 8.0% [LS–LS] pairs.

These results obtained in applied field clearly prove that the ST in the dinuclear compounds under study proceeds via [HS–HS] $\leftrightarrow$ [HS–LS] $\leftrightarrow$ [LS–LS]. Simultaneous ST in both iron centers of the [HS–HS] pairs, leading directly to [LS–LS] pairs, apparently can be excluded, at least in the systems discussed above. This is surprising in view of the fact that these dinuclear complexes are centrosymmetric, that is, the two metal centers have identical surroundings and therefore, experience the same ligand field strength and consequently, thermal ST is expected to set in simultaneously in both centers. In other dinuclear iron(II) complexes, however, thermally induced direct ST from [HS–HS] to [LS–LS] pairs does occur and, indeed, has been observed by Mössbauer measurements [30, 31].

An interesting Mössbauer study has been reported on the dinuclear SCO complex $[Fe_2^{II}(PMAT)_2](BF_4)_4\cdot$DMF (PMAT: 4-amino-3,5-bis{[(2-pyridylmethyl) amino]methyl}-4H-1,2,4-triazole), where thermal ST occurs from [HS–HS] to the stable endproduct [HS–LS] [32]. The molecular structure and magnetic behavior of this complex was reported earlier by Brooker et al. [33, 34] (Fig. 8.15). At ca. 225 K, the complex undergoes a sharp half ST from the HS state, $^5T_2$, to a state containing 50% HS and 50% LS, $^1A_1$, isomers. The single-crystal structural analysis

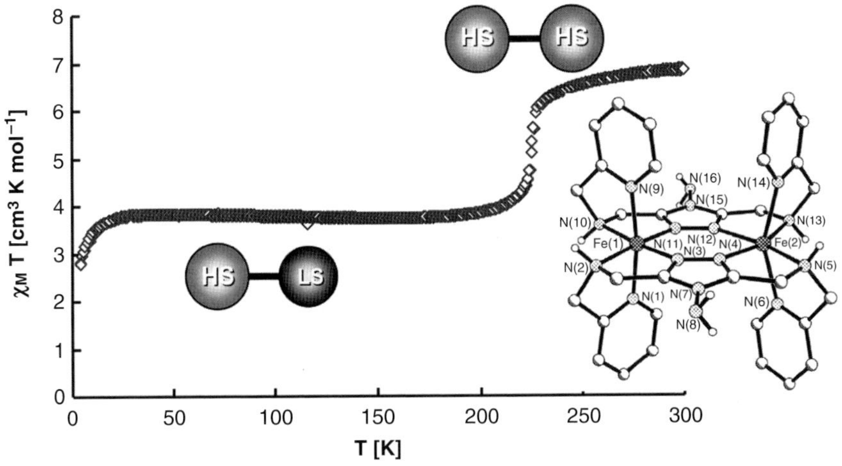

Fig. 8.15 Molecular structure and magnetic properties, $\chi_M T$ versus T, of $[Fe_2^{II}(PMAT)_2](BF_4)_4{\cdot}DMF$. Only half of the Fe(II) sites undergo thermal SCO from HS to LS (from [33, 34])

at 123 K has identified that the dinuclear complex contains one Fe(II) in the LS state and one in the HS state, the latter with Fe–N bond distances about 10% longer than those in the LS state. The sharp decrease of the magnetic susceptibility below 30 K is due to the ZFS effect at the 50% HS isomers. The inset of Fig. 8.15 illustrates (1) that the LS site has decreased in volume (by up to 5%) as compared to the HS site and (2) that the HS site (dark gray) in the LS–HS pair has suffered a slight structural distortion during ST from HS–HS to LS–HS. This situation is in full accord with the Mössbauer data obtained from $[Fe_2^{II}(PMAT)_2](BF_4)_4{\cdot}DMF$ [32]. Mössbauer spectra were recorded as a function of temperature between 298 and 4.2 K (Fig. 8.16). At 298 K, the only quadrupole doublet, $\delta = 0.96$ mm s^{-1} and $\Delta E_Q = 2.34$ mm s^{-1} is characteristic of Fe(II) in the HS state and is consistent with the presence of only [HS–HS] pairs. The slight intensity asymmetry is due to texture; therefore, all further measurements were carried out with the sample mounted at the magic angle of 54° with respect to the γ-radiation, which results in symmetric intensities of the quadrupole doublet. As the temperature is decreased to 240 K, two new doublets (black and dark gray) appear at the expense of the original HS doublet (light gray). The doublet (black) with $\delta = 0.36$ mm s^{-1}, $\Delta E_Q = 0.26$ mm s^{-1} and area fraction $A = 10.3\%$, arises from the LS site in [HS–LS] pairs, which are formed as a result of thermal HS to LS transition. The doublet (dark gray) with $\delta = 0.96$ mm s^{-1}, $\Delta E_Q = 1.66$ mm s^{-1} and area fraction $A = 10.3\%$, is unambiguously due to the HS site in these HS–LS pairs. The doublet arising from the original [HS–HS] pairs (light gray) now yields $\delta = 0.92$ mm s^{-1} and a slightly larger $\Delta E_Q = 2.65$ mm s^{-1} (compared to the value obtained at higher temperature) and area fraction $A \sim 80\%$. By further lowering the temperature, the doublet (light gray) arising from original [HS–HS] pairs continues to lose intensity in favor of the new doublets (dark gray and black) with equal area fractions. In the plateau region,

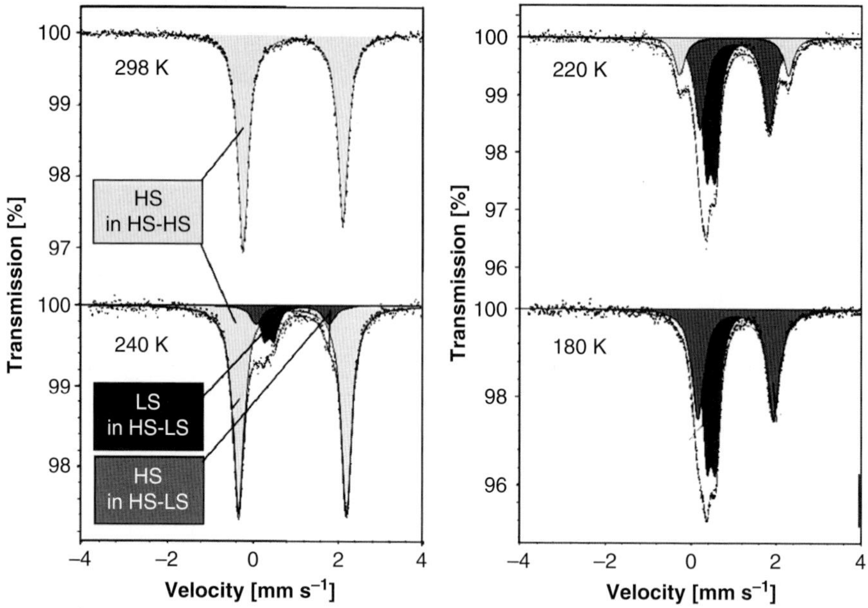

Fig. 8.16 ^{57}Fe Mössbauer spectra of $[Fe_2^{II}(PMAT)_2](BF_4)_4 \cdot DMF$ at selected temperatures. At 298 K, the only quadrupole doublet is characteristic of iron(II) in the HS state. SCO from HS to LS occurs at one Fe(II) site of the dinuclear complex at ca. 225 K. The second Fe(II) site remains in the HS state, but "feels" the spin state conversion of the neighboring atom by local distortions communicated through the rigid bridging ligand, giving rise to a new quadrupole doublet (*dark gray*), i.e., HS in [HS–LS], in the Mössbauer spectrum. The intensity ratio of the resonance signals of HS in [HS–LS] to that of LS (black) in [HS–LS] is close to 1:1 at all temperatures (from [32])

at 180 K and below, the half ST is complete and the doublet (light gray) from original [HS–HS] pairs has vanished.

Clearly, one of the two HS sites in the [HS–HS] pair undergoes thermal SCO generating a [HS–LS] pair. The unique feature is the concomitant appearance of the new HS quadrupole doublet, with a smaller ΔE_Q than the original HS doublet. This is clearly due to local structural distortion at the HS site induced by the newly formed LS state and communicated via the bridging ligand. This distortion at the iron center affects the electric field gradient (EFG) at the ^{57}Fe nucleus and causes a reduction of the ΔE_Q value of the new HS quadrupole doublet. The formation of [HS–LS] intermediates is favored by rigid bridging ligands in these dinuclear systems.

8.1.5 Spin Crossover in a Trinuclear Iron(II) Complex

Thermal ST occurs in the trinuclear iron(II) complex $[Fe_3(iptrz)_6(H_2O)_6](Trifl)_6$ (with iptrz = 4-isopropyl-1,2,4-triazole and Trifl = trifluoromethanesulfonate) only at the

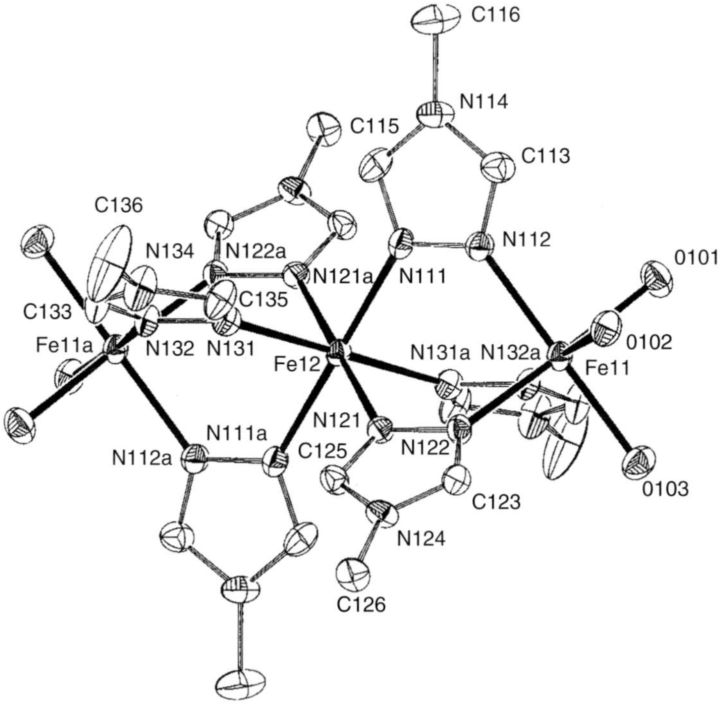

Fig. 8.17 Projection showing the molecular structure of $[Fe_3(iptrz)_6(H_2O)_6](Trifl)_6$ (from [35])

central iron(II) site, whereas the outer two iron(II) sites remain in the HS state at all temperatures under study [35]. A simplified version of the molecular structure is shown in Fig. 8.17. The central iron(II) site is coordinated to six nitrogen atoms belonging to three 1,2,4-triazole units on either side forming the bridges to the outer two iron(II) sites, which are capped by three water molecules on either site. The central site has a FeN_6 core and the appropriate ligand field strength to meet the thermal SCO conditions. The outer two sites have FeN_3O_3 cores and thus possess a weaker ligand field strength, obviously too weak for thermal SCO to occur. Figure 8.18 shows the Mössbauer spectra of this complex recorded at the indicated temperatures. At 300 K, the two quadrupole doublets, that is, HS_c (red) and HS_o (yellow), are characteristic for ferrous iron in the HS state and represent the central site and the two outer sites, respectively. The area fractions of the two doublets deviate from the ratio 1:2. The reason is that the outer Fe(II) sites with their FeN_3O_3 cores experience a relatively weak ligand field strength and are, therefore, more weakly bonded to their donor atoms than the central Fe(II) site with its FeN_6 core and stronger ligand field strength. Consequently, the Lamb– Mössbauer factor of the outer Fe(II) sites is lower than that of the central Fe(II) site and, thus, the corresponding intensities of the Mössbauer resonance lines do not reflect the

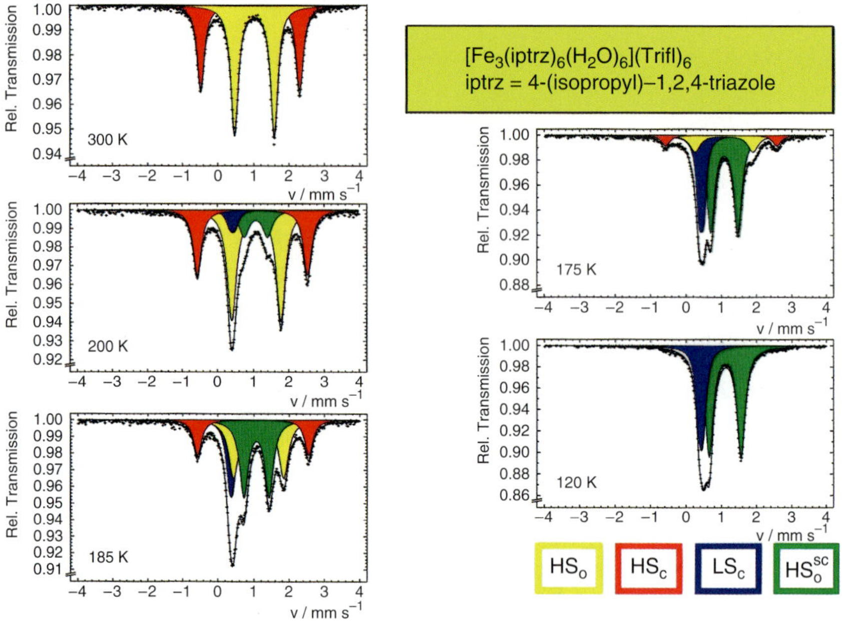

Fig. 8.18 ^{57}Fe Mössbauer spectra of the trinuclear complex $[Fe_3(iptrz)_6(H_2O)_6](Trifl)_6$ recorded at the indicated temperatures. Only the central Fe(II) site undergoes thermally induced ST, HS_c*(red)* → LS_c*(blue)*. The outer two Fe(II) sites, which remain in HS state at all temperatures, "feel" the ST occurring at the central iron site through a molecular distortion, the latter giving rise to a new quadrupole doublet, $HS_o \rightarrow HS_o^{SCO}$ (from [35])

actual concentrations; those of the HS_c doublet are overweighted in comparison with the HS_o doublet.

Lowering temperature to 200 K, the thermally induced ST at the central Fe(II) site sets in with the appearance of two new doublets in the Mössbauer spectrum: (1) the poorly resolved doublet termed LS_c refers to central Fe(II) sites now being in the LS state, and (2) the doublet termed HS_o^{SCO} arises from the outer Fe(II) sites of those trinuclear molecules, the central Fe(II) site of which has undergone HS → LS conversion (a similar situation was discussed above for the dinuclear complex $[Fe_2^{II}(PMAT)_2](BF_4)_4{\cdot}DMF$ also giving rise to a new HS doublet). The area fraction ratio of LS_c:HS_o^{SCO} derived from the Mössbauer spectra is nearly 1:2 at all temperatures under study; deviations are again due to different Lamb–Mössbauer factors as discussed earlier. The spectrum measured at 120 K can be decomposed into two doublets arising from LS_c and HS_o^{SCO}, respectively. Thus, thermal ST at the central Fe(II) site is complete at this temperature. The signals referring to HS_c and HS_o vanish in favor of LS_c and HS_o^{SCO}. In conclusion, the presented example demonstrates that the Mössbauer technique is an excellent tool to follow the selective spin switching in polynuclear systems.

8.1.6 Spin Crossover in Metallomesogens

Metallomesogens belong to a class of metal-containing compounds which exhibit liquid–crystal properties. The possibility of combining the properties of fluidity, ease of processability, one- or two-dimensional order, etc., with the properties associated with metal atoms, for example, color, paramagnetism, electron-rich environment, is the main objective of progress in this field. In addition, the possibility of tuning the physical (mesomorphic, optical, and magnetic) properties of metallomesogens is significantly extended, since the organic ligand of these systems can be varied. Liquid crystalline materials in which a spin crossover center is incorporated into the mesogenic organic skeleton constitute a separate class of compounds for which an interplay of structural transition and liquid crystallinity is expected. This may lead to advantages in practical applications, for example, processing SCO materials in the form of thin films, enhancement of ST signals, switching and sensing in different temperature regimes, and the achievement of photo and thermochromism in metal-containing liquid crystals. The change of color in coexistence with liquid crystallinity is certainly a phenomenon which is of interest in the field of material sciences.

Galyametdinov et al. [36] reported on temperature-dependent Mössbauer and magnetic susceptibility measurements of an iron(III) compound which exhibits liquid crystalline properties above – and thermal ST below – room temperature. This was the first example of SCO in metallomesogens. Later, different families of iron(II) and cobalt(II) systems were also investigated [37–43]. The question whether the solid–liquid crystal phase transition provokes the spin-state change in spin crossover metallomesogens has been addressed in several series of iron(II) systems employing a variety of physical measurements [43]. In all these studies, ^{57}Fe Mössbauer spectroscopy has been extremely helpful, for example, in controlling the completeness of ST in both the high and low temperature regions where $\chi_M T$ data are often not reliable due to calibration difficulties. Also, one can unambiguously decide whether a significant decrease of the $\chi_M T$ versus T plot toward lower temperatures is due to SCO or ZFS. An example is the study of the one-dimensional triazole-based compound $[Fe(C_{10}\text{-tba})_3](4\text{-}MeC_6H_4SO_3)_2 \cdot sH_2O$, with C_{10}-tba = 3,5-bis(decyloxy)-*N*-(4H-1,2,4-triazol-4-yl)benzamide, $s = 1$ or 0 [44, 45]. This system exhibits thermal SCO with a concomitant change of color between white (HS state) and purple (LS state). The magnetic properties of the pristine compound ($s = 1$) in the form of a $\chi_M T$ versus T plot are depicted in Fig. 8.19. At 300 K, the value of $\chi_M T = 0.20$ cm^3 K mol^{-1} indicates that this complex is apparently in the LS state. The Mössbauer spectrum recorded at 4.2 K is in agreement with the magnetic data, that is, the HS population is 4.8% and the LS population 95.2% (Fig. 8.20a). Upon heating, $\chi_M T$ increases abruptly within a few degrees, reaching the value of 3.74 cm^3 K mol^{-1} at 342 K. This clearly shows that ST from $S = 0$ to $S = 2$ has occurred. The thermogravimetric analysis (TGA) of this system has shown that dehydration takes place in the same temperature region where SCO occurs. The magnetic susceptibility of the dehydrated compound ($s = 0$)

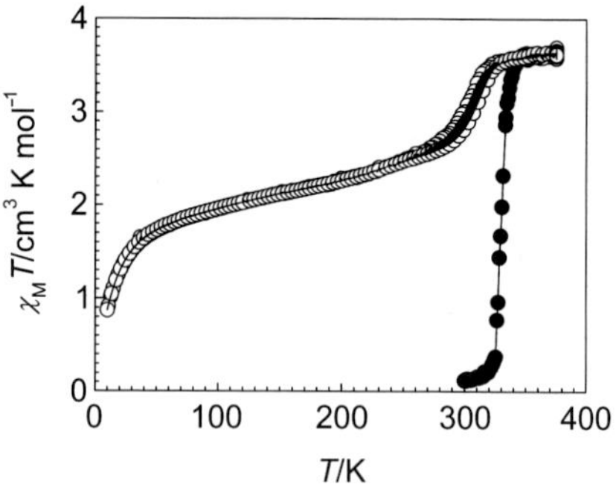

Fig. 8.19 Magnetic properties in the form of $\chi_M T$ versus T of the complexes [Fe(C_{10}–tba)$_3$](4-$MeC_6H_4SO_3$)$_2$·sH_2O, $s = 1$ (*filled circle*) and $s = 0$ (*open circle*). Both complexes exhibit thermal SCO (from [44, 45])

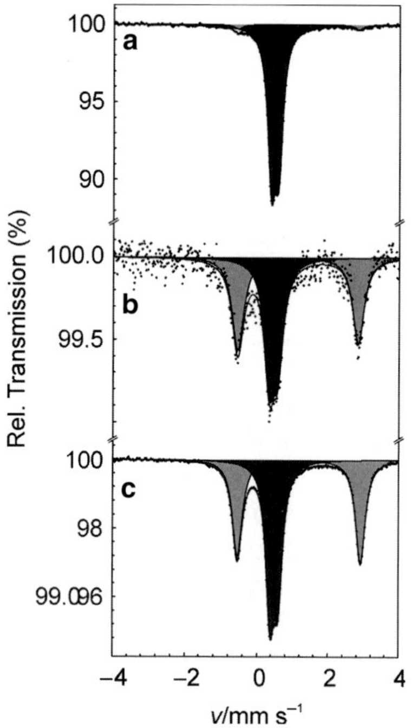

Fig. 8.20 Mössbauer spectra of [Fe(C_{10}–tba)$_3$](4-$MeC_6H_4SO_3$)$_2$· sH_2O, $s = 1$ at 4.2 K (**a**) and of $s = 0$ at 200 K (**b**) and at 4.2 K (**c**) (from [44, 45])

was recorded in a temperature loop, that is, from 375 K down to 10 K and then up again to 375 K. The dehydrated complex reveals incomplete and abrupt SCO, accompanied by hysteresis and color change (from purple in the LS state to white in the HS state) in the temperature region of 250–300 K. Around 50% of Fe(II) sites change spin state, as can be inferred from the value of $\chi_M T$ at 200 K. The Mössbauer spectrum recorded at 200 K show that 49.3% of the iron is in the LS state and 50.7% in the HS state. The further decrease of the $\chi_M T$ value below 100 K, particularly below 50 K, is due to the ZFS of those iron ions which remain in the HS state even at very low temperature, as derived from the Mössbauer spectrum at 4.2 K (HS population is 48.1%, LS population is 51.9%) (Fig. 8.20c).

8.1.7 Effect of Nuclear Decay: Mössbauer Emission Spectroscopy

The nuclear decay of radioactive atoms embedded in a host is known to lead to various chemical and physical "after effects" such as redox processes, bond rupture, and the formation of metastable states [46]. A very successful way of investigating such after effects in solid material exploits the Mössbauer effect and has been termed "Mössbauer Emission Spectroscopy" (MES) or "Mössbauer source experiments" [47, 48]. For instance, the electron capture (EC) decay of ^{57}Co to ^{57}Fe, denoted ^{57}Co(EC)^{57}Fe, in cobalt- or iron-containing compounds has been widely explored. In such MES experiments, the compound under study is usually labeled with ^{57}Co and then used as the Mössbauer source versus a single-line absorber material such as $K_4[Fe(CN)_6]$. The recorded spectrum yields information on the chemical state of the nucleogenic ^{57}Fe at ca. 10^{-7} s, which is approximately the lifetime of the 14.4 keV metastable nuclear state of ^{57}Fe after nuclear decay.

The MES technique has been applied to ^{57}Co-labeled cobalt(II) and iron(II) coordination compounds with strong, intermediate, and weak ligand fields. Among the most interesting results are the formation of metastable HS states of ^{57}Fe(II) in strong and intermediate ligand fields as exemplified by the following two cases: (1) $[Fe(phen)_3](ClO_4)_2$ (phen = 1,10-phenanthroline), a typical LS compound at room temperature and below, and (2) $[Fe(phen)_2(NCS)_2]$, a typical SCO system. Mössbauer emission spectra of $[^{57}Co/Co(phen)_3](ClO_4)_2$ as a function of temperature versus $K_4[Fe(CN)_6$ as absorber (which was kept at 295 K) are depicted in Fig. 8.21 and are compared with ^{57}Fe absorption spectra of $[^{57}Fe/Co(phen)_3]$ $(ClO_4)_2$ as a function of temperature versus ^{57}Co/Rh as source (which was kept at 295 K). The main component in the emission spectra of $[^{57}Co/Co(phen)_3](ClO_4)_2$ refers to $[^{57}Fe(phen)_3](ClO_4)_2$. At temperatures down to ca. 200 K, the spin state is LS, but below ca. 200 K, new signals, clearly indicative of ^{57}Fe(II) in the HS state, appear with increasing intensity at the expense of the ^{57}Fe(II)-LS signals. Thus, even in the case of a strong ligand field environment of the iron, such as $(phen)_3$, where Fe(II)-LS is the thermodynamically stable state, the metastable state ^{57}Fe(II)-HS can be detected in the low-temperature regime with lifetimes of the order of 10^{-7} s [48]. Interestingly, there is no significant difference if one uses the paramagnetic $[Co(phen)_3]^{2+}$ or the diamagnetic $[Fe(phen)_3]^{2+}$ compound as the Mössbauer source; the only important consideration is the immediate ligand cage of ^{57}Co or ^{57}Fe, respectively, after decay. If the ^{57}Co ion decays in a somewhat weaker ligand field such as $(phen)_2(NCS)_2$, which is known to induce thermal SCO, LS $\leftrightarrow$ HS, at the iron(II) center, the nucleogenic ^{57}Fe(II) manifests itself in the HS state at all temperatures down to 4.2 K [49]. This is shown in Fig. 8.22. Similar results have also been observed with other ^{57}Co-labeled iron(II) SCO systems [50].

The essential information from these Mössbauer source experiments is that, at a given temperature, the probability of "trapping" the HS state of the nucleogenic ^{57}Fe(II) within the Mössbauer time window is significantly greater in SCO compounds with intermediate ligand field strength than in those with strong ligand field. The discovery of the LIESST phenomenon (generating and trapping of metastable

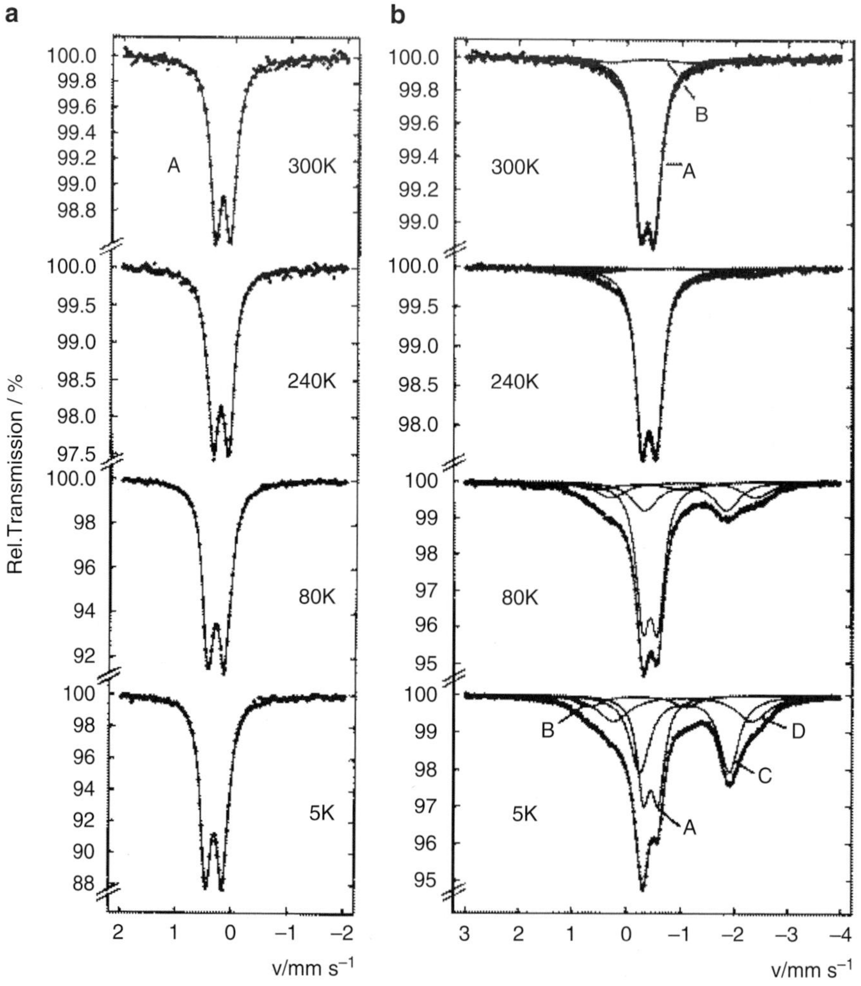

Fig. 8.21 (**a**, *left*) ^{57}Fe Mössbauer absorption spectra of [^{57}Fe/Co(phen)$_3$](ClO$_4$)$_2$ as a function of temperature versus ^{57}Co/Rh as source (which was kept at 295 K). (**b**, *right*) Time-integral ^{57}Fe Mössbauer emission spectra of [^{57}Co/Co(phen)$_3$](ClO$_4$)$_2$ as source as a function of temperature versus K$_4$[Fe(CN)$_6$ as absorber (which was kept at 295 K). Assignment: A, Fe(II)-LS; B, Fe(III)-LS; C, Fe(II)-HS1; D, Fe(II)-HS2. In (**a**) the source was moved relative to the absorber; in (**b**) the absorber was moved relative to the fixed source which was mounted in the cryostat. For direct comparison, the sign of the velocities (x-axis) must be changed either in (**a**) or in (**b**) (from [48, 50])

HS states by irradiation with light) has contributed to our understanding of the mechanism of the "Nuclear Decay Induced Excited Spin State Trapping (NIESST)," where the nuclear decay process ^{57}Co(EC)^{57}Fe may be regarded as an "internal" source for molecular excitation.

The mechanism of NIESST is essentially the same as for LIESST (see Fig. 8.10), except for the initial excitation step. The electronic structure of the nucleogenic

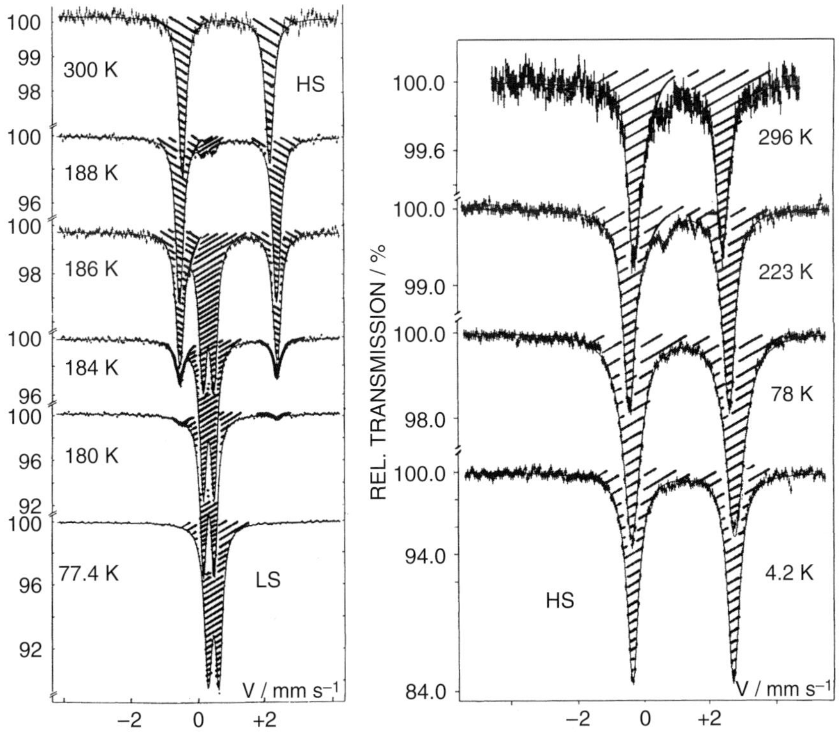

Fig. 8.22 *Left*: ^{57}Fe Mössbauer absorption spectra of [Fe(phen)$_2$(NCS)$_2$] as a function of temperature versus ^{57}Co/Rh as source (which was kept at 295 K). *Right*: Time-integral ^{57}Fe Mössbauer emission spectra of [^{57}Co/Co(phen)$_2$(NCS)$_2$] as source as a function of temperature versus K$_4$[Fe (CN)$_6$ as absorber (which was kept at 295 K). Assignment: The quadrupole doublet with the larger quadrupole splitting (ca. 3 mm s^{-1}) refers to the Fe(II)-HS state, and that with the smaller quadrupole splitting (ca. 0.5 mm s^{-1}) to the Fe(II)-LS state (from [43])

^{57}Fe(II) is still highly excited within the Mössbauer time window ($\sim 10^{-7}$ s after nuclear decay). The fast ($\ll 10^{-7}$ s) spin-allowed transitions to the $^1A_{1g}$ ground state as well as fast intersystem crossing processes ($\ll 10^{-7}$ s) are favored by spin-orbit coupling (SOC) as in the LIESST mechanism, feeding the metastable $^5T_{2g}$ state. The branching ratio for the two relaxation pathways (to either $^1A_{1g}$ or $^5T_{2g}$) depends on the ligand field strength of the system, in analogy to the observed correlation of the lifetime of the metastable LIESST state and the energy difference ΔE^0_{HL}, which also depends on the ligand field strength ("reduced energy-gap law").

Measurements of the lifetimes of NIESST states by time-resolved MES [51] and of LIESST states by time-resolved optical spectroscopy [52] on the very same system (a single crystal of [Fe$_x$Mn$_{1-x}$(bpy)$_3$]$^{2+}$ (bpy = bipyridine)) gave similar results. This supports the suggestion that the mechanisms for LIESST and NIESST relaxation are very similar, at least for the low-energy regime. The NIESST effect was also studied in Co(II) SCO compounds, viz. [^{57}Co/Co(terpy)$_2$]$X_2 \cdot n$H$_2$O (X = [ClO$_4$]$^-$, n = ½, X = Cl$^-$, n = 5), where terpy is the tridentate ligand terpyridine

[53]. The perchlorate salt shows thermal SCO with $T_{1/2}$ around 200 K and a high-spin fraction of nearly 100% at room temperature, whereas the chloride salt possesses a somewhat stronger ligand field giving rise to thermal ST at much higher temperatures (the HS fraction starts to develop around 200 K, reaches ca. 20% at 320 K and would be expected to increase further) [53]. Conventional Mössbauer absorption measurements were performed on the corresponding systems doped with 5% Fe(II) which was found to be in the LS ($S = 0$) state at all temperatures under study [53]. The emission spectra of the ^{57}Co-labeled cobalt complexes were measured using a home-made resonance detector which operates as conversion-electron detector with count rates 10–20 times higher than those of a conventional detector. At room temperature, the nucleogenic ^{57}Fe ions were found to have relaxed to the stable $^1A_1$ LS ground state. On lowering the temperature, a doublet from a metastable Fe(II)-HS state appears in the spectra with increasing intensities. The perchlorate derivative with the weaker ligand field strength shows, at comparable temperatures, a considerably higher amount of the Fe(II)-HS fraction than the chloride derivative with the stronger ligand field. The emission spectra recorded at 100 K displayed in Fig. 8.23 demonstrate this effect very clearly. For a comparative study on a related system with even weaker ligand-field strength at the central metal site, temperature-dependent NIESST experiments were carried out with the ^{57}Co-doped high-spin compound $[Mn(terpy)_2](ClO_4)_2{\cdot}1/2H_2O$. As expected, the ^{57}Fe (II)-HS fraction derived from the emission spectra of the manganese host with the weakest ligand field is the highest one among the three systems Mn/ClO_4, Co/ClO_4 and Co/Cl at comparable temperatures [53]. It can be seen that ΔE^0_{HL}, as defined in the relaxation scheme in Fig. 8.10, takes on increasing values in the order

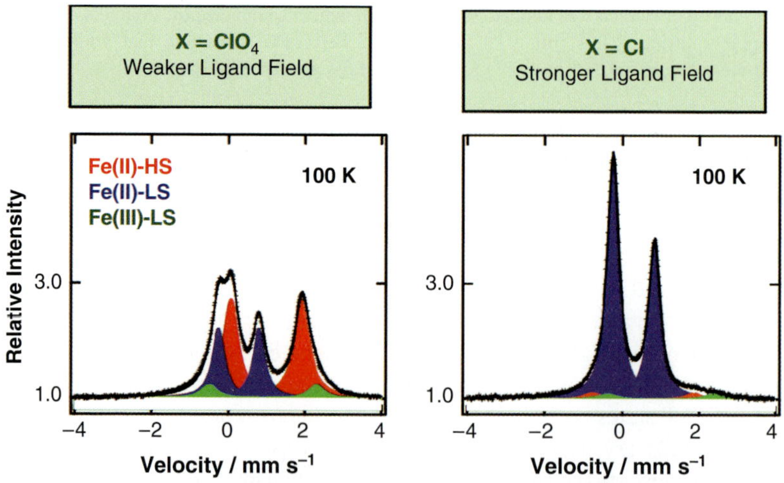

Fig. 8.23 Mössbauer emission spectra of $[^{57}Co/Co(terpy)_2]X_2{\cdot}nH_2O$ ($X = ClO_4^-$, $n = ½$; $X = Cl^-$, $n = 5$) as source material versus $K_4[Fe(CN)_6]$ as absorber (which was kept at 298 K) recorded at 100 K with a conversion-electron detector. *Left*: $X = ClO_4$, $n = ½$. *Right*: $X = Cl$, $n = 5$ (from [53])

Mn/ClO$_4$ < Co/ClO$_4$ < Co/Cl corresponding to the increasing ligand field strength in the same order. The ^{57}Fe(II)-HS fraction, at a given temperature, decreases in the reverse order, in full agreement with the "reduced energy gap law".

8.2 ^{57}Fe Mössbauer Spectroscopy: Unusual Spin and Valence States

By far the most utilized Mössbauer isotope is ^{57}Fe, particularly in (bio)inorganic chemistry. Most iron compounds are found in the oxidation states iron(II) and iron (III), either with low-spin or high-spin electron configuration. The literature on the application of ^{57}Fe Mössbauer spectroscopy in this field of research has been reviewed in several textbooks, which are referenced in Chap. 1. The present chapter is intended as a survey of the Mössbauer studies on iron compounds with less common, nevertheless increasingly interesting, valence and spin states.

8.2.1 Iron(III) with Intermediate Spin, S = *3/2*

The spin-quartet ground state ($S = 3/2$) of trivalent iron (configuration $3d^5$) is often regarded as exotic and marginal, probably because in crystal field theory (CFT), it is explicitly excluded as the ground state for octahedral complexes [54]. However, Gibson et al. reported in 1959 on a square-pyramidal ferric porphyrin chloride complex with EPR signals at $g_{\|} = 2$, $g_{\perp} = 3.8$, which Griffith, shortly after, explained by the presence of a ground state with $S = 3/2$ [55]. Some years later, it became obvious from the syntheses and magnetic studies of a series of iron(III) dithiocarbamato–halide complexes that the spin-quartet ground state prevails for several classes of ferric compounds because of their quasi-square-pyramidal geometry [56]. This subject has become a focus of interest during the recent decades partly because biological compounds such as ferricytochrome c' from photosynthetic and denitrifying bacteria [57–62], horseradish peroxidase [60, 63] and the fully oxidized form of cytochrome c oxidase [64] exhibit ground-state properties with spin-admixture, $S = (3/2, 5/2)$. New interest in intermediate-spin complexes arose because several types of the ligands that afford $S = 3/2$ for ferric iron are redox-active and hold promise for versatile new chemistry. Among these dithiolenes are presently experiencing a revival in interest in basic inorganic chemistry and material sciences [65].

8.2.1.1 Ligand Field Considerations

In a simple crystal-field description, a spin-quartet ground state ($S = 3/2$) of iron(III) is energetically stabilized over the spin-sextet ($S = 5/2$) when the energy separation

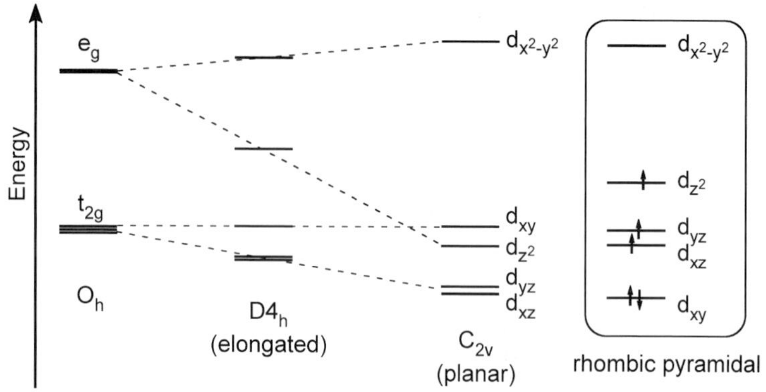

Fig. 8.24 Energy splitting of *d*-orbitals in an octahedral crystal field and the effect of increasing elongation of the coordination polyhedron along *z*. The box on the right presents the situation encountered with rhombic monopyramidal iron(III) complexes with $S = 3/2$ [66–68]

of the *d*-orbitals is such that one orbital lies at higher energy than the other four by a larger amount as compared to Coulomb repulsion and SOC. The five valence *d*-electrons then occupy only four orbitals (Fig. 8.24). Such an arrangement is afforded in C_{2v} symmetry, when the $d_{x^2-y^2}$ orbital is greatly destabilized by strong-field equatorial ligands and the d_{z^2} orbital remains low in energy because of a weak axial ligand [69]. Electrostatic arguments suggest that the ground-state configuration is $(d_{xy})^2\ (d_{yz}, d_{xz})^2\ (d_{z^2})^1$ with spin $S = 3/2$ according to Hund's rule [70]. Correspondingly, most ferric intermediate-spin complexes have quasi square-pyramidal symmetry with four strong ligands in the basal plane and a weak fifth one in the apical position; a similar orbital pattern arises for four-coordinate planar complexes.[1]

The single-electron crystal-field picture may be a crude over-simplification because Coulomb repulsion and SOC can lead to a situation where the ground term is a composite mixture of $S = 1/2$, 3/2, and 5/2 states that does not derive from a single configuration [71, 72]; the corresponding physical description is obtained from proper quantum chemical calculations [73–75].

8.2.1.2 Mössbauer Parameters

Intermediate spin occurs mainly for square- and rhombic-pyramidal five-coordinate iron complexes and also for planar four-coordinate and for asymmetric six-coordinate

[1]A priori also a linear iron(III) molecule would show a "one-over-four" orbital pattern. Alternatively, a "four-over-one" arrangement could also afford $S = 3/2$ for the d^5 configuration, but this cannot arise from only strong σ-interaction and would need competing strong π-interaction. Neither option has been observed so far.

complexes. The examples with pyramidal symmetry may be further divided into (1) complexes containing 4-sulfur- or 2-sulfur/2-nitrogen donor sets in equatorial positions and (2) complexes with macrocyclic ligands with four nitrogen atoms occupying the equatorial positions. Porphyrinates which also have a N_4-donor set will be described separately because they show a broad variety of phenomena arising from $S = (3/2, 5/2)$ spin admixture.

Square- and Rhombic-Pyramidal Complexes

S_4X Ligand Sphere

Ferric dithiocarbamato complexes have been intensely studied by Mössbauer spectroscopy since the first report on the synthesis and the unusual magnetic properties of iron(III) halide complexes of the type [Fe^{III}(dtc-R_2)$_2$ X], where dtc-R_2 = S_2CNR_2 which is a bis(*N,N′*-dialkyl-dithiocarbamato) ligand with methyl- or ethyl-groups etc. substituting for R, and X = Cl^-, Br^-, I^- [56]. Throughout this series, iron(III) has four equatorial sulfur-ligands and the halide is in the apical position. The room temperature effective magnetic moments range from 3.98 to 4.01μ_B, consistent with three unpaired electrons [56, 76]. The first Mössbauer data, including magnetic spectra recorded at 1.2 K, were reported in 1967 by Wickman for the diethyl chloride derivative, [Fe^{III}(dtc-Et_2)$_2$*Cl*] (**1**) [70, 77, 78], and later for the isopropyl and other analogs [79–81]. The magnetic properties of the diethyl-chloride derivatives and the 4-morphylyl iodide and bromide derivatives, [Fe(dtc-C_4H_8O)$_2$*I*] (**2**) and [Fe(dtc-C_4H_8O)$_2$*Br*] (**3**), respectively, have been intensely investigated by ^{57}Fe- and ^{129}I-Mössbauer spectroscopy [82] because the solids exhibit ferromagnetic ordering at liquid helium temperatures [78, 83, 84] and paramagnetic relaxation in the vicinity of the magnetic transition [77, 85]. The Mössbauer spectra of ferric dithiocarbamate complexes, in general, are characterized by relatively large isomer shifts ($\delta \approx 0.50$ mm s^{-1}) and large quadrupole splittings, ΔE_Q, close to +3 mm s^{-1} (Table 8.1). Exchange of the apical halide by thiocyanate, SCN^-, like in [Fe^{III}(dtc-Et_2)$_2$*SCN*] (**4**), has only a minor effect on the Mössbauer parameters [86].

Bis(dithiooxalato)ferric complexes, [Fe^{III}(dto)$_2$X]$^{2-}$ with dto = $S_2C_2O_2$ and X = I^-, Br^-, Cl^-, respectively (compounds **5–7**), have been studied by magnetic susceptibility, EPR and applied-field Mössbauer spectroscopy, [67]. The isomer shifts at 4.2 K are in the range $\delta \approx 0.25$–0.30 mm s^{-1} and the quadrupole splittings are larger than +3.25 mm s^{-1} (Table 8.1).

Coordination compounds of dianionic dithiolene ($S_2C_2{\cdot}R_2$) and benzene-1,2-dithiolene (bdt = ($S_2C_6H_4$) and their derivatives have been studied since the 1960s by Mössbauer spectroscopy [87] and other techniques. Nevertheless, many aspects of their electronic structure remained uncertain for a long time. The five-coordinate ferric complexes with two equatorial dithiolene ligands exhibit intermediate spin and show the Mössbauer parameters $\delta = 0.25$–0.38 mm s^{-1} and $\Delta E_Q = 1.6$–3.2 mm s^{-1}. For example, [Fe^{III}(mnt)$_2$*py*]$^{2-}$ with two mnt ligands (=$S_2C_2(CN)_2$) and an

Table 8.1 Mössbauer parameters of five- and six-coordinate ferric compounds with $S_{Fe} = 3/2$

Complex	δ^a (mm s^{-1})	$\Delta E_Q{}^b$ (mm s^{-1})	η^c	$\mathbf{A}/g_N\mu_N{}^d$ (T)	T (K)	Ref.
1 [Fe^{III} (dtc-Et$_2$)$_2$*Cl*]	0.50	+2.70	0.2	$\|22.0\|^{e,f}$	4.2	[77, 78]
2 [Fe^{III} (dtc-C$_4$H$_8$O)$_2$*I*]	0.48	+2.80	0.1	$\|22.4\|^{e,f}$	4.2	[84, 85]
3 [Fe^{III} (dtc-C$_4$H$_8$O)$_2$*Br*]	0.47	+2.72	0.2	$\|26.8\|^{e,f}$	4.2	[84, 85]
4 [Fe^{III}(dtc-Et$_2$)$_2$*SCN*]	0.42	2.65	–	–	77	[86]
5 [Fe^{III}(dto)$_2$Fe*I*]$^{2-}$	0.30	+3.33	0.1	$\|18.2\|^f$	4.2	[67]
6 [Fe^{III} (dto)$_2$*Br*]$^{2-}$	0.29	+3.25	0.1	$\|21.0\|^f$	4.2	[67]
7 [Fe^{III} (dto)$_2$*Cl*]$^{2-}$	0.25	+3.60	0.2	$\|22.5\|^f$	4.2	[67]
8 [Fe^{III}(mnt)$_2$(*idzm*)]	0.36	$\|2.64\|$	–	–	77	[88]
9 [Fe^{III}(^{Bu}bdt$^\bullet$) (^{Bu}bdt) (PMe$_3$)]	0.12^g	+3.05	0.7	−25.9, −17.7, +0.3	4.2	[92]
10 [Fe(^{Bu}bdt$^\bullet$)$_2$(PMe$_3$)]$^+$	0.29^h	+2.57	0.5	−18.7, −19.6, +1.0	4.2	[92]
11 [Fe^{III} (^{Bu}bdt)$_2$(^{t}Bu-py)]$^-$	0.34	+2.94	0	−19.1, −32.2, +1.5	4.2	[92]
12 [(^{SS}salen)Fe^{III}Cl]	0.38	$\|3.13\|$	–	–	4.2	[93]
13 [Fe(^{Me}iTSC$^\bullet$)$_2$(SCH$_3$)]	0.06	$\|2.32\|$	–	–	80	[97]
14 [Fe(^{Me}iTSC$^\bullet$)$_2$Cl]	0.20	$\|2.21\|$	–	–	80	[97]
15 [Fe(^{N,S}IBSQ$^\bullet$)$_2$I]	0.12	+3.18	0.5	−7.4, −16.1, +0.6	4.2	[94]
16 [Fe(^{N,O}IBSQ$^\bullet$)$_2$I]	0.24	+2.80	0.3	−12.4, −20.3, +1.3	4.2	[94]
17 [Fe^{III}(^{N,N}IBSQ$^\bullet$)$_2$*I*]	0.15	+3.03	0.2	−9.5, −28.3, +1.8	4.2	[98]
18 [Fe^{III}(^{N,N}IBSQ$^\bullet$)$_2$Ph$_2$Im]	0.11	+2.26			77	[98]
19 [Et$_4$N]$_2$[FeCl(η^4-MAC*)]	0.25	+3.60	0.2	−16.6, −22.0, +5.0	4.2	[100]
20 [(L-N$_4$)Fe*I*]	0.19	+3.56	0	−12.7, −12.7, +0.5	4.2	[68]
21 [Fe(TMCP)(EtOH)$_2$]ClO$_4$	0.35	$\|3.79\|$	–	–	77	[101]
22 [Fe(OETPP)(THF)$_2$]ClO$_4$	0.50	$\|3.50\|$	–	–	80	[102]
23 [Fe(L-N$_4$Me$_2$)(bdt)]ClO$_4$	0.32	$\|2.22\|$	–		200	[103]
24 [Fe^{III}(^{Bu}bdt$^\bullet$)(^{Bu}bdt) (PMe$_3$)$_2$]	0.17	+1.46	0.1	+1.3, −9.9, −30.6	4.2	[92]

^{a}Isomer shift versus α-iron at RT
^{b}Quadrupole splitting
^{c}Asymmetry parameter
^{d}Hyperfine-coupling tensor components in Tesla for the intrinsic spin of iron, $S_{Fe} = 3/2$
^{e}Estimated at 1.2 K due to magnetic self-ordering
^{f}Isotropic part
^{g}0.11 mm s^{-1} at 80 K
^{h}0.25 mm s^{-1} at 80 K

apical (*N*-coordinating) pyridine yields $\delta = 0.33$ mm s^{-1} at 77 K and $|\Delta E_Q| = 2.41$ mm s^{-1} [87], and the [Fe^{III} (mnt)$_2$(*idzm*)]·2dmf derivative (**8**), where idzm represents the bulky *N*-coordinating pyridyl ligand, exhibits $\delta = 0.36$ mm s^{-1}, $|\Delta E_Q| = 2.64$ mm s^{-1} at 77 K [88]. A usually steep drop in μ_{eff} was observed for (**8**) below 10 K, in conjunction with the formation of a new Mössbauer subspectrum, which was interpreted as a $S = 1/2 \leftrightarrow S = 3/2$ ST occurring at $T_{1/2} \sim 3$ K.

Many of the problems and misconceptions occurring for dithiolene compounds are related to the fact that the ligands are redox-active and can be oxidized to monoanionic radicals. Typical examples for this phenomenon are the mono and diradical complexes [Fe^{III} (^{Bu}bdt$^\bullet$)(^{Bu}bdt)(PMe$_3$)] (**9**) and [Fe^{III} (^{Bu}bdt$^\bullet$)$_2$(PMe$_3$)]$^+$ (**10**) for which ^{Bu}bdt and ^{Bu}bdt$^\bullet$ are *tert*-butyl-dithiolene and its one-electron oxidized form. Originally, these and other bdt derivatives had been described as iron(IV) and iron(V) compounds [89–91]. Using EPR, Mössbauer and susceptibility measurements, it was however, shown, that iron stays in the ferric intermediate-spin state in all cases, whereas the ligand is oxidized to bdt$^{\bullet-}$ [92]. The ligand radicals ($S' = 1/2$) interact antiferromagnetically with the spin of the central ferric ion ($S_{Fe} = 3/2$, $J \gg kT$) and thus, affords a well-isolated ground state with total spin $S_t = 1$ for **9** and $S_t = 1/2$ for **10**. The quartet ground state of iron has been directly observed for [Fe^{III} (^{Bu}bdt)$_2$(^{t}Bu-py)]$^-$ (**11**), which contains the dithiolene ligands in their closed-shell form without an interacting radical spin [92]. All of the complexes **9**–**11** exhibit a single doublet in their zero-field Mössbauer spectra, with isomer shifts ranging from 0.12 to 0.34 mm s^{-1} (Table 8.1). Interestingly, the isomer shift increases slightly upon oxidation **9** → **10**, in contrast to the situation normally expected for a metal-centered oxidation.

S_2N_2X Ligand Sphere

Mixed sulfur and nitrogen donors as encountered with iminothiophenolate ligands also yield intermediate spin ferric complexes. A detailed spectroscopic analysis was necessary to unravel the electronic structure of the bis(*o*-iminothiobenzosemiquinonate) iron(III) complex [Fe(^{N,S}IBSQ$^\bullet$)$_2$I] (**15**), because this is again a diradical complex with spin $S_t = 1/2$, like in **10**. The ground state here also results from strong antiferromagnetic coupling of the two ligand radicals ($S_{rad} = 1/2$) to $S_{Fe} = 3/2$ of the ferric ion [94]. Magnetically split Mössbauer spectra revealed two negative and one positive components of the intrinsic hyperfine tensor of iron ($A/g_N\mu_N(S_{Fe} = 3/2) = (-7.38, -16.05, +0.60)$ T), similar to the situation found for the complexes **9**–**11**. The isomer shift of $\delta = 0.12$ mm s^{-1}, and the large positive quadrupole splitting ($\Delta E_Q = +3.18$ mm s^{-1}) corroborate the presence of an intermediate-spin ferric species in **15**.

O_2N_2X Ligand Sphere

Closely related to the above example is the *o*-iminobenzosemiquinonate compound ([Fe^{III}(^{N,O}IBSQ$^\bullet$)$_2$I], **16**), which to the best of our knowledge is the only example of a five-coordinate ferric intermediate-spin complex with oxygen donors in the equatorial plane. Like for its (S_2N_2) analog **15**, both ligands are oxidized π-radicals (monoanions) and the ground state of the molecule corresponds to $S_t = 1/2$. Magnetically split Mössbauer spectra revealed a larger intrinsic field at the iron nucleus ($A/g_N\mu_N(S_{Fe} = 3/2) = (-12.40, -20.32, +1.30)$ T) and a larger isomer shift value ($\delta = 0.24$ mm s^{-1}) for **16** than for **15**. This reflects the reduced covalency of the N_2O_2I donor set as compared to N_2S_2I. Interestingly, the spin state of Fe^{III} in the *chloride* analog of **16** is *high-spin*, whereas for the *bromide* analog, both spin isomers, $S_{Fe} = 3/2$ and $S_{Fe} = 5/2$, are present in the crystalline state in a 1:1 ratio [95]. Depending on the type of crystalline state, the bromide complex can also show a $S_{Fe} = 1/2 \leftrightarrow S_{Fe} = 3/2$ ST [96].

N_4X Ligand Sphere

The bis(thiosemicarbazide) compounds [Fe^{III}(^{Me}iTSC$^{\cdot}$)$_2$(SCH_3)] (**13**) and [Fe^{III} (^{Me}iTSC$^{\bullet}$)$_2$Cl] (**14**) were recently shown to be ferric intermediate-spin complexes with two ligand radicals (^{Me}iTSC$^{\bullet}$) [97]. The total spin is again $S_t = 1/2$ due to strong antiferromagnetic coupling. The equatorial ligand atoms coordinating to iron are nitrogen, whereas methylthiolate or chloride, respectively, occupy the apical positions. The low isomer shift of $\delta = 0.06$ mm s^{-1} of **13**, in contrast to $\delta = 0.20$ mm s^{-1} for **14**, is explained by an exceptionally strong covalent Fe–SCH_3 bond compared to the lower covalency of the Fe–Cl bond.

The neutral complexes ([Fe^{III}(^{N,N}IBSQ$^{\bullet}$)$_2$I], **17**) and ([Fe^{III}(^{N,N}IBSQ$^{\bullet}$)$_2$$Ph_2$Im], **18**) also possess iron in its ferric intermediate-spin state with the *N,N′*-coordinating equatorial ligands in their monoanionic π-radical form, *o*-diiminobenzosemiquinonate(1–) [98]. The ground state is $S_t = 1/2$ and the Mössbauer parameters are $\delta = 0.16$ and 0.11 mm s^{-1}, and $\Delta E_Q = 2.90$ and 2.26 mm s^{-1}, respectively, at 77 K. More examples of such complexes are found in [98].

The classical compound [(^{SS}salen)Fe^{III}Cl] (**12**) exhibits an effective magnetic moment of 3.90μ_B and the Mössbauer isomer shift and quadrupole splitting recorded at 4.2 K are $\delta = 0.38$ mm s^{-1} and $|\Delta E_Q| = 3.13$ mm s^{-1} [93]. Other five-coordinate intermediate-spin ferric complexes with four equatorial N donor atoms have macrocyclic ligands, such as [Fe^{III}(Ph_2[16]N_4)(SPh)], where [16]N_4 is a 16-member tetraaza macrocycle ligand ($\delta = 0.26$ mm s^{-1}, $\Delta E_Q = 1.93$ mm s^{-1} at 300 K, [99]). More recently, the electronic structure of [Et_4N]$_2$[Fe^{III}Cl(η^4-MAC*)]·CH_2Cl_2 (**19**) was investigated by comprehensive EPR- and applied-field Mössbauer spectroscopy [100] revealing the typical feature of two large negative magnetic hyperfine coupling components and one small positive component: $A/g_N\mu_N = (-16.6, -22.0, +5.0)$ T.

The intermediate-spin ground state of the ferric compounds published by Jäger and coworkers is also stabilized by a N_4-macrocyclic ligand, [N_4]$^{2-}$ which exist in different varieties of substitutions. The apical ligands are weakly coordinating halides or pseudohalides, such as iodide in the case of [Fe^{III}[N_4]I] (**20**) [68]. The electronic structure was elucidated by EPR, Mössbauer and DFT studies.

Six-Coordinate Complexes

Most six-coordinate iron(III) compounds with a spin-quartet ground state are highly nonplanar porphyrinates like the strongly ruffled chiroporphyrin [Fe (TMCP)(EtOH)$_2$]ClO_4 (**21**) [101] or the saddle-shaped compound [Fe(OETPP) (THF)$_2$]ClO_4 (**22**) [102]. The Mössbauer parameters for these two examples are $\delta = 0.35$ mm s^{-1}, $|\Delta E_Q| = 3.79$ mm s^{-1} at 77 K, and $\delta = 0.50$ mm s^{-1}, $|\Delta E_Q| = 3.50$ mm s^{-1} at 80 K, respectively. More information on porphyrinates is presented in the paragraph below dealing with spin admixture.

A remarkable nonporphyrin complex is the highly distorted compound [Fe(L-N_4Me_2)(bdt)]ClO_4·H_2O (**23**) [103] which undergoes a very gradual ST from a spin-doublet state persisting below 50 K to a spin-quartet state at ambient temperature. The X-band EPR recorded in MeCN/toluene displays a rhombic signal with $g^{\mathrm{eff}} = 5.50, 1.86, 1.40$ consistent with a large rhombicity parameter E/D for the quartet state. The intermediate-spin ferric species shows $\delta = 0.32$ mm s^{-1}, $|\Delta E_{\mathrm{Q}}| = 2.22$ mm s^{-1} at 200 K.

The octahedral iron complex [$\mathrm{Fe^{III}(^{Bu}bdt^{\bullet})(^{Bu}bdt)(PMe_3)_2}$] (**24**) plays a particular role among the quasi octahedral complexes because it has two strong trimethyl phosphine ligands and shows virtually D_2 symmetry without much tetragonal distortion of the iron coordination. The ground-state spin is $S_{\mathrm{t}} = 1$. An array of complementary spectroscopic data had been invoked to show that **24** contains an intermediate-spin ferric iron and one $\mathrm{bdt}^{\bullet-}$ radical and one dianion bdt^{2-} ligand (PMe_3 is neutral) [92]. The spin-triplet ground state arises from antiparallel coupling of the iron spin $S_{\mathrm{Fe}} = 3/2$ and the ligand-radical spin $S_{\mathrm{rad}} = 1/2$. The ground state exhibits large ZFS ($D_{\mathrm{t}} = +14.35$ cm^{-1}, $E/D = 0.02$) which, converted to local values of iron, yields $D_{\mathrm{Fe}} = +9.57$ cm^{-1}. Applied-field Mössbauer spectra reveal an anisotropic A tensor with a very large negative A_{zz} component and rather small A_{xx} and A_{yy} components: $A/g_{\mathrm{N}}\mu_{\mathrm{N}}(S_{\mathrm{Fe}} = 3/2) = (+1.25, -9.88, -30.58)$ T. The Mössbauer parameters $\delta = 0.17$ mm s^{-1}, $\Delta E_{\mathrm{Q}} = +1.46$ mm s^{-1} and $\eta = 0.1$ resemble those of the rhombic planar compounds, though ΔE_{Q} is rather small.

Common Features and Electronic Structures

Intermediate-spin ferric complexes have been found for rhombic pyramidal and distorted octahedral complexes containing strong equatorial and weak axial ligands. The equatorial ligands comprise: thiocarbamates, thiooxolates, thiosalen, thiosemicarbazides, benzenedithiolates, iminothiobenzosemiquinonates, and [N_4]-macrocycles, porphyrines and tetraazaporphyrines, which mostly have sulfur and nitrogen as coordinating atoms. Among the axial ligands are halides, $(SbF_6)^-$, $(ClO_4)^-$, H_2O, THF, Et_2O, halogenocarborane $(CB_{11}H_6X_6)^-$ and carborane $(CB_{11}H_{12})^-$ monoanions or, in a few cases, pyridines and thiolates.

Mössbauer isomer shifts are found between 0.06 and 0.50 mm s^{-1} at 80 K, with a clear statistical prevalence for $\delta \approx 0.3$ mm s^{-1}. A typical feature of $S_{\mathrm{Fe}} = 3/2$ compounds is a large quadrupole splitting up to 4.38 mm s^{-1} with positive V_{zz} (main component of the EFG). Moreover, the rhombic-pyramidal compounds are characterized by anisotropic magnetic hyperfine tensors with relatively large A_{xx} and A_{yy} components of about -18 to -26 T, and a small or even slightly positive A_{zz} component. ZFS of the spin quartet is usually large, on the order 2–20 cm^{-1}, with either a positive or a negative sign of D.

In a crystal-field picture, the electronic structure of iron in the five-coordinate compounds is usually best represented by a $(d_{xy})^2(d_{yz}, d_{xz})^2\,(d_{z^2})^1$ configuration [66, 70], as convincingly borne out by spin-unrestricted DFT calculations on the "Jäger compound" **20** [68]. The intermediate spin configuration with an empty $d_{x^2-y^2}$ orbital in the CF model, however, has a vanishing valence contribution to the

EFG. Early extended Hückel calculations suggested that the large quadrupole splitting of the spin-quartet ground state is entirely caused by covalency effects. The recent DFT molecular-orbital calculations by Grodzicki et al. [68] corroborate this conclusion and specify particularly the covalent contribution of the otherwise, in the crystal-field picture, completely empty $d_{x^2-y^2}$ orbital as the origin of a large quadrupole splitting of **20**. The shape of the magnetic orbitals, $(d_{yz}, d_{xz})^2\ (d_{z^2})^1$, renders a large spin-dipole contribution caused by the corresponding elongated spin-density distribution as the origin of the A-anisotropy of the five-coordinate ferric spin-quartet compounds.

8.2.1.3 Spin Admixture $S = (5/2, 3/2)$ in Porphyrinates: A Special Case?

The ground state of five-coordinate *high-spin* iron(III)-porphyrinates sometimes cannot be properly described by a pure $S = 5/2$ state. The classical examples are, as mentioned above, ferricytochrome c′ and horseradish-peroxidase [60, 63]; a recent review on corresponding synthetic porphyrinates is found in [104]. These ferric compounds display magnetic moments between 5.8 and $4\mu_B$ at room temperature [64, 69, 105–111], curved Curie–Weiss plots [107, 112], very low EPR effective $g_\perp$ values of 4.2–5.8 and large ZFS of more than 10 cm^{-1} [107, 109, 110, 113, 114], large temperature changes of the NMR chemical shift of the pyrrole protons [109, 112, 115, 116], and large Mössbauer quadrupole splittings $\Delta E_Q = 2.2$–4.1 mm s^{-1} with isomer shifts $\delta = 0.38$–0.43 mm s^{-1} at 4.2 K [69, 104, 107, 109, 110, 113, 114, 117, 118]. All these properties are distinct from those of $S = 1/2$ and $S = 5/2$ iron(III) porphyrinates and indicate that the spin state of the iron(III) is a coherent superposition (not a thermal mixture) of the $S = 3/2$ and $S = 5/2$ states.

Maltempo adapted the theoretical approach worked out by Harris [71] to explain the unusual spectroscopic data by a simplified model using quantum-mechanical mixing of the excited spin-quartet state ($^4A_2$) into the sextet ground state ($^6A_1$), caused by SOC [57, 59]. The energy gap between these states necessary to explain this observation is of the order of the SOC constant (200–400 cm^{-1}). The situation, in general, results from the coordination of weak-field axial ligand(s) that cause the porphyrinate core to contract, thereby destabilizing the $d_{x^2-y^2}$ orbital to the point where the $S = 3/2$ spin state becomes lowest in energy. *The actual ground state of iron(III) porphyrinates can be on either side of this crossover, that is, either largely $S = 3/2$ or $S = 5/2$* [114].

Spin-admixed ($S = 3/2, 5/2$) iron(III) porphyrinates are mostly observed when weak-field counter-anions are coordinated as axial ligands, such as ClO_4^-, $B_{11}CH_{12}^-$, SbF_6^-, BF_4^-, PF_6^-, $C(CN)_3^-$, or $SO_3CF_3^-$ [104, 116] (Table 8.2). The degree of spin admixture, however, also depends on the 2,6-substituents of the phenyl rings of tetraphenyl porpyhrins [116]. Spin admixture has also been observed in six-coordinate complexes such as iron(III)octaethylporphyrinate *bis*-ligated by 3,5-dichloropyridine [113, 118] or 3-chloropyridine [119] and in iron(III) tetraazaporphyrins. Additionally, the spin-admixed state exists in some five-coordinate iron(III) phthalocyanines [120]. The first truly four-coordinate ferric heme

Table 8.2 Mössbauer parameters of some five- and the first truly four-coordinate ferric porphyrinates with admixed spin $S = (3/2, 5/2)$

Compound[a]	δ^{b} (mm s^{-1})	$\Delta E_Q{}^{c}$ (mm s^{-1})	Temp. (K)	% ($S = 3/2$)[d]	Ref.
Fe(TPP)($B_{11}CH_{12}$)(C_7H_8)	0.33	4.12	4.2	92	[111]
Fe(TPP)(ClO_4) (0.5*m*-xylene)	0.38	3.50	4.2	65	[107]
Fe(TPP)($FSbF_5$)·C_6H_5F	0.39	4.29	4.2–77	98	[123]
Fe(OETPP)Cl	0.35	0.92	280	4–10	[124]
Fe(OEP)ClO_4	0.40	3.54	4.2	100	[125]
Fe(TPP)(CF_3SO_3)	–	2.39	200	30	[126]
[Fe(TipsiPP)]$^+$[$CB_{11}H_6Br_6$]$^{-e}$	0.33	5.16	6	100	[121]

[a]*TPP* tetraphenyl porphyrin, *OETPP* 2,3,7,8,12,13,17,18-octaethyl-5,10,15,20-tetraphenylporphyrin, *OEP* octaethylporphyrin, *TipsiPP* 5,10,15,20-tetrakis(2′,6′-bis(triisopropylsiloxy)phenyl)-porphyrin (in short: bis-pocket siloxyl porphyrin)
[b]Isomer shift versus α-iron at RT
[c]Quadrupole splitting
[d]Derived from EPR *g*-values
[e]Iron in [Fe(TipsiPP)]$^+$ is four-coordinated

with the extremely hindered *bis*-pocket siloxyl porphyrin (HTipsiPP) shows, as expected, *pure intermediate spin* $S = 3/2$ [121].

An interesting example of a spin-admixed nonheme iron(III) complex with $S = (3/2, 5/2)$ ground state is the organometallic anion $[Fe^{III}(C_6Cl_5)_4]^-$ which has four pentachloro phenyl ligands in tetrahedrally distorted planar symmetry [122].

8.2.2 *Iron(II) with Intermediate Spin, S = 1*

8.2.2.1 Square-Planar Iron(II) Compounds

The planar complex of iron(II)-phthalocyanine, FePc, was the first example of an iron(II) compound ($3d^6$ configuration) for which a $S = 1$ ground state has been established. Phthalocyanine is a macrocyclic ligand with a planar N_4 set of donor atoms resembling the core of a porphyrin. The magnetic susceptibility of FePc has been measured by Klemm [127] who reported a value of $3.96\mu_B$ (Bohr magnetons) at room temperature, and later by Lever who found $3.85\mu_B$ for a highly purified powder of FePc [128]. Both values are considerably larger than the spin-only value of $2.83\mu_B$ for $S = 1$ ($\mu_{eff} = g \cdot [S(S+1)]^{1/2}$) but, nevertheless, the temperature-dependence of the magnetic data could be readily explained by the presence of a spin-triplet ground state, $S = 1$. Strong SOC partly restores the orbital momentum and yields large ZFS (see Chap. 4.7) as well as a large (average) g-factor for the Zeeman interaction; the corresponding spin Hamiltonian parameters are found in the range of $D = 64$–70 cm^{-1} and $g_{av} = 2.61$–2.74 [129, 130]. Essentially, the magnetic properties of FePc can be reasonably well explained in a crystal field (CF) model by mixing of the spin-triplet ground state with excited spin-triplet and spin-quintet states under the influence of SOC. Probably the best CF description invokes an energetically split ^3E ground state, according essentially to a $(d_{xy})^2$, $(d_{xz}, d_{yz})^3$, $(d_{z^2})^1$, $(d_{x^2-y^2})^0$ ground-state configuration [129–131]. Apparently, the $d_{x^2-y^2}$ orbital is destabilized by strong in-plane σ-interaction affording a one-over-four arrangement of valence d-orbitals as expected for a planar compound.

Metal complexes of phthalocyanine still attract much attention, mostly because of a wide range of applications, such as catalysts, dyes, optical switches, but more recently also because of their magnetic properties, particularly the iron complex FePc [132]. The "flat" molecules can aggregate in the solid and form chains of the so-called herringbone type with various properties. FePc crystallizes in an α- [133] and a β-form [134], which may explain many of the magnetic properties reported for FePc. The α-form which is most interesting for applications shows spontaneous magnetic ordering at low temperatures with a remarkably strong internal field of 66.2 T at the Mössbauer nucleus at 1.3 K due to sizable intermolecular spin coupling (a record value for an $S = 1$ system) [132]. In contrast, the β-form is paramagnetic and shows vanishing magnetization below 10 K because of the large (positive) ZFS of the triplet, providing a $M_s = 0$ ground state [129, 130].

A remarkable number of Mössbauer studies have been published since the first spectra reported in 1966 [135], most of them performed on the β-form when not specified differently [131, 132, 136–139]. Also, high pressure has been applied [140] and thin films were prepared [141]. Because of the ambiguity concerning the crystalline phase, the values of the hyperfine parameters show some dispersion. The isomer shift, $\delta = 0.4$–0.6 mm s^{-1}, is found in between the typical values known for high-spin iron(II) and low-spin iron(II). The quadrupole splitting is large, $\Delta E_Q = 2.4$–3.0 mm s^{-1} (Table 8.3), as one might expect because of the unusual non-cubic symmetry. Applied-field measurements revealed positive V_{zz}.

Interestingly, a CF model invoked for the basic interpretation of the magnetic properties can explain reasonably well the quadrupole splitting, including the weak temperature dependence. (Note, the valence contribution to the EFG as it can be inferred for the basic $(d_{xy})^2$, $(d_{xz}, d_{yz})^3$, $(d_{z^2})^1$, $(d_{x^2-y^2})^0$ configuration of the ^3E ground state has its main component along the x-axis of the crystal field: $V_{xx,\mathrm{val}} = +4/7e\langle r^{-3}\rangle$, $V_{yy,\mathrm{val}} = V_{zz,\mathrm{val}} = -2/7e\langle r^{-3}\rangle$, as can be inferred from Table 4.3). However, a major contribution from covalent bonding has been noted. Recent quantum chemical calculations on intermediate-spin phthalocyanines are found in [142].

Four-coordinate iron(II)-porphyrins also show an intermediate-spin ($S = 1$) ground state. The classical examples are the tetraphenyl, TPP, and octaethyl, OEP, derivatives, whose effective magnetic moments are about $4.2\mu_B$ at room temperature [143]. The magnetic properties of these planar complexes (with S_4 symmetry due to ruffling of the N_4-ligand plane for Fe(TPP)), studied by magnetic susceptibility, paramagnetic NMR [144, 145], and applied-field Mössbauer measurements [146–149], are compatible with a $^3A_{2g}$ ground stare in a crystal field model arising from a $(d_{xy})^2$, $(d_{z^2})^2$, $(d_{xz}, d_{yz})^2$ configuration, mixed by SOC with an excited ^3E_g state, $(d_{xy})^2$, $(d_{z^2})^1$, $(d_{xz}, d_{yz})^3$. Other suggestions for the ground state are ^3E_g and $^3B_{2g}$ [150]. More recent Mössbauer studies were performed with iron(II) octaethyltetrazaporphyrin, Fe(OETAP), which shows magnetic ordering in the

Table 8.3 Mössbauer parameters of iron(II) complexes with intermediate spin (S = 1)

Compounds	Temp. (K)	δ (mm s^{-1})	ΔE_Q (mm s^{-1})	Ref.
Fe(Pc)	4	0.49	+2.70	[136]
Fe(Pc)	77	0.51	+2.69	[136]
Fe(Pc)	293	0.40	+2.62	[136]
Fe(Pc), α-phase	4.2	0.46	+2.52	[132]
Fe(TPP)	4.2	0.52	+1.51	[148]
α-Fe(OEP)	4.2	0.59	+1.60	[147]
Fe(OETAP)	4.2	0.3	+3.09	[151]
porphyrinogen[a]	4.2	0.35	2.34	[152]
Fe(octaaza[14]annulene)[b]	78	0.19	+4.13	[156]
$(AsPh_4)_2$[Fe(II)bdt_2]	4.2	0.44	+1.16	[157]
$(N(C_2H_5))_2$[Fe(II)bdt_2]	4.2	0.45	+1.21	[158]

[a]A porphyrinogen is a reduced tetrapyrrole parent compound of a porphyrin
[b]The true electronic structure might be Fe(III) with intermediate spin (S_{Fe} = 3/2), antiferromagnetically coupled to a ligand radical (S' = 1/2)

solid with a large internal field of 62.4 T at the Mössbauer nucleus [151]. Iron(II)-porphyrinogens represent another group of quasi square-planar iron(II) complexes with $S = 1$ ground state [152].

Both Fe(II)(TPP) and Fe(II)(OEP) have positive electric quadrupole splitting without significant temperature dependence which, however, cannot be satisfactorily explained within the crystal field model [117]. Spin-restricted and spin-unrestricted X_α multiple scattering calculations revealed large asymmetry in the population of the valence orbitals and appreciable 4p contributions to the EFG [153] which then was further specified by ab initio and DFT calculations [154, 155].

Four-coordinate iron(II) complexes of porphyrin-derivatives, so-called porphyrinogens, show similar properties as the square planar porphyrins [152], whereas the macrocyclic N_4-donar ligand octaaza[14]annulene affords a planar iron(II) complex with a particularly low isomer shift, $\delta = 0.19$ mm s^{-1}. Since this value is significantly lower than the typical values for planar iron(II) complexes given in Table 8.3, one may presume that the ligand is not innocent and has oxidized the metal center so that the true electronic structure is better described by iron(III) intermediate-spin (S = 3/2) antiferromagnetically coupled to a ligand radical located on the one-electron reduced ligand in its trianion form.

Four-coordinate, planar iron(II)–dithiolate complexes also exhibit intermediate spin. The first example described was the tetraphenylarsonium salt of the square-planar bis(benzene-1,2-dithiolate)iron(II) dianion, $(AsPh_4)_2$[Fe(II)bdt_2], which showed $\delta = 0.44$ mm s^{-1} and $\Delta E_Q = 1.16$ mm s^{-1} at 4.2 K [157]. The electronic structure of a different salt was explored in depth by DFT calculations, magnetic susceptibility, MCD measurements, far-infra red spectroscopy and applied-field Mössbauer spectroscopy [158].

The four-coordinate iron(II) complex of cycloheptatrienylidene is a rare example of a fully reversible singlet ($S = 0$ at 6 K) to triplet ($S = 1$ at 293 K) transition in the slow relaxation regime [159].

Six-Coordinate Iron(II) Complexes

König and others published in the 1970s an impressive series of studies on six-coordinate iron(II)-bis-phenanthroline complexes [160–164] for which they inferred $S = 1$ from thorough magnetic susceptibility and applied-field Mössbauer measurements. Criteria for the stabilization of the triplet ground state for six-coordinate compounds with tetragonal (D_{4h}) and trigonal (D_{3d}) symmetry were obtained from LFT analyses [163]. The molecular structures, however, were not known because the materials could not be crystallized.

Elaborate studies finally revealed that the composition of the solid material in reality was not Fe(II)-(phenanthroline)$_2$-X, as presumed previously, where X is a dianionic ligand like oxalate or malonate, but a so-called mixed-valence "double-salt" of $[Fe(II)(phen)_3]_2$ and $[Fe(III)(dianion)_3]$ with (1/2-dianion)*xH$_2$O [164, 165]. The iron(II) compounds are low-spin ($S = 0$) and the iron(III) species are high-spin ($S = 5/2$), and the superposition of the magnetic moments of both centers accounts for the "effective" triplet signal. Most annoying, the Mössbauer spectra of the paramagnetic iron(III) species turned out to be broadened beyond recognition because of intermediate-spin relaxation. Their contribution was completely missed and the Mössbauer spectra, reported as those of iron(II) with $S = 1$, are, in reality, the spectra of the low-spin iron(II) species only ($\delta \sim 0.3$ mm s^{-1}, $\Delta E_Q \sim 0.25$ mm s^{-1} [162]). It took the authors almost a decade to uncover the true nature of these systems (which might be a lesson for every spectroscopist). In conclusion, we can state that, partly in contrast with textbook knowledge (e.g., [16] in Chap. 1), iron(II) with intermediate spin $S = 1$ has been unambiguously identified only for four-coordinate (mainly planar) compounds.

8.2.3 Iron in the High Oxidation States IV–VI

Iron centers that are more electron-deficient than iron(III) compounds are used for efficient and highly specific oxidation reactions in, for example, heme and nonheme enzymes [166–172]. Most iron(IV)-complexes found in biological reaction cycles possess terminal or bridging oxo groups as is known from a large number of structural and spectroscopic investigations. With the exception of iron(IV)-nitrido groups, nonoxo iron(IV) centers very rarely take part in such reactions.

The enzymatic reactions of peroxidases and oxygenases involve a two-electron oxidation of iron(III) and the formation of highly reactive $[Fe{=}O]^{3+}$ species with a formal oxidation state of +V. Direct (spectroscopic) evidence of the formation of a genuine iron(V) compound is elusive because of the short life times of the reactive intermediates [173, 174]. These species have been safely inferred from enzymatic considerations as the active oxidants for several oxidation reactions catalyzed by nonheme iron centers with "innocent," that is, redox-inactive, ligands [175]. This conclusion is different from those known for heme peroxidases and oxygenases which show a separation and delocalization of the two oxidation equivalents formally stored on one $[Fe{=}O]^{3+}$ unit. Upon oxidation, one of the two electrons is removed from iron(III) and one from the porphyrin ligand forming an iron(IV)-oxo species and a porphyrin (ligand) radical [176–178]. The reaction intermediate, called compound-I, was initially discovered because of its unusual and intense green color [179] appearing after a two-electron oxidation [180, 181]. In the next step of the reaction cycle of peroxidases, the red so-called compound-II is formed by a one-electron back-reduction of the radical, leaving the iron(IV)-oxo group in the ferryl state [182]. In some cases of modified heme proteins, when a redox-active amino acid like tyrosine is close to the iron, another organic radical (instead of the porphyrin radical) can be formed in conjunction with the iron(IV)–oxo group [183]. Although examples of oxidation states higher than iron(V) are elusive in biological systems, the whole range of high-valent iron sites have been intensely studied with synthetic compounds, as will be shown below.

Iron(IV), (V), and (VI) centers are also found in solid-state materials, the coordination chemistry of which is of considerable interest because unusual structures and remarkable electronic and catalytic properties are encountered [184]. Reviews of the corresponding Mössbauer properties are found in [185–187].

8.2.3.1 Crystal-Field Ground States

High-valent iron can occur in a wide variety of electronic configurations. Figure 8.25 (a-c, e-i) presents a summary of the corresponding one-electron crystal-field states for the $3d^4$, $3d^3$, and $3d^2$ electron configurations, allocated to HS and LS states in distorted octahedral and tetrahedral symmetry. Part d, in addition, depicts the case of "low–low-spin" iron(IV) found in some trigonal

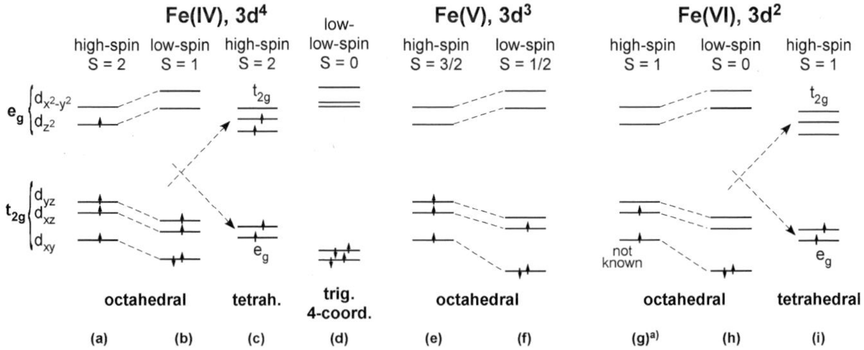

Fig. 8.25 Schematic presentation of the one-electron crystal-field states for the $3d^4$, $3d^3$, and $3d^2$ electron configurations of iron(IV), (V), and (VI). [a]Case (**g**) has not been observed yet

compounds. We note that in solid-state materials, only HS states are observed (parts a, c, e, and i).

8.2.3.2 Iron(IV) Oxides

The chemistry of iron(IV) in solid-state materials and minerals is restricted to that of oxides, since other systems such as iron(IV)-halides are not stable [186]. Iron(IV) oxides are easy to handle because they are usually stable in air, but they often have a substoichiometric composition, with oxygen vacancies contributing to varying degrees. Moreover, the samples may contain different amounts of iron(III) in addition to the intended iron(IV) oxide, a complication which may obscure the Mössbauer data [185]. Even iron(V) was found in iron(IV) oxides due to temperature-dependent charge disproportionation [188, 189].

The Mössbauer parameters of several iron(IV) compounds are summarized in Table 8.4. The calcium, barium, and strontium perovskites $CaFeO_3$ [188], $BaFeO_3$ [190] and $SrFeO_3$ [189, 191], together with their various modifications containing mixed metal ions and different oxygen content, are probably the best investigated systems among the iron(IV) oxides [185]. In perovskites, iron(IV) is coordinated by six oxygen atoms yielding spin $S = 2$, arising from the $t_{2g}{}^3e_g{}^1$ configuration (Fig. 8.25, case a). The isomer shift (relative to α-iron) is about +0.1 mm s^{-1} at 4.2 K and the quadrupole splitting is practically zero, as expected for iron in a cubic lattice site [185]. Octahedral oxoferrates, such as $Sr_{0.5}La_{1.5-}Li_{0.5}Fe_{0.5}O_4$, have isomer shifts at 4.2 K in the range of −0.12 to −0.08 mm s^{-1} [185].

More recently, tetrahedral $Fe(IV)O_4{}^{4-}$ anions have been found in the barium-rich iron oxides Ba_2FeO_4 and Ba_3FeO_5. The fragment is high spin, $S = 2$, according to the tetrahedral coordination (Fig. 8.25, case c) and yields an isomer shift of about −0.15 mm s^{-1} at 80 K and −0.24 mm s^{-1} at room temperature and a quadrupole splitting of 0.3–0.4 mm s^{-1} [187].

8.2.3.3 Iron(IV) in Metalloproteins and Coordination Compounds

The archetypical and spectroscopically best investigated example of a heme peroxidase is horseradish peroxidase (HRP). Moss et al. have shown by Mössbauer spectroscopy that compound-I and compound-II of HRP both possess the same oxidation state of iron [182]. Later, the corresponding iron(IV) and particularly the iron(IV)-radical states have been studied in great detail, revealing much insight into the electronic interaction of both paramagnetic sites, the iron and the porphyrin radical [63]. The iron(IV)-oxo group is in the LS state, $S = 1$, throughout in all heme systems. The isomer shifts and quadrupole splittings for HRP and chloroperoxidase (CPO), are summarized in Table 8.4.

Table 8.4 Mössbauer parameters of iron(IV) compounds[a]

Compound	CN[b]	Spin[c]	Temp. (K)	δ (mm s^{-1})	ΔE_Q (mm s^{-1})	Ref.
(1) Solid-state oxides						
$SrFeO_3$	6	2	4.2/300	0.11/0.07	<l.w.[d]	[191]
$Sr_{0.5}La_{1.5}Li_{0.5}Fe_{0.5}O_4$	6	2	4.2	−0.08	<l.w.[d]	[192]
Ba_2FeO_4 / Ba_3FeO_5	4	2	85/78	−0.15/−0.14	0.36/0.39	[187]
(2) Fe(IV) hemes						
HRP-II	6	1	4.2	0.03	+1.61	[63]
HRP-I	6	1	4.2	0.08	+1.25	[193]
CPO-I	6	1	4.2	0.14	+1.02	[194]
$[(TMP)Fe{=}O]^+$	6	1	4.2	0.08	+1.62	[195, 196]
$[TPP(2,6\text{-}Cl)Fe{=}O]^+$	5?	1	4.2	0.08	+1.8	[196]
$[(TMP)Fe^{IV}(OCH_3)_2]$	6	1	4.2	−0.025	2.10	[197]
Catalase (high pH, $Fe^{IV}{=}O$)	6	1	4.2	0.03	2.29	[198]
Catalase (low pH, $Fe^{IV}{=}OH$)	6	1	4.2	0.07	1.47	[198]
$(FeTPP)_2N$	5?	1?	131	0.18	1.08	[199]
$(FeTPP)_2C$	5?	1?	131	0.10	1.88	[199]
$(FeTPP)_2CCl_2$	5?	1?	131	0.10	2.28	[199]
(3) Nonheme compounds						
$[(cyc\text{-}ac)(ACN)Fe{=}O]^+$	5?	1	4.2	0.01	+1.37	[200]
$[Fe{=}O(TMC)(NCCH_3)](OTf)_2$	6	1	4.2	0.17	+1.24	[201]
$[FeIV(O)(TPA)]^{2+}$	5/6?	1	4.2	0.01	+0.92	[202]
TauD, intermediate **J**	6	2	4.2	0.30	−0.88	[203]
$[(H_2O)_5Fe^{IV}{=}O]^{2+}$	6	2	4.2	0.38	−0.33	[204]
$[Fe^{IV}(O)(TMG_3tren)]^{2+}$	5	2	4.2	0.09	−0.29	[205]
RNR, R2 intermediate **X**	6	2	4.2	0.26	−0.6	[206]
MMO, intermediate **Q**	6	2	4.2	0.17	0.53	[207]
$[([9]aneN_3)Fe^{III}(O)$ $(CH_3CO_2)_2Fe^{IV}([9]aneN_3)]^{3+}$	6	2 35%, 1 65%	4.2	−0.10, 0.05	+0.86, +1.14	[208]
$L(cat)[Fe^{III}{-}N{\equiv}Fe^{IV}](cat)L$	6	1	77	0.09	+0.81	[209, 210]
$L(cat)\,[Fe^{IV}{=}N{=}Fe^{IV}](cat)L]^+$	6	1	77	0.04	+1.55	[209, 210]
$[PhBP^{R}{}_3]Fe{\equiv}N$, R = *i*Pr, CH_2Cy	4	0	140	−0.34	6.01	[211]
$[(TIMEN^{mes})Fe{\equiv}N]^+$	4	0	77	−0.27	6.04	[212]
$[Fe(NTs)(N_4Py)]^{2+}$	6	1	4.2	0.02	+0.98	[213]
$[Fe(O)(N_4Py)]^{2+}$	6	1	4.2	−0.04	+0.93	[213]
(4) Nonoxo, nonnitrido compounds						
$[Fe(oetpp)C_6H_5)]^+$	5	1	4.2	0.13	+3.23	[214]
$[Fe^{IV}Cl(\eta^4\text{-}MAC^*)]^-$	5	2	4.2	−0.02	+0.89	[100]
$[(Me_3cyclam\text{-}acetate)Fe^{IV}Cl]^{2+}$	6	1	80	0.08	+2.40	[215]
$[(Me_3cyclam\text{-}acetate)Fe^{IV}F]^{2+}$	6	1	80	0.02	+2.43	[215]
$[(Me_3cyclam\text{-}acetate)Fe^{IV}N_3]^{2+}$	6	1	80	0.11	+1.92	[216]
$[N_3N']FeCN$	5	0	180	−0.22	3.28	[217]

[a] *HRP* horseradish peroxidase, *CPO* chloroperoxidase, *TMP* chloro-5,10,15,20-tetra(mesityl)porphyrin, *TPP(2,6-Cl)* tetra(2,6-dichlorophenyl)porphyrin, *cyc-ac* cyclam-acetate = 1,4,8,11-tetramethyl-1,4,8,11-tetraaza-cyclotetradecane-1-acetate, ACN = $(NCCH_3)$, TMG_3tren = $N[CH_2CH_2N{=}C(NMe_2)_2]_3$, which is based on the (tren)-backbone (=tris(2-ethylamino)amine) and has a set of three superbasic tetramethylguanidine (=TMG) N-donor atoms, *RNR* ribonucleotide reductase, *MMO* methane monooxygenase, L = 1,4,7-trimethyl-l,4,7-triazacyclononane, cat = (tetrachlorocatechol)$^{4-}$, $PhBP^R$ = tris(phosphino)borate, R = isopropyl, $TIMEN^R$ = tris[2-(3-aryl-imidazol-2-ylidene)ethyl]amine, R = mesityl (mes), η^4-MAC* = 1,4,8,11-tetraaza-13,13diethyl-2,2,5,5,7,7,10,10-octamethyl-3,6,9,12,14 pentaoxocyclotetradecane, $[N_3N']^{3-}$ = $[(t\text{-}BuMe_2\,SiNCH_2CH_2N)]^{3-}$

[b] Coordination number

[c] Spin of the iron(IV) site

[d] Less than the natural line width; materials are magnetic and exhibit a six-line spectrum at low temperatures without appreciable quadrupole interaction

A typical feature of the Mössbauer spectra of five- or six-coordinate iron(IV) with an axial oxo group (or a OCH_3, a nitrido or a imido group) is a low isomer shift ($+0.1 \pm 0.15$ mm s^{-1}), a large and positive quadrupole splitting (1–2 mm s^{-1}), an anisotropic hyperfine coupling tensor with moderately large values for $A_{xx}/g_N\mu_N$ and $A_{yy}/g_N\mu_N$ (-16 to -23 T) and a rather small value for $A_{zz}/g_N\mu_N$ (0 to -10 T) and also a large positive ZFS ($D = 5$–35 cm^{-1}). In many cases, the applied-field Mössbauer spectra of iron(IV) hemes and model compounds could be simulated by using a correlation of the electronic g- and D-values which is obtained from a perturbation treatment of ligand field theory [194, 218]. This is, however, not generally justified because the perturbation model neglects the so-called spin-flip contributions to the ZFS [219] (see Chap. 5).

Heme Iron(IV) Oxo Centers

The interesting electronic and catalytic properties of the heme enzymes have stimulated intense studies on synthetic porphyrin compounds with a large variety of possible modifications aiming for the preparation of systematic spectroscopic or functional probes. Some typical examples of synthetic iron(IV)–porphyrin models are summarized in Table 8.4. Interestingly, the one-electron oxidation of an iron (III) porphyrin can also be ligand-based, forming an porphyrin-cation radical that is exchange-coupled to iron(III). At least three types of iron(III)–porphyrin-radical complexes have been observed; high-spin iron(III) antiferromagnetically or ferromagnetically coupled to the radical affording total spin $S_t = 2$ and $S_t = 3$ ground states, respectively [220–222], and low-spin iron(III) antiferromagnetically coupled to the radical [223]. Experiments with different metal ions and porphyrin-ligand systems indicate that the energy of the metal d- and porphyrin π-orbitals controls the type of oxidation which is either metal- or ligand-based. It appeared that alkoxides, but not methanol, perchlorate or imidazole, are sufficiently strong π-donating ligands to stabilize iron(IV) in porphyrins with respect to iron(III)-porphyrin-cation radicals [197]. The iron(IV) complex $[(TMP)Fe^{IV}(OCH_3)_2]$ is included in Table 8.4 as an example.

Moreover, it has been recognized that protonation of the $Fe^{IV}{=}O$ group induces an increase of the quadrupole splitting with respect to the nonprotonated form. As a typical example, we mention the low-pH and high-pH forms of catalase from *Proteus mirabilis* [198] in Table 8.4.

One of the amazing and still not perfectly understood differences in the electronic structures of the biomimetic models on one side and the biological systems on the other side is the strength and the sign of the exchange coupling between the ferryl iron ($S_{Fe} = 1$) and the porphyrin radical ($S_{rad} = 1/2$) in the respective compound-I states. The interaction always appears to be ferromagnetic for the model compounds ($S = 3/2$ ground state) irrespective of how distorted the respective porphyrins are, whereas the enzymes show very weak and anisotropic magnetic coupling (HRP-I) or appreciable antiferromagnetic coupling with $S = 1/2$ ground state (CPO). Reviews of applied-field Mössbauer investigations on the heme enzymes are found in [117, 224].

Nonheme Iron(IV) Oxo Centers

Iron(IV) also occurs as an intermediate in the reaction cycle of a large family of *nonheme* enzymes that have in their resting state a catalytic mononuclear iron(II) center for the activation of oxygen and the oxidation of their substrates [169, 175, 225, 226]. The iron is coordinated by three protein ligands, two histidine, and one aspartic or glutamic acid, with three further sites for additional coordination. This $(His)_2$–(Asp/Glu) motif is known as "facial triade" [227]. The first iron(IV)–oxo intermediate of this type was characterized by Mössbauer spectroscopy in a transient state of taurine:αKG dioxygenase (called TauD). The species was trapped by *rapid-freeze quench* technique. This intermediate, named **J**, has a ground state with spin $S = 2$ with large, almost axial ZFS ($D = 10.5\ \mathrm{cm}^{-1}$) and an isomer shift of $\delta = 0.30\ \mathrm{mm\ s}^{-1}$ and a quadrupole splitting of $\Delta E_Q = -0.88\ \mathrm{mm\ s}^{-1}$ [203, 228]. Spectroscopy-biased DFT studies on a number of possible model structures suggested for intermediate **J** either trigonal-bipyramidal or octahedral iron coordination [229]. In the meantime, the high-valent species has been thoroughly characterized for four enzymes, and in all instances, the iron(IV) center was found to be high-spin, $S = 2$ [203, 230–232].

The first nonheme iron(IV)–oxo complex, $[(\text{cyclam-acetato})Fe^{IV}{=}O]^+$, was generated by the ozonolysis of a $[(\text{cyclam-acetato})Fe(III)(O_3SCF)]^+$ precursor. Its Mössbauer study yielded $\delta = 0.01\ \mathrm{mm\ s}^{-1}$ and $\Delta E_Q = 1.37\ \mathrm{mm\ s}^{-1}$ [200], but the exact molecular structure could not be determined. Subsequently, it was found that the corresponding Fe(IV)-oxo complex with the ligand TMC = Me_4cyclam ($[Fe{=}O(TMC)(NCCH_3)](OTf)_2$) was sufficiently stable that it could be crystallized to obtain its molecular structure by X-ray diffraction [201]. Since then, the number of known iron(IV)–oxo compounds has increased [233–235], and some have also been investigated by Mössbauer spectroscopy [236–238]. In one case, where the high-valent state has been generated by the reaction of Fe(II) with hypochloride in CH_3CN, the terminal ligand of Fe(IV) is presumably a methoxy group ($\delta = 0.03\ \mathrm{mm\ s}^{-1}$ and $\Delta E_Q = 1.21\ \mathrm{mm\ s}^{-1}$ at 4.2 K) [239].

All but two of the known synthetic iron(IV)–oxo compounds are low-spin, $S = 1$ [202, 240]. The first example of an iron(IV)–oxo model compound with spin $S = 2$ was the quasioctahedral complex $[(H_2O)_5Fe^{IV}{=}O]^{2+}$ ($\delta = 0.38\ \mathrm{mm\ s}^{-1}$, $\Delta E_Q = 0.33\ \mathrm{mm\ s}^{-1}$) which was obtained by treating $[Fe^{II}(H_2O)_6]^{2+}$ with ozone in acidic aqueous solution [204]. The spin state of iron in this type of structure is determined by the energy gap between the $d_{x^2-y^2}$ and the d_{xy} orbitals [241]. The weak water ligands induce a sufficiently small gap being less than the spin paring energy and stabilizing the HS state (Fig. 8.25, case a).

The second example of a high-spin iron–oxo complex, $[Fe^{IV}(O)(TMG_3tren)]^{2+}$ (see Table 8.4) has been published only recently. In this compound, the HS state is afforded by the trigonal bipyramidal symmetry of the TMG_3tren ligand, causing degeneracy of the d_{xy} and $d_{x^2-y^2}$ orbitals [205]. The differences in the electronic structures of high-spin and low-spin compounds matter because of the consequences for the reactivity of the complex [229].

The R2 protein of class I ribonucleotide reductase (RNR R2) and methane monooxygenase (MMO) are typical examples of enzymes with a diiron center in the active site. Their catalytic functions rely on the formation of two key intermediates, **X** and **Q**, respectively, with highly oxidized iron. Intermediate **X** of RNR contains a mixed-valent Fe^{III}–O–Fe^{IV} unit [206, 242], whereas **Q** has a so-called Fe^{IV}–(μ-O_2)–Fe^{IV} diamond core [243]. Intermediate **X** is oxidized by one equivalent above the diferric resting R2 state, involving high-spin Fe(III) (d^5, $S = 5/2$) and high-spin Fe(IV) (d^4, $S = 2$). The Mössbauer parameters are $\delta = 0.56$ mm s^{-1} and $\Delta E_Q = -0.9$ mm s^{-1} for the ferric and $\delta = 0.26$ mm s^{-1} and $\Delta E_Q = -0.6$ mm s^{-1} for the ferryl site. Antiferromagnetic exchange produces the observed $S = 1/2$ ground state with distinct EPR and Mössbauer properties [206]. The homovalent cluster of the MMO intermediate **Q** contains two identical ferryl sites; their local spin is $S = 2$ and the Mössbauer parameters are $\delta = 0.17$ mm s^{-1} and $\Delta E_Q = 0.53$ mm s^{-1} [172, 207, 244].

High-Valent Iron Dimers

The two iron(III) sites in MMO and RNR-R2, are bridged by a (μ-oxo) group and two (μ-carboxylato) groups [245, 246]. Upon oxidation, the diiron core is converted into a localized mixed-valent iron(III)–iron(IV) dimer as characterized by high-resolution molecular-structure studies of biomimetic model compounds [208] (and references therein). Of particular interest were site-selective Mössbauer measurements on structurally related compounds which contain mixed-metal clusters of the type (Fe^{III}–Cr^{III}) and (Cr^{III}–Fe^{IV}). The g-values, ZFS, and exchange interaction parameters could be determined in detail and were related to structural features via DFT calculations [208]. Interestingly, the iron(IV) site coordinated by the neutral synthetic N_3-donating ligand ([9]aneN_3) and the (μ-oxo)(μ-carboxylato)$_2$ bridges exhibits thermal spin-equilibrium. Corresponding to this observation, the trication $[([9]aneN_3)\ Fe^{III}(O)(CH_3CO_2)_2\ Fe^{IV}([9]aneN_3)]^{3+}$ measured in frozen CH_3CN solution exhibits the superpositions of two EPR spectra and of two types of magnetic Mössbauer spectra, respectively. The superpositions are related with the two different spin states of iron(IV). In one case (ca. 35%), the iron(IV) site is in the HS state, $S = 2$, and in the other (ca. 65%), it is in the LS state, $S = 1$. The corresponding Mössbauer parameters are summarized in Table 8.4. Since the iron (III) site is in the HS state, $S = 5/2$, the total spin of the two antiferromagnetically coupled spin-isomers is $S_t = 1/2$ and $S_t = 3/2$, respectively.

High-valent iron also occurs in μ-nitrido bridged dimers with linear $[Fe^{III}–N{\equiv}Fe^{IV}]^{4+}$ and $[Fe^{IV}{=}N{=}Fe^{IV}]^{5+}$ cores [209, 210] (and references therein). Such compounds have been prepared first by thermolysis [247] or photolysis [248] of iron(III)-porphyrin complexes with an azide ligand, $(N_3)^-$. Mixed-valent iron-nitrido porphyrin dimers exhibit valence delocalization as can be inferred from the unique Mössbauer spectra showing isomer shifts in the range $\delta = 0.08$–0.18 mm s^{-1} [249, 250]). At the same time, the dimers show $S = 1/2$ ground state due to very strong antiferromagnetic spin coupling. This is somewhat mysterious because valence delocalization as the result of double-exchange interaction should stabilize HS states [251–253].

On the contrary, ligands such as 1,4,7-trimethyl-1,4,7-triazacyclononane (=Me_3[9]ane) yield localized valences for the $[Fe^{III}–N{\equiv}Fe^{IV}]^{4+}$ core with a total spin of $S_t = 3/2$, arising from strong antiferromagnetic coupling ($-J > 150$ cm^{-1}) of high-spin ($S = 5/2$) iron(III) and low-spin ($S = 1$) iron(IV); the Mössbauer parameters of the latter are $\delta = 0.09$ mm s^{-1} and $\Delta E_Q = 0.81$ mm s^{-1}. The homo-valent oxidation product with a $[Fe^{IV}{=}N{=}Fe^{IV}]^{5+}$ core is diamagnetic up to room temperature ($-J > 200$ cm^{-1}), showing Mössbauer parameters $\delta = 0.04$ mm s^{-1} and $\Delta E_Q = 1.55$ mm s^{-1} [209]. DFT calculations have provided insight into the electronic structure and exchange pathways within the diiron core of these compounds [210].

Mononuclear Iron(IV)-Nitrido and -Imido Compounds

A mononuclear iron-nitride group has recently been suggested a constituent of the cofactor catalyzing biological nitrogen reduction by nitrogenase [254]. In order to gain further insight into the structure of iron-nitride complexes and their reactivity, a number of model compounds with terminal Fe=NR and Fe≡N groups have been synthesized and spectroscopically characterized [211, 255–257]. An exciting achievement has been the preparation and characterization of distorted tetrahedral iron(IV)-nitrido complexes [257], in which the Fe=N unit is coordinated by a polydentate ligand with three soft donor groups, either three phosphine ligands in the case of $[PhBP_3^R]Fe{\equiv}N$ [211, 258, 259] or three *N*-heterocyclic carbene groups in the case of $[(TIMEN^R)Fe{\equiv}N]^+$ [212] and another variant [260]. In these complexes, because of the pseudotetrahedral symmetry with strong trigonal distortion at the iron site, the valence *d*-orbitals of the iron are widely split in a "three-over-two" pattern such that the four valence electrons are accommodated in the low-lying nonbonding *e*-orbitals, that is, d_{xy} and $d_{x^2-y^2}$ [260]. Therefore, these compounds are diamagnetic, $S = 0$ (the state which is called the "low–low-spin" state of the d^4 configuration). As a consequence of the preferential population of nonbonding *d*-orbitals rather than antibonding orbitals, the Fe–N distances are very short (151–155 pm, [212, 259]), the Fe–N vibration frequencies are high (ν(Fe≡N) $\approx$ 1,000 cm^{-1}), and the species have high reactivity toward nucleophiles [257]. The d^4 configuration causes an unprecedentedly large EFG at the Mössbauer nucleus. Although not determined, one may presume a positive sign for V_{zz} because of the huge (positive) valence contribution to the *EFG* expected for four electrons in the two "planar" orbitals d_{xy} and $d_{x^2-y^2}$. The observed quadrupole splitting is 6.01 mm s^{-1} for $[PhBP^{iPr}{}_3]Fe{\equiv}N$ [211] and 6.04 mm s^{-1} for $[(TIMEN^{mes})Fe{\equiv}N]^+$ [212], which are the largest ever observed for an iron compound.

The corresponding isomer shifts are $-0.31\ \mathrm{mm\ s^{-1}}$ (at 140 K) and $-0.27\ \mathrm{mm\ s^{-1}}$ (at 77 K), respectively.

In analogy to the iron(IV)-oxo unit, the iron(IV)-imido unit (Fe(IV)=NR) has been proposed as the reactive intermediate of isolobal amination reactions catalyzed by iron [213, 261]. A model compound for such a species is the tosylimido analog of a nonheme iron(IV)–oxo complex $[Fe(NTs)(N4Py)]^{2+}$ (N4Py = *N,N*-bis (2-pyridylmethyl)bis(2-pyridyl)methylamine, NTs = mesityl-*N*-tosylimide). In contrast to related porphyrin–imido compounds, which contain an iron(II) center, $S = 2$ ($\delta = 0.72\ \mathrm{mm\ s^{-1}}$, $\Delta E_Q = 1.51\ \mathrm{mm\ s^{-1}}$) [262], the Mössbauer parameters of this nonheme complex are indicative of an iron(IV) species ($\delta = 0.02\ \mathrm{mm\ s^{-1}}$, $\Delta E_Q = +0.98\ \mathrm{mm\ s^{-1}}$, $S = 1$, $D = 29\ \mathrm{cm^{-1}}$). The parameters are very close to those of the analogous iron(IV)-oxo complex $[Fe(O)(N_4Py)]^{2+}$ ($\delta = -0.04\ \mathrm{mm\ s^{-1}}$, $\Delta E_Q = +0.93\ \mathrm{mm\ s^{-1}}$, $S = 1$, $D = 22\ \mathrm{cm^{-1}}$) [213]. More examples of iron(IV)–imido compounds are reviewed in [257].

Nonoxo-, Nonnitrido-Iron(IV) Compounds

While most examples of genuine iron(IV) species utilize bridging or terminal oxo or nitrido ligands, high-valent complexes without these groups are quite rare. To our knowledge, the only well-characterized mononuclear Fe(IV)–*heme* complex without an oxo ligand is the phenyl–iron derivative, $[Fe(oetpp)C_6H_5)]SbCl_6$, generated from the phenyl-iron(III) complex $Fe(III)(oetpp)C_6H_5$ by oxidation with an organic radical, $[C_{12}H_8OS]SbCl_6$ (oetpp = 2,3,7,8,12,13,17,18-octaethyl-5,10,15,20-tetraphenylporphyrin) [214]. The compound exhibits an isomer shift of $\delta = 0.13\ \mathrm{mm\ s^{-1}}$ and a quadrupole splitting of $\Delta E_Q = +3.23\ \mathrm{mm\ s^{-1}}$. Applied-field measurements revealed large ZFS ($D \approx 30\ \mathrm{cm^{-1}}$) and axially symmetric magnetic hyperfine parameters that are consistent with iron(IV), $S = 1$, but definitively exclude the possible alternative of iron(III) and an oxidized porphyrin. A similar iron site has been recently observed for the reactive oxidized state of the enzyme MauG which utilizes two covalently bound c-type hemes to catalyze the synthesis of the so-called cofactor tryptophan–tryptophylquinone. It was proposed that the enzyme stores two oxidation equivalents in a highly oxidized pair of a heme Fe(IV)=O group ($\delta = 0.06\ \mathrm{mm\ s^{-1}}$, $\Delta E_Q = 1.70\ \mathrm{mm\ s^{-1}}$) and a heme Fe (IV) nonoxo species ($\delta = 0.17\ \mathrm{mm\ s^{-1}}$, $\Delta E_Q = 2.54\ \mathrm{mm\ s^{-1}}$) [263].

The first well-characterized mononuclear *nonheme* Fe(IV) complex without an oxo or nitrido ligand is $[Fe^{IV}Cl(\eta^4\text{-MAC*})]^-$ (H_4[MAC*] = 1,4,8,11-tetraaza-13,13diethyl-2,2,5,5,7,7,10,10-octamethyl-3,6,9,12,14-pentaoxocyclotetradecane) where iron is coordinated by a plane of four amide-nitrogen anions from a macrocyclic ligand and one axial chloride ligand [100, 264]. Its spin is $S = 2$ with small negative ZFS ($D = -2.6\ \mathrm{cm^{-1}}$) as inferred from EPR and applied-field Mössbauer measurements [100]. For a short review of high-valent iron complexes with the η^4-MAC* ligand, including Fe(IV) species with $S = 1$, see [265].

The iron(IV) state in $[(Me_3\text{cyclam-acetate})FeX]^{2+}$, X = Cl^-, N_3^-, F^-, [215, 216] is stabilized by the five-coordinating macrocyclic ligand $(Me_3\text{-cyclam-acetate})^-$

which is a variant of cyclam-acetate (mentioned earlier) with three additional methyl groups [215, 266, 267]. A full spectroscopic investigation of these compounds (and their related Fe(III) and Fe(II) derivatives) by applied-field Mössbauer studies and DFT calculations showed that the spin state of iron(IV) is $S = 1$ and that the energy of the single-electron valence orbitals increases in the order $d_{xy} < d_{xz} \approx d_{yz} < d_{x^2-y^2} < d_{z^2}$, as expected from ligand field theory [215]; the Mössbauer parameters are collected in Table 8.4. Note the increased isomer shift and reduced quadrupole splitting of the fluoro complex within the series which reflects the fact that the iron(IV)-fluoro group is an "isoelectronic twin" of the iron (IV)-oxo compounds listed above.

The first well characterized "low–low-spin" iron(IV) compound ($S = 0$) is the complex [N$_3$N′]FeCN ($\delta = -0.22$ mm s^{-1}, $\Delta E_Q = 3.28$ mm s^{-1}) [217], where [N$_3$N′]$^{3-}$ represents the trianionic triamido-amine ligand [(*t*-BuMe$_2$SiNCH$_2$CH$_2$N)]$^{3-}$. The three bulky amido substituents form a "pocket" and protect the metal site leaving only the apical position open for functional reactions. The "low–low-spin" state arises from trigonal bipyramidal ligand-field symmetry which provides a "three-over-two" orbital splitting with two energetically low-lying *e* orbitals.

Corroles and Other Noninnocent Ligands

There has been a longstanding debate in the literature on the electronic structure of chloro-iron corroles, especially for those containing the highly electron-withdrawing *meso*-tris(pentafluorophenyl)corrole (TPFC) ligand [268–270]. Two alternative electronic structures were proposed for this and for the related Fe (TDCC)Cl (TDCC = *meso*-tris(2,6-dichlorophenyl)corrole) complex, that is, a high-valent ferryl species ($S = 1$) chelated by a trianionic corrolato ligand (Fe (IV)Cor^{3-}) or an intermediate-spin ferric iron ($S = 3/2$) that is antiferromagnetically coupled to a dianionic π-radical corrole (Fe(III)Cor$^{2-\cdot}$) yielding an overall spin-triplet ground state ($S_t = 1$). A combined experimental (Mössbauer) and computational (DFT) investigation of two series of corrole-based iron complexes finally reached the following conclusion: "The electronic structures of Fe*X*Cor (X = F, Cl, Br, I, Cor = TPFC, TDCC) are best formulated as (Fe(III)Cor$^{2-\cdot}$), similar to chloro-iron corroles containing electron-rich corrole ligands" [271]. Antiferromagnetic exchange between Fe(III) ($S = 3/2$) and the corrole radical ($S' = 1/2$) afforded by a singly occupied Fe$-d_{z^2}$ and corrole a_{2u}-like π-orbitals, leads to the observed spin-triplet ground state, $S_t = 1$. The coupling constants exceed those of analogous porphyrin systems by a factor of 2–3. In the corroles, the combination of lower symmetry, extra negative charge and smaller cavity size (relative to the porphyrins) leads to exceptionally strong iron-corrole σ-bonds. Hence, the Fe$-d_{x^2-y^2}$-based molecular orbital is unavailable in the corrole complexes (contrary to the porphyrin case), and the local spin state is $S_{Fe} = 3/2$ in the corroles versus $S_{Fe} = 5/2$ in the porphyrins. Since these results clearly rule out the presence of iron(IV), corrole complexes cannot be included in the list of high-valent compounds assembled here.

Similarly, the stable densely colored oxidation products of many iron(III) compounds with ligands which are known or may be suspected to be noninnocent, that is, redox-active, such as dithiocarbamate (dtc), phenanthroline (phen), dithiocarbazonate and particularly dithiolate (mnt, bdt, etc.) and their various derivatives, have been assigned to iron(IV) species in the past [186]. However, recent work (still in progress) based on high-resolution structure determination, EPR, Mössbauer and XAS spectroscopy, magnetic susceptibility measurements and DFT calculations unambiguously shows that the electronic structure of such systems is actually much better described by iron(III) and a ligand radical, or even iron(II) and two ligand radicals [74, 92, 272–275]. Therefore, we refrain from reviewing such highly oxidized coordination compounds with noninnocent ligands in this section. Instead, some example from the previous list of falsely assigned "iron(IV) complexes" [186] are mentioned in the paragraph on iron(III) intermediate-spin complexes (see for instance, compounds **9**, **10** [92] in Table 8.1).

8.2.3.4 Iron(V) Compounds

Genuine iron(V) is a very rare oxidation state. In the preparation of iron oxides (and of other solid-state materials), the intended iron(V) disproportionates mostly into an iron(III) fraction and two parts of an iron(VI) fraction [276]. The only example of an iron(V) oxide for which the Mössbauer parameters are known [185] is La_2LiFeO_6. A low isomer shift of $\delta = -0.41$ mm s^{-1} was observed at room temperature with practically zero quadrupole splitting [277], which was taken as a proof that iron is accommodated in octahedral FeO_6 sites surrounded by six Li ions. Although repeatedly cited, it seems that the spectra have never been published, and the data must therefore be considered with care.

Iron-porphyrin complexes of biological as well as synthetic origin do not attain the iron(V) state, because the porphyrin is not innocent and takes one of the two oxidation equivalents with respect to the iron(III) state. Recently, however, an optically detected intermediate generated by laser flash photolysis of a mutant of cytochrome P-450 was tentatively assigned to an iron(V)-oxo species [278]. In contrast, a nonheme iron(V)-oxo unit has been proposed, since long to be the oxidant in the large family of Rieske dioxygenases [169]. However, only a single genuine iron(V)–oxo compound has been synthesized to date with the redox-innocent tetranionic macrocyclic tetraamido (TAML) ligand B* (B* is related to η^4-MAC mentioned above [265]). The ligand provides four exceptionally strong amide-N σ-donor groups, which are capable of stabilizing iron(V) when an iron(III) precursor complex is treated with an oxygen-transfer agent such as a peroxy acid forming $[B^*Fe^V(O)]^-$ [173]. The presence of a $3d^3$ configuration was verified using Mössbauer spectroscopy ($\delta = -0.42$ mm s^{-1} at 4.2 K (-0.46 mm s^{-1} at 140 K), $\Delta E_Q = 4.25$ mm s^{-1}), EPR spectroscopy ($S = 1/2$, $g = [1.99, 1.97, 1.74]$), K-edge XAS and EXAFS spectroscopy. The molecular structure could not be characterized because of the thermal instability of the compound, but spectroscopy-biased DFT calculations provide a convincingly detailed description of the perferryl state of the iron [173].

Resonance Raman studies yielded the first indication of the formation of an iron (V)-nitrido species obtained by the photolysis of a porphyrin-iron(III) azide precursor by the Raman laser [279–281]; the product, however, could never be obtained in an otherwise detectable amount. Further spectroscopic evidence for the formation of an iron(V) species was possible for the Fe(V)≡N group stabilized by the redox-innocent ligands of the cyclam type [282]. Comprehensive studies were performed on the complex $[(\text{cyclam-ac})\text{Fe}^{V}{\equiv}\text{N}]^{+}$ using K-edge XAS and EXAFS measurements, magnetic susceptibility, Mössbauer spectroscopy, and nuclear inelastic scattering in conjunction with DFT calculations [200, 255, 283]. These studies characterize the electronic structure of the compound as a low-spin $(d_{xy})^2(d_{xz}, d_{yz})^1$ system with a formal bond order for Fe≡N of 2.5, leaving significant radical character on the nitrogen atom [255, 257]. The electron configuration yields essentially an E ground state which is subjected to Jahn–Teller effect and the related perturbations of the magnetic properties. The Mössbauer parameters of $[(\text{cyclam-ac})\text{Fe}^{V}{\equiv}\text{N}]^{+}$ ($\delta = -0.02$ mm s^{-1}, $\Delta E_Q = 1.60$ mm s^{-1}, Table 8.5) are consistent with those found earlier for $[(\text{cyclam})\text{Fe}^{V}{\equiv}\text{N}]^{+}$ in CH_3CN solution which presumably provides a sixth ligand to the iron ($\delta = -0.04$ mm s^{-1}, $\Delta E_Q = 1.67$ mm s^{-1}) [282].

8.2.3.5 Iron(VI) Compounds

Basic ligand-field considerations predict the following energetic order of valence d-orbitals for six-coordinate iron–nitrido compounds with approximate C_{4v} symmetry: $d_{xy} < d_{xz} \approx d_{yz} < d_{x^2-y^2} < d_{z^2}$, whereby only the d_{xy} orbital is nonbonding

Table 8.5 Mössbauer parameters of iron(V) and iron(VI) compounds[a]

Compound	CN[b]	Spin[c]	Temp. (K)	δ (mm s^{-1})	ΔE_Q (mm s^{-1})	Ref.
Iron(V)						
$La_2LiFe^{V}O_6$	6	3/2	r.t.	−0.41	0	[277][d]
$[B{*}Fe^{V}(O)]^{-c}$	5	1/2	4.2	−0.42	4.25	[173]
$[(\text{cyclam-ac})\text{Fe}^{V}{\equiv}\text{N}]^{+}$	6	1/2	4.2	−0.02	1.60	[200, 255]
Iron(VI)						
$[(\text{Me}_3\text{cyclam-acetate})\text{Fe}^{VI}{\equiv}\text{N}]^{2+}$	6	0	4.2	−0.29	+1.53	[284]
K_2FeO_4	4	1	78	−0.85	0	[285]
$SrFeO_4$	4	1	78	−0.83	0	[285]
$BaFeO_4$	4	1	78	−0.81	0	[285]

[a]B* = a tetraamido macrocycle, cyclam-acetate = 1,4,8,11-tetramethyl-1,4,8,11-tetraaza-cyclotetradecane-1-acetate
[b]Coordination number
[c]Spin of the iron(IV) ion
[d]Spectra not published

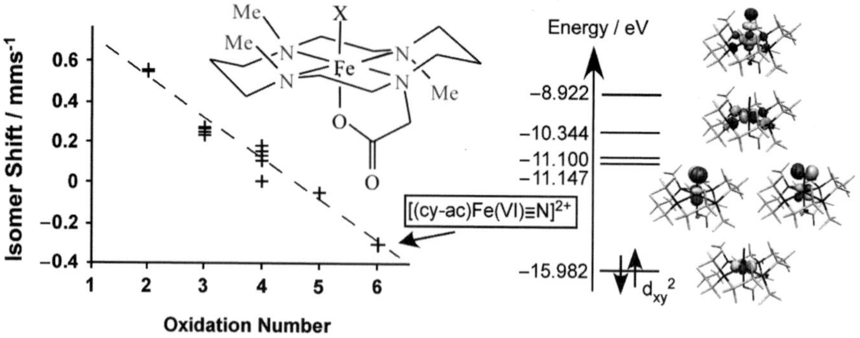

Fig. 8.26 (*Left*) Isomer-shift correlation diagram for a systematic series of [(Me$_3$cyclam-acetate)Fe(X)]$^{n+}$ complexes, where X is an azide (N_3^-) or a nitrido (≡N) group. (*Right*) Orbital scheme of [(Me$_3$cyclam-acetate)Fe^{VI}≡N]$^{2+}$

with respect to all ligands [257]. This suggests relatively high stability for a d^2 or a d^1 configuration. Accordingly, the first and only known molecular iron(VI) compound is a recently discovered Fe(VI)-nitrido species [284]. The bulky, innocent ligand (Me$_3$cyclam-acetate)$^-$ made it possible to synthesize [(Me$_3$cyclam-acetate)Fe^{VI}≡N]$^{2+}$ by low-temperature photolysis of the iron(IV)-azide precursor [(Me$_3$cyclam-acetate)Fe^{IV}N$_3$]$^{2+}$ in the sample cup of the Mössbauer spectrometer. The product is diamagnetic, $S = 0$, as predicted by the LF diagram, and the Fe≡N bond (157 pm obtained from EXAFS) is significantly shorter than that for the corresponding iron(V) compound (161 pm, [255]). The isomer shift is -0.29 mm s^{-1}, which is 0.40 mm s^{-1} lower than that of the corresponding iron(III)-azide starting material and 0.19 mm s^{-1} lower than that of the corresponding iron(V)–nitrido compound which was, together with the XAS preedge energies, the key argument for the assignment of the oxidation state (VI), see Fig. 8.26. The general trends in the electronic structure of such iron-nitrido systems as a function of the number of electrons has been elucidated in a recent DFT study of a series of hypothetical and real complexes [286].

The only other and long known iron(VI) species is the tetrahedral ferrate ion, FeO_4^{2-}, occurring in different salts such as K_2FeO_4, Cs_2FeO_4, or $SrFeO_4$ and $BaFeO_4$ [186]. The salts are rather unstable and represent strong oxidants used in chemistry and technical applications. According to its local tetrahedral symmetry, the iron(VI) is in high spin state, $S = 1$, and the isomer shifts takes unrivaled large negative values ($\delta = -0.80 - 0.85$ mm s^{-1} at 75 K, [287]). A brief review on other basic physical and chemical properties of ferrates are found in [186].

8.2.4 Iron in Low Oxidation States

Low-valent iron in the formal oxidation states (I) and (0) is mostly found in iron-carbonyl and metal–organic compounds which have been intensively studied since

Table 8.6 Isomer shift and quadrupole splitting of iron carbonyls and metal–organic compounds

Compound	Formal valency[a]	Temp. (K)	δ (mm s^{-1})	ΔE_Q (mm s^{-1})	Ref.
$Fe(CO)_5$	0	78	−0.09	2.57	[288]
$Fe_2(CO)_9$	0	78	0.10	0.54	[288]
$Fe_3(CO)_{12}$	0	78	0.04 0.10	0.13 (1×), 1.12 (2×)	[288]
$Na_2^+[Fe(CO)_4]^{2-}$	–II	80	−0.18	0	[17] in Chap. 1
$Et_4N^-[Fe(CO)_4H]^+$	I	80	−0.17	1.36	[17] in Chap. 1
I_2:$Fe(CO)_4$	II	4.2	0.14	0.38	[17] in Chap. 1
$LFe(CO)_4$[b]	0	4.2	−0.09 to +0.07	0.4–2	[17] in Chap. 1
Butadiene-$Fe(CO)_3$	0	4.2	0.03	1.46	[288], [17] in Chap. 1
$(Cp)_2Fe$[c]	II	80	0.53	2.37	[17] in Chap. 1
$(Cp)_2FeBr$	III		0.43	~0.2	[17] in Chap. 1
Substit. Fe-Cp compounds	II		0.51–0.55	2.0–2.4	[17] in Chap. 1

[a]Formal valence state of iron
[b]L is a two-electron donor ligand
[c]Cp = π-cyclopentadienyl

the early years of Mössbauer spectroscopy [17] in Chap. 1, [288]. We refrain from reviewing this field again, but rather present a cursory table of values to set out the possible impact of Mössbauer spectroscopy for the identification of valence states and structural features (Table 8.6).

Metal–organic bonds are short and covalent; therefore, the assignment of oxidation numbers to the central metal is problematic *per se*. The composition of the relevant molecular orbitals frequently approaches 50% (or more) admixture of ligand character into the metal *d*-orbitals. While a reasonable correlation of the isomer shift with the oxidation state could be found for iron(II) to iron(VI), particularly when the comparison is limited to the same spin state and to similar ligand systems, such trends are mostly elusive for low-valent low-spin iron. The major reason is the strong degree of π-backbonding for low-valent metal ions, which changes upon oxidation of the compound and compensates for the change in the number of valence 3*d* electrons. The classical examples for this behavior are $K_4[Fe(CN)_6]\cdot 3H_2O$ and $K_3[Fe(CN)_6]$ which have virtually the same isomer shift [289] and also the same metal–ligand bond lengths. Therefore, simple trends in isomer shifts for low-valent compounds (and particularly, low-spin) should be accepted only with caution.

8.2.4.1 Low-Valent Iron Porphyrins

In some electron-rich systems, the usual trend that the isomer shift increases with the number of valence electrons appears to be inverted. This is the case for the comparison of planar iron porphyrins in the formal oxidation states (I) and (0), which are obtained by one- and two-electron reduction of iron(II) porphyrins in

THF solution [290]. There had been much confusion in the past about the magnetic moment of the iron(I) species and the correct Mössbauer and NMR properties caused by impurities of the samples and possible unknown axial ligation. However, with clean crystallized samples with known molecular structure [291], the first reduction product of iron(II)(tetraphenylporyphrin), $[Fe(TPP)]^-$, can be clearly characterized as a quasiplanar iron(I) complex with spin $S = 1/2$ (EPR: $g_\perp = 2.28$, $g_\| = 1.92$). The isomer shift is very high for a porphyrin, $\delta = 0.65$ mm s^{-1}, and is about 0.13 mm s^{-1} more positive than that for planar Fe(II)(TPP) (Table 8.7), consistent with the high electron density of a low-spin d^7 system.

However, it was pointed out that two other observations are out of line with the iron(I) formulation and more consistent with an iron(II)-porphyrin *radical anion* [290]: (1) the low-intensity red-shifted Soret band in the UV–VIS spectrum with broad maxima in the α,β-region compared to, for instance, Fe(TPP) in THF, is typical of a porphyrin radical, and (2) the bond lengths of the porphyrin core indicate population of the (antibonding) LUMO of the ligand (i.e., the presence of an "extra" electron in the π-system). The presence of porphyrin radical character in the electronic ground state was also inferred from the paramagnetic NMR-shifts of the pyrrole protons at the meso and β-carbon atoms [291].

The second reduction step of Fe(II)(TPP) yields an extremely air-sensitive green product which can be assigned the formula $[Fe(I)(TPP)^{\cdot}]^{2-}$ because of the red shift of the Soret band in the UV–VIS spectrum. The pure material is diamagnetic ($S = 0$) but this does not allow one to distinguish between the three possible descriptions as an iron(0) d^8-porpyhrin, a spin-coupled $S = 1/2$ iron(I)-porphyrin

Table 8.7 Isomer shift and quadrupole splitting of some low-valent iron complexes

Compound	Val.[a]	Spin	Temp. (K)	δ (mm s^{-1})	ΔE_Q (mm s^{-1})	Ref.
$Fe(II)(TPP)(pip)_2$	II	$S = 0$	77	0.51	1.44	[146]
Fe(II)(TPP)	II	$S = 1$	77	0.52	1.51	[146]
$[Fe(TPP)]^-$	I	$S = 1/2$	77	0.65	2.23	[290, 291]
$[Fe(TPP)]^{2-}$	0	$S = 0$	77	0.48	1.29	[290, 291]
LFe(HCCPh)	I	$S = 3/2$	4.2	0.50	2.05[b]	[292]
$LFe^{II}Cl$	II	$S = 2$	4.2	0.74	−1.61	[293]
$LFe^{II}CH_3$	II	$S = 2$	4.2	0.48	+1.74	[293]
$L^{Me}Fe^{II}NNFe^{II}L^{Me}$	II	$S_{Fe} = 2$	4.2	0.62	1.41[b]	[294]
$\{[PhBP^{iPr}{}_3]Fe^I\}_2(\mu\text{-}N_2)$	I	$S_{Fe} = 3/2$	4.2	0.53	0.89	[211]
$[PhBP^{iPr}{}_3]Fe^IPMe_3$	I	$S = 3/2$	4.2	0.57	0.23	[211]
$[PhBP^{iPr}{}_3]Fe^{II}(dbabh)$[c]	II	$S = 2$	4.2	0.55	1.75	[211]

[a]Formal valence state of the iron

[b]The value is derived from a zero-field spectrum recorded at 150 K. ΔE_Q could not be determined at 4.2 K because the compound is in the limit of slow paramagnetic relaxation and the strong unquenched orbital moment forces the internal field into the direction of an "easy axis of magnetization." As a consequence, the quadrupole shift observed in the magnetically split spectra results only from the component of the *EFG* along the internal field and the orientation of the *EFG* is not readily known

[c]dbabh is a bulky N-coordinating amide

anion radical ($S' = 1/2$), or an iron(II) ($S = 1$) site coupled to a porphyrin dianionic radical. The Fe–N bond distances (196.8 pm), which are consistent with an LS state, support the suggestion of an iron(I)-radical formulation but do not readily rule out an iron(II)-diradical moiety.

The Mössbauer spectrum of the dianion was a surprise because the isomer shift ($\delta = 0.48$ mm s^{-1}) was substantially *more negative* than that of the monoanion, and resembles more that of Fe(II)(TPP) than that of $[Fe(I)TPP]^-$. Reed and Scheidt therefore suggested a resonance hybrid of iron(I) and iron(II) for the electronic structure of the dianion molecule instead of the formulation $[Fe(0)(TPP)]^{2-}$ [290, 291].

8.2.4.2 Three- and Four-Coordinate Low-Valent Iron Compounds

A quest for low-valent iron(I)-coordination compounds started in the early 2000s after it was demonstrated that the so-called Chatt-type nitrogen reduction cycle [295] can be mediated by a mononuclear molybdenum compound with low coordination number ([296] and references therein). Although the Chatt-cycle has been originally proposed for the Mo cofactor of the biological nitrogen fixation system, metal ions other than Mo were thought to play a role in binding nitrogen in nitrogenase and iron was certainly an interesting possibility. The nitrogenase enzyme system [297–299] needs to be loaded with electrons before substrate binding, and the Chatt-cycle for an iron analog implicates iron-oxidation states from (I) to (IV). The feasibility of a corresponding synthetic model compound is under intense exploration. Within this field of research, a number of low-coordinated iron(I) compounds have been crystallographically characterized but only a few Mössbauer spectra have been recorded.

The first Mössbauer spectra of an iron(I) compound were reported for LFe (HCCPh), L = $[HC(C[^t Bu]N\text{-}[2,6\text{-diisopropylphenyl}])_2]^-$ [292]. The iron is coordinated to the two nitrogen atoms of the β-diketiminate ligand L and to the C≡C bond of the phenylacetylene ligand via one of the π orbitals which renders the compound essentially three-coordinate. Various spectroscopic measurements, including EPR and applied-field Mössbauer spectroscopy, in conjunction with DFT calculations, showed clearly that the electronic structure of LFe(HCCPh) is best described by a $3d^7$ high-spin configuration with spin $S = 3/2$ (Table 8.7). Within the LF picture for the three-coordinate compound, the spin–orbit interaction in the space of the energetically quasi degenerate orbitals d_{z^2} and d_{yz} yields a consistent explanation for the remarkable magnetic properties, that is, the uniaxial hyperfine field of 68.8 T at the Mössbauer nucleus, the huge ZFS, $D \approx -100$ cm^{-1}, and the highly anisotropic effective g values of about nine and zero. For one direction, the orbital momentum of iron is almost unquenched. The electronic properties of LFe(HCCPh) can be compared with those of the related iron(II) compounds $LFe^{II}Cl$ and $LFe^{II}CH_3$ [293]. Note that the isomer shifts of the divalent iron compounds are more positive than those of the monovalent iron.

An interesting reaction product of the β-diketiminate iron complex is the N_2-bridged diiron complex $L^{Me}FeNNFeL^{Me}$ [294] which also contains a three-coordinate iron site with the formal oxidation number (I). The ground state with spin $S_t = 3$ exhibits a large orbital contribution to the magnetic moment and a uniaxial positive hyperfine field of +68.1 T. A comprehensive study [294], however, led to the conclusion that the compound is best described as an iron(II) dimer ($S_{Fe} = 2$) with a bridging N_2^{2-} group which, in analogy with the isoelectronic O_2 molecule, is in a spin-triplet state ($S_{N2} = 1$). With this interpretation, the total spin state $S_t = 3$ can be readily explained by antiferromagnetic coupling of the spin $S_{N2} = 1$ and the spins $S_{Fe,1} = 2$ and $S_{Fe,2} = 2$ of the two terminal iron(II) sites without invoking an exotic ferromagnetic coupling scheme between two high-spin iron(I) sites.

Weak ferromagnetic coupling ($J \approx 4\ \mathrm{cm}^{-1}$) has been found for the related dinitrogen complex $\{[PhBP^{iPr}{}_3]Fe^I\}_2(\mu\text{-}N_2)$ which appears to be more likely an iron(I) complex [211]. Both iron sites are four-coordinate by three phosphorous atoms from the ligand $[PhBP^{iPr}{}_3]$ [300] and one nitrogen atom from the bridging N_2. The Mössbauer parameters are summarized in Table 8.7, together with those of the monomeric monovalent "reference" complex $[PhBP^{iPr}{}_3]Fe^IPMe_3$.

8.2.4.3 Low-Spin Iron in the Active Site of Hydrogenases

Low-valent iron is an essential constituent of the active center of hydrogenases, the biological catalysts for the production of molecular hydrogen. Huge amounts of H_2 are produced in the anoxic biological habitats of the planet by microbial fermentation. Anaerobic microorganisms use most of it to reduce CO_2 to methane or sulfate to hydrogen sulfide [301]. These reactions and the sheer numbers of turnover have mesmerized scientists for a long time and have stimulated very active research in biology and bioinorganic chemistry. The literature in this rapidly developing field has been reviewed in many major articles ([301–304] and references therein). Here, we summarize the essential Mössbauer parameters of the different types of active clusters and their model compounds that have been synthesized to explore either the structural and spectroscopic or the functional features of the natural systems. Three types of hydrogenases are known, and these are labeled according to the composition of their active sites, [FeFe]-, [FeNi]-, and [Fe]-hydrogenases.

[FeFe]-Hydrogenases

The [FeFe]-hydrogenases have been known since the 1930s [305], and Warburg recognized sulfur-coordinate iron as an essential element of the enzyme [306]. The [FeFe]-hydrogenase from *Clostridium pasteurianum* accommodates 20 iron atoms organized in one [2Fe–2S] cluster, three [4Fe–4S] clusters and the so-called H-cluster which is a [4Fe–4S] cluster covalently linked to a dinuclear [FeFe] compound [307]. The iron atoms in the dinuclear center are bridged by CO and two sulfur atoms from a dithiol group. Only one of the two iron atoms is additionally bridged by a single cysteine sulfur to the attached [4Fe–4S] cluster, forming together the H-cluster. The other terminal ligands of the [FeFe] dimer are cyanide and CO [308] originally discovered in the vibration spectra of the protein [309].

Elaborate Mössbauer studies on enzyme samples obtained from two different microorganisms (which was impeded not only by the large number of iron sites but also by a large number of possible redox states and sample treatments) provided the electronic structure of the cyanide- and CO-ligated [FeFe] center and its interaction with the attached iron–sulfur cluster [310, 311]. The Mössbauer spectrum of the dinuclear center was recognized for the first time in a sample with reduced H clusters (H_{red}) showing a quadrupole doublet indicating low-spin iron (II) ($\delta = 0.08$ mm s^{-1}, $\Delta E_Q = 0.87$ mm s^{-1}) [310]. Later, different states of the [FeFe] center could be identified from the spectra of hydrogenase from a different organism, in three oxidized, one reduced, and a CO-treated state [311]. The iron sites are persistently low-spin and respond to redox titration only with minor shifts of δ and ΔE_Q, but they take part in the paramagnetic behavior of the H-cluster and show appreciable hyperfine coupling at redox potentials of -110 to -350 mV (SHE) and in the CO-reacted sample [311]. Isomer shifts and quadrupole splittings of the [FeFe] center are listed in Table 8.8, together with those of some model compounds.

[NiFe]-Hydrogenases

[NiFe]-hydrogenases are intensely studied enzymes [303, 304]. The first evidence of CO and CN^- groups as integral parts of the active center of a biological system was obtained from infrared spectroscopy for this class of hydrogenases [312]. Later, the first X-ray structure [313] revealed that these diatomic groups are coordinated to the iron atom of the [NiFe] center. Early Mössbauer investigations had revealed a spectral component with an unusually low isomer shift, $\delta = 0.05$–0.15 mm s^{-1} [314] which only later, in conjunction with the first model compound having a P–S_3–CO–CN coordination environment [315], has been safely assigned to the corresponding low-spin iron(II) center; the data are given in Table 8.8. Further conclusive Mössbauer work on the [NiFe] center is elusive, mostly because of the difficulties to identify and subtract the dominating and overlapping subspectra from FeS clusters and the tantalizing "plethora of so-called 'states' of the enzyme" [304] (at least 11!), some of which might be just artifacts of protein isolation and processing.

[Fe]-Hydrogenase

The third type of hydrogenases from some methanogenic archea, also referred to as iron–sulfur-cluster-free hydrogenase, or *Hmd,* has attracted interest in bioinorganic

Table 8.8 Isomer shift and quadrupole splitting of low-spin iron in the active site of hydrogenases and biomimetic model compounds

Compound	Temp. (K)	δ (mm s^{-1})	ΔE_Q (mm s^{-1})	Ref.
[FeFe]-hydrogenase and models				
H_{Red}, *Clostridium pasteurianum*	4.2	0.08	0.87	[310]
H_{Red}, *Desulfovibrio vulgaris*	4.2	0.13	0.85	[311]
H_{OX+1} [a]	4.2	0.16	1.09	[311]
$H_{OX-2.06}$ [a]	4.2	0.17	1.08	[311]
$H_{OX-2.10}$ [a]	4.2	0.13, 0.14	0.85, 0.67	[311]
H_{OX}-CO [a]	4.2	0.17, 0.13	0.70, 0.65	[311]
$[Fe^{II}(PS_3)(CO)(CN)]$	4.2	0.15	0.91	[315]
$[\{Fe_2S_3L\}(CN)(CO)_4]$ [b]	80	0.05, 0.11	0.92, 0.44	[316]
$[\{Fe_2S_3L\}(CO)_5]$ [c]	80	0.03, 0.12	0.94, 0.25	[316]
[NiFe]-hydrogenase and models				
State D, *Chromatium vinosum*	4.2	0.05–0.15	–	[314]
$[Ni(dsdm)(Fe^0(CO)_3)_2]$ [e]	80	0.07, 0.03	0.66, 0.57	[317]
$[Fe^0(CO)_3SMe]_2$	80	0.03	1.00/0.88 [d]	[317]
[Fe]-hydrogenase and models				
[Fe], native	80	0.06	+0.65	[318]
[Fe], cofactor	80	0.03	0.43	[318]
[Fe] + H2	80	0.06	0.65	[318]
[Fe] + CO	80	−0.03	−1.38	[318]
[Fe] + KCN	80	−0.001	−1.73	[318]
[Fe(*cis*-CO)$_2$(N-R)$_2$]	80	0.10	0.79	[319]
[Fe(*cis*-CO)$_2$(S-R)$_2$]	80	0.07	0.51	[320]
Fe(CO)$_2$(PPh$_3$)IL	50	0.10	0.48	[321]

[a] H clusters are termed according to one-electron oxidation and the highest EPR g value, or CO treatment, see [311]
[b] Compound **5a**
[c] Compound **4a** of [316]
[d] *Syn* and *anti*-configuration, respectively
[e] More examples of compounds of the type $[X(Fe(CO)_3)_2]$ are found in [317]

chemistry because it does not contain additional obscuring nickel or iron–sulfur clusters in its active site [301]. The abbreviation Hmd is related to the catalyzed reaction, namely, the reversible reduction of methenyltetrahydromethanopterin with H_2 to methylenetetrahydromethanopterin and a proton [322]. The mononuclear [Fe]-center has a square-pyramidal coordination polyhedron provided by the N-atom of an unusual guanylyl-pyridone cofactor in the apical position and a cysteine-sulfur atom, two CO molecules and an unknown ligand in the basal plane [323, 324]. The presence and geometric arrangement of the CO groups was discovered first by infrared spectroscopy which also revealed the reaction of the center with CO and the reversible formation of a presumably six-coordinate derivative with three CO ligands and alternatively, the bonding of three CN^- ligands [325]. These free forms, as well as a H_2-bound form and the quasiisolated iron center in a protein-free environment with only the cofactor and sulfur substituents, could be distinguished by Mössbauer spectroscopy (Table 8.8). The [Fe] center is clearly low-spin as can be inferred from the low isomer shifts, $\delta < 0.1$ mm s^{-1}, and it is diamagnetic as seen from the lack of EPR signals and from applied-field Mössbauer spectra [318]. As the total charge of the iron center is not known, the oxidation state could not be derived from the Mössbauer spectra since two oxidation states, Fe(II) and Fe(0), both show $S = 0$, and the isomer shifts are not conclusive; only Fe(I) is ruled out because it is paramagnetic. However, an inspection of the IR frequencies for a large number of complexes with a {Fe(*cis*-$CO)_2$} motif strongly supports an Fe(II) LS state [319].

8.3 Mobile Mössbauer Spectroscopy with MIMOS in Space and on Earth

Göstar Klingelhöfer* and Iris Fleischer*

8.3.1 Introduction

The NASA Mars Exploration Rovers (MER) (Fig. 8.27), Spirit and Opportunity, landed on the Red Planet in January 2004. The goal of the mission was to search for signatures of water possibly being present in the past. Both NASA Mars Exploration Rovers [326] carry the miniaturized Mössbauer spectrometer (MIMOS II) [327] as part of their scientific payload. The scientific basis for landing a Mössbauer spectrometer on Mars is extensively discussed by Knudsen et al. [328, 329]. The scientific objectives of the MIMOS II Mössbauer spectrometer investigation on Mars are (1) to identify its mineralogical composition and (2) to measure the relative abundance of iron-bearing phases (e.g., silicates, oxides, carbonates, phyllosilicates, hydroxyoxides, phosphates, sulfides, and sulfates), (3) to distinguish between magnetically ordered and paramagnetic phases and provide, from measurements at different temperatures, information on the size distribution of magnetic particles, and (4) to measure the distribution of Fe among its oxidation states (e.g., Fe^{2+}, Fe^{3+}, and Fe^{6+}). These data characterize the present state of Martian surface materials and provide constraints on climate history and weathering processes by which the surface evolved to its present state.

The MER mission was originally planned to last for three months on the Martian surface. At the time of writing (July 2010), both rovers and their instruments have spent more than 6 years exploring their landing sites and are still operational.

*Institut für Anorganische Chemie und Analytische Chemie, Johannes Gutenberg-Universität Mainz, Staudingerweg 9, 55099 Mainz, Germany; e-mail: klingel@uni-mainz.de.

Fig. 8.27 NASA Mars-Exploration-Rover artist view (courtesy NASA, JPL, Cornell). On the front side of the Rover, the robotic arm (IDD) carrying the Mössbauer spectrometer and other instruments can be seen

The mission success criteria for the rovers had been to drive more than 600 m, and the goal for the Mössbauer spectrometers had been to collect spectra from at least three different soil and rock samples at each landing site. To date, the rovers have covered distances of more than 7 km (Spirit) and more than 20 km (Opportunity), respectively. The total amount of scientific targets investigated by the Mössbauer spectrometers exceeds 300, the total number of integrations exceeds 600. A total of 14 unique Fe-bearing phases were identified up to now: Fe^{2+} in olivine, pyroxene, and ilmenite; Fe^{2+} and Fe^{3+} in magnetite and chromite; Fe^{3+} in nanophase ferric oxide (npOx), hematite, goethite, jarosite, an unassigned Fe^{3+} sulfate, and an unassigned Fe^{3+} phase associated with jarosite. Fe^{0} was identified in kamacite, an Fe–Ni alloy, and schreibersite ($(Fe,Ni)_3P$) present in Fe-meteorites. Both Mössbauer spectrometers remain operational and continue to return valuable scientific data [330–335].

Besides the extraterrestrial application of MIMOS, there are a number of terrestrial applications, such as the investigation of rock paintings, ancient artifacts, and environmental science, where the instrument has been applied successfully.

8.3.2 The Instrument MIMOS II

The instrument MIMOS II is extremely miniaturized compared to standard laboratory Mössbauer spectrometers and is optimized for low power consumption and high detection efficiency (see Sect. 3.3) and [326, 327, 336–339]. All components were selected to withstand high acceleration forces and shocks, temperature variations over the Martian diurnal cycle, and cosmic ray irradiation. Mössbauer measurements can be done during day and night covering the whole diurnal temperature variation between about −100°C (min) and about +10°C (max) of a Martian day [340–343].

To minimize experiment time a very strong $^{57}Co/Rh$ source was used, with an initial source strength of about 350 mCi at launch. Instrument internal calibration is accomplished by a second, less intense radioactive source mounted on the end of the velocity transducer opposite to the main source and in transmission measurement geometry with a reference sample. For further details, see the technical description in Sect. 3.3.

The MIMOS II Mössbauer spectrometer sensor head (see Sect. 3.3) is located at the end of the *I*nstrument *D*eployment *D*evice *IDD* (see Fig. 8.27) On Mars-Express Beagle-2, an *E*uropean *S*pace *A*gency (ESA) mission in 2003, the sensor head was also mounted on a robotic arm integrated to the *P*osition *A*djustable *W*orkbench (*PAW*) instrument assembly [344, 345]. The sensor head shown in Figs. 8.28 and 8.29 carries the electromechanical transducer with the main and reference $^{57}Co/Rh$ sources and detectors, a contact plate, and sensor. The contact plate and sensor are used in conjunction with the IDD to apply a small preload when it places the sensor head, holding it firmly against the target.

Because of the complexity of sample preparation, backscatter measurement geometry is the choice for an in situ planetary Mössbauer instrument [327]. No sample preparation is required, because the instrument is simply presented to the sample for analysis. Both 14.41 keV γ-rays and 6.4 keV Fe X-rays are detected simultaneously.

MIMOS II has three temperature sensors, one on the electronics board and two on the sensor head. One temperature sensor in the sensor head is mounted near the internal reference absorber, and the measured temperature is associated with the reference absorber and the internal volume of the sensor head. The other sensor is mounted outside the sensor head at the contact ring assembly. It gives the analysis temperature for the sample on the Martian surface. This temperature is used to route

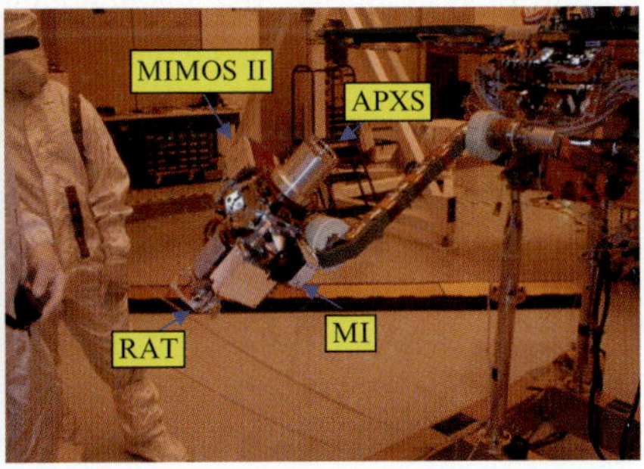

Fig. 8.28 External view of the MIMOS II sensor head without contact plate assembly (*left*); MIMOS II sensor head mounted on the robotic arm (IDD) of the Mars Exploration Rover. The IDD also carries the α-Particle-X-ray Spectrometer APXS, also from Mainz, Germany, for elemental analysis, the Microscope Imager MI for high resolution microscopic pictures (∼30 μm per pixel), and the RAT for sample preparation (brushing; grinding; drilling (<1 cm depth)). Picture taken at Kennedy-Space-Center KSC, Florida, USA

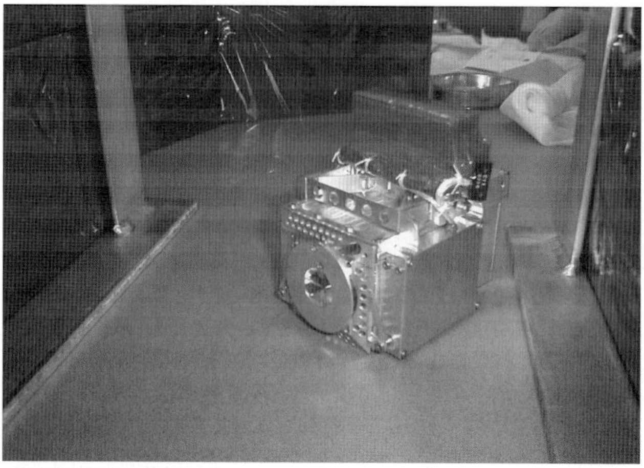

Fig. 8.29 The flight unit of the MIMOS II Mössbauer spectrometer sensor head (for the rover *Opportunity*), with the circular contact plate assembly (*front* side). The circular opening in the contact plate has a diameter of 15 mm, defining the field of view of the instrument

the Mössbauer data to the different temperature intervals (maximum of 13, with the temperature width software selectable) assigned in memory areas (for more details, see Sect. 8.3.3).

Sampling depth. In addition to no requirement for sample preparation, backscatter measurement geometry has another important advantage. Emission of internal conversion electrons, Auger electrons, and X-rays, which occur along with the recoilless emission and absorption of the 14.4 keV γ-ray of ^{57}Fe, can also be used for Mössbauer measurements. For ^{57}Fe, X-rays resulting from internal conversion have an energy of 6.4 keV. Because the penetration depth of radiation is inversely proportional to energy, the average depth from which 14.4 keV γ-rays emerge in emission measurements is greater than that for 6.4 keV X-rays. The importance of this difference in emission depths for an in situ Mössbauer spectrometer is that mineralogical variations that occur over the scale depths of the 14.4 and 6.4 keV radiations can be detected and characterized. Such a situation on Mars might arise for thin alteration rinds and dust coatings on the surfaces of otherwise unaltered rocks [346, 347].

Cosine smearing. Because instrument volume and experiment time must both be minimized for a planetary Mössbauer spectrometer, it is desirable in backscatter geometry to illuminate as much of the sample as possible with source radiation. However, this requirement at some point compromises the quality of the Mössbauer spectrum because of an effect known as "cosine smearing" [327, 348, 349] (see also Sects. 3.1.8 and 3.3). The effect on the Mössbauer spectrum is to increase the linewidth of Mössbauer peaks (which lowers the resolution) and shift their centers outward (affects the values of Mössbauer parameters). Therefore, the diameter of the source "γ-ray beam" incident on the sample, which is determined by a collimator, is a compromise between acceptable experiment time and acceptable velocity resolution. The distortion in peak shape resulting from cosine smearing can be accounted for mathematically in spectral fitting routines [327, 339, 348–350].

8.3.3 Examples

8.3.3.1 Mars-Exploration-Rover Mission

A MIMOS II instrument was mounted on the robotic arm of each of two identical rovers, called *Spirit* and *Opportunity,* which were launched separately in June 2003 (from Kennedy Space Center, Florida) and landed successfully on Mars in January 2004, *Spirit* in Gusev Crater and *Opportunity* in Meridiani Planum (opposite side of Mars). The primary objective of the Mars-Exploration-Rover (MER) science investigation is to explore with the *Athena* instrument payload two sites on the Martian surface where water may once have been present, and to assess past environmental conditions at those sites and their suitability for life [326, 327, 351, 352]. The rovers are ~1.5 m long, ~1.5 m high, and weigh ca. 185 kg each. The solar panels and a lithium-ion battery system provide a power of up to 900 W-h per Martian day. The rovers have been designed to cover a total distance during the mission of up to about 1 km taking photographs, performing remote sensing with their optical instruments, and recording Mössbauer and X-ray fluorescence (XRF) spectra of rock and soil on their way. Typical measuring times for Mössbauer spectra have been 2–4 h at mission beginning, with a source intensity of ~150 mCi at landing. Temperatures change between day and night from as high as about +10°C to as low as about −100°C, depending on the landing site and the Martian season. The robotic arm (see Fig. 8.30) carries the Mössbauer spectrometer MIMOS II, an α-Particle-X-ray-Spectrometer (APXS) for elemental analysis, developed at the Max Planck-Institute for Chemistry in Mainz and the University Mainz [353, 354], a *M*icroscopic *I*mager MI, and a *R*ock *A*brasion *T*ool RAT for polishing rocks and drilling holes of up to ~10 mm into rocks, with a diameter of 4.5 cm (see e.g., Fig. 8.31).

Gusev Crater. The Gusev Crater landing site, a flat-floored crater with a diameter of 160 km and of Noachian age, is located at about 14.5°S of the equator. Gusev was hypothesized to be the site of a former lake, filled by Ma'adim Vallis, one of the largest valley networks on Mars. *The first Mössbauer spectrum* ever recorded on the Martian surface was obtained on soil at Spirit's landing site on the plains in Gusev crater (see Fig. 8.32). It shows a basaltic signature dominated by the minerals olivine and pyroxene. This type of soil and dust were found to be globally distributed on Mars. Spectra obtained on soil at Opportunity's landing site in Meridiani Planum are almost identical to those recorded in Gusev crater.

The collection of spectra obtained at Spirit's landing site reveals various mineralogical signs of weathering. Spectra obtained on the basaltic rocks and soil on the plains show mainly olivine and pyroxene and small amounts of nonstoichiometric magnetite, and only comparably small amounts of weathering. An example

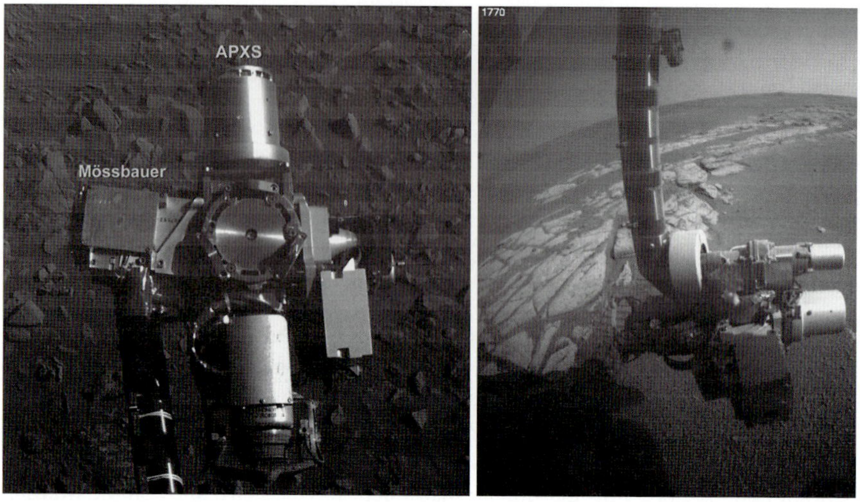

Fig. 8.30 The Instrument Deployment Device (IDD) above the surface of Mars, showing all the four in situ instruments: (*left*) the MIMOS II with its contact ring can be seen in the front; picture taken at Meridiani Planum, Mars; (*right*) MIMOS II is located on the *left* side; picture taken at Gusev Crater, Mars

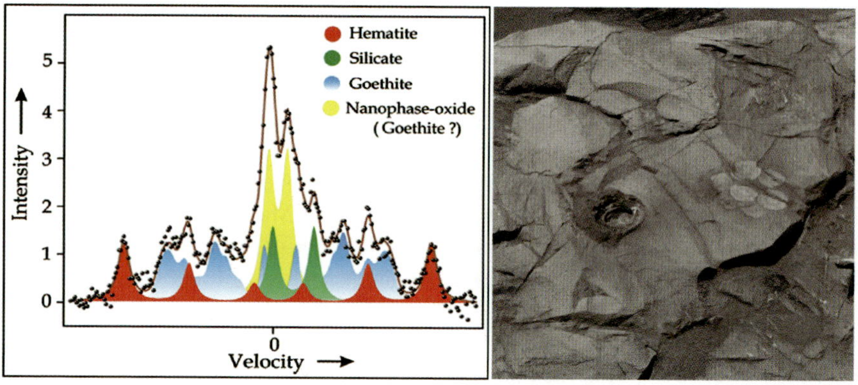

Fig. 8.31 (*Left*) In the Mössbauer spectrum taken in the Columbia Hills at a rock called *Clovis* the mineral goethite (GT) (α-FeOOH) could be identified. GT is a clear mineralogical evidence for aqueous processes on Mars. (*Right*) The rock *Clovis* is made out of rather soft material as indicated by the electric drill-current when drilling the $\sim$1 cm deep hole seen in the picture. Drill fines are of brownish color. The pattern to the right of the drill hole was made by brushing the dust off the surface by using the RAT

of such a rock named *Adirondack*, found close to Spirit's landing site, is shown in Fig. 8.33, together with its Mössbauer spectrum. The Fe mineralogy of this $\sim$50 cm wide rock is composed of the dominating mineral olivine, pyroxene, nanophase

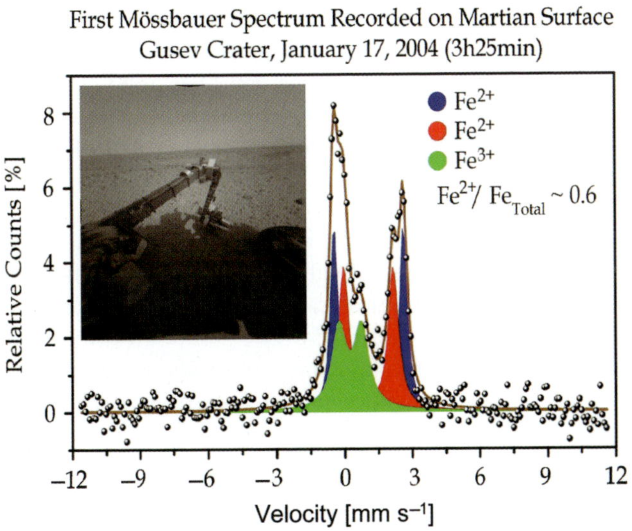

Fig. 8.32 The first Mössbauer Spectrum ever taken outside Earth, at Gusev Crater, Mars. It shows the basaltic composition of the plains at the landing site of the Rover *Spirit*

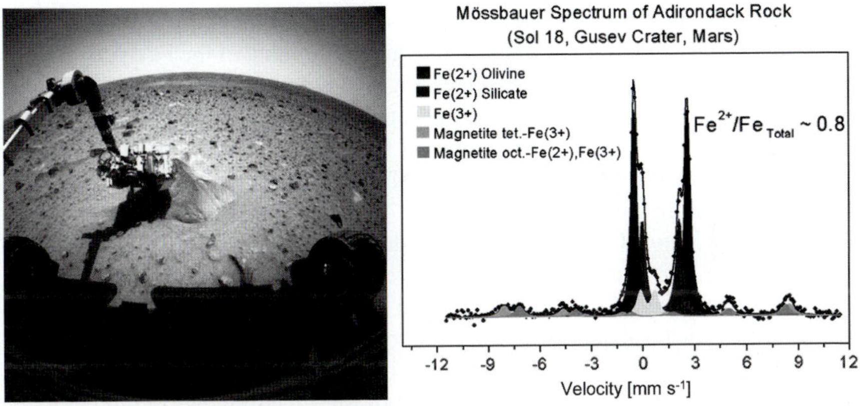

Fig. 8.33 *Left*: robotic arm with MIMOS II positioned on the rock *Adirondack*, as seen by the navigation camera of the rover; *Right*: Mössbauer Spectrum (14.4 keV; temperature range 220–280 K) of the rock Adirondack at Spirit landing side Gusev Crater, plains. The data were taken at the as-is dusty surface (not yet brushed). The spectrum shows an olivine-basalt composition, typical for soil and rocks in Gusev plains, consisting of the minerals olivine, pyroxene, an Fe^{3+} doublet, and nonstoichiometric magnetite

ferric oxide (npOx), and a small contribution of nonstoichiometric magnetite. The composition does not change within a depth of ~8 mm (drill hole obtained with the Rock Abrasion Tool (RAT)). On a similar rock, also in the plains, named *Mazatzal*, a dark surface layer was detected with the Microscopic Imager after the first of two

RAT grinding operations. This surface layer was removed except for a remnant in a second grind. Spectra – both 14.4 keV and 6.4 keV – were obtained on the undisturbed surface, on the brushed surface and after grinding. The sequence of spectra shows that nanophase Oxide (npOx) is enriched in the surface layer, while olivine is depleted. This is also apparent from a comparison of 14.4 keV spectra and 6.4 keV spectra [332, 346, 347]. The thickness of this surface layer was determined by Monte-Carlo (MC)-Simulation to about 10 μm. Our Monte Carlo simulation program [346, 347] takes into account all kinds of absorption processes in the sample as well as secondary effects of radiation scattering. For the MC-simulation, a simple model of the mineralogical sample composition was used, based on normative calculations by McSween [355].

The spectra of the rocks in the plains are very similar to the spectra obtained on the soil (see above). The ubiquitous presence of olivine in soil suggests that physical rather than chemical weathering processes currently dominate at Gusev crater.

On the contrary, thoroughly weathered rocks were encountered in the Columbia Hills, about 2.5 km away from the Spirit landing site. The Mössbauer signature is characteristic of highly altered rocks. Spectra obtained on these samples show larger amounts of nanophase ferric oxides and hematite. In spectra obtained on ~20 rock samples, the mineral goethite (α-FeOOH) was identified (see Fig. 8.31), a clear indicator for aqueous weathering processes in the Columbia hills in the past [332, 356–358]. In particular, one of the rocks, named Clovis (Fig. 8.31), which was found in the West Spur region, contains the highest amount of the Fe oxyhydroxide goethite (GT) of about 40% in area (see Fig. 8.31) found so far in the Columbia Hills. A detailed analysis of these data indicates a particle size of ~10 nm. The rock Clovis also contains a significant amount of hematite. The behavior of hematite is complex because the temperature of the Morin transition (~260 K) lies within diurnal temperature variations on Mars.

Meridiani Planum. Opportunity touched down on 24 January 2004 in the eastern portion of the Meridiani Planum landing ellipse in an impact crater 20 m in diameter named Eagle Crater. The Meridiani Planum landing site is the top stratum of a layered sequene about 600 m thick that overlies the Noachian cratered terrain. Orbital data indicated the presence of significant amounts of the mineral hematite, an indicator for water activities. The surface of Meridiani Planum explored by the Opportunity rover can be described as a flat plain of sulfur-rich outcrop that is mostly covered by thin superficial deposits of aeolian basaltic sand and dust, and lag deposits of hard Fe-rich spherules (and fragments thereof) that weathered from the outcrop and small unidentified rock fragments and cobbles, and meteorites. Surface expressions of the outcrop occur at impact craters (e.g., Eagle (Figs. 8.34 and 8.35), Fram, Endurance, Erebus, Victoria (see Fig. 8.36) craters) and occasionally in troughs between ripple crests [330, 334, 335, 358].

Mössbauer spectra measured by the Opportunity rover at the Meridiani Planum landing site (see Fig. 8.35) revealed four mineralogical components in Meridiani Planum at Eagle crater: jarosite- and hematite-rich outcrop (see Fig. 8.34), hematite-rich soil, olivine-bearing basaltic soil, and a variety of rock fragments such as

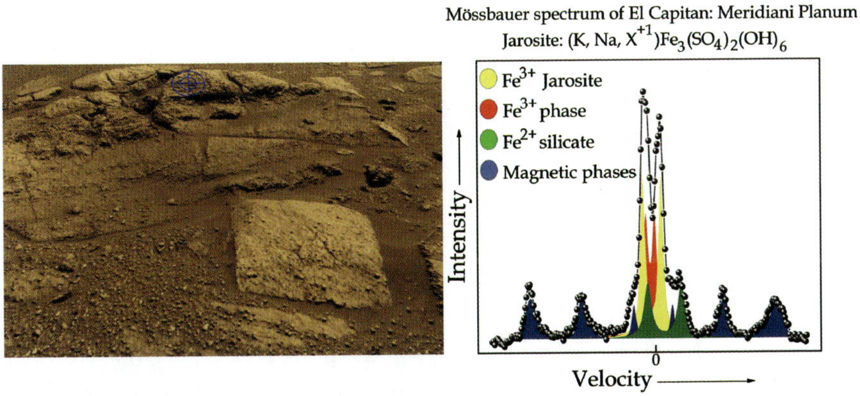

Fig. 8.34 (*Left*): outcrop rocks found at the crater wall of Eagle Crater, where the rover Opportunity landed on 24 January 2004. Clearly, the sedimentary structure is seen. (*Right*): in the spectrum, taken on sol 33 (sol = Martian day) of the mission, the mineral Jarosite, an Fe^{3+}-sulfate, could be identified at the Meridiani Planum landing site. It forms only under aqueous conditions at low pH ($<\sim$3–4) and is therefore clear mineralogical evidence for aqueous processes on Mars

meteorites and impact breccias. Spherules (Fig. 8.37), interpreted to be concretions, are hematite-rich and dispersed throughout the outcrop.

The same Fe-sulfate jarosite containing material was found everywhere along the several (more than 20) kilometer-long driveway of Opportunity to the south, in particular, at craters Eagle, Fram, Endurance, and Victoria, suggesting that the whole area is covered with this sedimentary jarositic, hematite and Fe-silicate (olivine; pyroxene) containing material. The mineral jarosite ((K,Na)$Fe_3(SO_4)_2$ $(OH)_6$) contains hydroxyl and is thus direct mineralogical evidence for aqueous, acid-sulfate alteration processes on early Mars. Because jarosite is a hydroxide sulfate mineral, its presence at Meridiani Planum is mineralogical evidence for aqueous processes on Mars, probably under acid-sulfate conditions as it forms only at pH-values below $\sim$3–4.

Hematite in the soil is concentrated in spherules and their fragments, which are abundant on nearly all soil surfaces. Several trenches excavated using the rover wheels showed that the subsurface is dominated by basaltic sand, with a much lower abundance of spherules than at the surface. Olivine-bearing basaltic soil is present throughout the region. At several locations along the rover's traverse, sulfate-rich bedrock outcrops are covered by no more than a meter or so of soil.

Meteorites on Mars. Meridiani Planum is the first *Iron meteorite* discovered on the surface of another planet, at the landing site of the Mars Exploration rover Opportunity [359]. Its maximum dimension is $\sim$30 cm (Fig. 8.38). Meteorites on the surface of solar system bodies can provide natural experiments for monitoring weathering processes. On Mars, aqueous alteration processes and physical alteration by Aeolian abrasion, for example, may have shaped the surface of the meteorite, which therefore has been investigated intensively by the MER instruments. Observations at mid-infrared wavelengths with the Mini-TES

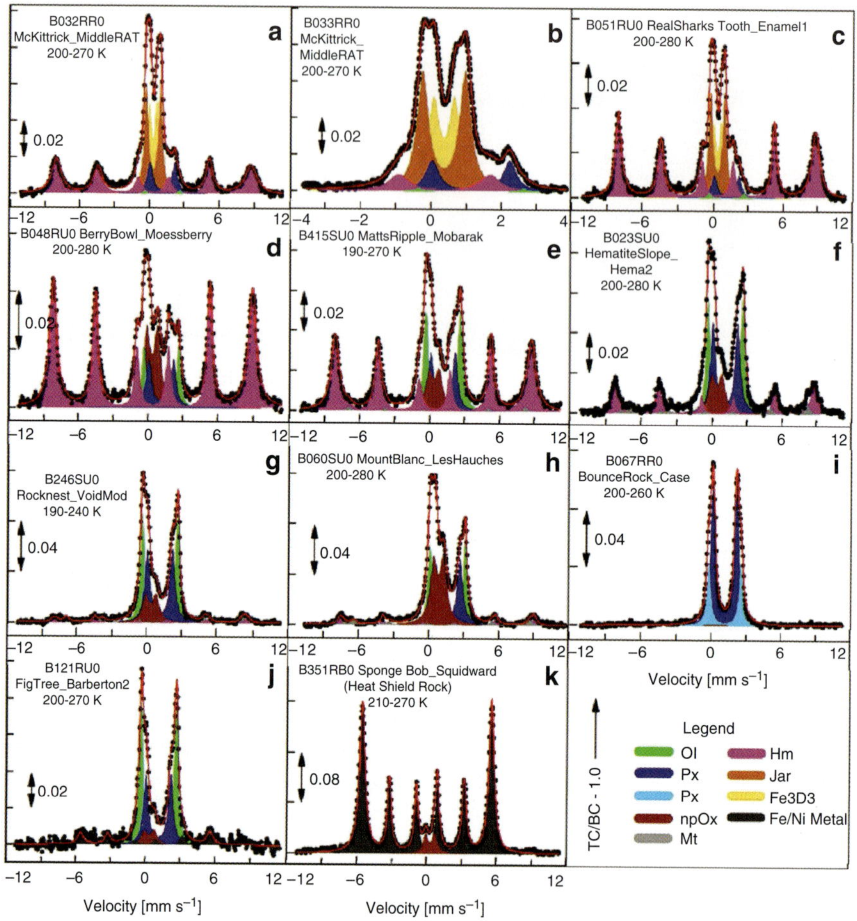

Fig. 8.35 Overview on Meridiani Planum Mössbauer mineralogy. The large variability in mineral composition at this landing site can be seen. Shown are representative Mössbauer spectra. Spectra are the sum over all temperature windows within the indicated temperature ranges. The computed fit and component subspectra (Lorentzian lineshapes) from least-squares analyses are shown by the solid line and the solid shapes, respectively. Full (**a**) and reduced (**b**) velocity Mössbauer spectra for interior Burns outcrop exposed by RAT grinding show that hematite, jarosite, and Fe3D3 (acronym for unidentified Fe^{3+} phase; see [331–334, 341, 346, 352] are the major Fe-bearing phases. Mössbauer spectrum for a rind or crust (**c**) of outcrop material that has an increased contribution of hematite relative to jarosite plus Fe3D3. Mössbauer spectra of two soils (**d** and **e**) have high concentrations of hematite. The spectrum (**e**) is typical for hematite lag deposits at ripple crests. The soil spectra (**f**)–(**h**) [334, 341] are basaltic in nature and have olivine, pyroxene, and *nanophase ferric oxide* as major Fe-bearing phases. The soil target named *MountBlanc_LesHauches* (**h**) is considered to be enriched in martian dust. Mössbauer spectra in (**i**–**k**) are for three single-occurrence rocks: BounceRock (**i**) is monomineralic pyroxene. Barberton (**j**) contains kamacite (iron–nickel metal), and Heat Shield Rock is nearly monomineralic kamacite identified as an Fe–Ni meteorite (see below). TC = total counts and BC = baseline counts (From Morris et al. [334])

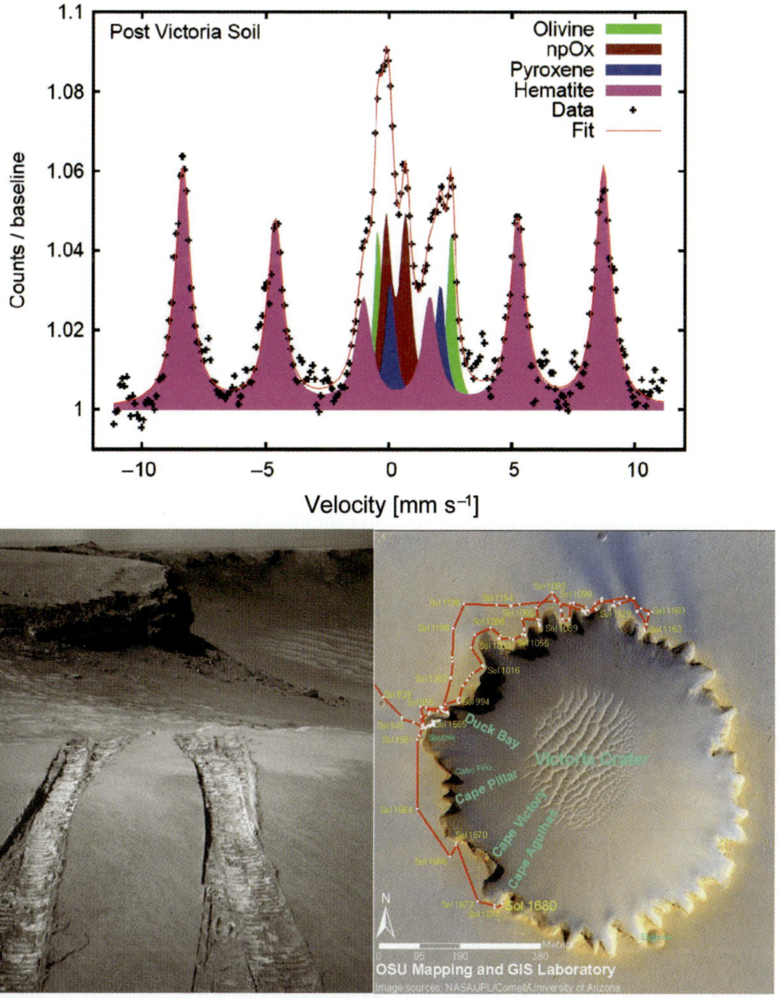

Fig. 8.36 *Left*: Spectrum of the soil close to the crater rim where *Opportunity* entered and exited the crater. The basaltic soil is unusually high in hematite (but no indication of significant contribution from hematitic spherules). *Middle*: rover tracks. *Right*: ~750 m diameter (~75 m deep) eroded impact crater *Victoria Crater*, formed in sulfate-rich sedimentary rocks. Image acquired by the Mars Reconnaissance Orbiter High-Resolution Science Experiment camera (Hirise). The *red line* is the drive path of *Opportunity* exploring the crater. (Courtesy NASA, JPL, ASU, Cornell University)

(Thermal Emission Spectrometer) instrument indicated the metallic nature of the rock [340]. Observations made with the panoramic camera and the microscopic image revealed that the surface of the rock is covered with pits interpreted as regmaglypts and indicate the presence of a coating on the surface. The α-Particle-X-ray spectrometer (APXS) and the Mössbauer spectrometer were used to investigate the undisturbed and the brushed surface of the rock. Based on the Ni and Ge

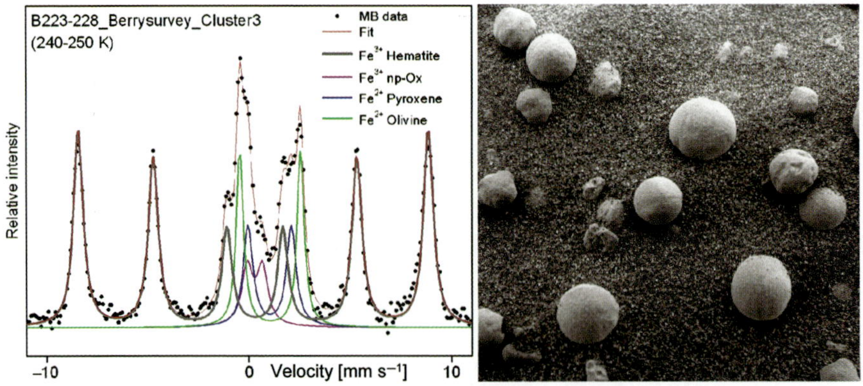

Fig. 8.37 *Left*: spectrum of an accumulation of hematite rich spherules (*Blueberries)* on top of basaltic soil (Sol 223–228 of the mission; 1 Sol = 1 Martian day). The spectrum is dominated by the hematite signal. Estimations based on area ratios (blueberries/soil) and APXS data indicate that the blueberries as composed mainly of hematite. *Right*: MI picture (3×3 cm^2) of hematitic spherules (*blueberries*) on basaltic soil at Meridiani Planum

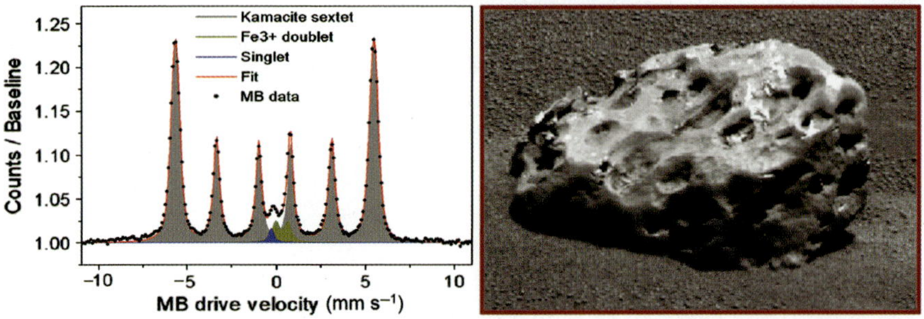

Fig. 8.38 (*Left*): The Mössbauer spectrum of the rock called "Heat Shield rock," clearly shows with high intensity the mineral Kamacite, an Fe–Ni alloy with about 6–7% Ni; (*Right*): The iron–nickel meteorite "Meridiani Planum" (originally called "Heat Shield Rock") at Opportunity landing site, close to the crater Endurance. The meteorite is about 30 cm across (Courtesy NASA, JPL, Cornell University)

contents derived by APXS, Meridiani Planum was classified as an iron meteorite of the IAB complex. The brushed meteorite surface was found to be enriched in P, S and Cl in comparison to Martian soil.

Mössbauer Spectral Analysis and Analog Measurements. Mössbauer spectra were obtained in the temperature range between 200 and 270 K and in two different energy windows (14.4 and 6.4 keV), which provide depth selective information about a sample [346]. To compensate for low counting statistics due to limited integration time, all available spectra were summed for the integrations on the undisturbed and brushed surface, respectively. In addition to kamacite (α-(Fe,Ni)) ($\sim$85%) and small amounts of ferric oxide (see Fig. 8.38), all spectra exhibit features indicative for an additional mineral phase. Based on analog measurements performed with a specimen of the Mundrabilla Iron meteorite, the additional spectral features can be attributed to schreibersite ($(Fe,Ni)_3P$) [360]. Mineral inclusions of schreibersite in Meridiani Planum are also consistent with an enrichment of P observed with the APXS [360].

Besides this iron meteorite, there have been four other rocks identified to be probably of meteoritic origin. These centimeter-sized pebbles, named *Barberton*, *Santa Catarina*, *Santorini* and *Kasos*, show troilite and/or kamacite signatures in the corresponding Mössbauer spectra [359]. The range of Fe oxidation states suggests the presence of a fusion crust. The four cobbles have a very similar chemical composition determined by the APXS, and therefore they may be fragments of the same impactor that created *Victoria Crater* [361].

Victoria Crater. After a journey of more than 950 Sols (1 Sol = 1 Martian day ~24 h and 37 min) or more than 2.5 Earth years (~1.3 Martian years), Opportunity arrived at the 750 m diameter crater Victoria (Fig. 8.36) [335], after traveling across the plains of Meridiani Planum more than 10 km to the south from its landing site in Eagle Crater. The rover has explored this eroded impact crater for more than 750 Sols (more than 2 Earth years or 1 Martian year), descending into the crater on Sol 1293 to begin in situ physical and chemical/mineralogical observations. After exiting the crater on Sol 1634, Opportunity performed more imaging and in situ investigations to the southwest of *Duck Bay,* name of a part of the crater rim, before leaving Victoria heading south toward a 20 km diameter crater named Endeavor.

The outline of Victoria Crater is serrated, with sharp and steep promontories separated by rounded alcoves (Fig. 8.36). The crater formed in sulfate-rich sedimentary rocks, and is surrounded by a smooth terrain that extends about one crater diameter from the rim. On the crater floor is a dune field. There are no perched ejecta blocks preserved on the smooth terrain around the crater rim, probably planed off by Aeolian abrasion. The Mössbauer mineralogy of the sedimentary rocks at the crater rim and inside the crater itself is nearly the same as at Eagle crater landing site and Endurance crater, both about 6–8 km away [335].

The soil close to the crater rim, at the exit point of Opportunity, shows a high hematite content in the Mössbauer spectrum (Fig. 8.36), which can be attributed to the presence of hematite spherules *(blueberries).*

8.3.3.2 Terrestrial Applications

In Situ Monitoring of Airborne Particles. The identification of air pollution sources is one of the main goals of any monitoring and controlling of the air quality in an industrialized urban region. In the metropolitan region of Vitória (MRV), Espírito Santo, Brazil, monitoring and analysis of iron-bearing materials in airborne particles in atmospheric aerosols was performed using MIMOS II [362]. The MRV covers an area of 1,461 km^2. This region has nearly 1.3 million inhabitants, houses a large industrial complex, is undergoing a rapid expansion and has to bear an intense traffic. During the second half of the 1980s, particulate matter (TSP) concentration above 80 μg m^{-3} was recorded. These concentrations did not comply with the Brazilian air quality standards. High pollution concentration was also associated with a period of low pluviometric precipitation, strong winds, vegetation burnings, and intense industrial activities. A MIMOS II instrument was installed inside an airborne particle sampler. MIMOS II operated very stable in field, and the performance was not affected by the vibrations of the air pump system. The obtained results are in good agreement with the results recorded in the laboratory. Hematite was identified as a predominant phase in the suspended particles from MRV. As subordinate phases geothite (GT), pyrite, iron-containing silicates and, sometimes, an ultrafine Fe^{3+} phase were detected. The hematite comes mostly from the industrial plants producing iron ore pellets and from soil, GT from soil and pyrite from handling and storing coal in the industrial area. Ultrafine particles are a result of strongly weathered tropical soils and industrial emissions. Eventually, magnetite was also found. Magnetite is related to steelwork plants and silicates stem from soil and civil constructions.

In Situ Monitoring of Soils Below the Surface. Iron in soil is one of the five most important chemical elements in abundance. It plays a major role in biogeochemical cycles as electron donor and acceptor in oxido-reduction reactions, the main source of energy for life. Iron minerals found in soils are iron oxides (most abundant), clay minerals, and less commonly, iron sulfides and carbonates. Soil genesis and soil properties are influenced by iron in minerals and in soil solution [363]. In order to monitor in situ transformations of iron minerals in soils, a special experimental setup for subsurface in-field applications was developed at the University of Mainz [363, 364]. A modified version of the instrument MIMOS II is mounted on a movable platform inside a 2 m long and 20 cm diameter Plexiglas tube, which has been installed in hydromorphic soil in the field. Positioning of the sensor head to different depths is done by a stepping motor arrangement with an accuracy of about 1 mm. Also, dedicated software for running measurement sequences (e.g., different depth positions at different times etc.) was developed. The setup can work autonomously up to several months in the field.

Experimental results confirm that the fougerite mineral found in hydromorphic soils is Fe(II)/Fe(III) hydroxycarbonate green rust. With a total iron concentration of about 4% in the bulk soil, spectra were obtained after 1–2 days of count accumulation. Calibrations were done in the laboratory. The example spectra demonstrate that the instrument can acquire Mössbauer spectra with reasonable statistics even with low-intensity sources. The spectra display three quadrupole doublets (D1, D2, and D3) associated with the fougerite mineral encountered in gley soils and responsible for their bluish-green color where D1 and D2 with a large quadrupole splitting are attributed to Fe(II) state and D3 to Fe(III). Intensities of the peaks that are found to be 50, 10, and 40% from computer fitting are consistent with the Fe(II)/Fe(III) hydroxycarbonate green rust where stoichiometry intensity ratios are 1/2:1/6:1/3 [363].

Archaeological Artifacts. A Lekythos Greek vase (500 years B.C.) was analyzed by backscattering Mössbauer spectroscopy [365, 366]. This Lekythos vase has three black human figures with small red-painted details painted on a yellow-fired clay

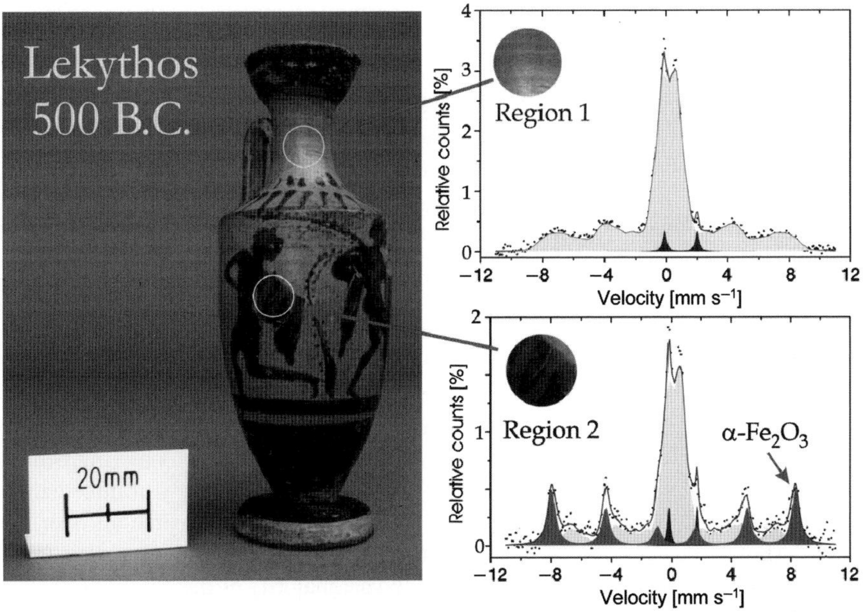

Fig. 8.39 Picture of the Lekythos vase (500 years B.C.), and two examples of spectra obtained with the portable spectrometer MIMOS II at the areas indicated in the figure above. The circular field of view of the instrument was set to about 1 cm diameter

(Fig. 8.39). The Lekythos vase was built as an oil or perfume jar having an ellipsoidal body, narrow neck, flanged mouth, curved handle, and a narrow base terminating in a foot. It was used chiefly for ointments and religious (funerary) ceremonies. This is substantiated by the fact that Lekythos have been found in and around tombs and excavated from ancient homes. Mössbauer backscattering spectra were recorded at room temperature with MIMOS II in three different regions of the Lekythos vase (Fig. 8.39). One was on nonpainted surfaces and the other was on the painted surfaces (only black-painted, and black with red details). Free painted surface shows a broad spectrum that can be associated with poor-crystallized iron oxides produced during the firing clay process. The painted surfaces show, in addition to the characteristic nonpainted area, a well-defined sextet. The Mössbauer parameters of this sextet correspond to well-crystallized hematite. The Mössbauer spectrum taken over the red painted details shows no significant difference from the nonpainted surface. Therefore, the red details are presumably iron-free.

The usefulness of this nondestructive technique may go further than pigment characterization, and it is proposed as an additional method for checking the authenticity of ancient archaeological artifacts.

Rock Paintings in the Field. The manufacturing and composition of paints from prehistoric periods are of great archaeological interest because it is possible to deduce aspects of the ancient cultures based on their abilities to produce works of arts. Many pigments used in ancient paintings are iron-based compounds. In most cases, for obvious reasons, it is impossible to remove samples, and therefore in situ methods have to be applied.

The impressive paintings in Santana do Riacho, associated with the recent development of the spectrometer MIMOS II, prompted the authors to perform, for the first time, nondestructive in situ Mössbauer measurements on ancient rock paintings in Brazil (near Belo Horizonte, Minas Gerais) [366–369]. We have used a prototype of the instrument to investigate the iron oxide composition of rock paintings in the field. The instrument was mounted on a tripod, and the power was supplied by battery (see Fig. 8.40a). Spectra were collected during 24–48 h ($\sim$30 mCi ^{57}Co/Rh source), with an average daily temperature of 27–32°C and 18–22°C during night.

The archaeological site of Santana do Riacho is located at about 90 km north-eastern of Belo Horizonte, in the "Serra do Cipó" region. The natural shelter of the archaeological site where the rock paintings were made is approximately 80 m long and 5 m high. Excavations of all sorts of materials (bones, stones, pigmented materials, etc.) have been performed over the years and a vast body of information was collected. Hundreds of prehistorical paintings are clearly visible all over the wall, but so far, no conclusive proof exists about the nature of the materials (e.g., paints) that were used in these drawings. The vast majority of the paintings are red

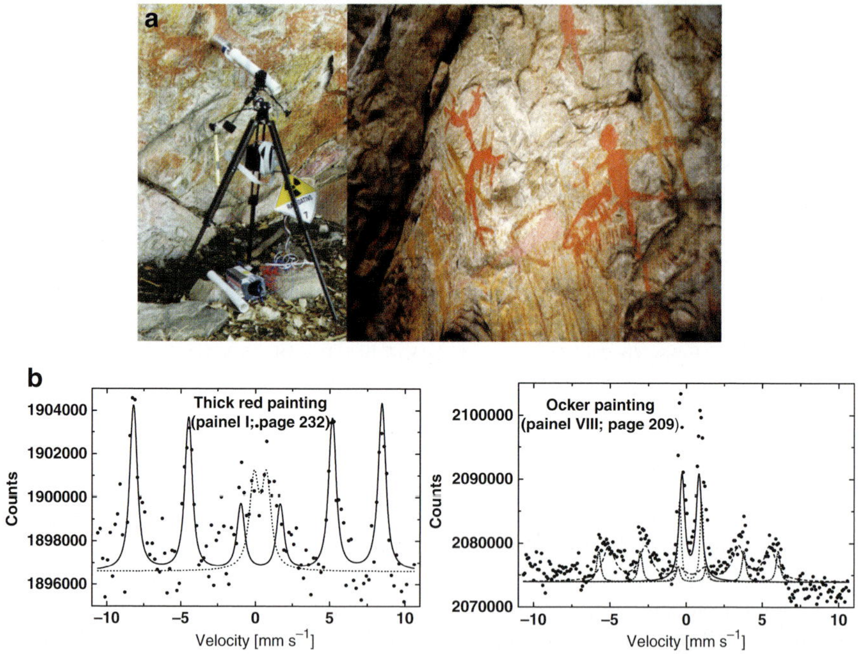

Fig. 8.40 (**a**). The MIMOS II instrument is mounted on a tripod and positioned to the picture of interest (*left*); part of the wall showing several distinct paintings with different colors (*right*). (**b**). Mössbauer spectra of (1) the thick red painting located in *panel* I, page 232, and (2) the ochre painting located in *panel* VIII, page 209 of the book about the archaeological site of Santana do Riacho, Brazil [368, 369]

(70%) and yellow (20%), but there are others in orange, black, brown, and ochre. Iron oxides are most probably the pigments that were used. The paintings were selected based on their colors and apparent thickness, and six spots were measured. The analysis of the data suggests that hematite and GT seem to be the pigments responsible for the red and yellow colors (see Fig. 8.40b), but no conclusion was yet found for an ochre painting. Possibly, it is related to the mineral jarosite.

8.3.3.3 The Advanced Instrument MIMOS IIa

The Mars-Exploration-Rover mission has demonstrated that Mössbauer spectroscopy is an important tool for the in situ exploration of extraterrestrial bodies and the study of Fe-bearing samples. MIMOS II is part of the scientific payload of future space missions, in particular, Phobos Grunt (Russian Space Agency; 2011) and ExoMars (European Space Agency (ESA) and NASA; 2018). In comparison to MIMOS II, the instrument under development for ExoMars will use Si-Drift detectors (SDD) which also allows high-resolution XRF analysis in parallel to Mössbauer spectroscopy [366, 370, 371]. The new design of the improved version of the MIMOS II instrument is lighter than the MER instrument. A new ring detector system of four SDDs [370] will replace the four Si-PIN detectors of the current version greatly improving the energy resolution and the area fill factor around the collimator. The SDDs with a sensitive area of $2 \times 45\ mm^2$ each exhibit an energy resolution <280 eV at room temperature approaching the theoretical limit of ~130 eV at moderate cooling (about −20°C). This results in an increase of the signal to noise ratio (SNR) (or sensitivity) of more than a factor of 10 (Fig. 8.41). In addition to the Mössbauer data, SDD allows the simultaneous acquisition of the XRF spectrum, thus providing data on the sample's elemental composition (Fig. 8.41). For more details, see Sect. 3.3.

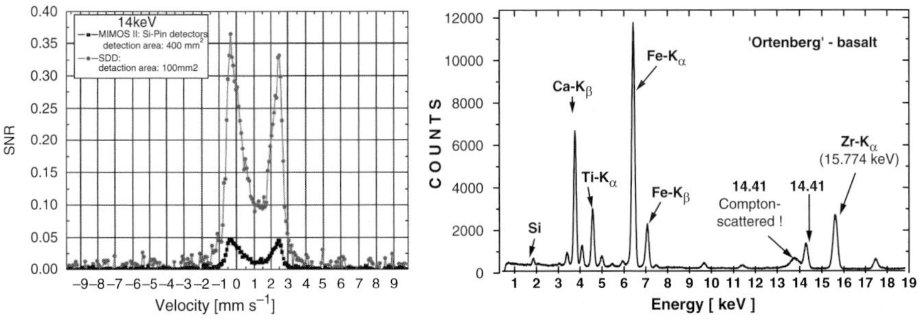

Fig. 8.41 *Left*: Comparison of SNR of 14.4 keV Mössbauer spectra, taken with a Si-PIN detector system (MER instrument; four diodes) and with a SDD detector system (advanced MIMOS instrument; only one diode chip); *Right*: X-ray spectrum of a basalt (Ortenberg basalt; see [366, 371], taken with a high resolution Si-drift detector system at ambient pressure (1 atm), demonstrating the XRF capability of the "advanced MIMOS" instrument. Excitation source: $^{57}Co/Rh$ Mössbauer source. Energy resolution: about 170 eV at +10°C (SDD temperature)

8.3.4 Conclusions and Outlook

The miniaturized Mössbauer instruments have proven as part of the NASA Mars Exploration Rover 2003 mission that Mössbauer spectroscopy is a powerful tool for planetary exploration, including our planet Earth. For the advanced model of MIMOS II, the new detector technologies and electronic components increase sensitivity and performance significantly. In combination with the high-energy resolution of the SDD, it will be possible to perform XRF analysis in parallel to Mössbauer spectroscopy. In addition to the Fe-mineralogy, information on the sample's elemental composition will be obtained.

The instrument MIMOS II will be part of the upcoming ESA–NASA space missions ExoMars in 2018, and the Russian Space Agency sample return mission Phobos Grunt scheduled for launch in 2011 to visit the Mars moon Phobos.

The miniaturized Mössbauer spectrometer MIMOS II has been used already in several terrestrial applications which would not have been possible before. A number of other possible terrestrial applications, for example, in the field, in industry, and fundamental research, are under consideration. With the new generation of the Mössbauer spectrometer MIMOS II, the method itself can be applied to numerous new fields in research, environmental science, planetary science, and many other fields. Because of this reason, Mössbauer spectroscopy may become a more widely used method than it is today.

References

References to Sect. 8.1

1. König, E.: Coord. Chem. Rev. **3**, 471 (1968)
2. Goodwin, H.A.: Coord. Chem. Rev. **18**, 293 (1976)
3. Gütlich, P.: Struct. Bond. (Berlin) **44**, 83 (1981)
4. Gütlich, P., Hauser, A., Spiering, H.: Angew. Chem. Int. Ed. **33**, 20 (1994)
5. Gütlich, P., Goodwin, H.A., (eds.): Top. Curr. Chem. 233–235 (2004)
6. Gütlich, P., Garcia, Y., Spiering, H.: In: Miller, J.S., Drillon, M. (eds.) Magnetism: Molecules to Materials IV, p. 271. Wiley-VCH, Weinheim (2003)
7. Dézsi, I., Molnar, B., Tarnozci, T., Tompa, K.: J. Inorg. Nucl. Chem. **29**, 2486 (1967)
8. Jesson, J.P., Weiher, J.F.: J. Chem. Phys. **46**, 1995 (1967)
9. Jesson, J.P., Weiher, J.F., Trofimenko, S.: J. Chem. Phys. **48**, 2058 (1968)
10. Long, G.J., Grandjean, F., Reger, D.L.: Top. Curr. Chem. **233**, 91 (2004)
11. Sorai, M., Ensling, J., Gütlich, P.: Chem. Phys. **18**, 199 (1976)
12. Gütlich, P., Link, R., Steinhäuser, H.G.: Inorg. Chem. **17**, 2509 (1978)
13. Spiering, H., Meissner, E., Köppen, H., Müller, E.W., Gütlich, P.: Chem. Phys. **68**, 65 (1982)
14. Sorai, M., Ensling, J., Hasselbach, K.M., Gütlich, P.: Chem. Phys. **20**, 197 (1977)
15. Gütlich, P., Köppen, H., Steinhäuser, H.G.: Chem. Phys. Lett. **74**, 475 (1980)
16. Köppen, H., Müller, E.W., Köhler, C.P., Spiering, H., Meissner, E., Gütlich, P.: Chem. Phys. Lett. **91**, 348 (1982)

17. Drickamer, H.G.: Angew. Chem. **86**, 61 (1974)
18. Meissner, E., Köppen, H., Spiering, H., Gütlich, P.: Chem. Phys. Lett. **95**, 163 (1983)
19. Köhler, C.P., Jakobi, R., Meissner, E., Wiehl, L., Spiering, H., Gütlich, P.: J. Phys. Chem. Solids **51**, 239 (1990)
20. Decurtins, S., Gütlich, P., Köhler, C.P., Spiering, H., Hauser, A.: Chem. Phys. Lett. **105**, 1 (1984)
21. Decurtins, S., Gütlich, P., Hasselbach, K.M., Hauser, A., Spiering, H.: Inorg. Chem. **24**, 2174 (1985)
22. Hauser, A.: Chem. Phys. Lett. **124**, 543 (1986)
23. Poganiuch, P., Decurtins, S., Gütlich, P.: J. Am. Chem. Soc. **112**, 3270 (1990)
24. Real, J.A., Gaspar, A.B., Muñoz, M.C., Gütlich, P., Ksenofontov, V., Spiering, H.: Top. Curr. Chem. **233**, 167 (2004)
25. Real, J.A., Zarembowitch, J., Kahn, O., Solans, X.: Inorg. Chem. **26**, 2939 (1987)
26. Real, J.A., Bolvin, H., Bousseksou, A., Dworkin, A., Kahn, O., Varret, F., Zarembowitch, J.: J. Am. Chem. Soc. **114**, 4650 (1992)
27. Ksenofontov, V., Spiering, H., Reiman, S., Garcia, Y., Gaspar, A.B., Moliner, N., Real, J.A., Gütlich, P.: Chem. Phys. Lett. **348**, 381 (2001)
28. Zimmermann, R., Ritter, G., Spiering, H.: Chem. Phys. **4**, 133 (1974)
29. Zimmermann, R., Ritter, G., Spiering, H., Nagy, D.L.: J. Phys. **35**, C6 (1974)
30. Yoneda, K., Nakano, K., Fujioka, J., Yamada, K., Suzuki, T., Fuyuhiro, A., Kawata, S., Kaizaki, S.: Polyhedron **24**, 2437 (2005)
31. Yoneda, K., Adachi, K., Hayami, S., Maeda, Y., Katada, M., Fuyuhiro, A., Kawata, S., Kaizaki, S.: Chem. Commun., 45 (2006)
32. Grunert, C.M., Reiman, S., Spiering, H., Kitchen, J.A., Brooker, S., Gütlich, P.: Angew. Chem. Int. Ed. **47**, 2997 (2008)
33. Klingele, M.H., Moubaraki, B., Cashion, J.D., Murray, K.S., Brooker, S.: Chem. Commun., 987 (2005)
34. Klingele, M.H., Moubaraki, B., Murray, K.S., Brooker, S.: Chem. Eur. J. **11**, 6962 (2005)
35. Kolnaar, J.J.A., van Dijk, G., Kooijman, H., Spek, A.L., Ksenofontov, V., Gütlich, P., Haasnoot, J.G., Reedijk, J.: Inorg. Chem. **36**, 2433 (1997)
36. Galyametdinov, Y., Ksenofontov, V., Prosvirin, A., Ovchinnikov, I., Ivanova, G., Gütlich, P., Haase, W.: Angew. Chem. Int. Ed. **40**, 4269 (2001)
37. Fujigaya, T., Jiang, D.L., Aida, T.: J. Am. Chem. Soc. **125**, 14690 (2003)
38. Hayami, S., Danjobara, K., Inoue, K., Ogawa, Y., Matsumoto, N., Maeda, Y.: Adv. Mater. **16**, 869 (2004)
39. Hayami, S., Moriyama, R., Shuto, A., Maeda, Y., Ohta, K., Inoue, K.: Inorg. Chem. **46**, 7692 (2007)
40. Hayami, S., Motokawa, N., Shuto, A., Masuhara, N., Someya, T., Ogawa, Y., Inoue, K., Maeda, Y.: Inorg. Chem. **46**, 1789 (2007)
41. Seredyuk, M., Gaspar, A.B., Ksenofontov, V., Galyametdinov, Y., Kusz, J., Gütlich, P.: Adv. Funct. Mater. **18**, 2089 (2008)
42. Seredyuk, M., Gaspar, A.B., Ksenofontov, V., Galyametdinov, Y., Verdaguer, M., Villain, F., Gütlich, P.: Inorg. Chem. **47**, 10232 (2008)
43. Seredyuk, M., Gaspar, A.B., Ksenofontov, V., Galyametdinov, Y., Kusz, J., Gütlich, P.: J. Am. Chem. Soc. **130**, 1431 (2008)
44. Seredyuk, M., Gaspar, A.B., Ksenofontov, V., Reiman, S., Galyametdinov, Y., Haase, W., Rentschler, E., Gütlich, P.: Chem. Mater. **18**, 2513 (2006)
45. Seredyuk, M., Gaspar, A.B., Ksenofontov, V., Reiman, S., Galyametdinov, Y., Haase, W., Rentschler, E., Gütlich, P.: Hyperfine Interact. **166**, 385 (2006)
46. Matsuura, T. (ed.): Hot Atom Chemistry. Kodansha, Tokyo (1984)
47. Sano, H., Gütlich, P.: Hot Atom Chemistry. Kodansha, Tokyo (1984) (and references therein)

48. Ensling, J., Fitzsimmons, B.W., Gütlich, P., Hasselbach, K.M.: Angew. Chem. Int. Ed. **9**, 637 (1970)
49. Ensling, J., Gütlich, P., Hasselbach, K.M., Fitzsimmons, B.W.: Chem. Phys. Lett. **42**, 232 (1976)
50. Gütlich, P.: Top. Curr. Chem. **234**, 231 (2004)
51. Deisenroth, S., Hauser, A., Spiering, H., Gütlich, P.: Hyperfine Interact. **93**, 1573 (1994)
52. Hauser, A.: Chem. Phys. Lett. **173**, 507 (1990)
53. Oshio, A., Spiering, H., Ksenofontov, V., Renz, F., Gütlich, P.: Inorg. Chem. **40**, 1143 (2001)

References to Sect. 8.2

54. Griffith, J.S.: J. Inorg. Nucl. Chem. **2**, 1–10 (1956)
55. Griffith, J.S.: Discuss. Faraday Soc. **26**, 8–86 (1959)
56. Hoskins, B.F., Martin, R.L., White, A.H.: Nature **211**, 627–628 (1966)
57. Maltempo, M.M.: J. Chem. Phys. **61**, 2540–2547 (1974)
58. Maltempo, M.M., Moss, T.H., Cusanovich, M.A.: Biochim. Biophys. Acta **342**, 290–305 (1974)
59. Maltempo, M.M., Moss, T.H.: Q. Rev. Biophys. **9**, 181–215 (1976)
60. Maltempo, M.M., Ohlsson, P.I., Paul, K.G., Petersson, L., Ehrenberg, A.: Biochemistry **18**, 2935–2941 (1979)
61. LaMar, G.N., Jackson, J.T., Dugad, L.B., Cusanovich, M.A.: J. Biol. Chem. **265**, 16173–16180 (1990)
62. Fujii, S., Yoshimura, T., Kamada, H., Yamaguchi, K., Suzuki, S., Takakwa, S.: Biochim. Biophys. Acta – Protein Struct. Mol. Enzym. **1251**, 161 (1995)
63. Schulz, C.E., Rutter, R., Sage, J.T., Debrunner, P.G., Hager, L.P.: Biochemistry **23**, 4743–4754 (1984)
64. Schauer, C.K., Akabori, K., Elliott, C.M., Anderson, O.P.: J. Am. Chem. Soc. **106**, 1127–1128 (1984)
65. Karlin, K.D.: Dithiolene Chemistry: Synthesis, Properties, and Applications, vol. 52. Wiley, New York (2004)
66. De Vries, J.L.K.F., Keijzers, C.P., De Boer, E.: Inorg. Chem. **11**, 1343–1348 (1972)
67. Niarchos, D., Kostikas, A., Simopoulos, A., Coucouvanis, D., Piltingsrud, D., Coffman, R.E.: J. Chem. Phys. **69**, 4411–4418 (1978)
68. Keutel, H., Käpplinger, I., Jäger, E.-G., Grodzicki, M., Schünemann, V., Trautwein, A.X.: Inorg. Chem **38**, 2320–2327 (1999)
69. Dolphin, D.H., Sams, J.R., Tsin, T.B.: Inorg. Chem. **16**, 711–713 (1977)
70. Wickman, H.H., Trozzolo, A.M., Williams, H.J., Hull, G.W., Merritt, F.R.: Phys. Rev. **155**, 563 (1967)
71. Harris, G.: Theor. Chim. Acta **10**, 119–154 (1968)
72. Harris, G.: Theor. Chim. Acta. **10**, 155–180 (1968)
73. Patra, A.K., Bill, E., Bothe, E., Chlopek, K., Neese, F., Weyhermüller, T., Stobie, K., Ward, M.D., McCleverty, J.A., Wieghardt, K.: Inorg. Chem. **45**, 7877–7890 (2006)
74. Chlopek, K., Muresan, N., Neese, F., Wieghardt, K.: Chem. Eur. J. **13**, 8391–8403 (2007)
75. Ray, K., George, S.D., Solomon, E.I., Wieghardt, K., Neese, F.: Chem. Eur. J. **13**, 2783–2797 (2007)
76. Ganguli, P., Marathe, V.R., Mitra, S.: Inorg. Chem. **14**, 970–973 (1975)
77. Wickman, H.H., Wagner, C.F.: J. Chem. Phys. **51**, 435–444 (1969)
78. Chapps, G.E., McCann, S.W., Wickman, H.H., Sherwood, R.C.: J. Chem. Phys. **60**, 990–997 (1974)
79. Wickman, H.H., Merritt, F.R.: Chem. Phys. Lett. **1**, 117–118 (1967)

80. Wickman, H.H., Trozzolo, A.M., Williams, H.J., Hull, G.W., Merritt, F.R. Bell Telephone Research Reports, Murray Hill, NJ. (1967)
81. Wickman, H.H., Trozzolo, A.M.: Inorg. Chem. **7**, 63–68 (1968)
82. Petridis, D., Simopoulos, A., Kostikas, A., Pasternak, M.: J. Chem. Phys. **65**, 3139–3145 (1976)
83. Wickman, H.H.: J. Chem. Phys. **56**, 976–982 (1972)
84. Grow, J.M., Hopkins, T.E., Robbins, G.L., Silverthorn, W.E., Wickman, H.H.: J. Chem. Phys. **67**, 5275–5281 (1977)
85. Grow, J.M., Robbins, G.L., Wickman, H.H.: J. Chem. Phys. **67**, 5282–5290 (1977)
86. De Vries, J.L.K.F., Trooster, J.M., De Boer, E.: Inorg. Chem. **10**, 81–85 (1971)
87. Birchall, T., Greenwood, N.N.: J. Chem. Soc. A, 286–291 (1969)
88. Fettouhi, M., Morsy, M., Waheed, A., Golhen, S., Ouahab, L., Sutter, J.-P., Kahn, O., Menendez, N., Varret, F.: Inorg. Chem. **38**, 4910–4912 (1999)
89. Sellmann, D., Geck, M., Knoch, F., Ritter, G., Dengler, J.: J. Am. Chem. Soc. **113**, 3819 (1991)
90. Sellmann, D., Emig, S., Heinemann, F.W.: Angew. Chem. Int. Ed. Engl. **36**, 1734–1736 (1997)
91. Sellmann, D., Emig, S., Heinemann, F.W., Knoch, F.: Angew. Chem. Int. Ed. Engl. **36**, 1201–1203 (1997)
92. Ray, K., Bill, E., Weyhermüller, T., Wieghardt, K.: J. Am. Chem. Soc. **127**, 5641–5654 (2005)
93. Fallon, G.D., Gatehouse, B.M., Marini, P.J., Murray, K.S., West, B.O.: J. Chem. Soc. Dalton Trans., 2733–2739 (1984)
94. Ghosh, P., Bill, E., Weyhermüller, T., Wieghardt, K.: J. Am. Chem. Soc. **125**, 3967–3979 (2003)
95. Chun, H., Weyhermüller, T., Bill, E., Wieghardt, K.: Angew. Chem. Int. Ed. **40**, 2489 (2001)
96. Chun, H.P., Bill, E., Weyhermüller, T., Wieghardt, K.: Inorg. Chem. **42**, 5612–5620 (2003)
97. Blanchard, S., Bill, E., Weyhermüller, T., Wieghardt, K.: Inorg. Chem. **43**, 2324–2329 (2004)
98. Chlopek, K., Bill, E., Weyhermüller, T., Wieghardt, K.: Inorg. Chem. **44**, 7087–7098 (2005)
99. Koch, S., Holm, R.H., Frankel, R.B.: J. Am. Chem. Soc. **97**, 6714–6723 (2002)
100. Kostka, K.L., Fox, B.G., Hendrich, M.P., Collins, T.J., Rickard, C.E.F., Wright, L.J., Münck, E.: J. Am. Chem. Soc. **115**, 6746–6757 (1993)
101. Simonato, J.-P., Pecaut, J., Le Pape, L., Oddou, J.-L., Jeandey, C., Shang, M., Scheidt, W.R., Wojaczynski, J., Wolowiec, S., Latos-Grazynski, L., Marchon, J.-C.: Inorg. Chem. **39**, 3978–3987 (2000)
102. Ikeue, T., Ohgo, Y., Yamaguchi, T., Takahashi, M., Takeda, M., Nakamura, M.: Angew. Chem. Int. Ed. **40**, 2617–2620 (2001)
103. Koch, W.O., Schünemann, V., Gerdan, M., Trautwein, A.X., Krüger, H.-J.: Chem. Eur. J. **4**, 686–691 (1998)
104. Weiss, R., Gold, A., Terner, J.: Chem. Rev. **106**, 2550–2579 (2006)
105. Ogoshi, H., Watanabe, E., Yoshida, Z.: Chem. Lett., 989–992 (1973)
106. Kastner, M.E., Scheidt, W.R., Mashiko, T., Reed, C.A.: J. Am. Chem. Soc. **100**, 666–667 (1978)
107. Reed, C.A., Mashiko, T., Bentley, S.P., Kastner, M.E., Scheidt, W.R., Spartalian, K., Lang, G.: J. Am. Chem. Soc. **101**, 2948–2958 (1979)
108. Masuda, H., Taga, T., Osaki, K., Sugimoto, H., Yoshida, Z., Ogoshi, H.: Inorg. Chem. **19**, 950–955 (1980)
109. Boersma, A.D., Goff, H.M.: Inorg. Chem. **21**, 581–586 (1982)
110. Toney, G.E., Gold, A., Savrin, J., Terhaar, L.W., Sangaiah, R., Hatfield, W.E.: Inorg. Chem. **23**, 4350–4352 (1984)
111. Gupta, G.P., Lang, G., Lee, Y.J., Scheidt, W.R., Shelly, K., Reed, C.A.: Inorg. Chem. **26**, 3022–3030 (1987)

112. Dugad, L.B., Mitra, S.: Proc. Indian Acad. Sci. Chem. Sci. **93**, 295–311 (1984)
113. Scheidt, W.R., Osvath, S.R., Lee, Y.J., Reed, C.A., Shaevitz, B., Gupta, G.P.: Inorg. Chem. **28**, 1591–1595 (1989)
114. Gismelseed, A., Bominaar, E.L., Bill, E., Trautwein, A.X., Winkler, H., Nasri, H., Doppelt, P., Mandon, D., Fischer, J., Weiss, R.: Inorg. Chem. **29**, 2741–2749 (1990)
115. Goff, H., Shimomura, E.: J. Am. Chem. Soc. **102**, 31–37 (1980)
116. Nesset, M.J.M., Cai, S., Shokhireva, T.K., Shokhirev, N.V., Jacobson, S.E., Jayaraj, K., Gold, A., Walker, F.A.: Inorg. Chem. **39**, 532–540 (2000)
117. Debrunner, P.G.: Mössbauer spectroscopy of iron porphyrins. In: Lever, A.B.P., Gray, H.B. (eds.) Iron Porphyrins Part III, pp. 137–234. VCH, Weinheim (1989)
118. Kintner, E.T., Dawson, J.H.: Inorg. Chem. **30**, 4892–4897 (1991)
119. Gupta, G.P., Lang, G., Scheidt, W.R., Geiger, D.K., Reed, C.A.: J. Chem. Phys. **85**, 5212–5220 (1986)
120. Kennedy, B.J., Murray, K.S., Zwack, P.R., Homborg, H., Kalz, W.: Inorg Chem. **25**, 2539–2545 (1986)
121. Fang, M., Wilson, S.R., Suslick, K.S.: J. Am. Chem. Soc. **130**, 1134–1135 (2008)
122. Alonso, P.J., Arauzo, A.B., Forniés, J., Garcìa-Monforte, M.A., Martínez, A.M.J.I., Menjón, B., Rillo, C., Sáiz-Garitaonandia, J.J.: Angew. Chem. Int. Ed. **45**, 6707–6711 (2006)
123. Gupta, G.P., Lang, G., Reed, C.A., Shelly, K., Scheidt, W.R.: J. Chem. Phys. **86**, 5288–5293 (1987)
124. Schünemann, V., Gerdan, M., Trautwein, A.X., Haoudi, N., Mandon, D., Fischer, J., Weiss, R., Tabard, A., Guilard, R.: Angew. Chem. Int. Ed. **38**, 3181–3183 (1999)
125. Ogoshi, H., Sugimoto, H., Watanabe, E., Yoshida, Z., Maeda, Y., Sakai, H.: Bull. Chem. Soc. Jpn. **54**, 3414–3419 (1981)
126. Reed, C.A., Guiset, F.: J. Am. Chem. Soc. **118**, 3281–3282 (1996)
127. Klemm, L., Klemm, W.: J. Prakt. Chem. **143**, 82–89 (1935)
128. Lever, A.B.P.: J. Chem. Soc., 1821–1829 (1965)
129. Dale, B.W., Williams, R.J.P., Johnson, C.E., Thorp, T.L.: J. Chem. Phys. **49**, 3441–3444 (1968)
130. Barraclough, C.G., Martin, R.L., Mitra, S., Sherwood, R.C.: J. Chem. Phys. **53**, 1643–1648 (1970)
131. Labarta, A., Molins, E., Vinas, X., Tejada, J., Caubet, A., Alvarez, S.: J. Chem. Phys. **80**, 444–448 (1984)
132. Filoti, G., Kuz'min, M.D., Bartolome, J.: Phys. Rev. B Condens. Matter Mater. Phys. **74**, 134420 (2006)
133. Ercolani, C., Neri, C., Porta, P.: Inorg. Chim. Acta **1**, 415 (1967)
134. Linstead, R.P., Robertson, J.M.: J. Chem. Soc. **1936**, 1736 (1936)
135. Hudson, A., Whitefield, H.J.: Inorg. Chem. **6**, 1120–1123 (1966)
136. Dale, B.W., Williams, R.J.P., Edwards, P.R., Johnson, C.E.: J. Chem. Phys. **49**, 3445–3449 (1968)
137. Deszi, I., Balazs, A., Molnar, B., Grobchenko, V.D., Kukoshevich, L.I.: J. Inorg. Nucl. Chem. **31**, 1661 (1969)
138. Dale, B.W.: Mol. Phys. **28**, 503–511 (1974)
139. Bell, N.A., Brooks, J.S., Robinson, J.K., Thorpe, S.C.: J. Chem. Soc. Faraday Trans. **94**, 3155 (1998)
140. Grenoble, D.C., Drickamer, H.G.: J. Chem. Phys. **35**, 1624 (1971)
141. Silver, J., Luker, P., Houlton, A., Howe, S., Hey, P., Ahmet, M.T.: J. Mater. Chem. **2**, 849 (1992)
142. Tanaka, K., Elkaim, E., Li, L., Jue, Z.N., Coppens, P., Landrum, J.: J. Chem. Phys. **84**, 6969–6978 (1986)
143. Boyd, P.D.W., Buckingham, D.A., McMeeking, R.F., Mitra, S.: Inorg. Chem. **18**, 3585–3591 (1979)
144. Mispelter, J., Momenteau, M., Lhoste, J.M.: J. Chem. Phys. **72**, 1003–1012 (1980)

145. McGarvey, B.R.: Inorg. Chem. **27**, 4691–4698 (1988)
146. Collman, J.P., Hoard, J.L., Kim, N., Lang, G., Reed, C.A.: J. Am. Chem. Soc. **97**, 2676–2681 (1975)
147. Dolphin, D., Sams, J.R., Tsin, T.B., Wong, K.L.: J. Am. Chem. Soc. **98**, 6970–6975 (1976)
148. Lang, G., Spartalian, K., Reed, C.A., Collman, J.P.: J. Chem. Phys. **69**, 5424–5427 (1978)
149. Strauss, S.H., Silver, M.E., Long, K.M., Thompson, R.G., Hudgens, R.A., Spartalian, K., Ibers, J.A.: J. Am. Chem. Soc. **107**, 4207–4215 (2002)
150. Walker, F.A.: Proton NMR and EPR spectroscopy of paramagnetic metalloporphyrins. In: Kadish, K.M., Smith, K.M., Guilard, R. (eds.) Porphyrin Handbook, pp. 81–183. Academic, San Diego (2000)
151. Reiff, W.M., Frommen, C.M., Yee, G.T., Sellers, S.P.: Inorg. Chem. **39**, 2076–2079 (2000)
152. Bachmann, J., Nocera, D.G.: J. Am. Chem. Soc. **127**, 4730–4743 (2005)
153. Sontum, S.F., Case, D.A., Karplus, M.: J. Chem. Phys. **79**, 2881–2892 (1983)
154. Choe, Y.-K., Nakajima, T., Hirao, K., Lindh, R.: J. Chem. Phys. **111**, 3837–3845 (1999)
155. Liao, M.-S., Scheiner, S.: J. Chem. Phys. **116**, 3635–3645 (2002)
156. Reiff, W.M., Wong, H., Baldwin, J.E., Huff, J.: Inorg. Chim. Acta **25**, 91–96 (1977)
157. Sellmann, D., Geck, M., Moll, M.: J. Am. Chem. Soc. **113**, 5259–5264 (2002)
158. Ray, K., Begum, A., Weyhermüller, T., Piligkos, S., Slagereren, J.V., Neese, F., Wieghardt, K.: J. Am. Chem. Soc. **127**, 4403–4415 (2005)
159. Bill, E., Bender, U., Gonser, U., Trautwein, A.X., Jones, W.M., Sabin, J.D.: In: Proc. Ind. Natl. Sci. Acad. Phys. Sci., Spec. Vol. for the International Conference on the Applications of the Mössbauer Effect, pp. 669–671, New Delhi (1982)
160. König, E., Madeja, K.: Inorg. Chem. **7**, 1848–1855 (1968)
161. König, E., Kanellakopulos, B.: Chem. Phys. Lett. **12**, 485–488 (1972)
162. König, E., Ritter, G., Kanellakopulos, B.: J. Chem. Phys. **58**, 3001–3009 (1973)
163. König, E., Schnakig, R.: Theor. Chim. Acta **30**, 205–208 (1973)
164. König, E., Ritter, G., Goodwin, H.A.: Inorg. Chem. **20**, 3677–3682 (1981)
165. König, E., Ritter, G., Kanellakopulos, B.: Inorg. Chim. Acta **59**, 285–291 (1982)
166. Feig, A.L., Lippard, S.J.: Chem. Rev. **94**, 759–805 (1994)
167. Que, L., Watanabe, Y.: Science **292**, 651–653 (2001)
168. Neese, F., Slep, L.D.: Angew. Chem. Int. Ed. **42**, 2942–2945 (2003)
169. Costas, M., Mehn, M.P., Jensen, M.P., Que, L.: Chem. Rev. **104**, 939–986 (2004)
170. Visser, S.P.: Angew. Chem. Int. Ed. **45**, 1790–1793 (2006)
171. Kovaleva, E.G., Lipscomb, J.D.: Nat. Chem. Biol. **4**, 186–193 (2008)
172. Xue, G., Fiedler, A.T., Martinho, M., Münck, E., Que, L.: Proc. Natl. Acad. Sci. U. S. A. **105**, 20615–20620 (2008)
173. de Oliveira, F.T., Chanda, A., Banerjee, D., Shan, X., Mondal, S., Que Jr., L., Bominaar, E.L., Münck, E., Collins, T.J.: Science **315**, 835–838 (2007)
174. Yoon, J., Wilson, S.A., Jang, Y.K., Seo, M.S., Nehru, K., Hedman, B., Hodgson, K.O., Bill, E., Solomon, E.I., Nam, W.: Angew. Chem. Int. Ed. **48**, 1257–1260 (2009)
175. Que, L.: J. Biol. Inorg. Chem. **9**, 684–690 (2004)
176. Meunier, B.: In: McClaverty, J., Meyer, T.J. (eds.) Comprehensive Coordination Chemistry II, pp. 261–280. Elsevier, New York (2003)
177. Meunier, B., de Visser, S.P., Shaik, S.: Chem. Rev. **104**, 3947–3980 (2004)
178. Shaik, S., Kumar, D., de Visser, S.P., Altun, A., Thiel, W.: Chem. Rev. **105**, 2279–2328 (2005)
179. Theorell, H.: Enzymologia **10**, 250 (1941)
180. Chance, B.: Arch. Biochem. Biophys. **41**, 404–415 (1952)
181. George, P.: Biochem. J. **54**, 267–276 (1953)
182. Moss, T., Ehrenberg, A., Bearden, A.J.: Biochemistry **8**, 4159–4162 (1969)

183. Hersleth, H.-P., Uchida, T., Rohr, A.K., Teschner, T., Schünemann, V., Kitagawa, T., Trautwein, A.X., Görbitz, C.H., Andersson, K.K.: J. Biol. Chem. **282**, 23372–23386 (2007)
184. Weller, M.T., Hector, A.L.: Angew. Chem. Int. Ed. **39**, 4162–4163 (2000)
185. Menil, F.: J. Phys. Chem. Solids **46**, 763–789 (1985)
186. Russo, U., Long, G.J.: Mössbauer spectroscopic studies of the high oxidation states of iron. In: Long, G.J., Grandjean, F. (eds.) Mössbauer Spectroscopy Applied to Inorganic Chemistry, pp. 289–329. Plenum, New York (1989)
187. Delattre, J.L., Stacy, A.M., Young, V.G., Long, G.J., Hermann, R., Grandjean, F.: Inorg. Chem. **41**, 2834–2838 (2002)
188. Takeda, Y., Naka, S., Takano, M., Shinjo, T., Takada, T., Shimada, M.: Mater. Res. Bull. **13**, 61–66 (1978)
189. Takano, M., Okita, T., Nakayama, N., Bando, Y., Takeda, Y., Yamamoto, O., Goodenough, J.B.: J. Solid State Chem. **73**, 140–150 (1988)
190. Takeda, Y., Shimada, M., Kanamaru, F., Koizumi, M., Yamamoto, N.: Mater. Res. Bull. **9**, 537–543 (1974)
191. Kawasaki, S., Takano, M., Takeda, Y.: J. Solid State Chem. **121**, 174–180 (1996)
192. Demazeau, G., Pouchard, M., Chevreau, N., Thomas, M., Ménil, F., Hagenmuller, P.: Mater. Res. Bull. **16**, 689–696 (1981)
193. Schulz, C.E., Devaney, P.W., Winkler, H., Debrunner, P.G., Doan, N., Chiang, R., Rutter, R., Hager, L.P.: FEBS Lett. **103**, 102–105 (1979)
194. Rutter, R., Hager, L.P., Dhonau, H., Hendrich, M., Valentine, M., Debrunner, P.: Biochemistry **23**, 6809–6816 (1984)
195. Boso, B., Lang, G., McMurry, T.J., Groves, J.T.: J. Chem. Phys. **79**, 1122–1126 (1983)
196. Bill, E., Ding, X.Q., Bominaar, E.L., Trautwein, A.X., Winkler, H., Mandon, D., Weiss, R., Gold, A., Jayaraj, K., Hatfield, W.E., Kirk, M.L.: Eur. J. Biochem. **188**, 665–672 (1990)
197. Groves, J.T., Quinn, R., McMurry, T.J., Nakamura, M., Lang, G., Boso, B.: J. Am. Chem. Soc. **107**, 354–360 (1985)
198. Horner, O., Oddou, J.L., Mouesca, J.M., Jouve, H.M.: J. Inorg. Biochem. **100**, 477–479 (2006)
199. English, D.R., Hendrickson, D.N., Suslick, K.S.: Inorg. Chem. **22**, 368–370 (1983)
200. Grapperhaus, C.A., Mienert, B., Bill, E., Weyhermüller, T., Wieghardt, K.: Inorg. Chem. **39**, 5306–5317 (2000)
201. Rohde, J.-U., In, J.-H., Lim, M.H., Brennessel, W.W., Bukowski, M.R., Stubna, A., Münck, E., Nam, W., Que, L.: Science **299**, 1037–1039 (2003)
202. Lim, M.H., Rohde, J.U., Stubna, A., Bukowski, M.R., Costas, M., Ho, R.Y.N., Münck, E., Nam, W., Que, L.: Proc. Natl. Acad. Sci. U. S. A. **100**, 3665–3670 (2003)
203. Price, J.C., Barr, E.W., Tirupati, B., Bollinger, J.M., Krebs, C.: Biochemistry **42**, 7497–7508 (2003)
204. Pestovsky, O., Stoian, S., Bominaar, E.L., Shan, X.P., Münck, E., Que, L., Bakac, A.: Angew. Chem. Int. Ed. **44**, 6871–6874 (2005)
205. England, J., Martinho, M., Farquhar, E.R., Frisch, J.R., Bominaar, E.L., Münck, E., Que, L.: Angew. Chem. Int. Ed. **48**, 3622–3626 (2009)
206. Sturgeon, B.E., Burdi, D., Chen, S.X., Huynh, B.H., Edmondson, D.E., Stubbe, J., Hoffman, B.M.: J. Am. Chem. Soc. **118**, 7551–7557 (1996)
207. Lee, S.K., Fox, B.G., Froland, W.A., Lipscomb, J.D., Münck, E.: J. Am. Chem. Soc. **115**, 6450–6451 (2002)
208. Slep, L.D., Mijovilovich, A., Meyer-Klaucke, W., Weyhermüller, T., Bill, E., Bothe, E., Neese, F., Wieghardt, K.: J. Am. Chem. Soc. **125**, 15554–15570 (2003)
209. Jüstel, T., Weyhermüller, T., Wieghardt, K., Bill, E., Lengen, M., Trautwein, A.X., Hildebrandt, P.: Angew. Chem. Int. Ed. Engl. **34**, 669–672 (1995)
210. Jüstel, T., Müller, M., Weyhermüller, T., Kressl, C., Bill, E., Hildebrandt, P., Lengen, M., Grodzicki, M., Trautwein, A.X., Nuber, B., Wieghardt, K.: Chem. Eur. J. **5**, 793–810 (1999)

211. Hendrich, M.P., Gunderson, W., Behan, R.K., Green, M.T., Mehn, M.P., Betley, T.A., Lu, C. C., Peters, J.C.: Proc. Natl. Acad. Sci. U. S. A. **103**, 17107–17112 (2006)
212. Vogel, C., Heinemann, F.W., Sutter, J., Anthon, C., Meyer, K.: Angew. Chem. Int. Ed. **47**, 2681–2684 (2008)
213. Klinker, E.J., Jackson, T.A., Jensen, M.P., Stubna, A., Juh·sz, G., Bominaar, E.L., Münck, E., Que Jr., L.: Angew. Chem. Int. Ed. **45**, 7394–7397 (2006)
214. Bill, E., Schünemann, V., Trautwein, A.X., Weiss, R., Fischer, J., Tabard, A., Guilard, R.: Inorg. Chim. Acta **339**, 420–426 (2002)
215. Berry, J.F., Bill, E., Bothe, E., Neese, F., Wieghardt, K.: J. Am. Chem. Soc. **128**, 13515–13528 (2006)
216. Berry, J.F., Bill, E., Bothe, E., Weyhermüller, T., Wieghardt, K.: J. Am. Chem. Soc. **127**, 11550–11551 (2005)
217. Cummins, C.C., Schrock, R.R.: Inorg. Chem. **33**, 395–396 (1994)
218. Paulsen, H., Müther, M., Grodzicki, M., Trautwein, A.X., Bill, E.: Bull. Soc. Chim. Fr. **133**, 703–710 (1996)
219. Neese, F., Solomon, E.I.: Inorg. Chem. **37**, 6568–6582 (1998)
220. Gans, P., Buisson, G., Duee, E., Marchon, J.C., Erler, B.S., Scholz, W.F., Reed, C.A.: J. Am. Chem. Soc. **108**, 1223–1234 (1986)
221. Lang, G., Boso, B., Erler, B.S., Reed, C.A.: J. Chem. Phys. **84**, 2998–3004 (1986)
222. Erler, B.S., Scholz, W.F., Lee, Y.J., Scheidt, W.R., Reed, C.A.: J. Am. Chem. Soc. **109**, 2644–2652 (1987)
223. Goff, H.M., Phillippi, M.A.: J. Am. Chem. Soc. **105**, 7567–7571 (1983)
224. Schünemann, V., Winkler, H.: Rep. Prog. Phys. **63**, 263–353 (2000)
225. Solomon, E.I., Brunold, T.C., Davis, M.I., Kemsley, J.N., Lee, S.K., Lehnert, N., Neese, F., Skulan, A.J., Yang, Y.S., Zhou, J.: Chem. Rev. **100**, 235–349 (2000)
226. Hausinger, R.P.: Crit. Rev. Biochem. Mol. Biol. **39**, 21–68 (2004)
227. Koehntop, K.D., Emerson, J.P., Que, L.: J. Biol. Inorg. Chem. **10**, 87–93 (2005)
228. Krebs, C., Price, J.C., Baldwin, J., Saleh, L., Green, M.T., Bollinger, J.M.: Inorg. Chem. **44**, 742–757 (2005)
229. Sinnecker, S., Svensen, N., Barr, E.W., Ye, S., Bollinger, J.M., Neese, F., Krebs, C.: J. Am. Chem. Soc. **129**, 6168–6179 (2007)
230. Hoffart, L.M., Barr, E.W., Guyer, R.B., Bollinger, J.M., Krebs, C.: Proc. Natl. Acad. Sci. U. S. A. **103**, 14738–14743 (2006)
231. Eser, B.E., Barr, E.W., Frantorn, P.A., Saleh, L., Bollinger, J.M., Krebs, C., Fitzpatrick, P.F.: J. Am. Chem. Soc. **129**, 11334 (2007)
232. Galonic, D.P., Barr, E.W., Walsh, C.T., Bollinger, J.M., Krebs, C.: Nat. Chem. Biol. **3**, 113–116 (2007)
233. Oh, N.Y., Suh, Y., Park, M.J., Seo, M.S., Kim, J., Nam, W.: Angew. Chem. Int. Ed. **44**, 4235–4239 (2005)
234. Nam, W.: Acc. Chem. Res. **40**, 522–531 (2007)
235. Sastri, C.V., Lee, J., Oh, K., Lee, Y.J., Jackson, T.A., Ray, K., Hirao, H., Shin, W., Halfen, J.A., Kim, J., Que, L., Shaik, S., Nam, W.: Proc. Natl. Acad. Sci. U. S. A. **104**, 19181–19186 (2007)
236. Kaizer, J., Klinker, E.J., Oh, N.Y., Rohde, J.U., Song, W.J., Stubna, A., Kim, J., Münck, E., Nam, W., Que, L.: J. Am. Chem. Soc. **126**, 472–473 (2004)
237. Martinho, M., Banse, F., Bartoli, J.F., Mattioli, T.A., Battioni, P., Horner, O., Bourcier, S., Girerd, J.J.: Inorg. Chem. **44**, 9592–9596 (2005)
238. Krebs, C., Fujimori, D.G., Walsh, C.T., Bollinger, J.M.: Acc. Chem. Res. **40**, 484–492 (2007)
239. Balland, V., Charlot, M.F., Banse, F., Girerd, J.J., Mattioli, T.A., Bill, E., Bartoli, J.F., Battioni, P., Mansuy, D.: Eur. J. Inorg. Chem., 301–308 (2004)
240. Que, L.: Acc. Chem. Res. **40**, 493–500 (2007)
241. Decker, A., Rohde, J.U., Klinker, E.J., Wong, S.D., Que, L., Solomon, E.I.: J. Am. Chem. Soc. **129**, 15983–15996 (2007)

242. Riggs-Gelasco, P.J., Shu, L.J., Chen, S.X., Burdi, D., Huynh, B.H., Que, L., Stubbe, J.: J. Am. Chem. Soc. **120**, 849–860 (1998)
243. Shu, L.J., Nesheim, J.C., Kauffmann, K., Münck, E., Lipscomb, J.D., Que, L.: Science **275**, 515–518 (1997)
244. Martinho, M., Xue, G.Q., Fiedler, A.T., Que, L., Bominaar, E.L., Münck, E.: J. Am. Chem. Soc. **131**, 5823–5830 (2009)
245. Wallar, B.J., Lipscomb, J.D.: Chem. Rev. **96**, 2625–2657 (1996)
246. Lammers, M., Follmann, H.: Struct. Bond. **54**, 27–91 (1983)
247. Summerville, D.A., Cohen, I.A.: J. Am. Chem. Soc. **98**, 1747 (1976)
248. Buchler, J.W., Dreher, C.: Z. Naturforsch. **B 39**, 222–230 (1984)
249. Scheidt, W.R., Summerville, D.A., Cohen, A.: J. Am. Chem. Soc. **98**, 6623 (1976)
250. Ercolani, C., Hewage, S., Heucher, R., Rossi, G.: Inorg. Chem. **32**, 2975 (1993)
251. Ding, X.Q., Bill, E., Trautwein, A.X., Winkler, H., Kostikas, A., Papaefthymiou, V., Simopoulos, A., Beardwood, P., Gibson, J.F.: J. Chem. Phys. **99**, 6421–6428 (1993)
252. Gamelin, D.R., Bominaar, E.L., Kirk, M.L., Wieghardt, K., Solomon, E.I.: J. Am. Chem. Soc. **118**, 8085–8097 (1996)
253. Glaser, T., Beissel, T., Bill, E., Weyhermüller, T., Schünemann, V., Meyer-Klaucke, W., Trautwein, A.X., Wieghardt, K.: J. Am. Chem. Soc. **121**, 2193–2208 (1999)
254. Einsle, O., Tezcan, F.A., Andrade, S.L.A., Schmid, B., Yoshida, M., Howard, J.B., Rees, D.C.: Science **297**, 1696–1700 (2002)
255. Aliaga-Alcalde, M., George, S.D., Mienert, B., Bill, E., Wieghardt, K., Neese, F.: Angew. Chem. Int. Ed. **44**, 2908–2912 (2005)
256. Mehn, M.P., Peters, J.C.: J. Inorg. Biochem. **100**, 634–643 (2006)
257. Berry, J.F.: Comm. Inorg. Chem. **30**, 28–66 (2009)
258. Betley, T.A., Peters, J.C.: J. Am. Chem. Soc. **126**, 6252–6254 (2004)
259. Rohde, J.U., Betley, T.A., Jackson, T.A., Saouma, C.T., Peters, J.C., Que, L.: Inorg. Chem. **46**, 5720–5726 (2007)
260. Scepaniak, J.J., Fulton, M.D., Bontchev, R.P., Duesler, E.N., Kirk, M.L., Smith, J.M.: J. Am. Chem. Soc. **130**, 10515 (2008)
261. Jensen, M.P., Mehn, M.P., Que Jr., L.: Angew. Chem. Int. Ed. **42**, 4357 (2003)
262. Mahy, J.P., Battioni, P., Mansuy, D., Fisher, J., Weiss, R., Mispelter, J., Morgenstern-Badarau, I., Gans, P.: J. Am. Chem. Soc. **106**, 1699–1706 (1984)
263. Li, X.H., Fu, R., Lee, S.Y., Krebs, C., Davidson, V.L., Liu, A.M.: Proc. Natl. Acad. Sci. U. S. A. **105**, 8597–8600 (2008)
264. Collins, T.J., Kosta, K.L., Münck, E., Uffelmann, E.S.: J. Am. Chem. Soc. **112**, 5637 (1990)
265. Chanda, A., Popescu, D.-L., de Oliveira, F.T., Bominaar, E.L., Ryabov, A.D., Münck, E., Collins, T.J.: J. Inorg. Biochem. **100**, 606–619 (2006)
266. Berry, J.F., Bill, E., Garcia-Serres, R., Neese, F., Weyhermüller, T., Wieghardt, K.: Inorg. Chem. **45**, 2027–2037 (2006)
267. Song, Y.F., Berry, J.F., Bill, E., Bothe, E., Weyhermüller, T., Wieghardt, K.: Inorg. Chem. **46**, 2208–2219 (2007)
268. Cai, S., Licoccia, S., D'Ottavi, C., Paolesse, R., Nardis, S., Bulach, V., Zimmer, B., Shokhireva, T.K., Walker, F.A.: Inorg. Chim. Acta **339**, 171 (2002)
269. Ghosh, A., Steene, E.: J. Inorg. Biochem. **91**, 423 (2002)
270. Simkhovich, L., Goldberg, I., Gross, Z.: Inorg. Chem. **41**, 5433 (2002)
271. Ye, S., Tuttle, T., Bill, E., Simkhovich, L., Gross, Z., Thiel, W., Neese, F.: Chem. Eur. J. **14**, 10839–10851 (2008)
272. Muresan, N., Lu, C.C., Ghosh, M., Peters, J.C., Abe, M., Henling, L.M., Weyhermüller, T., Bill, E., Wieghardt, K.: Inorg. Chem. **47**, 4579–4590 (2008)
273. Roy, N., Sproules, S., Bill, E., Weyhermüller, T., Wieghardt, K.: Inorg. Chem. **47**, 10911–10920 (2008)
274. Khusniyarov, M.M., Weyhermüller, T., Bill, E., Wieghardt, K.: J. Am. Chem. Soc. **131**, 1208–1221 (2009)

275. Milsmann, C., Patra, G.K., Bill, E., Weyhermüller, T., George, S.D., Wieghardt, K.: Inorg. Chem. **48**, 7430–7445 (2009)
276. Levason, W., McAuliffe, C.A.: Coord. Chem. Rev. **12**, 151–184 (1974)
277. Demazeau, G., Buffat, B., Pouchard, M., Hagenmuller, P.: Z. Anorg. Allg. Chem. **491**, 60–66 (1982)
278. Newcomb, M., Zhang, R., Chandrasena, R.E.P., Halgrimson, J.A., Horner, J.H., Makris, T. M., Sligar, S.G.: J. Am. Chem. Soc. **128**, 4580–4581 (2006)
279. Wagner, W.-D., Nakamoto, K.: J. Am. Chem. Soc. **110**, 4044–4045 (1988)
280. Wagner, W.-D., Nakamoto, K.: J. Am. Chem. Soc. **111**, 1590–1589 (1989)
281. Dey, A., Ghosh, A.: J. Am. Chem. Soc. **124**, 3206–3207 (2002)
282. Meyer, K., Bill, E., Mienert, B., Weyhermüller, T., Wieghardt, K.: J. Am. Chem. Soc. **121**, 4859–4876 (1999)
283. Petrenko, T., George, S.D., Aliaga-Alcalde, N., Bill, E., Mienert, B., Xiao, Y., Guo, Y., Sturhahn, W., Cramer, S.P., Wieghardt, K., Neese, F.: J. Am. Chem. Soc. **129**, 11053–11060 (2007)
284. Berry, J.F., Bill, E., Bothe, E., George, S.D., Mienert, B., Neese, F., Wieghardt, K.: Science **312**, 1937–1941 (2006)
285. Herber, R.H., Johnson, D.: Inorg. Chem. **18**, 2786–2790 (1979)
286. Berry, J.F., George, S.D., Neese, F.: Phys. Chem. Chem. Phys. **10**, 4361–4374 (2008)
287. Shinjo, T., Ichida, T., Takada, T.: J. Phys. Soc. Jpn. **29**, 111–116 (1970)
288. Fluck, E.: 57-Fe: metal organic compounds. In: Goldanskii, V.I., Herber, R.H. (eds.) Chemical Applicaitons of Mössbauer Spectroscopy. Academic, New York (1968)
289. Kerler, W., Neuwirth, W.: Z. Physik. **167**, 176–193 (1962)
290. Reed, C.A.: Iron(I) and iron(IV) porphyrins. In: Kadish, K.M. (ed.) Electrochemical and Spectrochemical Studies of Biological Redox Components, pp. 333–356. American Chemical Society (1982)
291. Mashiko, T., Reed, C.A., Haller, K.J., Scheidt, W.R.: Inorg. Chem. **23**, 3192–3196 (1984)
292. Stoian, S.A., Yu, Y., Smith, J.M., Holland, P.L., Bominaar, E.L., Münck, E.: Inorg. Chem. **44**, 4915–4922 (2005)
293. Andres, H., Bominaar, E.L., Smith, J.M., Eckert, N.A., Holland, P.L., Münck, E.: J. Am. Chem. Soc. **124**, 3012–3025 (2002)
294. Stoian, S.A., Vela, J., Smith, J.M., Sadique, A.R., Holland, P.L., Münck, E., Bominaar, E.L.: J. Am. Chem. Soc. **128**, 10181–10192 (2006)
295. Chatt, J., Dilworth, J.R., Richards, R.L.: Chem. Rev. **78**, 589–625 (1978)
296. Yandulov, D.V., Schrock, R.R.: Science **301**, 76–78 (2003)
297. Burgess, B.K., Lowe, D.J.: Chem. Rev. **96**, 2983–3011 (1996)
298. Howard, J.B., Rees, D.C.: Chem. Rev. **96**, 2965–2982 (1996)
299. Seefeldt, L.C., Dean, D.R.: Acc. Chem. Res. **30**, 589–625 (1997)
300. Brown, S.D., Betley, T.A., Peters, J.C.: J. Am. Chem. Soc. **125**, 322–323 (2003)
301. Shima, S., Thauer, R.K.: Chem. Rec. **7**, 37–46 (2007)
302. Capon, J.F., Gloaguen, F., Petillon, F.Y., Schollhammer, P., Talarmin, J.: Coord. Chem. Rev. **253**, 1476–1494 (2009)
303. Ogata, H., Lubitz, W., Higuchi, Y.: Dalton Trans., 7577–7587 (2009)
304. Tard, C., Pickett, C.J.: Chem. Rev. **109**, 2245–2274 (2009)
305. Kemper, W., Kubowitz, F.: Biochem. Z. **265**, 245 (1933)
306. Warburg, O.: Schwermetalle als Wirkungsgruppen von Fermenten. Verlag Dr. Werner Sänger, Berlin (1946)
307. Peters, J.W., Lanzilotta, W.N., Lemon, B.J., Seefeldt, L.C.: Science **282**, 1853–1858 (1998)
308. Nicolet, Y., de Lacey, A.L., Vernede, X., Fernandez, V.M., Hatchikian, E.C., Fontecilla-Camps, J.C.: J. Am. Chem. Soc. **123**, 1596–1601 (2001)
309. Pierik, A.J., Hulstein, M., Hagen, W.R., Albracht, S.P.J.: Eur. J. Biochem. **258**, 572–578 (1998)
310. Popescu, C.V., Münck, E.: J. Am. Chem. Soc. **121**, 7877–7884 (1999)

311. Pereira, A.S., Tavares, P., Moura, I., Moura, J.J.G., Huynh, B.H.: J. Am. Chem. Soc. **123**, 2771–2782 (2001)
312. Bagley, K.A., Duin, E.C., Roseboom, W., Albracht, S.P.J., Woodruff, W.H.: Biochemistry **34**, 5527–5535 (1995)
313. Volbeda, A., Charon, M.-H., Piras, C., Hatchikian, E.C., Frey, M., Fontecilla-Camps, J.C.: Nature **373**, 580–587 (1995)
314. Surerus, K.K., Chen, M., Vanderzwaan, J.W., Rusnak, F.M., Kolk, M., Duin, E.C., Albracht, S.P.J., Münck, E.: Biochemistry **33**, 4980–4993 (1994)
315. Hsu, H.F., Koch, S.A., Popescu, C.V., Münck, E.: J. Am. Chem. Soc. **119**, 8371–8372 (1997)
316. Razavet, M., Davies, S.C., Hughes, D.L., Barclay, J.E., Evans, D.J., Fairhurst, S.A., Liu, X.M., Pickett, C.J.: Dalton Trans., 586–595 (2003)
317. Chalbot, M.U., Mills, A.M., Spek, A.L., Long, G.J., Bouwman, E.: Eur. J. Inorg. Chem., 453–457 (2003)
318. Shima, S., Lyon, E.J., Thauer, R.K., Mienert, B., Bill, E.: J. Am. Chem. Soc. **127**, 10430–10435 (2005)
319. Wang, X.F., Li, Z.M., Zeng, X.R., Luo, Q.Y., Evans, D.J., Pickett, C.J., Liu, X.M.: Chem. Commun., 3555–3557 (2008)
320. Mauro, A.E., Casagrande, O.L., Nogueira, V.M., Santos, R.H.A., Gambardella, M.T.P., Lechat, J.R., Filho, M.F.J.: Polyhedron **12**, 2370 (1993)
321. Obrist, B.V., Chen, D., Ahrens, A., Schünemann, V., Scopelliti, R., Hu, X.: Inorg. Chem. **48**, 3514–3516 (2009)
322. Zirngibl, C., Hedderich, R., Thauer, R.K.: FEBS Lett. **261**, 112 (1990)
323. Shima, S., Pilak, O., Vogt, S., Schick, M., Stagni, M.S., Meyer-Klaucke, W., Warkentin, E., Thauer, R.K., Ermler, U.: Science **321**, 572–575 (2008)
324. Hiromoto, T., Warkentin, E., Moll, J., Ermler, U., Shima, S.: Angew. Chem. Int. Ed. **48**, 6457–6460 (2009)
325. Lyon, E.J., Shima, S., Boecher, R., Thauer, R.K., Grevels, F.W., Bill, E., Roseboom, W., Albracht, S.P.J.: J. Am. Chem. Soc. **126**, 14239–14248 (2004)

References to Sect. 8.3

326. Squyres, S.W., Arvidson, R.E., Baumgartner, E.T., Bell III, J.F., Christensen, P.R., Gorevan, S., Herkenhoff, K.E., Klingelhöfer, G., Madsen, M.B., Morris, R.V., Rieder, R., Romero, R. A.: J. Geophys. Res. **108**(E12), 8062 (2003)
327. Klingelhöfer, G., Morris, R.V., Bernhardt, B., Rodionov, D., de Souza Jr., P.A., Squyres, S.W., Foh, J., Kankeleit, E., Bonnes, U., Gellert, R., Schröder, C., Linkin, S., Evlanov, E., Zubkov, B., Prilutski, O.: J. Geophys. Res. **108**(E12), 8067 (2003)
328. Knudsen, J.M., Madsen, M.B., Olsen, M., Vistisen, L., Koch, C.B., Moerup, S., Kankeleit, E., Klingelhöfer, G., Evlanov, E.N., Khromov, V.N., Mukhin, L.M., Prilutskii, O.F., Zubkov, B., Smirnov, G.V., Juchniewicz, J.: Hyperfine Interact. **68**, 83 (1992)
329. Knudsen, J.M.: Hyperfine Interact. **47**, 3 (1989)
330. Schröder, C., Klingelhöfer, G., Morris, R.V., Rodionov, D.S., Fleischer, I., Blumers, M.: Hyperfine Interact. **182**, 149 (2008)
331. Morris, R.V., Klingelhöfer, G., Schröder, C., Fleischer, I., Ming, D.W., Yen, A.S., Gellert, R., Arvidson, R.E., Rodionov, D.S., Crumpler, L.S., Clark, B.C., Cohen, B.A., McCoy, T.J., Mittlefehldt, D.W., Shmidt, M.E., de Souza, P.A., Squyres, S.W.: J. Geophys. Res. **113**, E12S42 (2008)
332. Morris, R.V., Klingelhöfer, G., Schröder, C., Rodionov, D.S., Yen, A., de Souza Jr., P.A., Ming, D.W., Wdowiak, T., Gellert, R., Bernhardt, B., Evlanov, E.N., Zubkov, B., Foh, J.,

Bonnes, U., Kankeleit, E., Gütlich, P., Renz, F., Squyres, S.W., Arvidson, R.E.: J. Geophys. Res. **111**, E02S13 (2006)
333. Klingelhöfer, G., Morris, R.V., Bernhardt, B., Schröder, C., Rodionov, D.S., de Souza Jr., P.A., Yen, A., Gellert, R., Evlanov, E.N., Zubkov, B., Foh, J., Bonnes, U., Kankeleit, E., Gütlich, P., Ming, D.W., Renz, F., Wdowiak, T., Squyres, S.W., Arvidson, R.E.: Science **306**, 1740 (2004)
334. Morris, R.V., Klingelhöfer, G., Schröder, C., Rodionov, D.S., Yen, A., Ming, D.W., De Souza Jr., P.A., Wdowiak, T., Fleischer, I., Gellert, R., Bernhardt, B., Bonnes, U., Cohen, B.A., Evlanov, E.N., Foh, J., Gütlich, P., Kankeleit, E., McCoy, T., Mittlefehldt, D.W., Renz, F., Schmidt, M.E., Zubkov, B., Squyres, S.W., Arvidson, R.E.: J. Geophys. Res. **111**, E12S15 (2006)
335. Squyres, S.W., Knoll, A.H., Arvidson, R.E., Ashley, J.W., Bell III, J.F., Calvin, W.M., Christensen, P.R., Clark, B.C., Cohen, B.A., De Souza Jr., P.A., Edgar, L., Farrand, W.H., Fleischer, I., Gellert, R., Golombek, M.P., Grant, J., Grotzinger, J., Hayes, A., Herkenhoff, K.E., Johnson, J.R., Jolliff, B., Klingelhöfer, G., Knudson, A., Li, R., McCoy, T.J., McLennan, S.M., Ming, D.W., Mittlefehldt, D.W., Morris, R.V., Rice Jr., J.W., Schröder, C., Sullivan, R.J., Yen, A., Yingst, R.A.: Science **324**, 1058 (2009)
336. Klingelhöfer, G.: Hyperfine Interact. **113**, 369 (1998)
337. Klingelhöfer, G.: In: Garcia, M., Marco, J.F., Plazaola, F. (eds.) Industrial Applications of the Mössbauer Effect, p. 369. American Institute of Physics, New York (2005)
338. Klingelhöfer, G., Fegley Jr., B., Morris, R.V., Kankeleit, E., Held, P., Evlanov, E., Priloutskii, O.: Planet. Space Sci. **44**, 1277–1288 (1996)
339. Klingelhöfer, G., Held, P., Teucher, R., Schlichting, F., Foh, J., Kankeleit, E.: Hyperfine Interact. **95**, 305 (1995)
340. Bell, J. (ed.): The Martian Surface. Cambridge University Press, Cambridge (2008)
341. Morris, R.V., Klingelhöfer, G.: In: Bell, J. (ed.) The Martian Surface, p. 339. Cambridge University Press, Cambridge (2008)
342. Kieffer, H.H., Jakosky, B.M., Snyder, C.W., Mathews, M.S. (eds.): Mars. University of Arizona Press, Tucson (2008)
343. Barlow, N. (ed.): Mars: An Introduction to its Interior, Surface and Atmosphere. Cambridge University Press, Cambridge (2008)
344. Gibson, E.K., Pillinger, C.T., Wright, I.P., Morgan, G.H., Yau, D., Stewart, J.L.C., Leese, M.R., Praine, I.J., Sheridan, S., Morse, A.D., Barber, S.J., Ebert, S., Goesmann, F., Roll, R., Rosenbauer, H., Sims, M.R.: LPSC **35**, 1845 (2004)
345. Pillinger, C.: Space is a Funny Place. Barnstorm Productions, Venice (2007)
346. Fleischer, I., Klingelhöfer, G., Schröder, C., Morris, R.V., Hahn, M., Rodionov, D., Gellert, R., de Souza, P.A.: J. Geophys. Res. **113**, E06S21 (2008)
347. Fleischer, I., Klingelhöfer, G., Schröder, S., Rodionov, D.: Hyperfine Interact. **186**, 193 (2008)
348. Klingelhöfer, G., Imkeller, U., Kankeleit, E., Stahl, B.: Hyperfine Interact. **71**, 1445 (1992)
349. Riesenmann, R., Steger, J., Kostiner, E.: Nucl. Instrum. Method. **72**, 109 (1969)
350. Held, P.: MIMOS II – Ein miniaturisiertes Mößbauerspektrometer in Rückstreugeometrie zur mineralogischen Analyse der Marsoberfläche. Ph.D. thesis, Institute of Nuclear Physics, TU Darmstadt (1997)
351. Squyres, S.W., Arvidson, R.E., Bell III, J.F., Brückner, J., Cabrol, N.A., Calvin, W., Carr, M. H., Christensen, P.R., Clark, B.C., Crumpler, L., Des Marais, D.J., d'Uston, C., Economu, T., Farmer, J., Farrand, W., Folkner, W., Golombek, M., Gorevan, S., Grant, J.A., Greely, R., Grotzinger, J., Haskin, L., Herkenhoff, K.E., Hviid, S., Johnson, J., Klingelhöfer, G., Knoll, A.H., Landis, G., Lemmon, M., Li, R., Madsen, M.B., Malin, M.C., McLennan, S. M., McSween, H.Y., Ming, D.W., Moersch, J., Morris, R.V., Parker, T., Rice Jr., J.W., Richter, L., Rieder, R., Sims, M., Smith, P., Soderblom, L.A., Sullivan, R., Wänke, H., Wdowiak, T., Wolff, M., Yen, A.: Science **306**, 1698 (2004)
352. Squyres, S.W., Arvidson, R.E., Bell III, J.F., Brückner, J., Cabrol, N.A., Calvin, W., Carr, M. H., Christensen, P.R., Clark, B.C., Crumpler, L., Des Marais, D.J., d'Uston, C., Economu, T., Farmer, J., Farrand, W., Folkner, W., Golombek, M., Gorevan, S., Grant, J.A., Greely, R.,

Grotzinger, J., Haskin, L., Herkenhoff, K.E., Hviid, S., Johnson, J., Klingelhöfer, G., Knoll, A.H., Landis, G., Lemmon, M., Li, R., Madsen, M.B., Malin, M.C., McLennan, S.M., McSween, H.Y., Ming, D.W., Moersch, J., Morris, R.V., Parker, T., Rice Jr., J.W., Richter, L., Rieder, R., Sims, M., Smith, M., Smith, P., Soderblom, L.A., Sullivan, R., Wänke, H., Wdowiak, T., Wolff, M., Yen, A.: Science **305**, 794 (2004)

353. Rieder, R., Gellert, R., Brückner, J., Clark, B.C., Dreibus, G., d'Uston, C., Economou, T., Klingelhöfer, G., Lugmeier, G.W., Wänke, H., Yen, A., Zipfel, J., Squyres, S.W., Athena Science Team: Lunar Planet. Sci. **35**, 2172 (2004)
354. Rieder, R., Gellert, R., Brückner, J., Klingelhöfer, G., Dreibus, G., Yen, A., Squyres, S.W.: J. Geophys. Res. **108**(E12), 8066 (2003)
355. McSween, H., Wyatt, M.B., Gellert, R., Bell III, J.F., Morris, R.V., Herkenhoff, K.E., Crumpler, L.S., Milam, K.A., Stockstill, K.R., Tornabene, L., Arvidson, R.E., Bartlett, P., Blaney, D., Cabrol, N.A., Christensen, P.R., Clark, B.C., Crisp, J.A., Des Marais, D.J., Economou, T., Farmer, J.D., Farrand, W., Ghosh, A., Golombek, M., Gorevan, S., Greeley, R., Hamilton, V.E., Johnson, J.R., Joliff, B.L., Klingelhöfer, G., Knudson, A.T., McLennan, S., Ming, D., Moersch, J.E., Rieder, R., Ruff, S.W., Schroeder, C., de Souza Jr., P.A., Squyres, S.W., Wänke, H., Wang, A., Yen, A., Zipfel, J.: J. Geophys. Res. **111**, E02S10 (2006)
356. Klingelhöfer, G., DeGrave, E., Morris, R.V., Van Alboom, A., De Resende, V.G., De Souza Jr., P.A., Rodionov, D., Schröder, C., Ming, D.W., Yen, A.: Hyperfine Interact. **166**, 549 (2005)
357. Van Cromphaut, C., de Resende, V.G., De Grave, E., Van Alboom, A., Vandenberghe, R.E., Klingelhöfer, G.: Geochim. Cosmochim. Acta **71**, 4814 (2007)
358. Klingelhöfer, G., Morris, R.V., de Souza Jr., P.A., Rodionov, D., Schröder, C.: Hyperfine Interact. **170**, 169 (2006)
359. Schröder, C., Rodionov, D.S., McCoy, T.J., Joliff, B.L., Gellert, R., Nittler, L.R., Farrand, W.H., Johnson, J.R., Ruff, S.W., Ashley, J.W., Mittlefehldt, D.W., Herkenhoff, K.E., Fleischer, I., Haldemann, A.F.S., Klingelhöfer, G., Ming, D.W., Morris, R.V., de Souza Jr., P.A., Squyres, S. W., Weitz, C., Yen, A.S., Zipfel, J., Economu, T.: J. Geophys. Res. **113**, E06S22 (2008)
360. Fleischer, I., Schröder, C., Klingelhöfer, G., Gellert, R.: Meteorit. Planet. Sci. Suppl., 5238 (2009)
361. Schröder, C., Ashley, J.W., Chapman, M.G., Cohen, B.A., Farrand, W.H., Fleischer, I., Gellert, R., Herkenhoff, K.E., Johnson, J.R., Jolliff, B.L., Joseph, J., Klingelhöfer, G., Morris, R.V., Squyres, S.W., Wright, S.P., Athena Science Team: LPSC **40**, 1665 (2009)
362. de Souza Jr., P.A., Klingelhöfer, G., Morimoto, T.: Hyperfine Interact. **C 5**, 487 (2002)
363. Feder, F., Trolard, F., Klingelhöfer, G., Bourrie, G.: Geochim. Cosmochim. Acta **69**, 4463 (2005)
364. Rodionov, D., Klingelhöfer, G., Bernhardt, B., Schröder, C., Blumers, M., Kane, S., Trolard, F., Bourrie, G., Genin, J.-M.: Hyperfine Interact. **167**, 869 (2006)
365. de Souza Jr., P.A., Bernhardt, B., Klingelhöfer, G., Gütlich, P.: Hyperfine Interact. **151/152**, 125–130 (2003)
366. Klingelhöfer, G.: In: Miglierini, M., Petridis, D. (eds.) Mössbauer Spectroscopy in Materials Science, p. 413. Kluwer Academic, Amsterdam (1999)
367. Klingelhöfer, G., da Costa, G.M., Prous, A., Bernhardt, B.: Hyperfine Interact. **C 5**, 423–426 (2002)
368. Prous, A., Malta, I.M.: Santana do Riacho, Tomo I. Arquivos do Museu de História Natural **12**, 384 (1991)
369. Prous, A.: Santana do Riacho, Tomo I. Arquivos do Museu de História Natural **13/14**, 417 (1992)
370. Strüder, L., Lechner, P., Leutenegger, P.: Naturwissenschaften **11**, 539–543 (1998)
371. Blumers, M., Bernhardt, B., Lechner, P., Klingelhöfer, G., d'Uston, C., Soltau, H., Strüder, L., Eckhardt, R., Brückner, J., Henkel, H., Girones Lopez, J., Maul, J.: Nucl. Instr. Meth. A, (2010), doi:10.1016/j.nima.2010.04.07

Chapter 9
Nuclear Resonance Scattering Using Synchrotron Radiation (Mössbauer Spectroscopy in the Time Domain)

9.1 Introduction

Conventional Mössbauer spectroscopy (MS) can be considered as "spectroscopy in the energy domain." It has been widely used since its discovery in 1958 [1]. Nuclear resonant forward scattering (NFS) of synchrotron radiation has been successfully employed as a time-differential technique since 1991 [2]. Another related technique, nuclear inelastic scattering (NIS) of synchrotron radiation [3], can be regarded as an extension of conventional, energy-resolved MS (in the range 10^{-9} to 10^{-7} eV) to energies on the order of molecular vibrations (in the range 10^{-3} to 10^{-1} eV). So far only a few "Mössbauer" stations for NFS and NIS measurements have become available in synchrotron laboratories, i.e., in Germany, France, Japan, and the USA.

The use of synchrotron radiation overcomes some of the limitations of the conventional technique. The high brilliance of up to 10^{20} photons s^{-1} mm^{-2} $mrad^{-2}$/0.1% bandwidth of energy, and the extremely collimated synchrotron beam lead to a large flux of photons through a very small cross section (0.1–1 mm^2). This allows measurements with samples of small volume if isotopically enriched (with the relevant Mössbauer isotope, e.g., ^{57}Fe). Measurements that were described earlier [4] and that require a polarized Mössbauer source now become experimentally more feasible by making use of the polarization of the synchrotron radiation. Additionally, the energy can be tuned over a wide range. This facilitates measurements with those Mössbauer nuclei for which conventional sources are available but with life times that are too short for most experimental purposes, e.g., 99 min for ^{61}Co $\rightarrow$ ^{61}Ni and 78 h for ^{67}Ga $\rightarrow$ ^{67}Zn.

NIS of synchrotron radiation yields details of the dynamics of Mössbauer nuclei, while conventional MS yields only limited information in this respect (comprised in the Lamb–Mössbauer factor f). NIS shows some similarity with Resonance Raman- and IR-spectroscopy. The major difference is that, instead of an electronic resonance (Raman and IR), a nuclear resonance is employed (NIS). NIS is site-selective, i.e., only those molecular vibrations that contribute to the overall

P. Gütlich et al., *Mössbauer Spectroscopy and Transition Metal Chemistry*,
DOI 10.1007/978-3-540-88428-6_9, © Springer-Verlag Berlin Heidelberg 2011

mean-square displacement (msd) of the Mössbauer nucleus with a nonzero fraction are NIS visible. Thus, the measured peaks in the NIS spectrum can be directly interpreted in terms of vibrational modes since electronic properties, such as polarization, do not influence their intensity.

Some of the basic features of nuclear resonance scattering using synchrotron radiation (NFS and NIS) and of conventional MS are compared in Table 9.1.

9.2 Instrumentation

A typical setup for NFS and NIS measurements as installed, for example, at the European Synchrotron Radiation Facility (ESRF) [5] is shown in Fig. 9.1.

A monochromatic beam of X-rays with about 1 eV bandwidth is produced by the standard beamline equipment, the undulator and the high-heat-load premonochromator being the most important parts among them. Further monochromatization down to approximately the millielectronvolt bandwidth is achieved with the high-resolution monochromator. The width of a band of a millielectronvolt, however, is much more than the inherent linewidth of the ^{57}Fe γ-radiation, $\Gamma \sim 10^{-9}$ eV, or the full range of hyperfine-split Mössbauer lines, $\Delta E_{\mathrm{M}} \sim 10^{-7}$ eV. Yet, NFS is detectable because the coherent excitation of the nuclei is caused in the

Table 9.1 Comparison of basic features of NFS (NIS) and conventional MS

	NFS	MS
Sample cross section (mm^2)	0.1–1.0	10–100
Amount of ^{57}Fe[a] (μg)	~3	~3
Data acquisition time[b] (h)	1–5	24–120
Background[b] (%)	~1	~90
Data analysis[c]	Complex, visually controlled simulations necessary	Simple, least square fits using Lorentzians are routine

[a]NFS and MS are volume-sensitive; NIS is surface-sensitive, i.e., escape depth of the KX_{α}-radiation of Fe is ~100 nm
[b]For NIS: same value
[c]For NIS: simple, vibrational modes directly visible

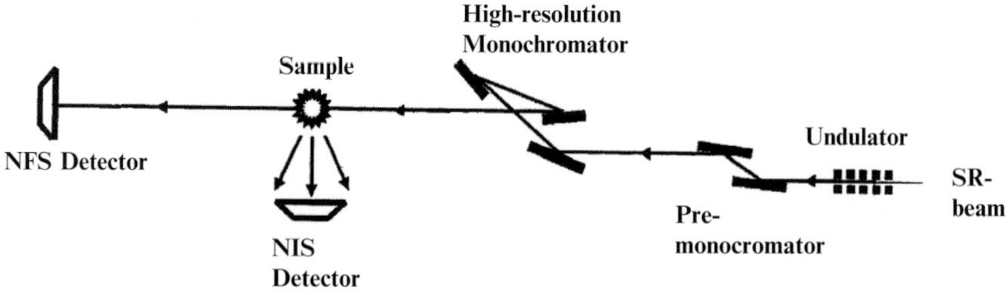

Fig. 9.1 Schematic representation of the experimental arrangement for nuclear resonant scattering, both for NIS and NFS

sample by at least some photons out of the relatively broad-band synchrotron radiation pulses.

The radiation that results from nuclear de-excitation has to be distinguished from the prompt radiation, i.e., from the photons that pass through the sample without interaction and from those that are scattered by the electrons. This is achieved if three characteristic times (Δt_1, Δt_2, and τ) are properly related:

(a) Synchrotron radiation is emitted by electron or positron bunches of $\Delta t_1 \sim 200$ ps length and adjustable temporal spacing of $\Delta t_2 \sim 2$–3,000 ns.
(b) Because electronic scattering of X-rays is fast compared to the slow nuclear scattering (the mean life time of the excited nuclear state of ^{57}Fe is $\tau = 141$ ns), one can separate the "prompt" electronic-scattering signal from the nuclear-scattering signal, which appears after successive nuclear decay as a time-delayed coherent electromagnetic radiation, by means of a fast detector system and by adjusting the spacing Δt_2 between the bunches relative to Δt_1 and τ, e.g., $\Delta t_1 \ll \tau$ and $\Delta t_2 \geq \tau$.

Nuclear scattering is counted by two avalanche photo diode (APD) detectors. The detector for NIS (Fig. 9.1) is located close to the sample. It counts the quanta scattered in a large solid angle. The detector for NFS is located far away from the sample. It counts the quanta scattered by the nuclei in the forward direction. These two detectors follow two qualitatively different processes of nuclear scattering:

(a) In the forward direction (NFS), the photons are scattered elastically; additionally, their radiation field in the sample is characterized by temporal and spatial coherence (see Sect. 9.3.2). The NFS spectrum is detected by monitoring the intensity of the radiation in the forward direction as a function of the delay time (with respect to the prompt signal).
(b) In the other case (NIS) when the photons are scattered inelastically, they acquire a certain phase shift and, therefore, they are no longer coherent with the incident radiation. If the phase shift is random for various nuclei, the scattering is spatially incoherent over the nuclear ensemble, and the scattered photons may be associated with some individual nuclei. The photons resulting from the de-excitation of individual nuclei are emitted in a large solid angle as spherical waves (neglecting polarization effects). Thus, the NIS detector monitors the energy spectrum of inelastic excitation. In addition to incoherent scattering of primary radiation, the detector may also collect delayed atomic fluorescent (KX_α) radiation resulting from internal conversion. This contribution is dominant in, for instance, ^{57}Fe, which is the most important Mössbauer isotope.

9.3 Nuclear Forward Scattering (NFS)

The primary parameters that can be extracted from conventional Mössbauer spectra are the Lamb–Mössbauer factor, *f*, as well as the various hyperfine parameters that provide information about the state of the electronic environment of the Mössbauer nuclei. NFS is in a similar way sensitive to the nuclear environment. Therefore time-differential NFS spectra also provide hyperfine parameters; in most cases, they are more precise than those obtained by conventional MS (examples are given in Figs. 9.22 and 9.23).

9.3.1 *Quadrupole Splitting: Theoretical Background*

In the case of resonance absorption of synchrotron radiation by an ^{57}Fe nucleus in a polycrystalline sample, the frequency dependence of the electric field of the forward scattered radiation, $R(\omega)$, takes a Lorentzian lineshape. In order to gain information about the time dependence of the transmitted radiation, the expression for $R(\omega)$ has to be Fourier-transformed into $R(t)$ [6].

With the two excited nuclear states E_1 and E_2 given in terms of the quadrupole splitting $\Delta E_Q = E_1 - E_2$ and taking $|R(t)|^2$, the time dependence of the intensity of the delayed radiation (quantum beats (QBs)) is given by:

$$I(t) \sim \exp\left(-\frac{t}{\tau}\right)\cos^2\left(\frac{\Delta E_Q t}{2\hbar}\right). \tag{9.1}$$

Equation (9.1) documents that quadrupole splittings ΔE_Q exhibit quantum-beat spectra with period $\hbar/2\pi\Delta E_Q$ superimposed over the time dependence of the nuclear decay $\exp(-t/\tau)$ with mean decay time $\tau = 141$ ns for ^{57}Fe. In Fig. 9.2, quadrupole splittings $\Delta E_Q = 0$ and 2 mm s^{-1} in the energy domain (conventional MS) are compared with those in the time domain (MS using synchrotron radiation) [7]. The QBs in the time domain spectrum for $\Delta E_Q = 2$ mm s^{-1} are the result of the interference between the radiation scattered by different nuclear resonances. Consequently, their frequencies correspond to the energetic differences between these resonances.

Equation (9.1) and therefore the spectra shown in Fig. 9.2 apply to "thin" nuclear scatterers, e.g., for effective thickness $t_{eff} < 1$. Using a cylindrical sample with a volume of 0.5 ml and a basal area of 0.5 cm^2, the situation $t_{eff} \sim 1$ corresponds to a concentration of about 1 mM in ^{57}Fe (yielding a quadrupole doublet with about 5% resonance effect in the energy domain). Samples with higher concentration, i.e., with higher effective thickness, $t_{eff} > 1$, exhibit a modulation of the forward scattering intensity (dynamical beats, see Fig. 9.3) in addition to the QBs.

9.3.2 *Effective Thickness, Lamb–Mössbauer Factor*

NFS is an elastic and coherent scattering process, i.e., it takes place without energy transfer to electronic or vibronic states and is delocalized over many nuclei. Owing to the temporal and spatial coherence of the radiation field in the sample,

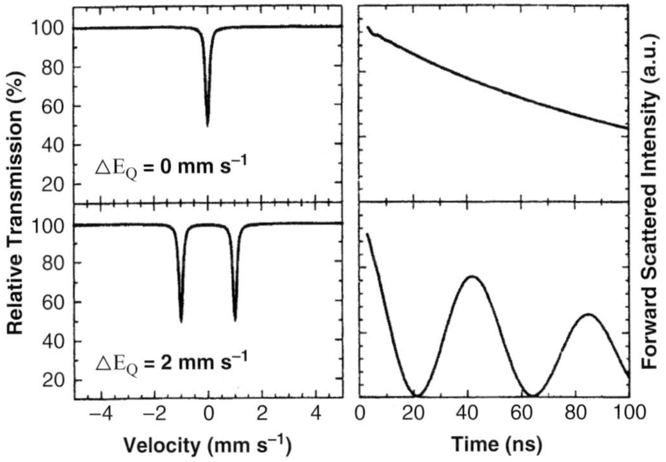

Fig. 9.2 Mössbauer spectra in the energy domain and in the time domain. Non-zero quadrupole splitting shows up in the time domain as quantum beats. (Taken from [7])

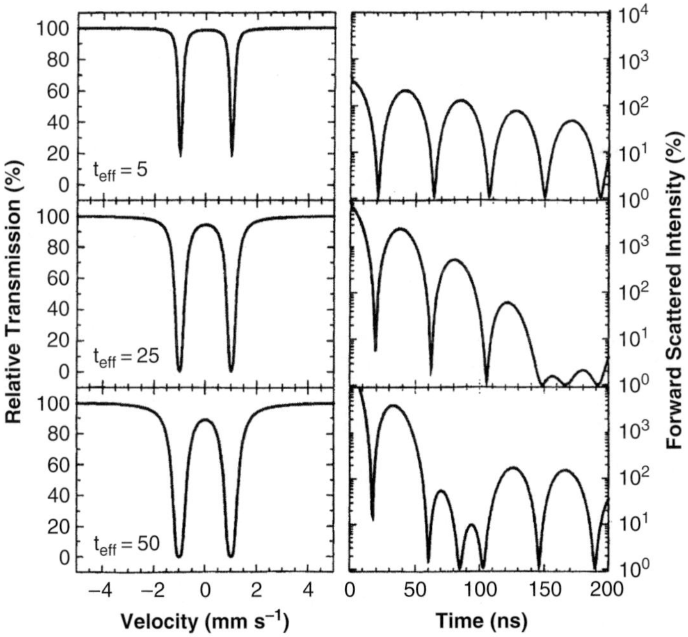

Fig. 9.3 Mössbauer spectra ($\Delta E_Q = 2$ mm s^{-1}) in the energy domain and in the time domain. High effective thickness t_{eff} appears in the energy domain as line broadening and in the time domain as dynamical beats which are superimposed over the quantum beats. (Taken from [7])

a characteristic pattern of so-called dynamical beats develops during the propagation through the sample. This pattern can be used to determine the effective thickness and thus the Lamb–Mössbauer factor of the sample in a much better way than in conventional MS. The time-differential intensity of NFS by a nuclear state of width $\Gamma = \hbar/\tau$ which is subject, for example, to an electric quadrupole splitting $|\Delta E_Q| = \hbar\Delta\omega$ with $|\Delta E_Q| \gg t_{\text{eff}}\Gamma$ can be approximated by [8].

$$I(t) \sim \frac{\Gamma t_{\text{eff}}}{\Delta E_\gamma t/\tau} \exp(-t/\tau - \sigma_{\text{el}} nd) \mathrm{J}_1^2 \sqrt{\frac{0.5 t_{\text{eff}} t}{\tau}} \cos^2\left(\frac{1}{2}\Delta\omega t + \frac{t_{\text{eff}}\Gamma}{8\hbar\Delta\omega}\right). \tag{9.2}$$

In (9.2), ΔE_γ is the bandwidth of the incoming radiation and σ_{el} is the electronic absorption cross section. The exponential decay is modulated by the square of a Bessel function of the first order (J_1), giving rise to the aforementioned dynamical beats. The positions of their minima and maxima (i.e., the slope of the envelope of the time-dependent intensity) can be determined with high accuracy and thus give precise information about the effective thickness of the sample,

$$t_{\text{eff}} = dnf\sigma_0, \tag{9.3}$$

which appears in the argument of J_1 (see Figs. 9.3 and 9.12a). Because the geometric thickness d of the sample, the number of Mössbauer nuclei/unit volume n, and the nuclear absorption cross section at resonance σ_0 are generally known, the Lamb–Mössbauer factor f can be determined without having to correct for the background and source properties as in conventional MS. Figure 9.3 shows the influence of t_{eff} upon Mössbauer spectra in the energy and time domains [7].

The mathematical description of the time-differential NFS intensity is, in many cases (e.g., in cases when frozen solutions are investigated), not as straightforward as it may appear in (9.2). The reason is that couplings between the various components of the delocalized radiation field in the sample have to be taken into account by an integration over all frequencies. This problem has been solved in different ways in a series of program packages, the most prominent of which are called CONUSS [9, 10], MOTIF [11, 12] and SYNFOS [13, 14].

A 1:1 superposition of two quadrupole splittings may be easily recognized by mere visual inspection of the spectrum in the energy domain; this, however, is not possible with the corresponding spectrum in the time domain. Two examples of this situation are provided in Fig. 9.4 [13]: practically the same beat pattern as for $t_{\text{eff}} = 10$, $\Delta E_Q = 4.2$ mm s^{-1} is obtained for a sample with $t_{\text{eff}} = 4$ and a 1:1 mixture of two species with quadrupole splittings 4.0 and 4.4 mm s^{-1}. The corresponding situation is obtained when comparing $t_{\text{eff}} = 40$ and $\Delta E_Q = 4.2$ mm s^{-1} with $t_{\text{eff}} = 4$ and $\Delta E_Q = 3.8$ and 4.6 mm s^{-1}. In both respective pairs of spectra (a,c) and (b,d), minima due to destructive interferences appear at about the same delay times. To resolve this ambiguity, it is necessary to perform measurements with different effective thicknesses, e.g., by recording temperature-dependent spectra of the sample, or to estimate t_{eff} separately.

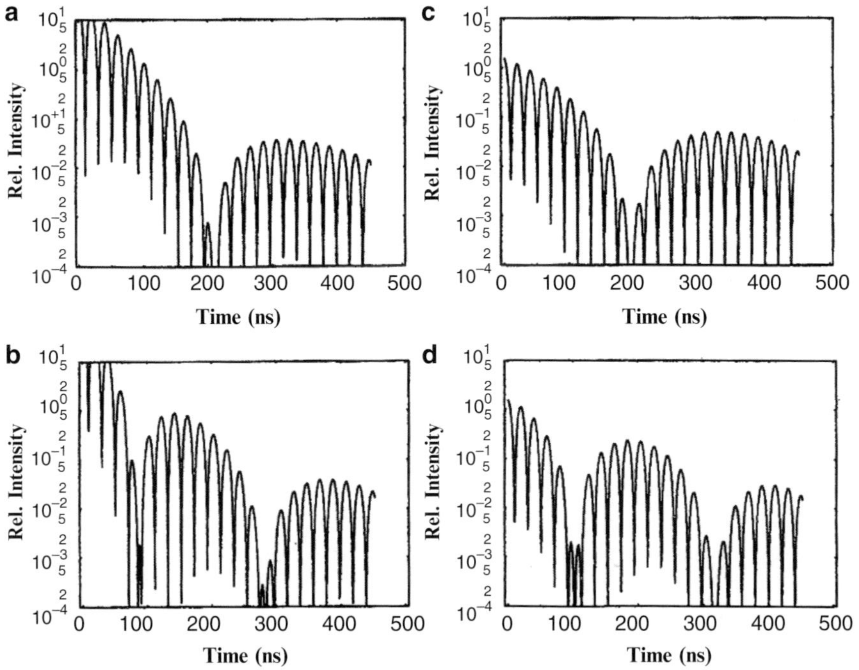

Fig. 9.4 Calculated time spectra for absorbers showing pure quadrupole splitting: (**a**) $\Delta E_Q = 4.2\ \text{mm s}^{-1}$, $t_{eff} = 10$, (**b**) $\Delta E_Q = 4.2\ \text{mm s}^{-1}$, $t_{eff} = 40$, (**c**) a 1:1 mixture of $\Delta E_Q = 4.0$ and $4.4\ \text{mm s}^{-1}$, $t_{eff} = 4$, and (**d**) a 1:1 mixture of $\Delta E_Q = 3.8$ and $4.6\ \text{mm s}^{-1}$, $t_{eff} = 4$. (Taken from [13])

This situation illustrates that special care has to be taken when analyzing NFS data. Nevertheless, high expectations rest on MS in the time domain especially for biological applications because it promises to be more sensitive to hyperfine parameters when the measurements are performed at high delay times and because of its applicability to samples of much smaller size.

9.4 NFS Applications

9.4.1 Polycrystalline Material Versus Frozen Solution (Example: "Picket-Fence" Porphyrin and Deoxymyoglobin)

Polycrystalline material and frozen solution differ in packing density and packing homogeneity. Polycrystalline material is normally characterized by high but inhomogeneous packing density, while for a frozen solution the reverse is observed,

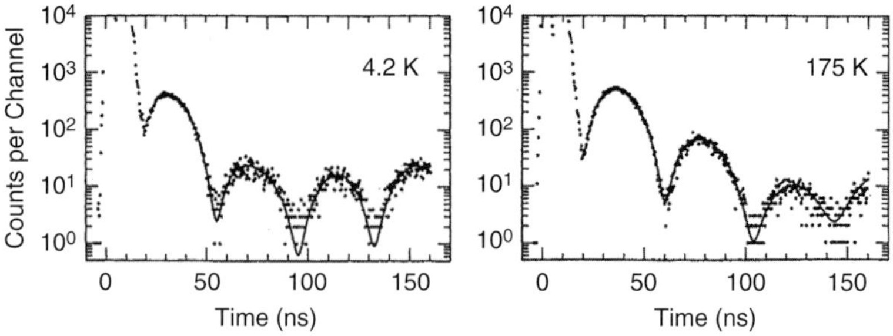

Fig. 9.5 NFS spectra of polycrystalline $FeO_2(SC_6HF_4)$ $(TP_{piv}P)$ in zero field. The fitted spectra (*solid lines*) were obtained with CONUSS [9, 10]. (Taken from [7])

i.e., packing density is low but it is homogeneous over the whole sample (if frozen adequately, for example by shock freezing).

The simulation of experimental NFS spectra of a polycrystalline sample of the "picket-fence" porphyrin $[FeO_2(SC_6HF_4)(TP_{piv}P)]$ at various temperatures (4.2 and 175 K, Fig. 9.5) [7] was performed with a temperature-dependent quadrupole splitting $\Delta E_Q(T)$ and with a rectangular effective thickness distribution ($t_{eff} \sim$ 34–74 at 4.2 K and 20–43 at 175 K). The physical background for the observation of ΔE_Q for this diamagnetic (ferrous low-spin) case is discussed together with that of oxymyoglobin (see Sect. 9.4.3). The distribution in t_{eff} indicates the sensitivity of the NFS method with respect to inhomogeneities of concentration over the sample. (Inhomogeneities can be made directly visible by scanning the highly collimated synchrotron beam over the sample area).

Frozen solution of deoxymyoglobin (Mb) has been the subject of an NFS investigation in the temperature range 3.2–230 K (Fig. 9.6) [15]. The synchrotron pulses were transmitted through the entrance window of the sample with a size of 12 mm · 2 mm (width · height). By scanning the sample area with a narrow beam (about 1 mm · 0.3 mm), the homogeneity of the effective thickness was determined as ±2.5%, which is more than ten times better than in the aforementioned case.

The envelope of the spectra is mainly determined by the effective thickness of the sample. For small effective thicknesses, the envelope can be approximated by an exponential decay of the measured intensity with time [13]. Therefore, on the semilogarithmic scale shown in Fig. 9.6, the slope of the envelope provides the effective thickness (see dashed line for $T = 3.2$ K). The slower decay with increasing temperature corresponds to a decrease in the effective thickness which is equivalent to an increase in the msd $\langle x^2 \rangle$ of iron with temperature

$$t_{eff} = Cf(T)$$

$$f(T) = \exp(-k^2 \langle x^2 \rangle), \tag{9.4}$$

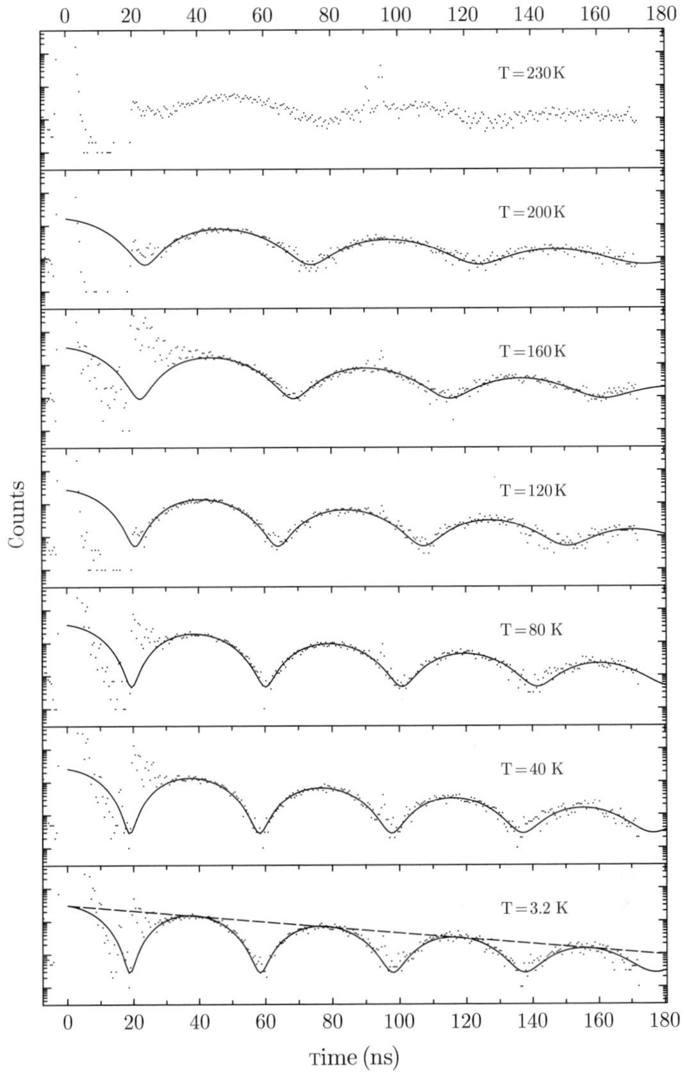

Fig. 9.6 Measured NFS spectra of deoxymyoglobin at the indicated temperatures. The *solid lines* are the simulations obtained with SYNFOS [13, 14] using the Debye model for the effective thickness as described in the text. Taken from [15]

with k representing the wave vector of the irradiation. Using the Debye model to simulate the temperature variation of the msd, the Lamb–Mössbauer factor can be approximated by:

$$f(T) = \exp\left\{-\frac{6E_\mathrm{R}}{k_\mathrm{B}\theta_\mathrm{D}}\left[\frac{1}{4}+\left(\frac{T}{\theta_\mathrm{D}}\right)^2\int_0^{\theta_\mathrm{D}}\frac{x}{\mathrm{e}^x-1}\mathrm{d}x\right]\right\}, \tag{9.5}$$

with $E_{R\gamma} = E^2/2mc^2$ being the recoil energy, m the mass of the iron atom, and k_B the Boltzmann constant. In this approach, only the Debye temperature θ_D is needed to describe the variation of the effective thickness with temperature.

All measured spectra in Fig. 9.6 were fitted simultaneously to derive the parameters θ_D and ΔE_Q. In contrast to ΔE_Q, which was allowed to vary with temperature, C and θ_D were kept constant over the whole temperature range. An initial value of C was obtained by a fit of the low-temperature spectrum recorded at $T = 3.2$ K.

The spectrum recorded at 230 K was discarded in the fit procedure because above 200 K the effective thickness decreases drastically because of a significant softening of protein-specific modes [16]. From the simultaneous fit of the spectra in the temperature range 3.2–200 K, the Debye temperature was determined as $\theta_D = 215$ K. ΔE_Q proved to be a temperature-dependent quantity, which is discussed later (see Sect. 9.4.2).

9.4.2 *Temperature-Dependent Quadrupole Splitting in Paramagnetic (S = 2) Iron Compounds (Example: Deoxymyoglobin)*

The heme iron in Mb is characterized by the ferrous high-spin (Fe^{II}, $S = 2$) state. The $3d^6$ configuration in this case exhibits, in a ligand-field picture, an electron distribution over e_g-type iron atomic orbitals (AOs), $d_{x^2-y^2}$ and d_{z^2}, and t_{2g}-type AOs, d_{xy}, d_{xz} and d_{yz}. For the hexa-coordinated iron in Mb, the t_{2g}-type AOs are lower in energy than the e_g-type AOs. The three t_{2g} AOs are moderately separated in energy because of a moderate electrostatic interaction of the heme iron with its direct coordination sphere. Within this electronic configuration (Fig. 9.7), a temperature-dependent Boltzmann population in the subspace of t_{2g} AOs is responsible for the temperature variation of ΔE_Q and, correspondingly, also for the QBs in the measured NFS spectra of Mb (Fig. 9.6). This behavior is in complete agreement with the results obtained from conventional Mössbauer studies of Mb [17–19].

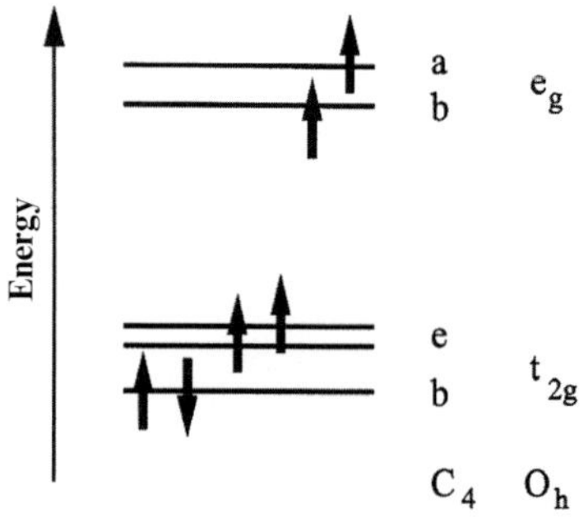

Fig. 9.7 Idealized $3d^6$, $S = 2$ electronic configuration for the heme iron in deoxymyoglobin

9.4.3 Dynamically Induced Temperature-Dependence of Quadrupole Splitting (Example: Oxymyoglobin)

The quadrupole splitting of ferrous low-spin systems (Fe^{II}, $S = 0$) is expected to be temperature independent. The reason is that the $3d^6$ configuration exhibits an energy separation between e_g- and t_{2g}-type iron AOs that is much larger than that typical of the ferrous high-spin case. Hence, the resulting low-spin $t_{2g}{}^6$ configuration (Fig. 9.8) is energetically unfavorable for temperature-dependent Boltzmann populations of empty excited e_g-type AOs. Nevertheless, oxymyoglobin (MbO_2) [20, 21] and related "picket-fence" porphyrin complexes [22, 23], which are truly diamagnetic as demonstrated by magnetic Mössbauer spectra [24, 25], do exhibit a pronounced temperature dependence of ΔE_Q.

Originally, this temperature variation of ΔE_Q was attributed to the dynamic distribution of the terminal oxygen of the FeO_2-moiety, as suggested by X-ray structural results for "picket-fence" porphyrins [26, 27]. This view is now supported by NFS studies which provide more information on dynamic processes in iron-containing molecules.

The NFS spectra of MbO_2 recorded at various temperatures are shown in Fig. 9.9 [15]. The spectra were fitted with ΔE_Q and t_{eff} treated as free parameters. The result of this analysis is that the values of $\Delta E_Q(T)$ agree with those obtained from earlier Mössbauer studies in the energy domain. However, the obtained t_{eff} values surprisingly increase with increasing temperature. The latter result cannot be correct because the Lamb–Mössbauer factor f and therefore t_{eff} grow only if $\langle x^2 \rangle$ decreases according to (9.4), and this should be the case only with decreasing temperature.

A physically more plausible mechanism for MbO_2 that influences the NFS spectra in the same way as an increase of t_{eff} is "relaxation effects." Both large t_{eff} and relaxation cause a fast decay of the NFS intensity with delay time. In the case of MbO_2, relaxation is represented by the dynamical structural disorder of the terminal oxygen (Fig. 9.10) which causes stochastic fluctuations of the electric field gradient (EFG) tensor. A consequence of this is a dephasing of originally coherent electromagnetic waves, thus affecting the quantum-beat pattern of the NFS spectra.

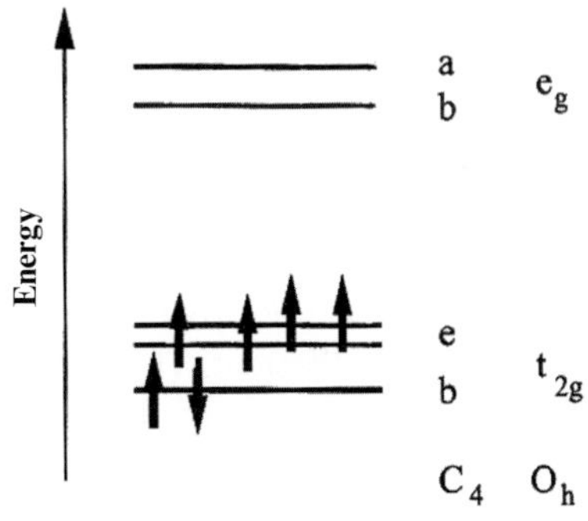

Fig. 9.8 Idealized $3d^6$, $S = 2$ electronic configuration for the heme iron in deoxymyoglobin

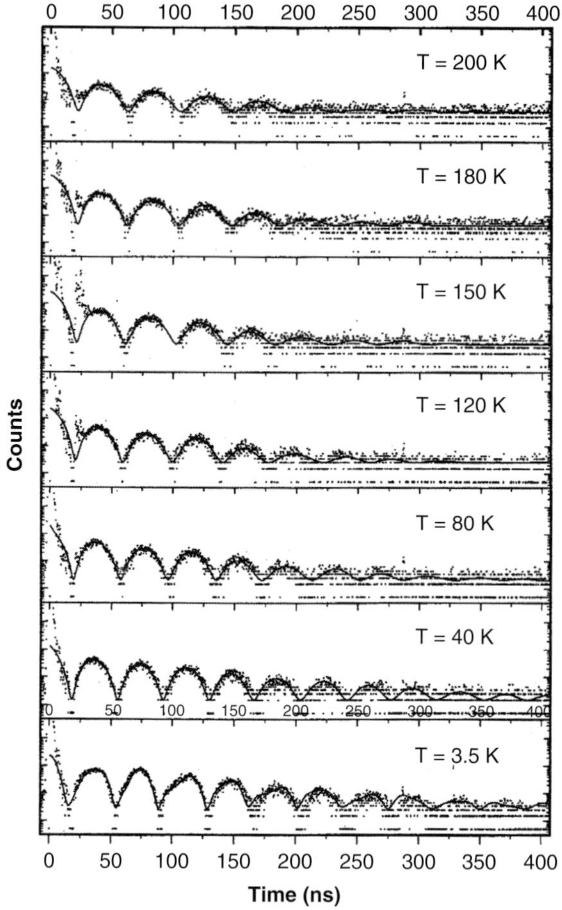

Fig. 9.9 NFS spectra of oxymyoglobin at various temperatures. (From [15])

The structural picture that was envisaged to represent the temperature-dependent fluctuations of the EFG tensor [15] is based on the X-ray structure of MbO_2 that exhibits a geometric disorder of FeO_2 with two different positions of the terminal O-atom [28]. Within this structure, the projection of the O–O bond on the heme plane is rotated by about 40° in position 2 compared to 1 (Fig. 9.10). Conventional Mössbauer studies of single crystals of MbO_2 have shown that the principal component V_{zz} of the EFG tensor lies in the heme plane and is oriented along the projection of the O–O bond onto this plane [29]. If the terminal O-atom is located in position 2, the EFG should be of the same magnitude as in position 1, but its orientation is different. The EFG fluctuates between positions 1 and 2 with a rate that depends on temperature.

In the final approach, all the NFS spectra of MbO_2 in the temperature range of 3.2–200 K (Fig. 9.9) were fitted simultaneously by:

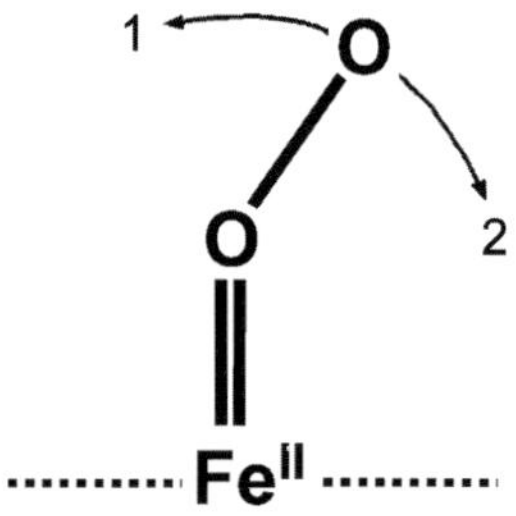

Fig. 9.10 Dynamic structural disorder of the terminal oxygen in oxymyoglobin between positions 1 and 2 which are related via a rotation by 40° about the heme normal

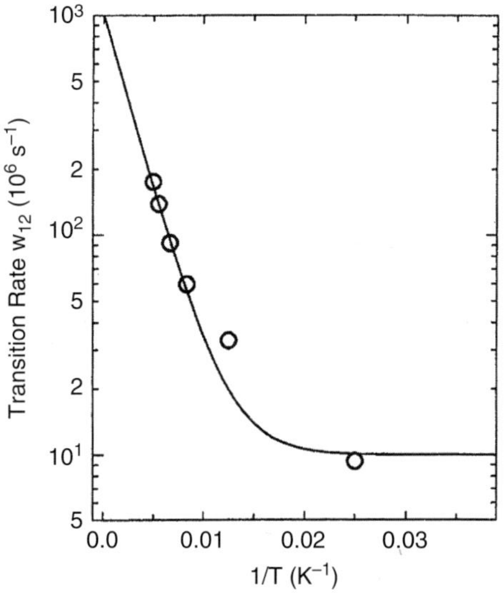

Fig. 9.11 Arrhenius plot of the EFG transition rates for $\theta_D = 215$ K. The *solid line* is a fit by the expression in (9.6). Taken from [15]

1. Applying the program package SYNFOS [30]
2. Using the Debye model (9.5) for the temperature dependence of t_{eff} (with $\theta_D = 215$ K)
3. Letting the EFG tensor fluctuate between positions 1 and 2 with a rate ω_{12}

The temperature-dependent fluctuation rates ω_{12} of the EFG, which result from this simultaneous fit, were interpreted in terms of low-temperature tunneling plus a temperature-dependent activation of the terminal O-atom from position 1 to 2 and vice versa, described by the expression

$$\omega_{12} = \omega_0 \exp\left(\frac{-E_A}{k_B T}\right) + \omega_T. \tag{9.6}$$

The fit of relaxation rates ω_{12} according to (9.6) (Fig. 9.11) yields $\omega_0 = 10^9\ \mathrm{s}^{-1}$, $\omega_T = 10^7\ \mathrm{s}^{-1}$, and an activation energy $E_A = 260 \pm 35\ \mathrm{cm}^{-1}$ [15].

9.4.4 Molecular Dynamics of a Sensor Molecule in Various Hosts (Example: Ferrocene (FC))

Rotational and translational motion of the resonant nuclei are probed by NFS via two effects:

1. Fluctuations of the EFG cause a dephasing of the originally coherent waves which affects the quantum-beat pattern (as described for MbO_2 in Sect. 9.4.3) and
2. Variations of the msd of the resonant nuclei affect the dynamical beat pattern

FC and substituted FCs have been used as detector molecules to investigate the dynamic properties of guest–host systems, including viscous liquids [31–35].

NFS spectra of the molecular glass former ferrocene/dibutylphthalate (FC/DBP) recorded at 170 and 202 K are shown in Fig. 9.12a [31]. It is clear that the pattern of the dynamical beats changes drastically within this relatively narrow temperature range. The analysis of these and other NFS spectra between ~100 and ~200 K provides f factors, the temperature dependence of which is shown in Fig. 9.12b [31]. Up to about 150 K, $f(T)$ follows the high-temperature approximation of the Debye model (straight line within the log scale in Fig. 9.12b), yielding a Debye temperature $\theta_D = 41$ K. For higher temperatures, a square-root term $f \sim \sqrt{(T_c - T)/T_c}$ was added to the regular Debye solution. This term is meant to describe the so-called "mode coupling theory (MCT) β process." The latter represents, in simple

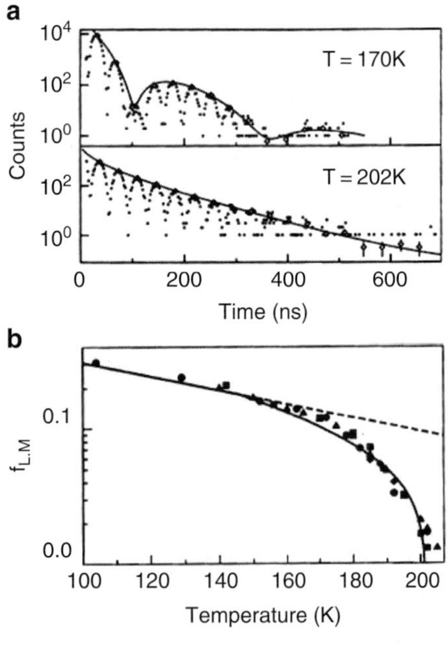

Fig. 9.12 (**a**) NFS spectra of FC/DBP with quantum beat and dynamical beat pattern. (**b**) Temperature-dependent f-factor. The solid line is a fit using the Debye model with $\theta_D = 41$ K below 150 K. Above, a square-root term $f \sim \sqrt{(T_c - T)/T_c}$ was added to account for the drastic decrease of f. At $T_c = 202$ K the glass-to-liquid transition occurs. (Taken from [31])

terms, the "local rattling of molecules in the cage formed by their neighbors" [31]. The quantitative analysis of $f(T)$, including the square-root term, yields the MCT crossover temperature $T_c = 202$ K at which the glass-to-liquid transition occurs.

In summary, the NFS investigation of FC/DBP reveals three temperature ranges in which the detector molecule FC exhibits different relaxation behavior. Up to ~150 K, it follows harmonic Debye relaxation ($\sim\exp(-t/\tau)^{\beta}$). Such a distribution of relaxation times is characteristic of the glassy state. The broader the distribution of relaxation times τ, the smaller β will be. In the present case, β takes values close to 0.5 [31] which is typical of polymers and many molecular glasses. Above the glass-to-liquid transition at $T_c = 202$ K, the msd of iron becomes so large that the f factor drops practically to zero.

The NFS investigation of FC/DBP in the pores of SiO_2 (with pore sizes of 25, 75 and 200 Å) [32] and under pressure (up to 4 kbar) [33] and of FC in the channels of molecular sieves $AlPO_4$-5 and SSZ-24 [34, 35] reveals that confinement leads to considerable slowing down of molecular motion.

By using NFS, information on both rotational and translational dynamics can be extracted. In many cases, it would be favorable to obtain separate information about either rotational or translational mobility of the sensor molecule. In this respect, two other nuclear scattering techniques using synchrotron radiation are of advantage. Synchrotron radiation-based perturbed angular correlations (SRPAC) yields direct and quantitative evidence for rotational dynamics (see Sect. 9.8). NIS monitors the relative influence of intra- and inter-molecular forces via the vibrational density of states (DOS) which can be influenced by the onset of molecular rotation (see Sect. 9.9.5).

9.4.5 *Temperature-Dependent Quadrupole Splitting and Lamb–Mössbauer Factor in Spin–Crossover Compounds (Example: [Fe^{II}(tpa)($NCS)_2$])*

The large family of thermally driven spin–crossover complexes allows an insight into the fundamental mechanisms governing metal coordination chemistry and, furthermore, provides promising materials for optical information storage and display devices [36, 37]. Therefore, these complexes were the subject of a thorough spectroscopic investigation [38], including Mössbauer studies in the energy domain and more recently also in the time domain [7, 39–44].

A series of NFS spectra of the spin–crossover complex [Fe(tpa)($NCS)_2$] were recorded over a wide temperature range [45]. A selection of spectra around the spin–crossover transition temperature is shown in Fig. 9.13. At 133 K, the regular quantum-beat structure reflects the quadrupole splitting from the pure high-spin (HS) phase, and the envelope of the spectrum represents the dynamical beating with a minimum around 200 ns. Below the transition, at 83 K, the QBs appear with lower frequency because of smaller ΔE_Q of the low-spin (LS) phase. Here the minima of

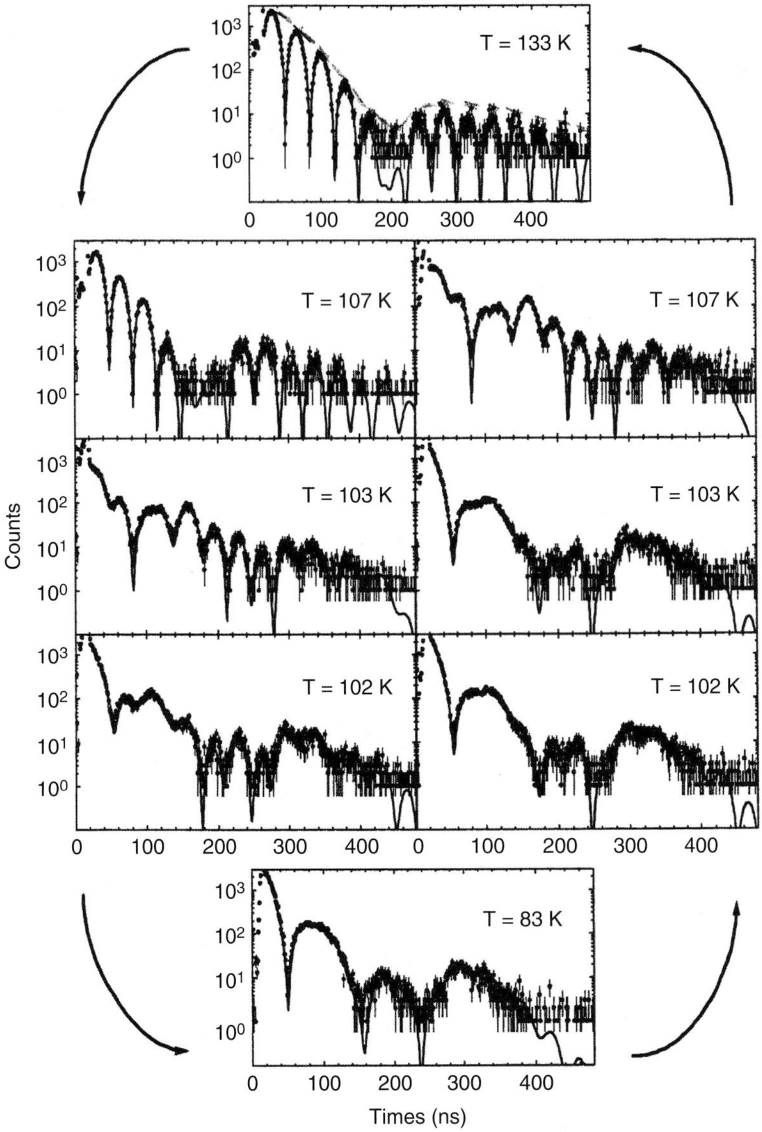

Fig. 9.13 Time evolution of the NFS intensity for various temperatures around the HS–LS transition of [Fe(tpa)(NCS)$_2$]. The measurements were performed at ID18, ESRF in hybrid-bunch mode. The *left-hand side* shows measurements in the transition region performed with decreasing temperature and the *right-hand side* with increasing temperature. (The spectral patterns at comparable temperatures do not match due to hysteresis in the spin-transition behavior). The points give the measured data and the curves are results from calculations performed with CONUSS [9, 10]. The *dashed line* drawn in the 133 K spectrum represents dynamical beating. (Taken from [41])

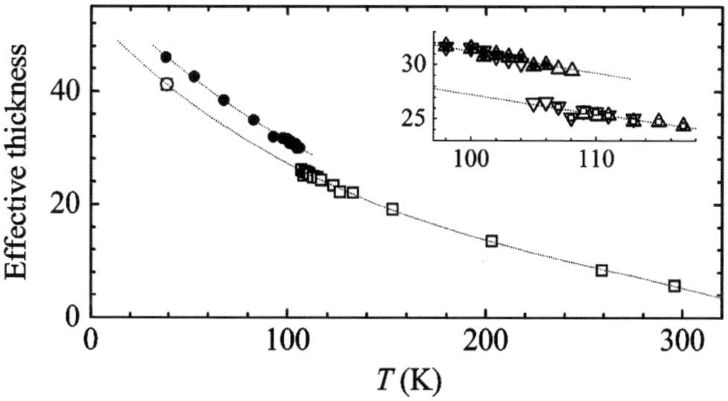

Fig. 9.14 Effective thickness derived from NFS measurements on [Fe(tpa)(NCS)$_2$]. The *open squares* (*solid circles*) denote measurements with a fraction of high-spin (low-spin) higher than 95%. The *open circle* at 40 K denotes the trapped high-spin isomer obtained after rapid cooling. The inset shows the step in the transition region: the upward directed triangles (*downward directed triangles*) denote measurements recorded with decreasing (increasing) temperature. The *lines* are guides to the eyes. (Taken from [41])

the dynamical beats are difficult to identify by eye. However, the computer analysis of this spectrum at 83 K reveals both ΔE_Q and f of the LS isomer unequivocally. The effective thickness t_{eff}, obtained from the NFS data (Fig. 9.14), grows with decreasing temperature. The step in the transition region is pronounced: on the branch measured with decreasing temperature, it changes from 26.5 (HS) to 30.1 (LS) and with increasing temperature from 29.4 (LS) to 25.4 (HS). This direct and accurate measure in the time domain of the discontinuous behavior of t_{eff} upon spin transition is superior to the less accurate results obtained with conventional Mössbauer measurements (in the energy domain) on similar compounds [46].

9.4.6 Coherent Versus Incoherent Superposition of Forward Scattered Radiation of High-Spin and Low-Spin Domains (Example: [Fe^{II}(tpa)(NCS)$_2$])

The form (gradual or abrupt transition) and width of the hysteresis in the temperature-dependent high-spin fraction (Fig. 9.15) were determined by the degree of cooperativity between the iron centers. In order to describe these features theoretically, it is necessary to know whether intermolecular interactions are long or short ranging and how domain growth develops (single or multidomain growth). By means of NFS, this question can, in principle, be addressed by making use of the limited transversal coherence length of the delayed radiation.

During an NFS experiment with a sample that contains more than one kind of scatterer (i.e., HS and LS isomer), the superposition of forward scattered waves could occur coherently or incoherently. Longitudinal scattering is always coherent, because there is no path-length difference for nuclei located along the X-ray beam.

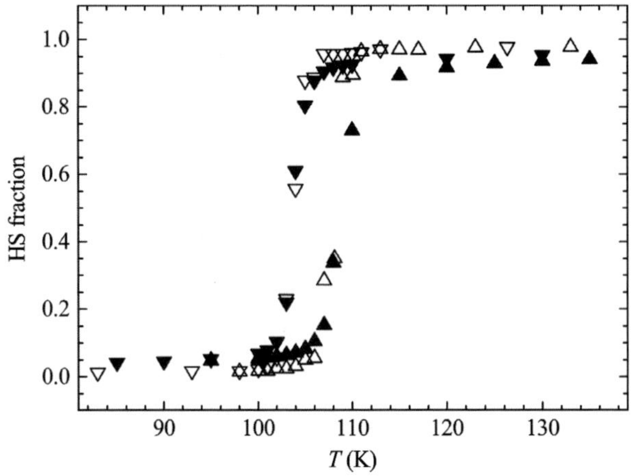

Fig. 9.15 HS fraction in the region of the transition temperature measured by NFS on [Fe(tpa)(NCS)$_2$]. The values derived during a temperature scan with increasing (decreasing) temperature are marked with *upward directed triangles* (*downward directed triangles*). (Taken from [41])

The transversal coherence length $L_{c,trans}$ is, however, limited. It is given by the expression:

$$\mathrm{L}_{c,trans} = \frac{\lambda}{2\pi} \cdot \frac{1}{\sigma_\theta} \text{with} \quad \sigma_\theta = \sqrt{\left(\frac{s_o}{S}\right)^2 + \left(\frac{\sigma_d}{D}\right)^2}, \tag{9.7}$$

where s_o is the diameter of the source and σ_d that of the detector, while S denotes the distance between the source and the scatterer and D that between the scatterer and the detector [47]. Discarding the contribution of s_o/S because of the large distance S, the transversal coherence length $L_{c,trans}$ is 3 nm, if for D 1 m and for σ_d 5 mm are assumed. During spin transition, the formation of single domains (either of HS or LS isomers) with a diameter $\gg$ 3nm would be manifested as an incoherent scattering process.

For the NFS spectrum of [Fe(tpa)(NCS)$_2$] recorded at 108 K, which exhibits a HS to LS ratio of about 1:1, a coherent and an incoherent superposition of the forward scattered radiation from 50% LS and 50% HS isomers was compared, each characterized by its corresponding QB pattern (Fig. 9.16) [42]. The experimental spectrum correlates much better with a purely coherent superposition of LS and HS contributions. However, this observation does not yield the unequivocal conclusion that the superposition is purely coherent, because in the 0.5 mm thick sample the longitudinal coherence predominates since many HS and LS domains lie along the forward scattering pathway. In order to arrive at a more conclusive result, the NFS measurement ought to be performed with a smaller ratio σ_d/D on a much thinner sample. Such an experiment would require a sample with 100% enriched ^{57}Fe and a much higher beam intensity.

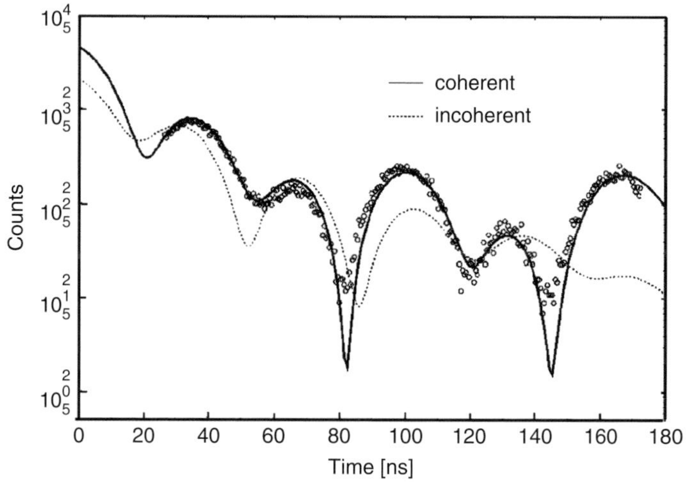

Fig. 9.16 Time-dependent NFS of $[Fe(tpa)(NCS)_2]$ recorded at 108 K. The *two curves* represent comparison of a coherent vs incoherent superposition of the scattering from 50 % LS and 50 % HS iron(II) characterized by their corresponding quantum beat pattern. The effective thickness of the sample was $t_{eff} = 18$. (Taken from [42])

9.4.7 Orientation-Dependent Line-Intensity Ratio and Lamb–Mössbauer Factor in Single Crystals (Example: $(CN_3H_6)_2[Fe(CN)_5NO]$)

Guanidinium nitroprusside (GNP), $(CN_3H_6)_2[Fe(CN)_5NO]$, has the advantage that the $[Fe(CN)_5NO]^{2-}$ anions are aligned parallel or antiparallel to each other (50%/ 50%) within the crystal. This situation allows the direct measurement of the anisotropy of the Lamb–Mössbauer factor f by appropriately orienting a GNP single crystal with respect to the synchrotron beam. By using the polarization characteristics of the radiation and the specific magnetic dipole transitions of oriented ^{57}Fe nuclei, one can visualize the different QB patterns for different crystal orientations.

NFS spectra recorded at 300 K for ***a***-cut and ***c***-cut crystals are shown in Fig. 9.17 [48]. The f factors for the two orientations were derived from the "speed-up" of the nuclear decay (i.e., from the slope of the time-dependent intensity in Fig. 9.17a and from the slope of the envelope in Fig. 9.17b). The factors obtained $f^{(a\text{-cut})} = 0.122$ (10) and $f^{(c\text{-cut})} = 0.206(10)$ exhibit significant anisotropic vibrational behavior of iron in GNP. This anisotropy in f is the reason for the observed asymmetry in the line intensity of the quadrupole doublet (in a conventional Mössbauer spectrum in the energy domain) of a powder sample of GNP caused by the Goldanskii–Karyagin effect [49].

An orientation-dependent asymmetry in line intensity of a quadrupole doublet is, in contrast to the aforementioned case, caused by the polarization response of

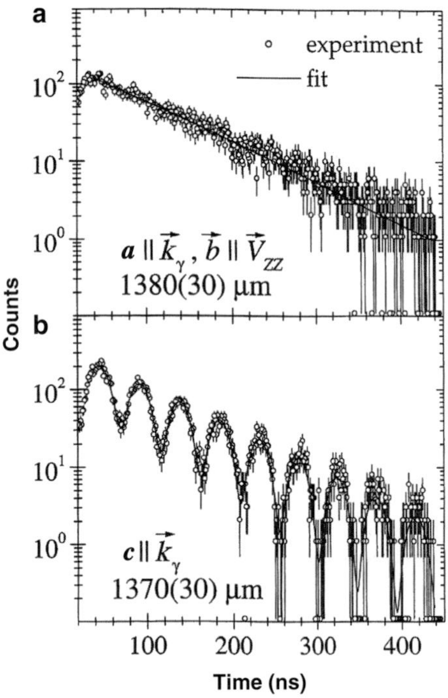

Fig. 9.17 NFS spectra of guanidinium nitroprusside single crystals recorded at room temperature with single-crystal orientations and thicknesses as indicated. The *solid lines* result from a least-squares fit using the CONUSS program [9, 10]. (Taken from [48])

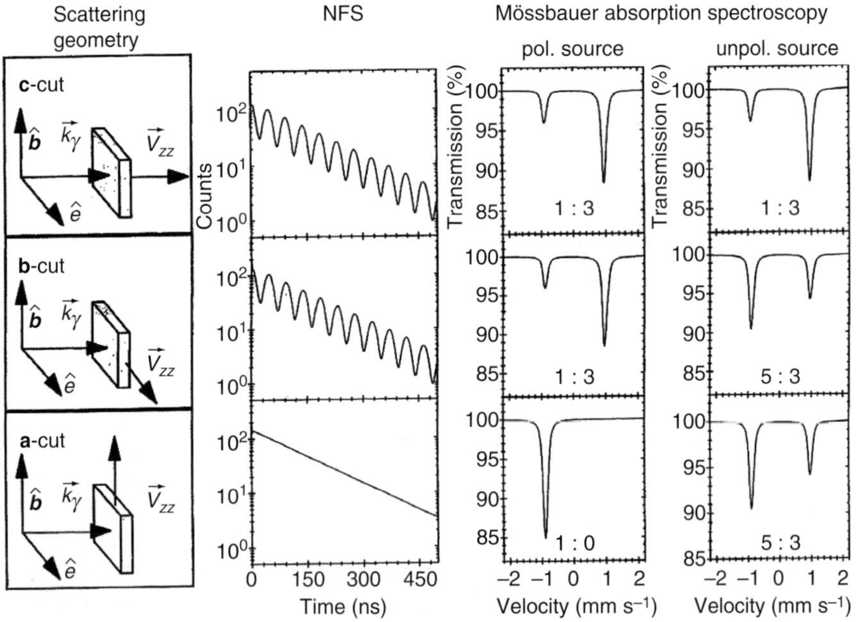

Fig. 9.18 Comparison of calculated time-domain spectra (NFS) and energy-domain spectra (Mössbauer absorption) for ***a***-, ***b***-, and ***c***-cut single crystals of guanidinium nitroprusside. For the calculations the approximation of complete alignment of V_{zz} parallel to the crystallographic ***c***-axis is used. The polarisation direction of the synchrotron beam is represented by ê. (Taken from [48])

oriented ^{57}Fe nuclei (as in single-crystalline GNP). Depending on the source, polarized or unpolarized, and on the orientation of a crystal, certain transitions are active, while others are silent [50]. Correspondingly, a single crystal of GNP exhibits a characteristic line-intensity pattern (in the energy domain) and quantum-beat pattern (in the time domain), depending on the orientation of the crystal and the polarization of the source (Fig. 9.18) [48].

9.5 Isomer Shift Derived from NFS (Including a Reference Scatterer)

The NFS spectra discussed in Sects. 9.3 and 9.4 are not influenced by the isomer shift δ. To detect δ of a (^{57}Fe-containing) scatterer S_1 relative to a reference scattering S_{ref} (also ^{57}Fe containing) requires a specific experimental setup: S_{ref} has to be placed into the synchrotron beam, either down- or up-stream of S_1. Preferably, S_{ref} is a single-line scatterer, e.g., stainless steel or $K_4[Fe(CN)_6]$. With this experimental setup, beating occurs between the resonances of S_1 and S_{ref}. In the case of the polycrystalline "picket-fence" porphyrin $[FeO_2(SC_6HF_4)$ $(TP_{piv}P)]$ at 4.2 K as scatterer S_1, and with $K_4[Fe(CN)_6]$ at room temperature as reference scatterer S_{ref}, three resonances are involved, i.e., the two quadrupole lines originating from S_1 and the single line from S_{ref} (inset in Fig. 9.19). A fit of the corresponding NFS spectrum (Fig. 9.19) with CONUSS [9, 10] yields an isomer shift of the porphyrin relative to stainless steel of $\delta = 0.30$ mm s^{-1} [25].

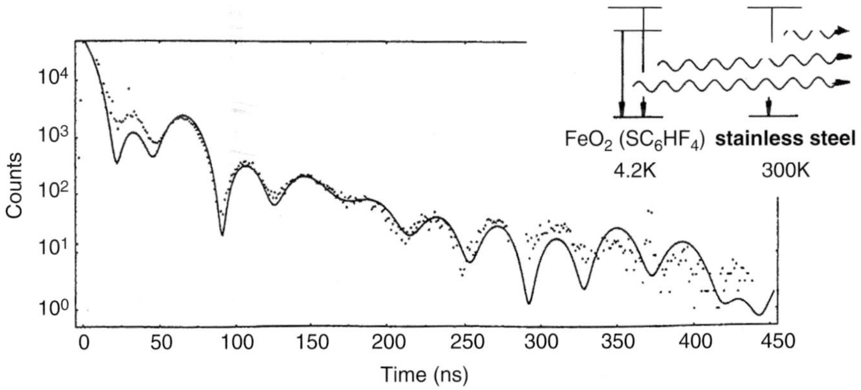

Fig. 9.19 Experimental NFS spectrum of $FeO_2(SC_6HF_4)$ $(TP_{piv}P)$ at 4.2 K together with stainless steel (SS) at 300 K as reference scatterer, the latter placed 30 cm downstream and outside the cryostat. A fit with the CONUSS program [9, 10] yields the value $\delta = 0.30$ mm s^{-1} as isomer shift of the porphyrin relative to stainless steel [25]

9.6 Magnetic Interaction Visualized by NFS

The investigation of magnetic interaction of diamagnetic or paramagnetic iron centers with applied magnetic fields is of interest in many fields of research. It benefits from NFS because additional external parameters (compared to conventional MS), like polarization and time structure of the probing radiation, can be introduced into the protocol of applied experimental conditions. In order to gain access to the information that can be obtained from time-resolved scattering experiments, especially on randomly oriented scatterers (frozen solution or polycrystalline material), theoretical approaches have been developed for a computer code by which NFS spectra can be simulated. The code that describes magnetic hyperfine splittings in Mössbauer nuclei in the framework of the spin-Hamiltonian formalism is SYNFOS [13, 14]; it has been successfully applied to several iron-containing systems [7, 13, 23, 25, 30, 51–54].

9.6.1 *Magnetic Interaction in a Diamagnetic Iron Complex (Example: [$FeO_2(SC_6HF_4)(TP_{piv}P)$])*

For the "picket-fence" porphyrin complex [$FeO_2(SC_6HF_4)(TP_{piv}P)$], which is a biomimetic model for oxymyoglobin, NFS spectra have been recorded at 4.2 K in applied fields (B = 4 T) pointing in different directions within a plane that is perpendicular to the synchrotron beam [25]. In Fig. 9.20 the angle θ represents the field orientation with respect to the polarization $\boldsymbol{\sigma}$ of the beam, with $\theta = 0$ corresponding to $\boldsymbol{B}\|\boldsymbol{\sigma}$. The iron in [$FeO_2(SC_6HF_4)(TP_{piv}P)$] is diamagnetic, and therefore the only magnetic contribution in the analysis of the measured spectra is the applied field. The field rotation in Fig. 9.20 shows that the NFS response on θ is clearly visible. The fit quality of the spectra, obtained either with CONUSS [9, 10] or with SYNFOS [13, 14], is comparable.

9.6.2 *Magnetic Hyperfine Interaction in Paramagnetic Iron Complexes (Examples: [$Fe(CH_3COO)(TP_{piv}P)$]$^-$ with S = 2 and [$TPPFe(NH_2PzH)_2$]Cl with S = 1/2)*

The application of magnetic fields to iron in the paramagnetic state, i.e., ferrous high-spin (Fe^{II}, S = 2), ferric low-spin (Fe, S = 1/2), ferric high-spin (Fe^{III}, S = 5/2), ferryl intermediate-spin (Fe^{IV}, S = 1), and ferryl high-spin (Fe^{IV}, S = 2), induces electronic spin-expectation values and consequently magnetic hyperfine interaction in the iron nuclei. The latter is represented by a complex beat structure in the NFS spectrum. Corresponding conventional magnetic Mössbauer spectra of polycrystalline iron complexes have been analyzed routinely by the spin-Hamiltonian

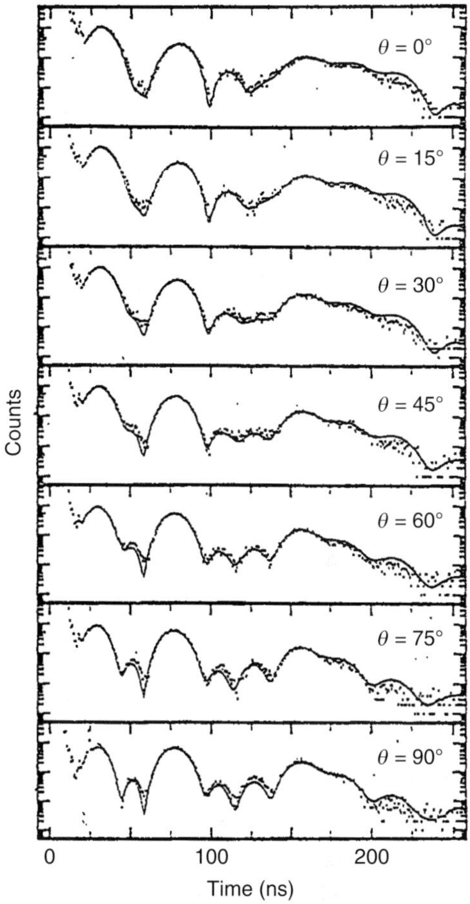

Fig. 9.20 Experimental NFS spectra of $FeO_2(SC_6HF_4)(TP_{piv}P)$ recorded at 4.2 K in a field of 4 T applied at different angles θ within a plane which is perpendicular to the synchrotron beam. $\theta = 0$ represents a field orientation parallel to the polarization of the beam. The *solid lines* are fits with the CONUSS program [9, 10]. (Taken from [25])

formalism [55]. It was challenging to apply this formalism to magnetic Mössbauer spectra in the time domain and to account for the complicated band of nuclear resonances in a randomly oriented paramagnetic sample. These resonances, via hyperfine interactions, are related to electronic spin-expectation values which depend on the polar and azimuthal angles θ and φ, respectively, of the applied field B with respect to the molecular frame of reference. The program package SYNFOS was designed to numerically evaluate NFS spectra for this general case [13].

The first test case was the ferrous high-spin state (Fe^{II}, S = 2) in the "picket-fence" porphyrin acetate complex $[Fe(CH_3COO)(TP_{piv}P)]^-$ [13, 23], which is a model for the prosthetic group termed P460 of the multiheme enzyme hydroxylamine oxidoreductase from the bacterium *Nitrosomonas europeae*. Both the "picket-fence" porphyrin and the protein P460 exhibit an extraordinarily large quadrupole splitting, as observed by conventional Mössbauer studies [56].

Figure 9.21 shows the measured and simulated NFS spectra of $[Fe(CH_3COO)(TP_{piv}P)]^-$. The solid lines are simulations with the SYNFOS program using S = 2

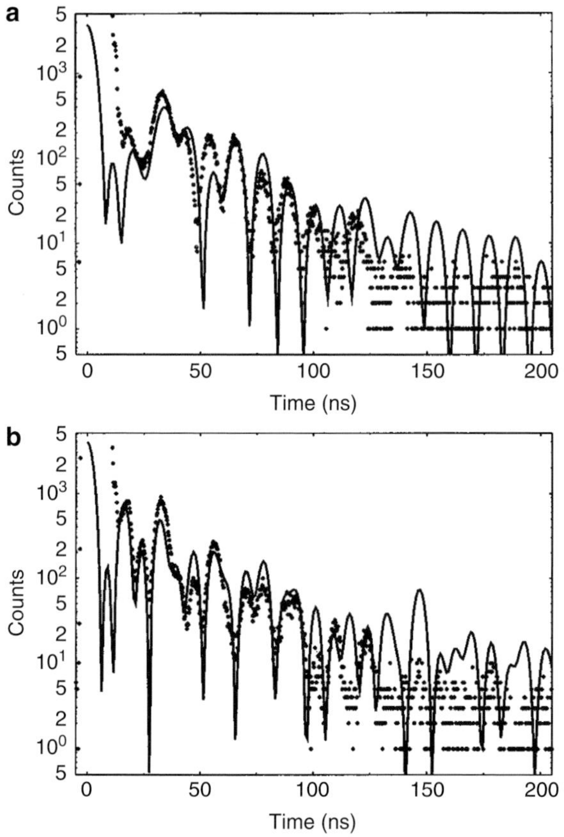

Fig. 9.21 NFS spectra of the paramagnetic "picket-fence" porphyrin complex $[^{57}Fe(CH_3COO)(TP_{piv}P)]^-$ obtained at 3.3 K in a field of 6.0 T applied (**a**) perpendicular to both the synchrotron beam and the polarization vector of the radiation and (**b**) perpendicular to the synchrotron beam but parallel to the polarization vector of the radiation. The *solid lines* are simulations with the SYNFOS program using $S = 2$ and parameters described in the text. (Taken from [13])

and zero-field splitting $D = -0.8\ cm^{-1}$, rhombicity parameter $E/D = 0$, magnetic hyperfine coupling tensor $A/g_n\beta_n = (-17, -17, -12)$ T, quadrupole splitting $\Delta E_Q = 4.25\ mm\ s^{-1}$, asymmetry parameter $\eta = 0$, and effective thickness $t_{eff} = 20$. These parameters have been obtained by conventional MS [56] and were used to test this first application of SYNFOS.

To visualize the sensitive response of the time-dependent forward scattering to small changes of hyperfine parameters, the time spectra for asymmetry parameters $\eta = 0.0$, 0.2, and 0.5 (keeping all other parameters the same) were simulated (Fig. 9.22). Comparable sensitivity was achieved for other hyperfine parameters, provided the measurements are performed to high enough delay times.

An additional application of SYNFOS to simulate the magnetic hyperfine pattern in time spectra is provided by the low-spin ferri-heme (Fe, S = 1/2) complex

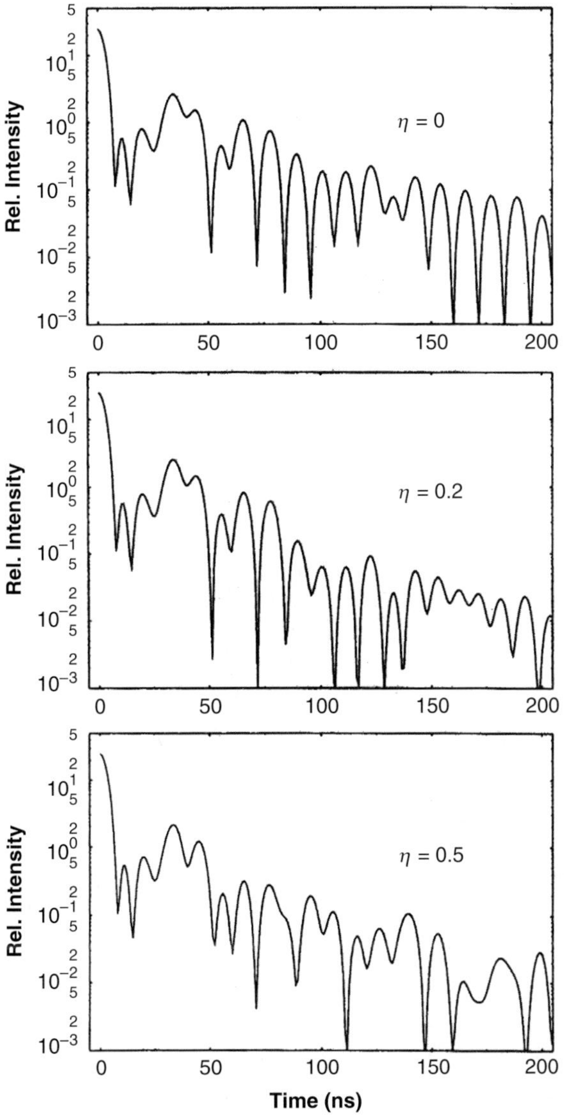

Fig. 9.22 NFS spectra calculated with SYNFOS for different values of the asymmetry parameter η. Other conditions are as described in Fig. 9.21a. (Taken from [13])

bis(3-aminopyrazole)tetraphenylporphyriniron(III) chloride, $[TPPFe(NH_2PzH)_2]Cl$ [57]. This complex is a model for cytochrome $P450_{cam}$ from *Pseudomonas putida* [58] and for chloroperoxidase from *Caldariomyces fumago* [59]. The NFS spectra were recorded at 4.2 K under applied field, varying in field strength and orientation with respect to the synchrotron beam (Fig. 9.23) [23, 51]. The solid lines obtained with SYNFOS originate from parameters that are slightly different from those resulting from conventional Mössbauer studies: i.e., $A_{xx} = (46.9 \pm 2.0)$ T,

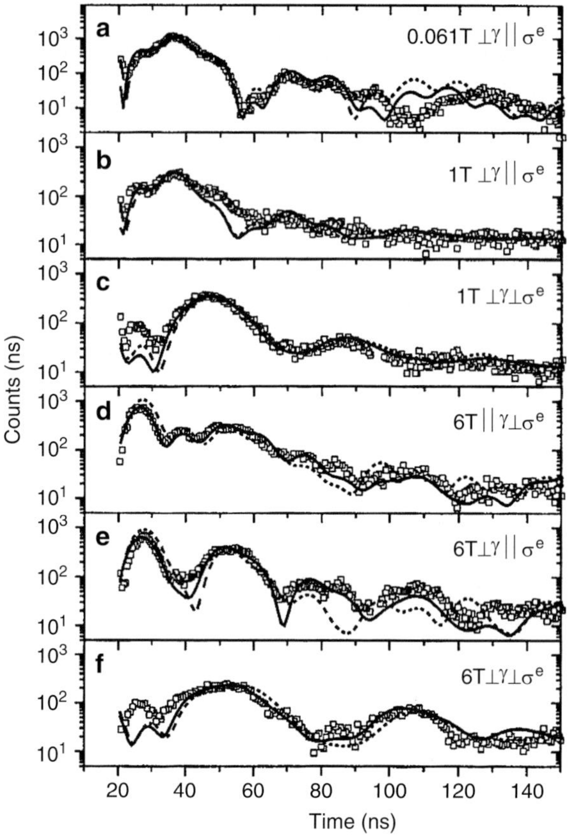

Fig. 9.23 NFS spectra of [TPPFe(NH_2PzH)$_2$]Cl recorded at 4.2 K in applied fields as indicated. The *solid lines* represent the best fit of NFS spectra using SYNFOS. The *dashed lines* are SYNFOS simulations with the parameters obtained from conventional Mössbauer studies. (Taken from [51])

A_{yy} = (9.5 ± 2.0) T, and A_{zz} = (13.6 ± 1.5) T (NFS) vs. A_{xx} = (45.6 ± 3.5) T, A_{yy} = (6.0 ± 5.5) T, and A_{zz} = (16.9 ± 2.5) T (conventional; dashed lines). This again demonstrates the sensitivity of NFS for hyperfine interactions in nuclear (^{57}Fe) scatterers. There can be no doubt that NFS benefits from experimental conditions such as polarization and time structure, and also from a beam diameter in the submillimeter range of the probing radiation.

However, when it comes to the simulation of NFS spectra from a polycrystalline paramagnetic system exposed to a magnetic field, it turns out that this is not a straightforward task, especially if no information is available from conventional Mössbauer studies. Our eyes are much better adjusted to energy-domain spectra and much less to their Fourier transform; therefore, a first guess of spin-Hamiltonian and hyperfine-interaction parameters is facilitated by recording conventional Mössbauer spectra.

9.6.3 Magnetic Hyperfine Interaction and Spin–Lattice Relaxation in Paramagnetic Iron Complexes (Examples: Ferric Low-Spin (Fe^{III}, S = 1/2) and Ferrous High-Spin (Fe^{II}, S = 2))

The spin state of a paramagnetic system with total spin S will lift its $(2S + 1)$-fold degeneracy under the influence of ligand fields (zero-field interaction) and applied fields (Zeeman interaction). The magnetic hyperfine field sensed by the iron nuclei is different for the $2S + 1$ spin states in magnitude and direction. Therefore, the absorption pattern of a particular iron nucleus for the incoming synchrotron radiation and consequently, the coherently scattered forward radiation depends on how the electronic states are occupied at a certain temperature.

When, however, phonons of appropriate energy are available, transitions between the various electronic states are induced (spin–lattice relaxation). If the relaxation rate is of the same order of magnitude as the magnetic hyperfine frequency, dephasing of the original coherently forward-scattered waves occurs and a "breakdown" of the quantum-beat pattern is observed in the NFS spectrum.

To visualize the effects induced by paramagnetic relaxation, the time-dependent forward-scattering intensity has been calculated by implementing the stochastic relaxation between spin states $|i\rangle$ and $|j\rangle$ into the SYNFOS program package [30]. The transition rates from $|i\rangle$ to $|j\rangle$ with $E_i > E_j$ are described by [53]

$$\omega_{i\to j} = \omega_o \frac{|E_i - E_j|^3/k_B^3}{\exp\{|E_i - E_j|/k_B T\} - 1}, \tag{9.8a}$$

and for the reverse transition by

$$\omega_{j\to i} = \omega_{i\to j} \exp\{|E_i - E_j|/k_B T\}. \tag{9.8b}$$

The scaling parameter ω_o in (9.8a) determines the strength of spin–phonon coupling.

A simple case is when a polycrystalline ferric low-spin system ($S = 1/2$), with effective thickness $t_{eff} = 20$ and values of $A_{x,y,z}/g_n\beta_n = -50$ T and $\Delta E_Q = 2$ mm s^{-1}, is exposed to an external field $B = 75$ mT. The transition rate from i (i.e., $S_z = -1/2$) to j (i.e., $S_z = +1/2$) is assumed to be given by (9.8a), and the rate of the reverse transition by (9.8b). NFS spectra for increasing relaxation rates (corresponding to increasing temperature) are shown in Fig. 9.24. As the orientation of the applied field was chosen perpendicular to direction $\boldsymbol{k}$ and the direction of polarization $\boldsymbol{\sigma}$ of the incoming beam, only the two $\Delta m = 0$ transitions are available. Therefore, at slow relaxation ($\omega = 0$ mm s^{-1}) essentially only one magnetic hyperfine frequency is observed, which is slightly modulated due to the powder average because this leads to different orientations of V_{zz} with respect to the effective magnetic field at the position of the nuclei. At fast relaxation ($\omega = 100$ mm s^{-1}), the effective

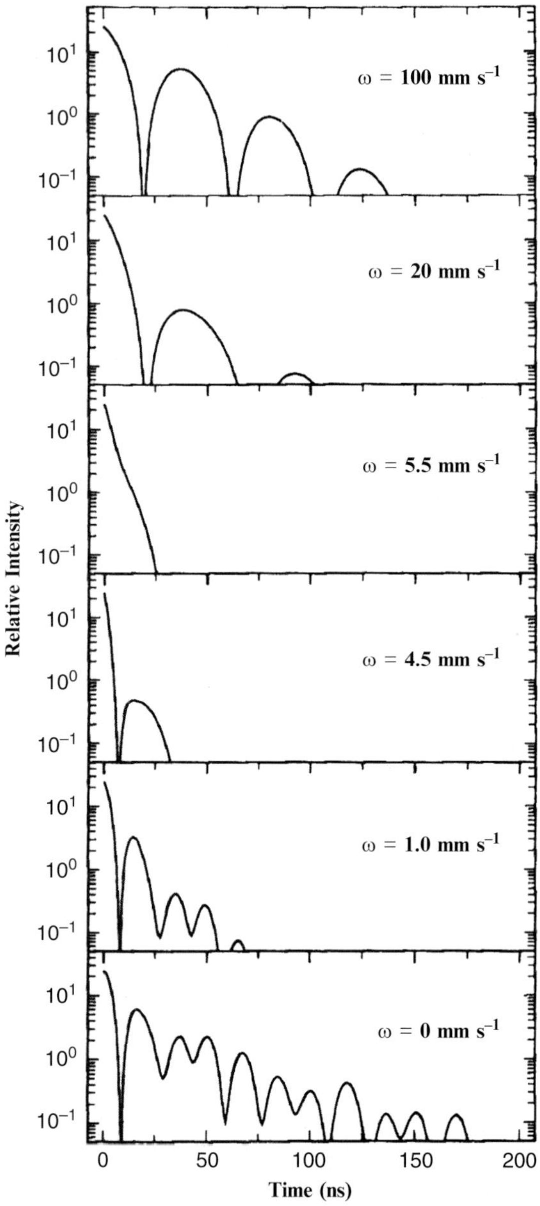

Fig. 9.24 Theoretical calculations of nuclear forward scattering for the relaxation rates as indicated for a system with electron spin $S = 1/2$, hyperfine parameters $A_{x,y,z}/g_n\beta_n = 50$ T, and $\Delta E_Q = 2$ mm s^{-1} in an external field of 75 mT applied perpendicular to $\boldsymbol{k}$ and $\boldsymbol{\sigma}$. The transition probabilities ω in ((9.8a) and (9.8b)) are expressed in units of mm s^{-1}, with 1 mm s^{-1} corresponding to $7.3 \cdot 10^7$ s^{-1}. (Taken from [30])

magnetic field corresponds to the very small applied field of 75 mT so only the quadrupole beats are visible. Around $\omega = 5.5$ mm s^{-1}, which corresponds approximately to the splitting between the two $\Delta m = 0$ resonances, complete dephasing of the at $\omega = 0$ mm s^{-1} and $\omega = 100$ mm s^{-1} coherently forward scattered waves occurs with the complete disappearance of the quantum-beat pattern in the NFS spectrum.

Figure 9.21 shows the NFS spectra of the ferrous high-spin (Fe^{II}, $S = 2$) complex $[Fe(CH_3COO)(TP_{piv}P)]^-$ at 3.3 K in an applied field of 6 T. This situation corresponds to the slow-relaxation limit ($\omega = 0$ mm s^{-1}). Figure 9.25 shows NFS spectra obtained from the same complex at various temperatures in a field of 4 T applied perpendicular to the wave vector $\boldsymbol{k}$ and to the direction of polarization (electric-field vector) $\boldsymbol{\sigma}$ of the incoming beam. The spectra recorded at 14 and 18 K clearly exhibit the progressive collapse of magnetic hyperfine splitting due to spin–lattice relaxation. The simulations with SYNFOS (solid lines in Fig. 9.25) were performed for all temperatures by means of one single value of 3.65×10^9 s^{-1} K^{-3} for the scaling parameter ω_o in (9.8a).

9.6.4 Superparamagnetic Relaxation (Example: Ferritin)

For many organisms, iron is an indispensable element. In order to prevent growth-limiting effects, a sufficient supply of iron must be guaranteed. Therefore, it is not surprising that nature has evolved systems for intracellular iron storage. One class of storage compounds is ferritins, which have been isolated from mammals, bacteria, and plants. Ferritins are composed of a protein shell harboring an iron-containing mineral core. The inner diameter of the protein shell is about 8 nm and can be filled to varying degrees with an inorganic "hydrous ferric oxide-phosphate" complex, $(FeOOH)_8(FeOPO_4H_2)$, up to a maximum of about 4,500 iron atoms in a single ferritin molecule [60].

Conventional Mössbauer studies (in the energy domain) of iron-rich ferritin show a well-resolved magnetic hyperfine pattern at 4.2 K that, by increasing temperature, changes into a quadrupole doublet [55, 61]. The transition temperature, which is also termed "blocking temperature" (T_B), depends on the size of the iron core. This behavior occurs in small particles of magnetically ordered materials and is called superparamagnetism. Figure 9.26 shows the temperature variation of the superparamagnetic relaxation as it is documented in time-domain spectra, recorded from ferritin of the bacterium *Streptomyces olivaceus* [7]. In the fast relaxation limit (100 K), the quadrupole interaction is visible as a quantum-beat spectrum with a single frequency. In the slow-relaxation limit (3.2 K), the full magnetic hyperfine interaction causes a complicated interference pattern. In the intermediate relaxation regime (about 7 K), the stochastic spin flips cause dephasing of the delayed electromagnetic radiation, yielding destructive interferences with concomitant loss of quantum-beat structure.

For a quantitative estimate of the particle size of the iron-containing mineral core, one has to make use of the relaxation rate f at which the magnetization

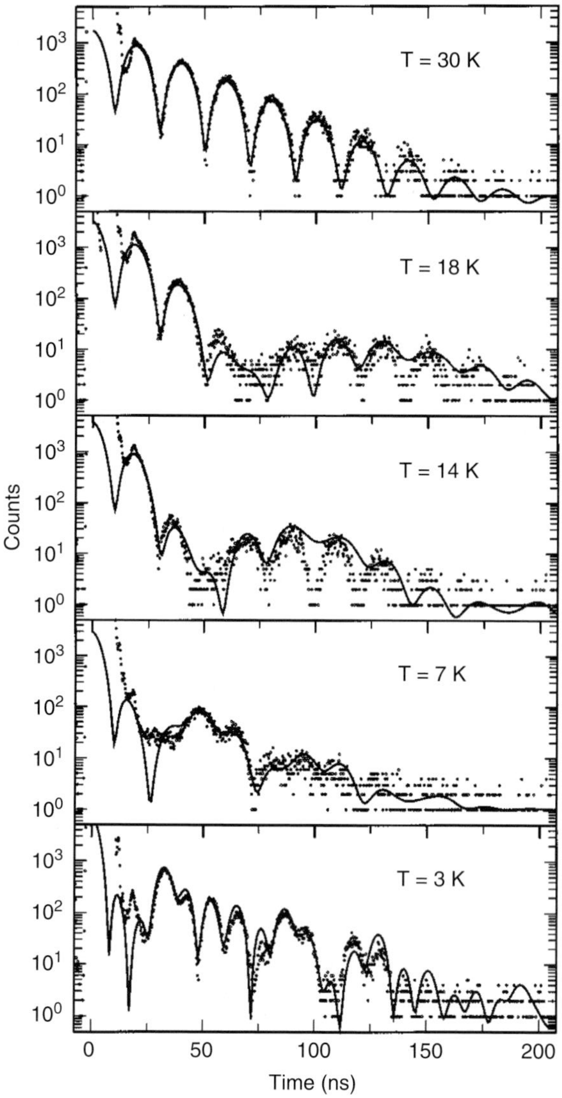

Fig. 9.25 Time dependence of the nuclear resonant forward scattering by the complex [Fe $(CH_3COO)(TP_{piv}P)]^-$ measured in an external field of 4 T applied perpendicular to $\boldsymbol{k}$ and $\boldsymbol{\sigma}$ at the temperatures indicated. (Taken from [30])

direction in each superparamagnetic particle changes between easy magnetic axes:

$$f = f_o \exp(-KV/k_B T). \tag{9.9}$$

In (9.9) K is the magnetic anisotropy constant, V is the volume of the particle, k_B is the Boltzmann constant, T is the temperature, and f_o is a constant of about 10^9 s^{-1}

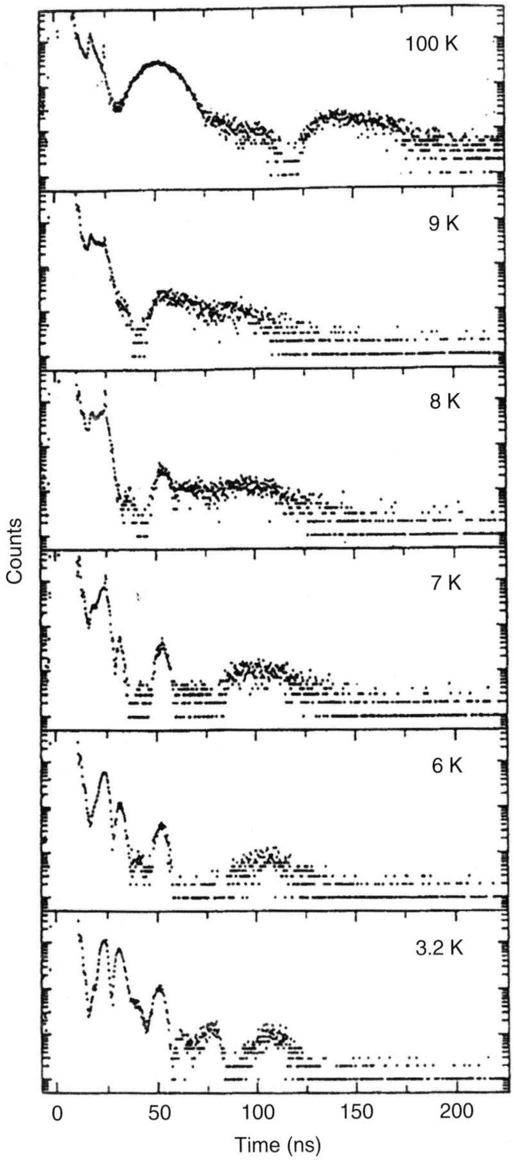

Fig. 9.26 NFS spectra from ferritin of the bacterium *Streptomyces olivaceus*. (Taken from [7])

[62]. Assuming that the anisotropy constant is the same for various ferritins [63], the relation between the diameter of the iron core and the blocking temperature is $d = (6.860T_B)^{1/3}$ (in Å). A quantitative determination of T_B, at which f is equal to the nuclear Larmor frequency (corresponding to the magnetic hyperfine field), yields about 8 K for T_B and about 4 nm for d, which is consistent with the analysis of energy-domain spectra [7]. For comparison, energy-domain spectra of human (instead of bacterium) ferritin have yielded 50 K for T_B and 7 nm for d [64].

9.6.5 High-Pressure Investigations of Magnetic Properties (Examples: Laves Phases and Iron Oxides)

Nuclear resonant scattering is extremely well suited for high-pressure studies because synchrotron radiation exhibits almost laser-like properties and can be collimated by optical devices such as mirrors and X-ray lenses, to spot sizes in the 10 μm range [65]. Pressures, applied with diamond-anvil cells (DAC) to samples of this size, reach values on the order of 100 GPa (1 Mbar). Such DAC-devices (Fig. 9.27) have been successfully used for low-temperature (down to 15 K) high-pressure (up to 105 GPa) studies of RFe_2 Laves phases (R = Y, Gd, Sc) [66, 67] as well as for high-temperature (up to 700 K) and high pressure (up to 80 GPa) studies of the same class of material (RFe_2, R = Y, Se, Lu) [65, 67], $FeBO_3$ (3.5–600 K, up to 55 GPa) [68], $BiFeO_3$ (295 K, up to 62 GPa) [69], and FeO and Fe_2O_3 (up to 2,500 K and 100 GPa) [70].

Temperatures as high as 2,500 K have been achieved by laser heating (LH). For such LHDAC experiments, the sample size was around 50–100 μm, the laser beam was focused to about 40 μm, and the synchrotron beam was microfocused to about 10 μm in diameter [70]. The photon-flux for the 14.4 keV (^{57}Fe) synchrotron radiation at the focusing spot was about 10^9 photons $s^{-1}$with a 1 meV energy bandwidth. This flux was reduced by a 5 mm path through diamond, via photo absorption, to 25% of its original value. For comparison: the flux of the 21.5 keV radiation of ^{151}Eu would be reduced to only 60%.

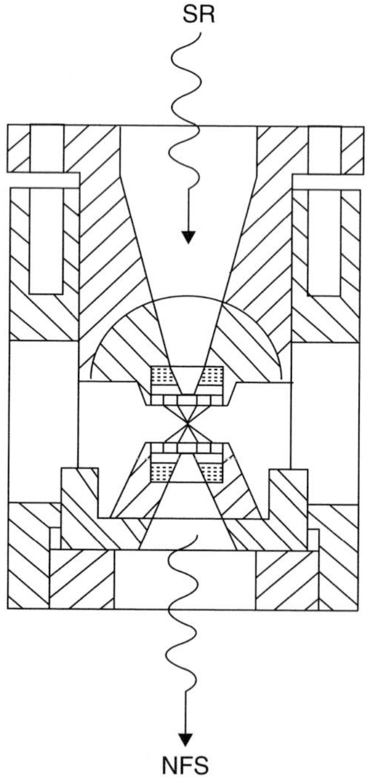

Fig. 9.27 Diamond-anvil cell used for NFS studies. The synchrotron radiation (SR) enters the cell along the diamond-anvil axis

The work of Wortmann et al. [65–67], Gavriliuk et al. [68, 69] and Sturhahn et al. [70] convincingly demonstrates the power of nuclear resonant scattering experiments with synchrotron radiation for high-pressure studies of magnetism and lattice dynamics. An illustrative example was presented at the Fifth Seeheim Workshop by Wortmann [65]: Fig. 9.28a shows NFS spectra of $LuFe_2$ at 295 and

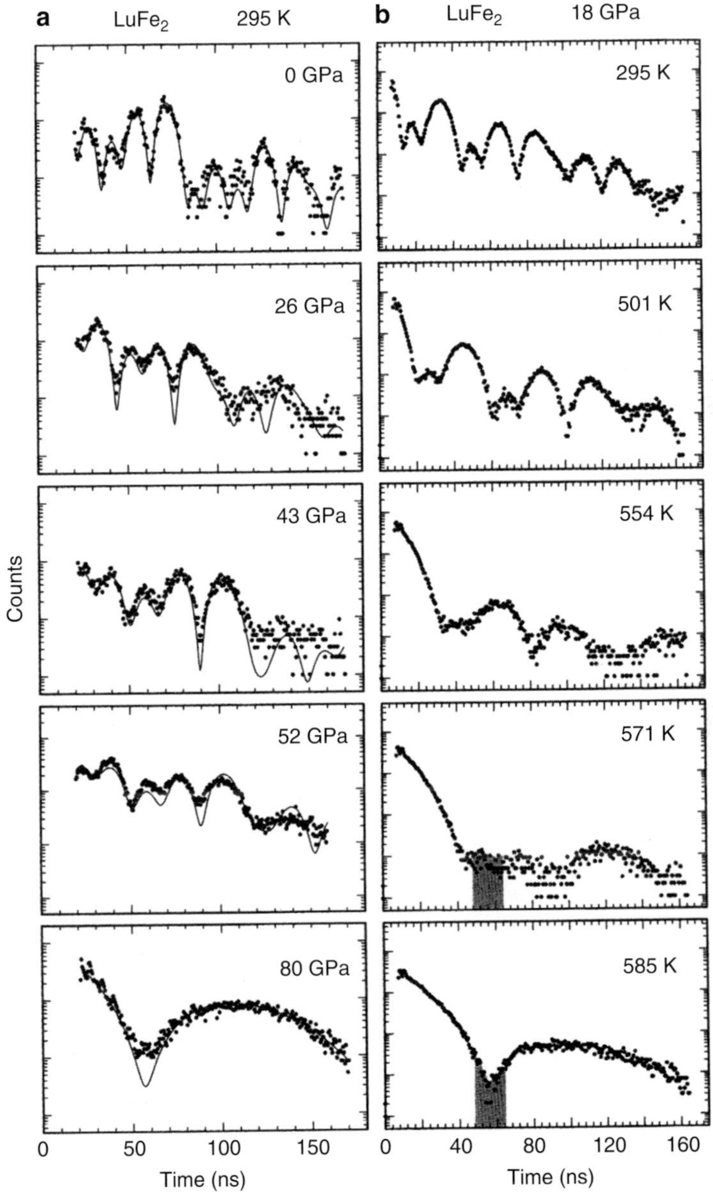

Fig. 9.28 NFS spectra for $LuFe_2$ at (**a**) 295 K and various pressures and (**b**) at 18 GPa and various temperatures. (Taken from [65])

various pressures up to 80 GPa. The spectra reveal a decrease of the magnetic hyperfine field B_{hf} with increasing pressure. At 80 GPa the NFS spectrum exhibits zero magnetism. From the pressure-dependent NFS measurements, the critical temperature $T_c = 295$ K (for magnetic ordering) correlates with an applied pressure of 75 GPa. A complementary series of measurements were performed by keeping the pressure at 18 GPa and varying the temperature between 295 and 585 K (Fig. 9.28b). Combining pressure- and temperature variations, the NFS spectra yielded information about T_c (Fig. 9.29a) and hyperfine field B_{hf} (Fig. 9.29b) depending on pressure. Independently, pressure-dependent XRD studies allowed a correlation of pressures with Fe–Fe distances, d(Fe–Fe) [71, 72]. For $LuFe_2$, the pressure, which corresponds to the relevant Fe–Fe distance in Fig. 9.29b, is given as a corresponding number (in GPa) in Fig. 9.29a.

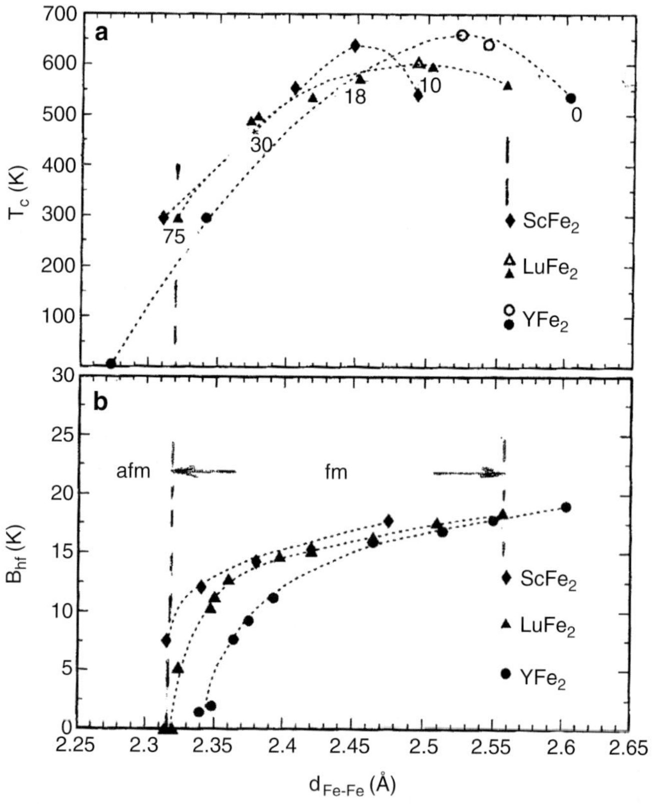

Fig. 9.29 (**a**) Magnetic ordering temperatures T_c and (**b**) magnetic hyperline fields B_{hf} (at 295 K) at various pressures as a function of the Fe–Fe distances $d_{Fe–Fe}$ obtained from NFS studies. The Fe–Fe distances were derived from XRD studies [71, 72]. The numbers in the upper panel for $LuFe_2$ indicate pressure values (in GPa) which correlate with $d_{Fe–Fe}$. For $LuFe_2$, the ferromagnetic (fm) – antiferromagnetic (afm) transition at 75 GPa is indicated. The *dotted lines* are guides to the eye. (Taken from [65])

The experimental results (for $LuFe_2$) are summarized as follows: the ferromagnetic ordering temperature T_c increases from 562 K at 0 GPa to 605 K at 10 GPa. Then it decreases to 573 K at 18 GPa to 496 K at 30 GPa and finally to 295 K at 75 GPa (Fig. 9.29a). This behavior is based on two competing contributions: (a) T_c increases with increasing pressure because the ferromagnetic exchange between the Fe band magnetic moments increases, however, (b) the band moments themselves sensitively depend on d(Fe–Fe), i.e., the band width increases with d^{-5}(Fe–Fe), causing a concomitant decrease of 3d electron density-of-states at the Fermi level and thus of band moment [73]. Consequently, at 295 K and 80 GPa, a complete disappearance of magnetic structure is observed in the NFS spectrum (Fig. 9.28a). This was assigned to a possible phase transition from ferromagnetic to antiferromagnetic interaction and finally to the entire loss of Fe 3d band moment [65].

9.7 NFS Visualized by the Nuclear Lighthouse Effect (NLE) (Example: Iron Foil)

A variant of NFS is the NLE, which was observed by Röhlsberger et al. [74]. It maps the nuclear time response of a resonantly scattering sample to an angular scale by rotating the sample with high speed (Fig. 9.30a). This way the collectively excited nuclear state as a whole is carried with the rotor and, therefore, the direction $\boldsymbol{k}(t)$ of the scattered radiation deviates from the direction of $\boldsymbol{k}_o$ by the rotation angle ϕ that has developed during the delay time t after excitation. The simple mapping $t \rightarrow \phi$ depends on Ω, the angular velocity of the rotor ($t = \phi/\Omega$). An angular window of 10 mrad corresponds, using a rotor with $\Omega = 10$ kHz as employed for investigating a ^{57}Fe-foil [75], to a time window of about 160 ns. The obtained NLE spectrum in this time window (Fig. 9.30b) corresponds to a NFS spectrum and therefore can be analyzed by the standard methods used for NFS (solid line in Fig. 9.30b).

Note that the delayed response in Fig. 9.30b extends to times beyond 170 ns and is almost twice the bunch separation of 96 ns used in this experiment. This means that the technique does not rely on any particular time structure of the synchrotron beam. An additional advantage of this technique is that it has the potential to overcome existing limits for time resolution. Such limits are inherent for NFS, for which the time resolution cannot be below the bunch length that is typically in the range of 50–200 ps. In the experimental setup described earlier (Fig. 9.30) [75], the time resolution was 150 ps. With angular velocities of up to 70 kHz (which seems to be a realistic goal), time resolutions of $\sim$20 ps can be expected. Even lower values may be reached if the beam is microfocused onto the detector plate. With spot sizes below 1 μm, as planned for PETRA III in Hamburg, combined with detectors that exhibit a comparable spatial resolution, time resolutions below 10 ps may be achieved. This would allow the study of isotopes with very short life times (e.g., rare earth nuclei with τ in the range 3–5 ns) and of short-period QBs that result from large hyperfine splittings (e.g., rare earth nuclei like ^{161}Dy with internal fields of

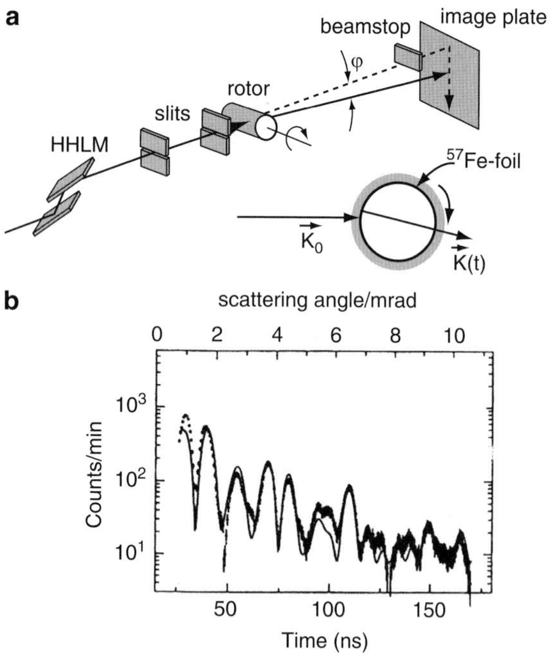

Fig. 9.30 (**a**) NLE setup to record the time response of a rotating resonantly scattering ^{57}Fe foil on an image plate. The direction of the radioactive decay $\boldsymbol{k}(t)$ deviates from the direction of $\boldsymbol{k}_o$ by an angle φ that has developed during the delay time t after excitation. HHLM = high heat-load monochromator. (**b**) NLE time spectrum of the radiation, which was scattered by a ^{57}Fe foil rotating at 10 kHz and which is obtained after background correction of the image plate data. The spectrum was obtained during a 1 min exposure. (Taken from [75])

several hundreds of Tesla and with expected quantum-beat periods below 100 ps). First successful NLE applications with isotopes with short life times have been reported for ^{149}Sm (22.5 keV, 10.3 ns) [76] and for ^{61}Ni (67.4 keV, 7.4 ns) [77].

NLE is far from being routinely applied. Additionally, at present, an unwanted source of background in NLE experiments is small-angle X-ray scattering (SAXS) from the rotor material. First experiments using single-crystalline sapphire rotors seem to be promising in suppressing SAXS [75]. Further, this source of background shifts to very small scattering angles with increasing photon energy.

In summary, the NLE technique offers a conceptually new approach to observe NFS. Existing limits for time resolution could be overcome by a microfocused synchrotron beam (as planned for PETRA III) and by detectors with high spatial resolution; and background from SAXS could be suppressed by employing high-energy transitions and crystalline sapphire as rotor material.

9.8 Synchrotron Radiation Based Perturbed Angular Correlation, SRPAC (Example: Whole-Molecule Rotation of FC)

SRPAC is a scattering variant of time-differential perturbed angular correlation (TDPAC). In TDPAC, an intermediate nuclear level is populated from "above" after the decay of a radioactive parent. If this nuclear level exhibits hyperfine interaction, for example quadrupole splitting, its decay into the ground state (according to its characteristic life time τ) is perturbed by the quadrupole precession time. This periodic perturbation becomes visible as an oscillating imprint on the measured decay-intensity $I(t)$ which decreases according to τ (Fig. 9.31a). The intensity $I(t)$ is recorded (conventionally in either 90° – or 180° – geometry) by an

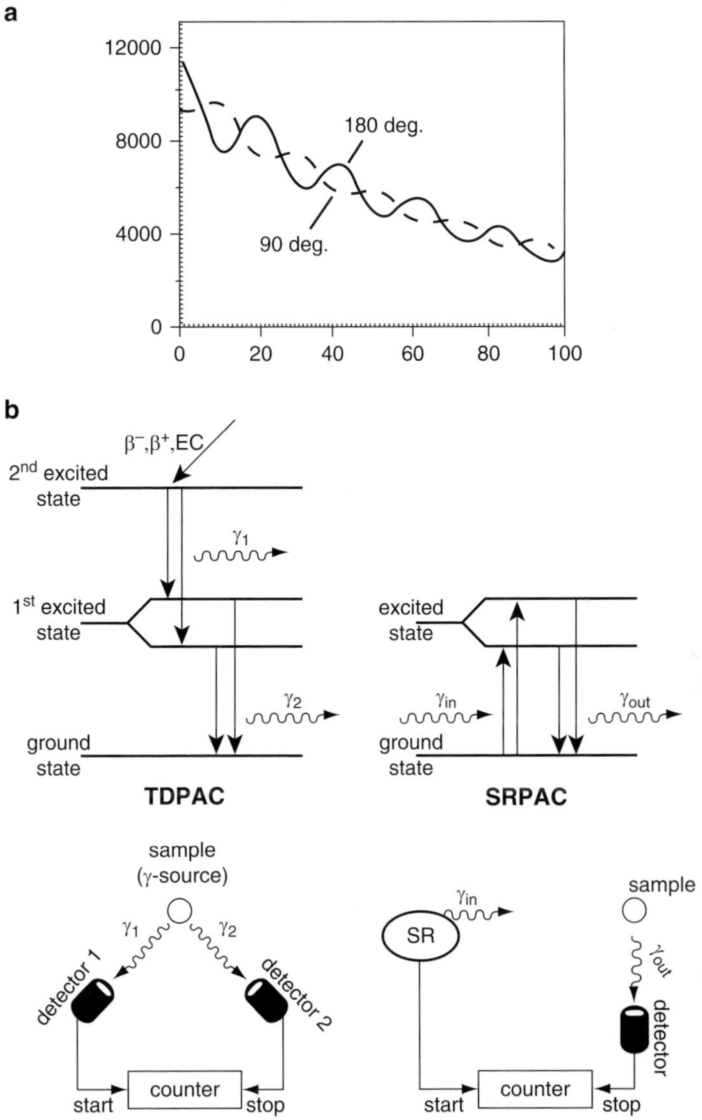

Fig. 9.31 (**a**) Coincidence spectrum of TDPAC with γ_1 as start- and γ_2 as stop signal. (**b**) Schematic representation of the principle and of the experimental setup for TDPAC and SRPAC

experimental setup that counts quanta γ_2 being emitted in coincidence with quanta γ_1 (Fig. 9.31b, left side).

In SRPAC, the intermediate nuclear level (excited state) is populated from "below" (from the groundstate) by the SR beam. In contrast to TDPAC, this procedure avoids any chemical or electronic after effects. Additionally, directional selection and timing by the first detector in TDPAC are replaced in SRPAC by the direction and the timing of the incident SR flash (Fig. 9.31b, right side). The hyperfine splitting of the excited state induces a QB pattern in SRPAC. If during the decay of the excited state (having a lifetime of 141.1 ns in the case of ^{57}Fe) the EFG is fluctuating in space – because, for example, the structure of the molecular complex under study is dynamically disordered – the QB pattern will fade depending on the geometry and timing of this relaxation process [78].

A specific feature of SRPAC is that the QB pattern remains unmodified by any translational motion of the Fe nuclei. This is due to the fact that in the incoherent channel (in 90 – geometry), the QB pattern is performed by the two interfering waves "γ_{out}" (Fig. 9.31b, right side) emitted by one and the same nucleus and not, as in NFS, by an ensemble of coherently emitting nuclei in the forward direction. Thus, SRPAC is independent of the Lamb–Mössbauer factor f and, therefore, allows monitoring of the dynamics of ^{57}Fe-containing complexes even in the liquid state. From NFS, as well as from conventional Mössbauer studies, information both on rotational and translational motion of Fe nuclei can be extracted, but it is not straightforward to unravel this information into its individual contributions. It is, therefore, advantageous that SRPAC yields direct and quantitative evidence for rotational dynamics only.

SRPAC has been applied to the study of the dynamics of the sensor molecule ferrocene (FC, which is diamagnetic and exhibits relatively large quadrupole splitting, $\Delta E_Q \sim 2.40$ mm s^{-1}) [78] and also of various substituted FCs like octamethyl ferrocene (OMF) and octamethyl ethinyl ferrocene (OMFA) [79–81] in various hosts. SRPAC studies of the molecular glass former DBP doped with 5% of FC enriched in ^{57}Fe have been performed in the temperature range from the glassy state up to the normal liquid state of the glass former. This was the first study of relaxational dynamics of a Mössbauer isotope in the regime of a low-viscosity liquid [78]. The temperature-dependent damping of the SRPAC signal shown in Fig. 9.32a describes the stochastic rotational relaxation of the EFG. Its theoretical analysis yielded relaxation rates λ in the frequency range from $\sim 1.5 \cdot 10^6$ s^{-1} at 180 K up to $\sim 4 \cdot 10^{10}$ s^{-1} at 328 K (Fig. 9.32b). At high temperatures, the relaxation rates exhibited a typical viscous-like temperature behavior, whereas below $\sim$210 K they follow an Arrhenius law (Fig. 9.32b) [78].

The SRPAC study of the dynamics of FC in the unidimensional channels of the molecular sieve SSZ-24 covered a wide temperature range from 140 to 505 K (to be compared with the bulk melting temperature of FC of 446 K and its bulk boiling temperature of 522 K) [81]. Figure 9.33a shows SRPAC signals up to 445 K. The relaxation manifested itself in a damping according to the increase of the relaxation rate λ with increasing temperature, in a slight (nearly invisible) change of the frequency and in the appearance of a sine term (at 445 K). A good fit of the

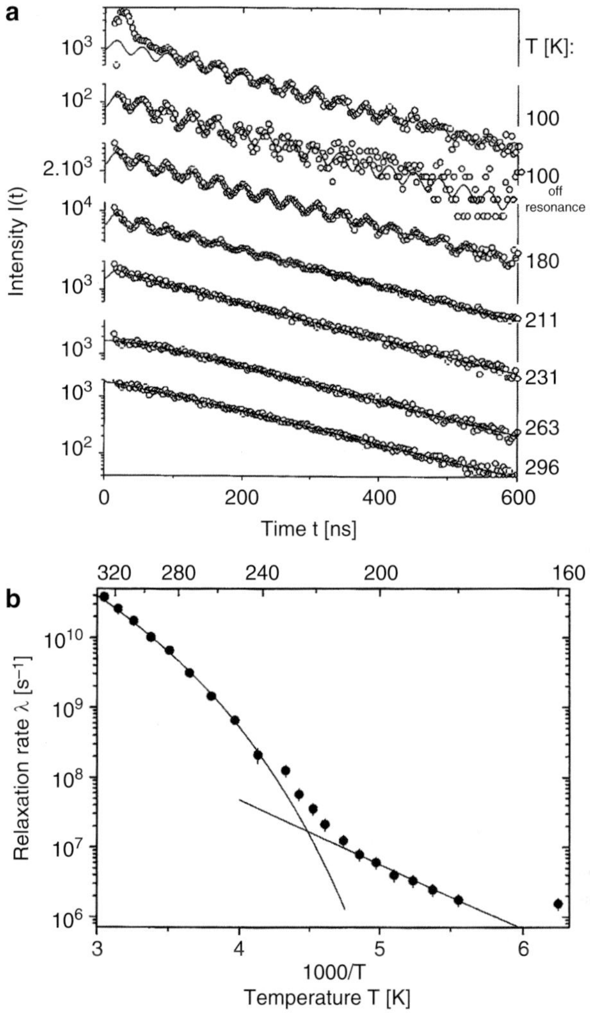

Fig. 9.32 (**a**) Time evolution of the SRPAC signal (in log scale) for several temperatures obtained from the molecular glass former DBP doped with ^{57}Fe-enriched ferrocene. (**b**) The relaxation rate λ (in log scale) as a function of inverse temperature 1,000 T^{-1} obtained from analyzing the SRPAC signals in (**a**). (Taken from [78])

SRPAC spectra of Fig. 9.33a with a model that allows random, single-activated jumps of the EFG on a cone (Fig. 9.33b) was possible over the entire temperature range. This "random jump cone" model follows an Arrhenius law with activation energy $E_a = (20.1 \pm 0.8)$ kJ mol^{-1} and frequency factor $A = (5.5 \pm 1.6) \cdot 10^{11}$ s^{-1} and it yields a cone opening angle of about 47° above 380 K [81].

In summary: SRPAC is a powerful tool that yields information about both rotation rates and rotation mechanisms. This method can be extended to molecules other than FC, provided they possess a finite EFG at the Fe center.

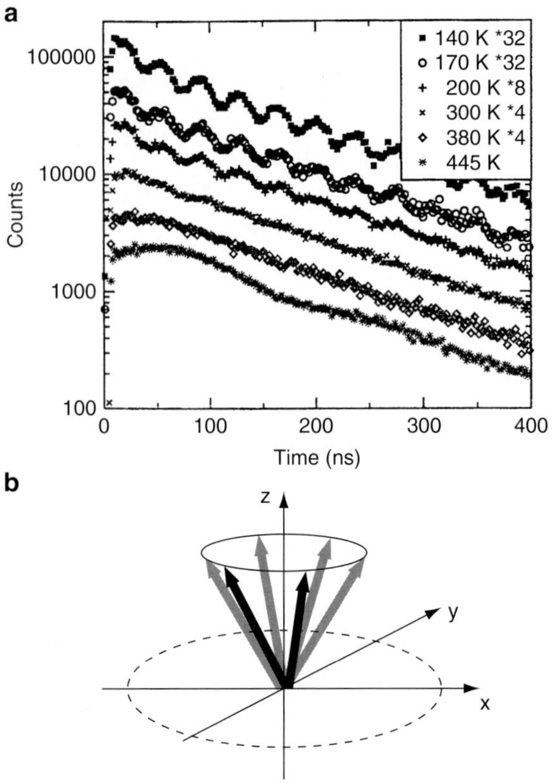

Fig. 9.33 (**a**) SRPAC time spectra of FC in SSZ-24 at different temperatures. For comparison the spectra at low temperatures are multiplied as indicated. (**b**) Model, which allows random jumps of the EFG on a cone. (Taken from [35])

9.9 Nuclear Inelastic Scattering

9.9.1 Phonon Creation and Annihilation

Nuclear absorption of incident X-rays (from the synchrotron beam) occurs elastically, provided their energy, E_γ, coincides precisely with the energy of the nuclear transition, E_0, of the Mössbauer isotope (elastic or zero-phonon peak at $E_\gamma = E_0$ in Fig. 9.34). Nuclear absorption may also proceed inelastically, by creation or annihilation of a phonon. This process causes inelastic sidebands in the energy spectrum around the central elastic peak (Fig. 9.34) and is termed nuclear inelastic scattering (NIS).

Complementary to other methods that constitute a basis for the investigation of molecular dynamics (Raman scattering, infrared absorption, and neutron scattering), NIS is a site- and isotope-selective technique. It yields the partial density of vibrational states (PDOS). The word "partial" refers to the selection of molecular vibrations in which the Mössbauer isotope takes part. The first NIS measurements were performed in 1995 to constitute the method and to investigate the PDOS of

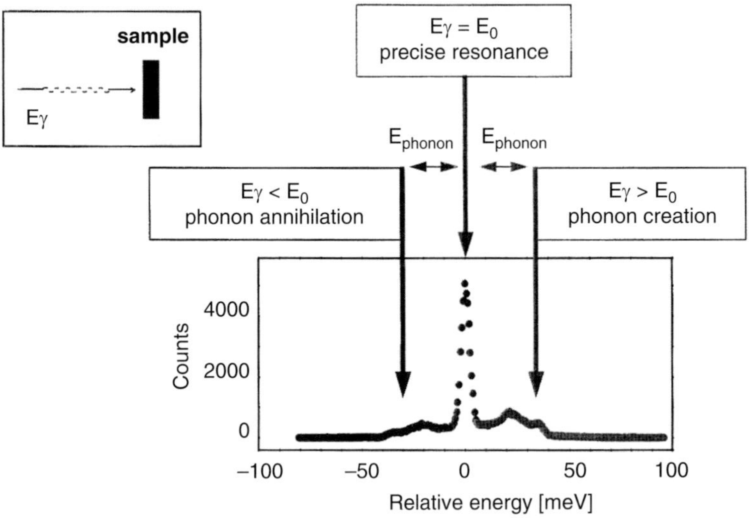

Fig. 9.34 Monitoring of inelastic excitations by nuclear resonant scattering. The sidebands of the excitation probability densities for phonon creation, $S(E)$, and for annihilation, $S(-E)$, are related by the Boltzmann factor, i.e., $S(-E) = S(E)\exp(-E/k_BT)$. This imbalance, known as "detailed balance," is an intrinsic feature of each NIS spectrum and allows the determination of the temperature T at which the spectrum was recorded

α-iron [82, 83]. Over the years, different groups have developed different preferences to address NIS:

- Phonon-assisted Mössbauer effect
- Phonon-assisted nuclear resonance absorption
- Atom-selective vibrational spectroscopy
- Nuclear resonance vibrational spectroscopy
- Nuclear resonance inelastic X-ray scattering

The main advantages of NIS, compared to Raman- and neutron-scattering and IR-spectroscopy, are site- and isotope selectivity, detectability independent of the aggregate state (solid, liquid, or even gaseous), and high accuracy with practically zero background. In its performance, NIS is competitive: the energy resolution of about 0.5 meV ($\sim$4 cm^{-1}) is better than that of many neutron spectrometers and it is only slightly inferior to that of IR ($\sim$1 cm^{-1}). NIS provides a precise quantitative description of molecular vibrations in which the resonant atom (Mössbauer isotope) takes part, i.e., besides the list of vibrational frequencies, the exact density of vibrational states for this selected (Mössbauer) site is obtained. Therefore, molecular dynamics calculations can be compared with experimental data not only via their eigen frequencies but also via their eigen vectors. These features simplify mode assignment tremendously. It should be noted, however, that NIS is not a routine spectroscopy; it requires enrichment of the probe with the respective Mössbauer isotope and it requires a research station like the ESRF in Grenoble, France, and therefore ought to be applied only to cases that are difficult to handle by other methods.

So far, NIS applications to the study of molecular dynamics have been performed mainly with ^{57}Fe-containing systems. NIS with ^{119}Sn has been used to investigate the dynamics of tin ions chelated by DNA [84]. The technique has also been applied to materials containing the isotopes ^{61}Ni, ^{83}Kr, ^{151}Eu, and ^{161}Dy [85]. Applications with these isotopes will hopefully become more routinely possible with the installation of PETRA III in Hamburg, Germany.

9.9.2 Data Analysis and DOS (Example: Hexacyanoferrate)

NIS aims at detecting the PDOS and especially its contributions from the individual molecular vibrational modes. The fraction $\tilde{g}_j(s) = |\boldsymbol{e}_j \boldsymbol{s}|^2$ of the PDOS that belongs to the *j*th vibrational mode in a single crystal depends (in the approximation of a single isolated molecule with Einstein-like vibrational modes) on the phonon polarization vector $\boldsymbol{e}_j$ and on the unit vector $\boldsymbol{k}/k$ along the direction of the synchrotron beam. Orientation dependence of the projected PDOS has been observed with anisotropic single crystals [86–88]. For randomly oriented powder, liquid, or glass – instead of $\tilde{g}_j(\boldsymbol{s})$ – the average over all crystallographic directions are:

$$g_j = \frac{1}{3} e_j^2 = \frac{1}{3} \left\{ (e_j)_x^2 + (e_j)_y^2 + (e_j)_z^2 \right\}.$$

Using normalization properties of the polarization vectors one obtains [83, 89]:

$$e_j^2 = \frac{\langle r_{j,R} \rangle^2 m_{\mathrm{R}}}{\Sigma_{k=1}^{N} \langle r_{j,k} \rangle^2 m_k}. \tag{9.10}$$

In (9.10), $\langle r_{j,\mathrm{R}} \rangle^2$ is the mean square displacement (msd) of the resonant atom R and $\langle r_{j,k} \rangle^2$ is that of the *k*th atom in the *j*th vibrational mode. From (9.10), it is clear that the contribution of an arbitrary vibrational mode to the PDOS is determined by the msd of the resonant atom in this very mode. The value e_j^2 gives the fraction of the vibrational energy in the *j*th mode deposited on the resonant atom and was therefore termed "mode composition factor" [89, 90].

The composition factor $e_{\mathrm{ac}}{}^2$ for the acoustic branch of the NIS spectrum is derived from (9.10) by assuming (in the approximation of total decoupling of inter- and intramolecular vibrations) that the msd $\langle r_k \rangle^2$ in acoustic modes are identical for all the atoms in the molecular crystal:

$$e_{\mathrm{ac}}^2 = \frac{m_{\mathrm{R}}}{M_{\Sigma}}. \tag{9.11}$$

The highest possible composition factor $e_{\max}{}^2$ for a given molecule is

$$\mathrm{e}_{\max}^2 = (\mathrm{M}_{\Sigma} - \mathrm{m}_{\mathrm{R}})/\mathrm{M}_{\Sigma}, \tag{9.12}$$

which corresponds to stretching of the resonant atom relative to the rest of the molecule.

Assuming that the resonant atom belongs to a rigid molecular fragment with mass M_1, which exhibits stretching against the rest of the molecule, the composition factor e^2_{str} for this stretching mode is

$$e^2_{str} = m_R(M_\Sigma - M_1)/M_1 M_\Sigma. \tag{9.13}$$

If a vibrational mode is not a pure stretching mode but contains bending contributions, the composition factor may deviate from e^2_{str} in both directions.

The PDOS of the hexacyanoferrate $(NH_4)_2Mg[Fe(CN)_6]$ (Fig. 9.35a) was analyzed employing these simple rules, i.e., (9.11–9.13) [89]. The anion $[Fe(CN)_6]^{4-}$ has ideal octahedral symmetry; therefore, one expects a pair of threefold degenerate stretching modes (Fig. 9.35b).

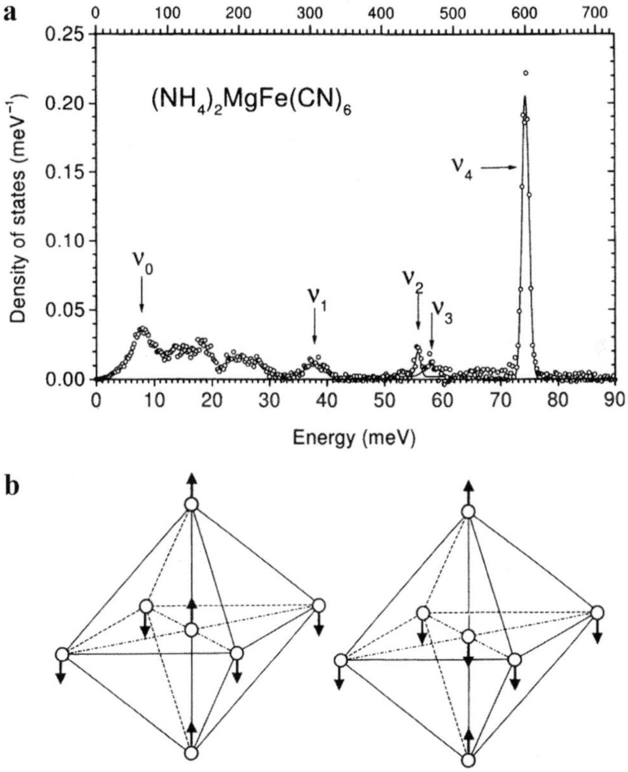

Fig. 9.35 (**a**) Measured PDOS of the resonant atom in $(NH_4)_2Mg^{57}Fe(CN)_6$ at 30 K. Taken from [89]. (**b**) The two threefold degenerated stretching modes of the octahedron $[Fe(CN)_6]^{4-}$ which have nonzero mean-square displacement of the central resonant atom

The assignment of acoustic modes, according to (9.11) is derived from the mass of the resonant iron atom (57 a.u.) and that of the whole anion $[Fe(CN)_6]^{4-}$ (273 a.u.), i.e., $e^2_{ac} = 0.21$. Integration of the PDOS in Fig. 9.35a reaches this value at 12.7 meV. Thus, the peak ν_o at 8 meV is attributed to acoustic modes.

The pronounced peak ν_4 at 74.3 meV has a composition factor $e^2 = 1.04$. Because of the aforementioned threefold degeneracy of stretching modes, the composition factor of one individual mode is $e^2 = 1.04/3 = 0.35$. According to (9.13), this value yields the mass $M_1 = 103$ a.u. for the corresponding molecular fragment which exhibits stretching against the rest of the cation. The iron atom and two CN groups define a fragment with mass $M_1 = 109$ a.u., which takes part in the threefold degenerate stretching mode shown on the left side in Fig. 9.35b. Thus, it is possible to assign the peak ν_4 at 74.3 meV to this specific stretch.

The other three degenerate stretching modes of hexacyanoferrate (right side in Fig. 9.35b) involve the iron atom together with four CN groups relative to the rest of the anion and have together, according to (9.13), the composition factor $e^2 = 0.44$. Integration over the PDOS in Fig. 9.35a yields much smaller composition factors for the resonances at ν_1, ν_2, and ν_3. This finding suggests that ν_1, ν_2 and ν_3 are not pure stretching modes but contain considerable contributions from bending modes [89]. Normal mode analysis confirms this qualitative assignment [91].

The simple rules of (9.11) and (9.12) were also successfully used to assign the stretching mode of the central iron atom in FC, $Fe(C_5H_5)_2$, relative to the rest of the molecule as well as the acoustic modes [89].

9.9.3 Data Analysis Using Absorption Probability Density (Example: Guanidinium Nitroprusside)

GNP, $(CN_3H_6)_2[Fe(CN)_5NO]$, may be considered as a calibration standard for NIS applications since nitroprusside complexes have been studied in detail over the past decades by a variety of experimental and theoretical methods. In addition, single crystals of GNP are well suited for the investigation of the anisotropy of molecular vibrations because the two nonequivalent NP anions, $[Fe(CN)_5NO]^{2-}$, in the unit cell of GNP have an almost antiparallel orientation.

Complementary to NIS measurements, NIS simulations from first principles provide information about the normal modes of a molecular complex. In the present example, geometry and vibrational modes of the NP anion in vacuo were calculated by applying density-functional theory (DFT). Deviations from C_{4v} symmetry along the Fe–N–O axis are minor ($<5°$ for the Fe–N–O bond angle and <1 pm for the different bond lengths), which facilitates the assignment of vibrational modes. NIS measurements and simulations were described for two different orientations of GNP, i.e., with surface normal of the single crystal either in the crystallographic ***a*** or in the ***c*** direction (***a***-cut or ***c***-cut) [86].

The NIS spectra of the ***a***-cut and ***c***-cut crystals of GNP are shown in Fig. 9.36a. They exhibit an intense and broad peak in the energy range below 30 meV which is almost twice as large as the energy resolution of the incident radiation (not shown). This behavior indicates that a considerable portion of this peak belongs to the inelastic part of the spectrum and, as shown in the following text, is due to vibrations of intra- as well as intermolecular nature. Apart from this low-energy peak, nonvanishing probability density is observed in the range between 30 and 90 meV. In this range, the measured probability density of the ***a***-cut crystal is significantly different from that of the ***c***-cut crystal.

It should be noted that there are different preferences for the presentation of NIS spectra. In Fig. 9.35, the NIS pattern of hexacyanoferrate, $(NH_4)_2Mg[Fe(CN)_6]$, is given as *partial* DOS $g(E)$ [89], while in Fig. 9.36 for GNP, $(CN_3H_6)_2[Fe(CN)_5NO]$, it is drawn as *probability density* $S(E)$ of nuclear inelastic absorption [86]. The projected probability density $S(E,\boldsymbol{k})$ can be simulated theoretically if the projected PDOS $g(E,\boldsymbol{k})$ is available. For molecular complexes, an approximation to simulate $S(E,\boldsymbol{k})$ outside the energy range of intermolecular vibrations, i.e., for $E \geq 15$ meV, is given by [86]:

$$S(E,\boldsymbol{k}) \sim e^{-\langle \boldsymbol{k}\boldsymbol{u}\rangle^2}\left[\delta(E) + k^2\Sigma_{j=1}^{L}\delta(E-E_j)\langle \boldsymbol{k}\boldsymbol{u}_j\rangle^2\right]. \tag{9.14}$$

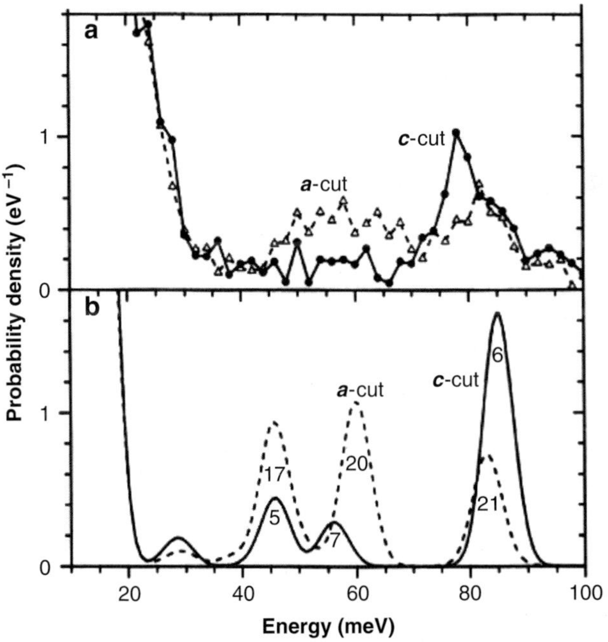

Fig. 9.36 (**a**) NIS spectra, measured at RT and (**b**) DFT simulations for ***a***-cut (*dashed lines*) and ***c***-cut (*solid lines*) single crystals of GNP. (Taken from [86])

The sum on the right side of (9.14) corresponds to $g(E,\boldsymbol{k})$ in a single crystal. The PDOS $g(E)$ for randomly oriented systems requires an averaging over all crystallographic directions.

The $L = 33$ normal modes of the NP anion [92] can be classified according to the five irreducible representations of the C_{4v} point group. In Table 9.2, these modes, their individual energies, and projected msd $\langle(\boldsymbol{k}\boldsymbol{u}_j)^2\rangle$, as derived from DFT calculations [86], are summarized together with measured energies, as obtained by IR- and Raman spectroscopy [93, 94]. The msd of the iron nucleus varies within the A_1 and E modes, while it remains zero for the other modes. Therefore, in principle, only A_1 and E modes are visible in the NIS spectra of GNP. Since the symmetry axis of the NP anion deviates slightly from the crystallographic $\boldsymbol{c}$ axis, only A_1 modes, which contribute to the msd of iron solely along this symmetry axis, are visible in the $\boldsymbol{c}$-cut NIS spectrum. E modes contribute to the msd perpendicular to this symmetry axis, and therefore are visible in the $\boldsymbol{a}$-cut NIS spectrum.

Table 9.2 Vibrational energies E_j and msd $\langle(\boldsymbol{k}\boldsymbol{u}_j)^2\rangle$ derived from DFT calculations for the nitroprusside anion *in vacuo* [86]

j	Sym. spec.	Primary contribution[a]	E_j (meV)		$\langle(\boldsymbol{k}\boldsymbol{u}_j)^2\rangle$ (pm^2)
			Exp.	Calc.	
1	A_1	C_{ax}–N s	269.9	266.1	0.00
2	A_1	C_{eq}–N s	268.8	265.2	0.00
3	A_1	N–O s	241.4	228.9	0.00
4	A_1	Fe–C_{eq} s	51.2	41.7	0.01
5	A_1	Fe–C_{ax} s	52.1	45.9	0.23
6	A_1	Fe–N s	81.5	84.8	0.75
7	A_1	Fe–C–N δ	57.4	55.9	0.14
8	A_1	C–Fe–C,N β,β'	12.6	14.3	5.02
9	A_2	Fe–C–N λ		38.2	
10	B_1	C–N_{eq} s	268.3	264.6	
11	B_1	Fe–C_{eq} s	50.0	40.7	
12	B_1	Fe–C–N δ	49.2	50.6	
13	B_1	C,N–Fe–C β,β'	13.0	9.4	
14	B_2	Fe–C–N λ	52.8	53.7	
15	B_2	C–Fe–C α	11.4	13.0	
16	E	C–N_{eq} s	266.3	264.1	0.00
17	E	Fe–C_{eq} s	56.3	45.5	0.50
18	E	Fe–C–N δ	39.7	38.3	0.04
19	E	Fe–C–N λ	53.9	51.0	0.05
20	E	Fe–C–N γ'	62.1	59.9	0.54
21	E	Fe–N–O γ	82.7	83.0	0.35
22	E	C,N–Fe–C β,β'	8.8	7.7	1.99
23	E	C–Fe–C β,β'	18.9	14.5	3.74
24	E	C,N–Fe–C β,β'	12.3	12.3	1.10

Measured energies for sodium nitroprusside are taken from [93]. For $\boldsymbol{c}$-cut ($\boldsymbol{k} = \boldsymbol{c}$) the A_1 modes and for $\boldsymbol{a}$-cut ($\boldsymbol{k} = \boldsymbol{a}$) the E modes are NIS active

[a] s denotes stretching modes and small Greek letters denote bending modes

The strongest contribution to the projected mean square displacement $\langle(\boldsymbol{ku})^2\rangle$ and therefore to the absorption probability $S(E)$ originates from C–Fe–C and N–Fe–C bending modes ($8A_1$, $22E$, $23E$, and $24E$ in Table 9.2). However, the energy range of these modes (8–15 meV) strongly overlaps with that of the acoustic modes (with composition factor $e_{\mathrm{ac}}^2 = 0.17$, $M_\Sigma = 337$, $m_{\mathrm{Fe}} = 57$) and therefore are not resolved in the NIS spectrum (Fig. 9.36). Other NIS peaks with significant msd and therefore individually visible in the energy range 30–90 meV are for the ***a***-cut crystal, the stretching mode Fe–C_{eq} ($17E$) and the bending modes Fe–C–N ($20E$) and Fe–N–O ($21E$) and, for the ***c***-cut crystal, the stretching modes Fe–N ($6A_1$) and Fe–C_{ax} ($5A_1$) and the bending mode Fe–C–N ($7A_1$). The anisotropy of the measured absorption probability density (Fig. 9.36a) in this energy range is convincingly reproduced by the calculated DFT results (Fig. 9.36b).

From the individual contributions of the A_1 modes to the msd along the ***c***-axis (~6 pm^2) and along the ***a***-axis (~8 pm^2), the corresponding calculated molecular Lamb–Mössbauer factors for the ***c***-cut crystal ($f_{\mathrm{LM,c}} = 0.90$) and for the ***a***-cut crystal ($f_{\mathrm{a}}^{\mathrm{calc}} = 0.87$) were derived. Comparison with the experimental f-factor, i.e., $f_{\mathrm{c}}^{\mathrm{exp}} = 0.20(1)$ and $f_{\mathrm{a}}^{\mathrm{exp}} = 0.12(1)$ [45], indicates that by far the largest part of the iron msd must be due to intermolecular vibrations (acoustic modes) of the nitroprusside anions and its counter ions. This behavior is reflected in the NIS spectrum of GNP by the considerable onset of absorption probability density below 30 meV in Fig. 9.36a.

NIS measurements (at 77 K) were also performed on a GNP single crystal (***c***-cut) (a) after the sample was illuminated with blue light (450 nm) at 50 K to populate metastable molecular states and (b) after warming up the illuminated crystal to 250 K at which the metastable states decay to the groundstate [94]. Comparison of measured and simulated (DFT) NIS spectra provided evidence for the isonitrosyl structure of the metastable state MS_{I} [95] which was proposed earlier by Carducci et al. in their pioneering X-ray study [96].

9.9.4 *Iron–Ligand Vibrations in Spin-Crossover Complexes*

9.9.4.1 Thermally Induced Spin Transition (Example: [Fe(tpa)(NCS)$_2$])

The driving force for the temperature-dependent spin crossover (SCO) is the entropy difference between the HS and the LS isomers which arises mainly from a shift of the vibrational frequencies when passing from the HS to the LS state [97–99]. This frequency shift has been studied by IR- and Raman-spectroscopy and recently also by NIS [23, 39, 87]. The NIS method is isotope (^{57}Fe) selective and, therefore, its focus is on iron–ligand bond-stretching vibrations which exhibit the most prominent contribution to the frequency shift upon SCO [87].

IR measurements on several SCO complexes with a central [FeN$_6$] octahedron indicate a remarkable increase of the energy of the Fe–N bond stretching mode from about 20–25 meV in the HS state to about 50–60 meV in the LS state.

For the [Fe(tpa)(NCS)$_2$] complex, energies of 59.5 and 65.7 meV were reported [100]. IR- and Raman spectra are rather complex in this energy region, making an unambiguous assignment of these modes difficult. The measured NIS spectra of the HS and LS isomers of [Fe(tpa)(NCS)$_2$] exhibit peaks at 30 meV (HS) and at 50 meV (LS) with a linewidth (12–15 meV) that is far more than the resolution of the high-resolution monochromator (6.4 meV) and the linewidth of the central peak (7.2 meV, not shown) (Fig. 9.37a) [41]. The large linewidth indicates that the two peaks (HS and LS) are superpositions of two or more individual peaks; this view is supported by results obtained from DFT calculations (*vide infra*).

Owing to the almost octahedral environment of the iron center, three out of six Fe–N stretching modes are invisible in NIS and IR spectra. Those modes that transform according to A_{1g} and E_g representations of the ideal octahedron do not contribute to the msd of the iron nucleus or to the variation of the electric dipole moment. Only the remaining three modes that transform according to T_{1u} representations can be observed in NIS- and IR spectra. These three modes, as obtained

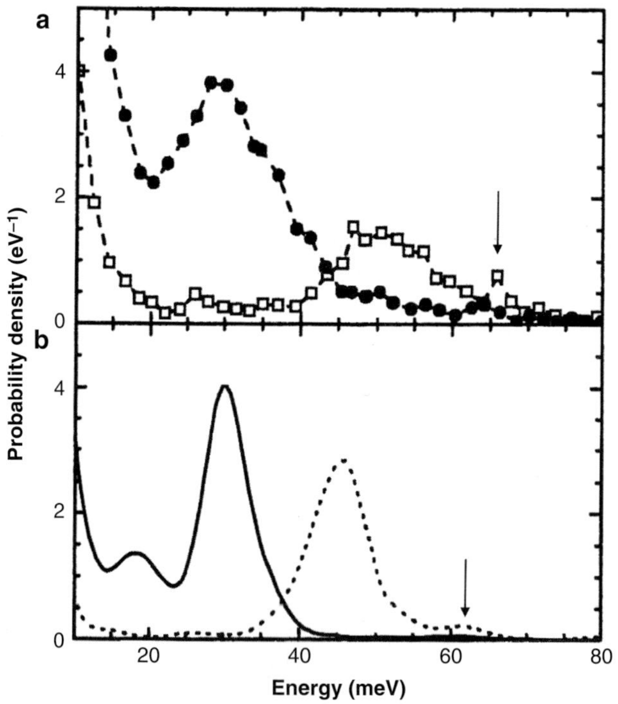

Fig. 9.37 (**a**) NIS spectra of the LS (*solid circles*) and HS (*open squares*) isomers of [Fe(tpa)(NCS)$_2$] measured with a high-resolution monochromator of 6.4 and 1.7 meV bandwidth at 293 and 13 K, respectively. The *dashed lines* are guides to the eye. The simulated NIS spectra (**b**) for the LS (*dotted line*) and for the HS isomer (*solid line*) are based on DFT calculations. (Taken from [41])

from DFT calculations, with energies of 29.1, 30.1, and 35.3 meV and 42.8, 46.6, and 52.6 meV give rise to prominent peaks in the simulated NIS spectra of both the LS and HS isomer of [Fe(tpa)(NCS)$_2$] (Fig. 9.37b) [41]. Considerable contributions to the calculated absorption probability also arise from N–Fe–N bending modes in the range from 3 to 20 meV. These modes cannot be identified in the experimental spectra because they are superimposed by modes originating from acoustic phonons.

The Fe–N bond stretching energies calculated for the LS isomer are in good agreement with the energies obtained from NIS; however, they are more than 10 meV smaller than the IR values given earlier. According to the DFT calculations, the low-intensity peak at 63 meV in the simulation, at 66 meV in the measured NIS spectrum (arrows in Fig. 9.37), and the line at 65.7 meV in the IR spectrum must be assigned to a mode that predominantly has a N–C–S bending character and to some extent also an Fe–N stretching character [41].

In molecular crystals, the Lamb–Mössbauer factor can be regarded (in good approximation) as a product of a molecular part, f^{mol}, and a lattice part, f^{lat} [101]:

$$f = f^{\mathrm{mol}} \cdot f^{\mathrm{lat}}. \tag{9.15}$$

The f factor derived from the NFS spectra (Fig. 9.9) comprises both the molecular and the lattice part. The contribution f^{mol} is obtained by integrating only over the "molecular part" in the NIS spectrum, i.e., after truncating its low-energy part (0–15 meV) which mainly accounts for the lattice contribution f^{lat}. The f values resulting from DFT calculations, from NFS spectra and from "truncated" NLS spectra for the HS and the LS isomers of [Fe(tpa)(NCS)$_2$], are listed in Table 9.3. At all temperatures the total f factor is considerably smaller than f^{mol}. According to (9.15), this is due to acoustic phonons which are represented by f^{lat}. The f (NFS) values decrease considerably with increasing temperature because of mode softening. This effect is not reflected in f^{mol} (NIS) values (Table 9.3) because soft modes have been neglected when f^{mol} was determined by truncating the low-energy part (0–15 meV) of the NIS spectra.

Table 9.3 Factorization of the Lamb–Mössbauer factor f_{LM} into a molecular part $f_{\mathrm{LM}}^{\mathrm{moll}}$ and into a lattice part $f_{\mathrm{LM}}^{\mathrm{lat}}$ at different temperatures

T (K)	Spin state	f_{LM}	$f_{\mathrm{LM}}^{\mathrm{mol}}$		$f_{\mathrm{LM}}^{\mathrm{lat}}$ [a]	
		NFS	NIS	DFT	NIS	DFT
34	LS	0.68	0.85	0.92	0.80	0.74
107	HS	0.38	0.86	0.75	0.44	0.51
200	HS	0.20	0.80	0.52	0.25	0.38

The factor f_{LM} was determined by NFS, and $f_{\mathrm{LM}}^{\mathrm{mol}}$ was derived by "truncated" NIS and by DFT calculations on [Fe(tpa)(NCS)$_2$]. Taken from [41]

[a]Derived from f_{LM} and $f_{\mathrm{LM}}^{\mathrm{mol}}$ according to (9.15)

9.9.4.2 Entropy Change Upon Transition (Example: [Fe(Phen)$_2$(NCS)$_2$])

The entropy difference ΔS_{tot} between the HS and the LS states of an iron(II) SCO complex is the driving force for thermally induced spin transition [97]. About one quarter of ΔS_{tot} is due to the multiplicity of the HS state, whereas the remaining three quarters are due to a shift of vibrational frequencies upon SCO. The part that arises from the spin multiplicity can easily be calculated. However, the vibrational contribution ΔS_{vib} is less readily accessible, either experimentally or theoretically, because the vibrational spectrum of a SCO complex, such as [Fe(phen)$_2$(NCS)$_2$] (with 147 normal modes for the free molecule) is rather complex. Therefore, a reasonably complete assignment of modes can be achieved only by a combination of complementary spectroscopic techniques in conjunction with appropriate calculations.

The vibrational modes of the LS and HS isomers of the SCO complex [Fe (phen)$_2$(NCS)$_2$] (phen = 1,10-phenanthroline) have been measured by NIS (Fig. 9.38a), IR- and Raman-spectroscopy, and the vibrational frequencies and normal modes were calculated by DFT methods [44]. The calculated difference $\Delta S_{vib} = 57$–70 J mol^{-1} K^{-1}, depending on the method) is in qualitative agreement with the experimentally derived values (20–36 J mol^{-1} K^{-1}).

Only the low-energy vibrational modes ($\sim$20% of the 147 modes) contribute to ΔS_{vib}. Furthermore, only the 15 modes of the central FeN_6 octahedron (six Fe–N stretching modes and nine N–Fe–N bending modes, marked by the letters *s* and *b*, respectively, in Fig. 9.38b) account for $\sim$75% of ΔS_{vib}.

In summary, the combined experimental (NIS, IR- and Raman-spectroscopy) and computational (DFT) approach has enabled the identification of the vibrational modes that contribute most to the entropic driving force for SCO transition.

9.9.5 Boson Peak, a Signature of Delocalized Collective Motions in Glasses (Example: FC as Sensor Molecule)

In molecular crystals or glasses, intermolecular interactions are mirrored directly in the low-energy part of the vibrational DOS. Slow processes, which give rise to a quasielastic broadening in the range of neV–μeV, cannot be resolved using NIS. However, faster processes, e.g., cage rattling processes in glasses and plastic crystals as well as the much-debated Boson peak, appear in the low-energy part of the DOS usually at energies well below the intramolecular modes [102].

If the center of mass of the molecular probe coincides with the Mössbauer nucleus, then the low-energy part of the spectrum monitors exclusively translational modes of the probe molecule thus providing a selective probe for fast translational processes on the lengthscale of several molecular diameters and larger. If, however, the center of mass does not coincide with the Mössbauer nucleus, then hindered rotations, i.e., librations, will contribute to the low-energy DOS. If

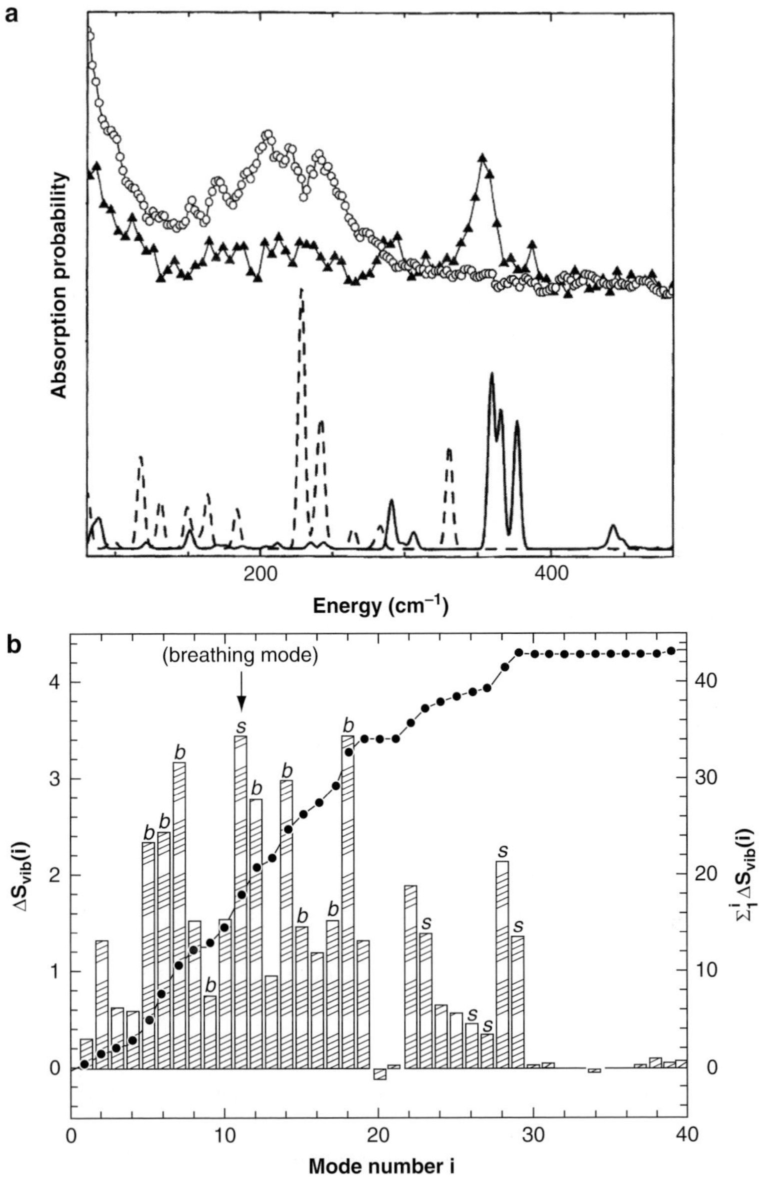

Fig. 9.38 (**a**) Measured (*upper panel*) and calculated (*lower panel*) NIS spectra of the HS (*open circle*, *dashed line*) and the LS (*filled triangle*, *solid line*) isomer (1 meV = 8.06 cm^{-1}) of [Fe $(phen)_2(NCS)_2$]. (**b**) Calculated contributions ΔS_{vib}(i) of individual modes i to the vibrational entropy difference (*bars*, *left axis*) and sum $\Sigma_1{}^i \Delta S_{vib}$ (i) of the contributions of modes 1 to *i* (*filled circle*, *right axis*). The 15 modes of an idealized octahedron (six Fe–N stretching modes and nine N–Fe–N bending modes) are marked by the letters *s* and *b*, respectively. (Taken from [44])

contributions from translational motions can be neglected, as is the case for plastic crystals, then a selective probe for vibrational dynamics on a local scale is available. It is therefore profitable to select an appropriate sensor molecule for NIS measurements to study the specific dynamical properties of hosts.

FC as sensor molecule has been used to investigate the low-energy mobility, i.e., the nature of the Boson peak and of the *trans*-Boson dynamics, of toluene, ethylbenzene, DBP and glycerol glasses [102]. The spectator nucleus Fe is at the center of mass of the sensor molecule FC. In this way, rotations are disregarded and one selects pure translational motions. Thus, the low-energy part of the measured NIS spectra represents the DOS, $g(E)$, of translational motions of the glass matrix (below about 15 meV in Fig. 9.39a).

The so-called Boson peak is visible as a hump in the reduced DOS, $g(E)/E^2$ (Fig. 9.39b), and is a measure of structural disorder, i.e., any deviation from the symmetry of the perfectly ordered crystal will lead to an excess vibrational contribution with respect to Debye behavior. The reduced DOS appears to be temperature-independent at low temperatures, becomes less pronounced with increasing temperature, and disappears at the glass–liquid transition. Thus, the significant part of modes constituting the Boson peak is clearly nonlocalized on FC. Instead, they represent the delocalized collective motions of the glasses with a correlation length of more than ~20 Å.

Beyond the Boson peak, the reduced DOS reveals for all studied glasses a temperature-independent precisely exponential behavior, $g(E)/E^2 \sim \exp(E/E_o)$ with the "decay" energies E_o correlating with the energies E_B of the Boson peak. This finding additionally supports the view that the low-energy dynamics of the glasses are indeed delocalized collective motions because local and quasilocal vibrations would be described in terms of a power law or a log–normal behavior [102].

9.9.6 Protein Dynamics Visualized by NIS

Conventional MS in the energy domain has contributed a lot to the understanding of the electronic ground state of iron centers in proteins and biomimetic models ([55], and references therein). However, the vibrational properties of these centers, which are thought to be related to their biological function, are much less studied. This is partly due to the fact that the vibrational states of the iron centers are masked by the vibrational states of the protein backbone and thus techniques such as Resonance Raman- or IR-spectroscopy do not provide a clear picture of the vibrational properties of these centers. A special feature of NIS is that it directly reveals the fraction of kinetic energy due to the Fe motion in a particular vibrational mode.

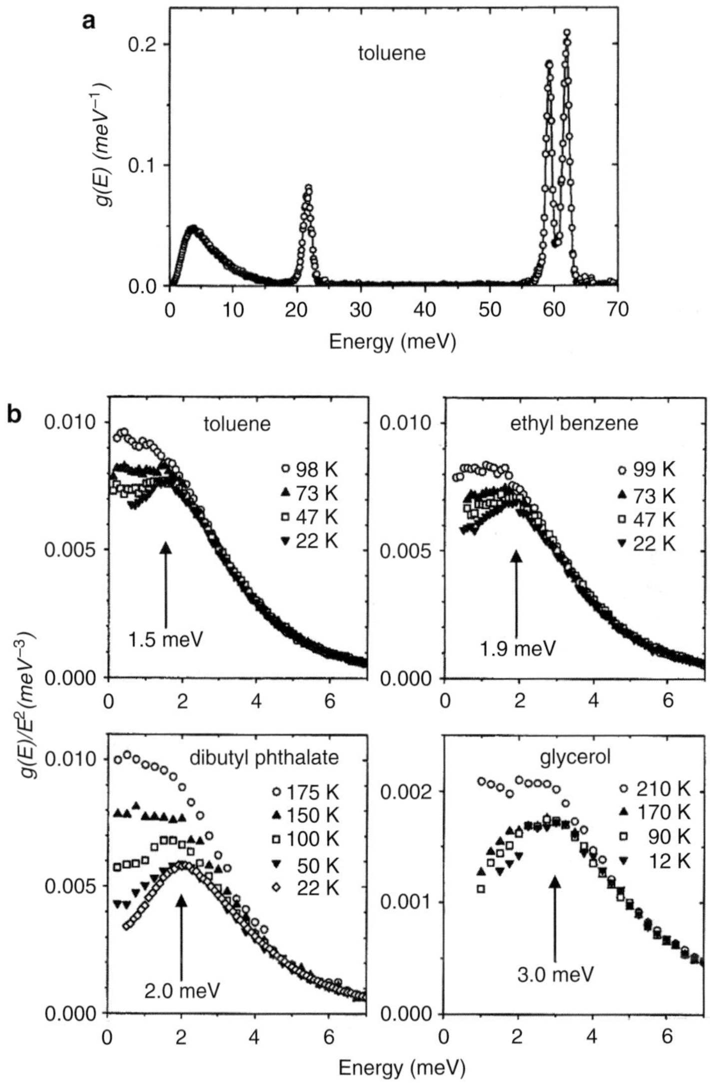

Fig. 9.39 (**a**) Density of states (DOS), $g(E)$, obtained from NIS at 22 K on ferrocene as sensor molecule in toluene glass. (**b**) Reduced DOS, $g(E)/E^2$, for various glasses. *Arrows* indicate the energy of the Boson peak. (Taken from [102])

9.9.6.1 Iron–Sulfur Proteins (Examples: FeS_4 – and Fe_4S_4 – Centers)

Iron–sulfur (Fe–S) proteins function as electron-transfer proteins in many living cells. They are involved in photosynthesis, cell respiration, as well as in nitrogen fixation. Most Fe–S proteins have single-iron (rubredoxins), or two-, three-, or four-iron (ferredoxins), or even seven/eight-iron (nitrogenases) centers.

NIS measurements have been performed on the rubredoxin (FeS_4) type mutant Rm 2–4 from *Pyrococcus abyssi* [103], on *Pyrococcus furiosus* rubredoxin [104], on Fe_2S_2 – and Fe_4S_4 – proteins and model compounds [105, 106], and on the P-cluster and FeMo-cofactor of nitrogenase [105, 107].

For example, Fig. 9.40 shows the NIS spectra of the oxidized and reduced FeS_4 centers of a rubredoxin mutant from *Pyrococcus abyssi* obtained at 25 K together with DFT simulations using different models for the Fe–S center [103]. The spectrum from the oxidized protein $Fe^{III}S_4$ ($S = 5/2$) reveals broad bands around 15–25 meV (121–202 cm^{-1}) and 42–48 meV (339–387 cm^{-1}) consistent with the results on rubredoxin from *Pyrococcus furiosus* [104].

These results confirm Resonance Raman studies of oxidized rubredoxin from *Desulfovibrio gigas*, which indicates that there are three bands (43.15, 45.01, and 46.62 meV; 348, 363, and 376 cm^{-1}) in the region where asymmetric Fe^{III}–S stretch modes are expected [108]. Resonance Raman data of the reduced rubredoxin

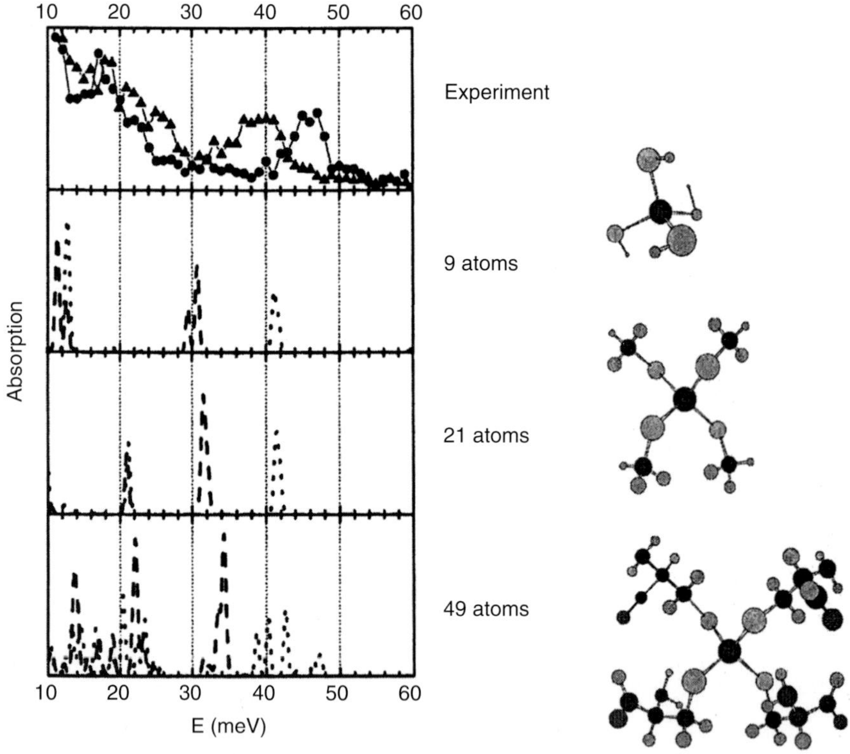

Fig. 9.40 NIS spectra of oxidized (*filled circle*) and reduced rubredoxin mutant Rm 2–4 (*filled triangle*) from *Pyrococcus abyssi* obtained at 25 K. The protein samples have been prepared with ^{57}Fe concentrations of about 10 mM. Theoretically calculated NIS spectra based on DFT calculations (B3LYP/CEP-3IG) of 9, 21 and 49 atoms are shown below. The *dotted lines* represent calculated NIS spectra for the oxidized $Fe^{III}S_4$ center and the *dashed lines* for the reduced $Fe^{II}S_4$ center. (Taken from [103])

with its iron in the $Fe^{II}S_4$ ($S = 2$) state could not be obtained. Thus, NIS is the method of choice to study the dynamical properties of $Fe^{II}S_4$ centers. Upon reduction (from the ferric to the ferrous state), the asymmetric stretch modes observed by NIS shift to lower energies (36–42 meV; 291–339 cm^{-1}). This can be rationalized by the fact that, upon reduction, the Fe^{II}–S bond lengths increase, which is in turn accompanied by a decrease in Fe^{II}–S binding energy and therefore by a decrease in the force constant of the Fe^{II}–S bond. This correlation between iron-oxidation state and Fe–S stretch-mode position is in line with DFT calculations for three different models [103] (Fig. 9.40), which have been performed on the basis of the crystal structure of the protein. The results for the model with 21 atoms are shown in Table 9.4. NIS-visible S–Fe–S bending modes are expected in the region 6–10 meV, Fe–S–C bending modes in the region 16–21 meV, and Fe–S stretching modes in the region 28–42 meV.

As illustrated in Fig. 9.40, progressively more complex models for the environment of Fe in oxidized or reduced rubredoxin produce better simulations of the NIS pattern. A simple $Fe(SCH_3)_4$ model (21 atoms) predicts a division of the spectrum into Fe–S stretch and S–Fe–S/Fe–S–C bend regions, but at least a model with 49 atoms is needed to reproduce the splitting of the stretch region and to capture some of the features between 10 and 30 meV. These results confirm the delocalization of the dynamic properties of the redox-active Fe site far beyond the immediate $Fe–S_4$ coordination sphere.

$[(n\text{-}Bu_4)N]_2[^{57}Fe_4S_4(SPh)_4]^{2-}$ was examined as a simple model for 4Fe ferredoxins and as a test to observe the effects in the PDOS originating from $^{32}S/^{36}S$ isotope replacement in the bridging positions (Fig. 9.41) [106]. Four major features were observed in the NIS spectrum of the ^{32}S variant of the complex: (a) a peak at about 430 cm^{-1}, (b) a band near 280 cm^{-1}, (c) a doublet at 267/290 cm^{-1}, and (d) a triplet at 137/148/157 cm^{-1}. With DFT calculations the modes were identified as (a) Fe–S^t stretch between iron and terminal sulfur atoms (S^t), (b, c) Fe–S^b stretch between iron and bridging sulfur atoms (S^b), and (d) S–Fe–S bend. Consistent with this classification, the peak at about 430 cm^{-1} (a) is invariant when substituting ^{36}S into the bridging position of the Fe_4S_4 cluster (Fig. 9.41). The modes that experience relatively large frequency change (down shift by about 7 cm^{-1}) upon ^{36}S substitution have significant Fe–S^b stretch contribution (b, c) [106]. Downshifts of

Table 9.4 Assignment of vibrational iron–ligand modes based on a DFT calculation on the model with 21 atoms

Character of vibrational modes	Energy (meV)	
	$Fe^{III}S_4$	$Fe^{II}S_4$
Four torsional modes	1–6	2–5
Two S–Fe–S bending modes (NIS vis.)	8–10	6–8
Seven torsional modes	10–14	9–13
Four Fe–S–C bending modes (NIS vis.)	17–21	16–21
Four Fe–S stretching modes (NIS vis.)	35–42	28–32

(Taken from [103])

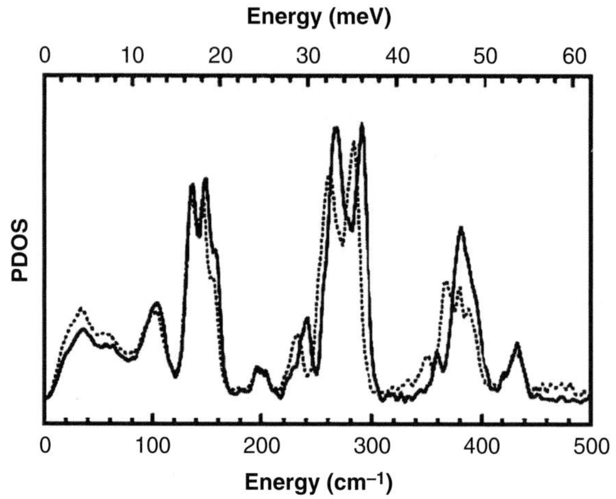

Fig. 9.41 Partial (vibrational) densities of states (PDOS) at 60 K for $[n\text{-}Bu_4]N_2[^{57}Fe_4S_4(SPh)_4]$ with ^{32}S (*solid line*) and ^{36}S (*dotted line*) substitution in the bridging position of the Fe_4S_4 cluster. (Taken from [106])

this amount are reproduced by an isolated ^{57}Fe–S harmonic oscillator when replacing ^{32}S by ^{36}S.

The catalytic site $MoFe_7S_9$ (FeMo-cofactor) of nitrogenase exhibits a complex NIS spectrum with a strong signal at about 190 cm^{-1}, where Fe_2S_2 – and Fe_4S_4 – clusters have weak spectral contributions. Comparison with NIS simulations on the basis of DFT and with experimental NIS results obtained for $[Fe_6N(CO)_{15}]^{3-}$ suggests that this strong intensity originates from cluster breathing modes whose frequency is raised by an interstitial atom (probably N) which causes extra rigidity of the FeMo-cofactor [107].

9.9.6.2 Heme Proteins (Examples: Deoxy-, CO- and Metmyoglobin)

Heme complexes and heme proteins have also been the subject of NIS studies. Of specific interest have been three features: the in-plane vibrations of iron, which have not been reported by Resonance Raman studies [108], the iron–imidazole stretch, which has not been identified in six-coordinated porphyrins before, and the heme-doming mode, which was assumed to be a soft mode.

The NIS investigation of heme complexes includes various forms of porphyrins (deuteroporphyrin IX, mesoporphyrin IX, protoporphyrin IX, tetraphenylporphyrin, octaethylporphyrin, and "picket fence" porphyrin) and their nitrosyl (NO) and carbonyl (CO) derivatives, and they have been the subject of a review provided by Scheidt et al. [109].

Sage et al. reported the complete vibrational spectrum of the iron site in deoxy- and CO-myoglobin [110]. The spectrum of photolyzed CO-myoglobin (frozen solution) resembles that of Mb. Because of high-resolution and reasonable statistics, they were able to identify three main regions: a Fe–N (histidine) (Fe–N_{His}) stretch at 234 cm^{-1} and Fe–N (pyrrole) (Fe–N_{pyr}) stretches at 251 and 267 cm^{-1}. Corresponding vibrations were reported for metmyoglobin by Achterhold et al. [111] (Fig. 9.42a): a vibration centered at about 180 cm^{-1} (22.3 meV; perpendicular to the heme plane) and a vibration centered at about 270 cm^{-1} (33.4 cm^{-1}; within the heme plane). Their identification as out-of-plane and in-plane vibrations was possible because the NIS studies of Achterhold et al. were performed on a single crystal of the protein. This experimental situation has the advantage that vibrational modes can be assigned by the projection of the vibrational amplitude onto the direction of the synchrotron beam (Fig. 9.42b).

For CO-myoglobin a Fe–CO stretch at 502 cm^{-1} and a Fe–C–O bend at 572 cm^{-1} has been observed [112]. The drastic increase of the out-of-plane stretch compared to deoxy- and metmyoglobin is due to the strong covalent Fe–CO bond. Raising the temperature from 50 to 110 K led to a broad resonance at around 25 cm^{-1} which has been assigned to the translational motion of the whole heme moiety.

The PDOS of the iron in deoxy- and CO-myoglobin and of myoglobin with different degrees of water content was also determined by Achterhold et al. [112, 113]. They found that the modes with an energy larger than 3 meV (24 cm^{-1}) are harmonic at physiologically relevant temperatures. Those below 3 meV exhibit a

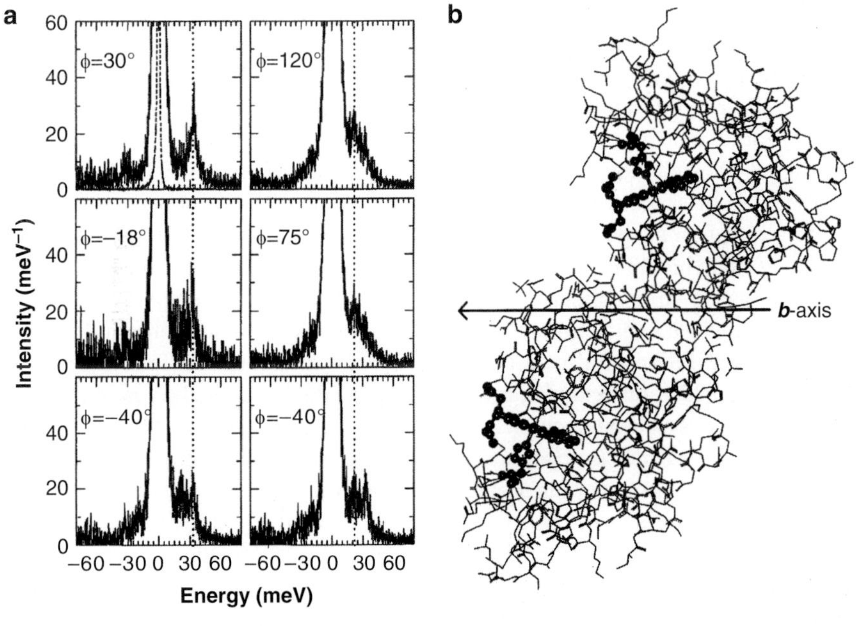

Fig. 9.42 (**a**) NIS spectra of a metmyoglobin single crystal with its ***b***-axis perpendicular to the synchrotron beam as a function of the rotation angle Φ around this axis. (**b**) Two metmyoglobin molecules in the unit cell together with *an arrow* indicating the crystallographic ***b***-axis are shown. $\Phi = 0°(90°)$ corresponds to an orientation of the synchrotron beam parallel (perpendicular) to the plane of this drawing. (Taken from [111])

temperature dependence that was interpreted as mode softening in the low-energy regime.

In summary, NIS provides an excellent tool for the study of the vibrational properties of iron centers in proteins. In spectroscopies like Resonance Raman and IR, the vibrational states of the iron centers are masked by those of the protein backbone. A specific feature of NIS is that it is an isotope-selective technique (e.g., for ^{57}Fe). Its focus is on the metal–ligand bond stretching and bending vibrations which exhibit the most prominent contributions to the mean square displacement of the metal atom.

9.10 Nuclear Resonance Scattering with Isotopes Other Than ^{57}Fe

Nuclear resonance scattering, delayed in time, is separated from the prompt synchrotron pulse by a fast detector. Additionally, the energy bandwidth of the incident radiation has to be decreased (to $\sim$100 eV) by appropriate monochromators; otherwise, the enormous intensity of the incident beam would lead to an overload of the detector. For low-energy nuclear resonances, this can be achieved using a monochromator with (a) high-order reflections of perfect silicon crystals which provide a narrow bandwidth (1–10 meV) and (b) high angular acceptance matching the angular divergence of the synchrotron beam [114]. For nuclear resonances beyond 30 keV, this approach causes problems because the angular acceptance of the high-order reflections becomes very small [115]. Backscattering with a sapphire crystal has allowed the observation of NFS for ^{121}Sb (37.13 keV) [116].

The NLE overcomes this shortcoming because here NFS is separated from the incident radiation in space (see Sect. 9.7) as demonstrated with ^{61}Ni (67.419 keV).

Extension to higher energies can also be achieved by using X-ray optics based on silicon crystals in combination with fast multielement detectors [117]. This approach provides sufficiently high angular acceptance using low-order reflections. High-quality NFS spectra have been obtained for a Ni foil and for NiO powder [117]. This application demonstrates that this procedure can also be applied to other high-energy Mössbauer transitions.

Potential Mössbauer isotopes for nuclear resonance scattering, which are within the spectral reach of synchrotron radiation sources, are summarized in Table 9.5 [118–120], and the synchrotron radiation sources which provide dedicated beam lines for specific Mössbauer isotopes are listed in Table 9.6 (adopted from [85]).

Looking ahead, new storage rings with enhanced brilliance such as PETRAIII, presently under construction at DESY, Hamburg, and PEPIII in its early design at SLAC, Stanford, will provide new scientific opportunities. In the next two decades, X-ray free-electron lasers (XFELs) with unique time structure, coherence, and a 5–6 orders of higher average brilliance are expected to revolutionize nuclear

Table 9.5 Potential Mössbauer isotopes for nuclear resonance scattering, which are within the spectral reach of currently available synchrotron radiation sources

Isotope	Energy E_0 (keV)	Half-lifetime[a] (ns)	Observed by various methods[b]
^{133}Cs	81	6.4	
^{161}Dy	25.655	28.2	NIS [118]
^{151}Eu	21.532	9.9	NFS [85]
^{57}Fe	14.413	97.8	NFS (see Sects. 9.4–9.6) NIS (see Sect. 9.9) NLE (see Sect. 9.7) SRPAC (see Sect. 9.8)
^{73}Ge	13.263	2,953	
^{129}I	27.770	16.8	
^{193}Ir	73	6.3	
^{40}K	29.560	4.25	
^{83}Kr	9.401	147	NIS [85]
^{145}Nd	67.100	67.1	
^{61}Ni	67.419	5.1	NFS [117], NLE [77]
^{121}Sb	37.1298	3.5	NFS [116, 119], NIS [119]
^{149}Sm	22.490	7.1	NLE [76]
^{119}Sn	23.870	17.75	NIS [84, 120]
^{125}Te	35.4931	1.48	NFS, NIS [119]
^{181}Ta	6.238	6,050	
^{169}Tm	8.401	4	
^{67}Zn	93.3	9,200	

[a]Half-lifetime $t_{1/2}$ and mean-lifetime τ are related by: $\tau = t_{1/2}/\ln 2$

[b]Nuclear inelastic scattering (NIS), nuclear forward scattering (NFS), nuclear lighthouse effect (NLE), synchrotron radiation-based perturbed angular correlation (SRPAC)

Table 9.6 List of synchrotron radiation sources which provide dedicated nuclear resonance beamlines for specific isotopes

Location	Isotopes
APS[a]	Dy, Eu, Fe, Kr, Sn
ESRF[b]	Dy, Eu, Fe, Ni, Sb, Sn, Ta, Te
HASYLAB[c]	Dy, Eu, Fe, Ni, Sn
KEK-AR[d]	Dy, Eu, Fe, Sn
SPRING-8[e]	Dy, Eu, Fe, K, Sn

[a]Advanced Photon Source, Argonne National Lab, USA

[b]European Synchrotron Radiation Facility, Grenoble, France

[c]Hamburg Synchrotron Radiation Laboratory, Germany

[d]National Laboratory for High Energy Physics, Accumulator Ring, Tsukuba, Japan

[e]Super Photon Ring, Hyogo, Japan

resonance applications in a major way. In an overview, Shenoy and Röhlsberger [121] have described the specific radiation characteristics of these new radiation sources and have provided a glimpse of scientific prospects and dreams in the nuclear resonance field associated with such sources.

References

1. Mössbauer, R.L.: Z. Phys. **151**, 124 (1958)
2. Hastings, J.B., Siddans, D.P., van Bürck, U., Hollatz, R., Bergmann, U.: Phys. Rev. Lett. **66**, 770 (1991)
3. Selo, M., Yoda, Y., Kikuta, S., Zhang, X.W., Ando, M.: Phys. Rev. Lett. **74**, 3828 (1995)
4. Gonser, U., Fischer, H.: In: Gonser, U. (ed.) Topics in Current Physics, Mössbauer Spectroscopy II: The Exotic Side of the Method. Springer, Berlin (1981)
5. Rüffer, R., Chumakov, A.J.: Hyperfine Interact. **589**, 97 (1996)
6. Hannon, J.P., Trammel, G.T.: In: Materlik, G., Sparks, C.J., Fischer, K. (eds.) Resonant Anomalous X-Ray Scattering, p. 565. Elsevier, New York (1994)
7. Grünsteudel, H., Haas, M., Leupold, O., Mandon, D., Matzanke, B.F., Meyer-Klaucke, W., Paulsen, H., Realo, E., Rüter, H.D., Trautwein, A.X., Weiss, R., Winkler, H.: Inorg. Chim. Acta **275/276**, 334 (1988)
8. Smirnov, G.V.: Hyperfine Interact. **97/98**, 551 (1996)
9. Sturhahn, W., Gerdau, E.: Phys. Rev. **B 49**, 9285 (1994)
10. Sturhahn, W.: Hyperfine Interact. **125**, 149 (2000)
11. Shvyd'ko, Yu V.: Phys. Rev. **B 59**, 9132 (1999)
12. Shvyd'ko, Yu V.: Hyperfine Interact. **25**, 173 (2000)
13. Haas, M., Realo, E., Winkler, H., Meyer-Klaucke, W., Trautwein, A.X., Leupold, O., Rüter, H.D.: Phys. Rev. **B 56**, 14082 (1997)
14. Haas, M., Realo, E., Winkler, H., Meyer-Klaucke, W., Trautwein, A.X.: Hyperfine Interact. **125**, 189 (2000)
15. Herta, C., Winkler, H., Benda, R., Haas, M., Trautwein, A.X.: Eur. Biophys. J. **31**, 478 (2002)
16. Keppler, C., Achterhold, K., Ostermann, K., van Bürck, U., Chumakov, A.F., Rüffer, R., Sturhahn, W., Alp, E.E., Parak, F.G.: Eur. Biophys. J. **29**, 146 (2000)
17. Eicher, H., Trautwein, A.X.: J. Chem. Phys. **50**, 2540 (1969)
18. Trautwein, A.X.: Struct. Bond. **20**, 101 (1974)
19. Eicher, H., Trautwein, A.X.: J. Chem. Phys. **52**, 932 (1970)
20. Lang, G., Marshall, W.: Proc. Phys. Soc. (Lond.) **87**, 3 (1966)
21. Kappler, H.M., Trautwein, A.X., Mayer, A., Vogel, H.: Nucl. Instrum. Methods **53**, 157 (1967)
22. Bill, E., Trautwein, A.X., Weiss, R., Winkler, H.: Focus MHL **3**, 91 (1986)
23. Paulsen, H., Schünemann, V., Trautwein, A.X., Winkler, H.: Coord. Chem. Rev. **249**, 255 (2005)
24. Debrunner, P.G.: In: Lever, A.B.P., Gray, H.B. (eds.) Iron Porphyrins (Phys. Bioinorg. Chem. Series Part 3), p. 139. VCH, New York (1989)
25. Trautwein, A.X., Winkler, H., Schwendy, S., Grünsteudel, H., Meyer-Klaucke, W., Leupold, O., Rüter, H.D., Gerdau, E., Haas, M., Realo, E., Mandon, D., Weiss, R.: Pure Appl. Chem. **70**, 917 (1998)
26. Montiel-Montoya, R., Bill, E., Trautwein, A.X., Winkler, H., Ricard, L., Schappacher, M., Weiss, R.: Hyperfine Interact. **29**, 1411 (1986)
27. Schappacher, M., Ricard, L., Fischer, J., Weiss, R., Bill, E., Montiel-Montoya, R., Winkler, H., Trautwein, A.X.: Eur. J. Biochem. **168**, 419 (1987)
28. Phillips, S.E.V.: J. Mol. Biol. **142**, 531 (1980)
29. Maeda, Y., Harami, T., Morita, Y., Trautwein, A.X., Gonser, U.: J. Chem. Phys. **75**, 36 (1981)
30. Haas, M., Realo, E., Winkler, H., Meyer-Klaucke, W., Trautwein, A.X., Leupold, O.: Phys. Rev. **B 61**, 4155 (2000)
31. Sergueev, I., Franz, H., Asthalter, T., Petry, W., van Bürck, U., Smirnov, G.V.: Phys. Rev. **B 66**, 184210 (2002)

32. Asthalter, T., Sergueev, I., Franz, H., Rüffer, R., Petry, W., Messel, K., Härter, P., Huwe, A.: Eur. Phys. J. **B 22**, 301 (2001)
33. Asthalter, T., Sergueev, I., Franz, H., Petry, W., Messel, K., Verbeni, R.: Hyperfine Interact. **5**, 29 (2003)
34. Asthalter, T., Garibay, J.V., Olszowka, V., Kornatowski, J.: J. Chem. Phys. **122**, 014508 (2005)
35. Asthalter, T.: Habilitation Thesis, University of Stuttgart (2007)
36. Gütlich, P., Hauser, A., Spiering, H.: Angew. Chem. **106**, 2109 (1994)
37. Gütlich, P., Hauser, A., Spiering, H.: Angew. Chem. **33**, 2024 (1994)
38. Gütlich, P., Goodwin, H.A.: Top. Curr. Chem. **233**, 1–47 (2004)
39. Grünsteudel, G., Paulsen, H., Meyer-Klaucke, W., Winkler, H., Trautwein, A.X., Grünsteudel, H.F., Baron, A.Q.R., Chumakov, A.I., Rüffer, R., Toftlund, H.: Hyperfine Interact. **113**, 311 (1988)
40. Grünsteudel, H., Paulsen, H., Winkler, H., Trautwein, A.X., Toftlund, H.: Hyperfine Interact. **123/124**, 841 (1999)
41. Paulsen, H., Grünsteudel, H., Meyer-Klaucke, W., Gerdan, M., Grünsteudel, H.F., Chumakov, A.I., Rüffer, R., Winkler, H., Toftlund, H., Trautwein, A.X.: Eur. Phys. J. **B 23**, 463 (2001)
42. Winkler, H., Chumakov, A.I., Trautwein, A.X.: Top. Curr. Chem. **235**, 137 (2004)
43. Böttger, L.H., Chumakov, A.I., Grunert, C.M., Gütlich, P., Kusz, J., Paulsen, H., Ponkratz, U., Rusanov, V., Rüffer, R., Trautwein, A.X., Wolny, J.A.: Chem. Phys. Lett **429**, 189 (2006)
44. Ronayne, K.L., Paulsen, H., Höfer, A., Denniss, A.C., Wolny, J.A., Chumakov, A.I., Schünemann, V., Winkler, H., Spiering, H., Bousseksou, A., Gütlich, P., Trautwein, A.X., McGarvey, J.: Phys. Chem. Chem. Phys. **8**, 4685 (2006)
45. Grünsteudel, H.: Ph.D. Thesis, University of Lübeck (1998)
46. Grünsteudel, H.F.: Diplom Thesis, University of Erlangen (1993)
47. Baron, A.Q.R., Chumakov, A.I., Grünsteudel, H.F., Grünsteudel, H., Niesen, L., Rüffer, R.: Phys. Rev. Lett. **77**, 4808 (1996)
48. Grünsteudel, H., Rusanov, V., Meyer-Klaucke, W., Trautwein, A.X.: Hyperfine Interact. **122**, 345 (1999)
49. Goldanskii, V.I., Makarov, E.F.: Chemical Applications of Mössbauer Spectroscopy, p. 102. Academic, New York (1968)
50. Gonser, U., Fischer, H.: In: Gonser, U. (ed.) Mössbauer Spectroscopy II, p. 99. Springer, Berlin (1981)
51. Benda, R., Herta, C., Schünemann, V., Winkler, H., Trautwein, A.X., Shvyd'ko, Y., Walker, F.A.: Hyperfine Interact. **C 5**, 269 (2002)
52. Grünsteudel, H., Meyer-Klaucke, W., Trautwein, A.X., Winkler, H., Leupold, O., Metge, J., Gerdau, E., Rüter, H.D., Baron, A.Q.R., Chumakov, A.I., Grünsteudel, H.F., Rüffer, R., Haas, M., Realo, E., Mandon, D., Weiss, R., Toftlund, H.: In: Trautwein, A.X. (ed.) Bioinorganic Chemistry, Transition Metals in Biology and Their Coordination Chemistry, p. 760. Wiley-VCA, Weinheim (1997)
53. Winkler, H., Meyer-Klaucke, W., Schwendy, S., Trautwein, A.X., Matzanke, B.F., Leupold, O., Rüter, H.D., Haas, M., Realo, E., Mandon, D., Weiss, R.: Hyperfine Interact. **113**, 443 (1998)
54. Herta, C., Winkler, H., Benda, R., Trautwein, A.X., Haas, M.: Hyperfine Interact. **C 5**, 245 (2002)
55. Trautwein, A.X., Bill, E., Bominaar, E.L., Winkler, H.: Struct. Bond. **78**, 1 (1991)
56. Bominaar, E.L., Ding, X.-Q., Gismelseed, A., Bill, E., Winkler, H., Trautwein, A.X., Nasri, H., Fischer, J., Weiss, R.: Inorg. Chem. **31**, 1845 (1992)
57. Schünemann, V., Raitsimring, A.M., Benda, R., Trautwein, A.X., Shokireva, T.K., Walker, F.A.: JBIC **4**, 708 (1999)
58. Scharrock, M., Debrunner, P.G., Schulz, C., Lipscomb, J.D., Marshall, V., Gunsalus, I.C.: Biochem. Biophys. Acta **420**, 8 (1976)

59. Champion, P.M., Münck, E., Debrunner, P.G., Hollenberg, P.F., Hager, L.P.: Biochemistry **12**, 426 (1973)
60. Harrison, P.M., Lilley, T.H.: In: Loehr, T.M. (ed.) Carriers and Proteins, p. 123. VCH, New York (1989)
61. Bauminger, E.R., Harrison, P.M., Hechel, D., Nowik, I., Treffry, A.: Biochem. Biophys. Acta **1118**, 48 (1991)
62. Morup, S., Dumesic, I.A., Topsoe, H.: In: Cohen, R.C. (ed.) Applications of Mössbauer Spectroscopy II, p. 1. Academic, New York (1980)
63. Bauminger, E.R., Nowik, I.: Hyperfine Interact. **50**, 484 (1989)
64. Bell, S.H., Weir, M.P., Dickson, D.P.E., Gibson, J.F., Sharp, G.A., Peters, T.J.: Biochem. Biophys. Acta **787**, 227 (1984)
65. Wortmann, G., Rupprecht, K., Giefers, H.: Hyperfine Interact. **144/145**, 103 (2002)
66. Lübbers, R., Wortmann, G., Grünsteudel, H.G.: Hyperfine Interact. **123/124**, 529 (1999)
67. Lübbers, R., Rupprecht, K., Wortmann, G.: Hyperfine Interact. **128**, 115 (2000)
68. Gafriliuk, A.G., Trjan, I.A., Lyubutin, I.S., Ovchinuikov, S.G., Sarkissian, V.A.: J. Exp. Theor. Phys. **100**, 688 (2005)
69. Gavriliuk, A.G., Struzhkin, V.V., Lyubutin, I.S., Hu, M.U., Mao, H.-K.: JEPT Lett. **82**, 224 (2005)
70. Zhao, J., Sturhahn, W., Liu, J.-F., Shen, G., Alp, E.E., Mao, H.-K.: HPR **24**, 447 (2004)
71. Friedmann, T.: Diploma Thesis, University of Paderborn (2001)
72. Reiß, G.: Ph.D. Thesis, University of Paderborn (2000)
73. Wagner, D., Wohlfahrt, F.: J. Phys. **F 11**, 2417 (1981)
74. Röhlsberger, R., Toellner, T.S., Sturhahn, W., Quast, K.W., Alp, E.E., Bernhard, A., Burkel, E., Leupold, O., Gerdau, E.: Phys. Rev. Lett. **84**, 1007 (2000)
75. Röhlsberger, R., Quast, K.W., Toellner, T.S., Lee, P.L., Sturhahn, W., Alp, E.E., Burkel, E.: Appl. Phys. Lett. **78**, 2970 (2001)
76. Röhlsberger, R., Quast, K.W., Toellner, T.S., Lee, P.L., Alp, E.E., Burkel, E.: Phys. Rev. Lett. **87**, 047601 (2001)
77. Roth, T., Leupold, O., Wille, H.C., Rüffer, R., Quast, K.W., Röhlsberger, R., Burkel, E.: Phys. Rev. **B 71**, 140401 (2005)
78. Sergueev, I., van Bürck, U., Chumakov, A.I., Asthalter, T., Smirnov, G.B., Franz, H., Rüffer, R., Petry, W.: Phys. Rev. **B 73**, 24203 (2006)
79. Asthalter, T., Sergueev, I., van Bürck, U., Dinnebier, R.: J. Phys. Chem. Solids **67**, 1416 (2006)
80. Asthalter, T., Sergueev, I., van Bürck, U.: J. Phys. Chem. Solids **66**, 2271 (2005)
81. Asthalter, T.: (private communication)
82. Seto, M., Yoda, Y., Kikuta, S., Zhang, S.W., Ando, M.: Phys. Rev. Lett. **74**, 3828 (1995)
83. Sturhahn, W., Toellner, T.S., Alp, E.E., Zhang, X., Ando, M., Yoda, Y., Kikuta, S., Seto, M., Kimball, C.W., Dabrowski, B.: Phys. Rev. Lett. **74**, 3832 (1995)
84. Barone, G., Böttger, L.H., Wolny, J.A., Paulsen, H., Trautwein, A.X., Silvestri, A., Sergueev, I., LaManna, G.: Hyperfine Interact. **165**, 299 (2005)
85. Alp, E., Sturhahn, W., Toellner, T.S., Zhao, Y., Hu, M., Brown, D.E.: Hyperfine Interact. **144/145**, 3 (2002)
86. Paulsen, H., Winkler, H., Trautwein, A.X., Grünsteudel, H., Rusanov, V., Toftlund, H.: Phys. Rev. **B 59**, 975 (1999)
87. Paulsen, H., Benda, R., Herta, C., Schünemann, V., Chumakov, A.I., Duelund, L., Winkler, H., Toftlund, H., Trautwein, A.X.: Phys. Rev. Lett. **86**, 1351 (2001)
88. Chumakov, A.I., Rüffer, R., Baron, A.Q.R., Grünsteudel, H., Grünsteudel, H.F., Krohn, V.G.: Phys. Rev. **B 56**, 10758 (1957)
89. Chumakov, A.I., Rüffer, R., Leupold, O., Sergueev, I.: Struct. Chem. **14**, 109 (2003)
90. Sage, J.T., Durbin, S.M., Sturhahn, W., Wharton, D.C., Champion, P.M., Hession, P., Sutter, J., Alp, E.E.: Phys. Rev. Lett. **86**, 4966 (2001)
91. Zakharieva-Pencheva, O., Dementiev, V.A.: J. Mol. Struct. **90**, 241 (1982)

92. In the molecular approximation used in (14) only the $L = 3N - 6$ (N is the number of atoms) discrete *intra*molecular vibrations of the molecular complex in vacuo are considered. In general these vibrations correspond to the L highest optical branches of the phonon spectrum. The *inter*molecular vibrations, which correspond to the three acoustical branches and to the three lowest optical branches are disregarded, i.e., the center of mass and – in case of small amplitudes – the inertial tensor of the complex are assumed to be fixed in space
93. Bates, J.B., Khanna, R.K.: Inorg. Chem. **9**, 1376 (1970)
94. Zakharieva, O., Woike, Th, Haussühl, S.: Spectrochim. Acta **A 51**, 447 (1995)
95. Paulsen, H., Rusanov, V., Benda, R., Herta, C., Schünemann, V., Janiak, Ch, Dorn, Th, Chumakov, A.I., Winkler, H., Trautwein, A.X.: J. Am. Chem. Soc. **124**, 3007 (2002)
96. Carducci, H.D., Pressprich, M.R., Coppens, P.J.: J. Am. Chem. Soc. **119**, 2669 (1997)
97. Sorai, M., Seki, S.: J. Phys. Chem. Solids **35**, 555 (1974)
98. Paulsen, H., Duelund, L., Winkler, H., Toftlund, H., Trautwein, A.X.: Inorg. Chem. **40**, 2201 (2001)
99. Paulsen, H., Trautwein, A.X.: Top. Curr. Chem. **235**, 197 (2004)
100. Højland, F.H., Toftlund, H., Yde-Andersen, S.: Acta Chem. Scand. **A 37**, 251 (1983)
101. Jung, J., Spiering, H., Yu, Z., Gütlich, Ph: Hyperfine Interact. **95**, 107 (1995)
102. Chumakov, A.I., Sergueev, I., van Bürck, U., Schirmacher, W., Asthalter, T., Rüffer, R., Leupold, O., Petry, W.: Phys. Rev. Lett. **92**, 245508 (2004)
103. Trautwein, A.X., Wegner, P., Winkler, H., Paulsen, H., Schünemann, V., Schmidt, C., Chumakov, A.I., Rüffer, R.: Hyperfine Interact. **165**, 295 (2005)
104. Xiao, Y., Wang, H., George, S.J., Smith, H.C., Adams, M.W.W., Jenney Jr., F.E., Sturhahn, W., Alp, E.E., Zhao, J., Yoda, Y., Dey, A., Salomon, E.I., Cramer, S.P.: J. Am. Chem. Soc. **127**, 14596 (2005)
105. Cramer, S.P., Xiao, Y., Wang, H., Guo, Y., Smith, M.C.: Hyperfine Interact. **170**, 47 (2006)
106. Xiao, Y., Koutmos, M., Case, D.A., Goucouvanis, D., Wang, H., Cramer, S.P.: Dalton Trans., 2192 (2006)
107. Xiao, Y., Fisher, K., Smith, M.C., Newton, W.E., Case, D.A., George, S.J., Wang, H., Sturhahn, W., Alp, E.E., Zhao, J., Yoda, Y., Cramer, S.P.: J. Am. Chem. Soc. **128**, 7608 (2006)
108. Czeruszewicz, R.S., LeGal, J., Mouro, I., Spiro, T.G.: Inorg. Chem. **25**, 696 (1986)
109. Scheidt, W.R., Durbin, S.M., Sage, J.T.: J. Inorg. Biochem. **99**, 60 (2005)
110. Sage, J.F., Durbin, S.M., Sturhahn, W., Wharten, D.C., Champion, P.M., Hession, P., Sutter, J., Alp, E.E.: Phys. Rev. Lett. **86**, 4966 (2001)
111. Achterhold, K., Parak, F.G.: J. Phys.: Condens. Matter **15**, S1683 (2003)
112. Achterhold, K., Sturhahn, W., Alp, E.E., Parak, F.G.: Hyperfine Interact. **141/142**, 3 (2002)
113. Achterhold, K., Keppler, C., Ostermann, A., van Bürck, U., Sturhahn, W., Alp, E.E., Parak, F.G.: Phys. Rev. **E 65**, 051916 (2002)
114. Toellner, T.: Hyperfine Interact. **125**, 3 (2000)
115. Gerdau, E.: Hyperfine Interact. **90**, 301 (1994)
116. Wille, H.C., Shvyd'ko, Y.V., Alp, E.E., Rüter, H.D., Leupold, O., Sergueev, I., Rüffer, R., Barla, A., Sanchez, J.P.: Europhys. Lett. **74**, 170 (2006)
117. Sergueev, I., Chumakov, A.I., Deschaux Beaume-Dang, T.H., Rüffer, R., Strohm, C., van Bürck, U.: Phys. Rev. Lett. **99**, 097601 (2007)
118. Kobayashi, H., Tsutsui, S., Baron, A.Q.R., Kunii, S.: Physica **B 359**, 974 (2005)
119. Wille, J.H.-C., Hermann, R.P., Sergueev, I., Leupold, O., van der Linden, P., Sales, B.C., Grandjean, F., Long, G.J., Rüffer, R., Shvyd'ko, YuV: Phys. Rev. **B 76**, 140301 (2007)
120. Polyakov, V.B., Mineev, S.D., Clayton, R.N., Hu, G., Mineev, K.S.: Geochim. Cosmochim. Acta **69**, 5531 (2005)
121. Shenoy, G.K., Röhlsberger, R.: Hyperfine Interact. **182**, 157 (2008)

Chapter 10
Appendices

Appendix A: Optimization of Sample Thickness

This paragraph presents a summary of the most relevant expressions provided by Long et al. [1] for the optimization of "thin" absorbers (effective thickness $t \ll 1$) with high mass absorption. The result of this work is used in Sect. 3.3.2 of the book. Following the approach of [1], we adopt for the signal-to-noise ratio:

$$\mathrm{SNR}(t') = \frac{N_S(t')}{\sqrt{\Delta N_\infty(t')^2 + \Delta N_0(t')^2}}. \tag{A.1}$$

The meaning of the various count numbers is illustrated in Fig. A.1 The value $N_S = N_\infty - N_0$ represents the signal amplitude, and N_0 and N_∞ are the total counts/channel in and off resonance, respectively. N_b is a nonresonant background contribution to the photons arriving at the detector by the scattering of high-energy radiation and X-ray fluorescence in the absorber, whereas $(1 - f_s)N_\infty - N_b$ is the nonresonant fraction of the Mössbauer radiation, and $N_{m,\infty} = f_s N_\infty - N_b$ describes the counts of the resonant Mössbauer radiation recorded by the detector.

A relative simple solution can be obtained for absorbers with low signal N_s, if the background contribution N_b can be neglected. With equation (2.32) for the fractional absorption in thin absorbers, i.e., $\varepsilon(t) \approx t/2$, the signal amplitude can be written as:

$$N_S(t') = N_{\infty,m}(t')f_s\varepsilon(t) = N_{\infty,m}(t')f_s t/2. \tag{A.2}$$

If we express the effective thickness t for the resonant absorption in terms of the total absorber thickness t in g cm^{-2}, a nuclear absorption coefficient can be defined in cm^2 g^{-1}, such that $t'\mu_n = t$. The signal amplitude is then given by:

$$N_S(t') = N_{\infty,m}(t')f_s t'\mu_n/2. \tag{A.3}$$

P. Gütlich et al., *Mössbauer Spectroscopy and Transition Metal Chemistry*,
DOI 10.1007/978-3-540-88428-6_10,

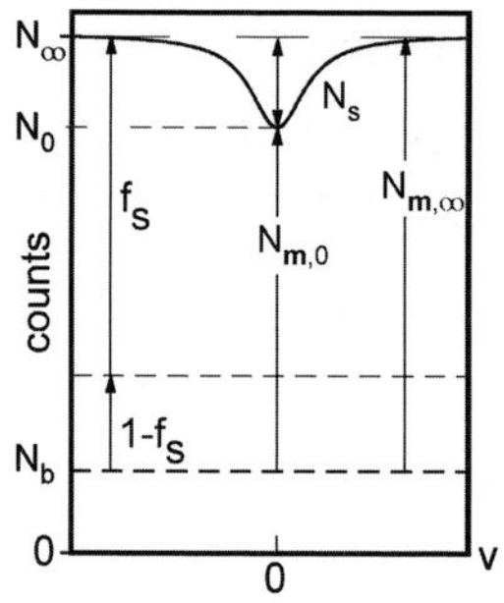

Fig. A.1 Contributions to a Mössbauer transmission spectrum. N_b is the nonresonant background from scattered high-energy γ-radiation and X-ray fluorescence in the source and the absorber

Since N_b is neglected in this case, $N_{m,\infty}$ is given by the attenuation of the incident counts $N(0)$ by mass absorption,

$$N_{\infty,m}(t') = N(0)e^{-t'\mu_e}, \tag{A.4}$$

so that the dependence of N_s on the absorber thickness t' is

$$N_s(t') \propto t'e^{-t'\mu_e}. \tag{A.5}$$

The denominator in (3.1) can be simplified because the statistical uncertainty of the baseline, ΔN_∞, is negligible in practice when the spectra are simulated with numerical line fit routines. The stochastic emission of γ-rays by the source leads to a Poisson distribution of counts with the width $\Delta N = \sqrt{N}$, and since N_s is small, the denominator of (3.1) can be written as:

$$\sqrt{N_0} \approx \sqrt{N_\infty} = \sqrt{N(0)e^{-t'\mu_e}}, \tag{A.6}$$

so that the sought signal-to-noise ratio is given by:

$$\mathrm{SNR}(t') \propto \frac{t'e^{-t'\mu_e}}{e^{-t'\mu_e/2}} = t'e^{-t'\mu_e/2}. \tag{A.7}$$

For a treatment of thick absorbers with $t > 1$, we refer to the expressions given in [1] and the references therein, but there may not be many cases relevant in practice.

References

1. Long, G.J., Cranshaw, T.E., Longworth, G.: In: Stevens, J.G., Stevens, V.E., White, R.M., Gibson, J.L. (eds.) Mössbauer Effect Reference and Data Journal, p. 42. Mössbauer Effect Data Center, North Carolina (1983)

Appendix B: Mass Absorption Coefficients

Table B.1 Mass absorption coefficients[a] for the 14.41-keV Mössbauer radiation of ^{57}Fe

Atom	No.	Mass (Da)	μ_e (cm^2 g^{-1})	Atom	No.	Mass (Da)	μ_e (cm^2 g^{-1})
H	1	1.008	0.387	Tc	43	98.91	35.5
He	2	4.003	0.194	Ru	44	101.07	37.5
				Rh	45	102.91	38
Li	3	6.941	0.23	Pd	46	106.4	41
Be	4	9.012	0.32	Ag	47	107.87	44
B	5	10.810	0.51	Cd	48	112.41	46
C	6	12.011	0.87				
N	7	14.007	1.4	In	49	114.82	49
O	8	15.999	2.2	Sn	50	118.69	54
F	9	18.998	2.7	Sb	51	121.75	56
Ne	10	20.179	4.0	Te	52	127.60	60
				I	53	126.91	61
Na	11	22.990	5.2	Xe	54	131.30	67
Mg	12	24.305	6.8				
Al	13	26.982	9.0	Cs	55	132.91	71
Si	14	28.086	12.4	Ba	56	137.33	75
P	15	30.974	14.2	La	57	138.91	76
S	16	32.062	17.0	Ce	58	140.12	77
Cl	17	35.453	20.0	Pr	59	140.10	81
Ar	18	39.948	21.0	Nd	60	144.24	88
				Pm	61	(145)	92
K	19	39.098	27.0	Sm	62	150.42	96
Ca	20	40.080	32.5				
Sc	21	44.956	33	Eu	63	151.96	100
Ti	22	47.900	42	Gd	64	157.25	104
V	23	50.941	43	Tb	65	158.93	107
Cr	24	51.996	48	Dy	66	162.50	110
M	25	54.938	53	Ho	67	164.93	115
Fe	26	55.847	64	Er	68	167.26	120
Co	27	58.933	66	Tm	69	168.93	127
Ni	28	58.700	75	Yb	70	173.04	133
Cu	29	63.546	82	Lu	71	174.97	139
Zn	30	65.380	92				
				Hf	72	178.49	145
Ga	31	69.72	97	Ta	73	180.95	150
Ge	32	72.59	102	W	74	183.85	155
As	33	74.92	126	Re	75	186.20	160
Se	34	78.96	110	Os	76	190.23	165
Br	35	79.90	130	Ir	77	192.22	172
Pb	82	207.2	120	Pt	78	195.09	178
Kr	36	83.80	126	Au	79	196.96	191
				Hg	80	200.59	165
Rb	37	85.47	21.0				
Sr	38	87.62	24.5	Tl	81	204.37	123
				Pb	82	207.2	120
Y	39	88.91	26.0	Bi	83	208.98	123
Zr	40	91.22	28.0	Po	84	(209)	126
Nb	41	92.91	29.5	At	85	(210)	140
Mo	42	95.94	34.0	Rn	86	(222)	60

(continued)

Table B.1 (continued)

Atom	No.	Mass (Da)	μ_e (cm^2 g^{-1})	Atom	No.	Mass (Da)	μ_e (cm^2 g^{-1})
Fr	87	(223)	62	Th	90	232.04	75
Ra	88	226.03	65	Pa	91	231.04	80
				U	92	238.03	85
Ac	89	227.03	70	Np	93	237.05	90

[a]Values taken from Long, G.J., Cranshaw, T.E., Longworth, G.: In: Stevens, J.G., Stevens, V.E., White, R.M., Gibson, J.L. (eds.) Mössbauer Effect Reference and Data Journal, p. 42. Mössbauer Effect Data Center, North Carolina (1983)

Appendix C: The Isomer Shift Calibration Constant

According to (4.22)–(4.24), the isomer shift is given in terms of Doppler velocity (mm s^{-1}) by the relation:

$$\delta_V = \alpha\left\{|\psi(o)|_A^2 - |\psi(o)|_S^2\right\}, \quad \text{where} \quad \alpha = \left(\frac{3Ze^2cR^2}{5\varepsilon_0 E_\gamma}\right)\frac{\delta R}{R}. \tag{C.1}$$

Inserting numerical values for the elementary charge e, speed of light c, and the electric constant ε_0, using

$$e = 1.602176 \cdot 10^{-19}\text{C}$$
$$c = 2.99792 \cdot 10^8 \text{ m s}^{-1}$$
$$\varepsilon_0 = 8.854187 \cdot 10^{-12} \text{ C V}^{-1} \text{ m}^{-1},$$

we obtain

$$\alpha = 6.1505 \cdot 10^{10}\, mm\ s^{-1} \cdot keV \cdot a.u. \cdot \left(\frac{ZeR^2}{E_\gamma}\right)\frac{\delta R}{R}. \tag{C.2}$$

The effective nuclear radius R has to be inserted here in atomic units (1 a.u. = 0.05291772 fm = $1a_0$, Bohr radius), the transition energy E_γ in keV, and the electron density $|\psi(o)|^2$ in (a.u.)$^{-3}$; Z is the dimensionless atomic number of the Mössbauer nuclide. The value of nuclear radius R can be estimated by using Elton's approximation

$$R \approx 1.123\ A^{1/3} + 2.352\ A^{-1/3} - 2.070\ A^{-1}\text{fm}, \tag{C.3}$$

where A is the mass number of the Mössbauer atom [1, 2]. The calibration constant α so far could not be determined without recourse to experimental data, because of the difficulties in calculating $\langle r^2\rangle\delta R/R$ from nuclear theory. Therefore, α is usually determined from the correlation of experimental isomer shifts and theoretically calculated values of $|\psi(o)|^2$ obtained by DFT methods. The relative change of the nuclear radius, $\delta R/R$, is an important parameter for the understanding of certain nuclear features [3], because it is closely related to the compressibility of nuclear matter. For Mössbauer isotopes, $\delta R/R$ can be estimated from α as obtained from isomer shifts and calculated charge densities. The value of the calibration constant given in Chap. 5 by Neese and Petrenko is $\alpha = 0.3666$. This was obtained by using nonrelativistic DFT for a large series of iron complexes with a wide range of isomer shifts.

Core electrons are highly relativistic and DFT methods may show systematic errors in calculating the charge density at the nucleus because of the inherent approximations. Fortunately, this does not hamper practical calculations of isomer shifts of unknown compounds, because only differences of $|\psi(o)|^2$ are involved. In practice, the reliability of the results depends more on the number of compounds used for calibration and how wide the spread of their isomer shift values was. The isomer shift scale for several Mössbauer isotopes has been calibrated by this approach, among which are ^{197}Au [1], ^{119}Sn [4], and ^{57}Fe [5–9]. For details on practical calculation of Mössbauer isomer shifts, see Chap. 5.

Finally, we mention that Filatov [10, 11] recently in an interesting new approach discussed the effect of finite nuclei in detail. He suggested calculating the isomer shift from the variation to the total electron energy in dependence of the nuclear charge extension.

References

1. Wdowik, U.D., Ruebenbauer, K.: Calibration of the isomer shift for the 77.34 keV transition in 197-Au using the full-potential linearized augmented plane-wave method. J. Chem. Phys. **129** (10), 104504 (2008)
2. Elton, L.R.B.: Nuclear Sizes. Oxford University Press, Oxford (1961)
3. Blaizot, J.P., Gogny, D., Grammaticos, B. Nucl. Phys. A **265**, 315 (1972)
4. Svane, A., Christensen, N.E., Rodriguez, C.O., Methfessel, M. Phys. Rev. B **55**, 12572 (1997)
5. Duff, K.J.: Phys. Rev. B **9**, 66 (1974)
6. Nieuwpoort, W.C., Post, D., van Duijnen, P.Th.: Phys. Rev. B **17**, 91 (1978)
7. Wdowik, U.D., Ruebenbauer, K.: Phys. Rev. B **76**, 155118 (2007)
8. Madsen, G.K.H., Blaha, P., Schwarz, K., Sjöstedt, E., Nordström, L.: Phys. Rev. B **64**, 195134 (2001)
9. Sinnecker, S., Neese, F.: Top. Curr. Chem. **268**, 47–83 (2007)
10. Filatov, M.: J. Chem. Phys. **127**, 084101 (2007)
11. Kurian, R., Filatov, M.: J. Chem. Theory Comput. **4**, 278 (2008)

Appendix D: Relativistic Corrections for the Mössbauer Isomer Shift

The expressions (4.22)–(4.23) found in chap. 4 for the isomer shift δ in nonrelativistic form may be applied to lighter elements up to iron without causing too much of an error. In heavier elements, however, the wave function ψ is subject to considerable modification by relativistic effects, particularly near the nucleus (remember that the spin–orbit coupling coefficient increases with Z^4!). Therefore, the electron density at the nucleus $|\psi(o)|^2$ will be modified as well and the aforementioned equations for the isomer shift require relativistic correction. This has been considered [1] in a somewhat restricted approach by using Dirac wave functions[1] and first-order perturbation theory; in this approximation the relativistic correction simply consists of a dimensionless factor $S'(Z)$, which is introduced in the above equations for δ,

$$\delta = (4\pi/5)Ze^2S'(Z)R^2(\delta R/R)\left\{|\psi(o)|_A^2 - |\psi(o)|_S^2\right\}. \tag{D.1}$$

Values of the "relativity factor" $S'(Z)$ for $Z = 1$–96 have been compiled by Shirley [1] and others [2, 3]. For example, $S'(Z) = 1.32$ for iron ($Z = 26$), 2.48 for tin ($i = 50$), and 19.4 for neptunium ($Z = 93$). The problem of relativistic corrections does not arise in Mössbauer effect studies, where one compares compounds of the same Mössbauer nuclide, because the relativity factor $S'(Z)$ is constant for all compounds of a given Mössbauer nuclide.

References

1. Shirley, D.A.: Rev. Mod. Phys. **36**, 339 (1964)
2. Trautwein, A., Harris, F.E., Freeman, A.J., Descaux, J.P.: Phys. Rev. B **11**, 4101 (1975)
3. Vries, J.L. K.F.d., Trooster, J.M., Ros, P.: J. Chem. Phys. **63**, 5256 (1975)

[1]A comparison between relativistic and nonrelativistic calculations shows that the correction factor $S'(Z)$ is slightly different for $1s$, $2s$, and $3s$ electrons of iron. One finds

for $3d^5$:	$S'_{1s}(Z) = 1.2619$	for $3d^6$:	$S'_{1s}(Z) = 1.2619$
	$S'_{2s}(Z) = 1.2998$		$S'_{2s}(Z) = 1.3002$
	$S'_{3s}(Z) = 1.3079$		$S'_{3s}(Z) = 1.3077$

These small differences are of great significance in the calculation of absolute electron densities.

Appendix E: An Introduction to Second-Order Doppler Shift

In order to elucidate the physical origin of second-order Doppler shift, δ_{SOD}, we consider the Mössbauer nucleus ^{57}Fe with mass M executing simple harmonic motion [1] (see Sect. 2.3). The equation of motion under isotropic and harmonic approximations can be written as

$$M\ddot{r} = -Kr, \tag{E.1}$$

where K is the force constant and r is the displacement of the ^{57}Fe atoms from their mean position at any instant. The acceleration experienced by the ^{57}Fe atom at a particular instant is given by:

$$\ddot{r} = -\frac{K}{M}r. \tag{E.2}$$

According to the “principle of equivalence” [2], this acceleration creates a gravitational field whose potential is defined as the work necessary to move a unit mass from the distance r to a point free of this force field, and may be given by

$$\phi = \int_r^0 \left(-\frac{K}{M}r\right)(-\mathrm{d}r) = -\frac{K}{2M}r^2. \tag{E.3}$$

Since the period of vibration in a crystal ($\approx 10^{-13}$ s) is much smaller than the lifetime of the excited state of the Mössbauer nucleus ($\approx 10^{-8}$ s), the vibrating nucleus only sees an average value of the potential

$$\langle\phi\rangle = -\frac{K}{2M}\langle r^2\rangle, \tag{E.4}$$

which causes a change [3] ΔE in the energy E_γ of the γ-ray emitted or absorbed by the vibrating nucleus: $\Delta E_\gamma = m_\gamma\langle\phi\rangle$. This (together with $E_\gamma = \hbar\omega_\gamma = m_\gamma c^2$) gives rise to a fractional energy shift

$$\delta_{SOD} = \frac{\Delta \mathrm{E}}{\mathrm{E}} = \frac{\langle\phi\rangle}{c^2} = \frac{\mathrm{K}}{2\mathrm{M}c^2}\langle r^2\rangle, \tag{E.5}$$

which is called *second-order Doppler shift*. Since the mean square displacement $\langle r^2\rangle$ of the vibrating nucleus is temperature-dependent, δ_{SOD} is also called *temperature shift*.

From a chemical point of view, the second-order Doppler shift is very interesting with respect to its simple relation connecting δ_{SOD}, the recoil-free fraction f, and the force constant K experienced by the Mössbauer nucleus in a simple harmonic potential. This relation follows from a combination of (E.5) and (2.14) [1]

$$\delta_{\mathrm{SOD}} = \frac{3\hbar^2}{2ME_\gamma^2} K \ln f. \tag{E.6}$$

Since it is difficult to derive δ_{SOD} directly from Mössbauer measurements, we substitute (E.6) into (4.25) and obtain

$$\delta_{\mathrm{total}} = \frac{3\hbar^2}{2ME_\gamma^2} K \ln f + \delta. \tag{E.7}$$

Because of the smallness of the variation of K and δ with temperature in comparison with that of ln f, the differentiation of (E.7) with respect to temperature gives

$$\frac{\partial \delta_{\mathrm{total}}}{\partial T} = \left(\frac{3\hbar^2}{2ME_\gamma^2} K \right) \frac{\partial \ln f}{\partial T} \tag{E.8}$$

$$\left(\frac{\Delta \delta_{\mathrm{total}}}{\Delta \ln f} \right)_T = \frac{3\hbar^2}{2ME_\gamma^2} K. \tag{E.9}$$

$\Delta\delta_{\mathrm{total}}$ and Δln f are the changes of the total energy shift δ_{total} and of ln f, respectively, corresponding to the variation in temperature T.

Taylor and Craig [4] have studied $\Delta\delta_{\mathrm{total}}$ and Δln f for ^{57}Fe in various host lattices over a wide temperature range. Taking the experimental values $\Delta\delta_{\mathrm{total}}/\Delta \ln f$ from their work, the force constants K experienced by the ^{57}Fe atoms in the different host lattices have been calculated by Gupta and Lal [1]. These calculations indicate that K(^{57}Fe) significantly depends on bonding properties and the chemical nature of the ligands which are coordinated to iron. Alternatively, the elastic properties of solids, including temperature dependence, are better described in the Debye model. This also connects the second-order Doppler shift, $\delta_{\mathrm{SOD}}(T)$, and the Lamb-Mössbauer factor, $f(T)$, using the Debye temperature θ_M and the effective Mössbauer mass M as crucial common parrameters instead of K (for details see Sects. 2.4 and 3.2.3, and for applications see Sects. 7.6.2 and 9.4.1).

References

1. Gupta, G.P., Lal, K.C.: Phys. Stat. Sol. **51**, 233 (1972)
2. Einstein, A.: Ann. Phys. (4. Folge) **35**, 898 (1911)
3. Lustig, H.: Am. J. Phys. **29**, 1 (1961)
4. Taylor, R.D., Craig, P.P.: Phys. Rev. **175**, 782 (1968)

Appendix F: Formal and Spectroscopic Oxidation States

Correlations of isomer shifts with the redox properties of iron compounds refer primarily to the oxidation number of the coordinated iron. According to the IUPAC regulations, the oxidation number of a metal ion in a coordination compound is defined by "the charge it would bear if all the ligands were removed along with the electron pairs that were shared with the central atom" [1] – that is, if the ligands were removed in their normal, closed-shell configuration [2]. The formal number is therefore not measurable and is thus without any physical meaning, but the definition provides a very useful system of classification for any possible transition of metal ion complexes with any number of electrons.

In contrast, it is accepted practice that referring to, for example, an iron(III) complex implies that this compound contains an iron ion with a d^5 high-, intermediate-, or low-spin electron configuration. Since n for a d^n configuration is, at least in principle, a measurable quantity,[2] it has been suggested [3] that an oxidation number n, which is derived from a known d^n configuration, should be specified as physical or *spectroscopic* oxidation number (state) [4–6].

In most cases, formal and spectroscopic oxidation numbers are identical as is exemplified for $[Fe(H_2O)_6]^{3+}$, where the high-spin d^5 iron ion possesses the oxidation state +III, both formally and spectroscopically. Discrepancies arise when redox-active ligands [3] are involved which can occur in more than one open or closed-shell configuration. This has been demonstrated most clearly, for example, for a complex of an iron ion with a d^5 configuration and an O-coordinated (neutral) phenoxyl radical (–O•–Ph) [7]. The complex was denoted as LFe(III)–O•–Ph, where *L* is a tri-anionic organic ligand. According to the aforementioned definition, the formal oxidation number of iron would be +IV, since a closed-shell phenolato *anion* would have to be removed. On the other hand, Mössbauer and resonance Raman spectra unequivocally prove the presence of a high-spin d^5 electron configuration at the metal ion and of a phenoxyl ligand, respectively [7]. Thus, the iron ion has a physical oxidation number of +III. Apparently, an Fe(III)-phenoxyl species has a distinctly different electronic structure than an Fe(IV)-phenolato complex.

The terms innocent and noninnocent ligands are widely used to emphasize the fact that some ligands do not necessarily possess a closed-shell configuration. These terms can be used meaningfully only in conjunction with the spectroscopic oxidation state of the metal ion. In this book, we generally refer to spectroscopic oxidation numbers, unless stated differently.

---

[2] In this context n is usually taken as a (positive) integer, as it would be used in a crystal field description, although a thorough quantum chemical calculation may yield a broken value. A d^n configuration, defined in this sense, can be determined from the Mössbauer isomer shift by comparison with similar complexes with known (spectroscopic) oxidation numbers. Another powerful method is X-ray absorption spectroscopy, by which K- and L-edge energies of the central metal are evaluated for the assignment of metal oxidation states.

The group of noninnocent ligands is continuously growing and comprises dithiolates, diimines, quinones, phenolates, and their derivatives with all combinations of coordinating N, O, and S atoms, and many others. Some examples are found in Sect. 8.2.

References

1. IUPAC. Compendium of Chemical Terminology, 2nd edn. Compiled by McNaught, A.D., Wilkinson, A.: Blackwell Scientific Publications, Oxford (1997). XML on-line corrected version: http://goldbook.iupac.org (2006-) created by Nic, M., Jirat, J., Kosata, B.; updates compiled by Jenkins, A. ISBN 0-9678550-9-8 (2006)
2. Hegedus, L.S.: Transition Metals in the Synthesis of Complex Organic Molecules. University Science Books, Mill Valley, California (1994)
3. Jörgensen, C.K.: Oxidation Numbers and Oxidation States. Springer, Berlin, Heidelberg, New York (1969)
4. Chaudhuri, P., Verani, C.N., Bill, E., Bothe, E., Weyhermüller, T., Wieghardt, K.: J. Am. Chem. Soc. **123**, 2213–2223 (2001)
5. Bill, E., Bothe, E., Chaudhuri, P., Chlopek, K., Herebian, D., Kokatam, S., Ray, K., Weyhermüller, T., Neese, F., Wieghardt, K.: Chem. Eur. J. **11**, 204–224 (2004)
6. Wikipedia Oxidation number. http://en.wikipedia.org/wiki/Oxidation_number
7. Snodin, M.D., Ould-Moussa, L., Wallmann, U., Lecomte, S., Bachler, V., Bill, E., Hummel, H., Weyhermüller, T., Hildebrandt, P., Wieghardt, K.: Chem. Eur. J. **5**, 2554–2565 (1999)

Appendix G: Spin-Hamiltonian Operator with Terms of Higher Order in S

Equations (4.69)–(4.72) do not give the most general form of the electronic spin-Hamiltonian operator possible for a transition metal ion with spin S. More extended representations with terms of higher order in S are found in textbooks [1] and EPR simulation programs (for example SOPHE [2, 3] or SIM [4, 5]. In general, these expressions result from the higher symmetry elements of the paramagnetic site in a crystal lattice [1]. However, such terms are found to be marginal in most experiments, unless D becomes unusually small as for iron(III) in a cubic environment such as in the $[Fe(III)(H_2O)_6]^{3+}$ complex, but such a situation is not often encountered for iron in most practical cases. Moreover, the corresponding parameters cannot be derived in a straightforward manner from orbital energies within LFT or MO calculations. Therefore, higher-order terms are usually neglected in Mössbauer spectroscopy [6–8]. If necessary, spin-Hamiltonian matrices for higher order terms are found for instance in [1].

Any set of energetically well-isolated levels can be described by an effective spin Hamiltonian operator by choosing S to match the corresponding number of levels. This can be just one isolated Kramers doublet of a high-spin multiplet if the zero-field splitting (ZFS) exceeds the Zeeman splitting of the "M_S"-levels, as in the example shown in Fig. 4.20 for the ^6S sextet of iron(III) high-spin. In this case, it is customary in EPR spectroscopy to attribute an effective spin $S^{eff} = 1/2$ to the resonant pair of levels, irrespective of its true electronic character [9]. The only parameter in the corresponding spin Hamiltonian is the *effective* g value of the effective Zeeman splitting, which then depends only on the rhombicity parameter E/D (in terms of the spin Hamiltonian for the physical spin $S = 5/2$). Since transition ions often show large ZFS and well-isolated Kramers doublets, corresponding plots of effective g values for the different Kramers doublets as a function of E/D are widely used for the interpretation of EPR spectra measured from high-spin systems [10].

References

1. Abragam, A., Bleaney, B.: Electron Paramagnetic Resonance of Transition Ions. Dover, New York (1986)
2. Wang, D., Hanson, G. R.: J. Magn. Res. A **117**, 1–8 (1995)
3. Hanson, G. R., Gates, K. E., Noble, C. J., Griffin, M., Mitchell, A., Benson, S.: J. Inorg. Biochem. **98**, 903–916 (2004)
4. Glerup, J., Weihe, H.: Acta Chem. Scand. **45**, 444–448 (1991)
5. Jacobsen, C.J.H., Pedersen, E., Villadsen, J., Weihe, H.: Inorg. Chem. **32**, 1216–1221 (1993)
6. Wickman, H.H., Klein, M.P., Shirley, D.A.: Phys. Rev. **152**, 345–357 (1966)
7. Debrunner, P.G.: In: Berliner, L.J., Reuben, J. (eds.) EMR of Paramagnetic Molecules, pp. 59–102. Plenum, New York (1993)
8. Münck, E.: In: Que, L. (ed.) Physical Methods in Bioinorganic Chemistry, pp. 287–319. University Science Books, Sausalito (2000)
9. Pilbrow, J.R.: Transition Ion Electron Paramagnetic Resonance, Clarendon, Oxford (1990)
10. Hagen, W.R.: Biomolecular EPR Spectroscopy. Boca Raton (2009)

Appendix H: Remark on Spin–Lattice Relaxation

Spin–lattice relaxation is related to the thermodynamics of solids with periodic crystal lattices. The decay of magnetization caused by spin–lattice relaxation is described by the relaxation time T_1 [1, 2]. The name was coined to describe the energy dissipation between the spin system of a paramagnetic center and the vibrational system of phonons existing in the lattice of the solid. But a very similar type of relaxation occurs also for amorphous solids like glasses or frozen solution, and also for liquid solutions. Since the spin operator $\vec{S}$ has no spatial components, the spin system is not "directly" coupled to the vibrational states of a molecule or a solid, but the (residual) orbital momentum of the paramagnetic center can mediate such a coupling. In an LFT picture, the orbital momentum $\langle \vec{L} \rangle$ of a state depends on the ligand field splitting of the electronic energy term, as well as on the strength of the spin–orbit coupling. Since the strength of the LF depends on the spatial arrangement of the ligands and their bond distances, $\langle \vec{L} \rangle$ fluctuates in conjunction with fluctuations of the ligand positions, which lead to energy dissipation. Therefore spin–lattice relaxation may be regarded as a spin–orbit–phonon relaxation. Accordingly, electronic configurations with relatively large orbital moments usually have strong T_1 relaxation, i.e., shorter T_1 values. For instance, low-spin iron(III) with a T ground state has shorter T_1 relaxation times than high-spin iron(III) with a ^6S ground state. It is also noteworthy that the relevant part of the vibrational spectrum of a (molecular) solid that controls spin–lattice relaxation is not the regime of the metal–ligand vibrations, such as the vibrations of CN groups in cyanides. These are measured by IR or Raman spectroscopy, presenting fingerprints of the molecular structure and the chemical bonds. Their energies are usually by far too high to be thermally excited ($<10^3$ cm^{-1}). Instead, much more important for spin–lattice relaxation are low-energy vibrations, which include large parts of a molecule and eventually also of its environment. In a solid-state picture, these are phonons with long wavelengths.

References

1. Orbach, R., Stapleton, H.J.: Electron spin–latice relaxation. In: Geschwind, S. (ed.) ElectronPramagnetic Resonance, pp. 212–216. Plenum, New York, London (1972)
2. Abragam, A., Bleaney, B.: Electron Paramagnetic Resonance of Transition Ions. Dover, New York (1986)

Appendix I: Physical Constants and Conversion Factors

"2006 CODATA recommended values," adapted from [1, 2] and http://physics.nist.gov/cuu/Constants/index.html

Frequently used fundamental physical constants

Speed of light in vacuum	c, c_0	299,792,458	m s^{-1} (exact)
Magnetic constant	μ_0	$4\pi \times 10^{-7}$	N A^{-2} (exact)
		$= 12.566370614... \times 10^{-7}$	N A^{-2}
Bohr magneton	μ_B	$9.27400915(23) \times 10^{-24}$	J T^{-1}
		$5.7883817555(79) \times 10^{-5}$	eV T^{-1}
	μ_B/h	13.99624604(35)	GHz T^{-1}
	μ_B/hc	0.466864515(12)	cm^{-1} T^{-1}
Nuclear magneton	μ_N	$5.05078324(13) \times 10^{-27}$	J T^{-1}
		$3.1524512326(45) \times 10^{-8}$	eV/T
	μ_N/hc	$2.542623616(64) \times 10^{-4}$	cm^{-1} T^{-1}
	μ_N/k	$3.6582637(64) \times 10^{-4}$	K T^{-1}
	μ_N/h	7.62259384(19)	MHz T^{-1}
Electron g factor	g_e	2.0023193043622(15)	
Electric constant ($1/\mu_0 c^2$)	ε_0	$8.854187817... \times 10^{-12}$	F m^{-1} (exact)
Planck constant	h	$6.62606896(33) \times 10^{-34}$	J s
		$4.13566733(10) \times 10^{-15}$	eV s
$h/2\pi$	$\hbar$	$1.054571628(53) \times 10^{-34}$	J s
		$6.58211899(16) \times 10^{-16}$	eV s
Elementary charge	e	$1.602176487(40) \times 10^{-19}$	C
Electron mass	m_e	$9.10938215(45) \times 10^{-31}$	kg
Proton mass	m_p	$1.672621637(83) \times 10^{-27}$	kg
Proton/electron mass	m_p/m_e	1836.15267247(80)	
Fine-structure constant ($e^2/4\pi\varepsilon_0\hbar c$)	α	$7.2973525376(50) \times 10^{-3}$	
Avogadro constant	N_A,L	$6.02214179(30) \times 10^{23}$	mol^{-1}
Molar gas constant	R	8.314472(15)	J mol^{-1} K^{-1}
Boltzmann constant R/N_A	k	$1.3806504(24) \times 10^{-23}$	J K^{-1}
	k	$8.617343(15) \times 10^{-5}$	eV K^{-1}
	k/h	$2.0836644(36) \times 10^{10}$	Hz K^{-1}
	k/hc	0.6950356(12)	cm^{-1} K^{-1}
Non-SI units accepted for use with the SI			
Electron volt: (e/C) J	eV	$1.602176487(40) \times 10^{-19}$	J
(Unified) atomic mass unit			
1 u $= m_u = 1/12$ m(^{12}C) $= 10^{-3}$ kg mol$^{-1}N_A$	u	$1.660538782(83) \times 10^{-27}$	kg

Conversion of Doppler velocity into γ-energy increments:

Energy increment caused by Doppler velocity, $\varepsilon_\gamma = E_0(v/c)$, for ^{57}Fe with $E_0 = 14.412497$ keV

1 mm s^{-1} ↔ 11.6248 MHz

1 mm s^{-1} ↔ 4.80766×10^{-8} eV

1 mm s^{-1} ↔ 3.8776×10^{-4} cm^{-1}

Magnetic flux density

1 T (Tesla) = 10^4 G (Gauss); 1 G ↔ 1 Oe (Oersted, magnetic field in vacuum)

Conversion Factors

Energy equivalents:

	Hartree	eV	cm^{-1}	kcal mol^{-1}	K	J	Hz
Hartree	1	27.21138386	219,474.63	627.503	315,775.	4.35974×10^{-19}	$6.57968 \times 10^{+15}$
eV	0.03674932254	1	8065.54	23.0609	11,604.5	1.60217×10^{-19}	$2.41798 \times 10^{+14}$
cm^{-1}	4.55633×10^{-6}	1.23981×10^{-4}	1	0.00285911	1.42879	1.98630×10^{-23}	$2.99793 \times 10^{+10}$
kcal mol^{-1}	0.00159362	0.0433634	349.757	1	503.228	6.95×10^{-21}	$1.04854 \times 10^{+13}$
K	3.16682×10^{-6}	8.61734×10^{-5}	0.6950356	0.00198717	1	1.38065×10^{-23}	$2.08366 \times 10^{+10}$
J	$2.29371 \times 10^{+17}$	$6.24151 \times 10^{+18}$	$5.03412 \times 10^{+22}$	$1.44 \times 10^{+20}$	$7.24296 \times 10^{+22}$	1	$1.50919 \times 10^{+33}$
Hz	1.51983×10^{-16}	4.13567×10^{-15}	3.33564×10^{-11}	9.53702×10^{-14}	4.79924×10^{-11}	6.62607×10^{-34}	1

Disclaimer: We cannot guarantee for the correctness of the above tables. When accuracy is very important, we recommend using instead the NIST website: http://physics.nist.gov/cuu/Constants/index.html

Conversion of magnetic hyperfine coupling constants A into internal fields for ^{57}Fe:

When the nuclear spin is decoupled from the electronic spin by application of an external field, the hyperfine coupling, according to (4.76), is defined as

$$\hat{H}_{\mathrm{hf}s} = \hat{\vec{I}} \cdot \overline{\overline{A}} \cdot \langle \vec{S} \rangle, \tag{I.1}$$

where $\overline{\overline{A}}$ is the hyperfine coupling tensor in units of energy and $\langle \vec{S} \rangle$ is the expectation value for the electron spin. By using the definition of the internal field according to (4.77)

$$\vec{B}^{\mathrm{int}} = -\overline{\overline{A}} \cdot \langle \vec{S} \rangle / g_N \mu_N = -\overline{\overline{A}}_T \langle \vec{S} \rangle, \tag{I.2}$$

where $A/g_{\mathrm{N}}\mu_{\mathrm{N}} = A_{\mathrm{T}}$ is the hyperfine coupling constant in units of Tesla, the corresponding magnetic splitting of the nuclear states (Fig. I.1) can be written in terms of the Zeeman interaction as

$$\hat{H}_{\mathrm{hfs}} = -g_N \mu_N \hat{\vec{I}} \cdot \vec{B}^{\mathrm{int}}. \tag{I.3}$$

Apparently, the equation holds

$$\hat{\vec{I}} \cdot \overline{\overline{A}} \cdot \langle \vec{S} \rangle = -g_N \mu_N \hat{\vec{I}} \cdot (-\overline{\overline{A}}_T \cdot \langle \vec{S} \rangle). \tag{I.4}$$

A comparison of both sides of (I.4) yields the sought relation for any component of the A tensor

$$A/\mathrm{energy} = g_N \mu_N (A_{\mathrm{T}}/\mathrm{field}). \tag{I.5}$$

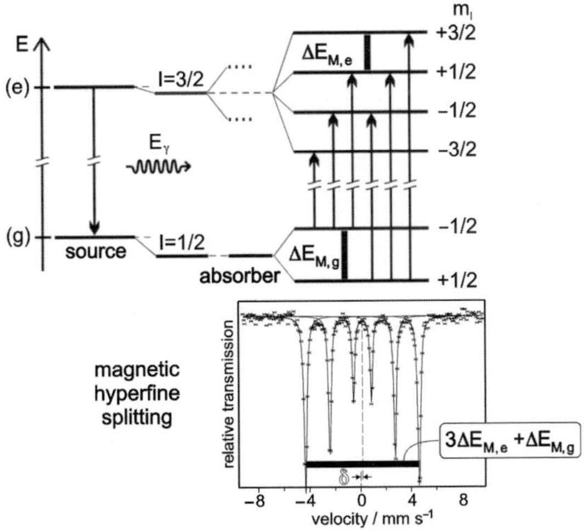

Fig. I.1 Nuclear Zeeman effect and magnetically split Mössbauer spectrum

Since in general the nuclear g factors are different for ground and excited states of a Mössbauer nucleus, the spin state must be quoted when giving numerical values for A in energy (which, however, is usually not necessary for NMR spectroscopy or other ground-state techniques). Thus, for a comparison of A values obtained from Mössbauer and NMR or ENDOR spectra, usually the ground state is considered.

The ground- and excited-state magnetic moments μ are tabulated as $\mu_g = 0.09062(3)$ n.m. (nuclear magnetons, μ_N) and $\mu_e = -0.1549$ n.m., respectively (see Table "*Properties of Isotopes Relevant to Mössbauer Spectroscopy*" provided by courtesy of Professor J. G. Stevens, Mössbauer Effect Data Center, cf. CD-ROM). Considering that nuclear magnetic moments are given by the relation $\mu = g_I \mu_N I$, the nuclear g factors for ^{57}Fe with $I_g = 1/2$ and $I_e = 3/2$ are $g_g = 0.09062 \times 2$ and $g_e = -0.1549 \times 2/3$. With these values and μ_N taken from the table seen earlier, we obtain for the ground state ($I = 1/2$) of ^{57}Fe,

$$A/\text{MHz} = 1.38152\, A_T/\text{Tesla} \quad \text{or} \quad A_T/\text{Tesla} = 0.723841\, A/\text{MHz}, \qquad (\text{I.6})$$

and for the excited state ($I = 3/2$) of ^{57}Fe,

$$A/\text{MHz} = -0.78716\, A_T/\text{Tesla} \quad \text{or} \quad A_T/\text{Tesla} = -1.27039\, A/\text{MHz}, \qquad (\text{I.7})$$

Note that the values of A_T are orders of magnitude larger than the A values given for the hyperfine splitting of an EPR spectrum (or ENDOR spectrum) in units of the field sweep necessary to obtain electron spin resonance. These are usually on the order of only 10^1 mT or less!

References

1. J. Mohr, P., Taylor, N., David, B.N., Newell, B.: Rev. Mod. Phys. **80**, 633–730 (2008)
2. Mohr, P.J., Taylor, B.N., Newell, D.B.: J. Phys. Chem. Ref. Data **37**, 1187–1284 (2008)

Index

P. Gütlich et al., *Mössbauer Spectroscopy and Transition Metal Chemistry*,
DOI 10.1007/978-3-540-88428-6, © Springer-Verlag Berlin Heidelberg 2011

O

P

Table 7.1 Nuclear data for Mössbauer transitions used in transition metal chemistry

Isotope	a (%)	E_γ (keV)	$t_{1/2}$ (ns)	2Γ (mm s^{-1})	Ig	Ie	α_T	E_R (10^{-3}eV)	σ_0 (10^{-20} cm^2)	MP	μ_g (n.m.)	μ_e (n.m.)	Q_g (barn)	Q_e (barn)	$\frac{\Delta\langle r^2\rangle}{\langle r^2\rangle}$ (10^{-4})	Popular precursor
^{57}Fe	2.14	14.413	97.81	0.1940	1/2$^-$	3/2$^-$	8.21	1.9563	255.75	M1	+ 0.090604	− 0.1547	–	+ 0.213	− 4.6	^{57}Co(EC, 270d)
														+ 0.16	− 30	
^{61}Ni	1.19	67.403	5.27	0.770	3/2$^-$	5/2$^-$	0.12	39.984	71.2	M1	− 0.74868	+ 0.481	+ 0.162	− 0.20	− 0.7	^{61}Co(β^-, 99m)
															− 4.8	
^{67}Zn	4.11	93.317	9,150	0.000320	5/2$^-$	3/2$^-$	0.89	69.766	4.96	E2	+ 0.8755	–	+ 0.16	–	–	^{67}Ga(EC, 78h)
^{99}Ru	12.72	89.36	20.5	0.1493	5/2$^+$	3/2$^+$	0.47^a	43.30	13.9	E2/M1 = 2.72	− 0.623	− 0.284	+ 0.12	+ 0.34	+ 28	^{99}Rh(EC, 16d)
^{101}Ru	17.1	127.22	0.59	3.676	5/2$^+$	3/2$^+$	0.16	86.04	8.687	E2/M1= 0.026	− 0.68	− 0.310	> 0	> 0	≈ + 24	^{101}Rh(EC, 3y)
^{177}Hf	18.5	112.97	0.5	4.843	7/2$^-$	9/2$^-$	3^a	38.72	5.99	M1/E2 = 0.24	+ 0.61	+ 1.06	+ 3	–	–	Coul. Excit.
^{178}Hf	27.14	93.17	1.5	1.96	0$^+$	2$^+$	4.6	26.18	25.2	E2	–	0.52	–	− 1.95	− 0.13	^{178}Ta(EC, β^+, 9.4m)
^{180}Hf	35.24	93.33	1.5	1.95	0$^+$	2$^+$	4.71^a	25.98	24.6	E2	–	0.63	–	− 1.96	− 0.11	^{180m}Hf(5.5h)
^{181}Ta	99.99	6.238	6,800	0.0064	7/2$^+$	9/2$^-$	46	0.115	168	E1	2.356	5.25	3.9	4.42	> 0	^{181}W(EC, 121.2d)
		136.25	0.04	50.2	7/2$^+$	9/2$^+$	1.76	55.05	5.97	E2/M1 = 0.4	–	–	–	–	–	–
^{180}W	0.14	103.7	1.27	2.08	0$^+$	2$^+$	3.44^a	32.01	25.7	E2	–	0.52	–	− 1.82	> 0	^{180m}Ta(β^-,8.1h)
^{182}W	26.41	100.10	1.37	2.09	0$^+$	2$^+$	3.85	29.55	25.2	E2	–	0.512	–	–	− 0.16	^{182}Ta(β^-, 115d)
^{183}W	14.4	46.48	0.183	34.4	1/2$^-$	3/2$^-$	40^a	5.54	6.3	E2/M1 = 0.077	0.117	− 0.62	–	–	–	^{183}Re(EC, 71d)
		99.08	0.692	3.99	1/2$^-$	5/2$^-$	4.12^a	28.79	8.2	E2	0.117	0.94	–	–	–	
^{184}W	30.64	111.19	1.28	1.92	0$^+$	2$^+$	2.61^a	36.07	27.4	E2	–	0.54	–	− 1.71	+ 0.13	^{184}Re(EC, 38d)
^{186}W	28.41	122.3	1.01	2.21	0$^+$	2$^+$	1.6^a	43.17	31	E2	–	0.134	–	− 1.63	+ 0.12	^{186}Re(EC, 90h)
^{186}Os	1.64	137.157	0.84	2.374	0$^+$	2$^+$	1.29^a	54.3	28.39	E2	–	0.562	–	− 1.5	–	^{186}Re(β^-, 90h)
^{188}Os	13.3	155.032	0.695	2.539	0$^+$	2$^+$	0.82^a	68.65	27.96	E2	–	0.58	–	− 1.36	–	^{188}Ir(EC, β^+, 41h)
^{190}Os	26.4	186.7	0.470	3.114	0$^+$	2$^+$	0.42^a	98.72	33.6	E2	–	0.63	–	− 1.18	–	^{190}Ir(EC, 12.1d)
^{189}Os	16.1	36.2	0.5	15.1	3/2$^-$	1/2$^-$	80^a	3.73	1.15	E2/M1 = 0.063	0.656	0.226	0.8	–	–	^{189}Ir(EC, 13.3h)
		69.6	1.64	2.4	3/2$^-$	5/2$^-$	8.0	13.75	8.4	E2/M1 = 0.687	–	0.977	–	− 0.62	–	^{189}Re(β^-, 24h)
		95.2	0.23	12.4	3/2$^-$	3/2$^-$	6.7	25.76	0.56	E2/M1 = 0.283	–	–	–	+ 0.8	–	
^{191}Ir	37.3	82.4	4.02	0.8258	3/2$^+$	1/2$^+$	10.7	19.09	1.540	E2/M1 = 0.64	+ 0.1440	+ 0.542	+ 1.5	–	–	^{191}Pt(EC, 3.0d)
		129.4	0.089	23.75	3/2$^+$	5/2$^-$	2.85^a	47.07	5.692	E2/M1 = 0.13	+ 0.1440	+ 0.58	+ 1.5	–	–	^{191}Os(β^-, 15d)
^{193}Ir	62.7	73.0	6.3	0.595	3/2$^+$	1/2$^+$	6.5	14.84	3.06	E2/M1 = 0.37	+ 0.1589	+ 0.470	+ 1.5	–	> 0	^{193}Os(β^-,31h)
		138.9	0.080	24.6	3/2$^+$	5/2$^-$	2.26	53.70	5.83	M1(+E2)	+ 0.1589	–	–	–	–	
^{195}Pt	33.8	98.9	170	16.28	1/2$^-$	3/2$^-$	7.2	26.90	6.11	M1(+E2)	0.606	− 0.62	–	–	$-1.6^{+4.4}_{-0.9}$	^{195}Au(EC, 183d)
		129.7	620	3.40	1/2$^-$	5/2$^-$	1.76	46.33	7.4	E2	–	0.90	–	–	9.0	
^{197}Au	100	77.345	1.879	1.882	3/2$^+$	1/2$^+$	4.3	16.3	3.86	E2/M1 = 0.316	0.1448	0.4163	0.594	–	3.0	^{197}Au(EC, 18h)
^{199}Hg	16.84	158.37	2.37	0.7276	1/2$^-$	5/2$^-$	≈0.77	67.7	8.264	E2	–	–	–	–	> 0 $\Delta\langle r^2\rangle$ $= 10^{-3}$fm^2	^{199}Au(β^-,3.15d)
^{201}Hg	13.22	32.19	<0.2	42.49	3/2$^-$	1/2$^-$	60^a	2.768	0.944	M1	− 0.556	–	+ 0.5	–	–	^{201}T1(EC, 73.5h)

[a]Value from theory